D0710711

Planar Transmission Line Structures

Edited by
Tatsuo Itoh
Professor of Electrical Engineering
The University of Texas at Austin

A volume in the IEEE PRESS Selected Reprint Series,
prepared under the sponsorship of the
IEEE Microwave Theory and Techniques Society

The Institute of Electrical and Electronics Engineers, Inc., New York

PRINTED IN THE UNITED STATES OF AMERICA

IEEE Order Number: PC02196

Library of Congress Cataloging-in-Publication Data

Planar transmission line structures.

(IEEE Press selected reprint series)
"Prepared under the sponsorship of the IEEE Microwave Theory and Techniques Society."
Includes bibliographies and indexes.
1. Microwave transmission lines. I. Itoh, Tatsuo.
II. IEEE Microwave Theory and Techniques Society.
TK7876.P55 1987 621.381'32 87-17277

ISBN 0-87942-232-7

Contents

Preface

TRANSMISSION lines are one of the most essential elements of microwave and millimeter-wave integrated circuits. Analysis, modeling, and design of transmission lines are important for any component and subsystem development. This book is concerned with those so-called planar transmission lines. Other classes of transmission lines such as dielectric waveguides are not treated in this volume.

Because of the importance of the subject, the planar transmission lines are a popular research subject and a large number of publications on this topic exist. This is exactly the reason for the publication of the present volume. Publications on the planar transmission lines are scattered throughout the literature, both in the U.S. and abroad. Beginners and students find it difficult to select representative articles on various aspects of planar transmission lines as well as for particular configurations. For working engineers, it is often not possible to spend a considerable amount of time on literature searches to access useful design and analysis information. This book is an attempt to alleviate such problems.

It is not an easy task to select papers on this proliferant subject. There are simply too many good and useful papers that must be left out in order to keep this volume a manageable size. In selecting the articles, the editor tried to include as much useful information as possible from the more recent publications. Several historically important papers, however, cannot be omitted from the list.

The largest portion of the book is devoted to the microstrip line as it is the most widely used in industry and, hence, the most extensively documented. In Part I, classical quasi-TEM (low frequency) analysis papers are included. Even though this approach is not rigorous, it is still useful in many designs and approximation treatments. Structures with isotropic, anisotropic, and magnetic substrates are treated in these papers. Part II contains papers dealing with microstrip lines based on the rigorous full-wave analysis. Several analytical and numerical techniques are treated. Many of the papers which are included here contain useful information for circuit and component designers. Part III has several representative papers on dielectric and conductor losses in the microstrip line. At higher frequencies, transmission lines with a suspended substrate are important as they may reduce the transmission loss and provide certain design flexibilities. Papers on this subject are in Part IV.

This book is not intended to cover passive components created from planar transmission lines. It is difficult to draw a boundary between the transmission line and the components. However, coupled transmission lines treated in Part V are considered useful enough to be included in this volume as they are used in many passive components.

Recent interests in computer aided design of microwave and millimeter-wave integrated circuits has enhanced the need for an accurate model and closed-form expressions for planar transmission lines in general and microstrip lines in particular. Part VI contains some of the papers dealing with the modeling aspect of microstrip lines. In connection with the microwave CAD, accurate characterizations of discontinuities in microstrip lines are increasingly important. Part VII includes some of the earlier papers based on quasi-static approaches. In addition, it contains more recent contributions which are based on more advanced techniques to characterize dynamic behaviors including the excitation of higher order modes and radiation. The latter subjects are rather young and substantial efforts are being expended at the present time by a number of researchers.

As more efforts are spent on monolithic integrated circuits for microwave and millimeter-waves, interest in coplanar waveguides has been renewed. Papers in Part VIII deal with quasi-TEM and dynamic characterizations of the coplanar waveguides as well as losses and empirical formulas.

Finlines and slot lines are treated in Part IX. For millimeter-wave applications, the past several years have seen considerable interest in these lines, particularly finlines. Analyses of propagation characteristics and attenuations for slot lines and finlines are the subjects in this part. Papers on empirical formulas and discontinuities are also included.

Finally, Part X is concerned with papers that deal with somewhat more exotic structures consisting of printed transmission lines exhibiting the ''slow-wave'' phenomena by virtue of dissipation loss in the layered substrate. These transmission lines have drawn some interest in connection with the monolithic integrated circuits in which the transmission lines are created on a semiconductor substrate.

Part I
Quasi-TEM Analysis of Microstrip Lines

THIS part deals with the quasi-TEM analysis method of microstrip lines. It is widely recognized that the microstrip line is the planar transmission line most extensively used and analyzed for microwave and millimer-wave integrated circuits. In its basic form, a microstrip line consists of a narrow conducting strip placed on a dielectric substrate which is in turn backed by a conducting ground plane. Because of its similarity to the printed circuit board, fabrication of integrated circuits has been expedited with the use of the microstrip line. Accessibility to the top surface is a convenient feature in circuit design, although in many practical applications, the microstrip is placed in a metal enclosure so that interference with an external space is eliminated. Furthermore, the size of the metal enclosure is usually small enough so that all higher order modes associated with the waveguide-like enclosure can be made below cutoff. Under such a condition, the microstrip line is operated in a single mode.

In spite of its simplicity in construction and practicality in function, the microstrip line belongs to a family of so-called inhomogeneously filled waveguides. This implies that no simple TEM or waveguide-type TE and TM modes exist independently. An accurate analysis is extremely difficult. In addition, no closed form expression satisfying all the physical conditions is known to exist. During the history of microstrip lines, a number of analytical and numerical solutions have been introduced. Although an accurate analysis requires the so-called hybrid mode solution of the full wave analysis, a quasi-TEM analysis is a good approximation at low frequencies. Under this approximation, the analysis of a microstrip line with a nonmagnetic material is reduced to that of a capacitance per unit length along its propagation direction. After this capacitance is calculated, the corresponding quantity of the microstrip less substrate, or air-filled microstrip, is computed. From these quantities, the characteristic impedance and the phase constant of the original microstrip line can be derived, though strictly speaking these values are valid only at dc. However, at lower microwave frequencies, these quantities are often useful, especially in light of the easy evaluation of this calculation process.

In this part, most of the papers are historical due to the primitive nature of the approximations introduced. The paper by Yamashita is a milestone; this paper is the prelude to a widely used spectral domain technique treated in later parts of this book. The paper by Wheeler is actually a revised version of his earlier works. Wheeler's work is based on a conformal mapping technique which can be applied exactly in many TEM transmission lines such as a stripline. In many applications, the substrate materials are non-isotropic and magnetic. Microstrip lines on these substrates are treated in the remaining papers in this chapter. The methods used are still based on the quasi-TEM approximations and hence, the results are valid at low microwave frequencies.

Variational Method for the Analysis of Microstrip-Like Transmission Lines

EIKICHI YAMASHITA, MEMBER, IEEE

Abstract—A theoretical method is presented by which microstrip-like transmission lines can be analyzed. These transmission lines are characterized by conducting strips, large ground planes, multi-dielectric-layer insulation, and planar geometry. The method is essentially based on a variational calculation of the line capacitance in the Fourier-transformed domain and on the charge density distribution as a trial function. A shielded double-layer microstrip line is analyzed by this method. Derived formulas for this structure are also applicable to simpler structures: a double-layer microstrip line, a shielded microstrip line, and a microstrip line. The calculated values of the line capacitance and the guide wavelength are compared with the measured values where possible. Oxide-layer effects on a silicon microstrip line and shielding effects on a sapphire microstrip line are also discussed based on this theory. The limitations and possible applications of this method are described.

I. Introduction

MICROSTRIP-LIKE transmission lines have been widely discussed recently in connection with the microwave integrated circuitry. These transmission lines are characterized by conducting strips, large ground planes, multi-dielectric-layer insulation, and planar geometry. It is, therefore, desirable to develop a general design theory which covers all transmission lines of this type.

The microstrip line is a simple structure mechanically and has been known for more than a decade. Yet it had not been analyzed with reasonable accuracy until a modified conformal mapping [1], a relaxation method [2], and a variational method [3] were investigated. The theoretical difficulty of this structure is attributed to the dielectric boundary conditions restricting electric fields. The above variational method for the microstrip line [3] treats the dielectric boundary conditions in a general way so that a multi-layer microstrip line can be analyzed without difficulty.

Manuscript received November 17, 1967; revised March 11, 1968. The work reported here was supported by U. S. Army Research Grant DA-G-646.

The author was with the Antenna Laboratory, Department of Electrical Engineering, University of Illinois, Urbana. He is now with the University of Electro-Communications, Tokyo, Japan.

In the microwave integrated circuit applications, the following modifications of the microstrip line are useful: a shielded (packaged) microstrip line, the oxide film coating of substrate [4], the glazing of alumina substrate [5], and an integrated circuit supported between two parallel ground planes. The last example is also a modification of an integrated circuit mounted in a coaxial structure [6].

This paper applies the above variational method [3] to a shielded double-layer microstrip line. Derived formulas for this structure are also applied to simpler structures, a double-layer microstrip line, a shielded microstrip line, and a microstrip line, by substituting appropriate structural parameters. The calculated values of the characteristic impedance and the guide wavelength are compared with the measured values where possible. Oxide-layer effects on a silicon-substrate microstrip line and shielding effects on a sapphire-substrate microstrip line are discussed based on this theory. Finally, possible applications and limitations of this method are described.

II. Shielded Double-Layer Microstrip Line

A transmission line as shown in Fig. 1 has the line capacitance C. Suppose all the dielectric layers are removed from this structure. The remaining conductor system has the line capacitance C_0, which is smaller than C. The theory of a distributed-parameter transmission line gives a relation between the wavelength of an unloaded line λ_0 and the guide wavelength of a capacitance-loaded line λ.

$$\lambda = \left(\frac{C_0}{C}\right)^{1/2} \lambda_0. \tag{1}$$

Reprinted from *IEEE Trans. Microwave Theory Tech.*, vol. MTT-16, no. 8, pp. 529–535, August 1968.

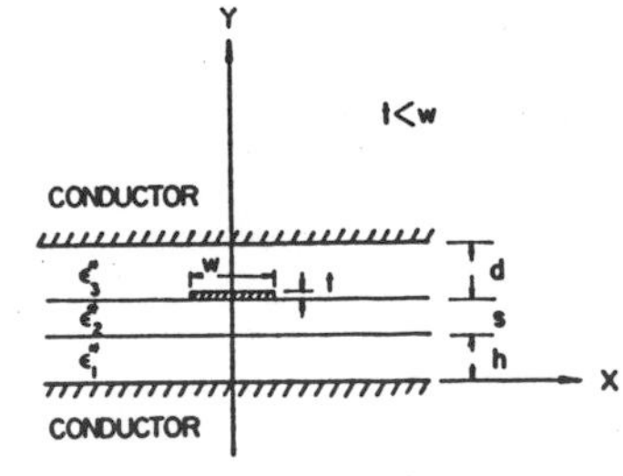

Fig. 1. Shielded double-layer microstrip line.

Similarly, it gives a relation between the characteristic impedance of the unloaded line Z_0 and the characteristic impedance of the capacitance-loaded line Z:

$$Z = \left(\frac{C_0}{C}\right)^{1/2} Z_0. \tag{2}$$

As is well known in the TEM transmission theory, λ_0 is identical to the free-space wavelength, and Z_0 is given by

$$Z_0 = \frac{1}{C_0 c} \tag{3}$$

where c is the velocity of light.

In this paper, the line capacitance is formulated as a variational integral in the Fourier-transformed coordinate. The processes of deriving the line capacitance formula are as follows.

1) Since the distributed capacitance is concerned here, the static theory can be used to calculate it. The potential distribution $\phi(x, y)$ in the cross-sectional area of the line is ruled by Poisson's equation

$$\nabla^2\phi(x, y) = -\frac{1}{\epsilon}\rho(x, y) \tag{4}$$

where $\rho(x, y)$ is the charge density distribution and ϵ is the permittivity.

2) A spacing of the dimension p is made between the strip conductor and the dielectric sheet.[1] Then the charge density on this infinitely thin strip is expressed as

$$\rho(x, y) = \delta(y - h - s - p)f(x) \tag{5}$$

where $\delta(y-h-s-p)$ is Dirac's delta function.

4) Then Poisson's equation is rewritten as

$$\left(\frac{d^2}{dy^2} - \beta^2\right)\tilde{\phi}(\beta, y) = 0 \quad (y \neq h, h+s, h+s+p). \tag{7}$$

The general solution of this differential equation is a linear combination of exp (βy) and exp $(-\beta y)$ in a bounded region. When d is taken as infinity, the solution takes a form of exp $(-|\beta| y)$ in the unbounded region.

5) The boundary conditions and continuity condition of this structure are given as follows:

$$\tilde{\phi}(\beta, 0) = 0 \tag{8a}$$

$$\tilde{\phi}(\beta, h + s + d) = 0 \tag{8b}$$

$$\tilde{\phi}(\beta, h + 0) = \tilde{\phi}(\beta, h - 0) \tag{8c}$$

$$\epsilon_2^* \frac{d}{dy}\tilde{\phi}(\beta, h + 0) = \epsilon_1^* \frac{d}{dy}\tilde{\phi}(\beta, h - 0) \tag{8d}$$

$$\tilde{\phi}(\beta, h + s + 0) = \tilde{\phi}(\beta, h + s - 0) \tag{8e}$$

$$\epsilon_3^* \frac{d}{dy}\tilde{\phi}(\beta, h + s + 0) = \epsilon_2^* \frac{d}{dy}\tilde{\phi}(\beta, h + s - 0) \tag{8f}$$

$$\tilde{\phi}(\beta, h + s + p + 0) = \tilde{\phi}(\beta, h + s + p - 0) \tag{8g}$$

$$\frac{d}{dy}\tilde{\phi}(\beta, h + s + p + 0) = \frac{d}{dy}\tilde{\phi}(\beta, h + s + p - 0) - \frac{1}{\epsilon_3^*\epsilon_0}\tilde{f}(\beta) \tag{8h}$$

where ϵ_1^*, ϵ_2^*, and ϵ_3^* are the relative dielectric constants. When more dielectric layers are involved, more boundary conditions have to be added in a similar fashion. It is noted that since the structure is symmetrical, $f(x)$ and $\tilde{f}(\beta)$ are also symmetrical.

6) Substituting these conditions to the general solution, one obtains a set of linear inhomogeneous simultaneous equations for the coefficients of potential functions. Now the spacing p is taken as zero so as to go back to the original structure. Then the solution of the potential distribution on the strip is

$$\tilde{\phi}(\beta, h + s) = \frac{1}{\epsilon_0}\tilde{f}(\beta)\tilde{g}(\beta) \tag{9}$$

where

$$\tilde{g}(\beta) = \frac{\epsilon_1^* \coth(|\beta| h) + \epsilon_2^* \coth(|\beta| s)}{|\beta| \{\epsilon_1^* \coth(|\beta| h)[\epsilon_3^* \coth(|\beta| d) + \epsilon_2^* \coth(|\beta| s)] + \epsilon_2^*[\epsilon_2^* + \epsilon_3^* \coth(|\beta| d)\coth(|\beta| s)]\}} \tag{10}$$

3) The Fourier transform is applied to all field quantities like

$$\tilde{f}(\beta) = \int_{-\infty}^{\infty} f(x)e^{j\beta x}dx. \tag{6}$$

[1] The original structure is expressed as the limit $p \to 0$ after applying boundary conditions. This artifice is to be convenient for separating the dielectric boundary condition and the continuity conditions at $y = h+s$, and, hence, for matching the number of equations to the number of unknown variables.

ϵ_0 is the permittivity of vacuum, and parameters in this expression are defined in Fig. 1.

7) The variational expression of the line capacitance in the x coordinate [7] is given by

$$\frac{1}{C} = \frac{1}{Q^2}\int_{-\infty}^{\infty} f(x)\phi(x, h + s)dx \tag{11}$$

where

$$Q \equiv \int_{-\infty}^{\infty} f(x)dx. \qquad (12)$$

The expression (11) is converted to the one in the β coordinate by using Parseval's formula [8]. The result is

$$\frac{1}{C} = \frac{1}{2\pi Q^2}\int_{-\infty}^{\infty} \tilde{f}(\beta)\tilde{\phi}(\beta, h + s)d\beta. \qquad (13)$$

The formula (13) is more convenient than (11), because the process of the inverse Fourier transform is unnecessary.

8) So far, only an infinitely thin strip has been considered. However, one can approximately take into account the strip thickness t by considering the potential distribution at $y=h+s+t$. Since there is a relation which is found in the above analytical processes

$$\tilde{\phi}(\beta, h + s + t) = \frac{\sin h(|\beta| d - |\beta| t)}{\sin h(|\beta| d)}\tilde{\phi}(\beta, h + s) \quad (14)$$

the line capacitance should be written in the form

$$\frac{1}{C} = \frac{1}{2\pi Q^2}\int_{-\infty}^{\infty} \tilde{f}(\beta)\tilde{\phi}(\beta, h + s)\tilde{h}(\beta)d\beta \qquad (15)$$

where

$$\tilde{h}(\beta) \equiv \frac{1}{2}\left\{1 + \frac{\sin h(|\beta| d - |\beta| t)}{\sin h(|\beta| d)}\right\}. \qquad (16)$$

The final form of the line capacitance including the strip thickness is, therefore,

$$\frac{1}{C} = \frac{1}{\pi Q^2 \epsilon_0}\int_{0}^{\infty} \{\tilde{f}(\beta)\}^2\tilde{g}(\beta)\tilde{h}(\beta)d\beta. \qquad (17)$$

9) The charge density distribution is properly chosen as discussed in the following section, and the line capacitance integral is numerically evaluated. C_0 is computed simply by substituting $\epsilon_1^* = \epsilon_2^* = \epsilon_3^* = 1$ in (17).

III. Charge Density Distribution

Since the charge density distribution $f(x)$ is still unknown, one has to find a reasonable trial function. The charge density on an infinitely thin conductor strip in free space has been calculated based on the electrostatic theory [9], and the solution indicates the rapid increase of the charge density near the edge of the strip, namely,

$$f(x) = \begin{cases} \left(1 - \left(\frac{2x}{w}\right)^2\right)^{-\frac{1}{2}} & \left(-\frac{w}{2} \leq x \leq \frac{w}{2}\right) \\ 0 & \text{(otherwise)} \end{cases} \qquad (18)$$

where w is the strip width. Dukes also observed a similar distribution in his experiment on a homogeneous strip line [10].

For the microstrip line, it would, therefore, be natural to assume a trial function of this type. It is an advantage of the variational expression that the accuracy of the calculated value is relatively insensitive to the choice of the trial function. Besides, the approximate line capacitance obtained by

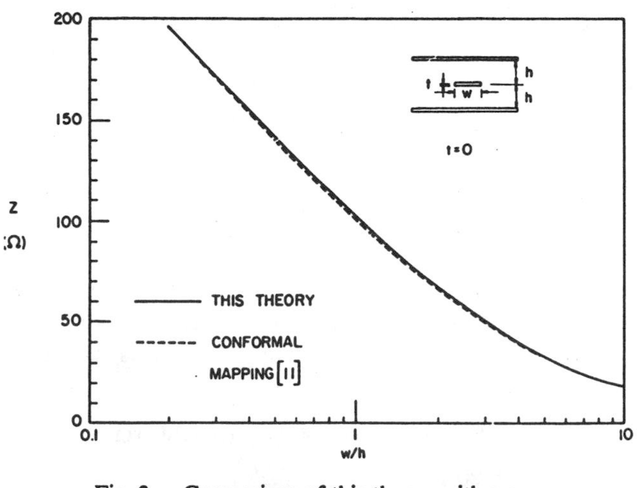

Fig. 2. Comparison of this theory with a conformal mapping in a special case.

this method is always smaller than the exact value. Therefore, one can use, as a criterion for choosing a trial function, that a trial function resulting in a larger line capacitance is a better trial function.

The charge distribution (18) is approximated by a series

$$f(x) = \begin{cases} A_0 + A_1|x| + A_2x^2 + \cdots & \left(-\frac{w}{2} \leq x \leq \frac{w}{2}\right) \\ 0 & \text{(otherwise.)} \end{cases} \qquad (19)$$

When a finite series is chosen, the coefficients of the series should be adjusted so as to obtain the maximum value of the line capacitance. The microstrip line calculation [3] shows that even a simple function, $f(x) = |x|$, gives the line capacitance with good accuracy. Taking many more terms than this trial function, one could calculate the capacitance more accurately, although more computing time would be necessary. A compromise between the computing time and the accuracy should be made in practice.

We tested a trial function,

$$f(x) = \begin{cases} 1 + A\left|\frac{2x}{w}\right|^3 & \left(-\frac{w}{2} \leq x \leq \frac{w}{2}\right) \\ 0 & \text{(otherwise)} \end{cases} \qquad (20)$$

for various structural parameters. As a result, it was found that the trial function (20) is slightly better than $f(x) = |x|$ when $A = 1$. Then the Fourier transform of $f(x)$ is readily given by

$$\frac{\tilde{f}(\beta)}{Q} = \frac{8}{5}\left\{\frac{\sin(\beta w/2)}{\beta w/2}\right\} + \frac{12}{5(\beta w/2)^2} \cdot \left\{\cos(\beta w/2) - \frac{2\sin(\beta w/2)}{\beta w/2} + \frac{\sin^2(\beta w/4)}{(\beta w/4)^2}\right\}. \qquad (21)$$

This is a rapidly decreasing function of βw. Consequently, the line capacitance integral converges fast.

A special case is considered for comparison with the other theory:

$$\epsilon_1^* = \epsilon_2^* = \epsilon_3^* = 1$$

$$d = h + s \qquad t = 0.$$

Then the structure in Fig. 1 becomes identical to a triplate strip line. The calculated values of the characteristic impedance by this theory and a conformal mapping [11] are shown in Fig. 2. The good agreement between these two theories supports this choice of the charge density distribution.

IV. Experiments on Shielded Double-Layer Microstrip Line

An experimental line was constructed which consists of a copper strip electroplated on a polystyrene sheet and two brass ground-planes as shown in Fig. 1. The parameters of the line were

$$\epsilon_1^* = \epsilon_3^* = 1 \qquad \epsilon_2^* = 2.55 \text{ (polystyrene)}$$
$$w = 5.20 \text{ mm} \qquad t = 0.20 \text{ mm}$$
$$s = 4.90 \text{ mm} \qquad h = 3.05 \text{ mm}.$$

The width of the ground plane was 200 mm. The height of the upper ground plane d was chosen as a variable in experiment, and the line capacitance at 1.5 MHz and the guide wavelength at 4 GHz were measured. The calculation of the line capacitance and the guide wavelength were carried out by the IBM 7094 computer. The computing time of the capacitance integral for one structure was less than ten seconds (GO TIME). Figs. 3 and 4 show the measured and calculated values. Agreement seems good enough to proceed with further investigation.

V. Shielded Microstrip Line

A shielded microstrip line, as shown in Fig. 5, corresponds to a special case ($s=0$) of the structure in Fig. 1. Accordingly, the formula of $\tilde{g}(\beta)$ is reduced to

$$\tilde{g}(\beta) = \frac{1}{|\beta|\{\epsilon_1^* \coth(|\beta| h) + \epsilon_3^* \coth(|\beta| d)\}}. \tag{22}$$

Substituting (16), (21), and (22) in (17), one can evaluate the line capacitance.

The shielding effects of a sapphire microstrip line [12] are considered here as an example case:

$$\epsilon_1^* = 9.9 \text{ (sapphire)} \qquad \epsilon_3^* = 1$$
$$d/h = 1, 2, 5$$
$$t/h = 0, 0.02$$
$$w/h = 0.1 \sim 10$$

The calculated values of the characteristic impedance and the guide wavelength are shown in Figs. 6, 7, 8, and 9. It is apparent from these results that the effect of shielding is almost negligible when d/h is more than 5. The effect is also larger for a wider strip.

VI. Double-Layer Microstrip Line

A double-layer microstrip line, as shown in Fig. 10, is a special case ($d\to\infty$, $\epsilon_3^*=1$) of the structure in Fig. 1. Therefore, $\tilde{g}(\beta)$ and $\tilde{h}(\beta)$ are reduced to:

$$\tilde{g}(\beta) = \frac{\epsilon_1^* \coth(|\beta| h) + \epsilon_2^* \coth(|\beta| s)}{|\beta|\{\epsilon_1^* \coth(|\beta| h)[1 + \epsilon_2^* \coth(|\beta| s)] + \epsilon_2^*[\epsilon_2^* + \coth(|\beta| s)]\}} \tag{23}$$

and

$$\tilde{h}(\beta) = \tfrac{1}{2}\{1 + \exp(-|\beta| t)\}. \tag{24}$$

The line capacitance is obtained by substituting (21), (23), and (24) in (17).

An experiment was carried out by removing the upper ground plane of the shielded double-layer microstrip line in Section IV. The height h was experimentally varied as multiple values of $h_0=3.05$ mm. The calculated and measured values are shown in Figs. 11 and 12.

Similarly, oxide-layer effects on a silicon microstrip line [4] can be analyzed. In this case

$$\epsilon_1^* = 12 \quad \text{(Si)}$$
$$\epsilon_2^* = 4.5 \quad \text{(SiO)} \quad [4].$$

The change of transmission properties are estimated as shown in Figs. 13, 14, 15, and 16. Two concrete examples are considered from these results:

1) $h = 125$ microns, $s = 1.25$ microns; $w = 12.5$ microns, $t = 2.5$ microns; $Z = 97.7$ ohms

2) $h = 125$ microns, $s = 0$; $w = 12.5$ microns, $t = 2.5$ microns; $Z = 92.0$ ohms

It is noted that the change in the characteristic impedance is about five percent increase even though the thickness of the added oxide layer is one percent of the substrate height. It is also expected that the glazing of alumina substrate [5] affects the transmission properties in a similar way.

VII. Microstrip Line

A microstrip line is a special case ($\epsilon_3^*=1$, $s=0$, $d\to\infty$) of the structure in Fig. 1. Therefore $\tilde{g}(\beta)$ is simplified as

$$\tilde{g}(\beta) = \frac{1}{|\beta|\{1 + \epsilon_1^* \coth(|\beta| h)\}}. \tag{25}$$

The line capacitance is obtained by substituting (21), (24), and (25) in (17). The calculated characteristic impedance and the guide wavelength of a sapphire microstrip line are shown, as an example, in Figs. 6, 7, 8, and 9, as indicated by $d/h\to\infty$.

VIII. Potential Distribution

The potential distribution can be evaluated by the inverse Fourier transform,

$$\phi(x, y) = \frac{1}{2\pi}\int_{-\infty}^{\infty} \tilde{\phi}(\beta, y)e^{-j\beta x}d\beta. \tag{26}$$

The potential distribution on the surface of the dielectric sheet gives information to be used in estimating the coupling between two adjacent microstrip lines. Applying (9) in (26),

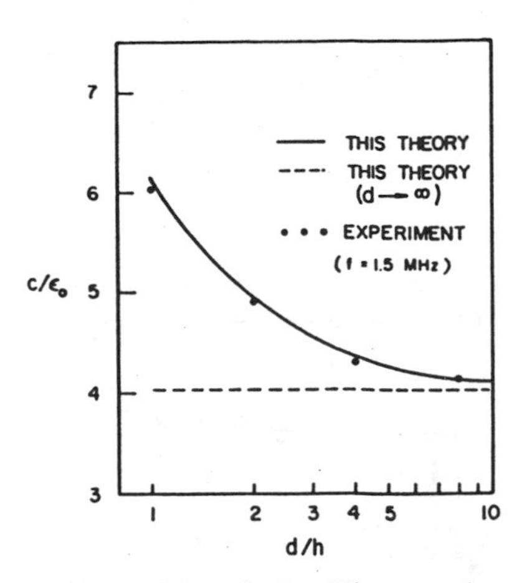

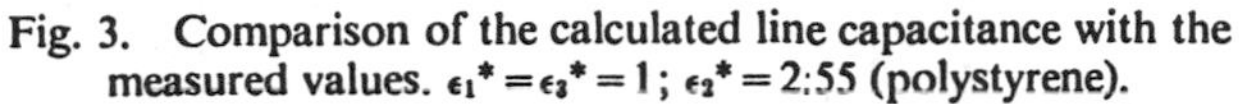
Fig. 3. Comparison of the calculated line capacitance with the measured values. $\epsilon_1^* = \epsilon_3^* = 1$; $\epsilon_2^* = 2.55$ (polystyrene).

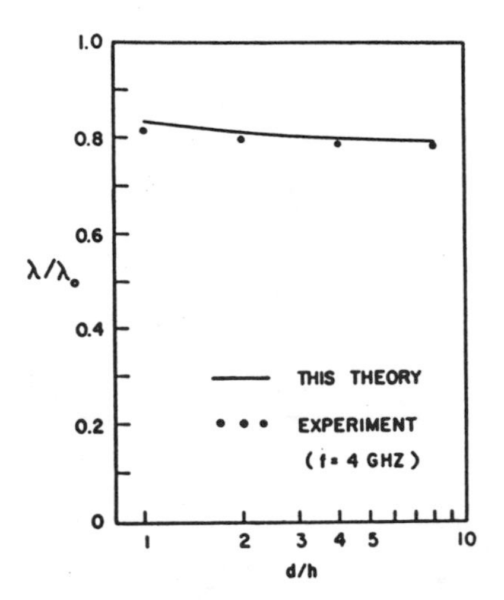

Fig. 4. Comparison of the calculated guide wavelength with the measured values. $\epsilon_1^* = \epsilon_3^* = 1$; $\epsilon_2^* = 2.55$ (polystyrene).

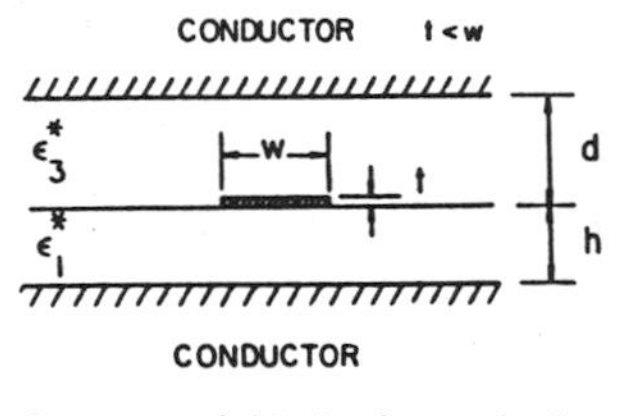

Fig. 5. Shielded microstrip line.

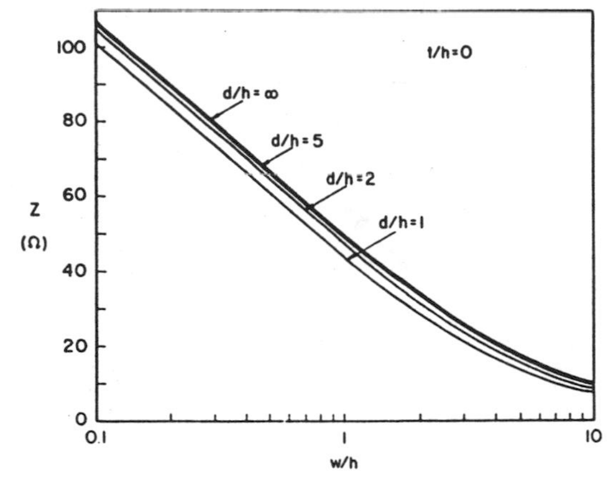

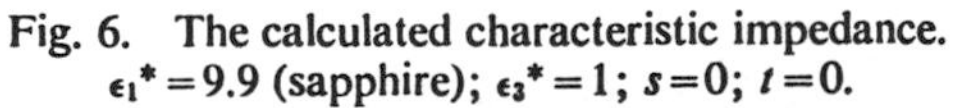
Fig. 6. The calculated characteristic impedance. $\epsilon_1^* = 9.9$ (sapphire); $\epsilon_3^* = 1$; $s = 0$; $t = 0$.

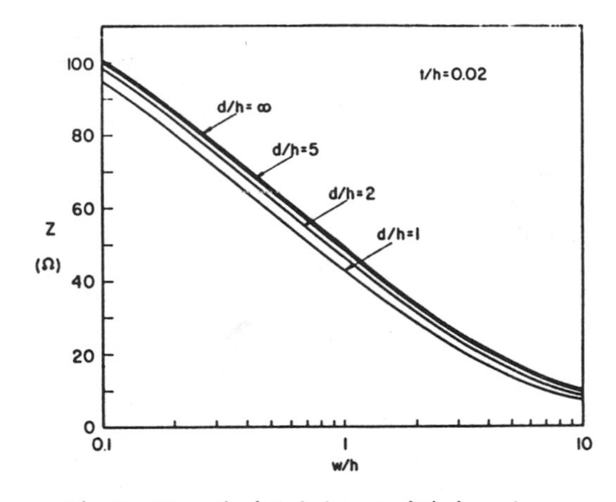

Fig. 7. The calculated characteristic impedance. $\epsilon_1^* = 9.9$ (sapphire); $\epsilon_3^* = 1$; $s = 0$; $t = 0.02h$.

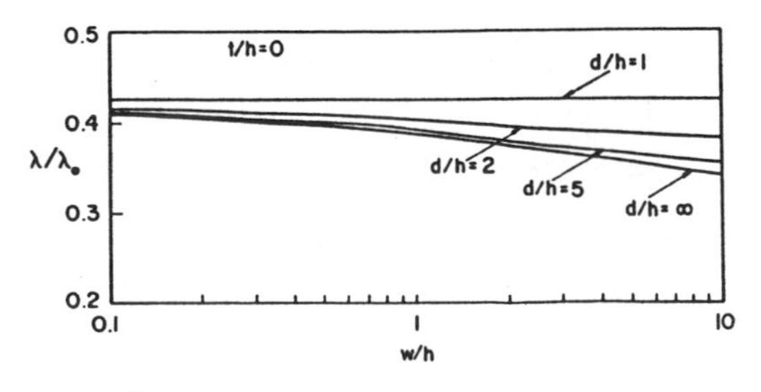

Fig. 8. The calculated guide wavelength. $\epsilon_1^* = 9.9$ (sapphire); $\epsilon_3^* = 1$; $s = 0$; $t = 0$.

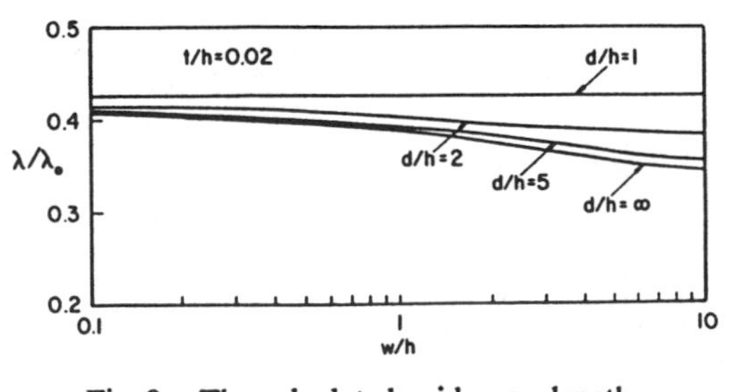

Fig. 9. The calculated guide wavelength. $\epsilon_1^* = 9.9$ (sapphire); $\epsilon_3^* = 1$; $s = 0$; $t = 0.02h$.

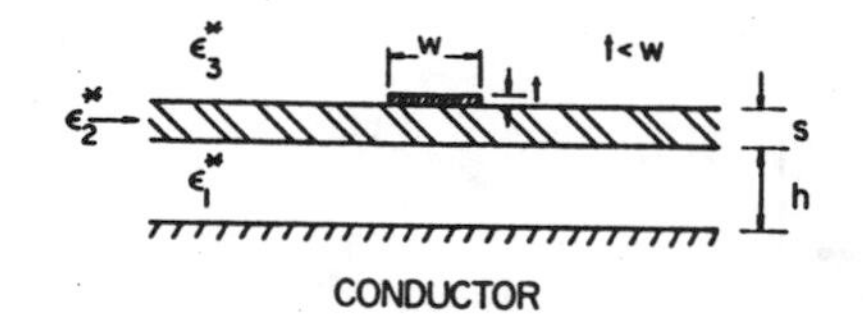

Fig. 10. Double-layer microstrip line.

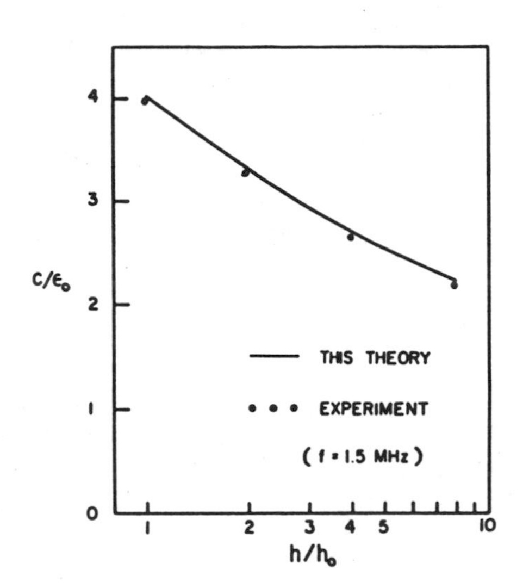

Fig. 11. Comparison of the calculated line capacitance and the measured values. $\epsilon_1^*=\epsilon_3^*=1$; $\epsilon_2^*=2.55$ (polystyrene); $d\rightarrow\infty$.

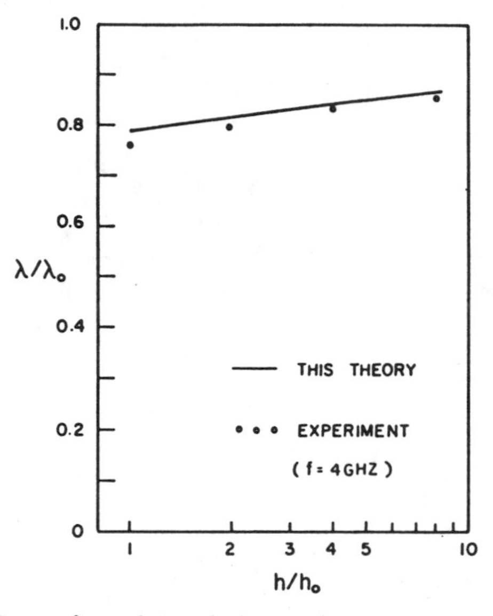

Fig. 12. Comparison of the calculated guide wavelength and the measured values. $\epsilon_1^*=\epsilon_3^*=1$; $\epsilon_2^*=2.55$ (polystyrene); $d\rightarrow\infty$.

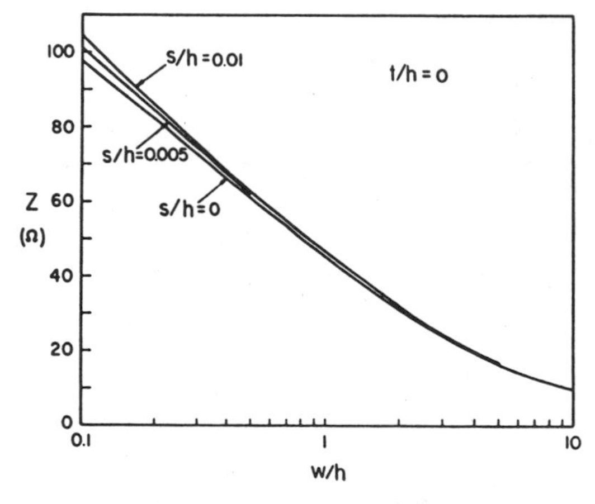

Fig. 13. The calculated characteristic impedance. $\epsilon_1^*=12$ (Si); $\epsilon_2^*=4.5$ (SiO); $\epsilon_3^*=1$; $d\rightarrow\infty$; $t=0$.

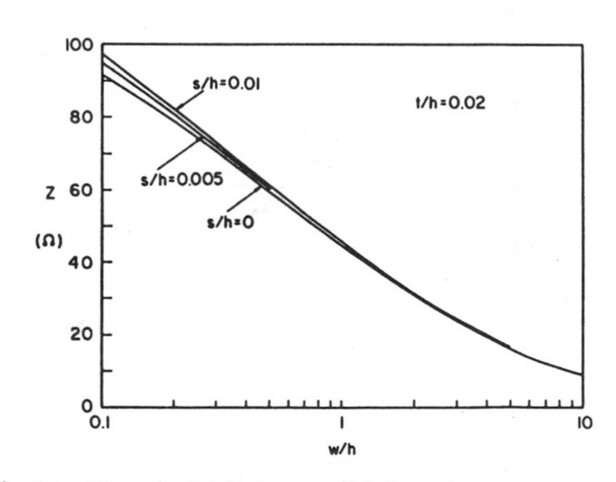

Fig. 14. The calculated characteristic impedance. $\epsilon_1^*=12$ (Si); $\epsilon_2^*=4.5$ (SiO); $\epsilon_3^*=1$; $d\rightarrow\infty$; $t=0.02h$.

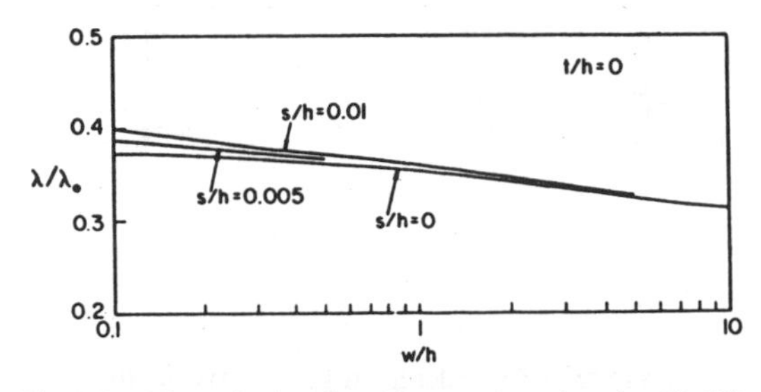

Fig. 15. The calculated guide wavelength. $\epsilon_1^*=12$ (Si); $\epsilon_2^*=4.5$ (SiO); $\epsilon_3^*=1$; $d\rightarrow\infty$; $t=0$.

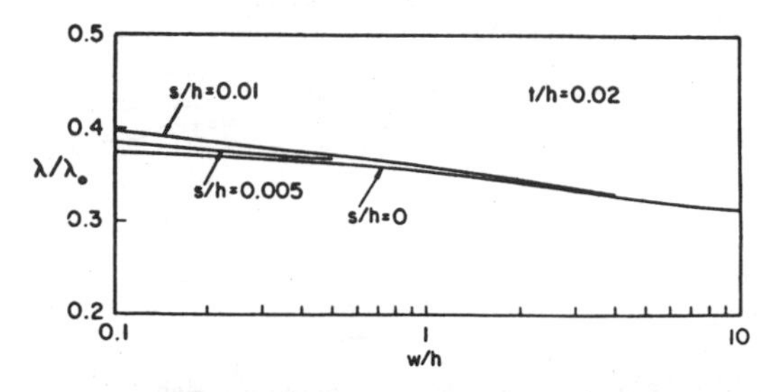

Fig. 16. The calculated guide wavelength. $\epsilon_1^*=12$ (Si); $\epsilon_2^*=4.5$ (SiO); $\epsilon_3^*=1$; $d\rightarrow\infty$; $t=0.02h$.

one obtains this potential:

$$\phi(x, h + s) = \frac{1}{2\pi\epsilon_0} \int_{-\infty}^{\infty} \tilde{f}(\beta)\tilde{g}(\beta)e^{-j\beta x} d\beta. \quad (27)$$

IX. Discussions

The variational method in the Fourier transformed coordinate was investigated and found to be useful to obtain basic design formulas for microstrip-like transmission lines. By this method, it is possible to take into account all the dielectric boundary conditions no matter how many planar boundaries exist in these lines. The characteristic impedance and the guide wavelength can be obtained for a wide range of structural parameters. The thickness of a thin strip can also be taken into account.

This approach is based on the calculation of the line capacitance by the static field theory, and, therefore, it is an approximation to the electromagnetic theory. One might recall that the conformal mapping and the relaxation method are also the static field theory. However, the analytical treatment of multiple boundaries may be easier by the variational method than by the modified conformal mapping. The computing time seems to be shorter by the variational method than by the relaxation method. At any rate, this variational method can be used within the following limitations.

1) The dielectic layer must be of low loss at the operating wavelength λ_0. This condition is written as $(2\pi\epsilon^* R_d/\eta_0) \gg \lambda_0$, where R_d is the resistivity and $\eta_0 = 120\pi$ ohms.
2) The static field approximation and the nonradiation assumption require $\lambda_0 \gg h+s$.
3) The thin strip approximation requires $t \ll h$ and $t \lesssim w$.

This method may also be applicable to other problems characterized by Poisson's equation, planar geometry, and multiple media. Examples of these problems are

1) the current distribution in multiple resistive media like a planar field-effect transistor,
2) the stationary temperature distribution in a microstrip-like structure with a strip heat source,
3) The wave velocity calculation of a traveling-wave-type electrooptical modulator using a microstrip-like transmission line.

Acknowledgment

The author thanks Prof. R. Mittra for his encouragement and advice, J. Welch for his experimental work.

References

[1] H. A. Wheeler, "Transmission-line properties of parallel strips separated by a dielectric sheet," *IEEE Trans. Microwave Theory and Techniques*, vol. MTT-13, pp. 172–185, March 1965.

[2] K. C. Wolters and P. L. Clar, "Microstrip transmission lines on high dielectric constant substrates for hybrid microwave integrated circuits," presented at the 1967 Internat'l Microwave Symp., Boston, Mass., May 1967, Session V-2.

[3] E. Yamashita and R. Mittra, "Variational method for the analysis of microstrip lines," *IEEE Trans. Microwave Theory and Techniques*, vol. MTT-16, pp. 251–256, April 1968.

[4] H. Guckel and P. A. Brennan, "Picosecond pulse response of interconnections in a common substrate monolithic system," presented at the Internat'l Solid-State Circuits Conf., (Philadelphia, Pa., February 1967), Session XI.

[5] B. T. Vincent, Jr., "Ceramic microstrip for microwave hybrid integrated circuitry," presented at the 1966 Internat'l Microwave Symp., Palo Alto, Calif., May 1966, Session V-2.

[6] H. C. Okean, "Integrated microwave tunnel diode device," presented at the 1966 Internat'l Microwave Symp., Palo Alto, Calif., May 1966, Session V-3.

[7] R. E. Collin, *Field Theory of Guided Waves.* New York: McGraw-Hill, 1960, p. 162.

[8] A. Papoulis, *The Fourier Integral and Its Applications.* New York: McGraw-Hill, 1962.

[9] E. Hallen, *Electromagnetic Theory.* London: Chapman & Hall, 1962, p. 64.

[10] J. M. C. Dukes, "An investigation into some fundamental properties of strip transmission lines with the aid of an electrolytic tank," *Proc. IEE* (London), vol. 103, pt. B, pp. 319–333, May 1956.

[11] S. B. Cohn, "Problems in strip transmission lines," *IRE Trans. Microwave Theory and Techniques*, vol. MTT-3, pp. 119–126, March 1955.

[12] M. Caulton, J. J. Hughes, and H. Sobol, "Measurements on the properties of microstrip transmission lines for microwave integrated circuits," *RCA Rev.*, vol. 27, pp. 377–391, September 1966.

Transmission-Line Properties of a Strip on a Dielectric Sheet on a Plane

HAROLD A. WHEELER, FELLOW, IEEE

Abstract—The subject is a strip line formed of a strip and a parallel ground plane separated by a dielectric sheet (commonly termed "microstrip"). Building on the author's earlier papers [1], [2], all the significant properties are formulated in explicit form for practical applications. This may mean synthesis and/or analysis. Each formula is a close approximation for all shape ratios, obtained by a gradual transition between theoretical forms for the extremes of narrow and wide strips. The effect of thickness is formulated to a second-order approximation. Then the result is subjected to numerical differentiation for simple evaluation of the magnetic-loss power factor from the skin depth.

The transition formulas are tested against derived formulas for overlapping narrow and wide ranges of shape. Some of these formulas are restated from the earlier derivations and others are derived herein. The latter include the second-order approximation for a narrow thin strip, and a close approximation for a narrow or wide square cross section in comparison with a circular cross section.

Graphs are given for practical purposes, showing the wave resistance and magnetic loss for a wide range of shape and dielectric. For numerical reading, the formulas are suited for programming on a digital pocket calculator.

I. Introduction

ONE FORM of strip line is naturally suited for the simplest fabrication in a printed circuit. It is the familiar type made of a dielectric sheet with a shield-plane conductor bonded on the bottom side and a pattern of strip lines on the top side.

The purpose of this paper is to present some improved formulas and graphs, including not only the wave resistance but also the losses. The effect of strip thickness is simply formulated to enable the evaluation of magnetic loss.

In the vernacular, this type of line is termed "microstrip," a term which is avoided in this scientific article because it is commonly used without a clear definition and is not self-descriptive. Apparently it was intended to be a short designation for "microwave strip line." The "microwave" description is ambiguous and only partially relevant. Furthermore it does not distinguish from the "sandwich" form of a microwave strip line.

Here also the descriptive term "wave resistance" is used in preference to the nondescriptive term "characteristic impedance."

The subject strip line may be described as half-shielded, by the ground plane on one side, as distinguished from the sandwich type, which is fully shielded, by ground planes on both sides. The half-shielding is adequate for some practical purposes, because the external field is relatively weak and does decay with distance.

Manuscript received October 29, 1976; revised February 23, 1977.
The author is with the Hazeltine Corporation, Greenlawn, NY 11740.

A peculiarity of the half-shielded line is the mixture of two different dielectrics. One is the material of the sheet between the strip and plane. The other is the air above the sheet. The simple rules of conformal mapping are restricted to a uniform dielectric or to some discrete boundaries that are different from the subject configuration. Various other approaches have been directed to this problem.

The first close approximation for this strip line with mixed dielectric was published by the author in 1965 [2]. It is based on some rigorous derivations for a thin strip by conformal mapping. These are supplemented by some logical concepts for interpolation between the extremes of dielectric. The uncertainties of interpolation are small enough to meet design requirements within practical tolerances. The result is a collection of formulas and charts which are complete for the wave resistance of a thin strip.

The loss power factor ($PF = 1/Q$) in a strip line has components of electric loss in the dielectric and magnetic loss in the conductor boundaries. These were not treated in the early paper but have been addressed by some other authors in the meantime.

In the frequency range where a strip line may have a length comparable with the wavelength, the magnetic loss is usually the dominant component. It is largely dependent on the strip thickness, so the formulas for a thin strip do not suffice. This loss PF can be evaluated from knowledge of the inductance of the line, which is independent of the dielectric. This evaluation can be made with the aid of the "incremental-inductance rule," published by the author in 1942 [3]. Other authors have applied this rule to the formulas of the early papers [13], [17] with the first-order thickness effect stated therein.

In the sandwich line, it has been simpler to evaluate its properties, for various reasons. First, the homogeneity of the dielectric avoids the problem of mixed dielectric, which is relevant for wave resistance. Second, the symmetry and two-sided shielding cause much greater decay of a field with distance. The symmetry simplifies the evaluation of the thickness effects, so those have been published, including the magnetic-loss PF [8]. These give an indication of trends in the subject line, but not quantitative values.

As in most of the previous articles, only the lowest mode of wave propagation in the line shall be considered, and, furthermore, only at frequencies so low that there is negligible interaction between the electric and magnetic fields. This is valid if the transverse dimensions are much less than half the wavelength in the dielectric. This mode may be termed the "quasi-TEM" mode, ignoring second-order effects of dispersion and surface-wave phenomena.

Reprinted from *IEEE Trans. Microwave Theory Tech.*, vol. MTT-25, no. 8, pp. 631–647, August 1977.

After the following list of symbols, the configuration will be defined and the scope of this article will be indicated.

II. Symbols

The units are MKS rationalized (meters, ohms, etc.).

k	=	dielectric constant of the sheet of material separating the strip and the ground plane.
k'	=	$1 + q(k-1)$ = effective dielectric constant of all space around the strip.
q	=	$(k'-1)/(k-1)$ = effective filling fraction of the dielectric material.
R_c	=	$377 = 120\pi$ = wave resistance of a square area of free space or air.
R	=	wave resistance of the transmission line formed by the strip and the ground plane (of perfect conductor) separated by a sheet of dielectric k.
R_1	=	R without dielectric ($k = 1$).
R_δ	=	R_1 subject to skin depth δ in a real conductor.
R/R_1	=	$1/\sqrt{k'} = \lambda_g/\lambda_0$ = speed ratio in mixed dielectric k' relative to free space or air.
w	=	width of the strip conductor.
h	=	height (separation) of the strip from the ground plane.
h	=	thickness of the dielectric sheet.
t	=	thickness of the strip conductor.
w'	=	effective width of a strip with some thickness.
w'	=	width of an equivalent thin strip ($t \to 0$).
Δw	=	$w' - w$ = width adjustment for thickness.
$\Delta w'$	=	width adjustment with mixed dielectric k'.
δ	=	skin depth in the conductor.
p	=	$1/Q$ = magnetic PF of the strip line.
p_k	=	electric PF of the dielectric material k.
p'	=	effective PF of mixed dielectric k'.
P	=	$p \div \delta/h = ph/\delta$ = normalized p.
α	=	rate of attenuation (nepers/meter).
λ_0	=	wavelength in free space or air.
λ_g	=	guide wavelength in mixed dielectric k'.
e	=	2.718 = base of natural logarithms.
$\exp x$	=	e^x = natural exponential function.
$\ln x$	=	$\log_e x$ = natural logarithm.
acosh x	=	anticosh $x = \cosh^{-1} x$.
asinh x	=	antisinh $x = \sinh^{-1} x$.
asin x	=	antisin $x = \sin^{-1} x$.

The following table translates some symbols from the author's earlier papers.

Here	[1] [2]
$w, h, t, \Delta w$	$2a, b, \Delta b, 2\Delta a$
R of 1 strip	R of 2 strips (twice as great)
(A-)(B-)	( )( ) formulas

III. A Strip Line on a Dielectric Sheet on a Plane

Fig. 1(a) shows the cross section of the subject line. It corresponds to the 1965 article [2] except for the translation to "practical" parameters. The latter are the wave resistance R of the asymmetric model (single strip and ground plane) and the descriptive dimensions w,h,t. Here the thickness is featured, and the equivalence between a practical strip and a wider theoretical thin strip (a perfect conductor with a thickness approaching zero). This equivalence is described in terms of the width adjustment Δw.

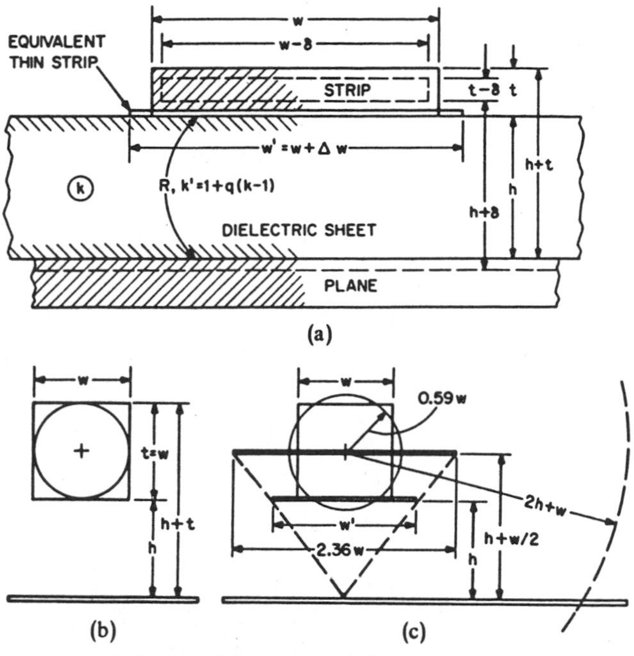

Fig. 1. A strip line parallel to a plane. (a) Rectangular cross section. (b) Cross-section square or inscribed circle. (c) Cross-section small square or equivalent circle.

For evaluation of the magnetic-loss PF, the skin effect is indicated in dashed lines. These boundaries are recessed by one half the skin depth ($\delta/2$) so they indicate the actual center of current. The actual boundary is the theoretical current center in a perfect conductor. The change between one and the other is involved in the computation of the magnetic PF. It is assumed that all conductive boundaries are nonmagnetic and have equal conductivity and skin depth.

As an extreme case of strip thickness, a square cross section is introduced, as shown in Fig. 1(b) and (c). These are related to a circular cross section in either of two ways, the inscribed circle (b) or the equivalent circle (c). Each is found to be helpful in some studies, mainly because the circle yields to simple exact formulation for comparison with an approximation for the square. Both will be used for reference.

IV. Scope

The thrust of this article is to enable explicit synthesis of a line to meet some specifications. This is achieved for various sequences. The wave resistance R is related to the dielectric k and the shape. On the other hand, the magnetic PF can be decreased by increasing the size, while the shape has a lesser effect. The PF is usually a tolerance rather than a requisite. The wave-speed ratio is taken not to be specified, but rather evaluated after synthesis of a design of a cross section.

Some graphs are introduced here, for reference in various sections. They present the relations needed for the purposes

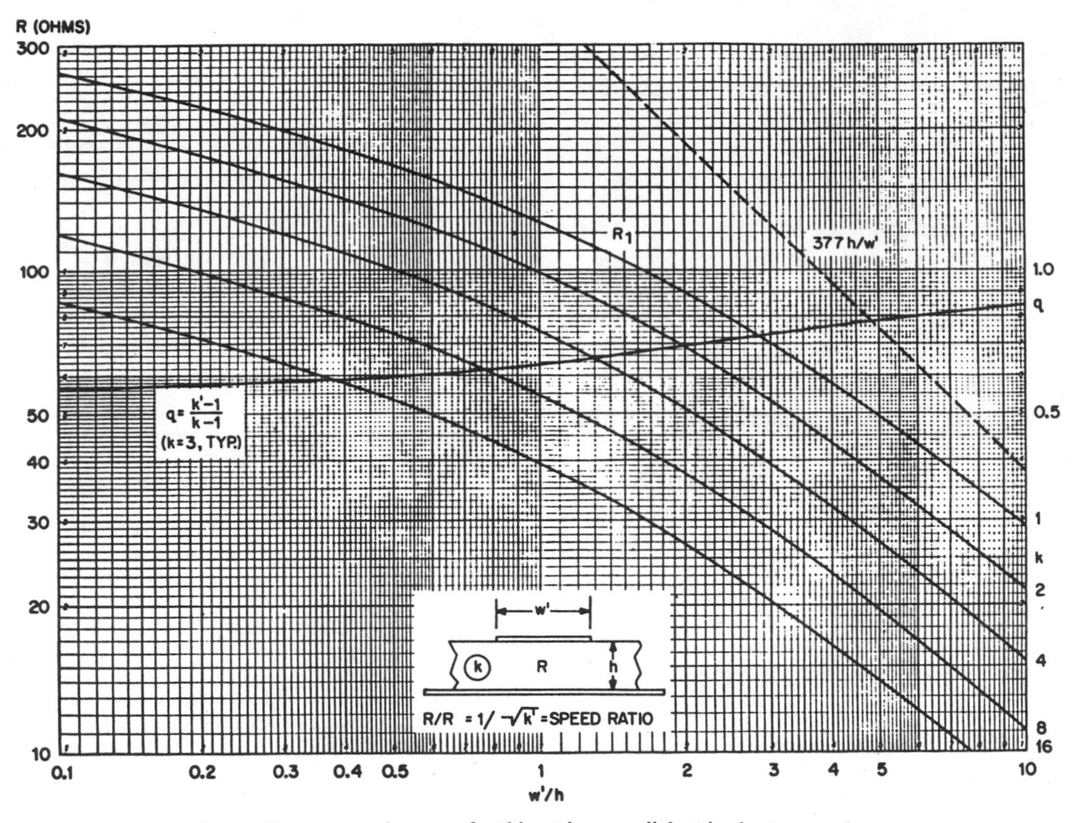

Fig. 2. The wave resistance of a thin strip on a dielectric sheet on a plane.

of practical design, and can be read close enough for ordinary purposes. The formulas to be given are intended as an alternative to the graphs, and also to give further insight into the relations. The formulas are designed for programming in a small digital calculator such as the HP-25 or HP-65.

Fig. 2 is a graph of the wave resistance of a thin strip, the same as previously published [2]. It is made with closest approximation by complete computation for overlapping ranges of narrow and wide strips. An alternative to this graph is the simple empirical formulas to be given for the entire range of width. The wave-speed ratio (relative to air or free space) for any width ratio is equal to the ratio of wave resistance with and without dielectric (R/R_1).

The effective filling fraction q of the dielectric is also graphed on Fig. 2 for a mean value of the dielectric constant ($k = 3$). It enables an alternative computation of the effective dielectric constant k' and the resulting speed ratio ($1/\sqrt{k'}$).

Fig. 3 is a graph of the thickness effect on the wave resistance without dielectric. The relative effect is less with dielectric, so the indicated effect is an upper bound. It is a small effect with respect to wave resistance but has a greater effect on the magnetic PF. This is generally similar to the first-order effect of thickness as previously stated [2] but is refined and extended to include the second-order effect in some degree.

Fig. 4 is a graph of the normalized magnetic PF ($P = p \div \delta/h$) as evaluated from the thickness effect. The magnetic PF is independent of the dielectric and its normalized value is independent of the size. The thickness parameter t/h is chosen as being a property of the laminate, specifically the thickness ratio of the conductive strip and the dielectric sheet.

New formulas are presented here in the main text without derivation. Most of them are empirical formulas providing a gradual transition between narrow and wide extremes. These are tested against the derived close approximations for overlapping narrow and wide ranges, which are reviewed in Appendix VI. Some derivations, not previously available, are given in Appendixes IV and V. Special emphasis is placed on some formulas which are "reversible" in the sense that a formula can be expressed explicitly in a simple form for either analysis or synthesis.

V. A Thin Strip Without Dielectric

The 1964 paper [1] gave the derivation for a wide thin strip without dielectric, and, incidentally, also gave formulas for a narrow thin strip. These together covered any width. Explicit formulas were given for both purposes, analysis and synthesis.

Recent studies yielded the discovery that the "narrow" formula could be put into a form which would also be asymptotic to the "wide" formula. This is accomplished while retaining its principal features for "narrow" approximation. Furthermore, this has been so arranged that the formula is "reversible." By this is meant that an explicit formula for either analysis or synthesis can be converted to an explicit formula for the other. This conversion is permitted no complication beyond the solution of a quadratic equation. The resulting formulas are empirical in the sense that they must be tested against derived formulas in the "wide" range and in the overlap of "wide" and "narrow." For

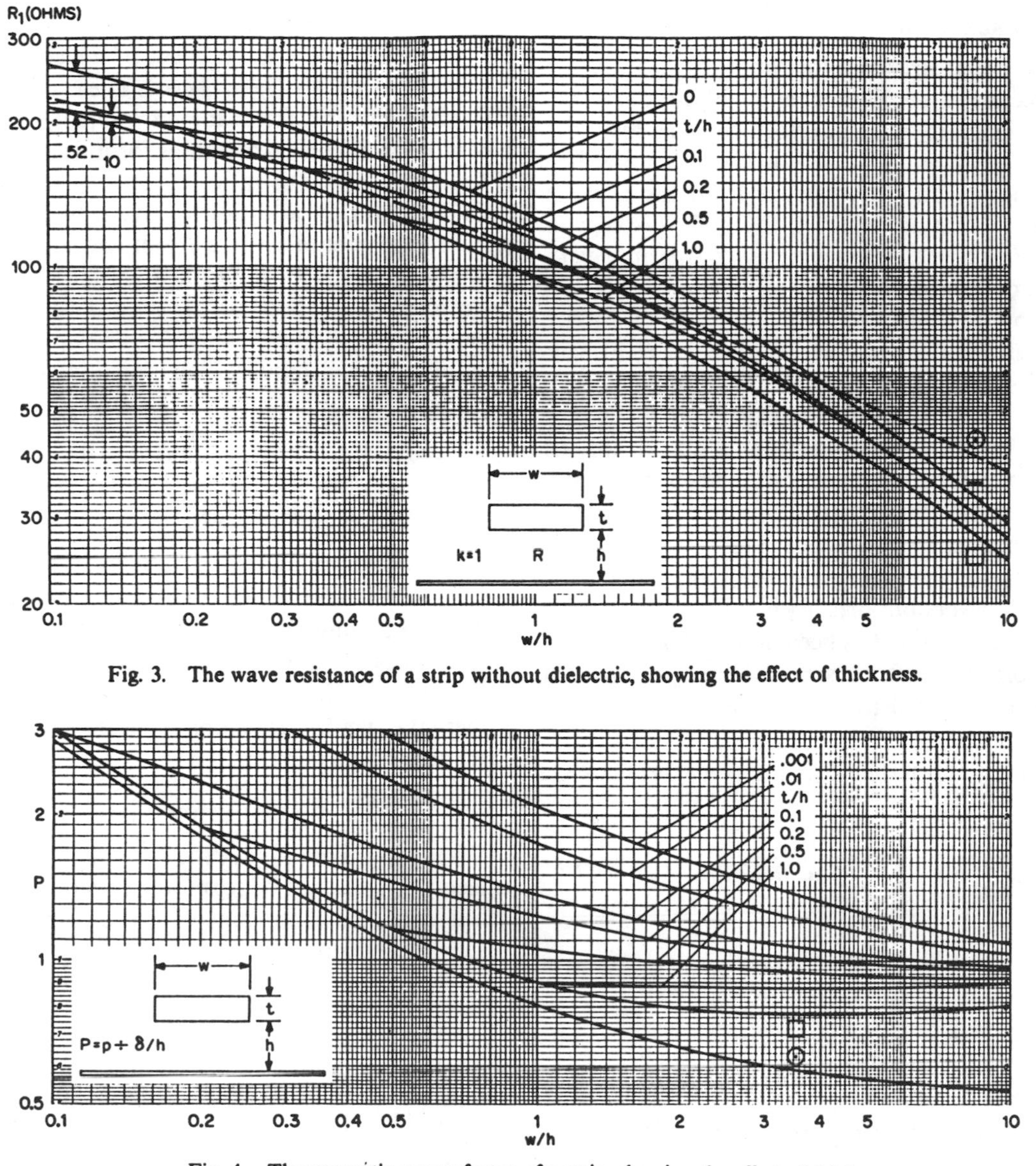

Fig. 3. The wave resistance of a strip without dielectric, showing the effect of thickness.

Fig. 4. The magnetic power factor of a strip, showing the effect of thickness.

the thin strip without dielectric, such derived formulas are the subject of the 1964 paper. The resulting formulas herein are based on "narrow" derivations which are relatively simple although the derivation of the second term has not been published before. It is based on a two-wire second-order approximation for a narrow thin strip. The overlap of the "narrow" and "wide" derivations is indicated, by stating the small error of either at the transition.

The new formulas are to be generalized for dielectric, but they are first given here in simplest form to show some features. Reversible formulas are here shown first for synthesis and then for analysis:

$$w'/h = 8\frac{\sqrt{(\exp R_1/30 - 1) + \pi^2/4}}{(\exp R_1/30 - 1)} \tag{1}$$

$$R_1 = 30 \ln \{1 + \tfrac{1}{2}(8h/w')[(8h/w') + \sqrt{(8h/w')^2 + \pi^2}]\} \tag{2}$$

where the error is $<0.01R_1$.

The following are the asymptotic forms for the narrow and wide extremes: narrow:

$$w'/h = 8(\exp - R_1/60)[1 + 1.73(\exp - R_1/60)^2]$$

$$R_1 = 60 \ln (8h/w' + 1.73(w'/8h)) \tag{3}$$

wide:

$$w'/h = 120\pi/R_1;\; R_1 = 120\pi(h/w'). \tag{4}$$

In the "narrow" formula, the second-order term has the proper form but its coefficient is compromised (1.73 instead of 2) to accomplish asymptotic "wide" behavior.

The asymptotic behavior at both extremes could be accomplished by any of several variants, yielding somewhat different behavior in the transition region. The form chosen was found to give close enough approximation with the minimum number of terms.

The form of these transitional approximate formulas shows some points of similarity to the exact formulas for a round wire near a plane, which are to be given here.

VI. Square or Circular Cross Section

As an extreme departure from the thin strip, a square or circular cross section is considered, still without dielectric. Fig. 1(b) shows a square or an inscribed circle as the cross section, with the description in the same terms as Fig. 1(a) ($t/w = 1$). It is noted that the distance from the plane is described by the separation height h, not by the distance to center (which is $h + t/2$). Hence it is compatible with separation by a dielectric sheet.

For the narrow case, simple formulations for a square wire and the equivalent round wire are known. See Appendix III. Fig. 1(c) shows this relation and the radius $(2h + w)$ of the outer circle equivalent to the plane.

For the wide case, the exact formula is known for the round wire but not for the square one. Therefore a close approximation for the square wire has been derived and is presented in Appendix IV.

For the round wire, the exact formula for any width ratio is known in simple reversible form. By modifying this form, a reversible empirical formula has been derived for the square shape. These formulas are presented here. (R_1 without dielectric is here simplified to R, because here there is no need for this distinction.)

For round wire without dielectric the exact formulas are as follows:

$$w/h = \frac{2}{\cosh R/60 - 1} = \frac{1}{(\sinh R/120)^2}$$

$$= \left(\frac{2}{\exp R/120 - \exp - R/120}\right)^2$$

$$= \frac{4}{\exp R/60 + \exp - R/60 - 2}$$

$$= \frac{4 \exp - R/60}{(1 - \exp - R/60)^2} = \frac{4 \exp R/60}{(\exp R/60 - 1)^2} \tag{5}$$

$$R = 60 \operatorname{acosh}(2h/w + 1) = 120 \operatorname{asinh} \sqrt{h/w}$$

$$= 60 \ln\left[(2h/w + 1) + \sqrt{(2h/w + 1)^2 - 1}\right]$$

$$= 120 \ln\left(\sqrt{h/w} + \sqrt{h/w + 1}\right). \tag{6}$$

For square wire without dielectric the approximate formulas are as follows:

$$w/h = \frac{1/0.59}{\exp R/60 - 0.2} \frac{2 + \exp - R/60}{1 - \exp - R/60}$$

$$= \frac{1}{0.118} \frac{2 + \exp - R/60}{5 \exp R/60 + \exp - R/60 - 6} \tag{7}$$

$$R = 60 \ln\left[\left(\frac{h}{0.59w} + 1.1\right) - 0.5 + \sqrt{\left(\frac{h}{0.59w} + 1.1\right)^2 - 1.05}\right]. \tag{8}$$

The relative error is < 0.025 or $< (0.005R + 0.5\,\Omega)$. If $R \to 0$, $w/h \to 381/R$ (near $377/R$). Each of these formulas is asymptotic in the first- and second-order terms for "narrow" and the first-order term for "wide."

These formulas are intended mainly for the magnetic-loss PF, for which there is no effect of dielectric, and only the "analysis" form (R of w/h) is used. The synthesis form (w/h of R) is shown mainly for academic interest, since it formed the basis for the empirical formulas for the square wire over the entire range of width ratio.

In Fig. 1(c), in addition to the equivalent circular and square cross sections, there are shown some equivalent thin strips. A round wire far from the plane has an equivalent concentric thin strip whose width is double the wire diameter ($2.36w$). If not so far from the plane, there is a thin strip of lesser width (w') which is equivalent by the following two tests:

a) height above the plane equal to that of the lower side of the square;
b) equal wave resistance.

The indicated geometric proportionality of the two strip widths is of interest in kind but not in degree, because their difference becomes substantial for a square so wide that the simple rules of equivalence are failing.

The lesser thin strip, compared with the square, determines the width adjustment here associated with the thickness of the square.

VII. A Thin Strip with Dielectric

The 1965 paper [2] gave the derivation for a thin strip with dielectric. Two sets of formulas covered wide and narrow strips, with close agreement in the transition region. The reversible formulas given above are here adapted to dielectric. Asymptotic behavior is achieved for the following conditions:

a) narrow strip, low-k and high-k extremes, with a logical interpolation therebetween;
b) wide strip, all k.

The resulting empirical formulas are found to track the derived formulas over the entire range of width and dielectric:

$$w'/h = 8 \frac{\sqrt{\left[\exp\left(\frac{R}{42.4}\sqrt{k+1}\right) - 1\right]\frac{7 + 4/k}{11} + \frac{1 + 1/k}{0.81}}}{\left[\exp\left(\frac{R}{42.4}\sqrt{k+1}\right) - 1\right]} \tag{9}$$

$$R = \frac{42.4}{\sqrt{k+1}} \ln\left\{1 + \left(\frac{4h}{w'}\right)\left[\left(\frac{14 + 8/k}{11}\right)\left(\frac{4h}{w'}\right) + \sqrt{\left(\frac{14 + 8/k}{11}\right)^2\left(\frac{4h}{w'}\right)^2 + \frac{1 + 1/k}{2}\pi^2}\right]\right\}. \tag{10}$$

The error is $< 0.02R$ (or $< 0.01R$ over most of range).

The analytic form gives R (for k) and R_1 (for $k = 1$), from which the speed ratio is

$$R/R_1 = 1/\sqrt{k'} < 1. \tag{11}$$

Therefore no other formula is needed for the speed ratio. The simpler formula (2) for R_1 may be used, but that is no advantage if the more general formula is recorded in a program for numerical computation.

While the effective filling fraction q [2] of the dielectric is not required in the procedures given here for design computations, it is a matter of some interest. Particularly, it is a factor in the electric-loss PF to be formulated. Schneider [14] has given an ingenious simple empirical formula, based on [2], which is close enough for practical purposes:

$$q = \frac{1}{2}\left(1 + \frac{1}{\sqrt{1 + 10h/w}}\right). \tag{12}$$

Compared with the derived formulas for narrow and wide ranges (for a mean value, $k = 3$) the departure is <0.02. It is a simple transition between the bounds $(\frac{1}{2},1)$. It lacks the shape that is peculiar to either extreme, which is contained in the derived formulas.

VIII. Strip Thickness and the Loss Power Factor

The earlier papers did not make any attempt to evaluate conductor loss, because it is not determined in the limit of a thin strip. However, there was given a width adjustment for the edge effect of a small thickness. From this adjustment, some other authors have formulated the losses to be expected, and their reduction by thickness [13], [17].

This subject has been reviewed. The width adjustment has been verified for small thicknesses of a narrow strip, and has been formulated more closely for a wide strip. A single formula is given here for the entire range of width. Also it is adapted to moderately large thicknesses (up to a square cross section for a narrow strip).

The loss PF (PF $= 1/Q$) of the magnetic field (bounded by the conductors) is evaluated by the rule proposed by the author in 1942 [3]. This "incremental-inductance rule" is based on differentiation of the inductance relative to the skin depth δ in the conductor boundaries, as indicated in Fig. 1. In a transmission line with perfect boundaries and no dielectric material, the inductance is proportional to the wave resistance. Only the relative change is significant, so the rule is here applied to the wave resistance R_1. This avoids the nuisance of magnetic units and surface resistivity.

A great simplification is now available by numerical differentiation. This was not available in the slide-rule computations of earlier days so analytical differentiation was necessary, however cumbersome. It was used by the other authors. It is no longer needed. What is needed is an analytic formula giving the wave resistance in terms of all dimensions but without dielectric.

The edge effect related to the strip thickness is here described in terms of the extra width Δw of a thin strip having equal wave resistance R_1 without dielectric. This is indicated in Fig. 1.

The first-order effect of a small thickness is given in the 1965 paper, for the extreme cases of narrow and wide strips. Three advances are here presented:

a) a refinement for the wide strip (Appendixes I and II),
b) a unified formula for the entire range of width,
c) a second approximation for greater thickness, within some restrictions.

The resulting formula is expressed in terms of the actual width w or the equivalent-thin-strip width w'. As mentioned above, these relations are based on free space, without dielectric:

$$\frac{\Delta w}{t} = \frac{1}{\pi} \ln \frac{4e}{\sqrt{\left(\frac{t}{h}\right)^2 + \left(\frac{1/\pi}{w/t + 1.10}\right)^2}} \tag{13}$$

or

$$\frac{1}{\pi} \ln \frac{4e}{\sqrt{\left(\frac{t}{h}\right)^2 + \left(\frac{1/\pi}{w'/t - 0.26}\right)^2}}. \tag{14}$$

This adjustment enables a width conversion either way between equivalent strips with or without thickness.

The development of this formula for the wide and intermediate regions has been enabled by complete computation of a few examples (Appendix II). These were accomplished by the technique of conformal mapping. Specifically, a few shapes (w,h,t) of rather small thickness were evaluated by numerical integration of the space gradient. This process is laborious and required some ingenuity near some bounds of integration.

Three examples so evaluated were sufficient to indicate two features implicit in this formula.

a) For a wide strip, the previous formula (1965) is refined in respect to its second-order effect. The ratio previously included as $2h/t$ is here changed to $4h/t$. The former ratio was based on unlimited width, and the change is an adaption to the limited width.
b) The "narrow" and "wide" formulas appear to be upper bounds, as would be expected. Furthermore, the quadratic sum of the two inverse ratios fits the sample points.

The adaptation of this formula for a greater thickness has been enabled by derivations for a square cross section. The extra numbers ($+1.10$ or -0.26) are chosen to match the square condition ($t = w$) for a narrow strip. The formula is a close approximation for moderate thicknesses ($t < h$) of a wide strip. (Another formula has been derived for a wide strip of square cross section, Appendix IV.)

For loss computation, the actual width and thickness (w,t) are converted to the width of an equivalent thin strip ($w' = w + \Delta w$). Then the thin-strip formula (R_1 of w') can be used for differentiation with respect to the actual dimensions (w,h,t).

As indicated in Fig. 1, each dimension is incremented by $\pm\delta$ and the same formula is used again to obtain R_δ. Then the (small) loss PF is computed by the incremental-inductance rule:

$$p = \frac{R_\delta - R_1}{R_\delta} = 1 - R_1/R_\delta = \ln R_\delta/R_1 \ll 1. \tag{15}$$

A normalized form for loss PF is proposed, which gives the effect of shape, independent of the size, frequency, and conductor material. It is normalized to the height h:

$$P = p \div (\delta/h) = p(h/\delta) \qquad p = P(\delta/h). \tag{16}$$

The reference (δ/h) is the nominal PF of a very wide strip.

In computing the normalized PF P, the value of the skin depth is immaterial if it is sufficiently small to approach the limiting behavior of the skin effect (which is usually of interest). Also it must not approach the sensitivity of the computer. In a computer giving ten decimal places, a fair compromise is $\delta/h = 0.0001$. Then the skin effect is well represented if all dimension ratios exceed 0.001.

For evaluation of a resonator made of a strip line, the loss PF (or dissipation factor or $1/Q$) is usually the most significant factor. The wave R is incidentally relevant in the circuit application of the resonator. The loss PF of the magnetic field is evaluated by the simplest formulas (R_1 and Δw without dielectric). For any shape, the value of P enables a computation of the size of the cross section to realize a value of p:

$$h = P\delta/p = P\delta Q. \tag{17}$$

The graphs in Fig. 4 show the loss PF in terms of P for a wide range of shapes. The common reference is the height h and the thickness ratio t/h because they may be fixed by a dielectric sheet and a conductive sheet bonded thereto.

For small thicknesses, the loss PF exceeds the reference value, as would be expected. Also the amount of excess is greater for lesser thickness, as a result of the current concentration at the edges. For example, reducing the thickness from square to $t/h = 0.02$ may double the PF (in the moderately narrow range).

An unexpected result is the loss PF being less than the reference value for a wide strip of substantial thickness. This happens because part of the magnetic energy is beyond the region bounded by the height. This part has boundaries further apart, and hence a lesser value of loss PF.

In Fig. 4, the two lowest curves give the loss PF for square and circular cross sections of the same width. It is less for the latter, the lower bound for the wide extreme being one half the reference value ($P \to \frac{1}{2}$). In the narrow region, it is less because of the following.

a) The two shapes are known to have equal skin resistance [9].
b) The circular shape has greater reactance. The proportionate wave resistance is greater by $60 \ln 1.18 = 10\,\Omega$; this is denoted, "the rule of 10 Ω."

If the thickness is comparable with the height, the relevant restriction may be the overall height $(h + t)$, perhaps for reasons of clearance space. Also the width may be restricted. Then the thickness ratio has an optimum value. This is found by minimizing a related normalized PF defined as follows:

$$P_{ht} = P\frac{h+t}{h} = P(1 + t/h) = (p/\delta)(h + t) \tag{18}$$

Square: w/h near 0.55, $\min P_{ht} = 1.65$ (19)

Circular: w/h near 0.50, $\min P_{ht} = 1.56$. (20)

Within specified bounds of the overall height and width (not less than the overall height), the optimum rectangular cross section is one which is bounded by these two dimensions and has a certain thickness ($t/h < 1$). The extreme optimum is a peculiar rounded shape bounded by these dimensions.

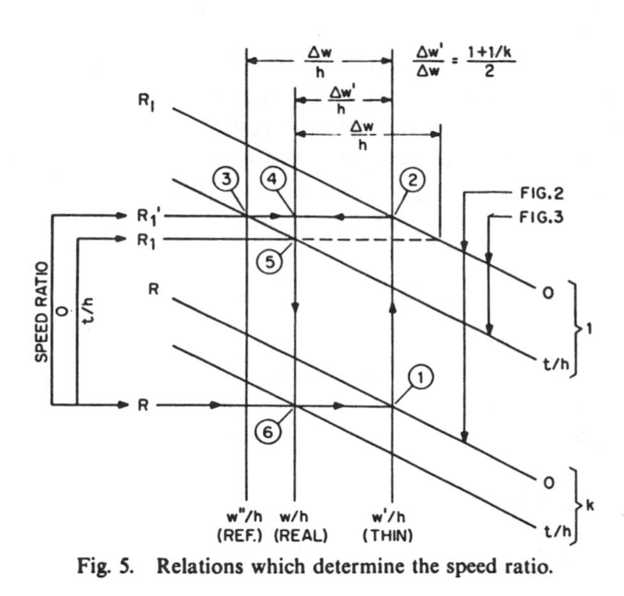

Fig. 5. Relations which determine the speed ratio.

IX. Strip Thickness with Dielectric

The effect of strip thickness is formulated above, but without the effect of dielectric. A width adjustment for thickness may be made in the synthesis for a specified wave resistance with dielectric.

The width of an equivalent thin strip is defined as one which is wider by the amount which yields an equal value of wave resistance. This involves both inductance and capacitance. The width adjustment for the former is independent of dielectric. That for the latter is less for a greater dielectric constant, because the thickness of the edge is somewhat spaced from the dielectric.

To approximate this effect, the width adjustment is divided in two equal parts, and one part is decreased by the factor $1/k$. The modified value becomes:

$$\Delta w' = \frac{1 + 1/k}{2}\Delta w, \qquad w = w' - \Delta w'. \tag{21}$$

The entire width adjustment Δw is effective for wave resistance without dielectric ($k = 1$) or for inductance alone. For capacitance alone, the entire width adjustment would be decreased by the factor $1/k$.

Fig. 5 shows the behavior of the width adjustment without or with dielectric. Especially it shows its graphic determination from Figs. 2 and 3. The full value of Δw is effective without dielectric, decreasing the wave resistance R_1 equally by decreasing inductance and by increasing capacitance. Its amount is represented by the horizontal separation of the upper pair of curves, both shown in Fig. 3. The effect of dielectric with a thin strip is represented by the separation of the upper curves (R_1,R) in the upper and lower pairs, both shown in Fig. 2. The reduced amount of width adjustment with dielectric is constructed and projected downward to give the lesser horizontal separation of the lower pair of curves.

The speed ratio for a thin strip is R/R_1 from Fig. 2. The interpolation for mixed dielectric, taking account of thickness, gives a greater speed ratio from this construction. The latter locates a point on the lower curve of the lower pair, not shown elsewhere.

As will be seen in a procedure and example to be given, the indicated numerical sequence yields all the quantities from the graphs in Figs. 2 and 3. Or they may be computed in this sequence by these formulas:

Sequence	Formulas
1	(9)
2	(2)
3	(14)
4	(21)
5	(13) (2)
6	(—)

In this sequence, 2–3 is the width adjustment downward from the upperbound for a thin strip. A parallel dashed line shows also the width adjustment upward from a strip with thickness. The former will be used in a synthesis procedure, the latter in analysis. The amount of the adjustment is designated alike in both ($\Delta w/h$) although it may differ slightly (too little for any practical significance).

X. ATTENUATION

The rate of attenuation with distance in a transmission line is simply expressed in terms of the average PF (magnetic p and electric p') and the wavelength λ_g in the line:

$$\alpha = \frac{p+p'}{2}\frac{2\pi}{\lambda_g} = \frac{p+p'}{2}\sqrt{k'}\frac{2\pi}{\lambda_0} = \frac{p+p'}{2}\frac{R_1}{R}\frac{2\pi}{\lambda_0} \text{ (Np/m).} \tag{22}$$

In words, this is the average PF (nepers) per radian length. The magnetic PF is evaluated by the skin effect, as described above.

The electric PF p' of the effective dielectric in the line (k') can be expressed in terms of the various parameters involved:

speed ratio:

$$\lambda_g/\lambda_0 = 1/\sqrt{k'} = R/R_1 \tag{23}$$

filling fraction:

$$q = \frac{k'-1}{k-1} = \frac{1}{k-1}[(R_1/R)^2 - 1] \tag{24}$$

electric PF:

$$p' = \frac{p_k}{1 + \dfrac{1/q - 1}{k}} \tag{25}$$

bounds:

$$p_k > p' > \left\{\begin{matrix} qp_k \\ \dfrac{p_k}{1+1/k} \end{matrix}\right\} > \tfrac{1}{2}p_k. \tag{26}$$

The electric PF p' is seen to be within $(\frac{3}{4} \pm \frac{1}{4})$ of the dielectric-material PF p_k, and usually it is nearer the upper bound. Therefore the electric PF is only slightly less than that of the material, so the complete formulation is not critical and may be unnecessary. If desired, it can be computed (as above) from R/R_1 and p_k.

Either attenuation or PF may be deduced from the other. However, it is preferable to evaluate the magnetic PF directly, because it is independent of the dielectric and the speed ratio. In a wide range of situations, it represents nearly all of the loss PF.

XI. PROCEDURES FOR COMPUTATION

The formulas are intended for useful applications, which may be theoretical and/or practical. As brought out in the earlier papers, "synthesis" and "analysis" are the alternative objectives, the former for practical design and the latter for evaluation of a configuration (the classical textbook approach). Both are needed here for a practical design to meet some specifications and tests. Therefore a few procedures and examples will be outlined to show the use of these formulas in arriving at a practical design.

The first few procedures start with the synthesis of a line to meet a specification of wave resistance. The subsequent evaluation of speed ratio and skin effect are inherently analysis, but the procedures build on the synthesis.

First Procedure: On a specified printed-circuit board, find the width for a 50-Ω line:

a) specify properties of a dielectric sheet with metal faces: $k = 2.5$, $h = 1$ mm, $t = 0.1$ mm;
b) specify wave resistance: $R = 50\ \Omega$;
c) width of thin strip by (9) or Fig. 2: $w'/h = 2.85$;
d) width adjustment (without dielectric) by (14) or Fig. 3: $\Delta w/h = 0.15$;
e) effect of dielectric by (21): $\Delta w'/h = 0.10$;
f) width by (21): $w/h = w'/h - \Delta w'/h = 2.75$; $w = 2.75$ mm.

Second Procedure: For the same line, evaluate the speed ratio, referring to Fig. 5:

g) no. 1 in sequence, c) above: $w'/h = 2.85$;
h) no. 2, find R_1' of thin strip by (2) or Fig. 2 or 3: $R_1' = 71.5$;
i) no. 3, d) above: $\Delta w/h = 0.15$, $w''/h = w'/h - \Delta w/h = 2.70$;
j) no. 4, e) above: $\Delta w'/h = 0.10$, $w/h = w'/h - \Delta w'/h = 2.75$;
k) no. 5, by Fig. 3: $R_1 = 71$, or can be computed by the "fourth procedure";
l) speed ratio $= R/R_1 = 50/71 = 0.70$.

Third Procedure: For the same line, evaluate the magnetic PF and the attenuation from this cause, referring to Fig. 4, Appendixes VII and VIII:

m) find the normalized PF by Fig. 4: $P = 1.10$; or it may be computed by (16) using a nominal small δ and numerical differentiation;
n) specify the frequency (or wavelength λ_0): $f = 1$ GHz, $\lambda_0 = 0.3$ m;

o) specify the conductivity (or material) of the metal boundaries: copper;
p) evaluate the skin depth by (62) or [7]: $\delta = 2.1$ μm;
q) compute the PF by (16): $p = 0.0023 = 2.3$ mil, $Q = 440$;
r) compute the attenuation rate from PF, speed ratio, etc., by (22): $\alpha = 0.034$ Np/m or 0.30 dB/m.

If used for a long line, the attenuation rate may be significant. If used for a resonator, the PF and speed ratio are relevant.

In the third procedure, if one is concerned with only one example (size, shape, materials, frequency) the actual skin depth may be used directly, then the procedure assumes this order: (n,o,p) (m,q,r).

A lesser PF may be required, or it may be desired to explore the compromise between the loss PF and the height and/or thickness. The first-order relation gives the PF inversely proportional to size (h,t,w). A closer evaluation may require complete computation of various examples, then interpolation.

The following example and procedure are modified to develop from analysis only. In particular, the width adjustment corresponds to the dashed line in Fig. 5.

Another Example: Design a resonator to be made of a square wire bonded to a printed-circuit board. Similarly lettered items refer to the foregoing procedures:

a) $k = 2.5$, $h = 1$ mm, $t = w$;
m) from Fig. 4: near-minimum $P = 0.8$ for $w/h = 2$, $w = t = 2$ mm;
p) $\delta = 2.1$ μm;
q) $p = 0.0017 = 1.7$ mil, $Q = 590$.

The speed ratio can be evaluated by the following procedure. It is found to be $50/67 = 0.75$.

Fourth Procedure: For any configuration, find the speed ratio. For a thin strip, see (11) and Fig. 2 for the simple rule. The following gives the effect of thickness:

a) specify configuration: $k = 2.5$, $w = 2.75$ mm, $h = 1$ mm, $t = 0.1$ mm, $w/h = 2.75$, $t/h = 0.1$;
b) width adjustment (without dielectric) by (14) or Fig. 3: $\Delta w/h = 0.15$, $w'/h = 2.90$;
c) wave resistance (without dielectric) by (2) or Fig. 2 or 3: $R_1 = 71$;
d) effect of dielectric by (21): $\Delta w'/h = 0.7$, $\Delta w/h = 0.10$, $w'/h = 2.85$;
e) wave resistance (with dielectric) by (10) or Fig. 2: $R = 50$;
f) speed ratio: $R/R_1 = 0.70$.

The speed ratio is slightly greater than that for a thin strip of the same width.

If resonance (small PF or high Q) is the principal objective (rather than wave resistance) a different procedure may be indicated. The following outline gives some relevant considerations.

a) Choose between a specified printed-circuit material (h,t) and the alternative of an attached thick strip (which may have a square or circular cross section). The latter offers a lesser PF.
b) If a thick strip is to be afforded, specify the bounds of the space (overall height and width, $h + t$ and w).
c) Specify whether the conductor (strip or whatever) is to be supported in contact with a dielectric sheet. If so, specify the height of the latter (h).
d) Subject to these restrictions, choose a cross section giving near-minimum P_{ht} (18).
e) If using a strip of small thickness t, a lesser PF is obtainable by greater width w and greater height h.
f) If using a square cross section in contact with a dielectric sheet $(t/h = w/h)$, the least PF is obtainable by a moderately wide shape (say w/h near 3).
g) If using a round wire in contact, a lesser PF is obtainable by greater width (diameter), but little reduction is obtainable beyond a moderate width (say w/h near 3).
h) If using a square or round wire with no need for contact, the least PF is obtainable by a width near one third the overall height.
i) If a rectangular space is specified, with the width not less than the overall height, the least PF obtainable with a rectangular cross section requires some thickness less than one third the overall height.

There is usually not available an explicit formula for the synthesis to realize a specified value of the loss PF. The graphs in Fig. 4 can be applied to this problem. Knowing the skin depth δ and specifying the material (h,t), a value of the loss PF p requires the P computed from (16). In Fig. 4, this value of P determines the shape ratio w/h and hence the width w. If this P is lower than a practical curve, the size (h,t) may be increased to permit a greater value of P.

XII. Conclusion

The transmission-line properties of a strip parallel to a plane, with or without an intervening dielectric sheet, are evaluated in simple formulas, each one adapted for all shape ratios. The formulas relating the width/height ratio with wave resistance are stated explicitly for both analysis and synthesis, with or without dielectric. The wave-speed ratio and the magnetic-loss PF are stated from the viewpoint of analysis, which is usually what is needed.

The advance over previous publications appears mainly in two areas:

a) a relation is expressed explicitly by a single simple formula for the entire range of the shape ratio;
b) the width adjustment for thickness is formulated and used for evaluation of the magnetic loss.

Each formula is an empirical relation obtained by designing a gradual transition between known simple formulas for both extremes of narrow and wide shapes.

All formulas are designed for ease of programming on a pocket calculator such as the HP-25 or HP-65. Particularly, the digital calculator enables the numerical differentiation (for loss evaluation) which is here used to realize a great simplification. While beyond the scope of this article, the writer would welcome inquiries relating to programs for the HP-25, some of which may be available on request.

The subject line, formed by a strip parallel to a plane, has presented problems of evaluation which are much more difficult than those of the strip between two planes (sandwich line). That configuration is symmetrical and the dielectric is homogeneous, so even the thickness effects have yielded to straightforward formulation [8]. The asymmetrical strip line is here formulated in a manner that is competitive, although necessarily involving mixed dielectric.

While there is always room for further progress, the graphs and formulas presented here are complete in that they offer the option of graphical or numerical reading for the all numerical values that may be needed for design purposes. Preliminary estimating is usually aided most by the graphs.

XIII. Acknowledgment

This study has been stimulated by the attempts of other authors, building on the writer's early papers. The stimulation has come partly from a perception of some deficiencies in progress, but more from an appreciation of the constructive efforts and interesting results of a few of the intervening workers. The final impetus was provided by the advent of the HP-25, which offered the computational power best suited for the "close support" essential to such a development.

Appendix I
Behavior of the Width Adjustment for Thickness

Formulas (13) and (14) for the width adjustment are based on some asymptotic relations and a transition therebetween. Asymptotic formulas for narrow and wide extremes were given in the early papers [1], [2]. Here a revision of the wide formula and an integrated formula with a simple form of transition were presented. This appendix is a graphical description of the behavior of this adjustment, for the purpose of visualizing the transition and some associated relations.

Fig. 6(a) shows a graph for a constant ratio of thickness/height (t/h). This may be the practical situation when designing for a printed circuit to be made by etching a conductive sheet bonded to a dielectric sheet. The width adjustment ratio $\pi\Delta w/t$ is plotted on the width ratio w/h. The scales are, respectively, linear and logarithmic, to give straight lines for the sloping graphs.

The normalized form for the width adjustment takes out the principal dependence on thickness, so one can see the variations of the coefficient which is dependent on shape. A higher value indicates a greater coefficient (responsive to thinness) but the amount of the adjustment is still nearly proportional to thickness.

There are two upper bounds (UB's) for this coefficient, based on the narrow and wide asymptotic behavior. The level upper line is based on the wide extreme, the edge-field pattern being influenced mainly by the proximity of the shield plane. The sloping lower line is based on the narrow extreme, the edge-field pattern being influenced mainly by the proximity of the two edges.

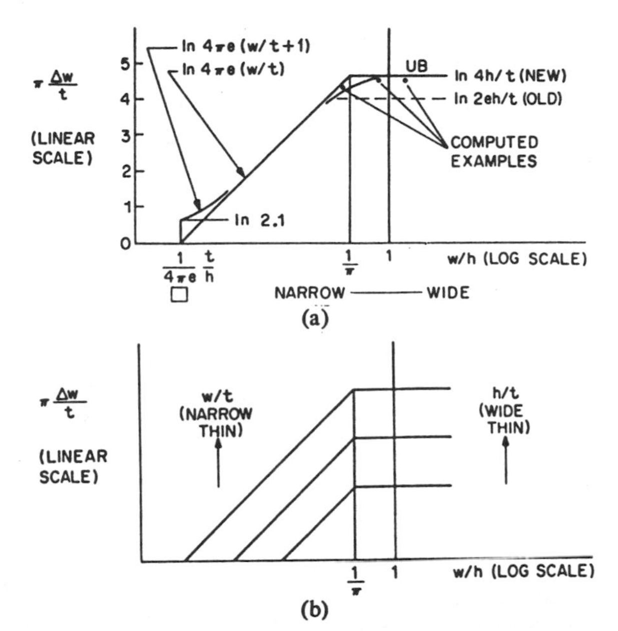

Fig. 6. The behavior of the width adjustment for thickness. (a) Transition between narrow and wide. (b) Family of transitions.

A smooth transition at the knee is provided by a quadrature combination in formula (13). This is validated by some computations to be described in Appendix II, indicated as three points on the curve. This validation requires a change in the wide formula, from the "old" in [2] to the "new" in (13). The computed points indicated that the level UB should be raised by a factor of two under the logarithm, as seen. (This factor is not exactly determined, but two appears to be the nearest and simplest number that might be indicated, and it may have an exact basis.) This is regarded as a refinement of the previous rule, whose derivation ignored the second-order interaction between the edges far apart. It is noted that the transition occurs in the vicinity of a width ratio somewhat less than unity ($w/h = 1/\pi$).

The asymptotic relations are based on the limiting condition of a thin strip. Formula (13) includes an adaptation ($w/t + 1.1$) which extends the close approximation to the square condition. This introduces another curved transition at the foot of the graph, raising the curve from the "square" point ($t/w = 1$). While beyond the present scope, it is noted that the curve has a minimum near the foot and approaches a higher level (π) in the narrow extreme ($w/t \ll 1$).

Fig. 6(b) is a diagram showing a family of such graphs. For greater thickness, the knee is closer to the foot of the graph, so the two curved transitions would merge as the thickness approaches the square shape. Then their separate descriptions become indefinite, so the validity of formula (13) is further tested on square shapes, as evaluated in Appendixes III and IV.

Appendix II
Small-Thickness Examples by Conformal Mapping

Formula (13) gives the width adjustment for thickness. It is an empirical transition between the narrow and wide

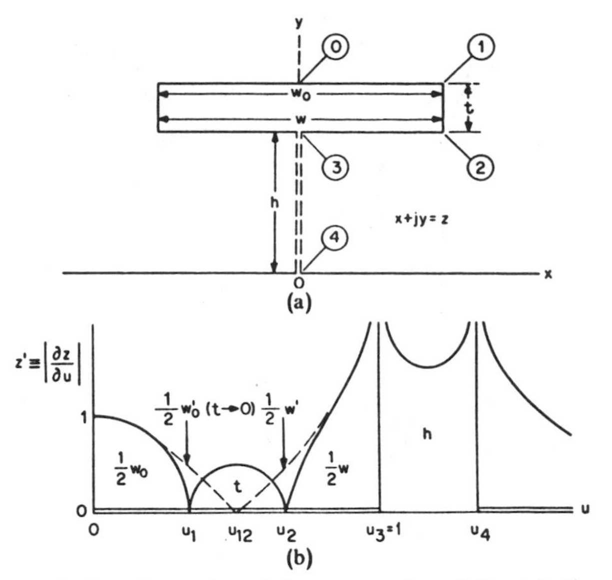

Fig. 7. Conformal mapping of the cross section of the strip line. (a) Contour in space. (b) Space gradient. (Each area equals dimension in space.)

extremes of the asymptotic behavior in the limit of a thin strip. The first-order relations for these extremes have been known [2] but not the behavior in the transition regions. Also there is found a second-order effect requiring a revision for the wide extreme.

The validation of this formula, especially in the transition region, is provided by some complete computations of a few examples by a procedure based on conformal mapping. The contour of the cross section is mapped on a straight line. The space gradient on this line is integrated to evaluate the dimensional ratios on the contour. Rather than implicit elliptic integrals, numerical integration is used. Even that is confronted by difficulties of integration where there is an infinite value and/or slope at either bound ($\infty, \sqrt{\infty}$, or $\sqrt{0}$).

Fig. 7 shows the essentials of the conformal mapping of the cross section of the strip line. The actual contour, Fig. 7(a), is described on the space plane ($x + jy = z$) in terms of the shape dimensions w,h,t whose ratios determine the properties. This contour is mapped on a straight line, Fig. 7(b). On the scale of this line u is graphed the space gradient (or inverse field gradient) on the contour. The area under the space gradient in each interval is equal to the dimension on the contour.

The space gradient is formulated by inspection, as follows:

$$Z' = |\partial z/\partial u| = \left| \frac{[1-(u/u_1)^2][1-(u/u_2)^2]}{[1-(u/u_3)^2][1-(u/u_4)^2]} \right|^{1/2}. \quad (27)$$

Only the area ratios are significant, so the scale is arbitrarily chosen for simplicity.

The analytic integration would involve elliptic integrals. There is a constraint that precludes an explicit solution. The upper and lower faces of the strip must have equal width ($w_0 = w$), to be realized by proportioning one of the critical values on the straight line.

Numerical integration is simple in concept and has been found useful in computing a few examples. Some special rules have been devised for closer convergence near the singular points which correspond to the angles of the contour. The result is a close approximation in cases where the singular points are not too closely spaced.

TABLE I
COMPUTED EXAMPLES

No.	1	2	3
R_1	188.5	133.7	88.5
u_4	1.414	1.1	1.01
u_{12}	0.7368	0.7774	0.8412
u_2	0.8867	0.8720	0.8988
u_1	0.5267	0.6520	0.7648
w'/h	0.348	0.879	2.020
w/h	0.237	0 .754	1.894
t/h	0.0755	0.0816	0.0838
$\Delta w/h$	0.111	0.125	0.126
$\Delta w/t$	1.48	1.53	1.51
(14)	1.47	1.55	1.56
dif.	+.01	-.02	-.05

The wave resistance is determined by the gaps in the straight line, both sides of center. For the upper half-plane,

$$R = \tfrac{1}{2} R_c \frac{K'(k)}{K(k)} = \frac{\left(\dfrac{1 + 1.14k^2}{\pi} \ln \dfrac{16}{k^2}\right)^{1-k^2}}{\left(\dfrac{1 + 1.14(1-k^2)}{\pi} \ln \dfrac{16}{1-k^2}\right)^{k^2}} \quad (28)$$

in which $k = 1/u_4$.

The latter (empirical) formula has a relative error < 0.005. It has the correct center value, skew symmetry, and asymptotic behavior at both extremes. A closer simple formula for a wide strip is

$$R = \frac{60\pi^2}{\ln \dfrac{8}{\ln 1/k}} = \frac{60\pi^2}{\ln \dfrac{8}{\ln u_4}}. \quad (29)$$

If $1/k = u_4 < 1.4$, the relative error is $< 0.001R$.

Three examples have been computed. They are summarized in Table I, numbered in order of increasing width. The first is a critical shape ($R_1 = 377/2$) while the others are chosen to give a range of widths in the transition region. These three examples have comparable values of the thickness ratio (t/h near 0.08). This ratio is small enough to be representative of small thicknesses ($t/h \ll 1$ and $w/h \ll 1$). Its value is the basis for the graph in Fig. 6(a), and the three points are plotted in relation to the curve of formula (13). The close agreement is regarded as confirmation of that formula (for small thicknesses), especially in the transition region which does not have a clear theoretical basis. This result was the objective of the complete computation of these few examples.

APPENDIX III
EQUIVALENT SQUARE AND CIRCULAR CROSS SECTIONS

The extreme thickness of a narrow strip line is taken to be a square cross section ($t/w = 1$). Therefore the formula for width adjustment contains a constant which assures a close approximation up to this thickness. Its derivation is based on the relations among three equivalent concentric cross sections, the square, the circle, and the thin strip, shown to scale in Fig. 1(c). Their dimensional ratios are such as to give equal values of capacitance and inductance (assuming a small skin depth) and the resulting wave resistance (all in free space).

Starting with the square (of width $= w$), the equivalent circle has a diameter which is greater in the ratio:

$$\sqrt{9/2\pi}\,\frac{\Gamma(5/4)}{\Gamma(7/4)} = 1.1803 = 1/0.8472. \tag{30}$$

The equivalent strip has a width which is double this diameter.

In Fig. 1(c), the large dashed arc is the circular boundary equivalent to the ground plane (radius $= 2h + w$).

Based on these equivalents, a narrow strip of square cross section has the following wave resistance:

$$R = 60 \ln \frac{2h + w}{0.59w} = 60 \ln 1.70(2h/w + 1) = 60 \ln 8h/w' \tag{31}$$

in which

$$w' = \frac{2.36w}{1 + w/2h}.$$

In terms of width and thickness, the corresponding adjustment is

$$\Delta w = w' - w = w\left(\frac{2.36}{1 + t/2h} - 1\right) = w'\left(1 - \frac{1 + t/2h}{2.36}\right). \tag{32}$$

In formula (13) or (14) for Δw, the constant $+1.1$ or -0.26 is inserted to give the correct value for a narrow strip of square cross section.

From another viewpoint, Fig. 1(b) shows cross sections of a square and an inscribed circle (having equal width). In the narrow case, these have wave resistances differing by

$$60 \ln 1.18 = 9.93 \text{ (near } 10\ \Omega). \tag{33}$$

This is denoted, the "10-Ω rule" for these two cross sections. They are known to have equal skin resistances [9] so the loss PF of the circular wire is less in the inverse ratio of its greater reactance and wave resistance.

Appendix IV
Wide Square Cross Section

The wide square cross section is here evaluated in simple terms by invoking a variety of techniques in four regions of each of the two active quadrants. These regions are described in Fig. 8(a). Each is to be evaluated first in terms of normalized capacitance C, which is the simplest concept for the boundaries involved. The dimensions are referred to the height ($h = 1$).

The first region, 1, is taken to be filled with uniform field:

$$C_1 = w/2. \tag{34}$$

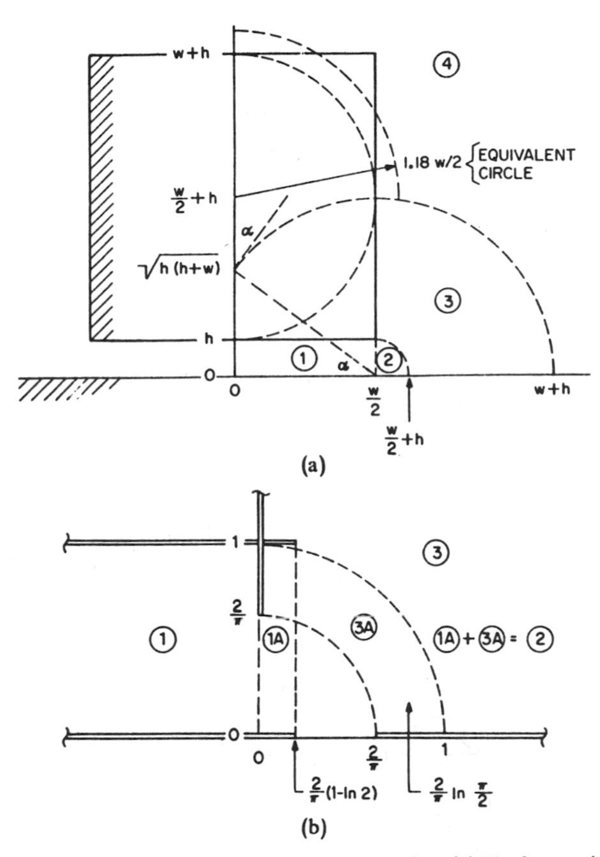

Fig. 8. Derivation for a wide square cross section. (a) The four regions in one quadrant. (b) Analysis of the corner region.

The third region, 3, is taken to be filled with logarithmic field, which is described by radial lines and concentric circles:

$$C_3 = \frac{2}{\pi} \ln (1 + w/2). \tag{35}$$

The second region, 2, is the excess in the transition between 1 and 3, taking into account the nearby distortion in both of those regions. The upper and lower boundaries (corner and straight line) are mapped on parallel straight lines. Then the relative displacement of far points evaluates a "stretch" which represents the excess in the transition. This is divided in two parts for the two directions from the corner. Fig. 8(b) shows this result diagrammatically. Region 2 is represented by 1A and 3A, the respective extensions of the adjacent regions. The validity of this viewpoint resides in the fact that the distortion from the transition decays rapidly in either direction, and also tends to average out. The resulting value of the transition region is

$$C_2 = \frac{2}{\pi}(1 - \ln 2) + \frac{2}{\pi} \ln \frac{\pi}{2}$$
$$= \frac{2}{\pi}\left(1 - \ln \frac{4}{\pi}\right) = 0.483. \tag{36}$$

The fourth region, 4, is closely related to an inscribed circle, as shown. The region around the inscribed circle would contribute

$$(C_4) = \frac{\alpha}{\pi} \frac{\pi}{\operatorname{acosh} (1 + 2/w)} = \frac{\operatorname{asin} \dfrac{1}{\sqrt{w+1}}}{\operatorname{asinh} \sqrt{1/w}} < 1. \tag{37}$$

For ease of computation, this ratio can be approximated by

$$(C_4) = \left(\frac{1}{1 + 2/w}\right)^{1/3} < 1. \tag{38}$$

Between the inscribed circle and the equivalent circle (of 1.18 times the radius) the nominal capacitance is that of one quadrant:

$$[C_4] = \frac{\pi/2}{\ln 1.18} = 9.49 = 1/0.105. \tag{39}$$

This is used to increase (C_3) to approximate the capacitance of the square in this quadrant:

$$C_4 = \frac{1}{1/(C_4) - 1/[C_4]} < 1.12. \tag{40}$$

The resulting wave resistance of the two quadrants (restoring w to w/h) is

$$R = \frac{377}{2C_1 + 2C_2 + 2C_3 + 2C_4}$$

$$= \frac{377}{w/h + 0.966 + \frac{4}{\pi}\ln(1 + w/2h) + \frac{2}{(1 + 2h/w)^{1/3} - 0.1}}. \tag{41}$$

This formula is best for a wide square cross section. The best for narrow is formula (31) for the equivalent round wire shown in Fig. 1(c). Their effective overlap is indicated by their close values for a transition shape ($w/h = 1$):

1) wide (41) above: 94.91;
2) narrow (31): 95.11 (close lower bound);
3) all (13) (2): 95.32.

The intermediate value is believed to be the closest approximation for this case.

Appendix V
Narrow Thin Strip

For a narrow thin strip (without dielectric) there is here derived the second-order approximation stated without proof in the early papers [1], [2]. It is based on a pair of small wires equivalent to the strip. It forms the basis for the simple formulas (1), (2) for any width.

Fig. 9(a) shows a single round wire of unit radius and its known equivalent thin strip whose width is 4 units. It is described on the z plane. It is to be transformed to another plane, $z' = \sqrt{z}$.

This transformation is here performed about one end of the strip cross section, and the result is seen in Fig. 9(b). An equal strip survives but the wire becomes a pair of smaller wires. This pair provides a second-order approximation to the far field of the strip. (This simple equivalence has not been seen by the author in any of the many published exercises in conformal mapping.) It is noted that the smaller wires are not strictly circular in cross section, but that is irrelevant in the use of the concept herein.

Fig. 9(c) shows the thin strip (or equivalent pair of wires) and its image in a ground plane. From this geometry, the mean distance between one pair of current centers and the other pair is increased to

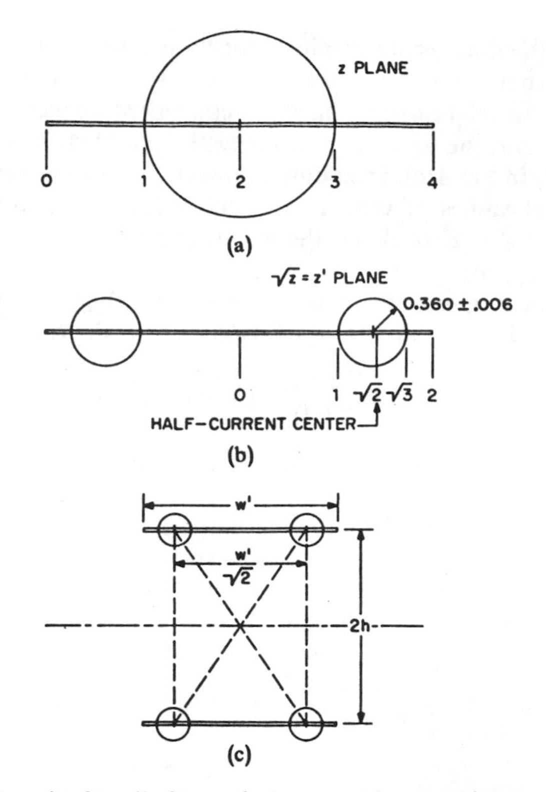

Fig. 9. The pair of small wires equivalent to a thin strip. (a) Thin strip and equivalent round wire. (b) Thin strip and equivalent pair of round wires. (c) Thin strip and its image in a ground plane.

$$2h' = \sqrt{2h\sqrt{(2h)^2 + w^2/2}} = 2h[1 + \tfrac{1}{8}(w/h)^2]^{1/4}$$
$$= 2h[1 + \tfrac{1}{32}(w/h)^2 - \cdots]. \tag{42}$$

The narrow-strip formula becomes

$$R = 60 \ln 8h'/w$$
$$= 30 \ln [(8h/w)\sqrt{(8h/w)^2 + 8}]. \tag{43}$$

This is a reversible formula, giving the following for synthesis:

$$w/h = 8\frac{\sqrt{\sqrt{\exp R/15 + 16} + 4}}{\exp R/30}. \tag{44}$$

For a narrow strip, the 4 term is of the second order ($\exp R/30$) and the 16 is of the fourth order ($\exp R/15$) relative to the first order ($\exp R/60$).

A modification of the above formula gives a simpler form which has a linear slope for $R \to 0$, while retaining the second-order approximation:

$$w/h = 8\frac{\sqrt{\exp R/30 + 2}}{\exp R/30 - 1}. \tag{45}$$

In the limit, $w/h \to (\sqrt{3})240/R = 416/R$.

This is not far from the desired $377/R$. The latter result can be obtained by substituting a slightly lesser value ($\pi^2/4 - 1 = 1.467$) for the constant 2. Asymptotic behavior for a wide strip is then realized at the cost of a slight deficiency in the second-order term. The result is the simple formula (1) giving a close approximation for any width.

The two preceding formulas are extremely close (the relative difference is <0.0002) in the narrow range ($w/h < 2$). The first term of error is proportional to $(w/h)^3$. Their average has a relative error <0.003 if $w/h < 4$. The second formula is closer for greater widths, and therefore probably for all widths. The corresponding formula for analysis is (2) except change π^2 to 12.

This exercise is a striking example of the technique of higher order approximation and its application to obtain a simple and versatile empirical formula with support from various theoretical relations.

Appendix VI
Previously Derived Formulas

Any empirical formula must be validated by comparison with derived formulas. These are typically more complicated and/or restricted as to the range of the width ratio.

Here are some such formulas for a thin strip, selected from the earlier papers, (A-#) referring to the first [1] and (B-#) referring to the second [2]. They are stated in a form that is convenient for computation, to provide a comparison test for the more recent empirical formulas covering the entire range of the width ratio. They are converted to the dimensions used herein (w, h, etc.).

Every one of the derived formulas is essentially the first few terms of a series converging in the extreme of a narrow or wide shape (and small thickness). Therefore a "narrow" or "wide" identity is necessary. The transition between the two occurs for a shape which may be "borderline" for close approximation by either. Hopefully the two kinds will overlap to give a coverage for all shapes. As stated in the early papers and as supported by more recent studies, the transition occurs near $w/h = 1$. The more sophisticated formulas give substantial overlap.

For a narrow thin strip without dielectric, the second-order approximation was not supported by a derivation. One is given here in Appendix VI. Formula (45) and the corresponding modification of (2) are presented as the closest approximation known to date. It provides overlap of the wide range. The relative error is $<0.003R$ if $w/h < 2$.

For a wide thin strip without dielectric, the first paper yields a remarkably close approximation with overlap of the narrow range. The synthesis form is an explicit formulation. Specify

$$R_1 < 60\pi = 188$$

(A-1), (A-45)
$$d' = \frac{\pi}{2} R_c/R_1 = 592/R_1 > \pi \tag{46}$$

(A-67)
$$d = d' + (2d')^2 \exp - (2d') > \pi \tag{47}$$

(A-10)
$$c = \sqrt{(d-1)^2 - 1} = \sqrt{d(2d-1)} \tag{48}$$

(A-68)
$$w/h = \frac{2}{\pi}[c - \operatorname{acosh}(d-1)]$$
$$= \frac{2}{\pi}[c - \ln(c + d - 1)] > 0.3. \tag{49}$$

The relative error is $<0.001R_1$.

The two preceding approximations give a large overlap. They are closest near $R_1 = 126$ or $w/h = 1$, where the relative difference is $0.0005R_1$. For the graphs in Fig. 2, the computation of any one point is made with the formula judged to be the closer of the two; if so, its relative error is less than this amount.

A comparison of these two formulas can be made in explicit form by the following sequence:

1) wide (49): w/h from R_{1w};
2) narrow (2) (modified): R_{1n} from w/h;
3) ratio: $R_{1n}/R_{1w} = 1 +$ relative difference. (50)

The relative difference of R_1 is the significant comparison.

A thin strip with dielectric likewise has different formulations for narrow and wide. The effective dielectric constant k' depends on the shape w/h and on the dielectric k. One sequence can be used for explicit formulations in any case:

1) specify: R_1;
2) compute (45): w/h;
3) specify: k;
4) compute (53), (57), (52): q, k', $R = R_1/\sqrt{k'}$;
5) graph: R for k, w/h. (51)

The "effective filling fraction" [2], defined as follows, depends mainly on the shape and less on the dielectric:

$$q = \frac{k'-1}{k-1}, \qquad k' = 1 + q(k-1). \tag{52}$$

Because it has only second-order dependence on the dielectric, a simple formula for a mean value of k is sufficient for practical purposes. A mean value ($k = 3$) is chosen because it places the effective dielectric k' midway between the extremes (for $1 < k < \infty$) and within the midrange of practical values. Some formulas will be stated for this mean value, with a supplemental term which may be ignored, having a factor $(1/k - \frac{1}{3})$. It is graphed in Fig. 2 in terms of w/h directly and R_1 indirectly.

The shape dependence of the filling fraction was derived in terms of the wave resistance without dielectric (R_1) and is most simply expressed in those terms. This R_1 and the actual shape w/h are related by various formulas. The filling fraction is here expressed in very simple form from the previous derivations for narrow and wide.

For a narrow thin strip with dielectric, the effective dielectric constant is formulated as follows. The shape is introduced in terms of the wave resistance without dielectric (R_1):

(B-32), (B-44)
$$q = \frac{1}{2} + \frac{30}{R_1}\left(\ln\frac{\pi}{2} + \frac{1}{k}\ln\frac{4}{\pi}\right)$$
$$= \frac{1}{2} + \frac{16}{R_1}\left[1 + 0.453\left(\frac{1}{k} - \frac{1}{3}\right)\right] \tag{53}$$

(B-32), (B-45)
$$k' = \frac{k+1}{2} + \frac{60}{R_1}\frac{k-1}{2}\left(\ln\frac{\pi}{2} + \frac{1}{k}\ln\frac{4}{\pi}\right). \quad (54)$$

The relative error is <0.01 of k' if R' is >70; w/h is <3; q is <0.72.

For a wide thin strip with dielectric, the effective dielectric constant is formulated in terms of parameters defined above and here:

(B-8)
$$d = \frac{\pi}{2}R_c/R_1 = 592/R_1 > \pi, \qquad R_1 < 188$$

(A-14), (A-16), (B-16)
$$s' = 0.732[\text{acosh}\,(d-1) - \text{acosh}\,(0.358d + 0.598)] \quad (55)$$

(B-25)
$$s'' = \ln 4 - 1 - 1/(2d-1) = 0.386 - 1/(2d-1) \quad (56)$$

(B-4)
$$q = 1 - \frac{1}{d}\left[\text{acosh}\,(d-1) - s'' + \frac{s''-s'}{k}\right]. \quad (57)$$

The relative error is <0.01 (estimated).

For the mean case ($k = 3$), this result is approximated very closely by the simple formula

$$q = 1 - \frac{R_1}{592}\ln\frac{710}{R_1}. \quad (58)$$

The overlap between the two simple formulas (53) and (58) occurs near $R_1 = 100$ or $w/h = 1.5$.

The narrow and wide simple formulas for the mean case can be integrated and supplemented by an adjustment for any k, as follows:

$$q = \frac{1}{7}\left[1 + \frac{6}{1 + \frac{R_1}{507}\ln\left(\frac{710}{R_1} + 1\right)}\right] - \frac{0.2}{\frac{R_1}{220} + \frac{220}{R_1}} + \left(\frac{1}{k} - \frac{1}{3}\right)\frac{0.05}{\frac{R_1}{90} + \frac{90}{R_1}}. \quad (59)$$

The relative error is <0.01 of k'. The first term is a transition between the narrow and wide extremes. The second term is a very close adjustment for the intermediate range. The last term is negligible in the practical effect on k', so it serves mainly to indicate the weakness of the dependence on k.

One simple example is here reviewed in Table II as a test of various derivations for a thin strip without dielectric. It is a shape ($w/h = 1$) which is in the region of transition between narrow and wide approximations. The wave resistance R_1 is based on free space (120π). The items are listed in order of increasing error from the first. The derivation is described with respect to its development from the extreme of narrow and/or wide strip. The following notes give further comments.

TABLE II
COMPARISON OF FORMULAS FOR ONE EXAMPLE

Identification	Derivation	R_1 (ohms)	Error
(S) [14]	unrestricted	126.553	0
(W.1) (43)	narrow	126.533	-.020
(W.2) (2) (modified)	narrow	126.528	-.025
(W.3) [1] (A-68)	wide	126.473	-.042
(W.4) [1] (A-66)	narrow	126.641	+.088
(W.5) (2)	narrow-wide	126.310	-.243
(K) [16]	wide	127.857	+1.304
(W.6) [1] (A-71)	wide	124.424	-2.129

(S) Schneider's example is derived rigorously from elliptic integrals and is taken to be "exact" for purposes of comparison. Its relation to the other items tends to confirm its validity.

(W.1) This is the derivation based on the pair of wires equivalent to a narrow thin strip. It is the closest approximation (the relative error is <0.0002).

(W.2) This is similar to (W.1) but modified to a form suitable for matching the wide extreme. It is the reverse of formula (45).

(W.3) (W.4) These are the closest approximations given in the 1964 paper. They bracket the correct value within a relative difference of ± 0.0007. Their computation is much easier than (S).

(W.5) This is the only item providing a rather close approximation over the entire range of shape.

(K) Kaden's "wide" formula is an approximation to his derivation from elliptic integrals. The error (about 1 percent) indicates that this shape is "borderline" for his approximation. It is comparable with (W.6) in its explicit form and in simplicity, and gives a closer approximation.

This concludes a summary of the earlier formulas, and some more recent, as required for the above procedure. They are adequate for a set of close computations for a thin strip with any dielectric. These may be used for checking any empirical formula such as those proposed herein. They are used for the graphs in Fig. 2.

APPENDIX VII
COMPUTATION OF LOSS BY NUMERICAL DIFFERENTIATION

In practical applications of a strip line, the PF of conductor loss is usually determined by the skin effect. Some simple rules are applicable if the skin depth δ is much less than the least transverse dimension. One is the "incremental-inductance rule" stated by the author [3]. It relates the skin loss with the inductance, by a formula based on differentiation.

In a transmission line made of perfect conductors, the wave resistance without dielectric (R_1) is uniquely related to the inductance, so that formula may be used instead. Then the loss PF of the skin effect may be expressed as follows:

$$p = \Delta L/L = \frac{R_\delta - R_1}{R_\delta} = 1 - R_1/R_\delta \ll 1. \tag{60}$$

The incremental-inductance rule is here represented by the relative increment of inductance ($\Delta L/L$) that would be caused by removing a thickness ($\delta/2$) from the face of every conductor bounding the field. The wave resistance (R_1) for a perfect conductor would be increased in the same ratio if the boundary were modified in the same manner. Then this change (from R_1 to R_δ) is used to compute the PF.

The elegant application of the incremental-inductance rule was stated in terms of the analytic differentiation of an inductance formula in terms of a simple continuous function. In its more general application, such a formula may not be available. The rule is still useful if the inductance variation is formulated continuously over a range of dimensions. Such a formula can be subjected to analytic differentiation, but the resulting expression may be much more complicated than the inductance formula from which it is derived. This has been the experience of some workers who have taken this approach in evaluating the loss PF in a strip line [13], [17].

In the meantime, the advent of the digital calculator has opened up a new opportunity for the differentiation of a complicated formula. It enables a close approximation of the derivative by computing a small finite difference. The basic formula is used twice, which requires little more effort in a programmable calculator. The versatility of this procedure reaches a peak in the Hewlett-Packard pocket models, HP-25 and HP-65. The convenience and availability of the HP-25 provided the author with the tools and incentive to prepare this paper.

Having stated the objective of numerical differentiation by finite differences, the programming is routine. Fig. 10 shows the flow chart of one such program. It serves to bring out the application of some features available in the HP-25. It includes as the principal subroutine, some formula for R_1 in terms of the transverse dimensions. After incrementing the dimensions in accord with the skin effect, this subroutine is traversed a second time for R_δ. The relative increase is interpreted as the PF in direct or normalized form. The following features of the program are notable.

The skin depth δ may be evaluated and then utilized to give the PF p for any example. The more versatile normalized PF P may be obtained by arbitrarily choosing a small difference (say $\delta = 0.0001$). Then the result approaches the analytic derivative. The value of the difference cancels out in the normalized form. The number of decimal places in the small difference may be somewhat less than one half the number available in the computation. The skin depth or small difference is entered once in one register (R4) where it need not be renewed unless a change is desired.

The dimensions w,h,t are entered in assigned registers R0,R1,R2. For the second computation, each dimension is incremented by $\pm\delta$. Each dimension is between two conductor faces so the removal of $\delta/2$ on each face requires that

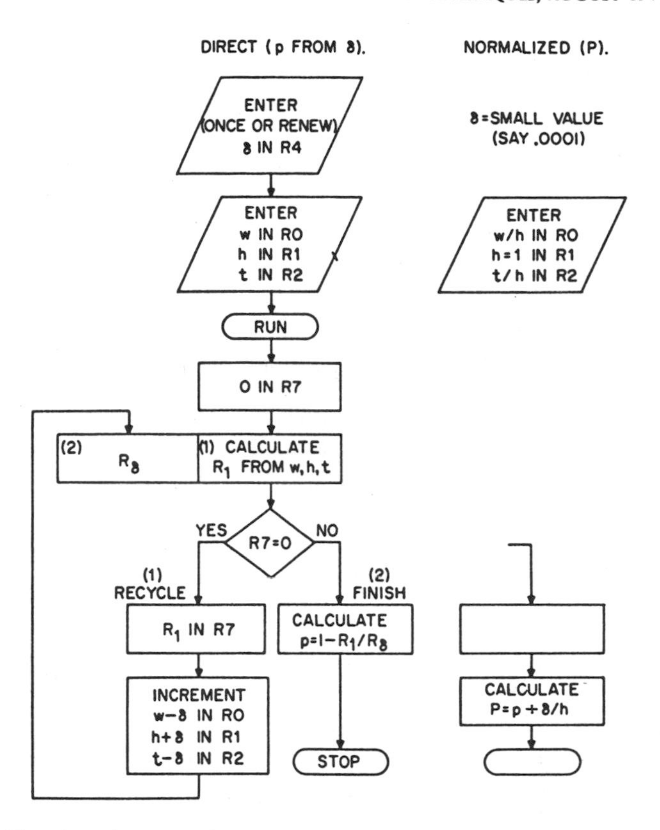

Fig. 10. Flow chart for computing the loss power factor by numerical differentiation.

the dimension be changed by δ, either increased or decreased. The increment is entered by arithmetic in the register.

Conditional branching is required at the end of each execution of the R_1 subroutine. One register (R7) is vacated until the end of the first execution, then occupied by R_1, which signals the end of the second. This serves also to retain R_1 for comparison with R_δ.

If the program storage is inadequate for the principal subroutine and also the transitional subroutines, one or more of the latter is easily performed manually.

In Fig. 10, the right-hand column of notes describe the program changes for the normalized form.

Appendix VIII
Formulas for the Skin Depth

Here are some formulas for the skin depth in nonmagnetic conductors [3], [7]:

$$\delta = \sqrt{\frac{2}{\omega\mu_0\sigma}} = \frac{1}{\sqrt{\pi f \mu_0 \sigma}} = \sqrt{\frac{\lambda G_0}{\pi\sigma}} = \sqrt{\frac{\lambda\rho}{\pi R_0}}$$

$$= \frac{1}{2\pi}\sqrt{\frac{\lambda}{30\sigma}} = \frac{1}{2\pi}\sqrt{\frac{\lambda\rho}{30}}$$

$$= 0.0291\sqrt{\lambda/\sigma} = \frac{1}{34.4}\sqrt{\lambda/\sigma}. \tag{61}$$

In copper ($\sigma = 58$ Mmho/m):

$$\delta_c = 3.81\sqrt{\lambda}\ \mu\text{m} = \sqrt{\frac{4.36\ \text{kHz}}{f}}\ \text{mm}$$

$$= \sqrt{\frac{4.36\ \text{GHz}}{f}}\ \mu\text{m}$$

$$= \frac{66}{\sqrt{f}}\ \text{mm} = \frac{0.066}{\sqrt{f\,(\text{MHz})}}\ \text{mm}. \tag{62}$$

Symbols used in (61) and (62) are defined below.

ω	=	$2\pi f$ = radian frequency (radians/second).
λ	=	wavelength in free space (meters).
R_0	=	$1/G_0$ = wave resistance of a plane wave in a square area in free space (ohms).
μ_0	=	magnetivity (magnetic permeability) in free space (henries/meter).
σ	=	$1/p$ = conductivity in copper (mhos/meter).
R_0	=	$377 = 120\pi\ \Omega$.
μ_0	=	$0.4\pi = 1.257\ \mu$H/m.

Appendix IX
Recent Article on Wide Strip with Thickness

Subsequent to the preparation of this paper, the author has seen a recent article related to the subject [19]. W. H. Chang has described an ingenuous and powerful approximation based on conformal mapping. To that extent, it has something in common with the author's 1964 paper [1]. Some thickness is accommodated at the expense of some refinements in other respects. The result is a very useful approximation for wide strips with thickness. To yield this in analytic form is a major achievement. Moreover, it appears that his result may be closely bracketed by further appreciation of his approximation.

Relevant to the present paper, Chang gives a table of examples computed from his formula and also by a numerical approximation from W. J. Weeks [18]. The agreement is very close. Most of those examples fall within the range of validity of the present paper, formulas (1), (2), (13), (14) without dielectric. The agreement is well within $0.01R$. The formulas herein offer a close approximation for any width. They are based on a thin strip with width adjustment for thickness. Chang's formulas for a wide strip are remarkable for including the width and the thickness in one formula.

The small-thickness examples reported in Appendix II are in the range of marginal approximation by Chang, so they have not been compared.

References

[1] H. A. Wheeler, "Transmission-line properties of parallel wide strips by a conformal-mapping approximation," *IEEE Trans. Microwave Theory Tech.*, vol. MTT-12, pp. 280–289, May 1964.

[2] ——, "Transmission-line properties of parallel strips separated by a dielectric sheet," *IEEE Trans. Microwave Theory Tech.*, vol. MTT-13, pp. 172–185, Mar. 1965.

[3] ——, "Formulas for the skin effect," *Proc. IRE*, vol. 30, pp. 412–424, Sept. 1942. (Skin loss by the "incremental-inductance rule.")

[4] F. Oberhettinger and W. Magnus, *Applications of Elliptic Functions in Physics and Technology*. New York: Springer, 1949. (Wave resistance of coplanar strip, p. 63. Tables of K'/K, p. 114.)

[5] E. Weber, *Electromagnetic Fields—Theory and Applications—Mapping of Fields*. New York: Wiley, 1950. (Thickness, p. 347.)

[6] H. A. Wheeler, "Transmission-line impedance curves," *Proc. IRE*, vol. 38, pp. 1400–1403, Dec. 1950.

[7] ——, "Universal skin-effect chart for conducting materials," *Electronics*, vol. 25, no. 11, pp. 152–154, Nov. 1952.

[8] S. Cohn, "Problems in strip transmission lines," *IEEE Trans. Microwave Theory Tech.*, vol. MTT-3, pp. 119–126, Mar. 1955. (Sandwich line, thickness and loss.)

[9] H. A. Wheeler, "Skin resistance of a transmission-line conductor of polygon cross section," *Proc. IRE*, vol. 43, pp. 805–808, July 1955.

[10] D. S. Lerner (Wheeler Labs., Inc.), unpublished notes, Nov. 1963. (Very wide strip, width adjustment for thickness. Half-shielded type compared with sandwich type.)

[11] M. Caulton, J. J. Hughes, and H. Sobol, "Measurements on the properties of microstrip transmission lines for microwave integrated circuits," *RCA Rev.*, vol. 27, pp. 377–391, Sept. 1966.

[12] H. Sobol, "Extending IC technology to microwave equipment," *Electronics*, vol. 40, no. 6, pp. 112–124, Mar. 1967. (A simple empirical formula, remarkably close for a practical range.)

[13] R. A. Pucel, D. J. Masse, and C. P. Hartwig, "Losses in microstrip," *IEEE Trans. Microwave Theory Tech.*, vol. MTT-16, pp. 342–350, June 1968. (By analytic differentiation.)

[14] M. V. Schneider, "Microstrip lines for microwave integrated circuits," *Bell Syst. Tech. J.*, vol. 48, pp. 1421–1444, May 1969.

[15] ——, "Dielectric loss in integrated microwave circuits," *Bell Syst. Tech. J.*, vol. 48, pp. 2325–2332, Sept. 1969.

[16] H. Kaden, "Advances in microstrip theory," *Siemens Forsch. u. Entwickl. Ber.*, vol. 3, pp. 115–124, 1974. (Computation of wave resistance and loss for a wide strip of small thickness.)

[17] R. Mittra and T. Itoh, "Analysis of microstrip transmission lines," in *Advances in Microwaves*, vol. 8. New York: Academic Press, 1974, pp. 67–141. (Latest review, many references.)

[18] W. J. Weeks, "Calculation of coefficients of capacitance for multiconductor transmission lines," *IEEE Trans. Microwave Theory Tech.*, vol. MTT-18, pp. 35–43, Jan. 1970. (Numerical method for a rectangular cross section with a parallel plane.)

[19] W. H. Chang, "Analytical IC metal-line capacitance formulas," *IEEE Trans. Microwave Theory Tech.*, vol. MTT-24, pp. 608–611, Sept. 1976. (Conformal mapping approximation for a wide strip with thickness. Examples compared with numerical method of Weeks [18] using his program.)

Characteristics of Single and Coupled Microstrips on Anisotropic Substrates

NICOLAOS G. ALEXOPOULOS, MEMBER, IEEE, AND CLIFFORD M. KROWNE, MEMBER, IEEE

Abstract—**In this paper, the effect of an anisotropic substrate on the characteristics of covered microstrip is presented for single and coupled lines. The Green's function is obtained in integral and series form for an arbitrary anisotropic substrate. Computer programs based on the method of moments approach [1], [2] are employed and results are presented in graphical form for impedance Z, coupling constant K, and phase velocity v_p as functions of n_x/n_y (the ratio of the substrate indices of refraction). Z, K, and v_p are studied for various w/H, S/H, and B/H ratios where w is the line width (w_1 and w_2 for coupled lines), S is the separation between coupled lines, B is the separation between ground planes, and H is the substrate thickness.**

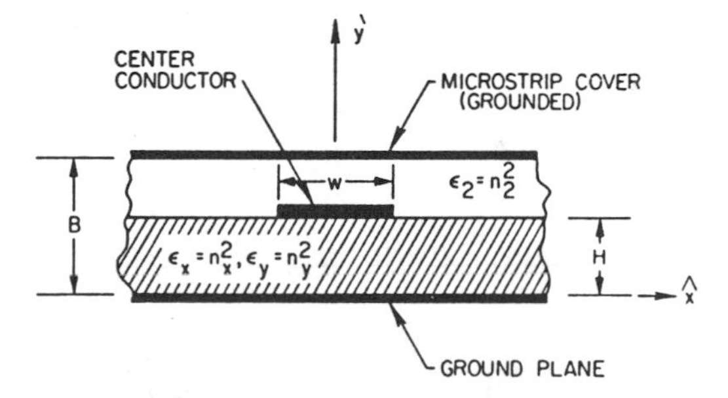

Fig. 1. Cross section of covered single microstrip geometry.

I. Introduction

EXTENSIVE results exist in the literature on the problem of microstrip lines on isotropic substrates, e.g., [1]–[13]. Therein, the Green's function of the problem is obtained either by image theory [7] or by a direct solution to the boundary value problem [13]. In most cases a quasi-static approach is presented, which necessitates solution to Laplace's equation for a given set of boundary conditions. A series of papers [8]–[12] presents solutions to the dispersion problem, again for isotropic substrates.

Recently [14], [15], the problem of anisotropic substrates was approached strictly from the numerical point of view. Specifically, the authors employed the method of finite differences to obtain the impedance characteristics of a single microstrip line over a single-crystal sapphire substrate. Since this crystal is uniaxial, the permittivity dyadic is strictly diagonal with a relative permittivity along the optical axis $\epsilon_y = 11.60$, while on the plane perpendicular to this axis $\epsilon_x = \epsilon_z = 9.40$. The authors proceeded to compute an equivalent isotropic relative permittivity ϵ_{req} which enables them to proceed with the computations of the microstrip characteristics.

In the present paper, the problem of the anisotropic substrate is approached from the boundary value point of view. The boundary value approach necessitates the introduction of a grounded cover, although this by no means limits the usefulness of the solution since B (see Fig. 1) can be allowed to recede to infinity. An image theory approach would be much more preferable, but there appears to be no prior references on how conductors image over anisotropic media. By employing a quasi-static approach, the Green's function is incorporated into two methods of moments computer programs. These programs provide solutions to the single and coupled microstrip problem by employing the usual methods for the computation of self and mutual capacitances, characteristic impedances, and phase velocities. The results are presented for various values of the relative permittivities in the x, y, and z directions, and they are shown in Figs. 2–4 and 6 and 7.

Manuscript received August 30, 1976; revised July 15, 1977. This work was supported jointly by the U.S. Office of Naval Research under Contract N00014-76C-0896 and by the Watkins–Johnson Company.

N. G. Alexopoulos is with the Department of Electrical Sciences and Engineering, University of California, Los Angeles, CA 90024.

C. M. Krowne is with the Watkins–Johnson Company, Palo Alto, CA 94304.

Reprinted from *IEEE Trans. Microwave Theory Tech.*, vol. MTT-26, no. 6, pp. 387–393, June 1978.

II. Green's Function for an Anisotropic Substrate

The geometry of Fig. 1 is considered where a charged line of ρ_l C/m is assumed parallel to the $\hat{z}$ axis at $y=H$. The space $H<y<B$ is assumed homogeneous and isotropic with a relative permittivity ϵ_2. For $0<y<H$ (the substrate region), the medium is anisotropic and oriented in such a manner that the relative permittivity is given by the following dyadic:

$$\bar{\bar{\epsilon}}=\begin{bmatrix} \epsilon_x \hat{x}\hat{x} & 0 & 0 \\ 0 & \epsilon_y \hat{y}\hat{y} & 0 \\ 0 & 0 & \epsilon_z \hat{z}\hat{z} \end{bmatrix}. \tag{1}$$

A quasi-static solution to the potential problem can be obtained by solving Laplace's equations in the two dielectric regions subject to the proper boundary conditions. One needs to solve in the anisotropic region the equation

$$\nabla \cdot \left(\bar{\bar{\epsilon}} \cdot \nabla \phi_1\right)=0. \tag{2}$$

Due to the infinite extent of the line in the z direction, the problem is two dimensional. Therefore, (2) yields

$$\epsilon_x \frac{\partial^2 \phi_1}{\partial x^2}+\epsilon_y \frac{\partial^2 \phi_1}{\partial y^2}=0 \tag{3}$$

which has a solution of the form

$$\phi_1(x,y)=\left[a_1(k) \cos\left(\frac{kx}{n_x}\right)+b_1(k) \sin\left(\frac{kx}{n_x}\right)\right] \cdot \left[c_1(k) \sinh\left(\frac{ky}{n_y}\right)+d_1(k) \cosh\left(\frac{ky}{n_y}\right)\right] \tag{4}$$

with k being a continuous variable. Because of the even symmetry in x and the fact that at the ground plane $\phi_1(x,0)=0$, it follows that the general solution of (3) can be written as

$$\phi_1(k)=\int_{-\infty}^{\infty} A_1(k) \cos\left(\frac{kx}{n_x}\right)\left\{\frac{\sinh(ky/n_y)}{\sinh(kH/n_y)}\right\} dk \tag{5}$$

where $n_x=(\epsilon_x)^{1/2}$ and $n_y=(\epsilon_y)^{1/2}$. Using Laplace's equation in the isotropic region ($H<y<B$), i.e.,

$$\nabla^2 \phi_2=0 \tag{6}$$

one must obtain $\phi_2(x,y)$ such that $\phi_2(x,B)=0$ at the top ground plane and

$$\phi_1(x,y)|_{y=H^-}=\phi_2(x,y)|_{y=H} \tag{7}$$

while satisfying the boundary condition

$$\hat{y} \cdot \left(\bar{D}_1-\bar{D}_2\right)=-\rho_l \delta(x). \tag{8}$$

Normalizing the x and y variables in (6) to the same constant n_x, a general solution in region 2 is obtained in the form

$$\phi_2(x,y)=\int_{-\infty}^{\infty} A_2(k) \cos\left(\frac{kx}{n_x}\right)\left\{\frac{\sinh(k(B-y)/n_x)}{\sinh(k(B-H)/n_x)}\right\} dk. \tag{9}$$

From the continuity of potentials at $y=H$, it follows that $A_1(k)=A_2(k)=A(k)$. By employing the boundary condition at $y=H$ which is expressed in (8), it follows that

$$\rho_l \delta(x)=\int_{-\infty}^{\infty} A(k) \cos\left(\frac{kx}{n_x}\right) \cdot \left\{n_y \coth\left(\frac{kH}{n_y}\right)+\frac{n_2^2}{n_x} \coth\left[\frac{k(B-H)}{n_x}\right]\right\} k\, dk. \tag{10}$$

Since

$$\int_{-\infty}^{\infty} \cos\left(\frac{kx}{n_x}\right) dk=2\pi\delta\left(\frac{x}{n_x}\right)=2\pi n_x \delta(x) \tag{11}$$

$A(k)$ is obtained from (10) in the form

$$A(k)=\frac{\rho_l}{2\pi\epsilon_0} \cdot \left\{\frac{1}{k}\left(\frac{1}{n_x n_y \coth(kH/n_y)+n_2^2 \coth[k(B-H)/n_x]}\right)\right\}. \tag{12}$$

Here ϵ_0 is the permittivity of a vacuum. The potential is now given by

$$\phi_1(x,y)=\frac{\rho_l}{2\pi\epsilon_0}\int_{-\infty}^{\infty} \frac{dk}{k} \frac{\cos(kx/n_x) \sinh(ky/n_y)}{\{n_x n_y \cosh(kH/n_y)+n_2^2 \sinh(kH/n_y) \coth[k(B-H)/n_x]\}} \tag{13}$$

for $0<y<H$, and by

$$\phi_2(x,y)=\frac{\rho_l}{2\pi\epsilon_0}\int_{-\infty}^{\infty} \frac{dk}{k} \frac{\cos(kx/n_x) \sinh[k(B-y)/n_x]}{\{n_x n_y \sinh(k(B-H)/n_x) \coth(kH/n_y)+n_2^2 \cosh(k(B-H)/n_x)\}} \tag{14}$$

for $H<y<B$.

If $\rho_l = 1$ and $y = H$, the Green's function for the microstrip over an anisotropic substrate is expressed simply by the equation

$$\phi(x,H) = \frac{1}{2\pi\epsilon_0} \int_{-\infty}^{\infty} \frac{dk}{k} \cdot \frac{\cos(kx/n_x)}{\{n_x n_y \coth(kH/n_y) + n_2^2 \coth(k(B-H)/n_x)\}}. \quad (15)$$

III. Numerical Analysis

In order to compute the characteristic impedance Z, phase velocity v_p, and coupling constant K parameters of the microstrip, a numerical approach based on the method of moments has been employed. Two computer programs have been developed to compute $\phi(x,H)$ in (15). The first one simply evaluates (15) using a numerical integration approach [16]. This entails separating the integration interval into two ranges: Δ to P and P to ∞. If P is chosen large enough, the coth functions will approach one and the integration can be represented as follows:

$$\phi(x,H) = \frac{1}{\pi\epsilon_0} \int_{\Delta}^{P} \frac{dk}{k} \cdot \frac{\cos(kx/n_x)}{\{n_x n_y \coth(kH/n_y) + n_2^2 \coth(k(B-H)/n_x)\}} + \frac{1}{\pi\epsilon_0(n_x n_y + n_2^2)} \int_{P}^{\infty} \frac{dk}{k} \cos\left(\frac{kx}{n_x}\right). \quad (16)$$

In order for (16) to approximate $\phi(x,H)$ well, $\Delta \to 0$ and P must be such that the coth arguments $\theta_a > 6$. The lower limit in the first integral is not set $\Delta \equiv 0$ because the coth functions blow up at $k=0$, preventing a correct numerical evaluation. The second integral in (16) can be calculated using the IBM scientific subroutine package program SICI.

The first integral can be determined using a Simpson integration routine. The cos argument increment $|\Delta\theta_c$ must be chosen small enough to approximate the cos oscillation behavior, $\Delta\theta_c < 5°$. This $\Delta\theta_c$ value specifies the k integration interval $\Delta k = n_x \Delta\theta_c / x$.

The other computer program computes $\phi(x,H)$ in (15) through a series expression (19) derived as follows. By analytic continuation, $\phi(x,H)$ in (15) can be rewritten as

$$\phi(x,H) = \frac{1}{2\pi\epsilon_0} \operatorname{Re} \oint_C \frac{dz}{z} \cdot \frac{\exp(i(x/n_x)z)}{[n_x n_y \coth(Hz/n_y) + n_2^2 \coth[((B-H)/n_x)z]]} \quad (17)$$

where z is a complex variable and C is the path of integration which extends from $-\infty$ to $+\infty$ along the Re (z) axis and closes on the upper half z plane. It can be easily demonstrated that $z=0$ is an ordinary point, and that there exists an infinite number of poles restricted to the Im (z) axis. By invoking the residue theorem, $\phi(x,H)$ can be rewritten as

$$\phi(x,H) = \frac{1}{\epsilon_0} \operatorname{Re} \sum_{p=1}^{\infty} iQ_p \quad (18)$$

where Q_p denotes the residue at the pth pole. In accordance with this approach, the residue series yields

$$\phi(x,H) = \frac{1}{\epsilon_0} \sum_{p=1}^{\infty} \cdot \frac{\exp(-(|x| r_p / n_x))}{r_p H[n_x \csc^2(H r_p / n_y) + n_2^2 (\nu / n_x) \csc^2((H\nu / n_x) r_p)]} \quad (19)$$

where $\nu = (B/H) - 1$ and r_p is the pth zero of the determinantal equation

$$\frac{\cot(Hr/n_y)}{\cot(H\nu r/n_x)} = -\frac{n_2^2}{n_x n_y}. \quad (20)$$

Of interest is the special case where $n_x = n_y = n_1$, i.e., that of isotropic substrate. In this case, the parameter ν is of importance in locating the roots of r_p properly. For example, if $\nu = 1$, (19) simplifies to

$$\phi(x,H) = \frac{2}{\epsilon_0} \sum_{p=1}^{\infty} \frac{\exp(-(|x|/2\pi) p\pi)}{p\pi[n_1^2 + n_2^2]} \quad (21)$$

which agrees with the result obtained for isotropic substrates by Farrar and Adams [2].

In order to verify the accuracy of our computer programs, extensive computations were performed based on the Green's function given by (16) and (19). Using (19), agreement within 5 percent was found with the results of [2] for all cases of B/H and w/H for $\epsilon_x = \epsilon_y = 9.6$, as well as with the results of [3] for alumina with $\epsilon_x = \epsilon_y = 9.9$ and $B/H = 6$. For example, the case $B/H = 6$, $w/H = 1.0$, and $\epsilon_x = \epsilon_y = 9.6$ yields $Z = 48.73$ Ω ($N = 20$ partitions/line using ANIGREEN 1) and 48.4 Ω from [2]. This is a discrepancy of 0.7 percent. (Note that $n_2 = 1$. This is true throughout our paper.) In addition, the special case of a single-crystal sapphire substrate with $\epsilon_x = 9.40$ and $\epsilon_y = 11.60$ was examined. In this latter case, our results agree within about 5 percent of the values obtained in [14] for Z versus w/H. Figs. 2(a) and (b) show a comparison of the characteristic impedance Z and normalized phase velocity v_p/c versus w/H for alumina substrates with $B/H = 6$ (c = velocity of light in a vacuum). Fig. 2(a) indicates that the values of the characteristic impedance Z are quite close for polycrystalline and single-crystal aluminas; in fact, for $w/H > 4.5$, the two curves coalesce. On the other hand, the two alumina materials exhibit divergent phase velocity characteristics as evidenced by Fig. 2(b). This latter observation underlines the fact that more precise documentation of anisotropic substrate properties is required. To this end, computations for microstrip over anisotropic substrates have been performed for both single (Section IV) and coupled (Section V) microstrip lines. The calculations in the next two sections are performed using (19) for loosely coupled or uncoupled lines (computer program ANIGREEN 1, see [17]) and (16) for tightly coupled lines (computer program ANIGREEN 2,

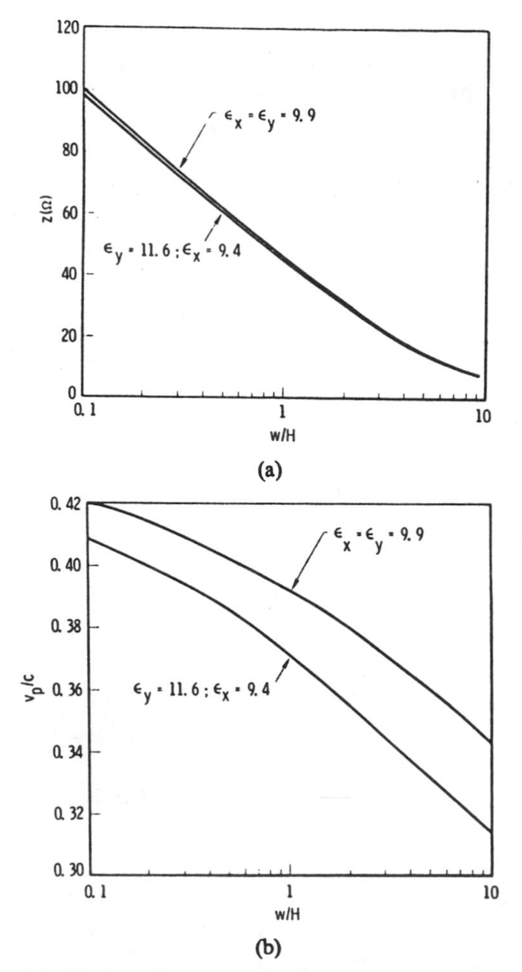

Fig. 2. (a) Single microstrip characteristic impedance Z versus $w/H(B/H=6)$ for alumina substrates. (b) Single microstrip phase velocity v_p/c (normalized to c) versus $w/H(B/H=6)$ for alumina substrates.

see [17]). Equation (16) yields coupled line parameters which have been found to agree within 5 percent of the alumina results found in [6]. Equation (16) also yields results within 5 percent of Bryant and Weiss' computer program MSTRIP [18]. (The computer program accuracy is increased by increasing the number of partitions into which the line is subdivided.)

IV. Single Microstrip Characteristics

Figs. 3(a), (b), and (c) show the behavior of Z versus n_x/n_y for $w/H=0.1$, 1, and 10, respectively, with $B/H=6$. In Fig. 3(a), the relative permittivity values are normalized to $\epsilon_y=9.9$, which is the relative permittivity usually employed in the literature for alumina substrates. It is observed that, as w/H increases in value from 0.1 to 10.0, the value of Z drops considerably for all n_x/n_y. This behavior parallels that found in isotropic substrates where, as w increases, a greater portion of the electric field lines are in the substrate as opposed to the air, increasing the effective dielectric constant and thereby decreasing Z. For a given w/H, Z decreases as n_x/n_y increases. This is due to an increase in the electric field lines in the x direction E_x with rising n_x or ϵ_x since n_y is fixed. The slope of the curves goes down with rising w/H since $E_y \gg E_x$.

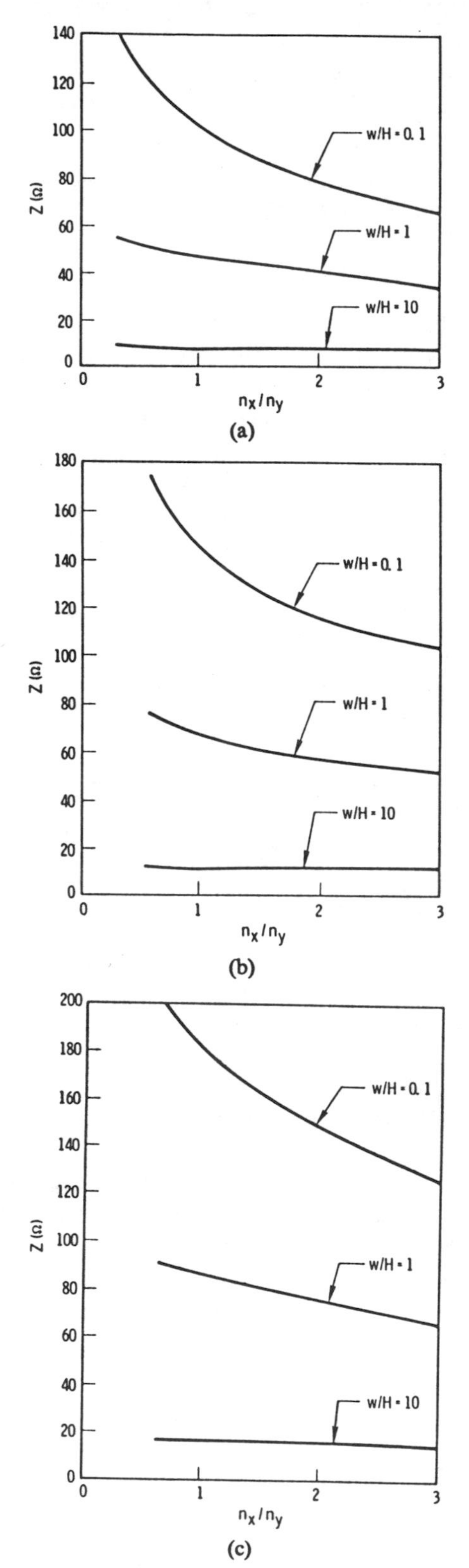

Fig. 3. (a) Z versus n_x/n_y for alumina-like substrates; $B/H=6$ (permittivity tensor normalized to $\epsilon_y=9.9$). (b) Z versus n_x/n_y for quartz-like substrates; $B/H=6$ (permittivity tensor normalized to $\epsilon_y=4.5$). (c) Z versus n_x/n_y for polystyrene-like substrates; $B/H=6$ (permittivity tensor normalized to $\epsilon_y=2.55$).

Figs. 3(b) and (c) show the characteristic impedance behavior for quartz-like (SiO material) and polystyrene-like substrates with the permittivity tensor elements being normalized to, respectively, $\epsilon_y=4.5$ and $\epsilon_y=2.55$. One notices that for a particular w/H value, Z increases in

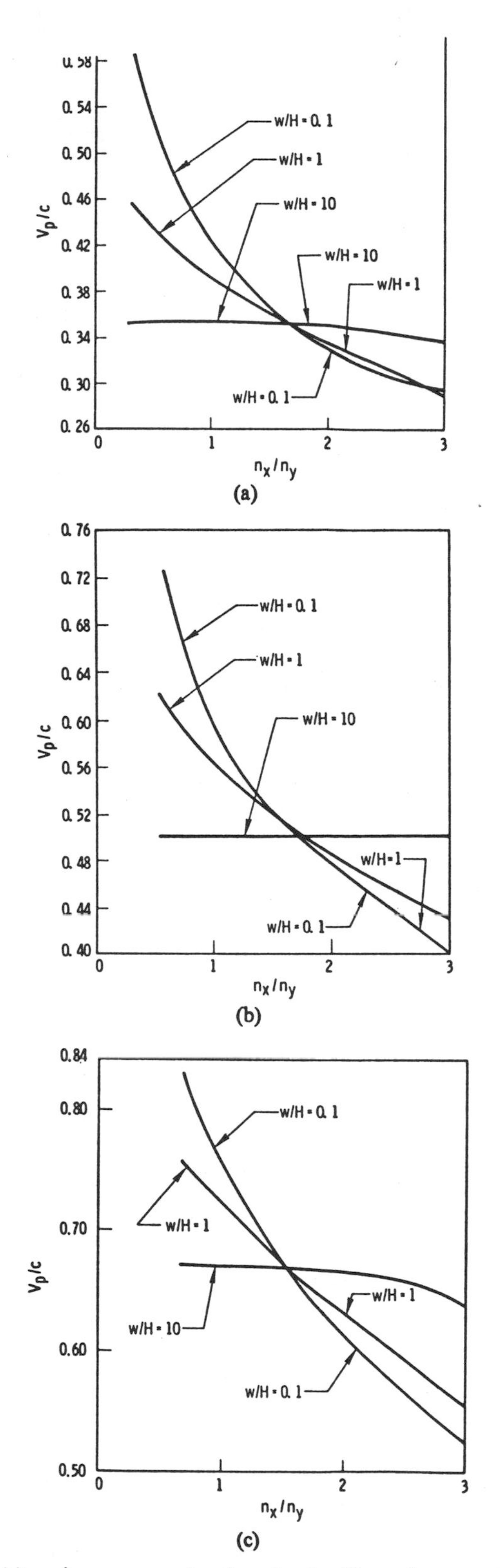

Fig. 4. (a) v_p/c versus n_x/n_y for alumina-like substrates; $B/H=6$ (permittivity tensor normalized to $\epsilon_y=9.9$). (b) v_p/c versus n_x/n_y for quartz-like substrates; $B/H=6$ (permittivity tensor normalized to $\epsilon_y=4.5$). (c) v_p/c versus n_x/n_y for polystyrene-like substrates; $B/H=6$ (permittivity tensor normalized to $\epsilon_y=2.55$).

going from Fig. 3(a) to 3(c) (with n_y getting smaller), because n_y controls E_y and therefore Z.

In Figs. 4(a), (b), and (c), the normalized phase velocity v_p/c is shown for $w/H=0.1$, 1, and 10 with $B/H=6.0$.

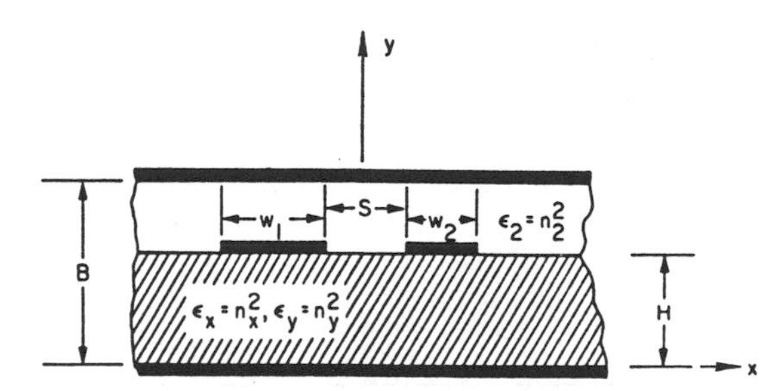

Fig. 5. Cross section of covered coupled microstrip lines geometry.

As expected, these curves show lower v_p/c values for all w/H as the normalizing parameter ϵ_y increases. Also, as w/H increases, it is clear that v_p/c changes very slowly as n_x/n_y increases and in fact, for quartz-like substrates, $v_p/c \cong 0.50$ when $n_x/n_y > 1$ ($w/H=10$).

V. Coupled Microstrip Characteristics

The characteristic impedance, phase velocity, and coupling constant parameters of coupled pairs of microstrip transmission lines are presented here for anisotropic substrates. The microstrip conductors are assumed to be parallel to the $\hat{z}$ direction, separated by a distance S and, in general, of unequal widths w_1, w_2, as shown in Fig. 5. The even- and odd-mode properties of the coupled microstrip have been obtained by employing the same approach which is customary in numerous quasi-static TEM studies of isotropic substrates, and therefore the analysis will not be reproduced here [6]. Our results are confined to alumina-like substrates [17]. Figs. 6(a) and (b) show the even- and odd-mode impedance and phase velocity plotted versus n_x/n_y. The results are for $B/H=20$, $w/H=1$, and $S/H=0.25$ with the permittivity tensor being normalized to $\epsilon_y=9.9$. As in the case of isotropic substrates, there exist large differences in characteristic impedance values between the even and odd modes for small S/H. This is evidenced in Fig. 6(a) where the even-mode impedance Z_e is larger than the odd-mode impedance Z_o by roughly 25 Ω for all values of n_x/n_y. On the other hand, contrary to the isotropic substrate results, the normalized phase velocity curves v_p/c for the even (v_{pe}) and odd (v_{po}) modes converge to $v_{po}/c \cong 0.334$ and $v_{pe}/c \cong 0.320$ values for $n_x/n_y \cong 2.0$. This indicates that $v_{po}-v_{pe}$ can be minimized for the proper value of n_x/n_y. Fig. 6(c) shows that the coupling K rises with n_x/n_y as expected, due to increasing E_x. K was calculated by employing the formula [19]

$$K=\frac{G}{\sqrt{AB}} \tag{22}$$

with

$$G=\frac{Z_{o1}^{-1}-Z_{e1}^{-1}}{2} =\frac{Z_{o2}^{-1}-Z_{e2}^{-1}}{2} \tag{23}$$

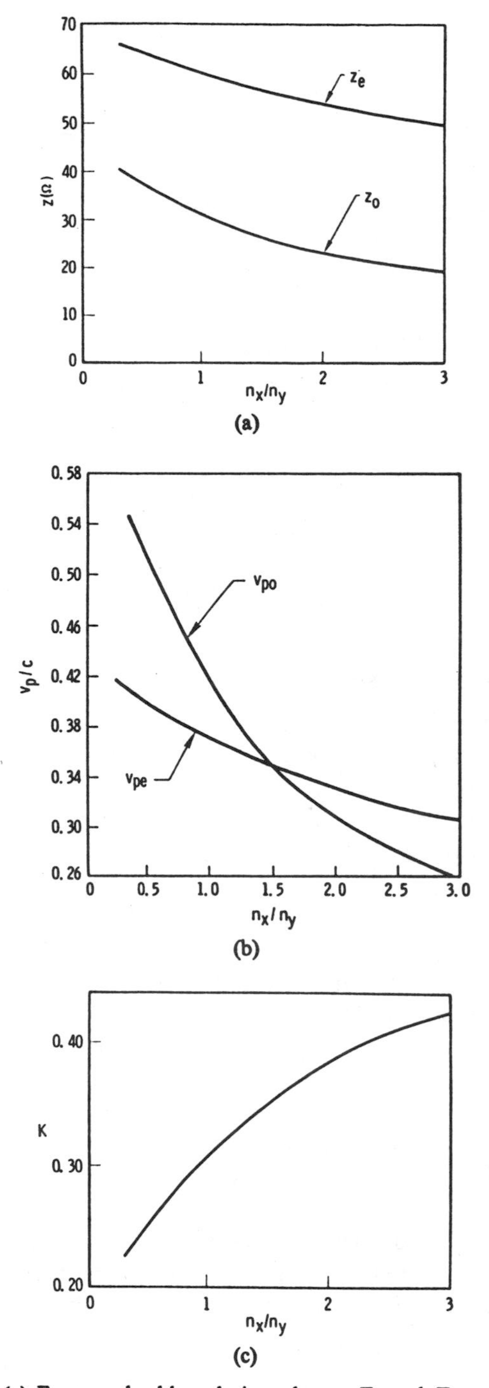

Fig. 6. (a) Even- and odd-mode impedances Z_e and Z_o versus n_x/n_y for parallel-coupled microstrip on an anisotropic alumina-like substrate ($\bar{\epsilon}$ normalized to $\epsilon_y=9.9$). $w/H=1$, $S/H=0.25$, and $B/H=20$. (b) v_{pe} and v_{po} versus n_x/n_y for parallel-coupled microstrip on an anisotropic alumina-like substrate ($\bar{\bar{\epsilon}}$ normalized to $\epsilon_y=9.9$). $w/H=1$, $S/H=0.25$, and $B/H=20$. (c) Coupling constant K versus n_x/n_y for parallel microstrip on an anisotropic alumina-like substrate ($\bar{\bar{\epsilon}}$ normalized to $\epsilon_y=9.9$). $w/H=1$, $S/H=0.25$, and $B/H=20$.

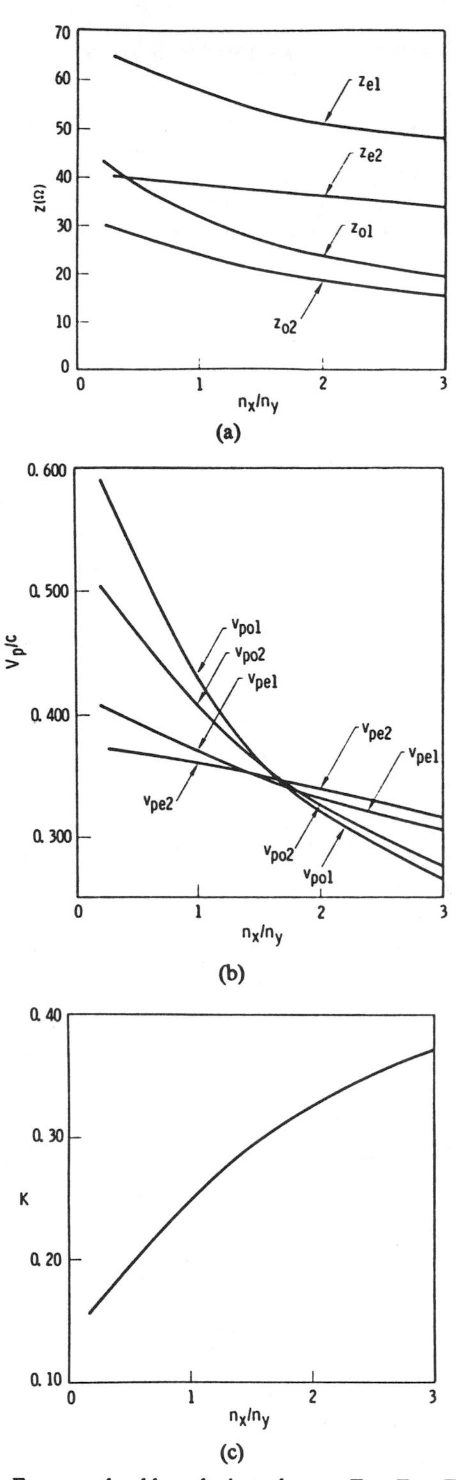

Fig. 7. (a) Even- and odd-mode impedances Z_{e1}, Z_{e2}, Z_{o1}, and Z_{o2} versus n_x/n_y for parallel-coupled microstrip on an anisotropic alumina-like substrate ($\bar{\epsilon}$ normalized to $\epsilon_y=9.9$). $w_2/H=1$, $w_2/H=2$, $S/H=0.25$, and $B/H=20$. (b) Even- and odd-mode phase velocities v_{e1}, v_{e2}, v_{o1}, and v_{o2} versus n_x/n_y for parallel-coupled microstrip on an anisotropic alumina-like substrate ($\bar{\bar{\epsilon}}$ normalized to 9.9). $w_1/H=1$, $w_2/H=2$, $S/H=0.25$, and $B/H=20$. (c) K versus n_x/n_y for parallel-coupled microstrip on an anisotropic alumina-like substrate ($\bar{\epsilon}$ normalized to $\epsilon_y=9.9$). w_1 substrate ($\bar{\epsilon}$ normalized to $\epsilon_y=9.9$). $w_1/H=1$, $w_2/H=2$, $S/H=0.25$, and $B/H=20$.

and

$$A=\frac{Z_{o1}^{-1}+Z_{e1}^{-1}}{2}$$

$$B=\frac{Z_{o2}^{-1}+Z_{e2}^{-1}}{2}. \qquad (24)$$

Figs. 7(a), (b), and (c) demonstrate the characteristics of coupled microstrip lines which are of unequal width; here, $w_1/H=1$ and $w_2/H=2$. $S/H=0.25$, $B/H=20$, and $\epsilon_y=$ 9.9, as we had in Figs. 6(a), (b), and (c). The unequal line widths cause the even- and odd-mode curves of Figs. 6(a), (b), and (c) to split up into Z_{o1}, Z_{o2}; Z_{e1}, Z_{e2}; v_{po1}, v_{po2}; and v_{pe1}, v_{pe2}. In Fig. 7(b), all phase velocities approach one another near $n_x/n_y=1.6$. They are within one percent of each other at this point. In general, for unequal but arbitrary line widths one does not expect the solution to yield a point where all v_p are equal or nearly equal.

VI. Discussion

Here we have demonstrated the effect of substrate anisotropy on the single and coupled microstrip line characteristics Z, v_p, and K, assuming a quasi-static TEM solution to hold. It is found for coupled lines that by properly choosing the substrate material, the difference between even- and odd-mode phase velocities can be significantly reduced compared to using isotropic or nearly isotropic substrates. This implies that the isolation and directivity of microstrip couplers can be markedly improved. One should also be able to improve the performance of microstrip filters.

An example of a substance which possesses enough anisotropy to make it a feasible substrate material to demonstrate an improvement in directivity of couplers due to a minimization in $v_{po} - v_{pe}$ is pyrolytic boron nitride (PBN). This crystal is currently being studied theoretically and experimentally at Watkins–Johnson Company [20].

Acknowledgment

C. M. Krowne would like to thank Dr. E. J. Crescenzi, Jr., for suggesting that he study the effect of anisotropy on microstrip waveguide electrical parameters, and R. C. Helvey for helping the author with a number of computational problems, both of Watkins–Johnson Company. The authors would also like to thank S. Kerner of the University of California, Los Angeles, for assisting with some of the numerical calculations.

References

[1] A. Farrar and A. T. Adams, "Characteristic impedance of microstrip by the method of moments," *IEEE Trans. Microwave Theory Tech.*, vol. MTT-18, pp. 65–66, Jan. 1970.
[2] ——, "A potential theory method for covered microstrip," *IEEE Trans. Microwave Theory Tech.* vol. MTT-21, pp. 494–496, July 1973.
[3] E. Yamashita, "Variational method for the analysis of microstrip-like transmission lines," *IEEE Trans. Microwave Theory Tech.*, vol. MTT-16, pp. 529–535, Aug. 1968.
[4] E. Yamashita and K. Atsuki, "Distributed capacitance of a thin-film electrooptic light modulator," *IEEE Trans. Microwave Theory Tech.*, vol. MTT-23, pp. 177–178, Jan. 1975.
[5] T. G. Bryant and J. A. Weiss, "Parameters of microstrip transmission lines and of coupled pairs of microstrip lines," *IEEE Trans. Microwave Theory Tech.*, vol. MTT-16, pp. 1021–1027, Dec. 1968.
[6] S. V. Judd, I. Whiteley, R. J. Clowes, and D. C. Rickard, "An analytical method for calculating microstrip transmission line parameters," *IEEE Trans. Microwave Theory Tech.*, vol. MTT-18, pp. 78–87, Feb. 1970.
[7] W. T. Weeks, "Calculation of coefficients of capacitance of multiconductor transmission lines in the presence of a dielectric interface," *IEEE Trans. Microwave Theory Tech.*, vol. MTT-19, pp. 35–43, Jan. 1970.
[8] R. Mittra and T. Itoh, "A new technique for the analysis of the dispersion characteristics of microstrip lines," *IEEE Trans. Microwave Theory Tech.*, vol. MTT-19, pp. 47–56, Jan. 1971.
[9] P. Daly, "Hybrid-mode analysis of microstrip by finite-element methods," *IEEE Trans. Microwave Theory Tech.*, vol. MTT-19, pp. 19–25, Jan. 1971.
[10] D. G. Corr and J. B. Davies, "Computer analysis of the fundamental and higher order modes in single and coupled microstrip," *IEEE Trans. Microwave Theory Tech.*, vol. MTT-20, pp. 669–678, Oct. 1972.
[11] M. K. Krage and G. I. Haddad, "Frequency-dependent characteristics of microstrip transmission lines," *IEEE Trans. Microwave Theory Tech.*, vol. MTT-20, pp. 678–688, Oct. 1972.
[12] E. J. Denlinger, "A frequency dependent solution for microstrip transmission lines," *IEEE Trans. Microwave Theory Tech.*, vol. MTT-19, pp. 30–39, Jan. 1971.
[13] M. K. Krage and G. I. Haddad, "Characteristics of coupled microstrip transmission lines—II: Evaluation of coupled-line parameters," *IEEE Trans. Microwave Theory Tech.*, vol. MTT-18, pp. 222–228, Apr. 1970.
[14] R. P. Owens, J. E. Aitken, and T. C. Edwards, "Quasi-static characteristics of microstrip on an anisotropic sapphire substrate," *IEEE Trans. Microwave Theory Tech.*, vol. MTT-24, pp. 499–505, Aug. 1976.
[15] T. Edwards and R. P. Owens, "2–18 GHz dispersion measurements on 10–100 Ω microstrip lines on sapphire," *IEEE Trans. Microwave Theory Tech.*, vol. MTT-24, pp. 506–513, Aug. 1976.
[16] J. A. Weiss and T. G. Bryant, "Dielectric Green's function for parameters of microstrip," *Electron. Lett.*, vol. 6, no. 15, pp. 462–463, July 1970.
[17] N. G. Alexopoulos, C. Krowne, and S. Kerner, "Dispersionless coupled microstrip over fused silica-like anisotropic substrates," *Electron. Lett.*, vol. 12, no. 22, pp. 579–580, Oct. 28, 1976.
[18] T. C. Bryant and J. A. Weiss, "MSTRIP (parameters of microstrip); Computer program description," *IEEE Trans. Microwave Theory Tech.*, vol. MTT-19, pp. 418–419, 1971.
[19] E. G. Cristal, "Coupled-transmission-line directional couplers with coupled lines of unequal characteristic impedances," *IEEE Trans. Microwave Theory Tech.*, vol. MTT-14, pp. 337–346, July 1966.
[20] C. M. Krowne, "Microstrip transmission lines on pyrolytic boron nitride," *Electron. Lett.*, vol. 12, no. 24, pp. 642–643, Nov. 25, 1976.

Microstrip Propagation on Magnetic Substrates—Part I: Design Theory

ROBERT A. PUCEL, SENIOR MEMBER, IEEE, AND DANIEL J. MASSÉ, MEMBER, IEEE

Abstract—**Formulas and graphs are presented for the effective relative permeability and the filling factors of magnetic substrates in microstrip. Both the propagation and the magnetic loss filling factors are included. In the calculation of these quantities, use was made of the filling factors for dielectric substrates obtained from Wheeler's analysis and a duality relationship between magnetic and dielectric substrates derived in this paper.**

I. Introduction

A LARGE BODY of design information for microstrip on dielectric substrates has been accumulated over the last few years [1]–[3]. Equivalent design data for magnetic substrates are incomplete. It is our purpose to present the missing data in a form most useful to the design engineer. Before proceeding, we shall review briefly some basic formulas for dielectric substrates.

A cross section of microstrip on a dielectric–magnetic substrate is shown in Fig. 1. Provided the frequency is not too high, this structure will propagate a wave which for all practical purposes is a transverse electromagnetic wave. If the dielectric constant k of the substrate is much greater than unity, most of the electric energy is confined to the dielectric region in the vicinity of the strip conductor and ground plane. However, because some of the electric field also fringes out into the air space above the strip conductor, the value of the effective dielectric constant k_{eff} entering into the calculation of the characteristic impedance and phase velocity is less than k, that is $1<k_{eff}<k$. In other words, the propagation "filling factor" for the dielectric, here denoted as q_d, and defined by Wheeler [1] as

$$q_d = \frac{k_{eff} - 1}{k - 1} \tag{1}$$

is less than unity. Both k_{eff} and q_d are functions of the dielectric constant k and the geometrical factor w/h, the ratio of the conductor strip width to substrate height. This functional dependence can be derived from Wheeler's paper.

If dielectric losses are present, the effective value of the dielectric loss tangent $\tan \delta_{eff}$ is also less than the loss tangent of the substrate $\tan \delta_d$ and can be expressed in the form [4], [5]

Manuscript received May 27, 1971; revised August 12, 1971.
The authors are with the Research Division, Raytheon Company, Waltham, Mass. 02154.

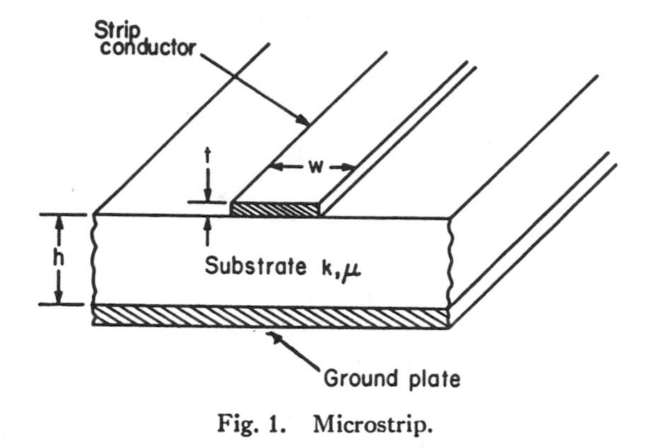

Fig. 1. Microstrip.

$$\tan \delta_{d_{eff}} = q_{d,l} \tan \delta_d \tag{2}$$

where $q_{d,l}$ is a filling factor for losses given by [5]

$$q_{d,l} = q_d \frac{k}{k_{eff}} = \frac{1 - k_{eff}^{-1}}{1 - k^{-1}}. \tag{3}$$

II. Magnetic Substrates

It would be convenient to have equivalent design formulas for substrates with magnetic properties. Fortunately, this information can be obtained from the above expressions by using a duality relationship for dielectric and magnetic substrates developed in the Appendix. This duality, based on an observation of Kaneki [6], allows one to calculate the functional dependence of the effective relative permeability μ_{eff} on w/h and the relative permeability μ of the substrate, once the functional dependence of k_{eff} on w/h and k is known. Thus the solution for the magnetic field distribution can be bypassed.

The duality relationship (which derives from the duality of k and $1/\mu$ in Maxwell's equations) is based on a TEM-mode approximation for the magnetic case, the same assumption as used for the dielectric case [1]–[3]. This relationship takes the form

$$\mu_{eff}(w/h, \mu) = \frac{1}{k_{eff}(w/h, \mu^{-1})}. \tag{4}$$

Note that the duality amounts to the conversions $k \rightarrow 1/\mu$ and $k_{eff} \rightarrow 1/\mu_{eff}$ in the formulas for the dielectric case. Equation (4) implies that one need not make a separate determination of the effective relative permeability if one has at hand tables or graphs of the effective dielectric constant.

Reprinted from *IEEE Trans. Microwave Theory Tech.*, vol. MTT-20, no. 5, pp. 304–308, May 1972.

It follows from (4) and (3) that a magnetic filling factor for propagation can be defined as

$$q_m = \frac{\mu_{eff}^{-1} - 1}{\mu^{-1} - 1}. \tag{5}$$

Note that $q_m(w/h, \mu) = q_d(w/h, \mu^{-1})$.

In like manner the expressions for the filling factor of the magnetic loss tangent $\tan \delta_m$ and the effective value of this loss tangent take the form

$$q_{m,l} = q_m \frac{\mu_{eff}}{\mu} = \frac{1 - \mu_{eff}}{1 - \mu} \tag{6}$$

or

$$\tan \delta_{m_{eff}} = q_{m,l} \tan \delta_m. \tag{7}$$

Our TEM assumption allows us to write simple formulas for the characteristic impedance Z_0, guide wavelength λ_g, and total substrate loss per wavelength $\alpha\lambda_g$ for microstrip on a substrate exhibiting both dielectric and magnetic properties. Thus we have

$$Z_0 = Z_0' \sqrt{\frac{\mu_{eff}}{k_{eff}}} \quad (\Omega) \tag{8}$$

$$\lambda_g = \lambda_0 / \sqrt{k_{eff}\mu_{eff}} \quad (\text{cm}) \tag{9}$$

$$\alpha\lambda_g = 27.3(\tan \delta_{d_{eff}} + \tan \delta_{m_{eff}}) \quad (\text{dB}). \tag{10}$$

The wavelength λ_0 corresponds to free space. Here Z_0' is the characteristic impedance when $\mu = k = 1$, which can be calculated exactly from the capacitance per unit length [7] for an air dielectric or from Wheeler's expressions letting $k=1$ [1]. The attenuation (10) of course only represents the substrate losses, to which must be added the contribution from conductor losses [5].

In Section III we shall present explicit formulas for μ_{eff} and, by way of review, for k_{eff} based on Wheeler's formulas.

III. Derivation of Formulas

Wheeler's analytic expressions [1] for the characteristic impedance of microstrip as a function of dielectric constant and the ratio w/h are given in piecewise form, one solution valid for $w/h \leq 2$, the other for $w/h \geq 2$. The two solutions do not join exactly at $w/h = 2$ (about a 5–10-percent error). His results may be graphically smoothed in this vicinity to join properly.

If we take Wheeler's expressions for Z_0, and set $k=1$ to give Z_0', the value for an air dielectric, then the effective value of dielectric constant may be obtained from

$$k_{eff} = \left[\frac{Z_0(w/h, 1)}{Z_0(w/h, k)}\right]^2. \tag{11}$$

For $w/h \leq 2$

$$k_{eff} = \frac{1 + k}{2}\left(\frac{A}{A - B}\right)^2 \tag{12a}$$

and for $w/h \geq 2$

$$k_{eff} = k\left(\frac{C - D}{C}\right)^2 \tag{12b}$$

where

$$A = \ln \frac{8h}{w} + \frac{1}{32}\left(\frac{w}{h}\right)^2 \tag{13a}$$

$$B = \frac{1}{2}\left(\frac{k - 1}{k + 1}\right)\left[\ln \frac{\pi}{2} + \frac{1}{k} \ln \frac{4}{\pi}\right] \tag{13b}$$

$$C = \frac{w}{2h} + \frac{1}{\pi}\left[\ln 2\pi\epsilon\left(\frac{w}{2h} + 0.94\right)\right] \tag{13c}$$

$$D = \frac{k - 1}{2\pi k}\left\{\ln \left[\frac{\pi\epsilon}{2}\left(\frac{w}{2h} + 0.94\right)\right] - \frac{1}{k} \ln \left(\frac{\epsilon\pi^2}{16}\right)\right\} \tag{13d}$$

from which the dielectric filling factors q_d and $q_{d,l}$, (1) and (3), respectively, may be computed.

The expressions for the effective relative permeability may be derived from the above by employing the duality relationship (4). Thus for $w/h \leq 2$

$$\mu_{eff} = \frac{2\mu}{1 + \mu}\left(\frac{A - B'}{A}\right)^2 \tag{14a}$$

and for $w/h \geq 2$

$$\mu_{eff} = \mu\left(\frac{C}{C - D'}\right)^2 \tag{14b}$$

where A and C are given by (13) and B' and D' are derived from B and D by letting $k \rightarrow \mu^{-1}$, that is,

$$B' = \frac{1}{2}\left(\frac{1 - \mu}{1 + \mu}\right)\left[\ln \frac{\pi}{2} + \mu \ln \frac{4}{\pi}\right] \tag{15a}$$

$$D' = \frac{1 - \mu}{2}\left\{\ln \left[\frac{\pi\epsilon}{2}\left(\frac{w}{2h} + 0.94\right)\right] - \mu \ln \left(\frac{\epsilon\pi^2}{16}\right)\right\}. \tag{15b}$$

The two filling factors q_m and $q_{m,l}$, (5) and (6), respectively, may be calculated by use of the expressions for μ_{eff} above.

Equations (14) and (15) together with (5)–(7) provide all the information necessary to design microstrip on magnetic substrates.

IV. Graphical Results

Since the purpose of this paper is to present design graphs for microstrip on ferrite and garnet substrates, our computations for μ_{eff}, q_m, and $q_{m,l}$ were made for values of magnetic constant less than unity, and indeed only for the practical range $0.4 < \mu < 1$. Because μ is less

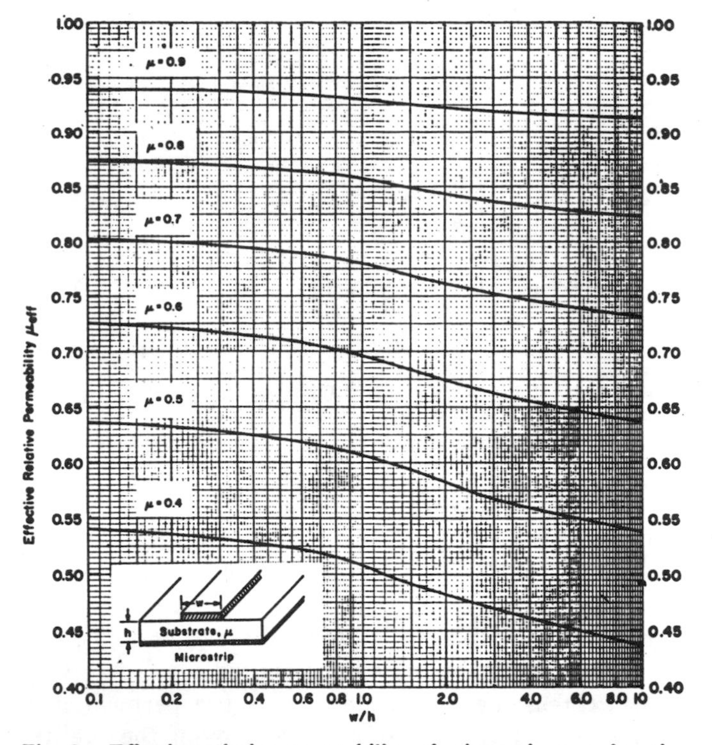

Fig. 2. Effective relative permeability of microstrip as a function of the relative permeability of the substrate and the geometrical parameter w/h.

than unity, the value for the air space above the microstrip, one should expect the effective value μ_{eff} to fall between μ and unity, that is $\mu<\mu_{eff}<1$. One can show that for $w/h\rightarrow 0$, $\mu_{eff}^{-1}\rightarrow\frac{1}{2}(1+\mu^{-1})$, and for $w/h\rightarrow\infty$, $\mu_{eff}\rightarrow\mu$.[1] In other words, μ_{eff} is bracketed in the range

$$\mu < \mu_{eff} < \frac{2\mu}{1+\mu} \cdot . \tag{16}$$

The curves of μ_{eff} in Fig. 2 illustrate this expected behavior.

The filling factors q_m and $q_{m,l}$ derived from μ are shown in Figs. 3 and 4. Because of the mild dependence of q_m on μ, only three curves were plotted to avoid crowding.

Observe in Fig. 4 that the loss filling factor becomes larger with increasing w/h, a reflection of the growing importance of the substrate. Using the limits on μ_{eff} derived above, one may show that $q_{m,l}$ falls in the range

$$\frac{1}{1+\mu} < q_{m,l} < 1. \tag{17}$$

Experimental verification of our design formulas are given in Part II of this paper [8], where we apply them to ferrite and garnet substrates, which, operated in certain biasing states, can be treated to a good approximation as reciprocal media.

[1] Wheeler shows that $\frac{1}{2}(k+1)<k_{eff}<k$, the lower limit applying to $w/h\rightarrow 0$, the upper to $w/h\rightarrow\infty$. Our results for the magnetic case are derived with the help of (4).

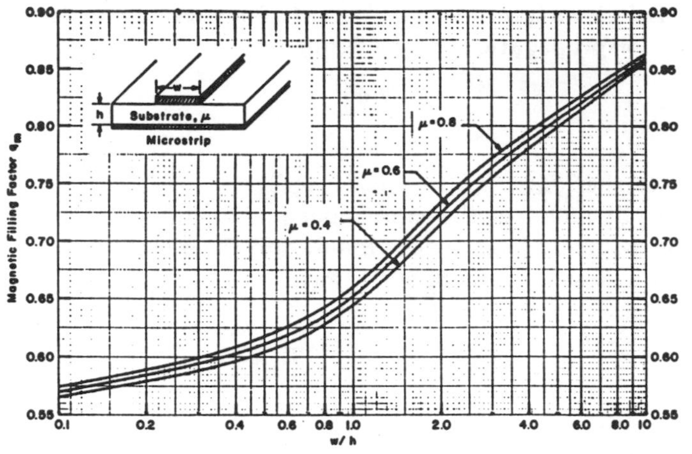

Fig. 3. Propagation magnetic filling factor of microstrip as a function of the relative permeability of the substrate and the geometrical parameter w/h.

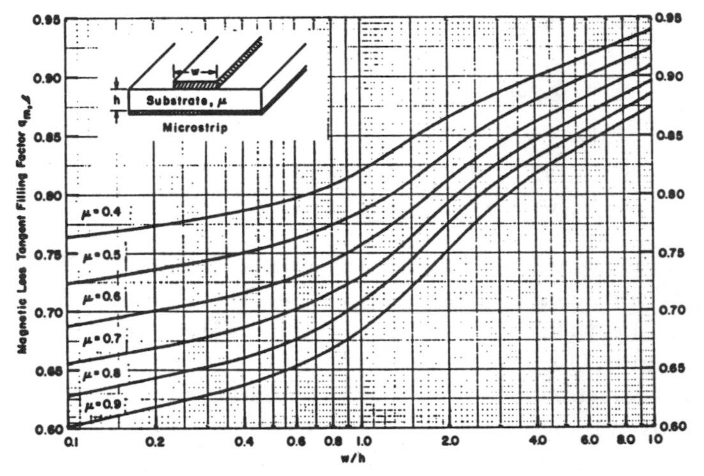

Fig. 4. Filling factor for magnetic loss tangent of microstrip as a function of the relative permeability of the substrate and the geometrical parameter w/h.

V. Summary

Design formulas and graphs were presented for the effective relative permeability and the propagation and attenuation filling factors of microstrip on magnetic substrates. The formulas were obtained by application of a duality relationship which exists between magnetic and dielectric substrates which circumvents the need for solution of the magnetic field distribution in microstrip.

Appendix

We wish to justify the duality relationship between the effective values of the dielectric constant and the relative permeability cited in (4), and establish the conditions under which it is valid.

Our point of departure is an enumeration of the assumptions for our analysis, namely, 1) TEM mode of propagation, 2) perfect conductors (infinite conductivity), and 3) isotropic, homogeneous, nongyromagnetic

TABLE I

RELATIONS PERTAINING TO SOLUTION FOR ELECTRIC AND MAGNETIC FIELDS

Condition	Electric Potential	Magnetic Potential
Potential function	$\psi_e(x, y, k)$	$\boldsymbol{k}\psi_m(x, y, \mu)$
Field vectors	$\boldsymbol{E} = -\nabla\psi_e$	$\boldsymbol{B} = \nabla \times (\boldsymbol{k}\psi_m) = -\boldsymbol{k} \times \nabla\psi_m$
	$\boldsymbol{D} = \epsilon_0 k(x, y)\boldsymbol{E}$	$\boldsymbol{H} = \mu_0^{-1}\mu^{-1}(x, y)\boldsymbol{B}$
Laplace's equation	$\nabla^2\psi_e = 0$	$\nabla^2\psi_m = 0$
Boundary conditions		
on conductor surfaces S_i, S_0	$k(x, y)\nabla\psi_e \cdot \boldsymbol{n} = -\sigma(s)/\epsilon_0$	$(\boldsymbol{k} \times \boldsymbol{n}) \cdot \nabla\psi_m = 0$
	$(\boldsymbol{k} \times \boldsymbol{n}) \cdot \nabla\psi_e = 0$	$\mu^{-1}(x, y)\nabla\psi_m \cdot \boldsymbol{n} = -\mu_0 j(s)$
on interface S_{ai}	$k\nabla\psi_{e,1} \cdot \boldsymbol{n} = \nabla\psi_{e,2} \cdot \boldsymbol{n}$	$(\boldsymbol{k} \times \boldsymbol{n}) \cdot \nabla\psi_{m,1} = (\boldsymbol{k} \times \boldsymbol{n}) \cdot \nabla\psi_{m,2}$
	$(\boldsymbol{k} \times \boldsymbol{n}) \cdot \nabla\psi_{e,1} = (\boldsymbol{k} \times \boldsymbol{n}) \cdot \nabla\psi_{e,2}$	$\mu^{-1}\nabla\psi_{m,1} \cdot \boldsymbol{n} = \nabla\psi_{m,2} \cdot \boldsymbol{n}$

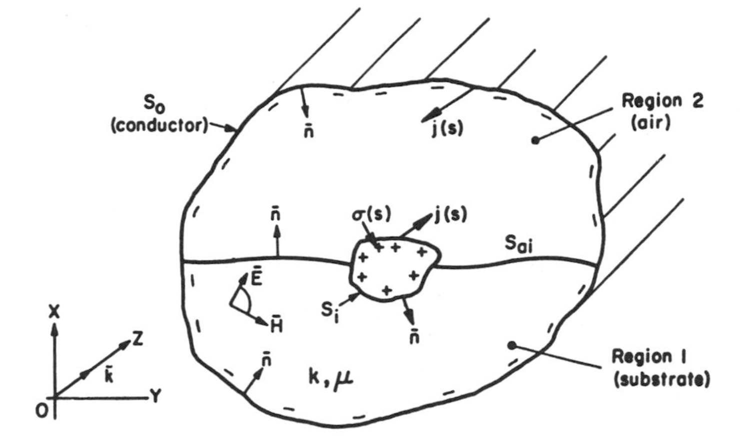

Fig. 5. Cross section of TEM structure relevant to derivation of duality relationship.

substrate. These assumptions are the ones used in all microstrip analyses; hence they are not restrictive for our purposes [1]–[3].

We assume for generality a cylindrical system of arbitrary cross section whose axis is along the Z direction, the assumed direction of propagation as illustrated in Fig. 5. Although a closed system is shown for convenience, our analysis also applies to open systems as well. The assumption of a closed outer conductor is not restrictive, since practical microstrip structures usually have an enclosure for shielding purposes.

With perfect conductors, the currents and charges, denoted by the surface densities $j(s)$ and $\sigma(s)$, reside on the conductor surfaces as shown in Fig. 5. Here s is a transverse surface coordinate on the conductors. For a TEM mode, the electric field $\boldsymbol{E}$ and the magnetic field $\boldsymbol{B} = \mu\boldsymbol{H}$ are in the x–y (transverse) plane. Their spatial distributions in this plane are solutions of a two-dimensional Laplace equation.

It is convenient to express $\boldsymbol{E}$ as the gradient of a scalar potential function ψ_e and $\boldsymbol{B}$ as the curl of a *vector* potential function $\boldsymbol{A}$ which is directed along the Z axis, the direction of the current; that is, $\boldsymbol{A}(x, y) = \boldsymbol{k}\psi_m$, where $\boldsymbol{k}$ is a unit vector along the Z axis. Thus

$$\boldsymbol{E} = -\nabla\psi_e(x, y, k) \tag{18}$$

$$\boldsymbol{B} = \nabla \times \boldsymbol{A} = -\boldsymbol{k} \times \nabla\psi_m(x, y, \mu). \tag{19}$$

Note that like $\boldsymbol{E}$, $\boldsymbol{B}$ is also proportional to the gradient of a scalar function (not to be confused with a scalar magnetic potential). Observe that ψ_e depends on the dielectric constant of the substrate k, but not on the relative permeability μ of the substrate. The converse is true for ψ_m. This is characteristic of a TEM solution. Since Laplace's equation is satisfied by the vector and magnetic potentials, then $\nabla^2\psi_e = 0$ and $\nabla^2\boldsymbol{A} = \boldsymbol{k}\nabla^2\psi_m = 0$ or $\nabla^2\psi_m = 0$. The solutions of these equations are determined by the geometry and the usual boundary conditions imposed on $\boldsymbol{E}$ and $\boldsymbol{B}$ at the conductor surfaces and at the substrate–air interface. These are summarized in Table I.

Perusal of Table I shows that ψ_e and ψ_m satisfy identical *sets* of boundary conditions provided the normal and tangential boundary conditions for $\boldsymbol{E}$ and $\boldsymbol{B}$ are interchanged (which is of no consequence to the solutions for ψ_e and ψ_m). Thus the *form* of the solution for ψ_m is identical to the *form* of the solution for ψ_e, if k is replaced by μ^{-1} and provided the surface densities $\sigma(s)$ and $j(s)$ are proportional. Assuming for a moment the latter to be true, then because of the linearity of the system, we may express ψ_e and ψ_m in the form

$$\psi_e(x, y, k) = \epsilon_0^{-1}QF(x, y, k) \tag{20}$$

$$\psi_m(x, y, \mu) = \mu_0 IF(x, y, \mu^{-1}) \tag{21}$$

where F is a scalar function satisfying Laplace's equation and $Q = \oint_{S_i}\sigma(s)\, ds$, $I = \int_{S_i} j(s)\, ds$ are the total charge/length and current on the conductors. Note that (20) and (21) imply that the magnetic field distribution can be obtained from a solution of an electrostatic problem. It is clear that (20) and (21) in conjunction with (18) and (19) establish the spatial orthogonality of the electric and magnetic fields.

In terms of (20) and (21) the effective dielectric constant k_{eff}, defined as the ratio of the stored electric energy per unit length with and without the substrate present ($k = 1$) at a specified charge Q, is expressible in the form

$$k_{eff} = K(g, k). \tag{22}$$

In a similar fashion, the magnitude of the effective relative permeability μ_{eff} equal to the ratio of the stored magnetic energy per unit length, with and without the substrate present ($\mu=1$) for a specified current is of the form

$$\frac{1}{\mu_{eff}} = K(g, \mu^{-1}) \tag{23}$$

where the energy density function K is given by

$$K(g, k) = \frac{\int_{\mathcal{A}} |\nabla F(x, y, 1)|^2 \, dx \, dy}{\int_{\mathcal{A}} k(x, y) \, |\nabla F(x, y, k)|^2 \, dx \, dy}. \tag{24}$$

Here $\mathcal{A}$ denotes the cross section of the propagating structure, excluding the conductors. The function $k(x, y)$ equals k in the substrate cross section, and unity in the air space above it. The parameter g is a geometrical factor, which equals w/h for the simple microstrip configuration of Fig. 1.

From (22) and (23) we obtain the interesting and useful duality relation

$$\mu_{eff}(g, \mu) = \frac{1}{k_{eff}(g, \mu^{-1})} \tag{25}$$

which was to be proven.

How realistic is the assumption of proportionality between $j(s)$ and $\sigma(s)$? For a system with a homogeneous cross section propagating a pure TEM mode, it is strictly correct. For a *non*homogeneous system, as we are considering here, $\sigma(s)$ and $j(s)$ *cannot* have identical distributions and this, *because* we postulate a TEM mode. Our reasoning is as follows. Suppose we have a TEM mode, and we assume $\sigma(s)$ and $j(s)$ to be proportional. Now consider a change in the dielectric constant of the substrate. Surely this will alter the charge distribution. By our assumption, the current distribution must also change, and so must the magnetic field distribution. But this cannot happen for a TEM mode, because the dielectric cannot affect the magnetic field.

Experience has shown that the magnetic field distribution, or more precisely, the inductance per unit length of microstrip, is *not* affected noticeably by the presence of a dielectric substrate. We can only conclude then that the charge and current distributions apparently do not deviate appreciably from proportionality and that the capacitance and inductance per unit length are not sensitive so much to the precise distribution of charge and current on the conductors, as they are to the geometrical configuration of the conductors and the substrate.

References

[1] H. A. Wheeler, "Transmission-line properties of parallel strips separated by a dielectric sheet," *IEEE Trans. Microwave Theory Tech.*, vol. MTT-13, pp. 172–186, Mar. 1965.

[2] H. E. Stinehelfer, Sr., "An accurate calculation of uniform microstrip transmission lines," *IEEE Trans. Microwave Theory Tech.*, vol. MTT-16, pp. 439–444, July 1968.

[3] E. Yamashita and R. Mittra, "Variational method for the analysis of microstrip lines," *IEEE Trans. Microwave Theory Tech.*, vol. MTT-16, pp. 251–256, Apr. 1968.

[4] J. D. Welch and J. J. Pratt, "Losses in microstrip transmission systems for integrated microwave circuits," *NEREM Rec. 8*, pp. 100–101, 1966.

[5] R. A. Pucel, D. Massé, and C. P. Hartwig, "Losses in microstrip," *IEEE Trans. Microwave Theory Tech.*, vol. MTT-16, pp. 342–350, June 1968; also "Correction to 'Losses in microstrip'," *IEEE Trans. Microwave Theory Tech.* (Corresp.), vol. MTT-16, p. 1064, Dec. 1968.

[6] T. Kaneki, "Analysis of linear microstrip using an arbitrary ferromagnetic substance as the substrate," *Electron. Lett.*, vol. 5, pp. 465–466, Sept. 18, 1969.

[7] H. B. Palmer, "The capacitance of a parallel-plate capacitor by the Schwarz–Christoffel transformation," *Elec. Eng.*, pp. 363–366, Mar. 1939.

[8] D. J. Massé and R. A. Pucel, "Microstrip propagation on magnetic substrates—Part II: Experiment," *IEEE Trans. Microwave Theory Tech.*, this issue, pp. 309–313.

Microstrip Propagation on Magnetic Substrates—Part II: Experiment

DANIEL J. MASSÉ, MEMBER, IEEE, AND ROBERT A. PUCEL, SENIOR MEMBER, IEEE

Abstract—**Experimental data taken on microstrip built on ferrite and garnet substrates are presented and compared with theoretical values calculated from formulas derived in a previous paper which were extended to gyromagnetic media. Good agreement has been obtained between experiment and theory. In particular the observed increase in wave attenuation at frequencies near ω_m is fully explained when the frequency dependence of the characteristic impedance is taken into account.**

I. Introduction

In Part I of this paper [1], we have shown how the duality principle permits one to calculate the effective permeability of a nongyromagnetic substrate once the expressions for a dielectric substrate are known. Thus it is now possible to predict the propagation characteristics of microstrip on a magnetic substrate as a function of the properties of the substrate material, and the geometrical parameters of the microstrip, against which experimental data may be compared.

In Part II we extend our design theory to gyromagnetic substrates, such as ferrites and garnets. Our analysis is supported by extensive experimental data. Particularly significant in our treatment is the fact that all losses observed on ferrite microstrip can now be accounted for and predicted. To our knowledge, this has not been done before.

Our measurements were taken for two bias conditions, namely for the substrate demagnetized and for the substrate latched, with the magnetization in the direction of propagation.

II. Permeability of Ferrite Substrates

The formulas presented in Part I are applicable, strictly speaking, only to substrate media exhibiting a scalar permeability. They must be modified to apply to gyromagnetic media, such as ferrites and garnets. For such substrates not only will the effective permeability depend on w/h, the ratio of strip conductor width to substrate height, but it also will be a function of the frequency of operation, as well as the magnitude and direction of the biasing field or magnetization.

For a TEM mode in microstrip the relative permeability of a gyromagnetic substrate biased along the direction of propagation, i.e., perpendicular to the transverse field, can be characterized by a two-dimensional tensor of the form,

$$\boldsymbol{\mu} = \begin{pmatrix} \mu & -j\kappa \\ j\kappa & \mu \end{pmatrix}. \tag{1}$$

We consider two biasing modes of operation that are of practical interest, namely, 1) the substrate demagnetized, and 2) the substrate partially magnetized. For this second case we shall be concerned only with the "latched" state, that is, the biasing state corresponding to remanent magnetization. We consider these two cases below.

A. Substrate Demagnetized

For this state, the off-diagonal tensor elements vanish, that is $\kappa=0$. Schlömann [2] has shown that for the demagnetized state the permeability tensor element $\mu=\mu_{\text{dem}}$ can be approximated by the simple expression

$$\mu_{\text{dem}} = \frac{1}{3}\left\{1 + 2\sqrt{1 - \left(\frac{\omega_m}{\omega}\right)^2}\right\} \tag{2}$$

where $\omega_m=\gamma(4\pi M_s)$, and ω is the operating frequency. Here γ is the gyromagnetic ratio and $4\pi M_s$ is the saturation magnetization. This expression also has been used by Denlinger in his analysis of microstrip [3].

B. Substrate Latched

Green and co-workers [4] at this laboratory, using their extensive experimental data, have proposed the following phenomenological expressions for the diagonal and off-diagonal elements of the tensor permeability for partially magnetized ferrites,

$$\mu = \mu_{\text{dem}} + (1 - \mu_{\text{dem}})\left(\frac{4\pi M}{4\pi M_s}\right)^{3/2} \tag{3}$$

$$\kappa = \frac{\gamma(4\pi M)}{\omega} \tag{4}$$

where $4\pi M$ is the magnetization. If the substrate is latched into remanence, the magnetization takes on its remanent value $4\pi M_r$.

Sandy, of this laboratory, extending his finite difference technique [5] for solving field distributions in microstrip to gyrotropic media [6], has obtained values for the inductance per unit length as a function of w/h, μ, and κ. By curve fitting to graphs of this inductance plotted against these parameters, he deduced the

Manuscript received May 27, 1971; revised August 12, 1971.
The authors are with the Research Division, Raytheon Company, Waltham, Mass. 02154.

Reprinted from *IEEE Trans. Microwave Theory Tech.*, vol. MTT-20, no. 5, pp. 309–313, May 1972.

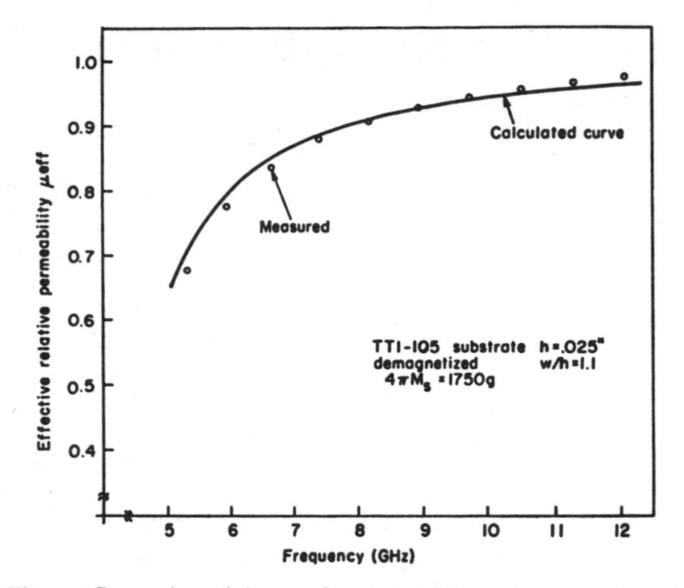

Fig. 1. Comparison of the experimental and theoretical values of the effective relative permeability for a demagnetized ferrite substrate.

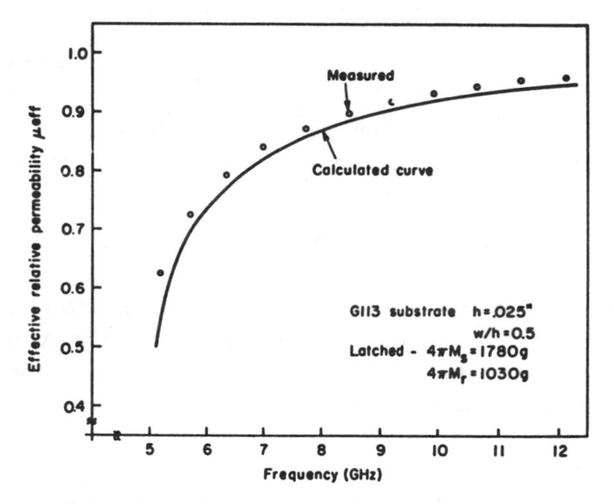

Fig. 2. Comparison of the experimental and theoretical values of the effective relative permeability for a latched garnet substrate.

following analytic approximation for the relative permeability $\mu=\mu_{\text{mag}}$ for the substrate,

$$\mu_{\text{mag}} = \frac{\mu^2-\kappa^2}{\mu} \quad \frac{1}{1-\frac{1}{7}\sqrt{\frac{h}{w}}\left(\frac{\kappa}{\mu}\right)^2 \ln\left(1+\frac{\mu}{\mu^2-\kappa^2}\right)} \tag{5}$$

where μ and κ are obtained from (3) and (4) above. The denominator term containing the ratio w/h is a first-order correction arising from the gyromagnetic nature of the substrate. It is a small perturbation of about 5 percent, depending on the frequency of operation. No physical basis is claimed for (5).

The expressions for μ_{dem} and for μ_{mag} for the two states of operation now can be used to evaluate the effective permeability and the magnetic filling factors.

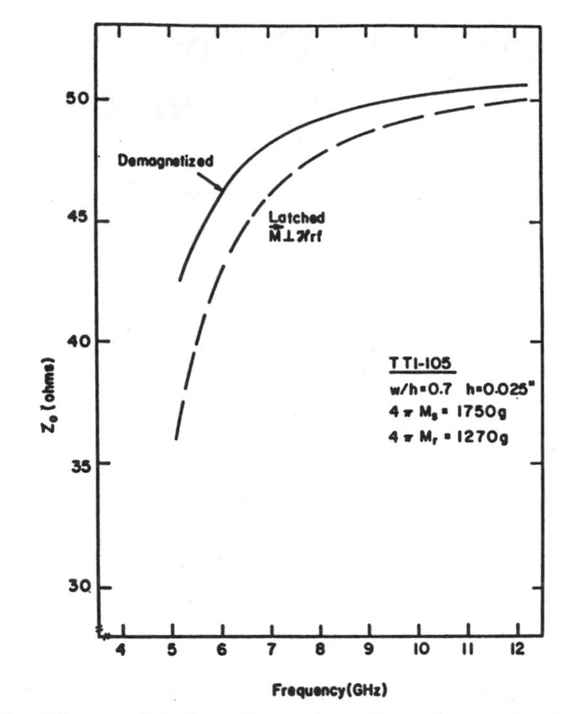

Fig. 3. Characteristic impedance of a microstrip on a ferrite substrate calculated from the measured values of relative permeability.

III. Experimental Results

The experimental data were obtained with ring resonators printed on various magnetic substrates.[1] Ring resonators, rather than open line sections, were used to reduce radiation losses to an insignificant amount [7]. From the observed resonant frequencies, the guide wavelength was determined. This provided information necessary for calculating the effective permeability. The losses were derived from the unloaded Q's of these resonance responses.

The guide wavelength corresponding to a particular resonant frequency f_0 was obtained from the formula

$$\lambda_g(f_0) = \frac{l}{n} \tag{6}$$

where l is the mean circumference of the ring and n is the mode number for that resonance, i.e., the number of wavelengths in the ring. By making l long enough to obtain many resonances in the frequency range of interest, we were able to plot a curve of λ_g versus f_0. To determine μ_{eff} from λ_g via (9) of Part I, it was first necessary for us to extract k_{eff}.

In principle k_{eff} can be calculated from Wheeler's analysis [8]. However, to do this, one must allow for the frequency dispersion observed with dielectric substrates [3], [9]. To include dispersion, we plotted against frequency the product $k_{\text{eff}}\,\mu_{\text{eff}}$ deduced from wavelength measurements. We then approximated $k_{\text{eff}}(f)$ by a linear

[1] Manufactured by Trans. Tech, Inc., Gaithersburg, Md.

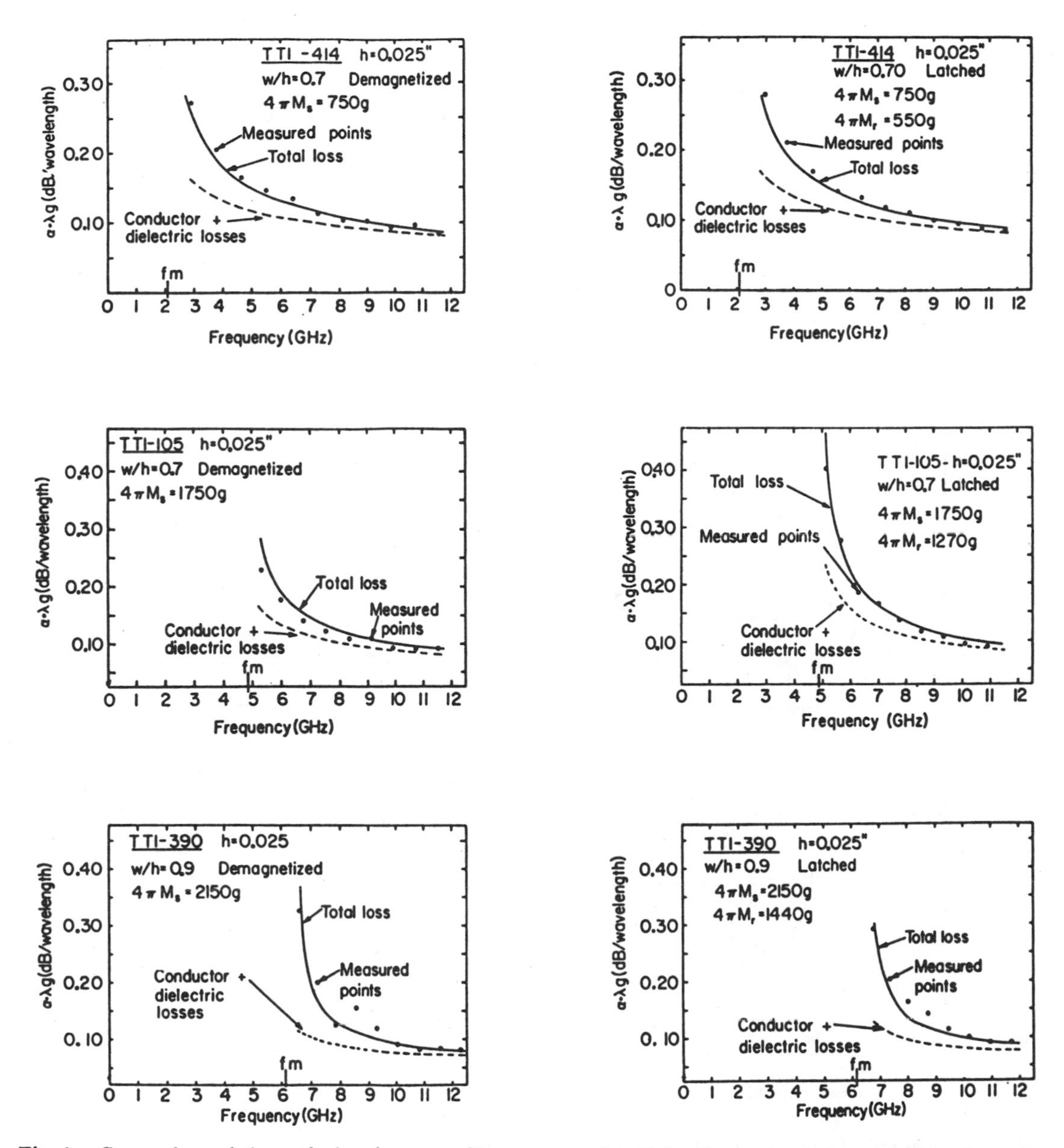

Fig. 4. Comparison of theoretical and measured loss per wavelength for three microstrips on ferrite substrates.

function of frequency having a positive slope which crossed the ordinate (zero-frequency) axis at the value of k_{eff} predicted by Wheeler's analysis. The slope of the line was adjusted to be asymptotic to the curve of k_{eff} μ_{eff} at the upper frequency end (where μ_{eff} approaches unity). This simple approximation for $k_{eff}(f)$ was satisfactory for the frequency range and microstrip dimensions considered. The effective permeability obtained by this method automatically reflects the frequency dependence of μ, as well as any possible additional dispersion, such as observed with k_{eff}. However, this additional contribution, if it did exist, was negligible in our experiments, since the deduced variation of μ_{eff} with frequency could be explained entirely by the frequency dependence of μ alone.

Figs. 1 and 2 compare measured and calculated values of the effective relative permeability for two different substrate materials and two values of w/h. Fig. 1 is for a demagnetized ferrite substrate and Fig. 2 is for a latched garnet substrate. It is evident that the agreement between theory and experiment is satisfactory.

With the values of μ_{eff} and k_{eff} obtained from these experiments, the characteristic impedance Z_0 was calculated from (8) of Part I. Fig. 3 illustrates the computed frequency dependence of Z_0 so obtained for a ferrite substrate in the demagnetized and latched states. The strong frequency variation, particularly at the low-frequency end, is a reflection of the frequency dependence of the permeability as ω approaches ω_m, characterized by (2)–(5). VSWR data, taken on a straight microstrip with good transitions, have confirmed our impedance calculations.

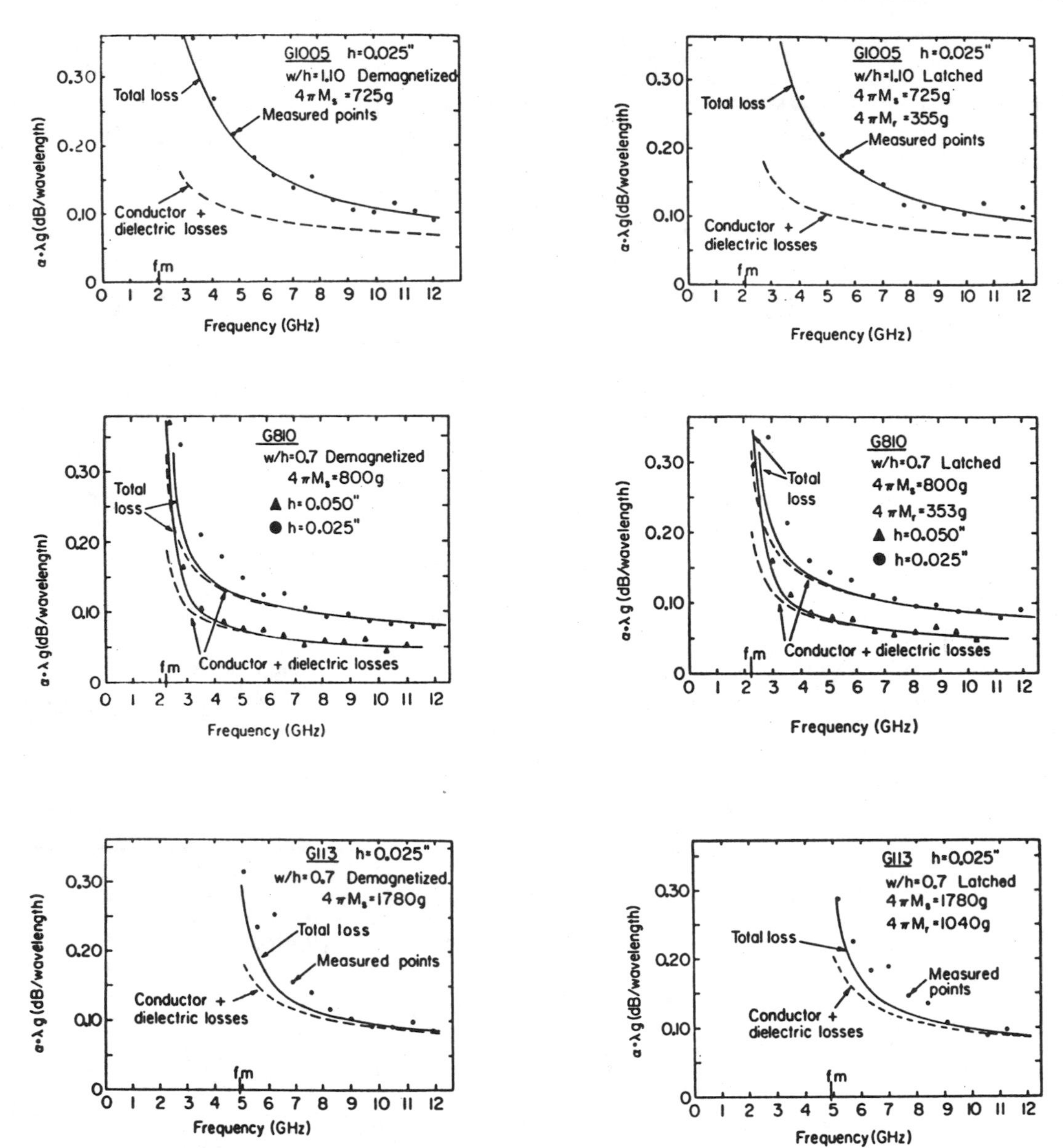

Fig. 5. Comparison of theoretical and measured loss per wavelength for three microstrips on garnet substrates.

The attenuation per wavelength was determined from measurements of the unloaded Q of the ring resonators [10]. The attenuation per wavelength, denoted here as $\alpha_T\lambda_g$, can be separated into a sum of three components, $\alpha_c\lambda_g$ due to skin losses in the strip conductor and the ground plane, $\alpha_d\lambda_g$ attributable to the dielectric losses of the substrate, and $\alpha_m\lambda_g$ arising from the magnetic losses of the substrate. The last two (substrate) components are given by (10) of Part I.

The magnetic loss tangents, $\tan \delta_m = \mu''/\mu'$, were calculated from the experimental data of Green *et al.* of this laboratory [4]. Here μ' and μ'' are the real and imaginary parts of the permeability. The dielectric loss tangents were obtained from manufacturers' data.

The conductor losses were obtained from the analysis of Pucel *et al.* [10] by correcting Z_0 and λ_g to account for the magnetic properties of the substrate in accordance with (8) and (9) of Part I. These losses were calculated for a gold metallization 5 μm thick, deposited on a substrate with a surface roughness of 20 μin rms.

We isolated the attenuation caused by magnetic losses by first calculating the attenuation attributable to the dielectric and conductor losses and comparing the sum of these two to the measured loss. We compared the difference with the computed magnetic losses and found good agreement.

Measurements were taken both on ferrite and garnet substrates,[2] for several w/h ratios and two substrate thicknesses. Both the demagnetized and the latched states were used. Figs. 4 and 5 summarize our calculated and measured results. The points represent measured data; the dashed lines, the calculated sum of conductor and dielectric attenuation; the solid line, calculated total attenuation.

[2] The substrates were annealed after machining.

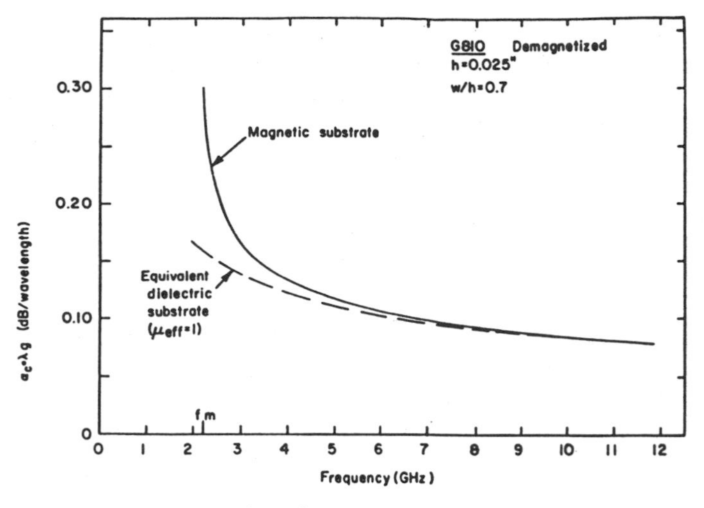

Fig. 6. Calculated conductor attenuation per wavelength for a dielectric and a garnet substrate.

The curves in Fig. 4 correspond to three ferrites with different values of saturation magnetization, hence ω_m. The agreement between the calculated and measured losses is very good in most cases. The significant increase in attenuation at the low-frequency end (as ω_m is approached) occurs for several reasons. First, magnetic losses (i.e., μ'') increase rapidly in the vicinity of ω_m. This can be seen from the difference in the solid and dashed curves. Second, and just as important, the contribution of the conductor losses increases rapidly as $\omega \rightarrow \omega_m$. This increase arises because the attenuation produced by skin losses varies inversely with the characteristic impedance Z_0 [10] and of course Z_0 drops rapidly as $\omega \rightarrow \omega_m$, as Fig. 3 shows.

The data in Fig. 5 apply to three garnets with different saturation magnetizations. Note that the data for G810 are for two different substrate thicknesses. The lower loss for the thicker substrate, of course, is attributable to the reduced copper losses, which vary inversely with thickness [10].

In general, agreement between theory and experiment is very good. This agreement shows that there are no "hidden" or anomalous sources of loss that are peculiar to magnetic substrates. Thus our accounting for the rapid increase in attenuation produced by skin losses as $\omega \rightarrow \omega_m$ removes the discrepancies between measured and calculated losses, which puzzled researchers in the past who failed to correct for the strong frequency dependence of Z_0. That the magnitude of this correction is not negligible is illustrated in Fig. 6 for one of the garnets. In this figure the conductor loss is plotted for the demagnetized garnet and for a material of identical dielectric properties but with μ_{eff} equal to 1 at all frequencies. The large increase in $\alpha_c \lambda_g$ when one approaches ω_m is shown clearly.

From the practical viewpoint the rapid rise in attenuation at low frequencies, which we now know cannot be reduced for a specific material, illustrates the importance of choosing a substrate with as small a magnetic loss (μ'') as possible, and emphasizes the necessity of operating as far from ω_m as is practical.

IV. Summary

We have proposed analytic formulas that adequately describe the microwave properties of microstrip on ferrite substrates. Good agreement is obtained between the calculated and measured properties. Our results show that wave attenuation on ferrite substrates can be predicted if proper account of the frequency dependence of the characteristic impedance is made.

Acknowledgment

The authors wish to thank Dr. C. Hartwig and Dr. F. Sandy for the many discussions. They also wish to thank the latter for allowing them to use his unpublished results; and Mrs. Y. Leighton, who fabricated the microstrip, and G. Flynn, who performed the microwave measurements.

References

[1] R. A. Pucel and D. J. Massé, "Microstrip propagation on magnetic substrates—Part I: Design theory," this issue, pp. 304–308.
[2] J. Green, E. Schlömann, F. Sandy, and J. Saunders, "Characterization of the microwave tensor permeability of partially magnetized materials," Semi-annual Rep. RADC-TR-69-93, Feb. 1969.
[3] E. J. Denlinger, "A frequency dependent solution for microstrip transmission lines," *IEEE Trans. Microwave Theory Tech.*, vol. MTT-19, pp. 30–39, Jan. 1971.
[4] J. Green, F. Sandy, and C. Patton, "Microwave properties of partially magnetized ferrites," Rome Air Development Center, Rome, N. Y., Final Rep. RADC-TR-68-312, Aug. 1968.
[5] F. Sandy and J. Sage, "Use of finite difference approximations to partial difference equations for problems having boundaries at infinity," *IEEE Trans. Microwave Theory Tech.* (Corresp.), vol. MTT-19, pp. 484–486, May 1971.
[6] F. Sandy, private communication, Jan. 1971.
[7] P. Troughton, "Measurement techniques in microstrip," *Electron. Lett.*, vol. 5, pp. 25–26, Jan. 23, 1969.
[8] H. A. Wheeler, "Transmission-line properties of parallel strips separated by a dielectric sheet," *IEEE Trans. Microwave Theory Tech.*, vol. MTT-13, pp. 172–185, Mar. 1965.
[9] C. P. Hartwig, D. Massé, and R. A. Pucel, "Frequency dependent behavior of microstrip," presented at the 1968 Int. Microwave Symp. (Detroit, Mich.), May 20–22, 1968.
[10] R. A. Pucel, D. Massé, and C. P. Hartwig, "Losses in microstrip," *IEEE Trans. Microwave Theory Tech.*, vol. MTT-16, pp. 342–350, June 1968; also "Correction to 'Losses in microstrip'," *IEEE Trans. Microwave Theory Tech.* (Corresp.), vol. MTT-16, p. 1064, Dec. 1968.

Part II
Dispersion Analysis of Microstrip Lines

THE topic of this part has been perhaps one of the most popular research subjects in the microwave community for the past decade or two. As described in the introduction to Part I, the quasi-TEM analysis, though simple, is not rigorous and hence becomes increasingly inaccurate as the operating frequency of the microstrip line is raised. This is because the quasi-TEM analysis neglects the hybrid nature of the guided modes in the microstrip line. Because of such a simple assumption, the quasi-TEM analysis cannot predict the dispersive nature of the microstrip line, even the dominant one. To correctly account for such a dynamic nature of the microstrip mode, a rigorous full-wave solution is needed which takes into account all possible field components and satisfies all the physically required boundary conditions in the structure.

Over the years, a number of analytical and numerical methods have been developed and applied for the microstrip line configurations. Many of them are reported in the papers in this part. They include the integral equation method (Yamashita and Atsuki), the method of lines (Schulz and Pregla), and the spectral domain method (Itoh; Knorr and Tufekcioglu). In the spectral domain method, one derives a coupled algebraic equation by taking a spatial Fourier transform of the coupled integral equations related to the current components on the strip and the tangential electric field components. This method has very attractive features from the numerical point of view as numerical efforts are reduced to a minimum. The method has been extensively studied and refined over the past decade. Although the spectral domain method appears to be the most popular technique for the analysis of microstrip lines, this method is not foolproof. For instance, at the frequencies where the conductor thickness is substantial with respect to the guide wavelength or other structural dimensions, this method is not a good choice. The nature of the formulation in the spectral domain method requires a conducting strip with an infinitesimal thickness. The method of lines is a hybrid algorithm which combines an analytical solution of the wave equation in the direction normal to the substrate and a discretized numerical process in the plane parallel to the substrate surface. In general, a method is more adaptable to a variety of structures if it contains more numerical features. Therefore, the spectral domain method is more restricted in use, whereas it is more numerically efficient in general.

A number of numerical methods can be applied to other planar transmission line structures described in the later parts. It is rather straightforward to extend the formulation of a microstrip line to structures containing multilayered configurations and multiple strips or slots.

Finally, it should be noted that the definition of the characteristic impedance in a microstrip line is somewhat arbitrary (Knorr and Tufekcioglu). There are three definitions possible; voltage-current, power-current and voltage-power. At the low frequency limit, all three converge to a unique value that can also be computed by the quasi-TEM approximation. Three definitions provide three different values at an arbitrary frequency. Choice of the definition is often a matter of specific application. In general, however, the power-current definition appears to be the most popular choice for the microstrip line.

A Rigorous Solution for Dispersive Microstrip

MASAHIRO HASHIMOTO, SENIOR MEMBER, IEEE

Abstract —Closed-form solutions are presented for the frequency-dependent characteristic impedance of microstrip as defined by the ratio of the electromagnetic power to the square of the electric current. The analysis uses the rigorous spectral-domain approach based on the charge-current formulation. Analytical expressions for the impedance solutions show that the frequency dispersion occurring in microstrip is characterized in terms of three different impedances. The characteristic impedance of a TEM line given in the limit as the frequency decreases is derived from one of these impedances, and the other two are involved in expressing the nature of dispersion to vanish in the limit. Conversely, as the frequency increases, these dispersive parts grow rapidly. Some comments are given in conjunction with previous works.

I. Introduction

SINCE A REAL microstrip line is not a TEM line, the problem of microstrip is treated as the problem of full-wave analysis. In the early stages, therefore, a large amount of attention was paid to evaluating, from Maxwell's equations, the frequency dispersion in microstrip. Recent concern of some people in the microwave community seems to have shifted to the subject of how the frequency dispersion can be characterized by a circuit-theory-based model. (The reader can find good introductions to current trends of microstrip in recent papers published in this TRANSACTIONS [1] or other related journals [2].)

The first important feature of the modeling mentioned above is that it will help us to explain the mechanism of dispersion by means of circuit description, just as a true TEM line is described in terms of circuit elements such as distributed line-capacitance, distributed line-inductance, and characteristic impedance. The second feature is that a certain extension of the fundamental concept of a TEM line may be possible. For the latter, however, we need to establish some other modeling that contains the influence of field excitation at terminals of microstrip. To do this, Getsinger [3] defines the "apparent characteristic impedance" on the basis of accurate measurements of the reflection loss in the transfer of power between the source and the stripline. Kuester, Chang, and Lewin [4] discuss the same problem from theoretical viewpoints, and conclude that if no definition can be found which has a sufficiently broad usefulness, one may have to bear certain possible definitions in mind. We must await further experimental evidence.

Nevertheless, whatever the results of measurements to follow, the significance for evaluating the characteristic impedance of dispersive microstrip remains unchanged. The main objective of this paper is to present analytical expressions for the characteristic impedance given by the ratio of the electromagnetic power flowing along the stripline to the square of the total longitudinal electric current. Unlike numerical procedures, lengthy calculations to obtain solutions are necessary, but the resulting expressions are simple. Although the paper does not claim to have given a new formulation, the closed-form expressions obtained for the characteristic impedance are new and rigorous. We shall begin with the known formulation for electric charge and electric currents on the strip.

II. Basic Equations

Fig. 1 shows a geometry of the open microstrip we wish to consider. The substrate material between a strip of zero thickness with width w and a ground plane is assumed to have magneto-dielectric properties. For the special case when the substrate is a dielectric as usual, we put $\mu_r = 1$ and $\epsilon_r > 1$. The electric charge sources and the electric current sources are induced over the upper and lower surfaces of the conducting strip. The surface charge density, given at a point x ($y = 0$) as the sum of the upper and lower charges, is denoted by ρ_s, and the surface current densities flowing at a point x toward the longitudinal direction (z-direction) of the stripline axis and the transverse direction (x-direction) are denoted by J_s and J_{st}, respectively. These are related by the continuity equation

$$\frac{\partial J_{st}}{\partial x} = j(\beta J_s - \omega \rho_s) \tag{1}$$

where β is the propagation constant and ω is the angular frequency. We note that the phase factor $e^{j(\omega t - \beta z)}$ is suppressed through the paper.

Since J_{st} stands for the sum of the upper and lower current densities on the strip, the edge condition for J_{st} is

$$J_{st}(\pm w/2) = 0. \tag{2}$$

The value of J_{st} may be considered to be rather small when narrow strip approximations are adopted, but neglecting this current results in the inaccurate solution which is unable to describe the whole nature of dispersive charac-

Manuscript received January 23, 1985; revised June 13, 1985. Portions of this work were presented at the Sino-Japanese Joint Meeting on Optical Fiber Science and Electromagnetic Theory, Beijin, China, May 16–19, 1985.

The author is with the Department of Applied Electronic Engineering, Osaka Electro-Communication University, Neyagawa, Osaka 572, Japan.

Reprinted from *IEEE Trans. Microwave Theory Tech.,* vol. MTT-33, no. 11, pp. 1131–1137, November 1985.

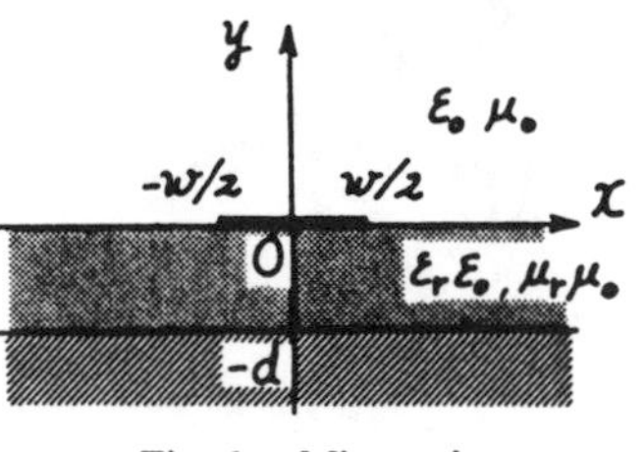

Fig. 1. Microstrip.

teristics. In particular, J_{st} plays an important role in expressing the characteristic impedance. To see this, we develop a rigorous theory based on the charge-current formulation.

Although a variety of approaches to obtain rigorous solutions are examined [5], the charge-current model appears to have a certain possibility of extending quasi-static approximations familiar to a quasi-TEM line. The mathematical formulation presented here was originated in 1971 by Fujiki, Hayashi, and Suzuki [6], and in 1972 independently by Itoh and Mittra [5], and refined later by Chang and Kuester [7]. Basic equations to be derived from the theory will be summarized as follows.

Integrating (1) over the strip and taking account of (2), we obtain

$$\beta I = \omega Q \tag{3}$$

where I and Q are, respectively, the total current and the total charge per unit length such that

$$I = \int_{-w/2}^{w/2} J_s \, dx \quad Q = \int_{-w/2}^{w/2} \rho_s \, dx. \tag{4}$$

Electromagnetic fields in air ($y > 0$) can be represented in terms of vector potential A and scalar potential ϕ by

$$\boldsymbol{E} = -j\omega \boldsymbol{A} - \nabla\phi \quad \boldsymbol{H} = \mu_0^{-1}\nabla \times \boldsymbol{A} \tag{5}$$

where A and ϕ obey

$$\frac{\partial^2\phi}{\partial x^2} + \frac{\partial^2\phi}{\partial y^2} + (k^2 - \beta^2)\phi = 0 \tag{6a}$$

$$\frac{\partial^2 \boldsymbol{A}}{\partial x^2} + \frac{\partial^2 \boldsymbol{A}}{\partial y^2} + (k^2 - \beta^2)\boldsymbol{A} = \boldsymbol{0} \tag{6b}$$

thus, satisfying the Lorentz gauge. Here, k is the wavenumber in air ($= \omega\sqrt{\epsilon_0\mu_0}$).

Because of the absence of y-directed currents on the strip, the components of vector A are A_x and A_z only; $A_y = 0$ everywhere. Thus

$$E_y = -\frac{\partial\phi}{\partial y}. \tag{7}$$

Differentiating (7) with respect to y, we obtain

$$\frac{\partial E_y}{\partial y} = -\frac{\partial^2\phi}{\partial y^2} = \left(\frac{\partial^2}{\partial x^2} + k^2 - \beta^2\right)\phi. \tag{8}$$

Finally, using Gauss' law and equating the term on the left-hand side to zero over the upper surface of the conducting strip ($y = +0$) as

$$\frac{\partial E_y}{\partial y} = -\frac{\partial E_x}{\partial x} + j\beta E_z = 0 \tag{9}$$

we find that ϕ satisfies a homogeneous differential equation of the second order. The symmetric solution which corresponds to the fundamental stripline mode is

$$\phi = A\cosh\left(\sqrt{\beta^2 - k^2}\,x\right) \tag{10}$$

where A is an arbitrary constant.

On the other hand, the scalar potential, as well as the vector potential, may be expressible in terms of ρ_s, J_s, and J_{st}. According to the literature [6],[7], these potentials are given on the strip by

$$\phi = \frac{1}{2\pi\epsilon_0}\int G_e(x - x')\rho_s(x')\,dx' \tag{11a}$$

$$A_x = \frac{\mu_0}{2\pi}\int G_h(x - x')J_{st}(x')\,dx' + \frac{1}{j\omega} \times \frac{1}{2\pi\epsilon_0}\int \frac{\partial}{\partial x}M(x - x')\rho_s(x')\,dx' \tag{11b}$$

$$A_z = \frac{\mu_0}{2\pi}\int G_h(x - x')J_s(x')\,dx' - \frac{\beta}{\omega} \times \frac{1}{2\pi\epsilon_0}\int M(x - x')\rho_s(x')\,dx' \tag{11c}$$

where

$$\int \cdot\, dx' \equiv \int_{-w/2}^{w/2} \cdot\, dx'$$

and $G_e(x)$, $G_h(x)$, and $M(x)$ are even functions of x, as listed in Appendix I. The tangential components of the electric vector are then

$$\begin{aligned} E_x &= -j\omega A_x - \frac{\partial\phi}{\partial x} \\ &= -j\omega \times \frac{\mu_0}{2\pi}\int G_h(x - x')J_{st}(x')\,dx' \\ &\quad - \frac{1}{2\pi\epsilon_0}\int \frac{\partial}{\partial x}[G_e(x - x') + M(x - x')]\rho_s(x')\,dx' \end{aligned} \tag{12a}$$

$$\begin{aligned} E_z &= -j\omega A_z + j\beta\phi \\ &= -j\omega \times \frac{\mu_0}{2\pi}\int G_h(x - x')J_s(x')\,dx' \\ &\quad + j\beta \times \frac{1}{2\pi\epsilon_0}\int [G_e(x - x') + M(x - x')]\rho_s(x')\,dx'. \end{aligned} \tag{12b}$$

Substituting (10) into the left-hand side of (11a) gives a Fredholm integral equation of the first kind, from which ρ_s is solvable. As stated in the theory by Fujiki, Hayashi, and Suzuki [6], letting $E_x = 0$ and $E_z = 0$ in (12) also give the integral equations of the same type. The solutions J_{st} and J_s are obtained using ρ_s previously obtained. The value of β can be determined from the edge condition (2). Such solutions are found to satisfy (1) or (3) exactly. In other words, the value of β can be calculated in a straightforward manner by (3), inserting J_s and ρ_s into (4). This is useful because we do not need to calculate J_{st}.

We start with these basic equations, which are rigorous to any structure of the open microstrip shown in Fig. 1.

III. Characteristic Impedance

For reasons discussed earlier [1]–[4] as to how we should define the characteristic impedance Z_0 for practical use in design applications, we assume[1]

$$Z_0 = \frac{2P}{II^*} \tag{13}$$

where P is the total average power in the z-direction

$$P = \frac{1}{2}\int \boldsymbol{E} \times \boldsymbol{H}^* \cdot d\boldsymbol{S}. \tag{14}$$

Since ρ_s, J_s, and J_{st} are assumed to have already been determined, it is possible to evaluate the electromagnetic fields in the air and substrate regions. Such fields can be described in terms of ρ_s and J_s. It follows that the power P can be described by the convolutions $J_s \times J_s$, $\rho_s \times \rho_s$, and $\rho_s \times J_s$. In fact, we have

$$P = P_{11} + P_{22} + P_{12} \tag{15}$$

where

$$P_{11} = \frac{1}{2}\iint z_{11}(x-x') \times \left[J_s(x')J_s^*(x) + J_{st}(x')J_{st}^*(x)\right] dx'\,dx \tag{16a}$$

$$P_{22} = \frac{1}{2}\iint z_{22}(x-x')\left(\frac{\omega}{\beta}\rho_s(x')\right)\left(\frac{\omega}{\beta}\rho_s^*(x)\right) dx'\,dx \tag{16b}$$

$$P_{12} = \iint z_{12}(x-x')J_s(x')\left(\frac{\omega}{\beta}\rho_s^*(x)\right) dx'\,dx \tag{16c}$$

and the functions $z_{11}(x-x')$, $z_{22}(x-x')$, and $z_{12}(x-x')$ are the "distributed mutual impedances" between the points x and x', as given in Appendix II. Note that the $J_{st} \times J_{st}$ term in P_{11} is derived by combining the three convolutions so as to use the relation (1).

If the "effective mutual impedances" Z_{ij} are defined as

$$P_{ij} = \frac{1}{2}Z_{ij}II^* \tag{17}$$

then

$$Z_0 = Z_{12} + Z_{11} + Z_{22}. \tag{18}$$

This is a rigorous expression for Z_0. We do not mention analytical details of the derivation outlined above so as not to become involved in mathematical complexities. Instead, we will show later another way to obtain the solutions, since the two solutions derived in different ways are in complete agreement.

In the static limit, P_{11} and P_{22} vanish, and P_{12} tends to the power of a TEM line

$$P_{12} \to \frac{1}{2}I\phi \qquad (M(x) \to 0). \tag{19}$$

[1] This subject is beyond the scope of the paper.

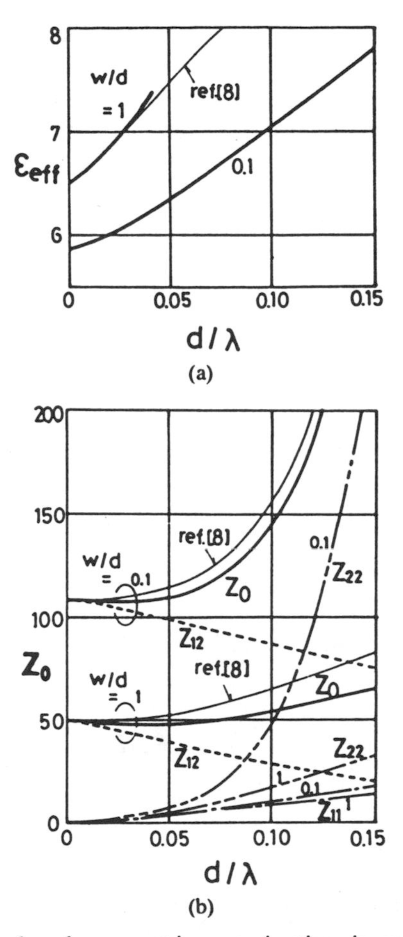

Fig. 2. Examples of narrow strip approximations in comparison with numerical solutions by Kowalski and Pregla [8] ($\epsilon_r = 9.7$, $\mu_r = 1$). (a) Effective dielectric constant ϵ_{eff} ($\equiv \beta^2/k^2$). (b) Characteristic impedance Z_0 ($= Z_{12} + Z_{11} + Z_{22}$).

This means that P_{11} and P_{22} are describing the "dispersive powers" due to dispersion under consideration.

To obtain an approximate solution valid for narrow strips, we use the fact that $z_{12}(x)$ possesses a logarithmic singularity at $x = 0$, whereas $z_{11}(x)$ and $z_{22}(x)$ are regular, and hence, set

$$Z_{11} \simeq z_{11}(0) \quad Z_{22} \simeq z_{22}(0). \tag{20}$$

For Z_{12} associated with P_{12}, we must perform double integration over the strip. However, calculations of the singular part of $z_{12}(x)$ give the static solution, which is reduced to the well-known formula for the characteristic impedance of a TEM line. The remaining terms are nonsingular and thus easy to obtain within the range of approximations (20). This procedure of calculation is proposed in [7]. Numerical examples for $\epsilon_r = 9.7$ and $\mu_r = 1$ are shown in Fig. 2. Curves in the two figures are plotted versus d/λ, where λ is the wavelength in air.

Fig. 2(a) is a test of the validity of the theory, because the result is the same with that in [7]. In Fig. 2(b), we plot curves of Z_{11}, Z_{22}, and Z_{12} for $w/d = 0.1$ and 1. It is important to note that the value of Z_{12} decreases with increasing the frequency and the others increase rapidly if the strip is narrow. The value of Z_0 calculated with the abovementioned approximations decreases a little, but soon begins to increase rapidly and reaches the reliable numerical solution of Kowalski and Pregla [8]. We therefore conclude that the dispersive nature of the characteristic impedance is mainly described by Z_{11} and Z_{22}, as shown in the figure.

IV. LINE INDUCTANCE AND LINE CAPACITANCE

Recalling that the functions $J_s(x)$ and $\rho_s(x)$ are even and real, whereas the function $J_{st}(x)$ is odd and imaginary, or

$$J_{st}^*(x) = -J_{st}(x) \tag{21}$$

we consider a lossy system of microstrip in the following. The power loss per unit length of the conducting strip can be calculated by

$$\frac{1}{2}R_l I^2 = \frac{1}{2}\int E_z J_s \, dx \tag{22a}$$

$$\frac{1}{2}R_t I^2 = -\frac{1}{2}\int E_x J_{st} \, dx \tag{22b}$$

$$\frac{1}{2}RI^2 = \frac{1}{2}(R_l + R_t)I^2 = \frac{1}{2}\int (E_z J_s - E_x J_{st}) \, dx \tag{22c}$$

where subscripts l and t denote "longitudinal component" and "transverse component," respectively. It should again be emphasized that the goal of this section is not to calculate a loss of the transmission system but to derive analytical expressions for the characteristic impedance. We will see this immediately.

Integrating by parts in (22c), we obtain

$$\begin{aligned}\frac{1}{2}RI^2 = &-j\omega \times \frac{1}{2}\left\{\frac{\mu_0}{2\pi}\iint G_h(x-x') \cdot [J_s(x')J_s(x) - J_{st}(x')J_{st}(x)]\, dx'\, dx \right.\\ &\left. - \frac{1}{2\pi\epsilon_0}\iint [G_e(x-x') + M(x-x')] \cdot \rho_s(x')\rho_s(x)\, dx'\, dx\right\}\\ = &-j\omega \times \frac{1}{2}\int (J_s A_z - J_{st}A_x - \rho_s \phi)\, dx.\end{aligned} \tag{23}$$

For lossy lines with complex β ($= \beta_0 - j\alpha$), $J_s(x)$, $J_{st}(x)$, and $\rho_s(x)$ are slightly deviated from the values in a lossless system, according to

$$\frac{\partial}{\partial x}(\delta J_{st}) = j(\delta\beta J_s + \beta\, \delta J_s - \omega\, \delta\rho_s). \tag{24}$$

In addition to these, we must calculate infinitesimal increments of $G_h(x)$, $G_e(x)$, and $M(x)$. Total loss $RI^2/2$ is given by the sum of these contributions. If, however, we undergo the increments of $J_s(x)$, $J_{st}(x)$, and $\rho_s(x)$, and if we ignore the effects of $G_h(x)$, $G_e(x)$, and $M(x)$, then

$$\delta \int (J_s A_z - J_{st} A_x - \rho_s \phi)\, dx = -\frac{2\,\delta\beta}{\omega} \times 2P_{12}. \tag{25}$$

This is a statement of variational principle for charge and currents. A proof of the theorem is given in Appendix III. Hence, the first variation of the integral on the right-hand side of (23) becomes

$$\begin{aligned}\frac{1}{2}RI^2 = &-j\omega \times \frac{1}{2}\left\{\frac{\mu_0}{2\pi}\iint \frac{\partial}{\partial\beta} G_h(x-x') \times [J_s(x')J_s(x) - J_{st}(x')J_{st}(x)]\, dx'\, dx \right.\\ &- \frac{1}{2\pi\epsilon_0}\iint \frac{\partial}{\partial\beta}[G_e(x-x') + M(x-x')] \cdot \rho_s(x')\rho_s(x)\, dx'\, dx\\ &\left. - \frac{2}{\omega} \times \frac{1}{2\pi\epsilon_0}\iint [G_e(x-x') + M(x-x')] \cdot \rho_s(x')J_s(x)\, dx'\, dx\right\}\delta\beta\end{aligned} \tag{26}$$

where the third term in { } corresponds to (25), and

$$\delta\beta = -j\alpha. \tag{27}$$

Another useful definition for α is

$$\alpha = \frac{R}{2Z_0}. \tag{28}$$

Using this formula, we have

$$\begin{aligned}\frac{1}{2}RI^2 &= Z_0 I^2 \times \alpha\\ &= 2(P_{11} + P_{22} + P_{12}) \times \alpha\\ &= (Z_{11}I^2 + Z_{22}I^2 + Z_{12}I^2) \times \alpha.\end{aligned} \tag{29}$$

Furthermore, comparing (26) with (29), we find

$$z_{11}(x) = -\frac{\omega}{2} \times \frac{\mu_0}{2\pi}\frac{\partial}{\partial\beta} G_h(x) \tag{30a}$$

$$z_{22}(x) = \frac{\omega}{2} \times \left(\frac{\beta}{\omega}\right)^2 \times \frac{1}{2\pi\epsilon_0}\frac{\partial}{\partial\beta}[G_e(x) + M(x)] \tag{30b}$$

$$z_{12}(x) = \frac{\beta}{\omega} \times \frac{1}{4\pi\epsilon_0}[G_e(x) + M(x)] \tag{30c}$$

which are identical with the previous results obtained in Appendix II.

Now, let us define line inductances and line capacitances as

$$L(\beta)=\frac{1}{I^2}\times\frac{\mu_0}{2\pi}\iint G_h(x-x')\times[J_s(x')J_s(x)-J_{st}(x')J_{st}(x)]\,dx'dx \quad (31a)$$

$$L_l(\beta)=\frac{1}{I^2}\times\frac{\mu_0}{2\pi}\iint G_h(x-x')J_s(x')J_s(x)\,dx'dx \quad (31b)$$

$$L_t(\beta)=\frac{1}{I^2}\times\frac{\mu_0}{2\pi}\iint G_h(x-x')[-J_{st}(x')J_{st}(x)]\,dx'dx \quad (31c)$$

$$\frac{1}{C(\beta)}=\frac{1}{Q^2}\times\frac{1}{2\pi\epsilon_0}\iint[G_e(x-x')+M(x-x')]\times\rho_s(x')\rho_s(x)\,dx'dx \quad (31d)$$

$$\frac{1}{C_l(\beta)}=\frac{1}{Q^2}\times\frac{1}{2\pi\epsilon_0}\iint[G_e(x-x')+M(x-x')]\times\rho_s(x')\left(\frac{\beta}{\omega}J_s(x)\right)dx'dx=\frac{1}{Q^2}\left(\frac{2\beta}{\omega}\right)P_{12} \quad (31e)$$

$$\frac{1}{C_t(\beta)}=\frac{1}{Q^2}\times\frac{1}{2\pi\epsilon_0}\iint\frac{\partial}{\partial x}[G_e(x-x')+M(x-x')]\times\rho_s(x')\left(\frac{J_{st}(x)}{j\omega}\right)dx'dx \quad (31f)$$

where

$$L(\beta)=L_l(\beta)+L_t(\beta) \quad (32a)$$

$$\frac{1}{C(\beta)}=\frac{1}{C_l(\beta)}+\frac{1}{C_t(\beta)}. \quad (32b)$$

We must be careful that the parameter β included in $J_s(x)$, $J_{st}(x)$, and $\rho_s(x)$ is not taken as a variable to calculate the circuit elements of (31). If this were done, the results which follow would be wrong.[2]

In terms of these circuit elements, (22) can be written

$$\frac{1}{2}RI^2=-\frac{1}{2}\left\{j\omega L(\beta)+\frac{\beta^2}{j\omega C(\beta)}\right\}I^2 \quad (33a)$$

$$\frac{1}{2}R_lI^2=-\frac{1}{2}\left\{j\omega L_l(\beta)+\frac{\beta^2}{j\omega C_l(\beta)}\right\}I^2 \quad (33b)$$

$$\frac{1}{2}R_tI^2=-\frac{1}{2}\left\{j\omega L_t(\beta)+\frac{\beta^2}{j\omega C_t(\beta)}\right\}I^2. \quad (33c)$$

Letting $R=R_l=R_t=0$ in (33) gives a set of dispersion equations for lossless lines

$$\beta=\omega\sqrt{L(\beta)C(\beta)} \quad (34a)$$

$$\beta=\beta_l(\beta)\equiv\omega\sqrt{L_l(\beta)C_l(\beta)} \quad (34b)$$

$$\beta=\beta_t(\beta)\equiv\omega\sqrt{L_t(\beta)C_t(\beta)}. \quad (34c)$$

If we want to determine the value of β, we can select one equation in (34) as a dispersion equation. These three conditions are incorporated in the theory so that if one of these is satisfied the others are satisfied too. A convenient choice may be (34a) or (34b), which is entirely valid even for pure-TEM and quasi-TEM modes. Note that, in [7], the value of β is determined from (34b).

The next step is to apply the above circuit description to the variational expression (26). The result is

$$\frac{1}{2}RI^2=-j\omega\times\frac{1}{2}\left\{\left(\frac{\partial L(\beta)}{\partial\beta}\right)I^2-\left(\frac{\partial}{\partial\beta}\frac{1}{C(\beta)}\right)Q^2-\left(\frac{2}{\beta}\right)\frac{Q^2}{C_t(\beta)}\right\}\delta\beta \quad (35)$$

or in the equivalent form

$$\frac{1}{2}RI^2=\frac{\alpha}{2}\left\{-\omega\frac{\partial L(\beta)}{\partial\beta}+\frac{\beta^2}{\omega}\frac{\partial}{\partial\beta}\frac{1}{C(\beta)}+\frac{2\beta}{\omega}\frac{1}{C_t(\beta)}\right\}I^2=\frac{\alpha}{2}\frac{\partial}{\partial\beta}\left\{\frac{\beta^2}{\omega C(\beta)}-\omega L(\beta)\right\}I^2-\frac{\alpha\beta}{\omega C_t(\beta)}I^2. \quad (36)$$

Hence, we have

$$Z_0=\frac{1}{2}\frac{\partial}{\partial\beta}\left\{\frac{\beta^2}{\omega C(\beta)}-\omega L(\beta)\right\}-\frac{\beta}{\omega C_t(\beta)}. \quad (37)$$

This is another rigorous expression for Z_0 with arbitrary parameters.

As the operating frequency decreases or the width of the conducting strip decreases, the transverse elements $L_t(\beta)$ and $1/C_t(\beta)$ described above become negligible, and therefore the theory provides the low-frequency operating solutions as given by Kuester, Chang, and Lewin [4]. Namely, if we replace $L(\beta)$ in (37) by $L_l(\beta)$ and $C(\beta)$ by $C_l(\beta)$ and neglect the last term, then we obtain their (*KCL*) solution. The accuracy of this class of approximation may, however, hold invalid over the entire (complex) β-plane, which will be used to determine the z-dependent field excited at an input terminal of microstrip by means of the spectral-domain method. The work presented in this section suggests further research that includes the investigation of the complex behavior of the transverse elements on the β-plane. The *KCL* solution for Z_0 behaves as an increasing function with increasing the frequency. This will be proved as follows, rewriting (37) with (34) as:

$$Z_0=\sqrt{\frac{L_l(\beta)}{C_l(\beta)}}-\sqrt{\frac{L_l(\beta)}{C_l(\beta)}}\frac{\partial\beta_l(\beta)}{\partial\beta}-\sqrt{\frac{L_t(\beta)}{C_t(\beta)}}\frac{\partial\beta_t(\beta)}{\partial\beta} \quad (38)$$

[2] Corrections should be made to these results. For example, in (37), the last term should be removed from the right side.

and neglecting the third term in (38). Note that the second term is the leading term which increases as the frequency increases. Note also that the first term becomes equal to Z_{12} because of

$$Z_{12} = \frac{\beta}{\omega C_l(\beta)} = \sqrt{\frac{L_l(\beta)}{C_l(\beta)}}. \tag{39}$$

We see that the increasing property of Z_0 can therefore be characterized in terms of the negative derivative $\partial\beta_l(\beta)/\partial\beta$.

In the case of $\omega \to 0$, $L(\beta)$ and $C(\beta)$ approach $L_l(\beta)$ and $C_l(\beta)$, respectively, and lastly these limiting values are to coincide with the values of the static elements by Vaynshteyn and Fialkovskiy [9].

V. Conclusion

A theory has been developed to obtain a rigorous solution for dispersive microstrip. Closed-form expressions for the characteristic impedance Z_0 have been derived. It is pointed out that the frequency dispersion of Z_0 in the graph is caused as a result of the negative slope of the curve $L(\beta)\times C(\beta)$ versus β. Since, in the previous theory, the transverse elements are ignored, the theory seems valid for limited use in the low-frequency range. The present theory holds valid at all frequencies and thus is applicable to strips with arbitrary width in the high-frequency operating regime, which are solved in [10].

Appendix I

Functions $G_h(x)$, $M(x)$, and $G_e(x)$ are as follows:

$$G_h(x) = 2\int_0^\infty \tilde{G}_h(\alpha)\cos(\alpha x)\,d\alpha \tag{A1a}$$

$$M(x) = 2\int_0^\infty \tilde{M}(\alpha)\cos(\alpha x)\,d\alpha \tag{A1b}$$

$$G_e(x) = 2\int_0^\infty \tilde{G}_e(\alpha)\cos(\alpha x)\,d\alpha \tag{A1c}$$

where

$$\tilde{G}_h(\alpha) = \frac{1}{\kappa_0 + \mu_r^{-1}\kappa_1\coth(\kappa_1 d)} \tag{A2}$$

$$\tilde{M}(\alpha) = \frac{(\epsilon_r\mu_r - 1)k^2}{(\mu_r\kappa_0 + \kappa_1\coth(\kappa_1 d))(\epsilon_r\kappa_0 + \kappa_1\tanh(\kappa_1 d))\kappa_0}$$

$$\tilde{G}_e(\alpha) = \frac{1}{\kappa_1 + \epsilon_r\kappa_0\coth(\kappa_1 d)} \times \left(\frac{\kappa_1}{\kappa_0}\right) \tag{A3}$$

$$= \tilde{G}_h(\alpha) - \frac{\alpha^2+\beta^2}{k^2}\tilde{M}(\alpha) \tag{A4}$$

and

$$\kappa_0 = \sqrt{\alpha^2+\beta^2-k^2} \tag{A5}$$

$$\kappa_1 = \sqrt{\alpha^2+\beta^2-\epsilon_r\mu_r k^2}. \tag{A6}$$

Appendix II

Mutual impedances between two points on the strip are defined as

$$z_{11}(x) = 2\int_0^\infty \tilde{z}_{11}(\alpha)\cos(\alpha x)\,d\alpha \tag{A7a}$$

$$z_{22}(x) = 2\int_0^\infty \tilde{z}_{22}(\alpha)\cos(\alpha x)\,d\alpha \tag{A7b}$$

$$z_{12}(x) = 2\int_0^\infty \tilde{z}_{12}(\alpha)\cos(\alpha x)\,d\alpha \tag{A7c}$$

where

$$\begin{aligned}\tilde{z}_{11}(\alpha) &= \frac{1}{4\pi}\left(\frac{k^2\beta}{\omega\epsilon_0}\right)\left[\frac{1}{\kappa_0} + \frac{1}{\mu_r\kappa_1}\right.\\ &\quad\left.\cdot\coth(\kappa_1 d) - \frac{d}{\mu_r\sinh^2(\kappa_1 d)}\right]\tilde{G}_h^2(\alpha)\\ &= -\frac{\omega}{2}\times\frac{\mu_0}{2\pi}\frac{\partial}{\partial\beta}\tilde{G}_h(\alpha)\end{aligned} \tag{A8}$$

$$\begin{aligned}\tilde{z}_{22}(\alpha) &= -\left(\frac{\beta}{k}\right)^2\tilde{z}_{11}(\alpha) + \frac{1}{4\pi}\left(\frac{\beta}{\omega\epsilon_0}\right)\left(\frac{\kappa_0\beta}{k}\right)^2\\ &\quad\cdot\left\{\frac{1}{\kappa_0}\left[\frac{1}{\epsilon_r\kappa_1}\tanh(\kappa_1 d)\right.\right.\\ &\quad\left.+\frac{d}{\epsilon_r\cosh^2(\kappa_1 d)}\right]\\ &\quad+\left[\frac{1}{\kappa_0} + \frac{1}{\mu_r\kappa_1}\coth(\kappa_1 d)\right.\\ &\quad\left.-\frac{d}{\mu_r\sinh^2(\kappa_1 d)}\right]\tilde{G}_h(\alpha)\\ &\quad-\left[\frac{1}{\kappa_0} + \frac{1}{\epsilon_r\kappa_1}\tanh(\kappa_1 d) + \frac{d}{\epsilon_r\cosh^2(\kappa_1 d)}\right]\\ &\quad\left.\cdot\tilde{G}_e(\alpha)\right\}\tilde{M}(\alpha)\\ &= \frac{\omega}{2}\times\left(\frac{\beta}{\omega}\right)^2\times\frac{1}{2\pi\epsilon_0}\frac{\partial}{\partial\beta}[\tilde{G}_e(\alpha) + \tilde{M}(\alpha)]\end{aligned} \tag{A9}$$

$$\tilde{z}_{12}(\alpha) = \frac{\beta}{\omega}\times\frac{1}{4\pi\epsilon_0}[\tilde{G}_e(\alpha) + \tilde{M}(\alpha)]. \tag{A10}$$

Appendix III

Calculate the first variation for charge and currents. Then

$$\begin{aligned}&\delta\int(J_s A_z - J_{st}A_x - \rho_s\phi)\,dx\\ &= \frac{\mu_0}{2\pi}\iint G_h(x-x')[J_s(x')2\,\delta J_s(x)\\ &\quad - J_{st}(x')2\,\delta J_{st}(x)]\,dx'\,dx\\ &\quad - \frac{1}{2\pi\epsilon_0}\iint[G_e(x-x') + M(x-x')]\\ &\quad\times\rho_s(x')2\,\delta\rho_s(x)\,dx'\,dx.\end{aligned} \tag{A11}$$

The $\delta\rho_s$ is given by

$$\delta\rho_s = \frac{\beta}{\omega}\,\delta J_s + \frac{\delta\beta}{\omega} J_s - \frac{1}{j\omega}\frac{\partial}{\partial x}(\delta J_{st}). \tag{A12}$$

Substituting this into (A11) and integrating by parts, the right-hand side becomes

$$-\frac{2\,\delta\beta}{\omega} \times \frac{1}{2\pi\epsilon_0} \iint [G_e(x-x') + M(x-x')]$$

$$\times\, \rho_s(x') J_s(x)\, dx'\, dx$$

$$-\frac{2}{j\omega}\int E_z \delta J_s\, dx + \frac{2}{j\omega}\int E_x \delta J_{st}\, dx.$$

The first double integral is found to be equal to $2P_{12}$, and the second and third integrals vanish because $E_z = 0$ and $E_x = 0$ on the strip.

References

[1] For example, see R. Bhartia and P. Pramanick, "A new microstrip dispersion model," *IEEE Trans. Microwave Theory Tech.*, vol. MTT-32, pp. 1379–1384, Oct. 1984.

[2] For example, see R. H. Jansen and M. Kirschning, "Arguments and an accurate model for the power-current formulation of microstrip characteristic impedance," *Arch. Elek. Übertragung.*, vol. 37, pp. 108–112, 1983.

[3] W. J. Getsinger, "Measurement and modeling of the apparent characteristic impedance of microstrip," *IEEE Trans. Microwave Theory Tech.*, vol. MTT-31, pp. 624–632, Aug. 1983.

[4] E. F. Kuester, D. C. Chang, and L. Lewin, "Frequency-dependent definitions of microstrip characteristic impedance," in *Dig. Int. URSI Symp. Electromagnetic Waves* (Munich), Aug. 26–29, 1980, pp. 335B1–3.

[5] T. Itoh and R. Mittra, "Analysis of microstrip transmission lines," Antenna Lab., Univ. Illinois, Urbana, IL, Sci. Rep. No. 72-5, June 1972.

[6] Y. Fujiki, Y. Hayashi, and M. Suzuki, "Analysis of strip transmission lines by iteration method," *Trans. Inst. Elec. Commun. Eng. Jpn*, vol. 55-B, pp. 212–219, May 1972.

[7] D. C. Chang and E. F. Kuester, "An analytic theory for narrow open microstrip," *Arch. Elek. Übertragung.*, vol. 33, pp. 199–206, 1979; see also Sci. Rep. No. 28, Univ. Colorado, Boulder, CO, May 1978.

[8] G. Kowalski and R. Pregla, "Dispersion characteristics of single and coupled microstrips," *Arch. Elek. Übertragung.*, vol. 26, pp. 276–280, 1972.

[9] L. A. Vaynshteyn and A. T. Fialkovskiy, "Modes in slotted and stripline waveguides: Variational method and simpler results," *Radio Eng. Electron. Phys.*, vol. 21, pp. 1–11, 1976.

[10] E. F. Kuester and D. C. Chang, "Theory of dispersion in microstrip of arbitrary width," *IEEE Trans. Microwave Theory Tech.*, vol. MTT-28, pp. 259–265, Mar. 1980; see also Sci. Rep. No. 35, Univ. Colorado, Boulder, CO, Sept. 1978.

A New Technique for the Analysis of the Dispersion Characteristics of Microstrip Lines

RAJ MITTRA, FELLOW, IEEE, AND TATSUO ITOH, MEMBER, IEEE

Abstract—Dispersion characteristics of shielded microstrip lines are investigated using a new technique. The method utilizes the well-known singular integral equation approach for deriving an alternate form of eigenvalue equation with superior convergence properties. It is shown that accurate numerical results may be obtained from this eigenvalue equation using only a 2×2 matrix equation. In comparison, the conventional formulation of the problem requires the use of matrices that are much larger in size. Aside from the numerical efficiency, the simplicity of the method makes it possible to conveniently extract higher order modal solutions for the propagation constants that affect the high-frequency application of microstrip lines.

Even though the derivation of the determinantal equation requires some intricate mathematical manipulations, the user may bypass these completely and use the final eigenvalue equation which is easily programmable on the computer.

I. Introduction

THE increasing use of microstrip lines at microwave frequencies has recently created considerable interest in the study of dispersion characteristics of these lines. Until quite recently, much of the work on the microstrip line was based on a TEM analysis [1]–[7]. This analysis is employed to calculate the static capacitance of the structure from which the characteristic impedance and the propagation wavenumber are subsequently derived. However, this analysis, which is necessarily approximate, is inadequate for estimating the dispersion properties of the line at higher frequencies. Recently it has been demonstrated by several authors [8]–[11] that a rigorous procedure based on hybrid mode analysis of the structure is required to study the high frequency performance of these lines. A number of different techniques have been employed for calculating the dispersion effects. For example, Hornsby and Gopinath [9]–[11] have applied the finite difference method as well as minimization techniques to derive the dispersion relations. On the other hand Zysman and Varon [8] have formulated the problem in terms of homogeneous, coupled, integral equations which were subsequently transformed into a matrix equation. In all of these approaches the size of the matrix that needs to be processed to determine the zeros of its determinant is usually quite large.

The purpose of this paper is to present a new formulation of the determinantal equation that allows one to extract accurate solutions for the dispersion characteristics of the line by seeking the roots of a 2×2 determinant. The accuracy of the solution is demonstrated by comparing the present results with those derived by other authors using matrices of the order of 40 or more. Aside from the numerical efficiency of the method, the simplicity of the determinantal equation allows one to readily predict and extract higher order modal solutions for the wavenumber β from a study of the functions involved in the equation. The knowledge of the higher order solutions is important since they obviously affect the usefulness of the microstrip line at high frequencies.

Manuscript received February 2, 1970; revised June 16, 1970. Part of this paper was presented at the 1970 International Microwave Symposium, Newport Beach, Calif., May 1970. This work was supported by U. S. Army Research Grant DA-ARO-G1103.

The authors are with the Antenna Laboratory, Department of Electrical Engineering, University of Illinois, Urbana, Ill. 61801.

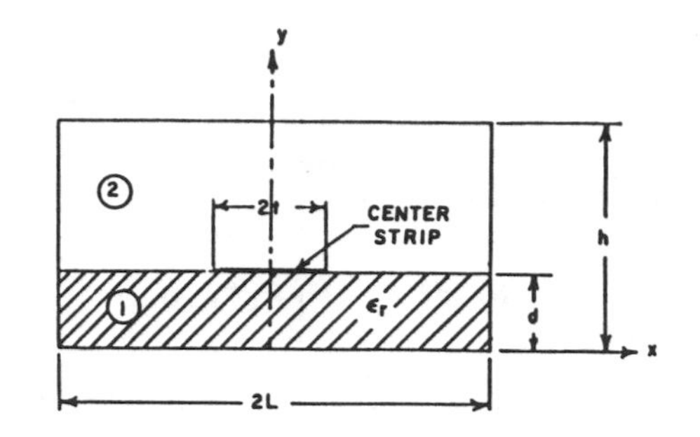

Fig. 1. Cross section of shielded microstrip lines.

II. Formulation

The cross section of the shielded microstrip line, to be analyzed in this paper, is shown in Fig. 1. It consists of a conducting strip placed on a dielectric substrate and a box-type enclosure of a perfectly conducting material. The initial steps in the formulation that parallel those employed by other authors (see, for instance, [8]), are as follows.

1) Write the electric and magnetic fields in subregions 1 and 2 in terms of hybrid modes, i.e., a superposition of TE and TM modes. The form of the representation is chosen such that the boundary conditions on the sidewalls are satisfied by the functions involved.

2) Apply the continuity conditions at the gap ($|x| > t$) and the boundary condition on the strip ($|x| < t$) at the interface $y=d$. Up to this point the formulation is strictly conventional.

3) The third step, which is the key one in the present formulation, involves the transformation of the original equations into an auxiliary set of equations that can be attacked by certain well-known methods useful for solving singular integral equations [12], [13]. The final result of the indicated manipulations is a determinantal equation which has a highly superior convergence

Reprinted from *IEEE Trans. Microwave Theory Tech.*, vol. MTT-19, no. 1, pp. 47–56, January 1971.

property as compared to the determinantal equation corresponding to the original set of equations.

4) The desired determinantal equation is studied for a suitable truncation size, and it is found that a 2×2 size is adequate for the range of parameters that is of interest. The resulting determinantal equation, which is rather simple in form, is then solved for the wavenumber β as a function of the free-space wavenumber k_0.

For the even symmetric case the TM and TE field components are derivable from scalar potentials $\psi^{(e)}$ and $\psi^{(h)}$. One can write TM fields as follows:

$$E_{zi} = j\frac{k_i^2 - \beta^2}{\beta}\psi_i^{(e)}e^{-j\beta z}$$

$$\bar{E}_{ti}^{(e)} = \nabla_t\psi_i^{(e)}e^{-j\beta z}$$

$$\bar{H}_{ti}^{(e)} = \frac{\omega\epsilon_i}{\beta}\hat{z}\times\nabla_t\psi_i^{(e)}e^{-j\beta z}. \tag{1}$$

TE fields may be written as follows:

$$H_{zi} = j\frac{k_i^2 - \beta^2}{\beta}\psi_i^{(h)}e^{-j\beta z}$$

$$\bar{E}_{ti}^{(h)} = -\frac{\omega\mu}{\beta}\hat{z}\times\nabla_t\psi_i^{(h)}e^{-j\beta z}$$

$$\bar{H}_{ti}^{(h)} = \nabla_t\psi_i^{(h)}e^{-j\beta z} \tag{2}$$

where β is the propagation constant in the z direction, the subscript t implies the transverse direction, $\hat{z}$ is the z-directed unit vector, and the superscripts (e) and (h) are associated with E(TM) and H(TE) fields, respectively. The subscript $i=1$, 2 designates the region 1 or 2, and

$$k_1 = \omega\sqrt{\epsilon_1\mu} = \omega\sqrt{\epsilon_r\epsilon_0\mu}; \qquad k_2 = \omega\sqrt{\epsilon_2\mu} = \omega\sqrt{\epsilon_0\mu} \tag{3}$$

where ω is the operating frequency, ϵ_0 and μ are the permittivity and permeability of a vacuum, and ϵ_r is the relative dielectric constant of the substrate. The scalar potentials $\psi_i^{(e)}$, $\psi_i^{(h)}$ satisfy the two-dimensional Helmholtz equations in regions 1 and 2 as well as the requirements that the tangential electric fields vanish on the metallic periphery of the box and that the tangential magnetic fields vanish at the plane of symmetry $x=0$. In view of the boundary conditions on the side walls it is appropriate to write

$$\psi_1^{(e)} = \sum_{n=1}^{\infty} A_n^{(e)} \sinh \alpha_n^{(1)}y \cos \hat{k}_n x \tag{4}$$

$$\psi_2^{(e)} = \sum_{n=1}^{\infty} B_n^{(e)} \sinh \alpha_n^{(2)}(h-y) \cos \hat{k}_n x \tag{5}$$

$$\psi_1^{(h)} = \sum_{n=1}^{\infty} A_n^{(h)} \cosh \alpha_n^{(1)}y \sin \hat{k}_n x \tag{6}$$

$$\psi_2^{(h)} = \sum_{n=1}^{\infty} B_n^{(h)} \cosh \alpha_n^{(2)}(h-y) \sin \hat{k}_n x \tag{7}$$

where $\hat{k}_n = (n-(1/2))\pi/L$, $\alpha_n^{(1)} = \sqrt{\hat{k}_n^2+\beta^2-\epsilon_r k_0^2}$, $\alpha_n^{(2)} = \sqrt{\hat{k}_n^2+\beta^2-k_0^2}$, with $k_0=\omega\sqrt{\epsilon_0\mu}$ the free-space wavenumber. The coefficients $A_n^{(e)}$, $A_n^{(h)}$, $B_n^{(e)}$, and $B_n^{(h)}$ are as yet unknowns. It is understood that the hyperbolic functions in (4)–(7) are to be replaced by trigonometric functions when $\alpha_n^{(1)}$ and $\alpha_n^{(2)}$ are imaginary.

The total fields derived from the potentials $\psi_i^{(e)}$ and $\psi_i^{(h)}$ must satisfy the interface conditions at $y=d$. Considering the symmetry with respect to the y axis, we have the following four conditions which are mutually independent:

1) $E_{z1} = E_{z2}, \quad 0 < x < L$
2) $E_{x1} = E_{x2}, \quad 0 < x < L$
3) a) $E_{z1} = 0, \quad 0 < x < t$
   b) $H_{x1} = H_{x2}, \quad t < x < L$
4) a) $E_{x1} = 0, \quad 0 < x < t$
   b) $H_{z1} = H_{z2}, \quad t < x < L.$

We impose these conditions on the field components derived from (1) and (2) using (4)–(7). Conditions 1) and 2) are used to express $B_n^{(e)}$ and $B_n^{(h)}$ in terms of $A_n^{(e)}$ and $A_n^{(h)}$. Next, the imposition of conditions 3) and 4) leads us to the following equations.

$$\sum_{n=1}^{\infty} \bar{A}_n^{(e)} \cos \hat{k}_n x = 0, \qquad 0 < x < t \tag{8}$$

$$\sum_{n=1}^{\infty} \bar{A}_n^{(e)}\hat{k}_n P_n(\beta) \cos \hat{k}_n x - \sum_{n=1}^{\infty} \bar{A}_n^{(h)}\hat{k}_n T_n(\beta) \cos \hat{k}_n x = 0, \qquad t < x < L \tag{9}$$

$$\sum_{n=1}^{\infty} \bar{A}_n^{(e)}\hat{k}_n \sin \hat{k}_n x - \sum_{n=1}^{\infty} \bar{A}_n^{(h)}\hat{k}_n \sin \hat{k}_n x = 0, \qquad 0 < x < t \tag{10}$$

$$\sum_{n=1}^{\infty} \bar{A}_n^{(e)}Q_n(\beta) \sin \hat{k}_n x - \sum_{n=1}^{\infty} \bar{A}_n^{(h)}W_n(\beta) \sin \hat{k}_n x = 0, \qquad t < x < L \tag{11}$$

where

$$\bar{A}_n^{(e)} = A_n^{(e)} \sinh \alpha_n^{(1)}d \tag{12}$$

$$\bar{A}_n^{(h)} = \frac{\omega\mu}{\beta}\frac{\alpha_n^{(1)}}{\hat{k}_n} A_n^{(h)} \sinh \alpha_n^{(1)}d \tag{13}$$

and $P_n(\beta)$, $T_n(\beta)$, $Q_n(\beta)$, and $W_n(\beta)$ are expressed as follows:

$$P_n(\beta) = \epsilon_r\frac{\alpha_n^{(1)}}{\hat{k}_n}\coth \alpha_n^{(1)}d + \frac{\epsilon_r - \bar{\beta}^2}{1-\bar{\beta}^2}\frac{\alpha_n^{(2)}}{\hat{k}_n}\coth \alpha_n^{(2)}(h-d) + \bar{\beta}^2\frac{\hat{k}_n}{\alpha_n^{(2)}}\frac{1-\epsilon_r}{1-\bar{\beta}^2}\coth \alpha_n^{(2)}(h-d) \tag{14}$$

$$T_n(\beta) = \bar{\beta}^2\left(\frac{\hat{k}_n}{\alpha_n^{(1)}} \coth \alpha_n^{(1)}d + \frac{\hat{k}_n}{\alpha_n^{(2)}} \coth \alpha_n^{(2)}(h-d)\right) \quad (15)$$

$$Q_n(\beta) = \frac{\hat{k}_n}{\alpha_n^{(2)}} \frac{1-\epsilon_r}{1-\bar{\beta}^2} \coth \alpha_n^{(2)}(h-d) \quad (16)$$

$$W_n(\beta) = \frac{\epsilon_r - \bar{\beta}^2}{1-\bar{\beta}^2} \frac{\hat{k}_n}{\alpha_n^{(1)}} \coth \alpha_n^{(1)}d + \frac{\hat{k}_n}{\alpha_n^{(2)}} \coth \alpha_n^{(2)}(h-d) \quad (17)$$

where $\bar{\beta} = \beta/k_0$ is the normalized propagation constant.

Equations (8)–(11) may be transformed into an infinite set of homogeneous simultaneous equations for $\bar{A}_n^{(e)}$ and $\bar{A}_n^{(h)}$ via the conventional technique of taking a scalar product with a complete set of functions appropriate for the various regions. The solution for $\bar{\beta}$ may then be determined by seeking the zeros of the determinant associated with the above matrix equation.

We will, however, depart from this conventional procedure and instead transform (8)–(11) into an auxiliary set of equations with rapid convergence properties.

III. Transformation to Auxiliary Equations

We will now outline the steps necessary for the transformation of (8)–(11) into an auxiliary set of equations. To this end, differentiate (8) with respect to x and substitute the resultant equation into (10). This yields

$$\sum_{n=1}^{\infty} A_n^{(e)}\hat{k}_n \sin \hat{k}_n x = 0, \qquad 0 < x < t \quad (18)$$

$$\sum_{n=1}^{\infty} A_n^{(h)}\hat{k}_n \sin \hat{k}_n x = 0, \qquad 0 < x < t. \quad (19)$$

Similarly, differentiating (11) with respect to x, we have, after some rearrangement,

$$\sum_{n=1}^{\infty} A_n^{(e)}\hat{k}_n \cos \hat{k}_n x = f(x), \qquad t < x < L \quad (20)$$

$$\sum_{n=1}^{\infty} A^{(h)}\hat{k}_n \cos \hat{k}_n x = g(x), \qquad t < x < L \quad (21)$$

where

$$f(x) = \sum_{m=1}^{\infty} (a_m A_m^{(e)} + b_m A_m^{(h)}) \cos \hat{k}_m x,$$

$$g(x) = \sum_{m=1}^{\infty} (c_m A_m^{(e)} + d_m A_m^{(h)}) \cos \hat{k}_m x$$

and

$$a_m = \hat{k}_m\left[1 - \frac{P_m(\beta)W(\beta) - T(\beta)Q_m(\beta)}{P(\beta)W(\beta) - T(\beta)Q(\beta)}\right] \quad (22)$$

$$b_m = \hat{k}_m \frac{T_m(\beta)W(\beta) - T(\beta)W_m(\beta)}{P(\beta)W(\beta) - T(\beta)Q(\beta)} \quad (23)$$

$$c_m = \hat{k}_m \frac{P(\beta)Q_m(\beta) - P_m(\beta)Q(\beta)}{P(\beta)W(\beta) - T(\beta)Q(\beta)} \quad (24)$$

$$d_m = \hat{k}_m\left[1 - \frac{P(\beta)W_m(\beta) - T_m(\beta)Q(\beta)}{P(\beta)W(\beta) - T(\beta)Q(\beta)}\right]. \quad (25)$$

The functions $P(\beta)$, $T(\beta)$, $Q(\beta)$, $W(\beta)$ are the asymptotic limits of $P_m(\beta)$, $T_m(\beta)$, $Q_m(\beta)$, $W_m(\beta)$ as $m \to \infty$. Explicitly,

$$P(\beta) = \epsilon_r + \frac{\epsilon_r - \bar{\beta}^2}{1-\bar{\beta}^2} + \bar{\beta}^2 \frac{1-\epsilon_r}{1-\bar{\beta}^2} \quad (26)$$

$$T(\beta) = 2\bar{\beta}^2 \quad (27)$$

$$Q(\beta) = \frac{1-\epsilon_r}{1-\bar{\beta}^2} \quad (28)$$

$$W(\beta) = \frac{\epsilon_r - \bar{\beta}^2}{1-\bar{\beta}^2} + 1. \quad (29)$$

Equations (18)–(21) are similar to those obtained in connection with the quasi-static formulation of the iris discontinuity problem in a waveguide [12], [13]. Their most important characteristic is that the pairs (18)–(20) and (19)–(21) are exactly invertable via a singular integral equation approach, meaning that it is possible to express the coefficients $\bar{A}_n^{(e)}$ and $\bar{A}_n^{(h)}$ in terms of integrals involving the functions in the right-hand side, viz., $f(x)$, $g(x)$. In the capacitive discontinuity problem the right-hand side is known and the unknown coefficients are determined in this manner. In the present case, however, the functions $f(x)$ and $g(x)$ themselves contain the unknowns $\bar{A}_n^{(e)}$ and $\bar{A}_n^{(h)}$ and the result of the inversion of (18)–(21) is a homogeneous set of equations, leading in turn to an eigenvalue equation. This equation is much more efficient for numerical solution than the conventional one derived by the moment method. The reason for this is the rapid decay of the weight coefficients a_m, b_m, c_m, and d_m appearing in $f(x)$ and $g(x)$. The asymptotic behavior of the coefficients a_m, b_m, c_m, and d_m can be proven as follows. Consider, for instance, (22) for a_m which contains difference terms of the type $P(\beta)-P_m(\beta)$ and $Q(\beta)-Q_m(\beta)$ in the numerator. For large m,

$$P(\beta) - P_m(\beta) = \epsilon_r\left\{1 - \frac{\alpha_m^{(1)}}{\hat{k}_m} \coth \alpha_m^{(1)}d\right\} + \epsilon_r\left\{1 - \frac{\alpha_m^{(2)}}{\hat{k}_m} \coth \alpha_m^{(2)}(h-d)\right\}$$
$$\sim \left\{e^{-((2m-1)\pi/L)d} + e^{-((2m-1)\pi/L)(h-d)}\right\} \cdot \left\{-2\epsilon_r + 0(m^{-2})\right\}.$$

The behavior of $Q(\beta)-Q_m(\beta)$ is similar, and it follows that a_m decays exponentially for large m.[1] Similar comment applies to other coefficients as well.

[1] This limiting procedure continues to remain valid even if the first few $\alpha_n^{(1)}$ and $\alpha_n^{(2)}$ are imaginary, as they would be for $\beta < k_0$. This is because all of the $\alpha_n^{(1)}$ and $\alpha_n^{(2)}$ are real for large n for which the asymptotic behavior is derived as shown. However, the convergence is relatively slower in this region.

It should be pointed out that the rate of decay of these coefficients becomes smaller as L becomes large. Thus, for very large L, it is more efficient to employ an alternative approach that regards the geometry as a perturbation of a structure with open side walls. Such a method is currently being investigated by the authors.

Following the standard method [12], [13] for solving the pair of equations of the type (18)–(20) and (19)–(21), we introduce the functions $F_{t_1}(x)$ and $F_{t_2}(x)$ via the equations

$$\sum_{n=1}^{\infty} \bar{A}_n^{(e)}\hat{k}_n \sin \hat{k}_n x = F_{t_1}(x), \qquad t < x < L \tag{30}$$

$$\sum_{n=1}^{\infty} A_n^{(h)}\hat{k}_n \sin \hat{k}_n x = F_{t_2}(x), \qquad t < x < L. \tag{31}$$

Then from (18), (30) and (19), (31), we have

$$\bar{A}_n^{(e)}\hat{k}_n = \frac{2}{L}\int_t^L F_{t_1}(x') \sin \hat{k}_n x' \, dx' \tag{32}$$

$$\bar{A}_n^{(h)}\hat{k}_n = \frac{2}{L}\int_t^L F_{t_2}(x') \sin \hat{k}_n x' \, dx'. \tag{33}$$

Substituting these into the left-hand side of (20) and (21) we have the following two integral equations:

$$\frac{2}{L}\int_t^L F_{t_1}(x')K(x, x') \, dx' = \sum_{m=1}^{\infty} (a_m\bar{A}_m^{(e)} + b_m\bar{A}_m^{(h)}) \cos \hat{k}_m x \tag{34}$$

$$\frac{2}{L}\int_t^L F_{t_2}(x')K(x, x') \, dx' = \sum_{m=1}^{\infty} (c_m\bar{A}_m^{(e)} + d_m\bar{A}_m^{(h)}) \cos \hat{k}_m x \tag{35}$$

where the kernel $K(x, x')$ is

$$K(x, x') = \sum_{n=1}^{\infty} \cos \hat{k}_n x \sin \hat{k}_n x'. \tag{36}$$

We can write the formal solutions of (34) and (35) as

$$F_{t_1}(x) = \frac{L}{2}\sum_{m=1}^{\infty} (a_m A_m^{(e)} + b_m A_m^{(h)}) f_m(x) + \bar{c}_1 F_h(x) \tag{37}$$

$$F_{t_2}(x) = \frac{L}{2}\sum_{m=1}^{\infty} (c_m A_m^{(e)} + d_m A_m^{(h)}) f_m(x) + \bar{c}_2 F_h(x) \tag{38}$$

where $f_m(x)$ is the solution of

$$\int_t^L f_m(x')K(x, x') \, dx' = \cos \hat{k}_m x, \qquad t < x < L \tag{39}$$

and $F_h(x)$ is the homogeneous solution, satisfying

$$\int_t^L F_h(x')K(x, x') \, dx' = 0, \qquad t < x < L \tag{40}$$

and $\bar{c}_1$, $\bar{c}_2$ are unknown constants yet to be determined.

Substituting (37) and (38) into (32) and (33) we obtain

$$\bar{A}_n^{(e)}\hat{k}_n = \sum_{m=1}^{\infty} (a_m A_m^{(e)} + b_m \bar{A}_m^{(h)}) D_{nm} + \bar{c}_1 K_n, \qquad n = 1, 2, \cdots \tag{41}$$

$$\bar{A}_n^{(h)}\hat{k}_n = \sum_{m=1}^{\infty} (c_m \bar{A}_m^{(e)} + d_m A_m^{(h)}) D_{nm} + \bar{c}_2 K_n, \qquad n = 1, 2, \cdots \tag{42}$$

where

$$D_{nm} = \int_t^L f_m(x) \sin \hat{k}_n x \, dx \tag{43}$$

$$K_n = \frac{2}{L}\int_t^L F_h(x) \sin \hat{k}_n x \, dx. \tag{44}$$

The integral equations, (39) and (40), can be inverted exactly using the singular integral equation technique [12]–[14] to give

$$f_m(x) = \frac{2}{L} \cos \frac{\pi x}{2L} \left[\sum_{q=1}^{m-1} P_{mq} \sin q\theta - P_{m0} \frac{\cos \theta}{\sin \theta} \right] \tag{45}$$

$$F_h(x) = \frac{\cos \dfrac{\pi x}{2L}}{\sin \theta}. \tag{46}$$

The details of the derivation are given in Appendix I where the relationship between θ and x, and the values of P_{mq} and P_{m0} are also given. The expressions for D_{nm} and K_n appearing in (43) and (44) are given in Appendix II.

We will now show that the unknowns $\bar{c}_1$ and $\bar{c}_2$ in (41) and (42) can also be expressed in terms of $\bar{A}_m^{(e)}$ and $\bar{A}_m^{(h)}$ and hence these equations may be viewed as a coupled infinite set of homogeneous linear equations for the unknowns $\bar{A}_n^{(e)}$ and $\bar{A}_n^{(h)}$.

To express $\bar{c}_1$ in terms of $\bar{A}_n^{(e)}$ and $\bar{A}_n^{(h)}$, we return to (8), which is a statement of the requirement that $E_z = 0$ on the strip. To enforce this condition we have to require

$$\int_t^L F_{t_1}(x) \, dx = E_z(L) - E_z(t) = 0.$$

Substituting for F_{t_1} from (37) in the preceding equations and solving for $\bar{c}_1$ we get

$$\bar{c}_1 = -\frac{\dfrac{L}{2}\displaystyle\int_t^L \sum_{m=1}^{\infty} (a_m A_m^{(e)} + b_m A_m^{(h)}) f_m(x) \, dx}{\displaystyle\int_t^L F_h(x) \, dx} = \sum_{m=1}^{\infty} (M_m \bar{A}_m^{(e)} + N_m A_m^{(h)}) \tag{47}$$

where

$$M_m = -\frac{a_m \sum_{q=1}^{m-1} P_{mq} I_q}{I_h} \tag{48}$$

$$N_m = -\frac{b_m \sum_{q=1}^{m-1} P_{mq} I_q}{I_h}. \tag{49}$$

The evaluations of I_q and I_h appearing in (48) and (49) are given in Appendix III. Notice that M_m and N_m are proportional to a_m and b_m, and they also decrease rapidly with increasing m.

Similarly, for expressing $\bar{c}_2$ in terms of $\bar{A}_n^{(e)}$ and $\bar{A}_n^{(h)}$, we return to (11), and use (32) and (33) to get

$$\sum_{n=1}^{\infty} \frac{2}{L} \int_t^L F_{t_1}(x') \frac{\sin \hat{k}_n x' \sin \hat{k}_n x}{\hat{k}_n} Q_n(\beta)\, dx - \sum_{n=1}^{\infty} \frac{2}{L} \int_t^L F_{t_2}(x') \frac{\sin \hat{k}_n x' \sin \hat{k}_n x}{\hat{k}_n} W_n(\beta)\, dx = 0, \qquad t < x < L. \tag{50}$$

This equation is true for all x in the range $t<x<L$ and we can set $x=t+0$ without loss of generality. Solving for $\bar{c}_2$ we now get, after some rearrangement

$$\bar{c}_2 = \sum_{m=1}^{\infty} (X_m \bar{A}_m^{(e)} + Y_m \bar{A}_m^{(h)}) \tag{51}$$

where

$$X_m = \frac{S_m - M_m Q I_g - (Q a_m - W c_m) E_m}{S - W I_g} \tag{52}$$

$$Y_m = \frac{S_m' - N_m Q I_g - (Q b_m - W d_m) E_m}{S - W I_g} \tag{53}$$

and

$$S = \sum_{n=1}^{\infty} \frac{\sin \hat{k}_n t}{\hat{k}_n} (W - W_n) K_n \tag{54}$$

$$S_m = \sum_{n=1}^{\infty} \frac{\sin \hat{k}_n t}{\hat{k}_n} \left[D_{nm} \{ (Q - Q_n) a_m - (W - W_n) c_m \} + K_n (Q - Q_n) M_m \right] \tag{55}$$

$$S_m' = \sum_{n=1}^{\infty} \frac{\sin \hat{k}_n t}{\hat{k}_n} \left[D_{nm} \{ (Q - Q_n) b_m - (W - W_n) d_m \} + K_n (Q - Q_n) N_m \right]. \tag{56}$$

The evaluation of I_g and the expression for E_m are given in Appendix IV. We observe once again that the infinite summations just mentioned are rapidly convergent due to the asymptotic decrease of their coefficients.

The final step involves the substitution of (47) and (51) into (41) and (42). This leads to the desired matrix equations

$$\sum_{m=1}^{\infty} (\hat{k}_p \delta_{pm} - a_m D_{pm} - M_m K_p) \bar{A}_m^{(e)} - \sum_{n=1}^{\infty} (b_n D_{pn} + N_n K_p) \bar{A}_n^{(h)} = 0, \qquad p = 1, 2, \cdots$$

$$\sum_{m=1}^{\infty} (-c_m D_{qm} - X_m K_q) \bar{A}_m^{(e)} + \sum_{n=1}^{\infty} (\hat{k}_q \delta_{qn} - d_n D_{qn} - Y_n K_q) \bar{A}_n^{(h)} = 0, \qquad q = 1, 2, \cdots \tag{57}$$

where δ_{pm} is the Kronecker delta.

The solution of the determinantal equation for (57) yields the desired values for β. Although (57) comprises a doubly infinite set of equations, we can truncate the associated matrix to a small size since a_m, b_m, c_m, d_m, as well as M_m, N_m, X_m and Y_m decrease extremely rapidly with m. It turns out that retaining only a single equation from each set is sufficient for accurate computation of numerical results, as is demonstrated in the following section. Using the fact that $D_{11}=0$ the associated determinantal equation of the truncated set may be explicitly written as

$$D(\beta) = (\hat{k}_1 - M_1(\beta) K_1)(\hat{k}_1 - Y_1(\beta) K_1) - N_1(\beta) X_1(\beta) K_1^2 = 0 \tag{58}$$

where the expressions for M_1, Y_1, N_1, and X_1 are given in (48), (53), (49), and (52).

Clearly, this equation is much easier to handle than the determinantal equation of a large order matrix that results from conventional processing of (8)–(11). Though the derivation of (58) requires advance analytical processing, this effort is more than compensated for by the numerical efficiency that results due to the simplicity of the characteristic equation. In addition, if one is only interested in computing the numerical results, he may use (58) as a starting point and evaluate M_1, K_1, Y_1, N_1, and X_1 from the complete expressions provided earlier in this section. He may thus completely avoid the task of following through the details of the solution of the singular integral equation that led to the final result.

IV. Numerical Computations

The roots of the characteristic equation (58) for the propagation constant β were calculated using a digital computer. The behavior of the function $D(\bar{\beta})$ as a function of $\bar{\beta}$ depends considerably on the structural and operating parameters of the microstrip line. Typical plots of $D(\bar{\beta})$ versus $\bar{\beta}^2$ are shown in Fig. 2 for a number of different choices for the structural parameters and k_0, the free-space wavenumber, and it is observed that there can be more than one zero crossing of $D(\bar{\beta})$, indicating

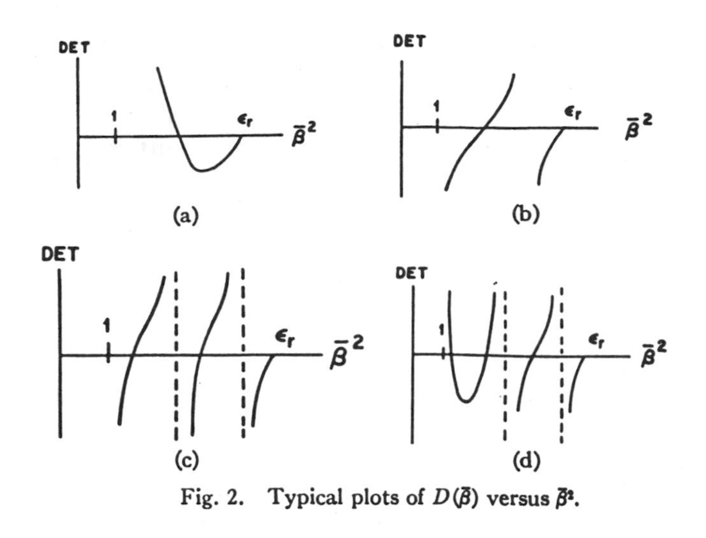

Fig. 2. Typical plots of $D(\bar{\beta})$ versus $\bar{\beta}^2$.

TABLE I
COMPARISON OF β(mm^{-1}) OBTAINED BY DIFFERENT SIZE MATRICES

Frequency (GHz)	2×2 Matrix	4×4 Matrix	Hornsby and Gopinath
10	0.530	0.531	0.55
20	1.10	1.115	1.17
30	1.71	1.74	1.77

Physical parameters
$\epsilon_r = 9.0$
$L = 1.75$ mm
$t = 0.5$ mm
$d = 0.5$ mm
$h = 2.0$ mm

the existence of higher order modes. As expected, the number of higher order propagating modes increases with increasing frequency.

Fig. 3 shows the dispersion relation computed for a choice of parameters identical to that used by Hornsby and Gopinath using a matrix size of the order of 100 ×100 [10]. Their results are also exhibited on the same diagram to facilitate comparison. Although the present method uses only a 2×2 matrix, the results compare quite favorably with those of Hornsby and Gopinath. Furthermore, as shown in Table I, doubling the size of the matrix introduced only very small differences in the result derived here. The curve shown here is for the lowest order symmetric mode which shows no low-frequency cutoff. Higher order modes are not present in the frequency range considered. The two dotted lines in Fig. 3 correspond to air-filled and dielectric-filled transmission lines. The quasi-TEM solution for open microstrip line is also shown for comparison [11]. The guide wavelength obtained by quasi-TEM analysis shows no frequency dependence. This is due to the fact that the propagation effect is neglected in such an analysis.

It should be remarked that a major factor contributing to the slow convergence in other methods, viz., the singular behavior of the fields at the edge of the strip, has been accounted for exactly in the present method, this being a well-known property of the singular integral equation approach.

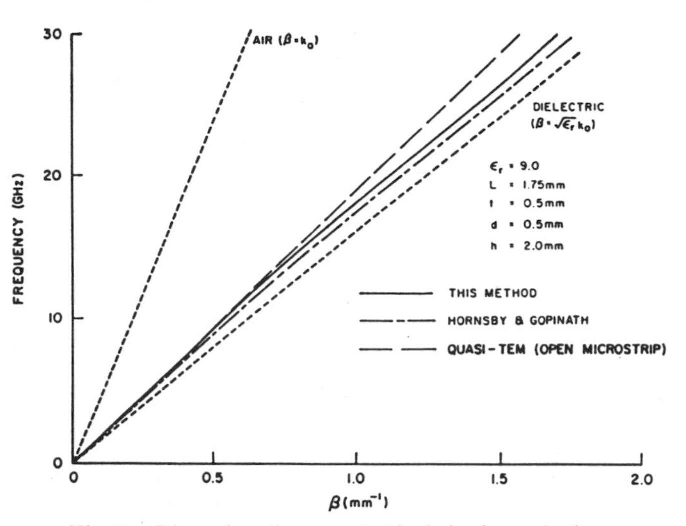

Fig. 3. Dispersion diagram of shielded microstrip line.

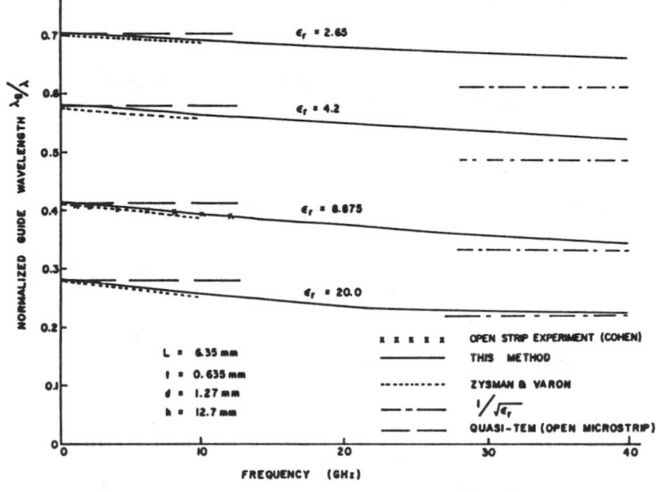

Fig. 4. Variation of guide wavelength with frequency.

The guide wavelength for the lowest order mode was also calculated by the present method for parameters investigated by Zysman and Varon [8]. (See Fig. 4). The agreement with their results, which were available for the range $0 < f < 10$ GHz, is again quite good. These results are also in good agreement with experimental data[2] shown in Fig. 4. Once again, the results for the (4×4) matrix were calculated and were found to be very close. They are therefore not plotted in Fig. 4.

From Figs. 3 and 5, it is evident that with increasing frequencies the dispersion curves for the lowest order mode become parallel to the dotted line corresponding to the dielectric-filled TEM transmission line. Thus for high frequencies the normalized guide wavelength for the lowest order mode approaches $1/\sqrt{\epsilon_r}$, the ratio of guide wavelength in the dielectric-filled TEM transmission line to that in an air-filled line (see Fig. 4).

Fig. 5 shows the dispersion characteristics for the four cases displayed in Fig. 4. The existence of a number

[2] Courtesy of J. Cohen and Dr. M. Gilden, Monsanto Company, St. Louis, Mo. (private communication). Zysman and Varon compared their theoretical results with experimental results reported by Hartwig *et al.* [15] and found good agreement.

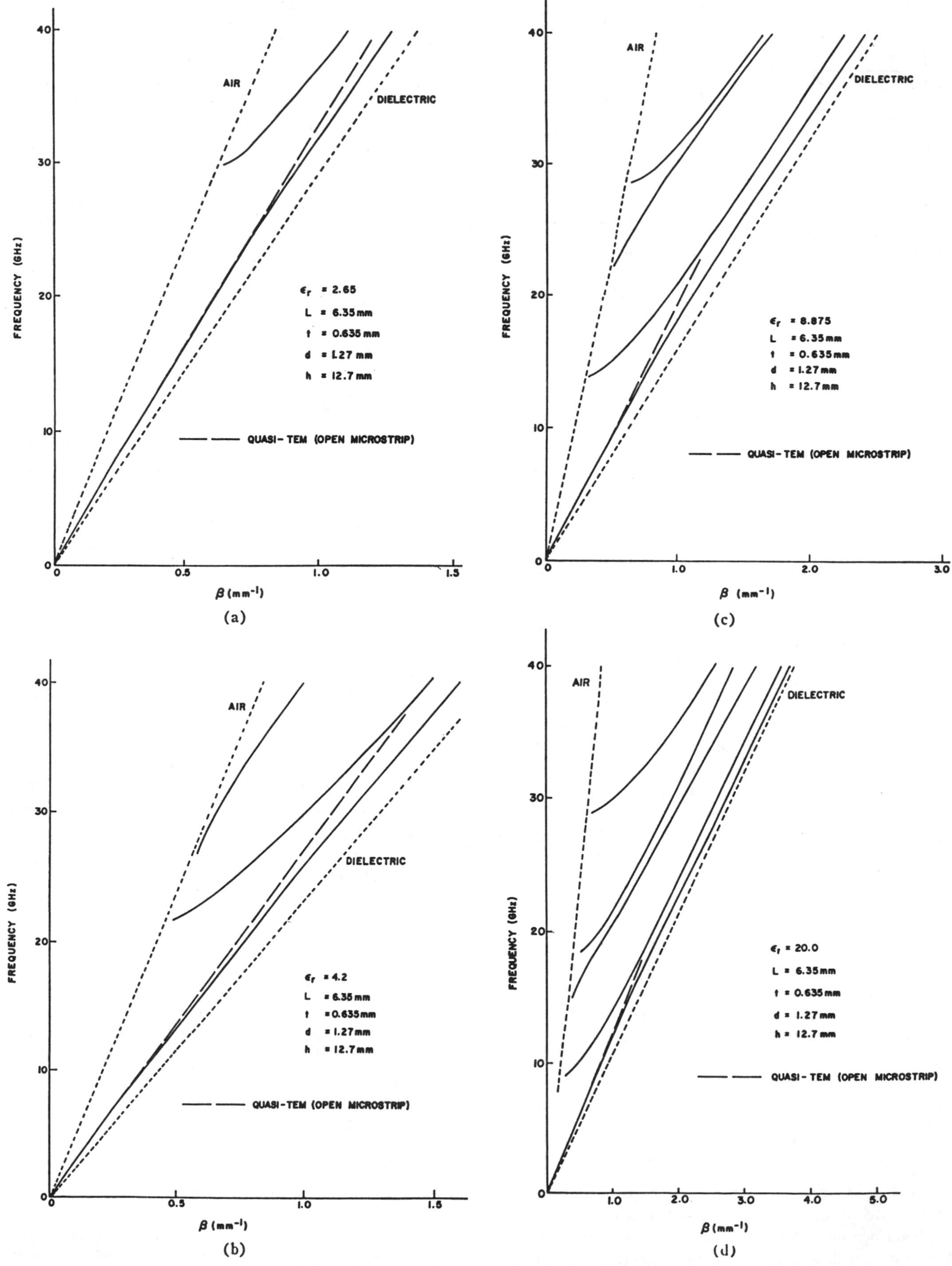

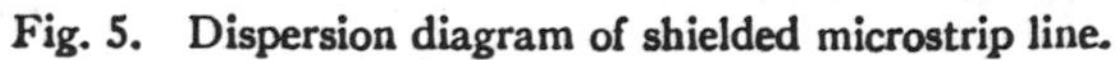
Fig. 5. Dispersion diagram of shielded microstrip line.

of higher order modes is evident in these plots. It is also seen that the frequency at which the first higher order mode begins to appear is lower for a guide filled with a dielectric substrate which has a larger dielectric constant. As pointed out in the introduction, the simplicity of the determinantal equation in the present analysis allows one to conveniently find these higher modes.

It may be worthwhile to comment on the applicability of the method in the fast wave region. As pointed out earlier, a few of the $\alpha_n^{(1)}$ and $\alpha_n^{(2)}$ now become imaginary though asymptotically $\alpha_n^{(1)}$ and $\alpha_n^{(2)}$ are still real for large n. Everything in the analysis goes through if the hyperbolic functions with α_n in their argument are replaced by corresponding trigonometric functions. However, the convergence of the method in this region is relatively slower than that in the slow wave region because the asymptotic values of the various functions of n are reached for larger n. Finally, it may be pertinent to quote the typical computational time using the method developed in the paper. Average time for calculating all roots for a single frequency was between 20 and 30 s on a CDC-G20, which is about seven to eight times slower than the IBM 360/75.

V. Conclusions

A new technique is presented for the calculation of dispersion characteristics of hybrid stripline modes. Though the derivation of the determinantal equation does require certain amount of analytical manipulation using well-known techniques, the final determinantal equation is rather simple. Thus the details of the derivation may be completely bypassed as those whose principal interest is to obtain numerical results. The computational effort in the present method is at least ten times smaller than the methods reported in the literature.

Another advantage is the ease with which the higher order modes can be located by studying the function $D(\bar{\beta})$ which has a rather simple form.

Appendix I

Solution of the Singular Integral Equations

We first consider the homogeneous integral equation

$$\int_t^L F_h(x')K(x, x')dx' = 0, \qquad t \leq x \leq L \tag{59}$$

with a singular kernel

$$K(x, x') = \sum_{n=1}^{\infty} \cos \hat{k}_n x \sin \hat{k}_n x' = \frac{\sin \frac{\pi x'}{2L} \cos \frac{\pi x}{2L}}{\cos \frac{\pi x}{L} - \cos \frac{\pi x'}{L}}. \tag{60}$$

The integral in (59) is to be interpreted as a Cauchy principal value. Let us introduce the Schwinger transformation [12]

$$\cos \frac{\pi x}{L} = \alpha_1 + \alpha_2 \cos \theta, \qquad t \leq x \leq L, \; 0 \leq \theta \leq \pi \tag{61}$$

$$\alpha_1 = \frac{1}{2}\left(\cos \frac{\pi t}{L} - 1\right), \qquad \alpha_2 = \frac{1}{2}\left(\cos \frac{\pi t}{L} + 1\right)$$

where α_1 and α_2 have been determined in the usual manner so as to transform the range t to L in x to 0 to π in θ. Introducing this change of variables x to θ in (59) we get, using (60) and (61)

$$\int_0^\pi \left[\frac{F_h(x')}{\cos \frac{\pi x'}{2L}}\right] \frac{\sin \theta' d\theta'}{\cos \theta - \cos \theta'} = 0. \tag{62}$$

Next, we quote the following identities from Tricomi [14]:

$$\int_0^\pi \frac{\cos nu \, du}{\cos u - \cos \alpha} = \pi \frac{\sin n\alpha}{\sin \alpha}, \qquad n = 0, 1, 2, \cdots \tag{63}$$

$$\int_0^\pi \frac{\sin (n+1)u \sin u \, du}{\cos u - \cos \alpha} = -\pi \cos (n+1)\alpha, \qquad n = 0, 1, 2, \cdots. \tag{64}$$

Using (64) in (62) we have

$$\frac{F_h(x')}{\cos \frac{\pi x'}{2L}} = \frac{1}{\sin \theta'}, \qquad \text{or } F_h(x') = \frac{\cos \frac{\pi x'}{2L}}{\sin \theta'}. \tag{65}$$

This concludes the derivation of the homogeneous solution.

The inhomogeneous solution is obtained in a similar manner. The equation

$$\int_t^L f_m(x')K(x, x') \, dx' = \cos \hat{k}_m x, \qquad t \leq x \leq L, \; m = 1, 2, \cdots \tag{66}$$

becomes, after effecting Schwinger transformation,

$$\frac{L}{2\pi} \int_0^\pi \left[\frac{f_m(x')}{\cos \frac{\pi x'}{2L}}\right] \frac{\sin \theta' \, d\theta'}{\cos \theta - \cos \theta'} = \frac{\cos \hat{k}_m x}{\cos \hat{k}_1 x}. \tag{67}$$

The right-hand side may be expressed as a function of θ as follows:

$$\frac{\cos \hat{k}_m x}{\cos \hat{k}_1 x} = \sum_{q=0}^{m-1} P_{mq} \cos q\theta \tag{68}$$

with

$P_{10} = 1$

$P_{20} = 2\alpha_1 - 1, \qquad P_{21} = 2\alpha_2$

$$P_{30} = 4\alpha_1^2 - 2\alpha_1 - 1 + 2\alpha_2^2, \qquad P_{31} = 8\alpha_1\alpha_2 - 2\alpha_2,$$
$$P_{32} = 2\alpha_2^2 \tag{69}$$

and so on.

Thus (67) may be rewritten as

$$\frac{L}{2\pi}\int_0^\pi \left[\frac{f_m(x')}{\cos\frac{\pi x'}{2L}}\right]\frac{\sin\theta'\, d\theta'}{\cos\theta - \cos\theta'} = \sum_{q=0}^{m-1} P_{mq}\cos q\theta. \tag{70}$$

To solve (70), we first represent the unknown function $f_m(x)$ as

$$\frac{f_m(x)}{\cos\frac{\pi x}{2L}} = \sum_{q=1}^{m-1} Q_{mq}\sin q\theta + Q_{mo}\frac{\cos\theta}{\sin\theta}. \tag{71}$$

Substituting (71) into (70) and using (63) and (64), we derive by comparing both sides of (70)

$$Q_{mq} = \frac{2}{L}P_{mq}, \qquad q = 1, 2, \cdots, (m-1)$$

$$Q_{mo} = -\frac{2}{L}P_{mo}.$$

This completes the derivation of the solution of the inhomogeneous equation for $f_m(x)$.

Appendix II

Calculations of K_n and D_{nm}

The values of K_n are calculated by using the expression for $F_h(x)$ given in (65), and evaluating the integrals

$$K_n = \frac{2}{L}\int_t^L F_h(x)\sin \hat{k}_n x\, dx \tag{72}$$

with

$$F_h = \frac{\cos \hat{k}_1 x}{\sin\theta}.$$

The first few K_n are given by

$$K_1 = \alpha_2, \qquad K_2 = \alpha_2(1 + 2\alpha_1),$$
$$K_3 = \alpha_2(4\alpha_1^2 + 2\alpha_1 - 1 + 2\alpha_2^2).$$

The D_{nm} may also be calculated in a similar manner. They are given by

$$D_{nm} = \int_t^L f_m(x)\sin \hat{k}_n x\, dx. \tag{73}$$

$f_m(x)$ given in (71) is substituted in (73). The first few values of D_{nm} are given by

$$D_{11} = 0, \qquad D_{12} = \alpha_2^2$$
$$D_{21} = -\alpha_2^2, \qquad D_{22} = 2\alpha_2^2.$$

Appendix III

Calculation of I_h and I_q

The integrals I_h and I_q are defined as

$$\int_t^L F_h(x)\, dx = \frac{\alpha_2 L}{\sqrt{2}\pi} I_h \tag{74}$$

$$\int_t^L f_m(x)\, dx = \frac{\sqrt{2}\alpha_2}{\pi}\sum_{q=0}^{m-1} P_{mq} I_q. \tag{75}$$

To evaluate I_h we consider the integral on the left-hand side of (74). After transformation we can write

$$\int_t^L F_h(x)\, dx = \frac{\alpha_2 L}{\sqrt{2}\pi}\int_0^\pi \frac{d\theta}{\sqrt{1 - \alpha_1 - \alpha_2\cos\theta}}. \tag{76}$$

Similarly, the integral in (75) becomes

$$\int_t^L f_m(x)\, dx = \frac{\alpha_2 L}{\sqrt{2}\pi}\int_0^\pi \left[\frac{f_m(x)}{\cos\frac{\pi x}{2L}}\right]\frac{\sin\theta\, d\theta}{\sqrt{1 - \alpha_1 - \alpha_2\cos\theta}}$$

$$= \sum_{q=0}^{m-1} I_{mq}. \tag{77}$$

Using (71) I_{mq} are expressed as

$$I_{mo} = -\frac{\sqrt{2}\alpha_2}{\pi}P_{mo}\int_0^\pi \frac{\cos\theta\, d\theta}{\sqrt{1 - \alpha_1 - \alpha_2\cos\theta}} \tag{78}$$

$$I_{mq} = \frac{\sqrt{2}\alpha_2}{\pi}P_{mq}\int_0^\pi \frac{\sin q\theta \sin\theta\, d\theta}{\sqrt{1 - \alpha_1 - \alpha_2\cos\theta}},$$
$$q = 1, 2, \cdots, (m-1). \tag{79}$$

Thus I_h and I_q defined in (74) and (75) take the form

$$I_h = \int_0^\pi \frac{d\theta}{\sqrt{1 - \alpha_1 - \alpha_2\cos\theta}} \tag{80}$$

$$I_q = \begin{cases} -\displaystyle\int_0^\pi \frac{\cos\theta\, d\theta}{\sqrt{1 - \alpha_1 - \alpha_2\cos\theta}}, & q = 0 \\ \displaystyle\int_0^\pi \frac{\sin q\theta \sin\theta\, d\theta}{\sqrt{1 - \alpha_1 - \alpha_2\cos\theta}}, & q = 1, 2, \cdots, (m-1). \end{cases} \tag{81}$$

Since the integrals appearing in (80) and (81) are well behaved, their calculation is easily accomplished by using one of the numerical quadrature subroutines.

Appendix IV

Calculation of I_g and E_m

In the process of derivation of (52) and (53) it is necessary to evaluate the integrals

$$I_g = \frac{2}{L}\int_t^L F_h(x')G(t, x')\, dx' \tag{82}$$

$$E_m = \int_t^L f_m(x')G(t, x')\, dx' \tag{83}$$

where

$$G(t, x') = \sum_{n=1}^{\infty} \frac{\sin k_n x \sin k_n x'}{k_n}\bigg|_{x=t}$$

$$= \frac{L}{2\pi} \ln \left\{ \frac{\sqrt{1 - \alpha_1 - \alpha_2 \cos \theta'} + \sqrt{1 - \alpha_1 - \alpha_2}}{\sqrt{1 - \alpha_1 - \alpha_2 \cos \theta'} - \sqrt{1 - \alpha_1 - \alpha_2}} \right\}. \quad (84)$$

This kernel has a logarithmic singularity at $x' = t$ or $\theta' = 0$. Using (84), we have

$$I_g = \frac{\alpha_2 L}{\sqrt{2}\pi^2} \int_0^{\pi} \ln \left\{ \frac{\sqrt{1 - \alpha_1 - \alpha_2 \cos \theta'} + \sqrt{1 - \alpha_1 - \alpha_2}}{\sqrt{1 - \alpha_1 - \alpha_2 \cos \theta'} - \sqrt{1 - \alpha_1 - \alpha_2}} \right\} \frac{d\theta'}{\sqrt{1 - \alpha_1 - \alpha_2 \cos \theta'}} \quad (85)$$

$$E_m = \frac{\alpha_2 L}{\sqrt{2}\pi^2} \left(\sum_{q=1}^{m-1} P_{mq} J_q + P_{mo} J_o \right) \quad (86)$$

where

$$J_o = - \int_0^{\pi} \ln \left\{ \frac{\sqrt{1 - \alpha_1 - \alpha_2 \cos \theta'} + \sqrt{1 - \alpha_1 - \alpha_2}}{\sqrt{1 - \alpha_1 - \alpha_2 \cos \theta'} - \sqrt{1 - \alpha_1 - \alpha_2}} \right\} \frac{\cos \theta' \, d\theta'}{\sqrt{1 - \alpha_1 - \alpha_2 \cos \theta'}} \quad (87)$$

$$J_q = \int_0^{\pi} \ln \left\{ \frac{\sqrt{1 - \alpha_1 - \alpha_2 \cos \theta'} + \sqrt{1 - \alpha_1 - \alpha_2}}{\sqrt{1 - \alpha_1 - \alpha_2 \cos \theta'} - \sqrt{1 - \alpha_1 - \alpha_2}} \right\} \frac{\sin q\theta' \sin \theta' \, d\theta'}{\sqrt{1 - \alpha_1 - \alpha_2 \cos \theta'}}. \quad (88)$$

For the numerical evaluation of I_g and J_o we divide the range of the integrals into two, say, $0 < \theta < \theta_p$, $\theta_p \ll 1$, and $\theta_p < \theta < \pi$. Next, we use a numerical quadrature subroutine for the integrals ranging from θ_p to π where integral is free of singularities. Since $\theta_p \ll 1$, we approximate for the integrals over the range $0 < \theta < \theta_p$, $\cos \theta'$ by one in the integrand except in the argument of logarithmic function. The integrals can then be evaluated analytically and adding the result to the integrals for θ_p to π gives us the desired final results for I_q and J_o.

REFERENCES

[1] H. A. Wheeler, "Transmission-line properties of parallel strips separated by a dielectric sheet," *IEEE Trans. Microwave Theory Tech.*, vol. MTT-13, Mar. 1965, pp. 172–185.
[2] H. E. Stinehelfer, Sr., "An accurate calculation of uniform microstrip transmission lines," *IEEE Trans. Microwave Theory Tech.*, vol. MTT-16, July 1968, pp. 439–444.
[3] E. Yamashita and R. Mittra, "Variational method for the analysis of microstrip lines," *IEEE Trans. Microwave Theory Tech.*, vol. MTT-16, Apr. 1968, pp. 251–256.
[4] E. Yamashita, "Variational method for the analysis of microstrip-like transmission lines," *IEEE Trans. Microwave Theory Tech.*, vol. MTT-16, Aug. 1968, pp. 529–535.
[5] R. Mittra and T. Itoh, "Charge and potential distributions in shielded striplines," *IEEE Trans. Microwave Theory Tech.*, vol. MTT-18, Mar. 1970, pp. 149–156.
[6] T. G. Bryant and J. A. Weiss, "Parameters of microstrip transmission lines and of coupled pairs of microstrip lines," *IEEE Trans. Microwave Theory Tech.*, vol. MTT-16, Dec. 1968, pp. 1021–1027.
[7] P. Silvester, "TEM wave properties of microstrip transmission lines," *Proc. Inst. Elec. Eng.* (London), vol. 115, Jan. 1968, pp. 43–48.
[8] G. I. Zysman and D. Varon, "Wave propagation in microstrip transmission lines," presented at the Int. Microwave Symposium, Dallas, Tex., May 1969, session MAM-I-1.
[9] J. S. Hornsby and A. Gopinath, "Fourier analysis of a dielectric-loaded waveguide with a microstrip line," *Electron. Lett.*, vol. 5, June 12, 1969, pp. 265–267.
[10] ——, "Numerical analysis of a dielectric-loaded waveguide with a microstrip line—Finite-difference methods," *IEEE Trans. Microwave Theory Tech.*, vol. MTT-17, Sept. 1969, pp. 684–690.
[11] ——, "Numerical analysis of a dielectric-loaded waveguide with a microstrip line—Part II: Fourier series methods," Computing Laboratory, University College of North Wales, Bangor, Caerns., United Kingdom, Tech. Rep., July 1969.
[12] L. Lewin, *Advanced Theory of Waveguides*. London: Iliffe, 1951.
[13] ——, "The use of singular integral equations in the solution of waveguide problems," in *Advances in Microwaves*, vol. 1, L. Young, Ed. New York: Academic Press, 1966, pp. 212–284.
[14] F. G. Tricomi, *Integral Equations*. New York: Interscience, 1957, pp. 161–217.
[15] C. P. Hartwig, D. Massé, and R. A. Pucel, "Frequency dependent behavior of microstrip," presented at the Int. Microwave Symposium, Detroit, Mich., May 1968, session 3.

Spectral Domain Immitance Approach for Dispersion Characteristics of Generalized Printed Transmission Lines

TATSUO ITOH, SENIOR MEMBER, IEEE

Abstract—**A simple method for formulating the dyadic Green's functions in the spectral domain is presented for generalized printed transmission lines which contain several dielectric layers and conductors appearing at several dielectric interfaces. The method is based on the transverse equivalent transmission line for a spectral wave and on a simple coordinate transformation. This formulation process is so simple that often it is accomplished almost by inspection of the physical cross-sectional structure of the transmission line. The method is applied to a new versatile transmission line, a microstrip-slot line, and some numerical results are presented.**

I. Introduction

A FEW YEARS AGO, a method called the spectral-domain technique was developed for efficient numerical analyses for various planar transmission lines and successfully applied to a number of structures [1], [2]. One difficulty in applying this technique is that a lengthy derivation process is required in the formulation stage, especially for the more complicated structures such as the one recently proposed by Aikawa [3], [4] in which more than one conductor are located at different dielectric interfaces. This paper presents a simple method for deriving the dyadic Green's functions (immittance functions) which is based on the transverse equivalent circuit concept as applied in the spectral domain in conjunction with a simple coordinate transformation rule. This technique is quite versatile and the formulation of the Green's function may be done almost by inspection in many structures. It is noted that symmetry in the structure is not required and that the analysis can be extended to finite circuit elements, such as the disk resonator.

In what follows, we first illustrate the formulation process for the microstrip line and subsequently extend it to a more general microstrip-slot structure. Numerical results for the microstrip-slot structure are also presented.

II. Illustration of the Formulation Process

To illustrate the formulation process, we will use a simple shielded microstrip line shown in Fig. 1. In conventional space-domain analysis [5], this structure may be analyzed by first formulating the following coupled homogeneous integral equations and then solving for the unknown propagation constant β:

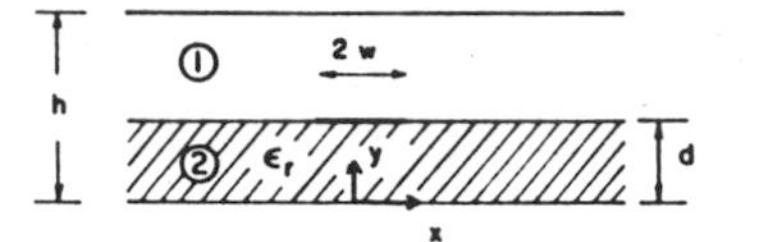

Fig. 1. Cross section of a microstrip line.

$$\int [Z_{zz}(x-x',d)J_z(x') + Z_{zx}(x-x',d)J_x(x')] \, dx' = 0 \tag{1}$$

$$\int [Z_{xz}(x-x',d)J_z(x') + Z_{xx}(x-x',d)J_x(x')] \, dx = 0 \tag{2}$$

where J_x and J_z are unknown current components on the strip and the Green's functions (impedance functions) Z_{zz}, etc., are functions of unknown β as well. The integration is over the strip, and (1) and (2) are valid on the strip. The left-hand sides of these equations give E_z and E_x components on the strip and, hence, are required to be zero to satisfy the boundary condition at the perfectly conducting strip. These equations may be solved provided that Z_{zz}, etc., are given. However, for the inhomogeneous structures, these quantities are not available in closed forms.

In the spectral domain formulation, we use Fourier transforms of (1) and (2) and deal with algebraic equations

$$\tilde{Z}_{zz}(\alpha,d)\tilde{J}_z(\alpha,d) + Z_{zx}(\alpha,d)\tilde{J}_x(\alpha,d) = \tilde{E}_z(\alpha,d) \tag{3}$$

$$\tilde{Z}_{xz}(\alpha,d)\tilde{J}_z(\alpha,d) + Z_{xx}(\alpha,d)\tilde{J}_x(\alpha,d) = \tilde{E}_x(\alpha,d) \tag{4}$$

instead of the convolution-type coupled integral equations (1) and (2). In (3) and (4), quantities with $\sim$ are Fourier transforms of corresponding quantities without $\sim$. The Fourier transform is defined as

$$\tilde{\phi}(\alpha) = \int_{-\infty}^{\infty} \phi(x) e^{j\alpha x} \, dx. \tag{5}$$

Notice that the right-hand sides of (3) and (4) are no longer zero because they are the Fourier transforms of E_z and E_x on the substrate surface which are obviously nonzero except on the strip. Hence, algebraic equations (3) and (4) contain four unknowns $\tilde{J}_z$, $\tilde{J}_x$, $\tilde{E}_z$, and $\tilde{E}_x$. However, $\tilde{E}_z$ and $\tilde{E}_x$ will be eliminated later in the solution process based on the Galerkin's procedure.

Manuscript received October 26, 1979; revised February 2, 1980. This work was supported in part by U.S. Army Research Office under Grant DAA29-78-G-0145.

The author is with the Department of Electrical Engineering, University of Texas, Austin, TX 78712.

Reprinted from *IEEE Trans. Microwave Theory Tech.*, vol. MTT-28, no. 7, pp. 733–736, July 1980.

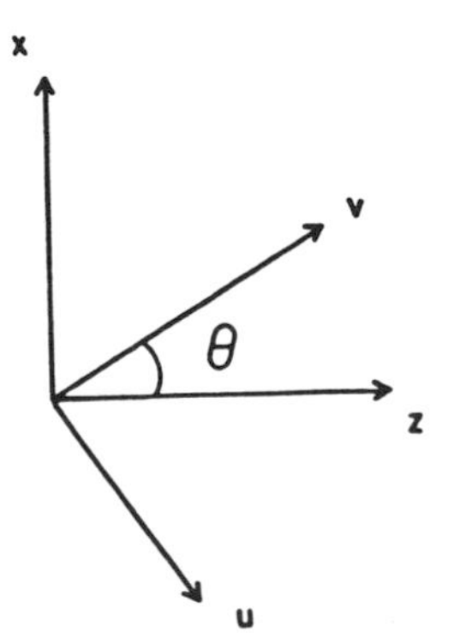

Fig. 2. Coordinate transformation.

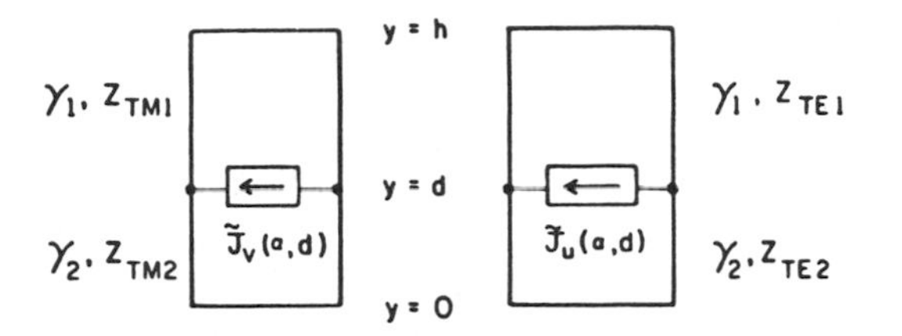

Fig. 3. Equivalent transmission lines for the microstrip line.

The closed forms of Green's impedance functions $\tilde{Z}_{zz}$, etc., can be derived by first writing the Fourier transforms of field components in each region in terms of superposition of TM-to-y and TE-to-y expressions by way of Maxwell's equations.

$$\tilde{E}_y(\alpha,y)=A^e\cosh\gamma_1 y, \qquad 0<y<d$$
$$=B^e\cosh\gamma_2(h-y), \qquad d<y<h \tag{6}$$

$$\tilde{H}_y(\alpha,y)=A^h\sinh\gamma_1 y, \qquad 0<y<d$$
$$=B^h\sinh\gamma_2(h-y), \qquad d<y<h \tag{7}$$

$$\gamma_1=\sqrt{\alpha^2+\beta^2-\epsilon_r k^2} \qquad \gamma_2=\sqrt{\alpha^2+\beta^2-k^2}. \tag{8}$$

Next, we match tangential (x and z) components at the interface and apply appropriate boundary conditions at the strip [1], [2]. By eliminating A^e, B^e, A^h, and B^h from these conditions, we obtain expressions for Green's impedance functions $\tilde{Z}_{zz}$, etc.

In the new formulation process we will make use of equivalent transmission lines in the y direction. To this end, we recognize that from

$$E_y(x,y)e^{-j\beta z}=\frac{1}{2\pi}\int_{-\infty}^{\infty}E_y(\alpha,y)e^{-j(\alpha x+\beta z)}\,d\alpha \tag{9}$$

all the field components are a superposition of inhomogeneous (in y) waves propagating in the direction of θ from the z axis where $\theta=\cos^{-1}(\beta/\xi)$, $\xi=\sqrt{\alpha^2+\beta^2}$. For each θ, waves may be decomposed into TM-to-y ($\tilde{E}_y,\tilde{E}_v,\tilde{H}_u$), and TE-to-$y$ ($\tilde{H}_y,\tilde{E}_u,\tilde{H}_v$) where the coordinates v and u are as shown in Fig. 2 and related with (x,z) via

$$u=z\sin\theta-x\cos\theta$$
$$v=z\cos\theta+x\sin\theta. \tag{10}$$

We recognize that $\tilde{J}_v$ current creates only the TM fields and $\tilde{J}_u$ the TE fields. Hence, we can draw equivalent circuits for the TM and TE fields as in Fig. 3. The characteristic admittances in each region are

$$Y_{TMi}=\frac{\tilde{H}_u}{\tilde{E}_v}=\frac{j\omega\epsilon_0\epsilon_i}{\gamma_i}, \qquad i=1,2 \tag{11}$$

$$Y_{TEi}=-\frac{\tilde{H}_v}{\tilde{E}_u}=\frac{\gamma_i}{j\omega\mu}, \qquad i=1,2 \tag{12}$$

where $\gamma_i=\sqrt{\alpha^2+\beta^2-\epsilon_i k^2}$ is the propagation constant in the y direction in the ith region. All the boundary conditions for the TE and TM waves are incorporated in the equivalent circuits. For instance, the ground planes at $y=0$ and h are represented by short circuits at respective places. The electric fields $\tilde{E}_v$ and $\tilde{E}_u$ are continuous at $y=d$ and are related to the currents via

$$\tilde{E}_v(\alpha,d)=\tilde{Z}^e(\alpha,d)\tilde{J}_v(\alpha,d) \tag{13}$$

$$\tilde{E}_u(\alpha,d)=\tilde{Z}^h(\alpha,d)\tilde{J}_u(\alpha,d). \tag{14}$$

$\tilde{Z}^e$ and $\tilde{Z}^h$ are the input impedances looking into the equivalent circuits at $y=d$ and are given by

$$\tilde{Z}^e(\alpha,d)=\frac{1}{Y_1^e+Y_2^e} \tag{15}$$

$$\tilde{Z}^h(\alpha,d)=\frac{1}{Y_1^h+Y_2^h} \tag{16}$$

where Y_1^e and Y_2^e are input admittances looking down and up at $y=d$ in the TM equivalent circuit and Y_1^h and Y_2^h are those in the TE circuit:

$$Y_1^e=Y_{TM1}\coth\gamma_1(h-d) \qquad Y_2^e=Y_{TM2}\coth\gamma_2 d \tag{17}$$

$$Y_1^h=Y_{TE1}\coth\gamma_1(h-d) \qquad Y_2^h=Y_{TE2}\coth\gamma_2 d. \tag{18}$$

The final step consists of the mapping from the (u,v) to (x,z) a coordinate system for the spectral wave corresponding to each θ given by α and β. Because of the coordinate transform (10), E_x and E_z are linear combinations of E_u and E_v. Similarly, J_x and J_z are superpositions of J_u and J_v. When these relations are used, the impedance matrix elements in (3) and (4) are found to be

$$\tilde{Z}_{zz}(\alpha,d)=N_z^2\tilde{Z}^e(\alpha,d)+N_x^2\tilde{Z}^h(\alpha,d) \tag{19}$$

$$\tilde{Z}_{zx}(\alpha,h)=\tilde{Z}_{xz}(\alpha,d)=N_xN_z\left[-\tilde{Z}^e(\alpha,d)+\tilde{Z}^h(\alpha,d)\right] \tag{20}$$

$$\tilde{Z}_{xx}(\alpha,d)=N_x^2\tilde{Z}^e(\alpha,d)+N_z^2\tilde{Z}^h(\alpha,d) \tag{21}$$

where

$$N_x=\frac{\alpha}{\sqrt{\alpha^2+\beta^2}}=\sin\theta \qquad N_z=\frac{\beta}{\sqrt{\alpha^2+\beta^2}}=\cos\theta. \tag{22}$$

Notice that $\tilde{Z}^e$ and $\tilde{Z}^h$ are functions of $\alpha^2+\beta^2$ and the ratio of α to β enters only through N_x and N_z.

It is easily shown that (19)–(21) are identical to those previously derived by means of boundary value problems imposed on the field expression [1], [2].

III. Extension to the Microstrip-Slot Structure

The method presented in the previous section may be extended to more complicated structures such as the microstrip-slot line structure in Fig. 4. This structure is

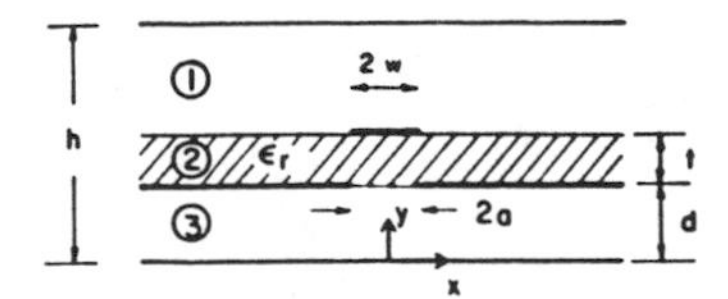

Fig. 4. Cross section of a microstrip-slot line.

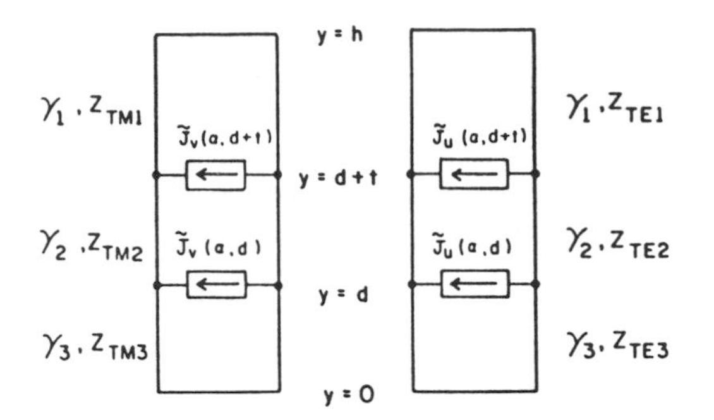

Fig. 5. Equivalent transmission lines for the microstrip-slot line.

believed to be useful in many microwave integrated circuit designs because there is an additional degree of freedom in design due to the existence of the slot [3], [4], [6]. The characteristic impedance and the propagation constant may be altered from those in the microstrip line by changing the slot width in the new structure.

By comparing the new structure with the microstrip line in Fig. 1, we may draw equivalent circuits in Fig. 5. From Fig. 5, we get

$$\tilde{E}_v(\alpha,d+t)=\tilde{Z}_{11}^e\tilde{J}_v(\alpha,d+t)+\tilde{Z}_{12}^e\tilde{J}_v(\alpha,d) \quad (23a)$$

$$\tilde{E}_u(\alpha,d+t)=\tilde{Z}_{11}\tilde{J}_u(\alpha,d+t)+\tilde{Z}_{12}^h\tilde{J}_u(\alpha,d) \quad (23b)$$

$$\tilde{E}_v(\alpha,d)=\tilde{Z}_{21}^e\tilde{J}_v(\alpha,d+t)+\tilde{Z}_{22}^e\tilde{J}_v(\alpha,d) \quad (24a)$$

$$\tilde{E}_u(\alpha,d)=\tilde{Z}_{21}^h\tilde{J}_u(\alpha,d+t)+\tilde{Z}_{22}\tilde{J}_u(\alpha,d) \quad (24b)$$

where $\tilde{Z}_{11}^e$ is the driving point input impedance at $y=d+t$ and $\tilde{Z}_{12}$ is the transfer impedance which expresses the contribution of the source at $y=d$ to the field at $y=d+t$. Other quantities may be similarly defined. Specifically

$$Z_{11}^e=\frac{1}{Y_1^e+Y_{2L}^e} \quad (25)$$

$$Y_1^e=Y_{\text{TM}1}\coth\gamma_1(h-d-t) \quad (26)$$

$$Y_{2L}^e=Y_{\text{TM}2}\frac{Y_{\text{TM}2}+Y_3^e\coth\gamma_2 t}{Y_3^e+Y_{\text{TM}2}\coth\gamma_2 t} \quad (27)$$

where

$$Y_3^e=Y_{\text{TM}3}\coth\gamma_3 d. \quad (28)$$

It is readily seen that Y_3^e and Y_{2L}^e are input impedances looking down at $y=d$ and $d+t$, respectively, while Y_1^e is the one looking upward at $y=d+t$. On the other hand

$$\tilde{Z}_{12}^e=\frac{1}{Y_3^e+Y_{2u}^e}\cdot\frac{Y_{\text{TM}2}/\sinh\gamma_2 t}{Y_1^e+Y_{\text{TM}2}\coth\gamma_2 t}. \quad (29)$$

Here

$$Y_{2u}^e=Y_{\text{TM}2}\frac{Y_{\text{TM}2}+Y_1^e\coth\gamma_2 t}{Y_1^e+Y_{\text{TM}2}\coth\gamma_2 t}$$

is the input admittance looking upward at $y=d$. We recognize that Z_{12}^e is the transfer impedance from Port 2 to Port 1 in the TM equivalent circuit. All other impedance coefficients in (23) and (24) may be similarly derived.

Impedance-matrix elements may be derived by the coordinate transform identical to the one used in the microstrip case. Some of the results are

$$\tilde{Z}_{zz}^{11}=N_z^2\tilde{Z}_{11}^e+N_x^2\tilde{Z}_{11}^h \quad (30)$$

$$\tilde{Z}_{zx}^{11}=N_zN_x(-\tilde{Z}_{11}^e+\tilde{Z}_{11}^h) \quad (31)$$

$$\tilde{Z}_{zz}^{12}=N_z^2\tilde{Z}_{12}^e+N_x^2\tilde{Z}_{12}^h. \quad (32)$$

The subscripts, say zx, indicate the direction of the field (E_z) caused by that of the contributing current (J_x). The superscripts, say 12, signify the relation between the interface where the field is observed (1) and the one where the current is present (2).

IV. Some Features of the Method

The method presented here is useful in solving many printed line problems. We will summarize the procedure for the formulation. 1) When the structure is given, we first draw TM and TE equivalent circuits. Each layer of dielectric medium is represented by different transmission lines and whenever conductors are present at particular interfaces, we place current sources at the junctions between transmission lines. At the ground planes, these transmission lines are shorted. 2) We derive driving point and transfer impedances from the equivalent circuits. 3) They are subsequently combined according to the sub- and superscript conventions described in the previous section, and we obtain the necessary impedance matrix elements.

The method has certain attractive features:

1) When the structures are modified, such changes are easily accommodated. For instance, when our structure has sidewalls, at say $x=\pm L$, to completely enclose the printed lines, all the procedures remain unchanged provided the discrete Fourier transform is used

$$\tilde{\phi}(\alpha)=\int_{-L}^{L}\phi(x)e^{j\alpha x}\,dx, \qquad \alpha=\frac{n\pi}{2L}. \quad (33)$$

On the other hand, when the top wall is removed, we only replace the shorted transmission line for the top-most layer with a semi-infinitely long one extending to $y\to+\infty$.

2) The formulation is independent of the number of strips and their relative location at each interface. Information on these parameters is used in the Galerkin's procedure to solve equations such as (3) and (4).

3) For some structures such as fin lines [8], it is more advantageous to use admittance matrix which provides the current on the fins due to the slot field. The formulation in this case almost parallels the present one. Instead of the current sources, we need to use voltage sources in the equivalent circuits.

4) It is easily shown that the method is applicable to finite structures such as microstrip resonators and antennas. Instead of (5), we need to use double Fourier transforms in x and z directions so that only the y dependence remains to allow the use of equivalent circuit concept.

5) Certain physical information is readily extracted. For instance, it is clear that denominators of typical impedance matrix elements give the transverse resonance equation when equated to zero. This implies that for certain spectral waves determined by α and β, surface wave poles may be encountered. How strongly the surface wave is excited, or if it is excited at all, is determined by the structure.

V. Numerical Example

Although the intention of this paper is to show the formulation process, the additional steps required to obtain numerical results are discussed for the sake of completeness. We computed dispersion characteristics of the microstrip-slot line with sidewalls at $x=\pm L$ by the present formulation followed by a Galerkin's procedure repeatedly used in the spectral-domain method.

In the previous section, the problem is formulated by using the impedance matrix with elements Z_{pq}^{ij}, $(i,j=1,2$ and $p,q=x,z)$ and we presumed that the current components on the conductors are unknown. It is more advantageous in numerical calculation if we choose the current components on the strip $\tilde{J}_z(\alpha,d+t)$ and $\tilde{J}_x(\alpha,d+t)$ and the aperture fields in the slot $\tilde{E}_z(\alpha,d)$ and $\tilde{E}_x(\alpha,d)$ for unknowns in the Galerkin's procedure. This is because the aperture field in the slot can be more accurately approximated than the current on the conductor at $y=d$ [4], [7]. To this end, we rearrange the impedance matrix equation to the one in which the above four unknown quantities are on the left-hand side. This modification can be readily accomplished. In the Galerkin's method, these unknowns are expressed in terms of known basis functions. Finally, we obtain homogeneous linear simultaneous equations as the right-hand side becomes identically zero by the inner product process [1], [2]. By equating the determinant to zero, we find the eigenvalue β.

There are two types of modes in the structure. One of them is a perturbed microstrip mode and another is a perturbed slot mode. For the perturbed microstrip quasi-TEM mode, we have computed dispersion relations by choosing only one basis function each for four unknowns. They are chosen such that appropriate edge conditions are satisfied at the edges of strip and slot. For instance, we can choose as the basis functions the Fourier transforms of

$$J_z(x,d+t)=\frac{1}{\sqrt{w^2-x^2}} \qquad J_x(x,d+t)=x\sqrt{w^2-x^2}$$

$$E_z(x,d)=\sqrt{a^2-x^2} \qquad E_x(x,d)=\frac{x}{\sqrt{a^2-x^2}}.$$

It is readily seen that Fourier transforms of these functions are analytically given in terms of Bessel functions. Fig. 6 shows some numerical examples of dispersion characteristics. The present results for a small slot width are compared with those of a shielded microstrip line [1]. It is clear that as the frequency increases, the presence of nonzero slot width becomes more significant. It is also seen that, as the slot width increases, the guide wavelength becomes larger because the effect for the free space below the slot is more pronounced. This suggests that the guide wavelength is adjustable by two means, one by changing the strip width and another by varying the slot width.

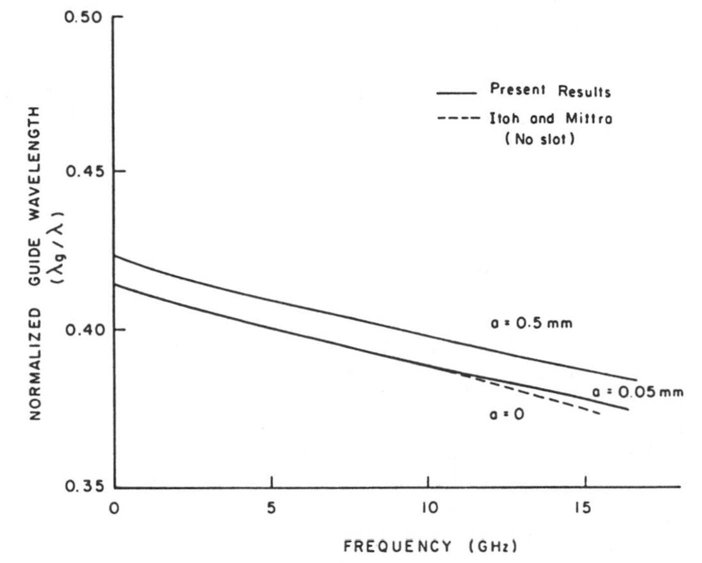

Fig. 6. Dispersion characteristics of microstrip-slot lines $L=6.35$ mm, $d=11.43$ mm, $t=1.27$ mm, $h=24.13$ mm, $w=0.635$ mm, $\epsilon_r=8.875$.

VI. Conclusions

We presented a simple method for formulating the eigenvalue problems for dispersion characteristics of general printed transmission lines. The method is intended to save considerable analytical labor for these types of problems. In addition, the method provides certain unique features. The method is applied to the problem of microstrip-slot line believed useful in microwave- and millimeter-wave integrated circuits. Numerical results are also presented.

References

[1] T. Itoh and R. Mittra, "A technique for computing dispersion characteristics of shielded microstrip lines," *IEEE Trans. Microwave Theory Tech.*, vol. MTT-22, pp. 896–898, Oct. 1974.

[2] T. Itoh, "Analysis of microstrip resonators," *IEEE Trans. Microwave Theory Tech.*, vol. MTT-22, pp. 946–952, Nov. 1974.

[3] M. Aikawa, "Microstrip line directional coupler with tight coupling and high directivity," *Electron. Commun. Jap.*, vol. J60-B, pp. 253–259, Apr. 1977.

[4] H. Ogawa and M. Aikawa, "Analysis of coupled microstrip-slot lines," *Electron. Commun. Jap.*, vol. J62-B, pp. 396–403, Apr. 1979.

[5] G. I. Zysman and D. Varon, "Wave propagation in microstrip transmission lines," presented at the Int. Microwave Symp. (Dallas, TX, May 1969, paper MAM-I-1.

[6] T. Itoh and A. S. Hebert, "A generalized spectral domain analysis for coupled suspended microstrip lines with tuning septums," *IEEE Trans. Microwave Theory Tech.*, vol. MTT-26, pp. 820–826, Oct. 1978.

[7] J. B. Davies and D. Mirshekar-Syahkal, "Spectral domain solution of arbitrary coupled transmission lines with multilayer substrate," *IEEE Trans. Microwave Theory Tech.*, vol. MTT-25, pp. 143–146, Feb. 1977.

[8] P. J. Meier, "Two new integrated circuit media with special advantages of millimeter wavelengths, presented at the 1972 IEEE G-MTT Int. Microwave Symp. (Arlington Heights, IL, May 1972).

Analysis of Microstrip-Like Transmission Lines by Nonuniform Discretization of Integral Equations

EIKICHI YAMASHITA, MEMBER, IEEE, AND KAZUHIKO ATSUKI

Abstract—**The nonuniform discretization of the integral equation on the tangential electromagnetic (EM) field on the boundary surface is proposed as a numerically efficient method to analyze the microstrip-like transmission lines. The calculated results of the propagation constant of the microstrip line based on this method are compared with other published analytical results. Various types of planar striplines are treated by the same formulas. The dominant and higher order modes of a shielded microstrip line are discussed and compared with the longitudinal-section electric (LSE) and linear synchronous motor (LSM) modes of a two-medium waveguide.**

I. Introduction

IN the early stage of microstrip-line analyses, the TEM approximation was effectively employed to calculate the line capacitance as a basic parameter of the inhomogeneous transmission line [1]–[4]. Though this approximation was useful in a wide range of frequencies, more rigorous analytical methods have been explored to find its theoretical limitations [5]–[8]. The dispersion characteristics of the microstrip line at high frequencies, for example, have been reported by many papers. However, published numerical values even on the same problem are not necessarily in good agreement.

This paper describes a straightforward and numerically effective method to compute characteristic values of microstrip-like transmission lines. The main features of this method are the formulation of integral equations for general structures in a form of Zysman and Varon [7] and the derivation of the solution by the nonuniform discretization of the integrals. Numerical results based on this method are compared with other data [8]–[11] and empirical formulas [12]–[14]. The treatments of coplanar striplines [15], [16], shielded slot lines [17], and microstrip lines are shown. The dominant and higher order modes of microstrip lines are also discussed compared with the LSE and LSM modes of a two-medium waveguide.

II. Formulation of Integral Equations

Fig. 1 shows a microstrip-like transmission line which contains three dielectric layers, multistrip conductors, and a metallic shield enclosure. The strip conductors are assumed to be negligibly thin and the line lossless.

A hybrid-mode analysis is apparently necessary in this inhomogeneous structure. When the scalar potentials for TM waves and TE waves are defined by $\psi^{(e)}$ and $\psi^{(h)}$, respectively, the electromagnetic (EM) fields of hybrid

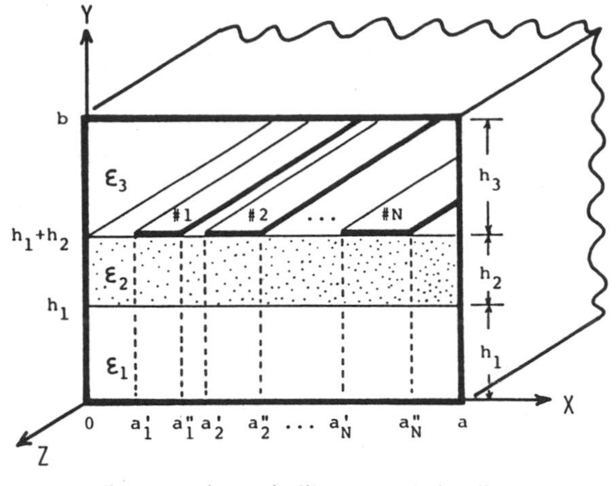

Fig. 1. Microstrip-like transmission line.

modes [18] are given by

$$E_{zi} = j\frac{k_i^2 - \beta^2}{\beta}\psi_i^{(e)}(x,y)e^{-j\beta z} \tag{1a}$$

$$H_{zi} = j\frac{k_i^2 - \beta^2}{\beta}\psi_i^{(h)}(x,y)e^{-j\beta z} \tag{1b}$$

$$E_{ti} = \nabla_t\psi_i^{(e)}(x,y)e^{-j\beta z} - \frac{\omega\mu_0}{\beta}a_z \times \nabla_t\psi_i^{(h)}(x,y)e^{-j\beta z} \tag{1c}$$

$$H_{ti} = \nabla_t\psi_i^{(h)}(x,y)e^{-j\beta z} + \frac{\omega\varepsilon_i}{\beta}a_z \times \nabla_t\psi_i^{(e)}(x,y)e^{-j\beta z}, \quad i = 1,2,3 \tag{1d}$$

where ω is the angular frequency, μ_0 is the magnetic permeability of a vacuum, ε_i is the permittivity in the ith medium ($i = 1,2,3$), k_i is $\omega\sqrt{\varepsilon_i\mu_0}$, the subscript t denotes the transverse direction, a_z is the z-directed unit vector, and β is the unknown phase constant of the hybrid mode.

The general solution of the wave equation in each dielectric layer can be obtained by the method of the separation of variables. After applying the boundary conditions on the surface of the shielding conductor to the general solution, one obtains the following form:

$$\psi_1^{(e)} = \sum_{n=1}^{\infty} A_n^{(e)} \sinh(\alpha_n^{(1)}y)\sin(a_n x) \tag{2a}$$

$$\psi_2^{(e)} = \sum_{n=1}^{\infty} \{B_n^{(e)} \sinh(\alpha_n^{(2)}y) + C_n^{(e)}\cosh(\alpha_n^{(2)}y)\} \cdot \sin(a_n x) \tag{2b}$$

Manuscript received August 18, 1975; revised October 27, 1975.
The authors are with the Department of Applied Electronics, University of Electro-Communications, Tokyo, Japan.

Reprinted from *IEEE Trans. Microwave Theory Tech.*, vol. MTT-24, no. 4, pp. 195–200, April 1976.

$$\psi_3^{(e)} = \sum_{n=1}^{\infty} D_n^{(e)} \sinh \{\alpha_n^{(3)}(b - y)\} \sin (a_n x) \tag{2c}$$

$$\psi_1^{(h)} = \sum_{n=0}^{\infty} A_n^{(h)} \cosh (\alpha_n^{(1)} y) \cos (a_n x) \tag{2d}$$

$$\psi_2^{(h)} = \sum_{n=0}^{\infty} \{B_n^{(h)} \cosh (\alpha_n^{(2)} y) + C_n^{(h)} \sinh (\alpha_n^{(2)} y)\} \cdot \cos (a_n x) \tag{2e}$$

$$\psi_3^{(h)} = \sum_{n=0}^{\infty} D_n^{(h)} \cosh \{\alpha_n^{(3)}(b - y)\} \cos (a_n x) \tag{2f}$$

where

$$a_n = \frac{n\pi}{a}$$

$$\alpha_n^{(i)} = \sqrt{a_n^2 + \beta^2 - k_i^2}, \qquad i = 1,2,3$$

and $A_n^{(e)}$, $B_n^{(e)}$, $C_n^{(e)}$, $D_n^{(e)}$, $A_n^{(h)}$, $B_n^{(h)}$, $C_n^{(h)}$, and $D_n^{(h)}$ are constants to be determined.

The boundary conditions of EM fields to connect these potentials are expressed as follows:

1) $y = h_1$

$$E_{z1} = E_{z2}, \qquad (0 \le x \le a) \tag{3a}$$

$$H_{z1} = H_{z2}, \qquad (0 \le x \le a) \tag{3b}$$

$$E_{x1} = E_{x2}, \qquad (0 \le x \le a) \tag{3c}$$

$$H_{x1} = H_{x2}, \qquad (0 \le x \le a). \tag{3d}$$

2) $y = h_1 + h_2$

$$E_{z2} = E_{z3}, \qquad (0 \le x \le a) \tag{4a}$$

$$E_{x2} = E_{x3}, \qquad (0 \le x \le a) \tag{4b}$$

$$E_{x2} = \begin{cases} f(x)e^{-j\beta z}, & (a_i'' \le x \le a_{i+1}', \; i = 0,1,\cdots,N, \text{ namely, on dielectrics}) \\ 0, & (a_i' \le x \le a_i'', \; i = 1,2,\cdots,N, \text{ namely, on conductors}) \end{cases} \tag{4c}$$

$$H_{x2} - H_{x3} = \begin{cases} 0, & (a_i'' \le x \le a_{i+1}', \; i = 0,1,\cdots,N) \\ g(x)e^{-j\beta z}, & (a_i' \le x \le a_i'', \; i = 1,2,\cdots,N) \end{cases} \tag{4d}$$

$$H_{z2} = H_{z3}, \qquad (a_i'' \le x \le a_{i+1}', \; i = 0,1,\cdots,N) \tag{4e}$$

$$E_{z2} = 0, \qquad (a_i' \le x \le a_i'', \; i = 1,2,\cdots,N). \tag{4f}$$

First, the EM fields in (1) are substituted for the boundary conditions in (3a)–(4d). By using the orthogonality of sinusoidal functions, the constants, $A_n^{(e)}$, $B_n^{(e)}$, $C_n^{(e)}$, $D_n^{(e)}$, $A_n^{(h)}$, $B_n^{(h)}$, $C_n^{(h)}$, $D_n^{(h)}$, can be derived as coefficients of the Fourier series expansion where, for simplicity, the following notations are defined:

$$F_n[f(\xi)] = \sum_{i=0}^{N} \int_{a_i''}^{a_{i+1}'} f(\xi) \cos (a_n \xi) \, d\xi \tag{5a}$$

$$G_n[g(\xi)] = \sum_{i=1}^{N} \int_{a_i'}^{a_i''} g(\xi) \sin (a_n \xi) \, d\xi. \tag{5b}$$

The rest of the conditions, (4e) and (4f), result in a set of homogeneous integral equations on the unknown functions, $f(\xi)$ $(a_i'' \le \xi \le a_{i+1}', \; i = 0,1,\cdots,N)$ and $g(\xi)$ $(a_i' \le \xi \le a_i'', \; i = 1,2,\cdots,N)$. Physically, $g(\xi)$ is proportional to the z directed electric currents on the boundary surface.

$$\sum_{n=0}^{\infty} \{P_n(\beta)F_n[f(\xi)] + R_n(\beta)G_n[g(\xi)]\} \cos (a_n x) = 0, \qquad (a_i'' \le x \le a_{i+1}', \; i = 0,1,\cdots,N) \tag{6a}$$

$$\sum_{n=0}^{\infty} \{R_n(\beta)F_n[f(\xi)] + Q_n(\beta)G_n[g(\xi)]\} \sin (a_n x) = 0, \qquad (a_i' \le x \le a_i'', \; i = 1,2,\cdots,N) \tag{6b}$$

where $P_n(\beta)$, $Q_n(\beta)$, and $R_n(\beta)$ are given in the Appendix.

III. Numerical Solutions by Nonuniform Discretization

The preceding simultaneous homogeneous integral equations (6) are numerically solved by the discretization of the integral regions, $\{0 \le \xi \le a_1', a_1' \le \xi \le a_1'', \cdots, a_N' \le \xi \le a_N'', a_N'' \le \xi \le a\}$, where the number of subregions are $\{M_0, M_1, \cdots, M_{2N}\}$, respectively, and the total number of subregions is M. When the functions to be solved, $f(\xi)$ and $g(\xi)$, are assumed to be constant in each subregion, the simultaneous integral equations (6) are rewritten as a set of simultaneous homogeneous linear equations with M variables. The determinant of these linear equations should vanish in order to have nontrivial solutions. The phase constant β can be calculated by solving this determinant equation.

The computation accuracy depends on the way of the discretization when the total number of subregions is fixed. We apply the nonuniform discretization method, which was developed in the TEM analysis of microstrip line [19], to the present case. There is a fact in the EM theory that fields vary very rapidly near strip-conductor edges. Therefore, our principle of the discretization is that the integral regions should be discretized more precisely when the considering point is closer to the edge. Fig. 2 shows an illustration of the uniform and nonuniform discretization methods. Fig. 3 shows the computation accuracy of the two discretization methods in the case of a shielded microstrip line. When the total number of subregions M is increased, the nonuniform discretization is seen to result in much faster convergence than the uniform discretization. The nonuniform discretization is employed throughout our work.

The numerical data obtained based on this method are compared with published data by other methods [8]–[11], in Fig. 4, and with those by empirical formulas [12]–[14], in Fig. 5. These curves indicate that our results are very close to those of Kowalski and Pregla (the mode-matching method) [11] in a wide range of frequencies. This agreement verifies the accuracy of the two methods. The empirical formulas in Fig. 5 seem to lose the accuracy at high frequencies.

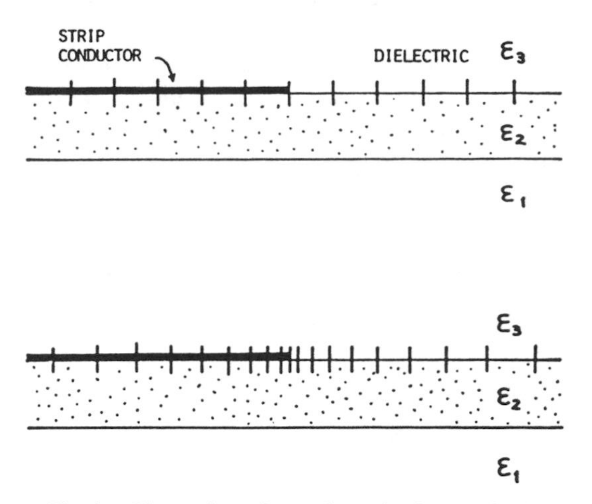

Fig. 2. Illustration of two discretization methods.

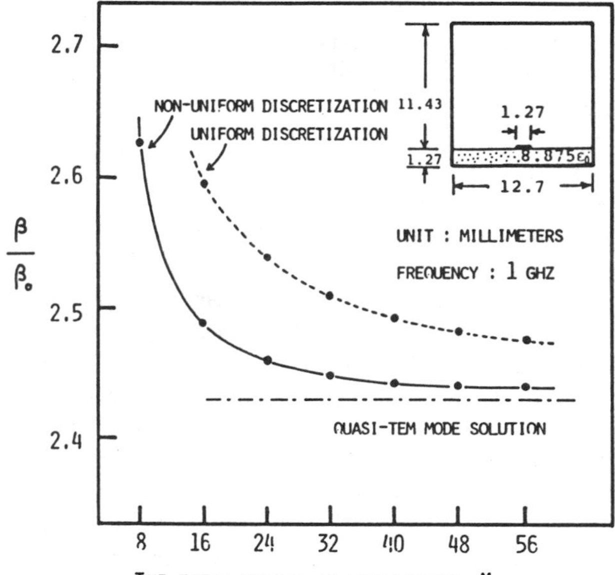

Fig. 3. The dependence of convergence on the way of the discretization.

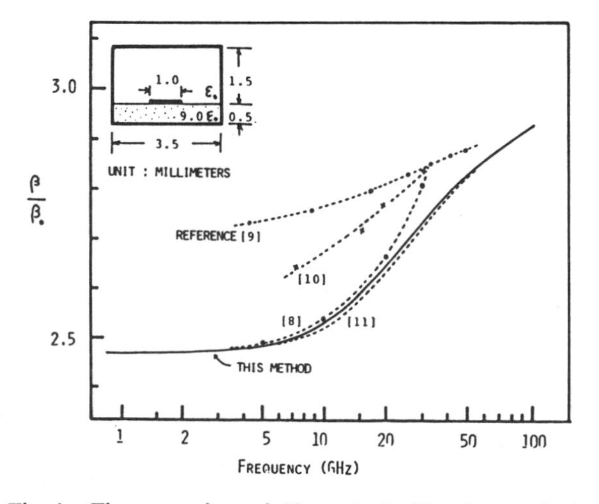

Fig. 4. The comparison of this method with other methods.

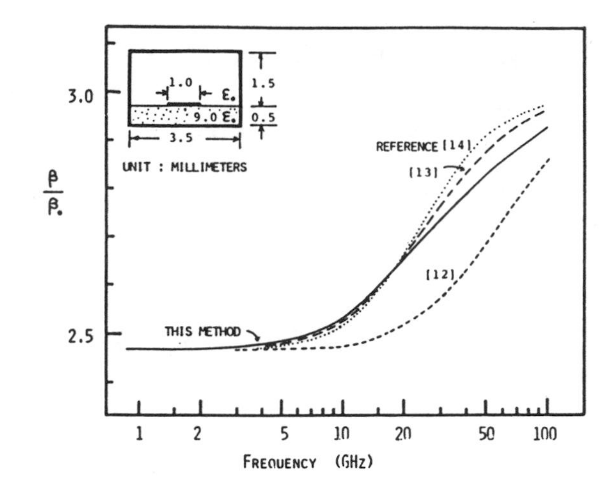

Fig. 5. The comparison of this method with empirical formulas.

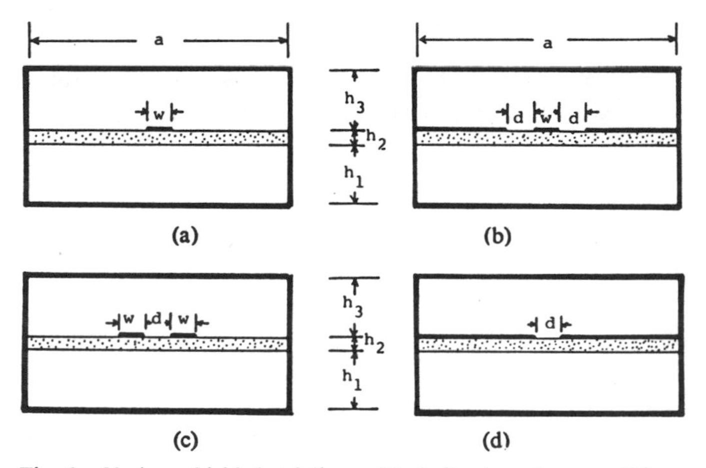

Fig. 6. Various shielded striplines with similar boundary conditions.

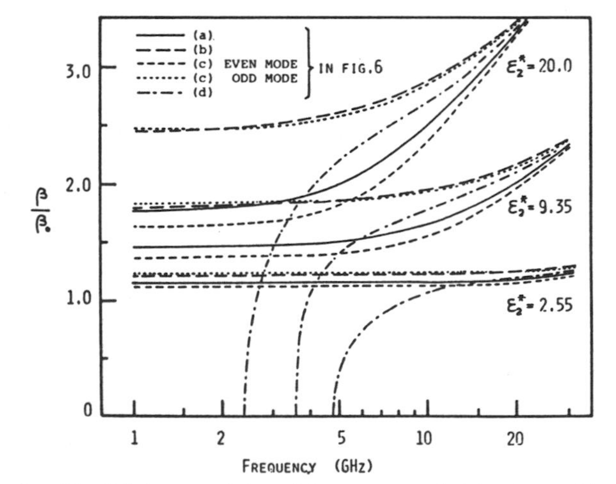

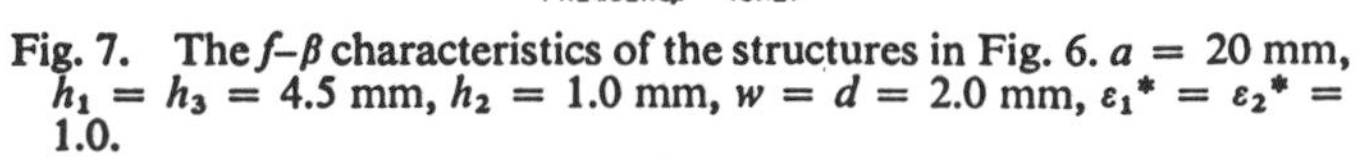

Fig. 7. The f–β characteristics of the structures in Fig. 6. $a = 20$ mm, $h_1 = h_3 = 4.5$ mm, $h_2 = 1.0$ mm, $w = d = 2.0$ mm, $\varepsilon_1^* = \varepsilon_2^* = 1.0$.

IV. Various Shielded Structures

The dispersion characteristics of the dominant mode of various shielded planar structures were investigated by the preceding method in a similar fashion. Fig. 6 shows the cross-sectional view of shielded transmission lines. Fig. 7 shows their characteristics for the cases, $\varepsilon_2^* = 2.55$,

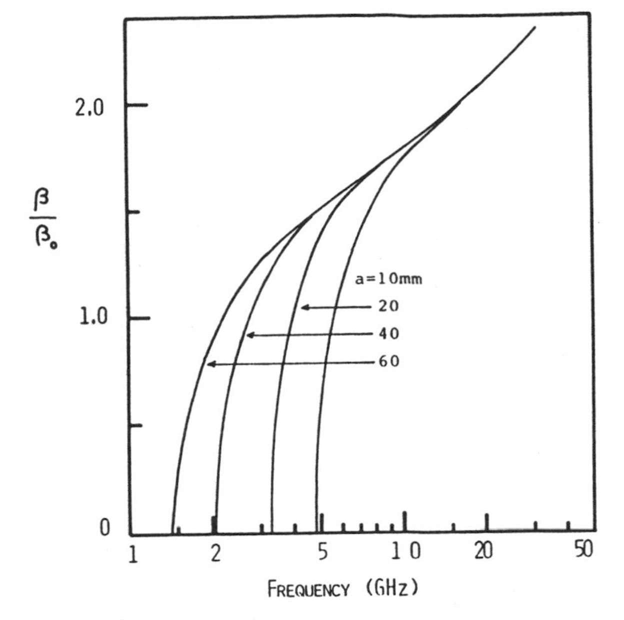

Fig. 8. The f–β characteristics of the shielded slot lines in Fig. 6(d). $h_1 = h_3 = 4.5$ mm, $h_2 = 1.0$ mm, $d = 2.0$ mm, $\varepsilon_1{}^* = \varepsilon_3{}^* = 1.0$, $\varepsilon_2{}^* = 9.35$.

$\varepsilon_2{}^* = 9.35$, and $\varepsilon_2{}^* = 20.0$. It is noted that the characteristics of the coplanar structure have a similarity to those of the odd-mode coupled microstrip line, and the slot line has a quite different nature from microstrip lines, though both are of the planar structure. Fig. 8 shows the dependence of the phase constant of the shielded slot lines in Fig. 6(d) on the frequency and the width of the shield conductor. It is seen that the phase constant is not much affected by the width in the high-frequency region since the EM field energy is concentrated around the slot at high frequencies.

V. Higher Order Modes of Shielded Microstrip Lines

This method was applied to the analysis of higher order modes of various structures. The cases of the shielded microstrip line, for example, are shown in Figs. 9–12. The longitudinal-section electric (LSE) and linear synchronous motor (LSM) modes of the two-medium waveguide in Fig. 9 are seen to be gradually changed to stripline modes when the center strip conductor appears and the strip width is increased.

By comparing Figs. 9 and 10, the following are observed.

1) The dominant mode close to the TEM solution appears in Fig. 10 because of the existence of the center conductor.

2) The degeneracies existing between even modes (solid lines) and between odd modes (dotted lines) in Fig. 9 have been resolved in Fig. 10. However, the degeneracies between even and odd modes in Fig. 9 have not been resolved in Fig. 10 because of the structural symmetry.

3) While the odd-mode characteristics in Fig. 9 are not much changed in Fig. 10, the even-mode characteristics are quite changed there.

When the characteristics of the lines with various strip widths are compared in Figs. 10–12, the following are observed with the increase of the width.

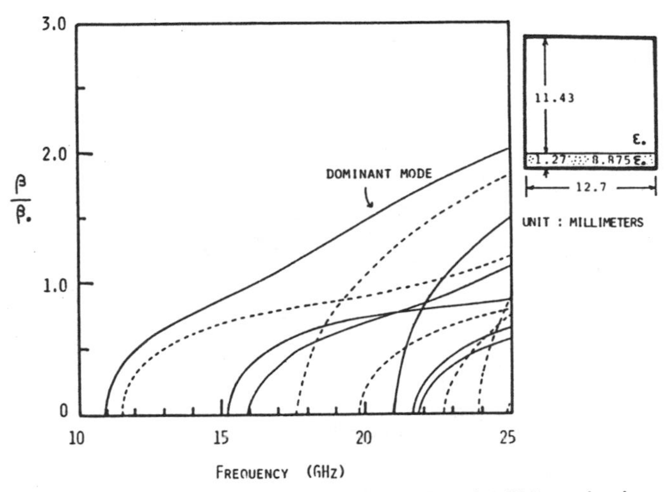

Fig. 9. The f–β characteristics of the LSE and LSM modes in a two-medium waveguide.

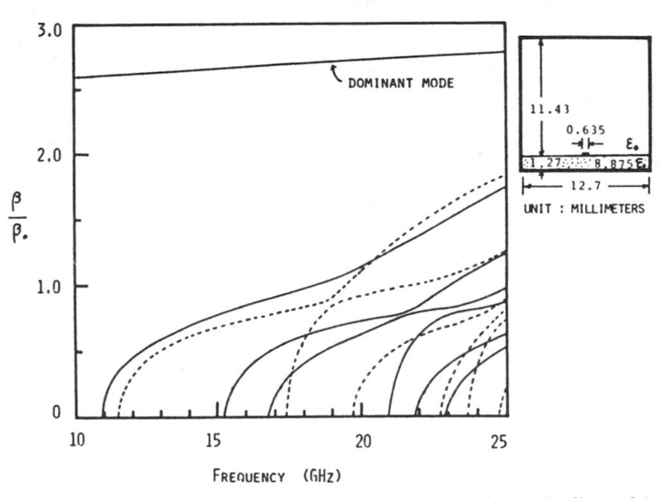

Fig. 10. The f–β characteristics of the shielded microstrip line with the strip width of 0.635 mm.

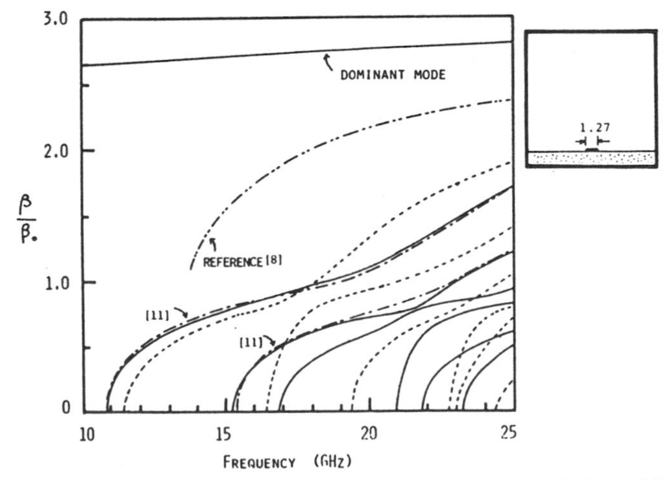

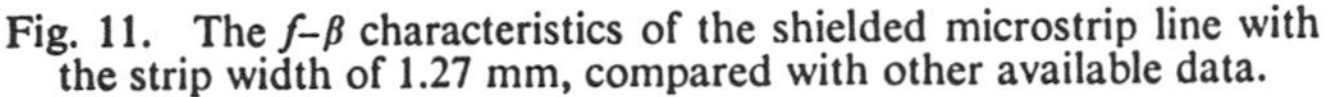

Fig. 11. The f–β characteristics of the shielded microstrip line with the strip width of 1.27 mm, compared with other available data.

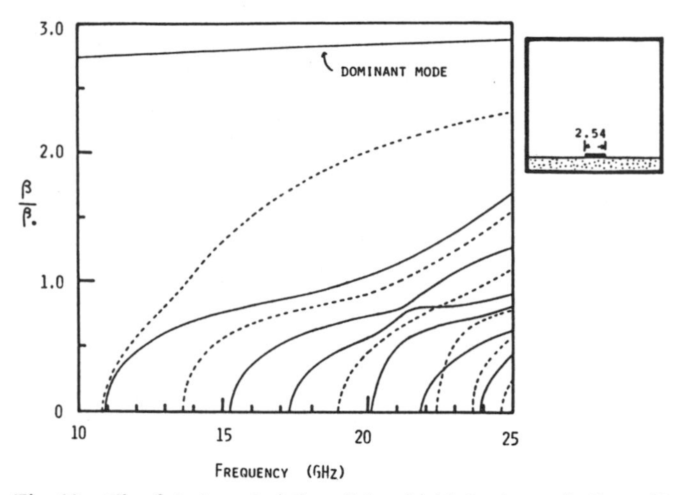

Fig. 12. The f-β characteristics of the shielded microstrip line with the strip width of 2.54 mm.

1) The phase constant of the dominant mode is gradually changed.

2) While the even-mode characteristics are not much changed, the odd-mode characteristics are changed rapidly.

3) The cutoff frequencies of the first- and the third-order mode of the shielded microstrip line are almost equal to the cutoff frequencies of the LSE and LSM modes of the corresponding two-medium waveguide.

4) The first higher order mode of the shielded microstrip line is the even mode when the strip is narrow, and the odd mode when the strip is wide.

Some of these numerical data are compared with other available data [8], [11] in Fig. 11. Again, our results are in good agreement with those of Kowalski and Pregla [11]. A small dip is noticed on the curve of the third higher order mode (solid line) at about 22 GHz in Fig. 11. The complexity of the phase constant curves for higher order modes such as this dip can be understood only by considering the resolution of waveguide-mode degeneracies as pointed out in the preceding.

VI. Conclusions

This paper described a method to analyze the microstrip-like transmission lines. The nonuniform discretization approach was found to be numerically effective in finding solutions of integral equations for the microstrip-like transmission lines. The results of computations in the case of the shielded microstrip line were very close to one of published data in a wide range of frequencies. The limitations of published empirical formulas at high frequencies were also recognized. Various planar transmission lines were compared to each other from the viewpoint of the f-β characteristics.

The higher order modes of the shielded microstrip lines were compared with those of the LSE and LSM modes. The observation on the similarity of cutoff frequencies between these modes may be useful to estimate the frequency range of the dominant mode of the microstrip line in a simple way. It was also found to be important to investigate the resolution process of waveguide-mode degeneracies due to the existence of a center strip conductor in order to explain the complex shape of the phase constant curves for higher order modes of microstrip lines.

Appendix

$P_n(\beta), R_n(\beta), Q_n(\beta)$ in (6) are given by

$$P_n(\beta) = \frac{\omega\mu_0}{\Delta_n}\Big[\alpha_n^{(1)}\alpha_n^{(2)}\Big\{\tanh^2(\alpha_n^{(1)}h_1) + \frac{\varepsilon_1}{\varepsilon_2}\tanh^2(\alpha_n^{(2)}h_2)\Big\} + \Big(\alpha_n^{(1)2} + \frac{\varepsilon_1}{\varepsilon_2}\alpha_n^{(2)2}\Big)\tanh(\alpha_n^{(1)}h_1)\tanh(\alpha_n^{(2)}h_2)\Big] \cdot \alpha_n^{(3)}\tanh(\alpha_n^{(3)}h_3)$$

$$R_n(\beta) = \frac{\beta a_n}{\Delta_n}\Big[\Big\{\alpha_n^{(2)}\tanh(\alpha_n^{(1)}h_1) + \frac{\varepsilon_1}{\varepsilon_2}\alpha_n^{(1)}\tanh(\alpha_n^{(2)}h_2) + \alpha_n^{(1)}\tanh^2(\alpha_n^{(1)}h_1)\tanh(\alpha_n^{(2)}h_2) + \alpha_n^{(2)} \cdot \Big(\frac{\varepsilon_1}{\varepsilon_2} + \frac{k_2^2 - k_1^2}{\alpha_n^{(2)2}}\Big)\tanh(\alpha_n^{(1)}h_1)\tanh^2(\alpha_n^{(2)}h_2)\Big\} \cdot \alpha_n^{(3)}\tanh(\alpha_n^{(3)}h_3) + \alpha_n^{(1)}\alpha_n^{(2)} \cdot \Big\{\tanh^2(\alpha_n^{(1)}h_1) + \frac{\varepsilon_1}{\varepsilon_2}\tanh^2(\alpha_n^{(2)}h_2)\Big\} + \Big(\alpha_n^{(1)2} + \frac{\varepsilon_1}{\varepsilon_2}\alpha_n^{(2)2}\Big)\tanh(\alpha_n^{(1)}h_1)\tanh(\alpha_n^{(2)}h_2)\Big]$$

$$\begin{aligned} Q_n(\beta) = \frac{1}{\Delta_n}\frac{k_2^2}{\omega\mu_0}\Big[\Big\{&\frac{\varepsilon_1}{\varepsilon_2}\alpha_n^{(1)}\alpha_n^{(2)} + \Big(\alpha_n^{(1)2} + \frac{\varepsilon_1}{\varepsilon_2}\alpha_n^{(2)2}\Big) \cdot \tanh(\alpha_n^{(1)}h_1)\tanh(\alpha_n^{(2)}h_2) + \alpha_n^{(1)}\alpha_n^{(2)} \cdot \tanh^2(\alpha_n^{(1)}h_1)\tanh^2(\alpha_n^{(2)}h_2)\Big\} \cdot \alpha_n^{(3)}\tanh(\alpha_n^{(3)}h_3) \\ &+ \frac{\varepsilon_1}{\varepsilon_2}\alpha_n^{(1)}\Big(\alpha_n^{(3)2} + \frac{\varepsilon_3}{\varepsilon_2}\alpha_n^{(2)2}\Big)\tanh(\alpha_n^{(2)}h_2) \\ &+ \alpha_n^{(2)}\Big(\frac{\varepsilon_1}{\varepsilon_2}\alpha_n^{(3)2} + \frac{\varepsilon_3}{\varepsilon_2}\alpha_n^{(1)2}\Big)\tanh(\alpha_n^{(1)}h_1) \\ &+ \alpha_n^{(2)}\Big(\frac{\alpha_n^{(1)2}}{\alpha_n^{(2)2}}\alpha_n^{(3)2} + \frac{\varepsilon_1\varepsilon_3}{\varepsilon_2^2}\alpha_n^{(2)2}\Big) \cdot \tanh(\alpha_n^{(1)}h_1)\tanh^2(\alpha_n^{(2)}h_2) \\ &+ \Big(\alpha_n^{(3)2} + \frac{\varepsilon_3}{\varepsilon_2}\alpha_n^{(2)2}\Big)\alpha_n^{(1)}\tanh^2(\alpha_n^{(1)}h_1) \cdot \tanh(\alpha_n^{(2)}h_2) + \frac{\varepsilon_3}{\varepsilon_2}\Big\{\alpha_n^{(1)}\alpha_n^{(2)}\tanh^2(\alpha_n^{(1)}h_1) \\ &+ \frac{\varepsilon_1}{\varepsilon_2}\alpha_n^{(1)}\alpha_n^{(2)}\tanh^2(\alpha_n^{(2)}h_2) + \Big(\alpha_n^{(1)2} + \frac{\varepsilon_1}{\varepsilon_2}\alpha_n^{(2)2}\Big)\tanh(\alpha_n^{(1)}h_1) \cdot \tanh(\alpha_n^{(2)}h_2)\Big\}\alpha_n^{(3)}\coth(\alpha_n^{(3)}h_3)\Big]\end{aligned}$$

$$\Delta_n = \Big[\alpha_n^{(2)}(k_1^2 - a_n^2)\tanh(\alpha_n^{(1)}h_1) + \frac{\varepsilon_1}{\varepsilon_2}\alpha_n^{(1)} \cdot (k_2^2 - a_n^2)\tanh(\alpha_n^{(2)}h_2) + \alpha_n^{(1)}(k_2^2 - a_n^2)\tanh^2(\alpha_n^{(1)}h_1)\tanh(\alpha_n^{(2)}h_2) + \alpha_n^{(2)}\left\{\frac{\varepsilon_1}{\varepsilon_2}(k_2^2 - a_n^2) + \frac{\beta^2}{\alpha_n^{(2)2}}(k_2^2 - k_1^2)\right\} \cdot \tanh(\alpha_n^{(1)}h_1)\tanh^2(\alpha_n^{(2)}h_2)\Big]\alpha_n^{(3)}\tanh(\alpha_n^{(3)}h_3) + (k_3^2 - a_n^2)\Big[\alpha_n^{(1)}\alpha_n^{(2)}\Big\{\tanh^2(\alpha_n^{(1)}h_1) + \frac{\varepsilon_1}{\varepsilon_2}\tanh^2(\alpha_n^{(2)}h_2)\Big\} + \left(\alpha_n^{(1)2} + \frac{\varepsilon_1}{\varepsilon_2}\alpha_n^{(2)2}\right) \cdot \tanh(\alpha_n^{(1)}h_1)\tanh(\alpha_n^{(2)}h_2)\Big].$$

ACKNOWLEDGMENT

The authors wish to thank Professor T. Okabe for his helpful advice.

REFERENCES

[1] H. A. Wheeler, "Transmission-line properties of parallel strips separated by a dielectric sheet," *IEEE Trans. Microwave Theory Tech.*, vol. MTT-13, pp. 172–185, Mar. 1965.
[2] H. E. Green, "The numerical solution of some important transmission-line problems," *IEEE Trans. Microwave Theory Tech. (Special Issue on Microwave Filters)*, vol. MTT-13, pp. 676–692, Sept. 1965.
[3] E. Yamashita and R. Mittra, "Variational method for the analysis of microstrip lines," *IEEE Trans. Microwave Theory Tech.*, vol. MTT-16, pp. 251–256, Apr. 1968.
[4] P. Silvester, "TEM wave properties of microstrip transmission lines," *Proc. Inst. Elec. Eng.*, vol. 115, pp. 43–48, Jan. 1968.
[5] P. Daly, "Hybrid-mode analysis of microstrip by finite-element methods," *IEEE Trans. Microwave Theory Tech.*, vol. MTT-19, pp. 19–25, Jan. 1971.
[6] E. J. Denlinger, "A frequency dependent solution for microstrip transmission lines," *IEEE Trans. Microwave Theory Tech.*, vol. MTT-19, pp. 30–39, Jan. 1971.
[7] G. I. Zysman and D. Varon, "Wave propagation in microstrip transmission lines," in *1969 Int. Microwave Symp. Dig.* (Dallas, TX), pp. 3–9.
[8] R. Mittra and T. Itoh, "A new technique for the analysis of the dispersion characteristics of microstrip lines," *IEEE Trans. Microwave Theory Tech.*, vol. MTT-19, pp. 47–56, Jan. 1971.
[9] J. S. Hornsby and A. Gopinath, "Fourier analysis of a dielectric loaded waveguide with a microstrip line," *Electron. Lett.*, vol. 5, pp. 265–267, June 1969.
[10] ——, "Numerical analysis of a dielectric-loaded waveguide with a microstrip line—Finite-difference method," *IEEE Trans. Microwave Theory Tech.*, vol. MTT-17, pp. 684–690, Sept. 1969.
[11] G. Kowalski and R. Pregla, "Dispersion characteristics of shielded microstrips with finite thickness," *Arch. Elek. Ubertragung*, vol. 107, pp. 163–170, Apr. 1971.
[12] M. V. Schneider, "Microstrip dispersion," *Proc. IEEE*, vol. 60, pp. 144–146, Jan. 1972.
[13] W. J. Getsinger, "Microstrip dispersion model," *IEEE Trans. Microwave Theory Tech.*, vol. MTT-21, pp. 34–39, Jan. 1973.
[14] H. J, Carlin, "A simplified circuit model for microstrip," *IEEE Trans. Microwave Theory Tech.*, vol. MTT-21, pp. 589–591, Sept. 1973.
[15] C. P. Wen, "Coplanar waveguide: A surface strip transmission line suitable for non-reciprocal gyromagnetic device applications," *IEEE Trans. Microwave Theory Tech. (1969 International Microwave Symp.)*, vol. MTT-17, pp. 1087–1090, Dec. 1969.
[16] A. Torisawa, "The analysis and experiment on parallel triple strip line," B.S. thesis, Univ. Electro-Communications, Tokyo, Japan, Mar. 1969.
[17] S. B. Cohn, "Slot line on a dielectric substrate," *IEEE Trans. Microwave Theory Tech.*, vol. MTT-17, pp. 768–778, Oct. 1969.
[18] R. F. Harrington, *Time-Harmonic Electromagnetic Fields.* New York: McGraw-Hill, 1961.
[19] K. Atsuki and E. Yamashita, "Analytical method for transmission lines with thick-strip conductor, multi-dielectric layers and shielding conductor," *J. Inst. Electron. Commun. Eng. Jap.*, vol. 53-B, pp. 322–328, June 1970.
[20] E. Yamashita, "Variational method for the analysis of microstrip-like transmission lines," *IEEE Trans. Microwave Theory Tech.*, vol. MTT-16, pp. 529–535, Aug. 1968.
[21] K. Atsuki and E. Yamashita, "Higher order modes of microstrip lines" (in Japanese), presented at the Inst. Electron. Commun. Eng. Jap. Technical Group Meeting on Microwaves, Feb. 1974, Paper MW73-116.
[22] ——, "Dispersion characteristics of strip lines" (in Japanese), presented at the Inst. Electron. Commun. Eng. Jap. Technical Group Meeting on Microwaves, June 1974, Paper MW74-18.
[23] D. G. Corr and J. B. Davies, "Computer analysis of the fundamental and higher order modes in single and coupled microstrip," *IEEE Trans. Microwave Theory Tech.*, vol. MTT-20, pp. 669–678, Oct. 1972.

Spectral-Domain Calculation of Microstrip Characteristic Impedance

JEFFREY B. KNORR, MEMBER, IEEE, AND AHMET TUFEKCIOGLU, STUDENT MEMBER, IEEE

***Abstract*—This paper presents a hybrid-mode solution for the characteristic impedance of microstrip on lossless dielectric substrate. A solution to the hybrid-mode equations is obtained by applying the method of moments in the Fourier transform domain. Numerical results are presented showing the frequency dependence of both wavelength and characteristic impedance for single and coupled strips. These results are compared with those of other investigators in the low-frequency range.**

I. INTRODUCTION

THE spectral-domain technique is a powerful, accurate, numerically efficient approach for analysis of planar transmission line structures. This technique was first suggested by Itoh and Mittra [1] and has been applied to calculate the dispersion characteristic of a single slot [1], the dispersion characteristic of a single microstrip [2], [3], and the resonant frequency of rectangular microstrip resonators [4] from which a calculation of microstrip open circuit end effect may also be obtained. The dispersion characteristic and characteristic impedances of coupled slots and coplanar strips have also been obtained using this approach [5].

Microstrip is a structure which has been studied by many investigators. There are numerous quasi-static analyses and a lesser number of frequency-dependent analyses which have been carried out. Among these frequency-dependent analyses is one by Krage and Haddad [6] which appears to be the only study to include an investigation of the frequency dependence of microstrip characteristic impedance. Results are presented for only a relatively low near-quasi-static frequency range ($\lambda > 0.1\lambda_d$), however.

The purpose of this paper is to present the results of a study of the frequency dependence of the characteristic impedance of microstrip using the spectral-domain approach. The method whereby the spectral-domain approach may be extended to calculate characteristic impedance will first be described. Numerical results showing the variation of characteristic impedance over a wide frequency range will then be presented, and it will be shown that these results converge to those of other investigators in the low-frequency range.

Manuscript received December 13, 1974; revised April 30, 1975. This work was supported in part by the Office of Naval Research through the Naval Postgraduate School Foundation Research Program. Computations were carried out at the W. R. Church Computer Center, Naval Postgraduate School.

J. B. Knorr is with the Department of Electrical Engineering, Naval Postgraduate School, Monterey, Calif. 93940.

A. Tufekcioglu was with the Naval Postgraduate School, Monterey, Calif. 93940. He is now with the Turkish Navy Tersane cad. 277/1, Aydin Blok, Gölcük, Turkey.

II. DISPERSION CHARACTERISTICS OF MICROSTRIP ON A LOSSLESS DIELECTRIC SUBSTRATE

To calculate the characteristic impedance of microstrip by the spectral-domain approach, it is first necessary to calculate the dispersion characteristic. The following discussion is included to provide an introduction to the method of analysis and to further reference the results of this study to those of other authors.

The spectral-domain dispersion analysis of microstrip is discussed in [2] and will be outlined only briefly here. With reference to Fig. 1, the microstrip field is expressed as a linear combination of TE and TM modes characterized by

$$E_{zi}(x,y,z) = k_{ci}^2\phi_i^e(x,y)e^{\gamma z} \tag{1a}$$

$$H_{zi}(x,y,z) = k_{ci}^2\phi_i^h(x,y)e^{\gamma z} \tag{1b}$$

where $k_{ci}^2 = \gamma^2 + k_i^2$, i denotes the appropriate region, and the ϕ_i are unknown scalar potential functions. Applying boundary conditions at $y = 0$ and $y = D$ leads to a set of boundary equations which still contain the $\phi_i(x,y)$. Although the ϕ_i are unknown, their Fourier transforms, $\Phi_i(\alpha,y)$, with respect to x can be found, and thus the boundary equations are transformed and the general solutions for the $\Phi_i(\alpha,y)$ are substituted. Extensive algebraic manipulation of the resulting equations leads to the coupled set

$$G_1(\alpha,\beta)\mathcal{J}_z(\alpha) + G_2(\alpha,\beta)\mathcal{J}_x(\alpha) = \mathcal{E}_z(\alpha) \tag{2a}$$

$$G_3(\alpha,\beta)\mathcal{J}_z(\alpha) + G_4(\alpha,\beta)\mathcal{J}_x(\alpha) = \mathcal{E}_x(\alpha) \tag{2b}$$

where α is the transform variable and $\mathcal{E}_i(\alpha)$ and $\mathcal{J}_i(\alpha)$ are the transforms of the electric field and the surface current at $y = D$. We next define the inner product

$$\langle A(\alpha),B(\alpha)\rangle = \int_{-\infty}^{+\infty} A(\alpha)B^*(\alpha)\,d\alpha \tag{3}$$

and take the inner product of (2a) and (2b) with weighting functions $W_i(\alpha)$. If we choose

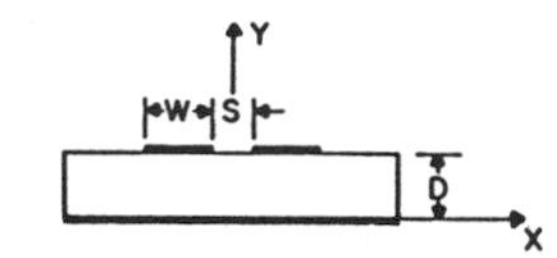

Fig. 1. Microstrip on a dielectric substrate.

Reprinted from *IEEE Trans. Microwave Theory Tech.*, vol. MTT-23, no. 9, pp. 725–728, September 1975.

$$W_1(\alpha) = \mathcal{J}_z(\alpha) \tag{4a}$$

$$W_2(\alpha) = \mathcal{J}_x(\alpha) \tag{4b}$$

we obtain

$$\langle G_1\mathcal{J}_x, \mathcal{J}_z\rangle + \langle G_2\mathcal{J}_z, \mathcal{J}_z\rangle = 0 \tag{5a}$$

$$\langle G_3\mathcal{J}_x, \mathcal{J}_z\rangle + \langle G_4\mathcal{J}_z, \mathcal{J}_z\rangle = 0. \tag{5b}$$

That the right-hand side of these equations is zero follows from Parseval's theorem since electric field and surface current at $y = D$ are orthogonal in the space domain.

Equations (5a) and (5b) are exact. The $G_i(\alpha,\beta)$ reflect substrate thickness, dielectric constant, and frequency while strip widths and current distributions determine the $\mathcal{J}$'s. A moment solution of (5) can be obtained by expanding $\mathcal{J}_x$ and $\mathcal{J}_z$ in a known set of basis functions and solving the resulting determinant. Various choices of bases have been considered by the authors and in [2]. Accurate results are obtained by neglecting transverse current ($\mathcal{J}_x(\alpha) \approx 0$) and assuming that longitudinal current is uniformly distributed. For simplicity and computational efficiency, the current has been assumed z directed and uniformly distributed in this study.

Fig. 2 shows the free space-to-microstrip wavelength ratio for a single strip and Fig. 3 shows the same ratio for the odd and even modes of coupled strips. Also shown are theoretical results published by Krage and Haddad [6], Wheeler [7], and Bryant and Weiss [8]. Where the present results overlap with those of [6], very good agreement is evident which tends to confirm the accuracy of both methods. The frequency-dependent analyses show an increasing free space-to-microstrip wavelength ratio with increasing frequency due to the relatively higher proportion of power in the dielectric. The inaccuracy of the quasi-static results at high frequencies is evident.

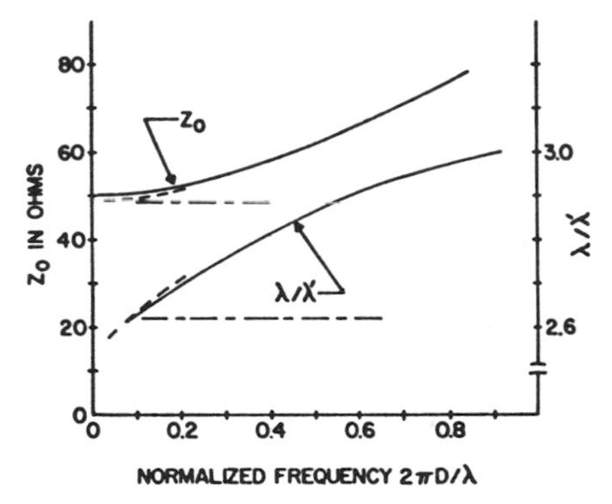

Fig. 2. Wavelength and characteristic impedance versus frequency for a single microstrip. $\epsilon_r = 10$, $W/D = 1$. Present method: ——; Krage and Haddad: ——; Wheeler: – —.

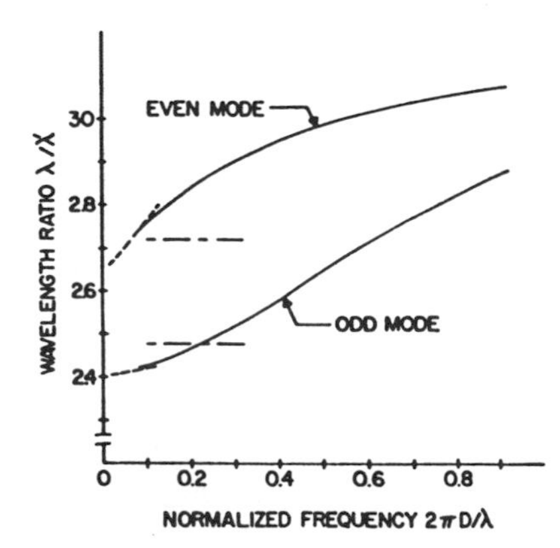

Fig. 3. Free space-to-microstrip wavelength ratio versus frequency for coupled microstrips. $\epsilon_r = 10$, $W/D = 1$, $S/D = 0.4$. Present method: ——; Krage and Haddad: ——; Bryant and Weiss: – —.

It is always desirable to compare theory with experiment, and this comparison appears in Figs. 4 and 5. In Fig. 4 the theoretical effective dielectric constant of a single microstrip is compared with data published by Getsinger [9]. In Fig. 5 the theoretical wavelength ratios for the odd and even modes of coupled microstrips are compared with data published by Gould and Tolboys [10]. In all cases the agreement between theory and experiment is better than 2 percent although the data from [10] show a constant offset. We cannot offer any explanation for this discrepancy. Getsinger [11] has obtained a somewhat better fit to these same data by using his approximate dispersion relation, but uses an empirical factor to do so.

III. CHARACTERISTIC IMPEDANCE OF MICROSTRIP ON A LOSSLESS DIELECTRIC SUBSTRATE

The extension of the spectral-domain technique to calculate the characteristic impedance of microstrip proceeds as follows. We again neglect transverse current and define

$$Z_{0i} = \frac{2P_{avg}}{I_{0z}^2} \tag{6a}$$

for a single strip or

$$Z_{0i} = \frac{P_{avg}}{I_{0z}^2} \tag{6b}$$

for coupled strips with reflection symmetry where I_{0z} is the total z-directed strip current. The average power is cal-

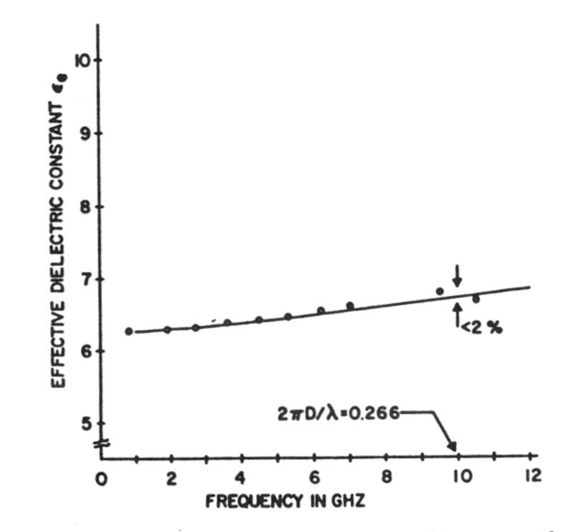

Fig. 4. Effective dielectric constant versus frequency for a single microstrip. $\epsilon_r = 10.185$, $W/D = 0.2$. $D = 0.050$ in. Present method: ——; Getsinger's data: O.

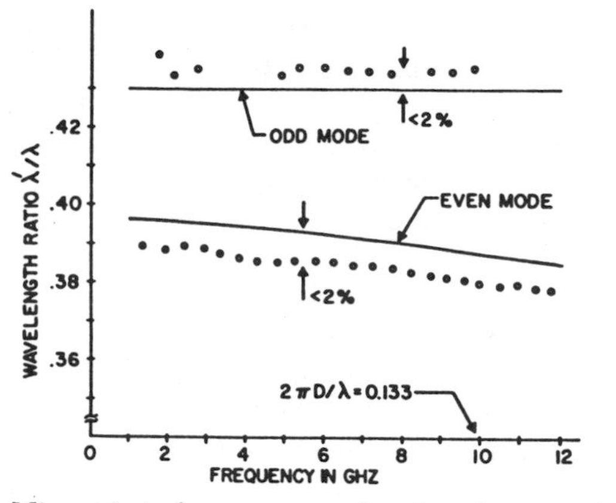

Fig. 5. Microstrip-to-free space wavelength ratio versus frequency for coupled microstrips. $\epsilon_r = 9.7$, $W/D = 0.3$, $D = 0.025$ in. Present method: ——; Gould and Tolboys' data: O.

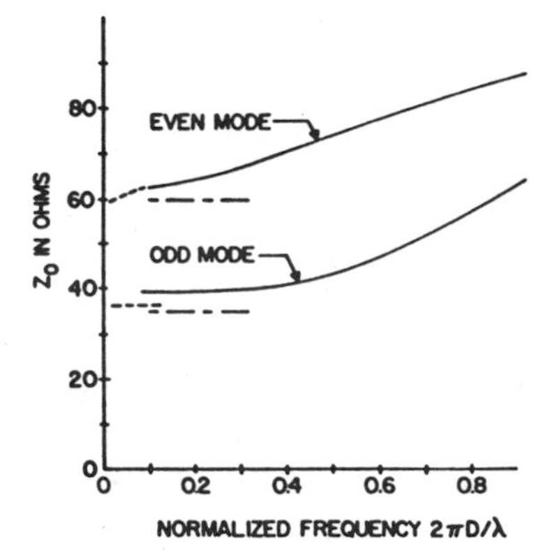

Fig. 6. Characteristic impedance versus frequency for coupled microstrips. $\epsilon_r = 10$, $W/D = 1$, $S/D = 0.4$. Present method: ——; Krage and Haddad: ——; Bryant and Weiss: – —.

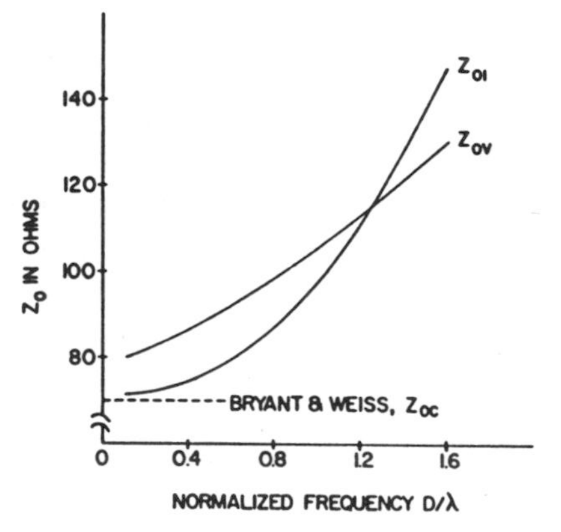

Fig. 7. Characteristic impedance versus frequency for a single microstrip using several definitions of impedance. $\epsilon_r = 9$, $W/D = 0.5$.

culated as

$$P_{\text{avg}} = \tfrac{1}{2}\,\text{Re} \iint_s (E_x H_y^* - E_y H_x^*)\, dx\, dy \tag{7}$$

where the transverse fields may be found from (1) and are thus given in terms of the unknown $\phi_i(x,y)$. Parseval's theorem may be applied, however, to obtain

$$P_{\text{avg}} = \frac{1}{4\pi}\,\text{Re} \int_{-\infty}^{+\infty}\int_{-\infty}^{+\infty} [\mathcal{E}_x(\alpha,y)\mathcal{H}_y^*(\alpha,y) - \mathcal{E}_y(\alpha,y)\mathcal{H}_x^*(\alpha,y)]\, dy\, d\alpha \tag{8}$$

where the script quantities denote the transforms of the fields and are given in terms of the $\Phi_i(\alpha,y)$. At this point the general solutions for the Φ_i may be substituted and integration with respect to y can be accomplished analytically. This leaves an equation of the form

$$P_{\text{avg}} = \frac{1}{4\pi} \int_{-\infty}^{+\infty} g(\alpha)\, d\alpha \tag{9}$$

which is evaluated numerically in each of the two regions.

Fig. 2 shows computed results for a single strip and Fig. 6 shows the characteristic impedances of the odd and even modes of coupled strips. The results again agree well with those from [6] where there is overlap. The increase of characteristic impedance with frequency, which appears to have been first discovered by Krage and Haddad, is verified by the present analysis.

There is some latitude in the definition of characteristic impedance for a structure such as microstrip. Another possible definition of characteristic impedance is given by

$$Z_{0v} = \frac{V^2(0)}{2P_{\text{avg}}} \tag{10}$$

for a single strip where $V(0)$ is given by

$$V(0) = -\int_0^D E_y(0,y)\, dy \tag{11}$$

and is the voltage between the center of the strip and the ground plane. It is interesting to compare the results obtained using (11) with those obtained using (6a). Fig. 7 shows results for a single strip. Also shown is the impedance from [8]. It is evident that the definition of impedance based upon strip current converges to the quasi-static characteristic impedance Z_{0c} which is defined in terms of static capacitance. In all probability, the reason that Z_{0v} does not converge to Z_{0c} is that $V(0)$ is sensitive to the assumed distribution of surface current while the total current I_{0z} used in (6a) is not. A better approximation (recall a uniform distribution was assumed) to the surface current such as $\mathcal{J}_z(x) = [(W/2)^2 - x^2]^{-1/2}, |x| < W/2$, would probably improve the result obtained using (10). Finally, it is to be noted that the geometric mean of the two curves in Fig. 7 gives the characteristic impedance defined by $V(0)/I_{0z} = (Z_{0v}Z_{0i})^{1/2}$.

IV. CONCLUSIONS

A spectral-domain technique for analysis of single and coupled microstrips has been described. It has been shown that this method can be successfully implemented computationally to accurately calculate characteristic impedance as well as wavelength as a function of frequency. The computational efficiency is quite good and, for a given frequency, the computations (wavelength and impedance) take about 2.5 s on the IBM 360.

The numerical results presented here have been shown to be in agreement with those of other investigators at low frequencies. Considerable departure from the quasi-static results has been shown to occur with increasing frequency, however. The analysis verifies the rise of characteristic impedance with frequency as predicted by Krage and Haddad [6].

REFERENCES

[1] T. Itoh and R. Mittra, "Dispersion characteristics of slot lines," *Electron. Lett.*, vol. 7, pp. 364–365, July 1971.

[2] ——, "Spectral-domain approach for calculating the dispersion characteristics of microstrip lines," *IEEE Trans. Microwave Theory Tech.* (Short Papers), vol. MTT-21, pp. 496–499, July 1973.

[3] ——, "A technique for computing dispersion characteristics of shielded microstrip lines," *IEEE Trans. Microwave Theory Tech.* (Short Papers), vol. MTT-22, pp. 896–898, Oct. 1974.

[4] T. Itoh, "Analysis of microstrip resonators," *IEEE Trans. Microwave Theory Tech.*, vol. MTT-22, pp. 946–952, Nov. 1974.

[5] J. B. Knorr and K.-D. Kuchler, "Analysis of coupled slots and coplanar strips on dielectric substrate," *IEEE Trans. Microwave Theory Tech.*, vol. MTT-23, pp. 541–548, July 1975.

[6] M. K. Krage and G. I. Haddad, "Frequency-dependent characteristics of microstrip transmission lines," *IEEE Trans. Microwave Theory Tech.*, vol. MTT-20, pp. 678–688, Oct. 1972.

[7] H. A. Wheeler, "Transmission-line properties of parallel strips separated by a dielectric sheet," *IEEE Trans. Microwave Theory Tech.*, vol. MTT-13, pp. 172–185, Mar. 1965.

[8] T. G. Bryant and J. A. Weiss, "Parameters of microstrip transmission lines and of coupled pairs of microstrip lines," *IEEE Trans. Microwave Theory Tech.* (*1968 Symposium Issue*), vol. MTT-16, pp. 1021–1027, Dec. 1968.

[9] W. J. Getsinger, "Microstrip dispersion model," *IEEE Trans. Microwave Theory Tech.*, vol. MTT-21, pp. 34–39, Jan. 1973.

[10] J. W. Gould and E. C. Tolboys, "Even- and odd-mode guide wavelengths of coupled lines in microstrip," *Electron. Lett.*, vol. 8, pp. 121–122, Mar. 1972.

[11] W. J. Getsinger, "Dispersion of parallel-coupled microstrip," *IEEE Trans. Microwave Theory Tech.* (Short Papers), vol. MTT-21, pp. 144–145, Mar. 1973.

Toward a Generalized Algorithm for the Modeling of the Dispersive Properties of Integrated Circuit Structures on Anisotropic Substrates

AKIFUMI NAKATANI, STUDENT MEMBER, IEEE, AND NICÓLAOS G. ALEXÓPOULOS, SENIOR MEMBER, IEEE

***Abstract*—A variety of substrate materials used in practice exhibit anisotropic behavior which is either inherent to the material or may be acquired during the manufacturing process. The development of highly accurate models for the propagation properties of integrated circuit structures necessitates careful accounting of substrate anisotropy. In this paper, an algorithm is developed which models structures such as microstrip, inverted microstrip, slotlines, and coplanar waveguides on anisotropic substrates. The model includes cases where the circuit has a cover or is enclosed in a rectangular shield.**

I. Introduction

Anisotropy is either an inherent property of a substrate material or may be acquired during the manufacturing process. Uniaxial crystalline substrates (such as, e.g., sapphire and quartz) belong to the former while fiber or ceramic impregnated plastics (such as, e.g., Epsilam-10) belong to the latter category. It has been clarified that when anisotropy is ignored in the development of design methods for single and coupled microstrip on Epsilam-10, an error is introduced which becomes significant for small linewidths and small line separation [1]. It follows that the development of highly accurate models for integrated circuit structures should account for substrate anisotropy. In addition, in certain applications of coupled line structures, anisotropy may prove beneficial in equalizing even–odd-mode phase velocities and reducing geometry tolerance sensitivity [2]. The dispersive behavior of microstrip [3]–[5], coupled slots [6], and coplanar waveguides [7] has been investigated to some extent including the case of microstrip discontinuities. This paper addresses the problem in a more generalized fashion in that a technique is developed which may encompass all useful structures with or without cover or with a rectangular shield. Furthermore, the algorithm discussed in this paper includes the possibility of multiple anisotropic layers, each of which is characterized by a diagonalized tensor permittivity $\bar{\bar{\epsilon}}$. The method consists of developing the LSE- and LSM-mode field solution to derive the Green's function for the structure under consideration. Subsequently, the Galerkin method is used to obtain the longitudinal and transverse current density or aperture electric-field components. In this manner, the dispersive behavior of several structures is obtained to show the versatility of the method. The error introduced when anisotropy is neglected is also discussed.

Manuscript received March 18, 1985; revised August 9, 1985. This work was supported in part by the National Science Foundation under Contract ECS 82-15408.

The authors are with the Electrical Engineering Department, University of California, Los Angeles, Los Angeles, CA 90024.

II. The Fourier Series Spectrum Method

The typical structures under consideration are shown in Fig. 1, where coupled microstrip and coplanar waveguide geometries are depicted with multiple anisotropic layers. The relative permittivity of the ith layer is given by

$$\bar{\bar{\epsilon}}^{(i)} = \begin{pmatrix} \epsilon_t^{(i)} & 0 & 0 \\ 0 & \epsilon_{yy}^{(i)} & 0 \\ 0 & 0 & \epsilon_t^{(i)} \end{pmatrix} \tag{1}$$

where $\epsilon_t^{(i)}$ is the relative permittivity on the xz-plane and, therefore, $\epsilon_t^{(i)} = \epsilon_{xx}^{(i)} = \epsilon_{zz}^{(i)}$. The sides of the rectangular shield may be chosen as electric or magnetic walls so as to simulate a closed or an open structure.

The dispersive properties of structures, such as those shown in Fig. 1, are obtained by solving Maxwell's equations in terms of LSE and LSM modes in the Fourier domain [5]. In this case, the finite even–odd-mode Fourier representation of the electromagnetic field is adopted by erecting a symmetric magnetic or electric wall at the plane $x = 0$. For the geometries of Fig. 1, the finite Fourier transforms are defined in each layer as

$$\tilde{\mathscr{E}}_y^{\mathrm{TM}}(k_n) = \int_0^a E_y^{\mathrm{TM}}(x) \begin{matrix}\cos\\ \sin\end{matrix}(k_n x)\, dx \tag{2}$$

and

$$\tilde{\mathscr{H}}_y^{\mathrm{TE}}(k_n) = \int_0^a H_y^{\mathrm{TE}}(x) \begin{matrix}\sin\\ \cos\end{matrix}(k_n x)\, dx \tag{3}$$

where the upper case corresponds to a magnetic and the lower case to an electric wall at $x = 0$. The choice of k_n also enforces a magnetic and electric wall at $x = a$, that is, at the side wall. The combination of (2) and (3) and k_n leads to an open or closed structure, when k_n is chosen as $n\pi/a$ or $(2n-1)\pi/2a$, respectively. If the lower case of (2)

Reprinted from *IEEE Trans. Microwave Theory Tech.*, vol. MTT-33, no. 12, pp. 1436–1441, December 1985.

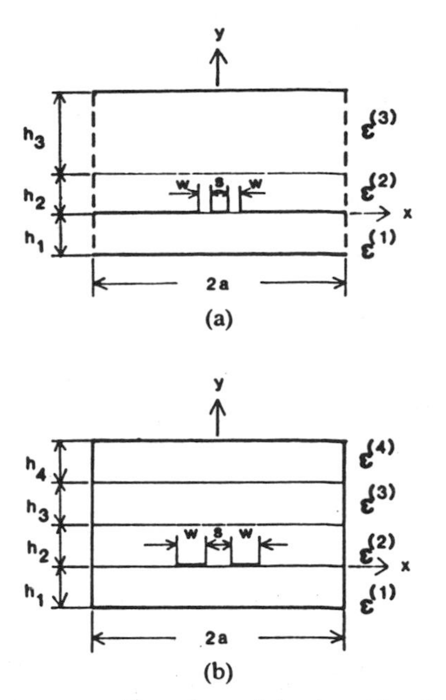

Fig. 1. Coplanar waveguide and coupled microstrip geometries on anisotropic layers. (a) Coplanar waveguide. (b) Coupled microstrip.

and (3) and $k_n = n\pi/a$ are chosen, then this corresponds to the odd-mode for the coupled microstrip lines or to a single microstrip line in the closed waveguide geometry. The other combination generates the case for an open side-wall geometry and it is useful for simulating an open structure for slot lines or other waveguiding systems.

LSE Modes

$$\tilde{\mathscr{E}}_x^{\mathrm{TE}} = \frac{\omega\mu_0\beta}{\beta^2+k_n^2}\tilde{\mathscr{H}}_y^{\mathrm{TE}} \tag{4}$$

$$\tilde{\mathscr{E}}_z^{\mathrm{TE}} = \pm\frac{j\omega\mu_0 k_n}{\beta^2+k_n^2}\tilde{\mathscr{H}}_y^{\mathrm{TE}} \tag{5}$$

$$\tilde{\mathscr{H}}_x^{\mathrm{TE}} = \pm\frac{k_n}{\beta^2+k_n^2}\frac{\partial\tilde{\mathscr{H}}_y^{\mathrm{TE}}}{\partial y} \tag{6}$$

and

$$\tilde{\mathscr{H}}_z^{\mathrm{TE}} = \frac{-j\beta}{\beta^2+k_n^2}\frac{\partial\tilde{\mathscr{H}}_y^{\mathrm{TE}}}{\partial y} \tag{7}$$

where $\tilde{\mathscr{H}}_y^{\mathrm{TE}}$ satisfies the wave equation

$$\frac{\partial^2\tilde{\mathscr{H}}_y^{\mathrm{TE}}}{\partial y^2} - \left[k_n^2+\beta^2-\epsilon_t k_0^2\right]\tilde{\mathscr{H}}_y^{\mathrm{TE}} = 0. \tag{8}$$

LSM Modes

$$\tilde{\mathscr{E}}_x^{\mathrm{TM}} = \pm\frac{k_n(\epsilon_{yy}/\epsilon_t)}{\beta^2+k_n^n}\frac{\partial\tilde{\mathscr{E}}_y^{\mathrm{TM}}}{\partial y} \tag{9}$$

$$\tilde{\mathscr{E}}_z^{\mathrm{TM}} = -\frac{j\beta(\epsilon_{yy}/\epsilon_t)}{\beta^2+k_n^2}\frac{\partial\tilde{\mathscr{E}}_y^{\mathrm{TM}}}{\partial y} \tag{10}$$

$$\tilde{\mathscr{H}}_x^{\mathrm{TM}} = -\frac{\epsilon_0\epsilon_{yy}\beta}{\beta^2+k_n^2}\tilde{\mathscr{E}}_y^{\mathrm{TM}} \tag{11}$$

and

$$\tilde{\mathscr{H}}_z^{\mathrm{TM}} = \pm\frac{j\omega\epsilon_0\epsilon_{yy}k_n}{\beta^2+k_n^2}\tilde{\mathscr{E}}_y^{\mathrm{TM}} \tag{12}$$

where $\tilde{\mathscr{E}}_y^{\mathrm{TM}}$ satisfies the wave equation

$$\frac{\partial^2\tilde{\mathscr{E}}_y^{\mathrm{TM}}}{\partial y^2} - \left(\frac{\epsilon_t}{\epsilon_{yy}}\right)\left[k_n^2+\beta^2-\epsilon_{yy}k_0^2\right]\tilde{\mathscr{E}}_y^{\mathrm{TM}} = 0. \tag{13}$$

The solution to the multiple boundary layer problem yields

$$\begin{bmatrix} j\tilde{\mathscr{J}}_x \\ -\tilde{\mathscr{J}}_z \end{bmatrix} = \begin{bmatrix} H_{11} & H_{12} \\ H_{21} & H_{22} \end{bmatrix}\begin{bmatrix} j\tilde{\mathscr{E}}_z \\ \tilde{\mathscr{E}}_x \end{bmatrix} \tag{14}$$

or inversely

$$\begin{bmatrix} j\tilde{\mathscr{E}}_z \\ \tilde{\mathscr{E}}_x \end{bmatrix} = \begin{bmatrix} G_{11} & G_{12} \\ G_{21} & G_{22} \end{bmatrix}\begin{bmatrix} j\tilde{\mathscr{J}}_x \\ -\tilde{\mathscr{J}}_z \end{bmatrix} \tag{15}$$

where $\tilde{\mathscr{J}}_x$ and $\tilde{\mathscr{J}}_z$ are finite Fourier transforms for the metallic strip current density components and $\tilde{\mathscr{E}}_z$ and $\tilde{\mathscr{E}}_x$ represent the finite Fourier transforms of the slot aperture electric-field components. The various terms involved in (16) and (17) are given by

$$H_{11} = H_{22} = \pm\left(\frac{\omega\epsilon_0}{k_0}\right)\frac{k_n\beta}{\gamma_n^2}\left[F_1(k_n,\beta)+F_2(k_n,\beta)\right] \tag{16}$$

$$H_{12} = \left(\frac{\omega\epsilon_0}{k_0}\right)\frac{1}{\gamma_n^2}\left[\beta^2F_1(k_n,\beta)-k_n^2F_2(k_n,\beta)\right] \tag{17}$$

$$H_{21} = \left(\frac{\omega\epsilon_0}{k_0}\right)\frac{1}{\gamma_n^2}\left[k_n^2F_1(k_n,\beta)-\beta^2F_2(k_n,\beta)\right] \tag{18}$$

$$G_{11} = H_{22}/\Delta \qquad G_{12} = -H_{12}/\Delta$$
$$G_{21} = -H_{21}/\Delta \quad \text{and} \quad G_{22} = H_{11}/\Delta \tag{19}$$

where

$$\Delta = H_{11}H_{22} - H_{12}H_{21}. \tag{20}$$

In addition

$$F_1(k_n,\beta) = \left[1/f_{t1} + \left(\alpha_t^{(2)^2}f_{t2}f_{t3}+1\right)/\left(f_{t2}+f_{t3}\right)\right] \tag{21}$$

$$F_2(k_n,\beta) = \left[1/g_{y1} + \left(g_{y2}g_{y3}+\alpha_y^{(2)^2}\right)/\left\{\alpha_y^{(2)^2}\left(g_{y2}+g_{y3}\right)\right\}\right] \tag{22}$$

$$\gamma_n^2 = \beta^2 + k_n^2 \tag{23}$$

$$\alpha_y^{(i)} = \sqrt{\left\{(\gamma_n/k_0)^2-\epsilon_y^{(i)}\right\}/\left(\epsilon_y^{(i)}\epsilon_t^{(i)}\right)} \tag{24}$$

$$\alpha_t^{(i)} = \sqrt{(\gamma_n/k_0)^2-\epsilon_t^{(i)}} \tag{25}$$

$$f_{ti} = \tanh\left(k_0\alpha_t^{(i)}h_i\right)/\alpha_t^{(i)} \tag{26}$$

$$g_{yi} = \alpha_y^{(i)}\tanh\left(\epsilon_t^{(i)}k_0\alpha_y^{(i)}h_i\right). \tag{27}$$

The choice between (16) and (17) depends on whether the structure of Fig. 1(a) or (b) is considered, respectively. Since duality applies, the expansion functions for $E_x(x)$ and $J_z(x)$ may be chosen alike. Similarly, this is true for

$E_x(x)$ and $J_x(x)$. For the former case, the chosen field or current distribution satisfies the edge condition, i.e., the expansion functions are selected to obey the Maxwell functional dependence across the aperture or across the metallic strip, while for the latter case, the chosen expansion functions are like sinusoids forced to zero value at the slot or metallic strip edges. In the application of the Galerkin method, a variety of expansion functions, such as Legendre polynomials, Chebyshev polynomials, and pulse functions, have been employed in conjunction with the Maxwell dependence to satisfy the edge condition, and to represent $E_x(x)$ on a slot or $J_z(x)$ on a metallic strip. These choices lead to solutions which involve special functions such as Bessel functions. Solutions in terms of special functions typically increase the required CPU time and therefore decrease the efficiency of the algorithm. To enhance the algorithm efficiency, a simple polynomial expansion is made for $E_x(x)$ or $J_z(x)$ and sine basis functions are chosen for the $E_z(x)$ or $J_x(x)$ distributions. These polynomial expansions introduce further numerical simplifications since simple recurrence relations may be developed, thereby contributing to further minimization of CPU computation time.

The aperture field or metallic strip current representations are chosen as

$$\begin{Bmatrix} E_x(x) \\ -J_z(x) \end{Bmatrix} = \sum_{k=1}^{N} C_k f_k(x) \tag{28}$$

and

$$\begin{Bmatrix} jE_z(x) \\ jJ_x(x) \end{Bmatrix} = \sum_{k=1}^{M} D_k g_k(x) \tag{29}$$

where

$$f_k(x) = \left(\frac{x - s/2 - w/2}{w/2} \right)^{k-1}$$

and

$$g_k(x) = \sin\left(k\pi \frac{x - s/2}{w/2} \right). \tag{30}$$

Application of the Galerkin method yields a system of $(N+M)\times(N+M)$ eigenequations to be solved. The dispersion relation for ϵ_{eff} as a function of β/k_0 is obtained by solving the determinant

$$\begin{bmatrix} X_{i,k}^{1,1} & X_{i,k}^{1,2} \\ X_{i,k}^{2,1} & X_{i,k}^{2,2} \end{bmatrix} = 0 \tag{31}$$

where each element is given for the magnetic wall case at the center by

$$\begin{bmatrix} X_{i,k}^{1,1} \\ X_{i,k}^{1,2} \\ X_{i,k}^{2,1} \\ X_{i,k}^{2,2} \end{bmatrix} = \sum_{n=1}^{\infty} \begin{bmatrix} \tilde{f}_i^s H_{11} \tilde{g}_k^c \\ \tilde{g}_i^c H_{12} \tilde{g}_k^c \\ \tilde{f}_i^s H_{21} \tilde{f}_k^s \\ \tilde{g}_i^c H_{22} \tilde{f}_k^s \end{bmatrix} \tag{32}$$

for slots, while for microstrip structures

$$\begin{bmatrix} X_{i,k}^{1,1} \\ X_{i,k}^{1,2} \\ X_{i,k}^{2,1} \\ X_{i,k}^{2,2} \end{bmatrix} = \sum_{n=1}^{\infty} \begin{bmatrix} \tilde{f}_i^c G_{11} \tilde{g}_k^s \\ \tilde{g}_i^s G_{12} \tilde{g}_k^s \\ \tilde{f}_i^c G_{21} \tilde{f}_k^c \\ \tilde{g}_i^s G_{22} \tilde{f}_k^c \end{bmatrix}. \tag{33}$$

The " ~ " symbol in (33) indicates the Fourier transformed basis functions given by (28) and (29). A recurrence relation can be easily derived for these Fourier transforms.

Thus, the recurrence relations of the $f_k^{(x)}$ basis functions are derived as

$$\begin{aligned} \tilde{f}_k^s(k_n) = {} & \frac{1}{k_n}\left[(-1)^{k-1}\cos(k_n s/2) - \cos\{k_n(w+s/2)\}\right] \\ & + \frac{(k-1)}{k_n^2 w/2}\left[\sin\{k_n(w+s/2)\} - (-1)^{k-2}\sin(k_n s/2)\right] \\ & - \frac{(k-1)(k-2)}{(k_n w/2)^2}\tilde{f}_{k-2}^s(k_n) \end{aligned} \tag{34}$$

for the sine Fourier transform and

$$\begin{aligned} \tilde{f}_k^c(k_n) = {} & \frac{1}{k_n}\left[\sin\{k_n(w+s/2)\} - (-1)^{k-1}\sin(k_n s/2)\right] \\ & - \frac{(k-1)}{k_n^2 w/2}\left[(-1)^{k-2}\cos(k_n s/2) - \cos\{k_n(w+s/2)\}\right] \\ & - \frac{(k-1)(k-2)}{(k_n w/2)^2}\tilde{f}_{k-2}^c(k_n) \end{aligned} \tag{35}$$

for the corresponding cosine Fourier transform.

It is observed that the choice of the $\{f_k(x)\}$ set yields recurrence relations in the Fourier domain which consist of simple sine and cosine functions, thereby contributing to the development of an efficient numerical algorithm. It is also interesting to note that a Maxwell distribution may be approximated readily with 6 or 7 basis function terms.

III. Characteristic Impedance

The characteristic impedance for microstrip is computed in this section by considering the definition $Z_0 = V(k_0)/I(k_0)$. For this calculation, the voltage and longitudinal current are given by

$$V(k_0) = -\int_{-h_1}^{0} E_y dy \tag{36}$$

and

$$I(k_0) = \sum_{n=1}^{N} \frac{2C_n}{n} \tag{37}$$

TABLE I

$2b/h_1$	h_2/h_1	s/h_1	w/h_1	Phase velocity	Ref (2)	Dynamic solution
6.2	0.8	1.6	0.58	v^e_{ph}	0.9629	0.9632
				v^o_{ph}	0.9646	0.9647
12.0	2.0	1.6	0.55	v^e_{ph}	0.9323	0.9324
				v^o_{ph}	0.9333	0.9333

respectively. This definition of Z_0 has been compared with the results reported for a single microstrip line on sapphire, and the agreement is excellent [8]. Also, a comparison with the computation for Z_0^e, Z_0^o for coupled microstrip lines on isotropic substrates [9], as computed by using the definition $Z_0 = 2P/I^2$, yields a 0.5-percent difference for higher frequencies and a 1.4-percent difference for lower frequencies for both the even- and odd-mode cases (see Fig. 6). This algorithm has also been tested against quasi-static calculations [1] with an agreement better than 0.1 percent.

IV. Numerical Results and Discussion

The results obtained with the algorithm developed in this article are within 0.5-percent convergence accuracy. Table I makes a comparison of the algorithm developed herein with the quasi-static results for coupled microstrip with cover on a sapphire substrate with an alumina overlay [2]. For this case, reference to Fig. 1(b) indicates that $\epsilon_{xx}^{(1)} = 9.4$, $\epsilon_{yy}^{(1)} = 11.6$, $\epsilon_{xx}^{(2)} = \epsilon_{yy}^{(2)} = 9.9$, while $\epsilon_{xx}^{(3)} = \epsilon_{xx}^{(4)} = \epsilon_{yy}^{(3)} = \epsilon_{yy}^{(4)} = 1$. To simulate a structure which is open on the sides, the dimension $2a$ has been chosen as 20. The results shown in Table I indicate agreement to within 0.05 percent between the two methods for both even- and odd-mode phase velocities. For these computations, the choice $N = 2$, $M = 10$, and $n_{sum} = 1000$ yields 0.05-percent convergence accuracy.

Fig. 2(a) demonstrates the dispersive behavior of λ_g/λ_0 for coupled slotlines on a saphire substrate. Comparison of this result with the dispersive behavior of coupled slots as obtained by the equivalent network method [6] indicates no visible discrepancy between the two methods. For this particular case $N = 6$, $M = 6$, $N_{sum} = 300$, and in order to simulate an open structure $2a = 20$ while $2b = 21$. In addition, it is observed that the dominant mode (odd mode) is obtained with electric side walls at $x = \pm a$, while the even mode results if the side walls at $x = \pm a$ are perfect magnetic conductors. Fig. 2(b) compares the ϵ_{eff} for even and odd modes for an Epsilam-10 substrate when anisotropy is accounted for or when it is neglected. It is clearly observed that with increasing frequency, the error when anisotropy is neglected increases. For the dominant mode, it is observed that it is 11.6 percent at low frequencies ($k_0 h_2 \doteq 0$) and it increases to 13.3 percent when $k_0 h_2 = 0.6$. This example demonstrates the significance of including substrate anisotropy in the development of highly accurate design algorithms for microwave and millimeter-wave circuits. Fig. 3 illustrates the dispersive behavior of

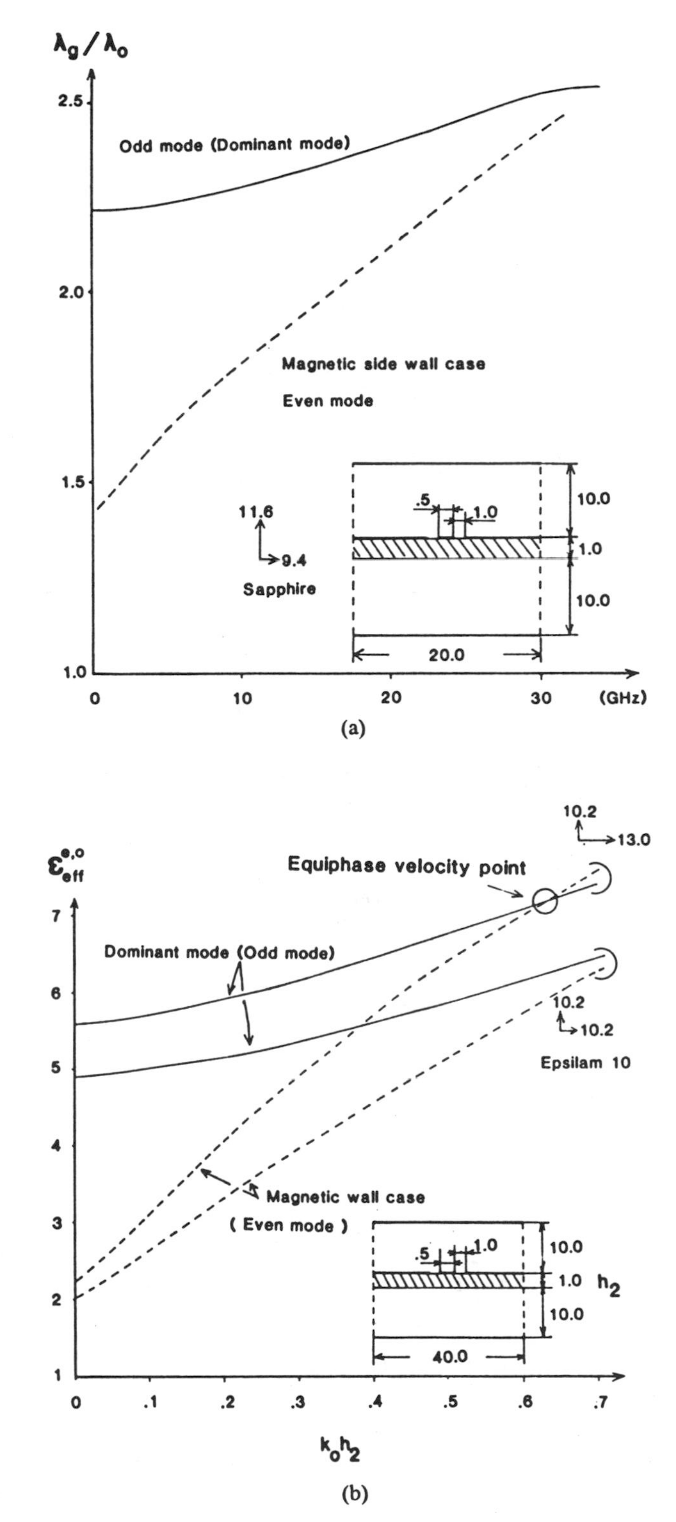

Fig. 2. (a) Dispersion characteristics of coupled slotlines on sapphire substrate. (b) Dispersion characteristics of coupled slotlines.

ϵ_{eff} for a slotline on sapphire. Of interest in this particular case are the waveguide modes shown in dashed lines. The first waveguide mode is generated by the substrate-filled portion of the waveguide, while the second, by the entire structure. The cutoff frequencies for these modes may be predicted by considering the ϵ_{eff}-filled waveguide mode and the transverse resonance method. The results are obtained as $k_{01} = 0.97$ with $\epsilon_{eff} = \sqrt{\epsilon_t \epsilon_y} = 10.44$. Also, the

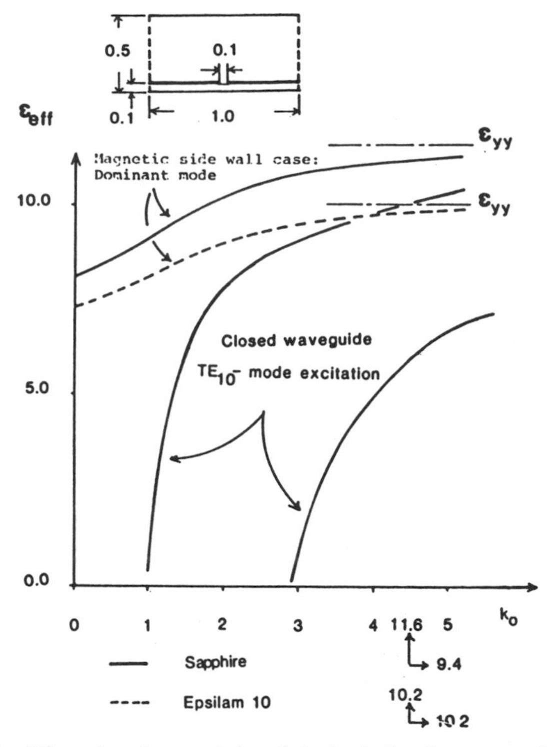

Fig. 3. Dispersion characteristics of single slotline for open and closed geometry.

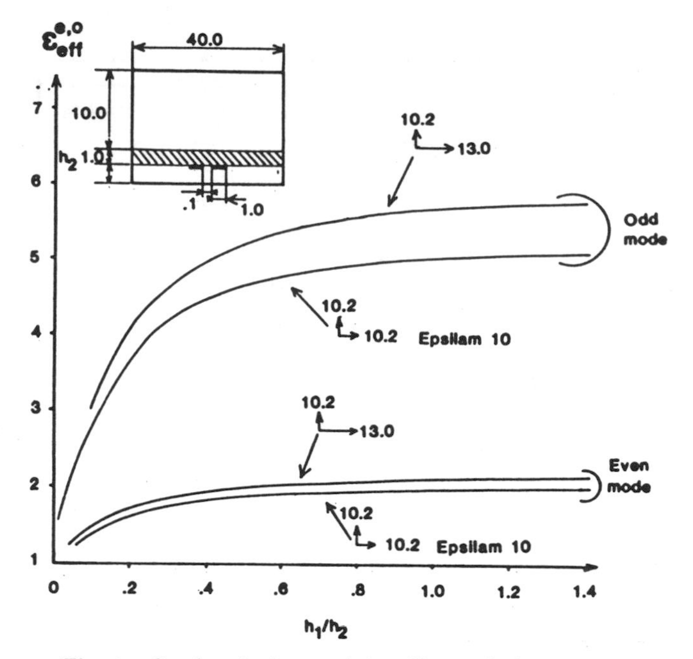

Fig. 4. Quasi-static characteristics of inverted microstrip lines.

TABLE II
C_k VALUES

Even mode	Odd mode
.2477	.0869
.0883	−.0494
.0518	.1895
−.2894	.1728
−.3373	−.7139
.5113	−.2935
1.0	1.0

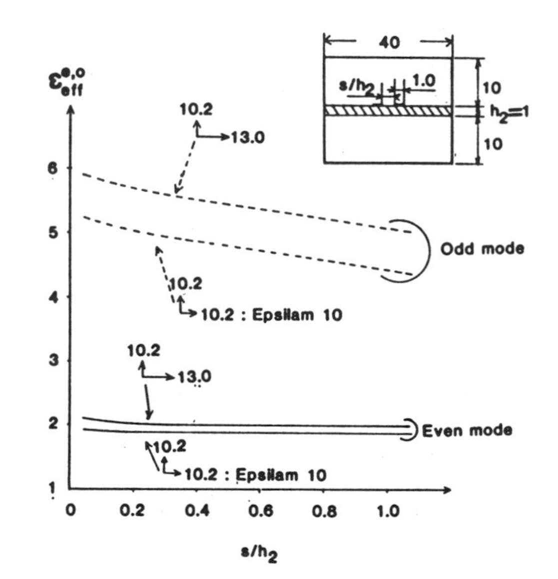

Fig. 5. Quasi-static properties of suspended microstrip lines.

transverse resonance method approximates k_{c2} here as

$$k_{c2} \sim \frac{\pi}{b}\left(\frac{\epsilon_{\text{eff}}(a-d)+d}{\epsilon_{\text{eff}}a}\right) \sim 2.84$$

which agrees with the computed result. In the above expression, a is the width, b is the height, and d is the substrate thickness.

The comparison to Epsilam 10 has also been shown for the magnetic wall case; an open structure is simulated. The dominant mode for this case exhibits a y-directed field component dominated by the ϵ_{yy} factor. The computed result shows the asymptotic behavior of the effective dielectric constant to the ϵ_{yy} component of the permittivity tensor.

Fig. 6 demonstrates the comparison to El-Sherbiny's results [8]. The effective dielectric constant and characteristic impedance show excellent agreement. The current distribution for low frequency is also shown where seven basis functions have been chosen here. The current density approximates the Maxwell current distribution with excellent agreement.

The polynomial coefficients C_k are found as shown in Table II. Table I demonstrates that this choice of a simple set of basis functions yields excellent results for the current distribution while at the same time it provides for a very efficient numerical algorithm.

Finally, in order to emphasize the versatility of this technique, the low-frequency behavior of ϵ_{eff} is demonstrated in Figs. 4 and 5 for inverted coupled microstrip and for suspended coupled microstrip lines on isotropic and anisotropic Epsilam-10 substrates.

V. Conclusion

This paper presents a generalized approach to analyze the dispersive properties of integrated circuit structures on multiple layers of anisotropic materials. The algorithm developed by this method includes as limiting cases geome-

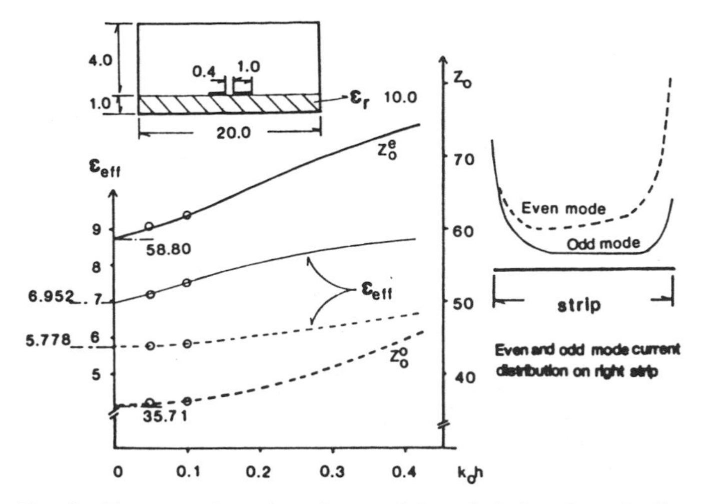

Fig. 6. Frequency-dependent characteristics of single microstrip line. —— Even mode. ---- Odd mode. ∘ Reference: Fig. 16 and Fig. 17 [9].

tries of integrated waveguiding structures with or without cover, as well as the case of a rectangular shield. The results presented in this paper are within 0.5-percent convergence accuracy. Finally, it has been demonstrated that neglecting substrate anisotropy leads to serious errors in the effort to derive an accurate integrated circuit model.

References

[1] N. G. Alexopoulos, "Integrated circuit structures on anisotropic substrates," *IEEE Trans. Microwave Theory Tech.*, vol. MTT-33, pp. 847–881, Oct 1985.

[2] N. G. Alexopoulos and C. M. Krowne, "Characteristics of single and coupled microstrips on anisotropic substrates," *IEEE Trans. Microwave Theory Tech.*, vol. MTT-26, pp. 387–393, June 1978.

[3] N. G. Alexopoulos and S. Maas, "Characteristics of microstrip directional couplers on anisotropic substrates," *IEEE Trans. Microwave Theory Tech.*, vol. MTT-30, pp. 1267–1270, Aug. 1982.

[4] J. E. Mariki, "Analysis of microstrip lines on inhomogeneous anisotropic substrates by the TLM technique," Ph.D. dissertation, Univ. of California, Los Angeles, 1978.

[5] K. Shibata and K. Hatori, "Dispersion characteristics of coupled microstrip with overlay on anisotropic dielectric substrate," *Electron. Lett.*, Jan. 1984.

[6] T. Kitazawa and Y. Hayashi, "Coupled slots on an anisotropic sapphire substrate," *IEEE Trans. Microwave Theory Tech.*, vol. MTT-29, pp. 1035–1040, Oct. 1981.

[7] T. Kitazawa and Y. Hayashi, "Quasi-static characteristics of coplanar waveguide on a sapphire substrate with its optical axis inclined," *IEEE Trans. Microwave Theory Tech.*, vol. MTT-30, pp. 920–922, June 1982.

[8] A. Abdel-Moniem El-Sherbiny, "Hybrid mode analysis of microstrip lines on anisotropic substrates," *IEEE Trans. Microwave Theory Tech.*, vol. MTT-29, pp. 1261–1265, Dec. 1981.

[9] K. Krage and I. Haddad, "Frequency-dependent characteristics of microstrip transmission lines," *IEEE Trans. Microwave Theory Tech.*, vol. MTT-20, pp. 678–688, Oct. 1972.

[10] T. Kitazawa and Y. Hayashi, "Propagation characteristics of striplines with multilayered anisotropic media," *IEEE Trans. Microwave Theory Tech.*, vol. MTT-31, pp. 429–433, June 1983.

A New Technique for the Analysis of the Dispersion Characteristics of Planar Waveguides

by Uwe Schulz* and Reinhold Pregla*

Dedicated to Professor Dr. rer. nat. Hans K. F. Severin on the occasion of his 60th birthday.

An efficient method for calculating the dispersion characteristics of planar waveguide structures is presented, of which the principle is known as the "method of lines" in mathematical literature. The wave equation is discretized in one direction, and the resulting differential-difference-equation can be solved analytically. As an example of application and as a test for the computed results single and coupled microstrip lines were calculated. The obtained results show good agreement with other available data.

Ein neues Verfahren zur Berechnung des Dispersionsverhaltens von planaren Wellenleitern

Ein leistungsfähiges Verfahren zur Berechnung der dispersiven Eigenschaften planarer Wellenleiterstrukturen wird beschrieben, dessen Grundprinzip in der mathematischen Literatur unter dem Namen „Methode der Geraden" bekannt ist. Die Wellengleichung wird in einer Richtung diskretisiert und die daraus entstehende Differential-Differenzengleichung analytisch gelöst. Als Beispiel wurde die einfache und verkoppelte Microstripleitung berechnet. Die erhaltenen Ergebnisse zeigen gute Übereinstimmung mit anderen vergleichbaren Daten.

1. Introduction

For hybrid mode analysis of planar waveguide structures, e.g. microstrip, slotline, coplanar-line, finline etc., several numerical methods exist. In spite of its simplicity the finite difference method is as yet not commonly applied to the more complicated structures, as it leads to a system of equations of very large size. Even improvements, like e.g. such of Corr and Davies [4], who apply a graduell mesh, could not help principally.

In this paper a method is presented, which is basicly a finite difference method too, but essentially more effective with respect to accuracy and computation time than the usual finite difference method, when applied to structures like those in Fig. 1.

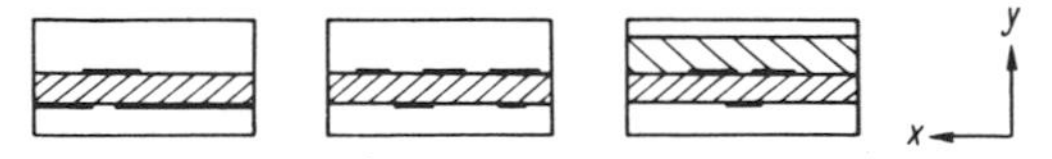

Fig. 1. Some planar structures.

The basic principle of this method is known as the "method of lines" in the mathematical literature, e.g. [1], [2]. To the best of the authors' knowledge this method has not been applied to the calculation of planar microwave structures up to the present. Merely Lennartsson's "network analogue method" [3], which has a certain similarity with the "method of lines", was applied to compute the quasi-TEM-parameters of microstrip structures.

* Dipl.-Ing. U. Schulz, Prof. Dr. R. Pregla, Fernuniversität, Lehrstuhl für Allgemeine und Theoretische Elektrotechnik, Frauenstuhlweg 31, D-5860 Iserlohn.

The "method of lines" is simple in concept — for a given system of partial differential equations all but one of the independent variables are discretized to obtain a system of ordinary differential equations. This semidiscrete procedure is apparently a very useful one in order to calculate planar waveguide structures like those in Fig. 1, since these consist of regions, which are homogeneous in one direction. Moreover, this method has no problem with the so called "relative convergence" as have e.g. the mode-matching- and Galerkin's method. This is an important advantage, especially when structures with more than one metallic strip are considered.

2. Method of Analysis

The electromagnetic field can be described by two scalar potentials $\psi^{(e)}$ and $\psi^{(h)}$, which have to satisfy Helmholtz' equation

$$\frac{\partial^2 \psi^{(e,h)}}{\partial x^2} + \frac{\partial^2 \psi^{(e,h)}}{\partial y^2} + (k^2 - \beta^2)\,\psi^{(e,h)} = 0 \quad (1)$$

and some boundary conditions within the considered region, where the dependence $e^{j(\omega t - \beta z)}$ has been assumed. The first step is to discretize the x-variable in the partial differential equation (1). For that purpose a family of N straight lines parallel to the y-axis is laid into the cross section. The distance between adjacent lines shall be constant and equal to h. Now, the potential ψ in eq. (1) can be replaced by a set $(\psi_1, \psi_2, \ldots, \psi_N)$ at the lines

$$x_i = x_0 + ih, \quad i = 1, 2, \ldots, N \quad (2)$$

and the derivatives with respect to x are replaced by finite differences. This procedure yields a system

Reprinted with permission from *AEU*, Band 34, Heft 4, pp. 169–173, 1980.

of N *coupled* ordinary differential equations

$$\frac{\partial^2 \psi_i}{\partial y^2} + \frac{1}{h^2}\left[\psi_{i-1}(y) - 2\psi_i(y) + \psi_{i+1}(y)\right] + + (k^2 - \beta^2)\,\psi_i(y) = 0, \quad i = 1, 2, \ldots, N. \tag{3}$$

When a column vector

$$\vec{\psi} = [\psi_1(y), \psi_2(y), \ldots, \psi_N(y)]^t \tag{4}$$

and the matrix

$$\boldsymbol{P} = \begin{pmatrix} p_1 & -1 & & & \\ -1 & 2 & -1 & & 0 \\ & \ddots & \ddots & \ddots & \\ 0 & & -1 & 2 & -1 \\ & & & -1 & p_2 \end{pmatrix} \tag{5}$$

are introduced. Eq. (3) can be expressed in matrix notation as

$$h^2 \frac{\partial^2 \vec{\psi}}{\partial y^2} - [\boldsymbol{P} - h^2(k^2 - \beta^2)\,\boldsymbol{I}]\,\vec{\psi} = \vec{0} \tag{6}$$

where $\boldsymbol{I}$ is the identity matrix. The lateral boundary conditions are already included in the matrix $\boldsymbol{P}$, i.e., in the numbers p_1 and p_2. The system of differential equations (6) can not be solved in the present form, because the equations are coupled. But it will be shown, that eq. (6) can be transformed into a system of uncoupled equations. As $\boldsymbol{P}$ is a real symmetric matrix, there exists an orthogonal matrix $\boldsymbol{T}$ such that

$$\boldsymbol{T}^t \boldsymbol{P} \boldsymbol{T} = \boldsymbol{\lambda} \tag{7}$$

is diagonal, where t denotes the transpose. The elements λ_i of the diagonal matrix $\boldsymbol{\lambda}$ are the eigenvalues of $\boldsymbol{P}$.

A transformed potential vector $\vec{U}$ is now introduced,

$$\boldsymbol{T}^t \vec{\psi} = \vec{U} \tag{8}$$

so that instead of eq. (6) one can write a system of N ordinary differential equations, which are now *uncoupled*:

$$h^2 \frac{\partial^2 U_i}{\partial y^2} - [\lambda_i - h^2(k^2 - \beta^2)]\,U_i = 0, \quad i = 1, 2, \ldots, N. \tag{9}$$

It is obvious, that eq. (9) can be solved analytically for each homogeneous region of the structure. The solution has a form similar to the line equations:

$$\begin{pmatrix} U_i(y_1) \\ h\dfrac{\partial U_i}{\partial y}(y_1) \end{pmatrix} = \begin{pmatrix} \cosh \dfrac{\varkappa_i(y_1 - y_2)}{h} & \dfrac{1}{\varkappa_i}\sinh \dfrac{\varkappa_i(y_1 - y_2)}{h} \\ \varkappa_i \sinh \dfrac{\varkappa_i(y_1 - y_2)}{h} & \cosh \dfrac{\varkappa_i(y_1 - y_2)}{h} \end{pmatrix} \begin{pmatrix} U_i(y_2) \\ h\dfrac{\partial U_i}{\partial y}(y_2) \end{pmatrix} \tag{10}$$

with

$$\varkappa_i = [\lambda_i - h^2(k^2 - \beta^2)]^{1/2}, \quad i = 1, 2, \ldots, N. \tag{11}$$

With the aid of eq. (10) the potential vector $\vec{\psi}$ can be transformed from one interface into another. So it is possible to replace the boundary value problem (1) by a linear algebraic system, the unknowns of which are only those potentials at the interfaces with metallic strips. Further, it is possible to reduce the order of the resulting matrix associated with the eigenvalue equation to the number of points on the strips or in the slots between the strips. Thus, the eigenvalue equation has a very low order.

3. The Lateral Boundary Conditions

It has been mentioned before, that the matrix $\boldsymbol{P}$ includes the lateral boundary conditions. The left boundary determines p_1, the right boundary p_2 (see eq. (5)). In Fig. 2 only the left boundary is considered. The consideration of the right boundary is equivalent. The Dirichlet condition (Fig. 2a) requires $\psi_0 = 0$. Inserting this in eq. (3) with $i = 1$ yields $p_1 = 2$. On the other hand, the Neumann-condition (Fig. 2b) requires $\psi_0 = \psi_1$. This condition can be satisfied with $p_1 = 1$.

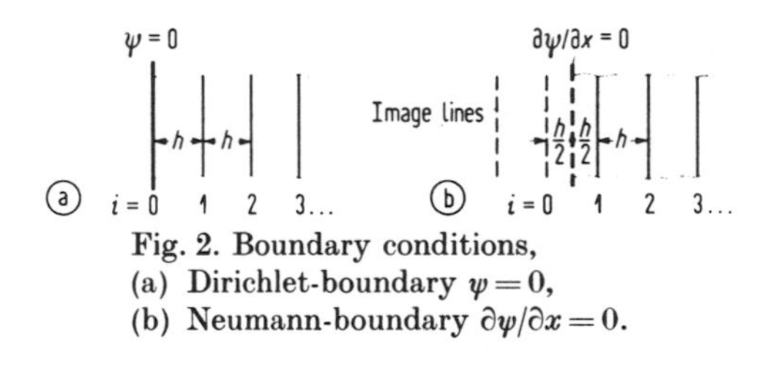

Fig. 2. Boundary conditions,
(a) Dirichlet-boundary $\psi = 0$,
(b) Neumann-boundary $\partial\psi/\partial x = 0$.

It should be mentioned, that this is not the only way to satisfy the boundary conditions. E.g., the Neumann-boundary can be laid at $i = 0$ too, but then the requirement $\psi_{-1} = \psi_1$ leads to a nonsymmetrical matrix $\boldsymbol{P}$, and the transformation matrix $\boldsymbol{T}$ is no longer orthogonal.

4. Discretisation of Hybrid Modes

It is well known that the hybrid field components can be expressed in terms of a superposition of the TE and TM fields, which are in turn derivable from the scalar potentials $\psi^{(e)} \sim E_z$ and $\psi^{(h)} \sim H_z$. Because of the complementary properties of the boundary conditions relative to $\psi^{(e)}$ and $\psi^{(h)}$ it is expedient to relatively shift the lines for $\psi^{(e)}$ and $\psi^{(h)}$ by half the discretisation distance. This is the only way to achieve that the matrix $\boldsymbol{P}$ will be symmetric for both $\psi^{(e)}$ and $\psi^{(h)}$. For instance, in Fig. 3 the left wall is a Neumann-boundary for $\psi^{(e)}$ and

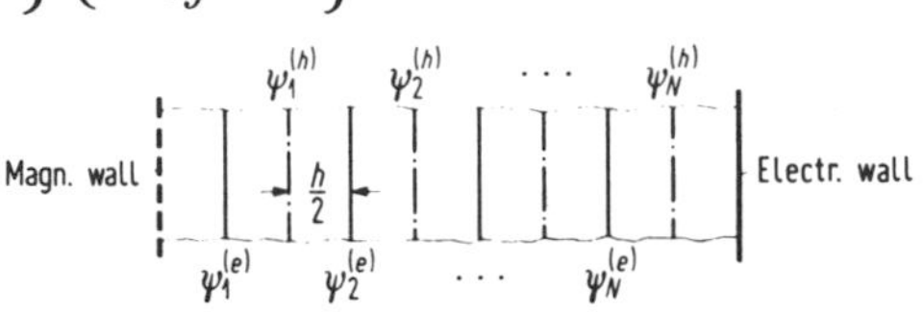

Fig. 3. Shifting of lines for $\psi^{(e)}$ and $\psi^{(h)}$.

a Dirichlet-boundary for $\psi^{(h)}$. At the right wall it is just the other way. The shifting of lines now guaranties a simple fitting of all the boundary conditions.

At dielectric interfaces the two potentials $\psi^{(e)}$ and $\psi^{(h)}$ are coupled by the continuity conditions of the tangential field components. This coupling is always such, that the derivative $\partial\psi_i^{(e)}/\partial y$ is related to the difference of the adjacent potentials $\psi_i^{(h)} - \psi_{i-1}^{(h)}$ and $\partial\psi_i^{(h)}/\partial y$ to the difference $\psi_{i+1}^{(e)} - \psi_i^{(e)}$ (see Fig. 3).

5. Edge Condition

Most of the considered waveguide structures contain metallic strips, which are assumed to have zero thickness. In [7] the discretisation error caused by the singularity due to the edge of such a strip is investigated. It was found there that this kind of discretisation error is negligibly small, if the field at the discrete lines satisfies the edge condition [5], [6]. This is approximately achieved, when the end of the strip exceeds the last $\psi^{(e)}$-line on the strip by $h/4$ and the last $\psi^{(h)}$-line by $\frac{3}{4}h$ (see Fig. 4). It is a further advantage of the shifting of the $\psi^{(e)}$- and $\psi^{(h)}$-lines, that these requirements can be satisfied simultaneously, so that the discretisation error caused by an edge can be compensated very well.

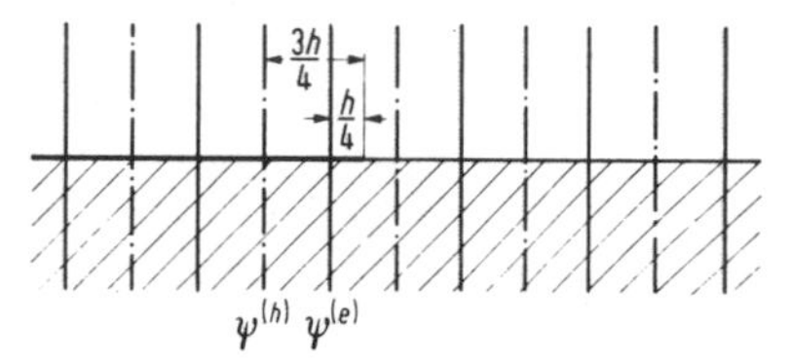

Fig. 4. To the edge condition.

Because of this, the number of lines in the considered structure and thus the order of the characteristic equation need not be very large, so that accurate results are obtained with rather low computational effort.

6. Example

For convenience, the method is demonstrated on a simple structure, the shielded microstrip line (Fig. 5). Because of symmetry only the half cross

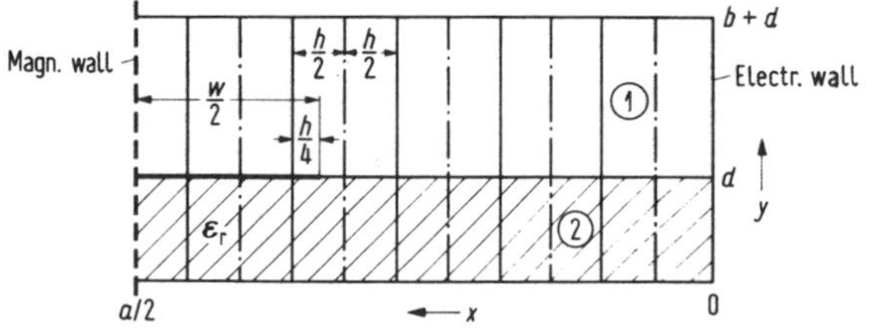

Fig. 5. Cross section of a shielded microstrip line.

section is considered. At the interface $y = d$ both potentials are related by the continuity conditions of the tangential field components. With

$$\psi^{(e)} = \frac{\mathrm{j}\,\omega\,\varepsilon}{k^2 - \beta^2} E_z \tag{12}$$

and

$$\psi^{(h)} = \frac{\mathrm{j}\,\omega\,\mu}{k^2 - \beta^2} H_z \tag{13}$$

the continuity conditions are:

$$\frac{\beta}{\omega\,\varepsilon_0}\frac{\partial}{\partial x}\left(\psi_{\mathrm{I}}^{(e)} - \frac{1}{\varepsilon_{\mathrm{r}}}\psi_{\mathrm{II}}^{(e)}\right) = \frac{\partial\psi_{\mathrm{II}}^{(h)}}{\partial y} - \frac{\partial\psi_{\mathrm{I}}^{(h)}}{\partial y}, \tag{14}$$

$$(k_0^2 - \beta^2)\,\psi_{\mathrm{I}}^{(e)} = \frac{1}{\varepsilon_{\mathrm{r}}}(\varepsilon_{\mathrm{r}}\,k_0^2 - \beta^2)\,\psi_{\mathrm{II}}^{(e)}, \tag{15}$$

$$\frac{\partial\psi_{\mathrm{I}}^{(e)}}{\partial y} - \frac{\partial\psi_{\mathrm{II}}^{(e)}}{\partial y} = \frac{\beta}{\omega\,\mu}\frac{\partial}{\partial x}(\psi_{\mathrm{I}}^{(h)} - \psi_{\mathrm{II}}^{(h)}) - J_z, \tag{16}$$

$$(k_0^2 - \beta^2)\,\psi_{\mathrm{I}}^{(h)} = (\varepsilon_{\mathrm{r}}\,k_0^2 - \beta^2)\,\psi_{\mathrm{II}}^{(h)} - \mathrm{j}\,\omega\,\mu\,J_x. \tag{17}$$

The subscripts indicate the subregions I and II. J_x and J_z are the current density distributions at $y = d$.

Now the operator $\partial/\partial x$ is replaced by the difference operator $\boldsymbol{D}$

$$\boldsymbol{D} = \begin{pmatrix} 1 & -1 & & \bigcirc \\ & \ddots & \ddots & \\ \bigcirc & & \ddots & \ddots \end{pmatrix} \tag{18}$$

with

$$\frac{\partial\psi^{(e)}}{\partial x} \to \frac{1}{h}\boldsymbol{D}\vec{\psi}^{(e)}.$$

and

$$\frac{\partial\psi^{(h)}}{\partial x} \to -\frac{1}{h}\boldsymbol{D}^{\mathrm{t}}\vec{\psi}^{(h)}. \tag{19}$$

The normal derivatives of $\psi^{(e,h)}$ at the interface are replaced by the following matrix expressions with the operators $\boldsymbol{G}^{(e,h)}$:

$$h\frac{\partial\psi_k^{(e)}}{\partial n} \to h\frac{\partial\vec{\psi}^{(e)}}{\partial n} = \boldsymbol{G}_k^{(e)}\vec{\psi}_k^{(e)}, \tag{20}$$

$$h\frac{\partial\psi_k^{(h)}}{\partial n} \to h\frac{\partial\vec{\psi}^{(h)}}{\partial n} = \boldsymbol{G}_k^{(h)}\vec{\psi}_k^{(h)}, \tag{21}$$

$$k = \mathrm{I}, \mathrm{II}$$

where $\partial n = -\partial y$ for $k = \mathrm{I}$ and $\partial n = \partial y$ for $k = \mathrm{II}$.

Eqs. (20) and (21) can be transformed into the diagonal form:

$$h\frac{\partial\vec{U}_k^{(e)}}{\partial n} = \boldsymbol{\gamma}_k^{(e)}\,\vec{U}_k^{(e)}, \tag{22}$$

$$h\frac{\partial\vec{U}_k^{(h)}}{\partial n} = \boldsymbol{\gamma}_k^{(h)}\,\vec{U}_k^{(h)}, \quad k = \mathrm{I}, \mathrm{II}. \tag{23}$$

The diagonal matrices $\boldsymbol{\gamma}_k^{(e,h)}$ are calculated analytically with the aid of eq. (10) and the known boundary conditions at $y = 0$ and $y = b$ (see Appendix B).

Now, eqs. (14) to (17) can be written in discretized form. After elimination of $\psi_{\mathrm{II}}^{(e)}$ and $\psi_{\mathrm{II}}^{(h)}$ with eqs. (15) and (17) and transformation with $\boldsymbol{T}^{(e)}$ and $\boldsymbol{T}^{(h)}$ respectively, remain

$$\frac{\beta}{\omega\,\varepsilon_0}(1 - \tau)\underbrace{\boldsymbol{T}^{(h)\mathrm{t}}\boldsymbol{D}\,\boldsymbol{T}^{(e)}}_{\boldsymbol{\delta}}\cdot\vec{U}_{\mathrm{I}}^{(e)} = (\boldsymbol{\gamma}_{\mathrm{I}}^{(h)} + \tau\,\boldsymbol{\gamma}_{\mathrm{II}}^{(h)})\cdot\vec{U}_{\mathrm{I}}^{(h)}, \tag{24}$$

$$-(\boldsymbol{\gamma}_{\mathrm{I}}^{(e)} + \varepsilon_{\mathrm{r}} \tau \boldsymbol{\gamma}_{\mathrm{II}}^{(e)}) \cdot \vec{U}_{\mathrm{I}}^{(e)} = \tag{25}$$
$$= \frac{-\beta}{\omega \mu} (1 - \tau) \underbrace{\boldsymbol{T}^{(e)\mathrm{t}} \boldsymbol{D}^{\mathrm{t}} \boldsymbol{T}^{(h)}}_{\boldsymbol{\delta}^{\mathrm{t}}} \cdot \vec{U}_{\mathrm{I}}^{(h)} - \boldsymbol{T}^{(e)\mathrm{t}} \vec{J}_z$$

with
$$\tau = \frac{1 - \varepsilon_{\mathrm{eff}}}{\varepsilon_{\mathrm{r}} - \varepsilon_{\mathrm{eff}}}. \tag{26}$$

Note, that the matrix product $\boldsymbol{T}^{(h)\mathrm{t}} \boldsymbol{D} \boldsymbol{T}^{(e)}$, denoted by $\boldsymbol{\delta}$, is a diagonal matrix, too. The elements δ_i are given by simple analytical expressions (see Appendix B).

Since the transverse current density is negligibly small, only the longitudinal current density distribution is considered in this example. Nevertheless, it is also possible to proceed without this restriction. From eqs. (24) and (25) results

$$\vec{U}_{\mathrm{I}}^{(e)} = \boldsymbol{\rho}\, \boldsymbol{T}^{(e)\mathrm{t}} \cdot \vec{J}_z \tag{27}$$

with the diagonal matrix

$$\boldsymbol{\rho} = [\boldsymbol{\gamma}_{\mathrm{I}}^{(e)} + \varepsilon_{\mathrm{r}} \tau \boldsymbol{\gamma}_{\mathrm{II}}^{(e)} - \varepsilon_{\mathrm{eff}} (1 - \tau)^2 \cdot \boldsymbol{\delta}^{\mathrm{t}} (\boldsymbol{\gamma}_{\mathrm{I}}^{(h)} + \tau \boldsymbol{\gamma}_{\mathrm{II}}^{(h)})^{-1} \boldsymbol{\delta}]^{-1}. \tag{28}$$

After inverse transformation of eq. (27) (see eq. (8)), for the potential vector $\vec{\psi}_{\mathrm{I}}^{(e)}$ is found

$$\vec{\psi}_{\mathrm{I}}^{(e)} = \boldsymbol{T}^{(e)} \boldsymbol{\rho}\, \boldsymbol{T}^{(e)\mathrm{t}} \cdot \vec{J}_z . \tag{29}$$

At last, the final boundary condition on the strip, which requires

$$\vec{\psi}_{\mathrm{I}}^{(e)} = 0 \quad \text{on the strip} \tag{30}$$

leads, with the additional relation

$$\vec{J}_z = \begin{cases} \vec{J}_{z\,\mathrm{red}} & \text{on the strip} \\ 0 & \text{elsewhere} \end{cases} \tag{31}$$

to the characteristic equation

$$(\boldsymbol{T}^{(e)} \boldsymbol{\delta}\, \boldsymbol{T}^{(e)\mathrm{t}})_{\mathrm{red}} \vec{J}_{z\,\mathrm{red}} = \vec{0} \tag{32}$$

where only the number of points on the strip determines the size of the matrix. The eigenvector $\vec{J}_{z\,\mathrm{red}}$ is the reduced part of the discretized current density vector.

Finally it should be emphasized, that except for the reduced characteristic equation (32) all operations can be carried out with diagonal matrices. This is an important property of the described method, and permits fast computation.

7. Numerical Results

The method described is so simple, that all numerical computations could be carried out on a HP 9825A desk calculator. Figs. 6 and 7 show the results for a single and a pair of coupled microstrip lines. A comparison with other results available [8], [9] shows good agreement. For the single microstrip line the total number of points at the interface is $N = 18$ and the number of points on the strip is 6. For the coupled lines 32 points were chosen at the interface, three points on each strip and two points in the slot. Of cause, these are the

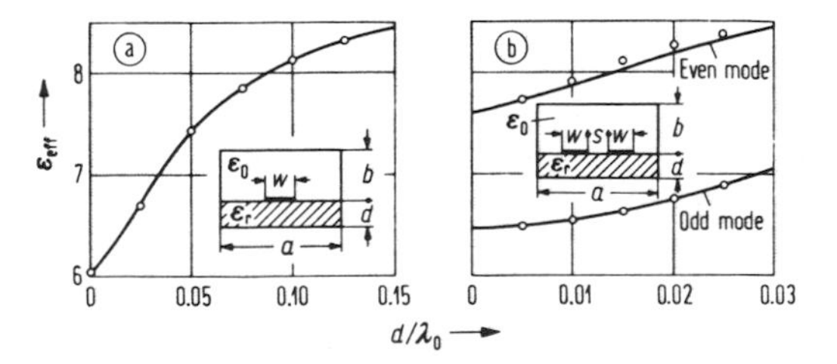

Fig. 6. (a) Effective dielectric constant of a single microstrip; $\varepsilon_{\mathrm{r}} = 9$, $w/d = 2$, $a/d = 7$, $b/d = 3$; o Kowalski, Pregla [8].
(b) Effective dielectric constant of a pair of coupled microstrips; $\varepsilon_{\mathrm{r}} = 10.2$, $w/d = 1.5$, $s/d = 1.5$, $a/d = 20$, $b/d = 19$; o Sharma, Bhat [9].

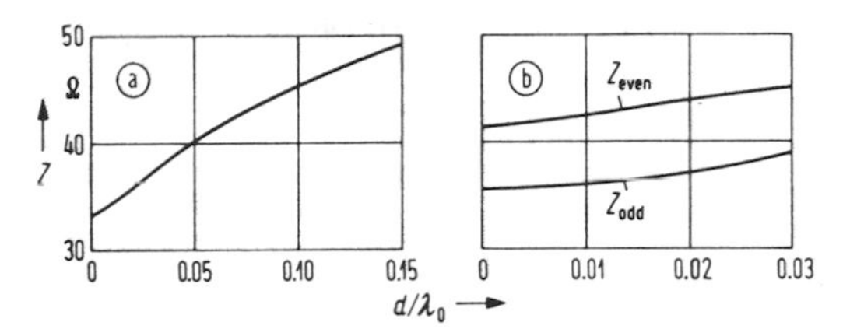

Fig. 7. (a) Characteristic impedance of a single microstrip; parameters as in Fig. 6a.
(b) Characteristic impedance of a pair of coupled microstrips; parameters as in Fig. 6b.

numbers for only one potential, say $\vec{\psi}^{(e)}$. Because of symmetry only one half of each structure has to be considered. Thus, in both cases the size of the matrix associated with the final characteristic equation has the order of only 3×3. Computing $\varepsilon_{\mathrm{eff}}$ and Z for one frequency on a HP 9825A takes between 0.5 and 1 minute. The accuracy of the results obtained (see Fig. 8) is better than 0.5%

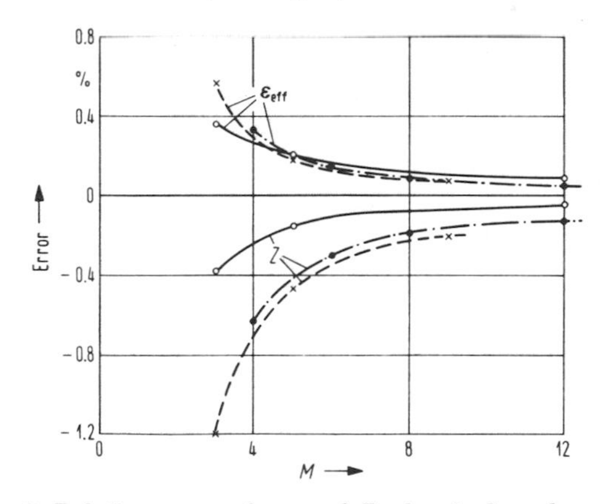

Fig. 8. Relative errors of $\varepsilon_{\mathrm{eff}}$ and Z of a single and a pair of coupled microstrip lines; parameters as in Fig. 6; —o— single line, --×-- even mode, —·—•—·— odd mode.

for $\varepsilon_{\mathrm{eff}}$ and about 1% for Z. This has been achieved with only three points on a strip! This is a consequence of the described line-shifting making it possible to compensate very well the discretisation error due to the edge of the strip.

(Received August 3rd, 1979.)

Appendix A

The matrix $\boldsymbol{P}$ and consequently the transformation matrix $\boldsymbol{T}$ (eq. (7)) depend on the lateral boundary conditions. The following table gives the matrix $\boldsymbol{T}$ and the eigenvalues λ_i for the various combinations of boundaries.

left boundary	right boundary	T_{ij}	λ_i
Dirichlet	Dirichlet	$\sqrt{\frac{2}{N+1}}\sin\frac{ij\pi}{N+1}$	$4\sin^2\frac{i\pi}{2(N+1)}$
Dirichlet	Neumann	$\sqrt{\frac{2}{N+1/2}}\sin\frac{i(j-1/2)\pi}{N+1/2}$	$4\sin^2\frac{(i-1/2)\pi}{2N+1}$
Neumann	Dirichlet	$\sqrt{\frac{2}{N+1/2}}\cos\frac{(i-1/2)(j-1/2)\pi}{N+1/2}$	$4\sin^2\frac{(i-1/2)\pi}{2N+1}$
Neumann	Neumann	$1/\sqrt{N},\quad j=1$ $\sqrt{\frac{2}{N}}\cos\frac{(i-1/2)(j-1)\pi}{N},\quad j>1$	$4\sin^2\frac{(i-1)\pi}{2N}$

Appendix B

The following formulas may be seen as a supplement to the considered example in section 6. The two transformation matrices $\boldsymbol{T}^{(e,h)}$ and the related eigenvalues λ_i are not given here, as they can be taken from Appendix A:

$$\boldsymbol{\Upsilon}_{\mathrm{I}}^{(e)} = \mathrm{diag}[\varkappa_i \coth(\varkappa_i b/h)], \tag{33}$$

$$\boldsymbol{\Upsilon}_{\mathrm{I}}^{(h)} = \mathrm{diag}[\varkappa_i \tanh(\varkappa_i b/h)], \tag{34}$$

$$\boldsymbol{\Upsilon}_{\mathrm{II}}^{(e)} = \mathrm{diag}[\eta_i \coth(\eta_i d/h)], \tag{35}$$

$$\boldsymbol{\Upsilon}_{\mathrm{II}}^{(h)} = \mathrm{diag}[\eta_i \tanh(\eta_i d/h)] \tag{36}$$

with

$$\varkappa_i = \left[4\sin^2\left(\frac{i-0.5}{2N+1}\pi\right) - h^2(k_0^2-\beta^2)\right]^{1/2} \tag{37}$$

and

$$\eta_i = \left[4\sin^2\left(\frac{i-0.5}{2N+1}\pi\right) - h^2(\varepsilon_r k_0^2-\beta^2)\right]^{1/2}, \tag{38}$$

$$\boldsymbol{\delta} = \mathrm{diag}\left[2\sin\left(\frac{i-0.5}{2N+1}\pi\right)\right]. \tag{39}$$

References

[1] Michlin, S. G. and Smolizki, Ch. L., Näherungsmethoden zur Lösung von Differential- und Integralgleichungen. B. G. Teubner-Verlag, Leipzig 1969, pp. 238–243.

[2] Ames, W. F., Numerical methods for partial differential equations. Academic Press, New York 1977, pp. 302–304.

[3] Lennartsson, B. L., A network analogue method for computing the TEM-characteristics of planar transmission lines. Transact. IEEE MTT-**20** [1972], 586–591.

[4] Corr, D. G. and Davies, J. B., Computer analysis of the fundamental and higher order modes in single and coupled microstrip. Transact. IEEE MTT-**20** [1972], 669–678.

[5] Meixner, J., Die Kantenbedingung in der Theorie der Beugung elektromagnetischer Wellen an vollkommen leitenden ebenen Schirmen. Ann. Phys. (Leipzig) **6** [1949], 2–9.

[6] Collin, R. E., Field theory of guided waves. McGraw-Hill Book Co., New York 1960, pp. 18–20.

[7] Schulz, U., On the edge condition with the method of lines in planar waveguides. AEÜ **34** [1980], 176–178.

[8] Kowalski, G. and Pregla, R., Dispersion characteristics of shielded microstrips with finite thickness. AEÜ **25** [1971], 193–196.

[9] Sharma, A. K. and Bhat, B., Dispersion characteristics of shielded coupled microstrip lines. AEÜ **32** [1978], 503–504.

Part III
Microstrip Lines Loss

IN an actual circuit design, particularly at high frequencies, attenuation along the transmission line is an important subject for investigation. In the case of the microstrip line, the wave attenuation is caused by the conductor loss, substrate loss, and radiation loss. In the case of nonmagnetic substrate, the substrate loss is commonly referred to as the dielectric loss and is caused by the dissipation factor of the material. If the substrate is a magnetic material, the substrate loss is caused by both the dielectric dissipation and the magnetic loss. These losses are dependent on the imaginary parts of the complex permittivity and the permeability. The conductor loss is caused by an imperfect metal used as a conductor. This loss is dependent on the skin depth or the surface resistivity of the conductor. Because the current density near the edges of the microstrip is quite high, the conductor loss is usually a predominant contribution to the wave attenuation.

In this part, papers included treat the formulation of the attenuation factor in a microstrip line. The analysis of the dielectric and conductor losses is undertaken essentially by a perturbation method. An interesting concept of the incremental inductance rule introduced by Wheeler is often used for conductor loss calculations (Pucel, Masse, and Hartwig; Denlinger). In this method, the change of the inductance of the strip due to a hypothetical reduction of the strip cross section by the amount of the skin depth is used for calculating the series resistance along the microstrip. From this information, the attenuation constant is obtained.

Although some evaluations of the radiation loss are reported in a number of earlier publications, such as the one by Denlinger in this part, the topic of radiation from the microstrip circuit has become an increasingly important subject in recent years. Therefore, papers treating this subject appear more extensively in the later part on end effects.

Losses in Microstrip

ROBERT A. PUCEL, SENIOR MEMBER, IEEE, DANIEL J. MASSÉ, MEMBER, IEEE, AND CURTIS P. HARTWIG, MEMBER, IEEE

Abstract—**Expressions are derived for the conductor loss in microstrip transmission lines. The formulas take into account the finite thickness of the strip conductor and apply to the mixed dielectric system. Good agreement with experimental data is obtained for rutile and alumina substrates.**

I. Introduction

THE INCREASING importance of miniature planar microwave integrated circuits has renewed interest on the part of microwave circuit designers in the various forms of planar strip transmission line systems. Of the many configurations possible, the open or "microstrip" geometry shown in Fig. 1 appears at present to be the most convenient and inexpensive system for batch processing of microwave integrated circuits. As illustrated, this transmission line consists of a narrow "strip" conductor of width w and thickness t separated from a conducting ground plane by an intervening supporting dielectric substrate of thickness h and width much greater than w. For maximum circuit size reduction, the dielectric constant of substrates now being used is of the order of ten or higher.

For design purposes, it is necessary to know how the characteristic impedance, phase velocity, and attenuation "constant" of the dominant microstrip mode depend on geometrical factors, on the electronic properties of the substrate and conductors, and on the frequency. Since this is a "mixed" dielectric system, the TEM mode cannot be supported. Nevertheless, Wheeler[1] recently has obtained simple expressions for the characteristic impedance of the dominant mode based on a TEM approximation which agrees to within five percent of the experimental data obtained with a time domain reflectometer at our Research Division and other laboratories.[1]

The attenuation constant of the dominant mode has not received much attention in recent years. Except for Welch and Pratt[3] who recently have treated the dielectric attenuation for the mixed dielectric system, no one, at least to our knowledge, has proposed any significant improvements to the expressions for conductor attenuation derived by Assadourian and Rimai[4] nearly fifteen years ago. These expressions, which predict losses higher than that observed,[5],[6] do not apply to the geometrical parameter ranges of most interest to designers of contemporary microwave integrated circuits. In fact, they are not applicable to these mixed dielectric systems, although they have been used for such purposes.

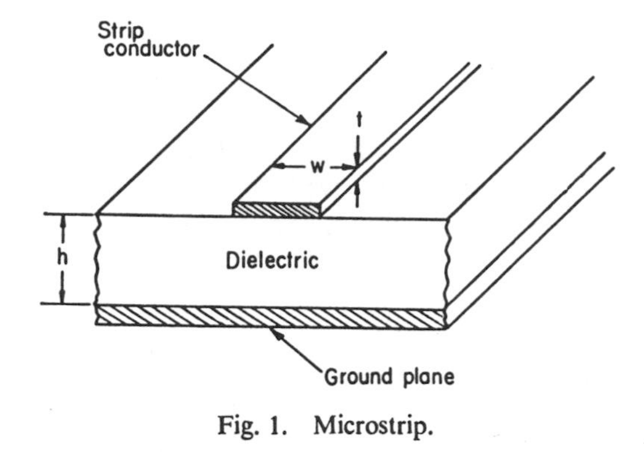

Fig. 1. Microstrip.

To help fill this gap, we propose approximate expressions for the conductor losses covering a wide range of geometrical parameters and applicable to the microstrip on a dielectric substrate. Our theory is compared with data taken on rutile and alumina microstrip. Excellent agreement is obtained in those cases where dielectric and conductor losses predominate.

II. Microstrip Losses

A. Introduction

We consider only nonmagnetic dielectric substrates. Thus, we may restrict our attention to two types of loss in the dominant microstrip mode, namely dielectric losses in the substrate and ohmic skin losses in the strip conductor and the ground plane.

Assuming these losses per unit length to be "small," we may represent them in terms of an attenuation factor α in the expression for transmitted power $P(z)$

$$P(z) = P_0 e^{-2\alpha z} \tag{1}$$

where z denotes a point along the direction of propagation parallel to the strip conductor and P_0 the transmitted power at an earlier point $z=0$. Letting $\alpha=\alpha_d+\alpha_c$, the sum of a dielectric attenuation factor α_d and an ohmic attenuation factor α_c, we get

$$\alpha = -\frac{dP/dz}{2P(z)} \approx \frac{P_c + P_d}{2P(z)} \ \frac{\text{nepers}}{\text{cm}} \tag{2a}$$

or

$$\alpha_d \approx \frac{P_d}{2P(z)} \ \frac{\text{nepers}}{\text{cm}} \tag{2b}$$

Manuscript received October 12, 1967; revised December 20, 1967, and February 26, 1968.

The authors are with the Research Division, Raytheon Company, Waltham, Mass.

[1] Such good agreement is not realized at microwave frequencies, however. We observe a frequency dependence of both the characteristic impedance and of the phase velocity of the order of 5 to 10 percent over the 1 to 6 GHz range, depending on the geometric parameter values and the dielectric constant. We have evidence indicating that this frequency dependence is caused by the lowest-order surface mode which is launched unintentionally.[2]

Reprinted from *IEEE Trans. Microwave Theory Tech.*, vol. MTT-16, no. 6, pp. 342–350, June 1968.

and

$$\alpha_c \approx \frac{P_c}{2P(z)} \frac{\text{nepers}}{\text{cm}}. \tag{2c}$$

Here P_d and P_c denote the average dielectric power loss and the average conductor power loss per unit length.

To compare one dielectric with another, it is more meaningful to use the loss per wavelength which can be obtained from (2b) and (2c) simply by multiplying them by the guide wavelength $\lambda_g = \lambda_0/\sqrt{k_e}$, where λ_0 is the free space wavelength, and k_e is the "effective" dielectric constant of the substrate. This latter quantity, which takes into account the dielectric "filling factor" is defined more precisely in Section II-B.

B. Dielectric Losses

Welch and Pratt[3] have analyzed the case of dielectric losses in a nonmagnetic mixed dielectric system. They have shown that their expressions account for most of the attenuation observed with a microstrip fabricated on a silicon substrate for which dielectric losses predominated.

We have used their expressions in conjunction with our own formulas for skin loss attenuation, derived herein, to predict the total attenuation to be expected for the rutile and alumina microstrip lines which were used in our experiments. For the convenience of the reader and for completeness, we summarize below the formulas derived by Welch and Pratt.

Following the approach of Wheeler[1] in determining the filling factor for the dielectric constant, Welch and Pratt derive a corresponding filling factor for the loss tangent tan δ, or equivalently, the conductivity σ of the dielectric substrate and obtain the following results for the case when the upper dielectric is air (assumed lossless):

$$\alpha_d \approx 27.3 \left(\frac{qk}{k_e}\right) \frac{\tan \delta}{\lambda_g} \quad \text{dB/cm} \tag{3a}$$

or

$$\approx 4.34 \frac{q}{\sqrt{k_e}} \sqrt{\frac{\mu_0}{\epsilon_0}} \, \sigma \quad \text{dB/cm} \tag{3b}$$

where we have used the factor $8.68 = 20 \log_{10} e$ to convert nepers to decibels. In these formulas, k denotes the dielectric constant of the substrate, and $k_e = 1 + q(k-1)$ where q is the dielectric filling factor as defined by Wheeler. The constants μ_0 and ϵ_0 denote the permeability and permittivity of free space. Equation (3a) is more convenient for nonconducting substrates, whereas (3b) is appropriate for substrates in which the conduction loss is the predominant component of loss.[2]

[2] Equations (3a) and (3b), strictly speaking, apply only if the loss tangent of the dielectric above the microstrip (in this case air) is equal to the loss tangent of the dielectric substrate. Equality of loss tangents is necessary to apply Wheeler's filling factor *unchanged*. However, for a low-loss dielectric substrate, tan $\delta \ll 1$, the assumption of a lossless upper dielectric introduces a negligible error.

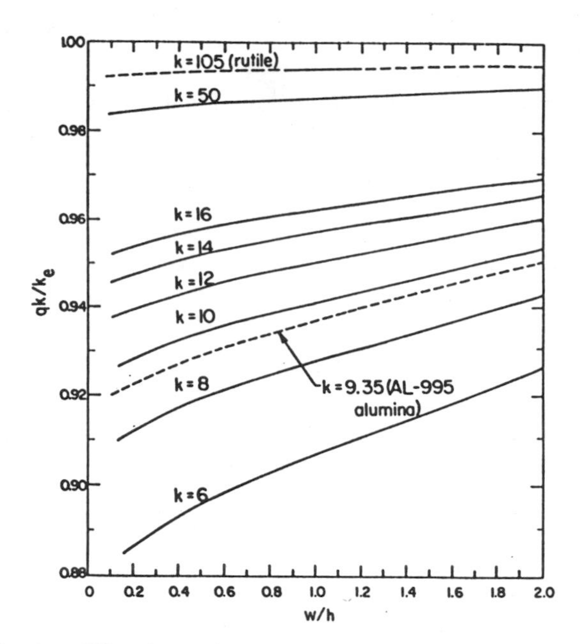

Fig. 2. Filling factor for loss tangent of microstrip substrate as a function of w/h.

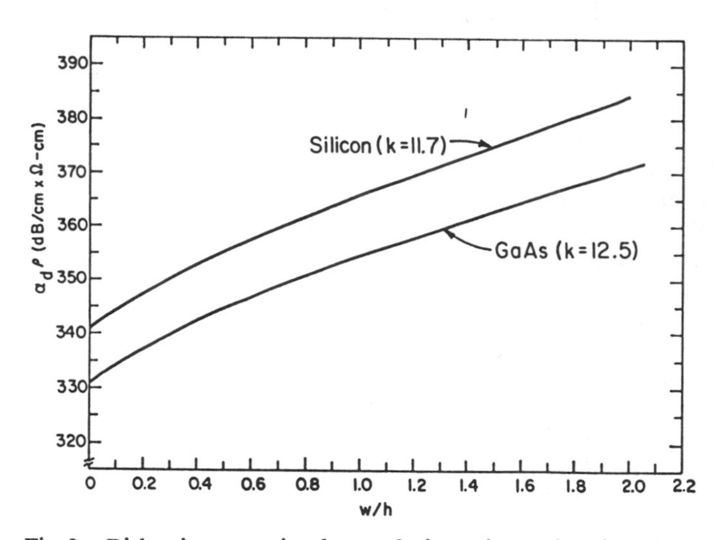

Fig. 3. Dielectric attenuation factor of microstrip as a function of w/h for silicon and gallium arsenide substrates.

Notice that if tan δ is frequency independent, the dielectric attenuation *per wavelength* is also frequency independent. On the other hand, if the substrate conductivity is frequency independent, as for a semiconductor, the dielectric attenuation *per unit length* is frequency independent.

Since q is a function of k and w/h, the filling factor for the loss tangent, qk/k_e, and for the conductivity, $q/\sqrt{k_e}$, are also functions of these quantities. To illustrate this dependence we have plotted in Fig. 2 the loss tangent filling factor against w/h for a range of dielectric constants suitable for microwave integrated circuits. In Fig. 3 we have graphed against w/h the product $\alpha_d\rho$ $(\rho = \sigma^{-1})$ for two semiconducting substrates which are being used for integrated microwave circuits, namely, silicon and gallium arsenide.

Observe that the loss tangent filling factor is of the order of unity and approaches this value as a limit as $k \rightarrow \infty$. For most practical purposes, this factor can be approximated by unity. The conductivity filling factor exhibits only a mild dependence on w/h which probably can be ignored in practice.

C. *Ohmic Losses*

1) *Introduction:* In a microstrip line over a low-loss dielectric substrate, the predominant sources of loss at microwave frequencies are the nonperfect conductors. The current density in the conductors is concentrated in a sheet approximately a skin depth deep inside the conductor surfaces exposed to the electric field. If the current distribution were known, one could compute the ohmic attenuation factor directly using the expression[7]

$$\alpha_c = \frac{R_{s1}}{2Z_0}\int_c \frac{|J_1|^2 dx}{|I|^2} + \frac{R_{s2}}{2Z_0}\int_{-\infty}^{\infty} \frac{|J_2|^2 dx}{|I|^2} \tag{4}$$

which is based on (2c). Here Z_0 denotes the characteristic impedance of the microstrip, $R_{s1}=(\pi f \mu_1 \rho_1)^{1/2}$ and R_{s2} $(\pi f \mu_2 \rho_2)^{1/2}$ the surface skin resistivity in ohms per square for the strip conductor and ground plane, respectively, $J_1(x)$ and $J_2(x)$ the corresponding surface current densities, and $|I|$ the magnitude of the total current per conductor. The quantities $\mu_{1,2}$ and $\rho_{1,2}$ represent the permeability and the bulk resistivity of the strip and ground conductors, respectively, and f denotes the operating frequency. The integral $\int_c$ implies integration of the surface current density around all surfaces of the strip conductor. Both the strip conductor thickness t and the ground plane thickness are assumed to be at least three or four skin depths thick.

The current density in the strip conductor and in the ground conductor is not uniform in the transverse plane because of the finite strip width. A sketch of the current distribution is shown in Fig. 4[5] for a strip of nonzero thickness t. The strip conductor contributes the major part of the skin loss. For an infinitesimally thin strip conductor, the current density diverges at the strip edges at such a rate that the skin loss is unbounded.[8]

Because of mathematical complexity, exact expressions for the current density for the practical case of a strip of nonzero thickness have never been derived.[9] Some workers[10],[11] have assumed for simplicity that the current distribution is uniform and equal to I/w in both conductors, and confined to the region $|x|<w/2$. With this assumption one obtains the formula

$$\alpha_c \approx \frac{8.68 R_s}{Z_0 w} \quad \text{dB/cm} \tag{5}$$

for the case $R_{s1}=R_{s2}$. This simple expression is valid only for arbitrarily large strip widths $w/h \rightarrow \infty$. Our experimental data shows that the above approximation overestimates the skin loss by 80 percent for $w/h<2$. This large discrepancy is understandable since the current density is far from uniform even for w/h as large as three.[5],[8] We believe that the apparent agreement with experiment claimed by some for this expression is probably caused by extraneous sources of loss.

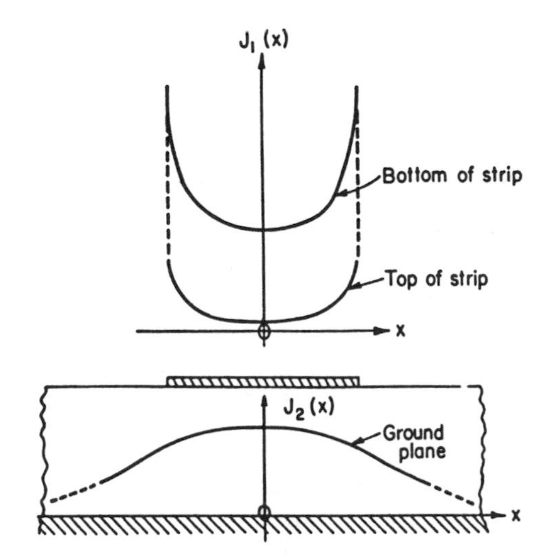

Fig. 4. Sketch of the current distribution on microstrip conductors.

Since the current distribution in the conductors is not known, we shall use another technique for calculating the skin losses, a technique based on the so-called "incremental inductance rule" which is also due to Wheeler.[12] This rule—really a formula—expresses the series skin resistance R per unit length in terms of that part of the total inductance per unit length which is attributable to the skin effect, that is, to the inductance produced by the magnetic field within the metallic conductors. The formula, of course, is as good as, and no better than, the assumptions and approximations used in the derivation of the inductance itself.

The formula is based on the known fact that the series skin impedance Z per unit length,

$$Z = R + jX \tag{6}$$

has a real part R (the desired quantity) which is equal to its imaginary part X, where

$$X = \omega L_i. \tag{7}$$

In most derivations of the inductance per unit length, the skin effect contribution L_i is neglected in comparison to the external inductance L per unit length, and indeed, is not even computed, since perfect conductors are assumed usually. In characteristic impedance calculations this approximation does not introduce serious error for low-loss lines.

Wheeler shows that L_i can be inferred from L as the incremental increase in L caused by an incremental recession of all metallic walls carrying a skin current. The amount of recession is equal to *half* the skin depth, δ, $\delta=(\rho/\pi\mu f)^{1/2}$. Naturally, this technique is useful only in those cases where an external inductance can be calculated—such as for TEM lines.[12] An assumption underlying this rule is that the radius of curvature and the thickness of the conductors exposed to the electric field be greater than the skin depth—preferably several skin depths.

From Wheeler we obtain

$$L_i = \sum_j \frac{\mu_j}{\mu_0}\frac{\partial L}{\partial n_j}\frac{\delta_j}{2}, \tag{8}$$

and

$$R = \frac{1}{\mu_0} \sum_j R_{sj} \frac{\partial L}{\partial n_j}, \tag{9}$$

where $\partial L/\partial n_j$ denotes the derivative of L with respect to the incremental recession of wall j, n_j the vector normal to this wall, and $R_{sj}=\omega\delta_j\mu_j/2$ the surface skin resistivity of wall j. Thus from (2c) and (4),

$$\alpha_c = \frac{|I|^2 R}{2|I|^2 Z_0} = \frac{1}{2\mu_0 Z_0} \sum_j R_{sj} \frac{\partial L}{\partial n_j}. \tag{10}$$

Equation (10) is the basis for our skin loss computations.

2) *Derivation of Skin Loss Expressions:* To apply (10) to the microstrip line, we assume that the inductance per unit length for the mixed dielectric case is the same as for the uniform dielectric case. This assumption implies that the stored magnetic energy per unit length is not affected by the presence of the dielectric. Apparently this is a reasonable assumption since it also underlies Wheeler's derivation for the characteristic impedance of the mixed dielectric system,[1] and has been verified experimentally at our laboratory.

This inductance can be obtained from Wheeler's impedance expressions by the formula

$$L = \sqrt{\mu_0 \epsilon_0} Z_0 \left(\frac{w'}{h}, \frac{t}{h}, k\right) \tag{11}$$

valid for TEM waves, if one lets $k=1$.

To take the thickness of the strip conductor into account, we use Wheeler's correction terms[1] which are added to the physical strip width w,

$$\Delta w = \frac{t}{\pi} \ln\left(\frac{4\pi w}{t} + 1\right) \qquad w/h \leq 1/2\pi \tag{12a}$$

$$(2t/h < w/h,\ 1/2\pi)$$

$$= \frac{t}{\pi} \ln\left(\frac{2h}{t} + 1\right) \qquad w/h \geq 1/2\pi \tag{12b}$$

thus,

$$w' = w + \Delta w. \tag{13}$$

Except for very narrow strip widths $w/h \ll 1$, these corrections can be neglected for impedance calculations, particularly if weighted by the factor $1/k$ as Wheeler suggests. However, for loss calculations, one must include them to account for the thickness dependence.

For the calculation of the inductance derivatives, it is helpful to refer to Fig. 5 which shows the surface recessions to be considered. We postulate that the conductor separation h is measured from the ground plane to the *bottom* of the strip conductor rather than to its horizontal center line. This assumption not only is consistent with the practical case of a film conductor on the surface of the dielectric, but also leads to the proper asymptotic expression for α_c as $w/h \to \infty$. We also assume that the strip corners are rounded sufficiently to insure a large enough radius of curvature, although in practice this may not be satisfied if the total thickness is only three or four skin depths.

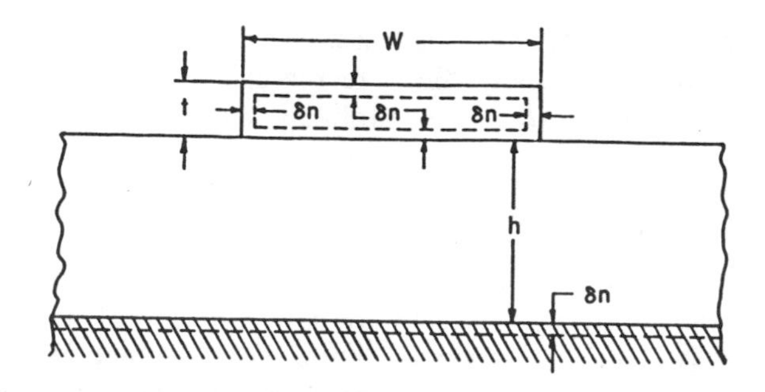

Fig. 5. Relevant to the derivation of conductor attenuation.

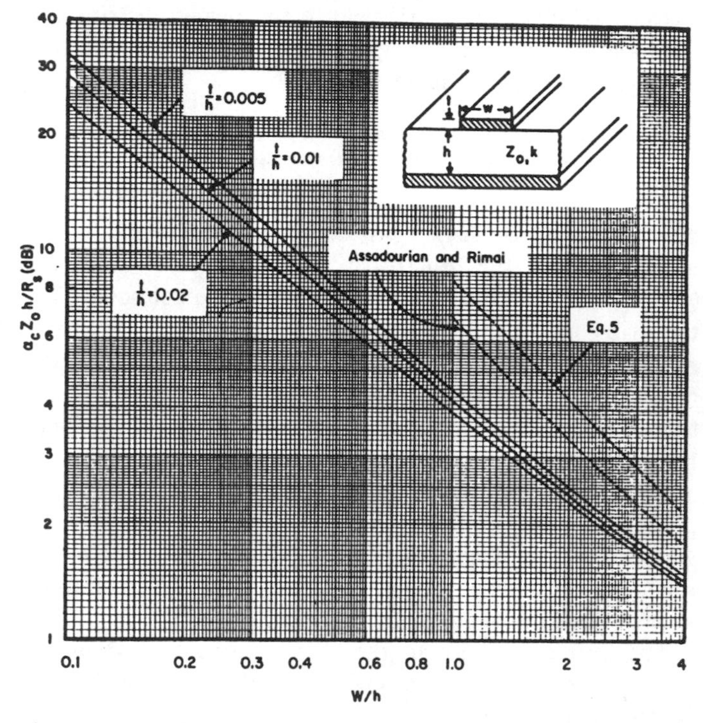

Fig. 6. Theoretical conductor attenuation factor of microstrip as a function of w/h.

We are on somewhat uncertain grounds, however, in deciding whether the recession of the bottom surface of the strip conductor can be counted towards a recession in t as well as an extension of h. We shall assume so here, but only experiment will decide this. By counting it as a recession of t, we do get a larger value for α_c which seems to agree better with our measured results.

In accordance with these assumptions and Fig. 5, we set

$\delta n_j = \delta h$ (recession of ground plane)
$= \delta h$ (recession of bottom of strip conductor)
$= -\delta w$ (recession of either edge of strip conductor)
$= -\delta t$ (recession of top and of bottom of strip conductor).

We obtain

$$\alpha_c = \frac{1}{2\mu_0 Z_0}\left[R_{s1}\left(\frac{\partial L}{\partial h} - 2\frac{\partial L}{\partial w} - 2\frac{\partial L}{\partial t}\right) + R_{s2}\frac{\partial L}{\partial h}\right]. \tag{14}$$

From Wheeler's approximate results[1] we get

$$L = \frac{\mu_0}{2\pi}\left[\ln\frac{8h}{w'} + \frac{1}{32}\left(\frac{w'}{h}\right)^2 + \cdots\right] \tag{15a}$$

for $w/h \leq 2$, and

$$L = \frac{\mu_0}{2} \frac{1}{\frac{w'}{2h} + \frac{1}{\pi}\ln\left[2\pi e\left(\frac{w'}{2h} + 0.94\right)\right]} \tag{15b}$$

for $w/h \geq 2$ where e is the Naperian base.

Assuming $R_{s1} = R_{s2} = R_s$, we obtain after a lengthy but straightforward calculation the results:

$w/h \leq 1/2\pi$:

$$\frac{\alpha_c Z_0 h}{R_s} = \frac{8.68}{2\pi}\left[1 - \left(\frac{w'}{4h}\right)^2\right]\left\{1 + \frac{h}{w'} + \frac{h}{\pi w'}\left[\ln\left(\frac{4\pi w}{t} + 1\right) - \frac{1 - t/w}{1 + t/4\pi w}\right]\right\}, \tag{16a}$$

$1/2\pi < w/h \leq 2$:

$$\frac{\alpha_c Z_0 h}{R_s} = \frac{8.68}{2\pi}\left[1 - \left(\frac{w'}{4h}\right)^2\right]\left\{1 + \frac{h}{w'} + \frac{h}{\pi w'}\left[\ln\left(\frac{2h}{t} + 1\right) - \frac{1 + t/h}{1 + t/2h}\right]\right\}, \tag{16b}$$

$2 \leq w/h$:

$$\frac{\alpha_c Z_0 h}{R_s} = \frac{8.68}{\left\{\frac{w'}{h} + \frac{2}{\pi}\ln\left[2\pi e\left(\frac{w'}{2h} + 0.94\right)\right]\right\}^2}\left[\frac{w'}{h} + \frac{w'/\pi h}{\frac{w'}{2h} + 0.94}\right] \cdot \left\{1 + \frac{h}{w'} + \frac{h}{\pi w'}\left[\ln\left(\frac{2h}{t} + 1\right) - \frac{1 + t/h}{1 + t/2h}\right]\right\} \tag{16c}$$

where α_c is in dB/cm. The joining error in the above expressions at $w/h = \frac{1}{2}\pi$ is less than 6 percent, at $w/h = 2$, less than 8 percent.

It is evident that the thickness terms become arbitrarily large as $t \rightarrow 0$, as they should. In the limit of wide strips $w/h \gg 1$, (16c) approaches the asymptotic value given by (5). For a fixed characteristic impedance, α_c decreases inversely with the substrate thickness h, and increases with the square root of frequency.

Values of α_c based on (16a–c) have been computed for a span of practical t/h ratios for the range $w/h < 4$. The results are plotted in Fig. 6. (The curves were smoothed graphically at the joining points.) It is evident that α_c varies little with t/h for the range used in the calculations.

For comparison purposes we have also plotted the attenuation factor based on Assadourian and Rimai's expressions as well as on the simplified expression (5). Observe that both of these predict higher losses than our expressions. Equation (5) yields attenuation values nearly 50 percent higher than our expressions for w/h as large as 4. The discrepancy is even greater for smaller w/h ratios or for larger t/h values.

3) *Experimental Results:* The substrates tested were mounted on a brass slab at either end of which was attached an ESCA[3] coaxial-to-strip line launcher with a shortened and narrowed-down center conductor tab. A photograph of the holder with a mounted substrate is shown in Fig. 7.

Our attenuation measurements were made with a resonance-Q technique on open-circuit terminated lines. This method, which is described in detail in the Appendix, is very sensitive to low losses. In this technique we capacitively couple the stripline launcher tab to the microstrip line via a small gap, adjust this gap for a match at the measurement frequency f_0, and then vary the frequency in either direction to determine the frequency separation Δf corresponding to the 3 dB points of absorbed power.

Assuming a lossless transition, we show that if $f_0/\Delta f$ is large (say 10 or greater), as is true for low loss lines, the attenuation factor in decibels per centimeter is given by the expression

$$\alpha \cong 4.34\left(\frac{\pi}{\lambda_g}\frac{\Delta f}{f_0} + \frac{1}{l}\ln R\right). \tag{17}$$

Here l denotes the linelength in centimeters, λ_g the measured guide wavelength, and R the magnitude of the (unknown) reflection coefficient at the open end of the line. By measuring

[3] Electronic Standards Corporation of America, Plainfield, N. J.

Δf for a series of linelengths, we separate α and R (both assumed to be invariant with l). The matching at the outset eliminates the unknown reactive effects of the transition.[4]

As a fair test of the theory we have chosen two substrates with vastly different dielectric constants, two substrate thicknesses, and three w/h ratios which were far apart. Measurements on many substrates were made for each w/h value. The substrates used were made from ceramic rutile ($k=105$) which was fabricated at our Research Division facilities, and a 99.5 percent alumina ($k=9.35$).[5] The dielectric slabs were cut to $\frac{3}{4}$ inch$\times 3\frac{1}{2}$ inch size and then diamond ground to the desired thickness, either 50 mils± 0.5 mils or 20 mils± 0.25 mils. The wafers were finish-ground to an rms surface roughness of 40 microinches with a series of silicon carbide powder slurries. The finished surfaces were metallized with copper in a sequence of plating steps to a layer approximately 0.4 mil thick. The copper was covered with a thin coating of gold to retard oxidation. The stripline patterns were then photoetched.

In our calculations we used $R_s=8.26\times 10^{-3}\sqrt{f}$ ohms as the surface resistivity of copper, where f is given in GHz; onto this value we added a correction term to take into account the surface roughness.[13] (This correction increased R_s by 13 percent at 1 GHz and 33 percent at 6 GHz.)

The dielectric loss tangent values used for rutile were obtained from measurements made at our laboratory on bulk material. This loss tangent is plotted as a function of frequency in Fig. 8. For the alumina we used a value of $\tan\delta=2.1\times 10^{-4}$, which was measured by us at 10 GHz. (This value is higher than the manufacturer's quoted value, 9×10^{-5}.) Any frequency dependence was neglected. Because of the low loss tangents of either substrate material, the calculated dielectric attenuation was less than 20 percent of the total attenuation. Therefore, the experimental data should provide a good test of our theory.

In Fig. 9(a)–(d), we compare our measured attenuation results for rutile with the calculated values. Because of the scatter in data points,[6] where possible, a curve was drawn through the points by "eye-ball" averaging. The agreement with theory is quite satisfactory. The sharp upturn in some of the data points between 5 and 6 GHz for the 50-mil thick substrates suggests that the TE_1 surface mode was excited and produced a loss through propagation of energy out of the edges of the substrate. This mode has a cutoff frequency[14] $f_{c0}=c/(4h)\sqrt{k-1}\approx 5.8$ GHz, where $c=3\times 10^{10}$ cm/s. Indeed, in and about this frequency range, radiation from the sides of the slab was observed.

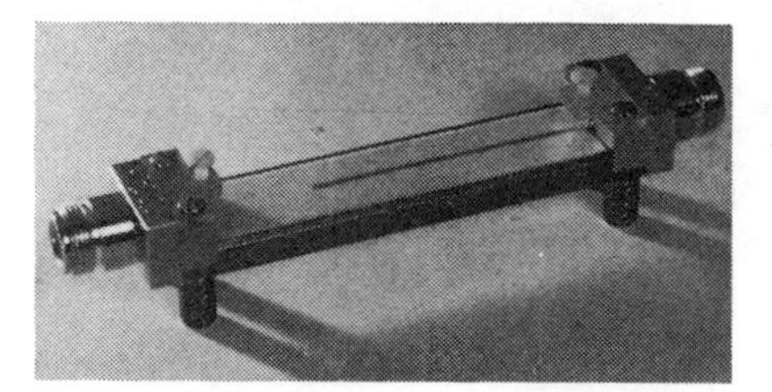

Fig. 7. Microstrip substrate in holder.

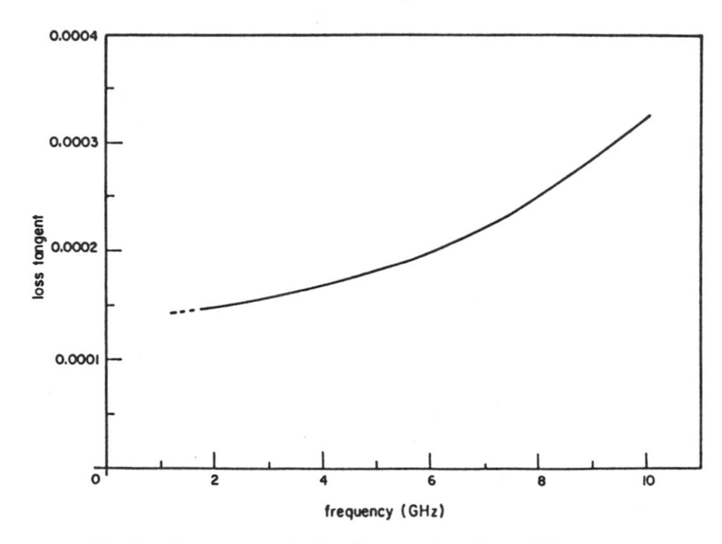

Fig. 8. Loss tangent of rutile as a function of frequency.

Fig. 10(a)–(d) illustrate our data for the alumina substrates. Good agreement with theory is evident in all but one case, namely that of Fig. 10(c) which corresponds to the thicker substrate and to the higher w/h ratio. It is possible that surface modes or higher-order microstrip modes[15] were responsible for the discrepancy. We must rule out the TE_1 surface mode, however, since its cutoff frequency is too high. On the other hand, the TM_0 surface mode has a zero cutoff frequency and a transverse electric field distribution similar to that for the dominant microstrip mode; therefore, it might have been launched rather easily with a wide strip.

One other possible explanation for the discrepancy was that the loss tangent between 1 and 6 GHz might have been considerably higher than the value measured at 10 GHz. One would expect a higher loss tangent to affect most that case which had the lowest copper losses, namely Fig. 10(c). While we could obtain considerably better agreement by choosing a higher loss tangent, we found that the value required was sufficiently high to disrupt the agreement with theory for the other three cases.

In comparing the attenuation per unit length for the two dielectric substrates, it might appear that alumina has a distinct advantage over low-loss rutile. However, this is not so. A meaningful comparison must be based on the loss per wavelength. On this basis, rutile and low-loss alumina are comparable.

[4] Incidentally, from these and other measurements we have concluded that the reflection coefficient at the end of an open line in microstrip for the dielectrics used can be considered equal to $+1$ for all practical purposes.

[5] AL-995 produced by Western Gold and Platinum Company, Belmont, Calif.

[6] The scatter was due to, among other things, ripple and asymmetries in some of the reflection coefficient versus frequency data, which made it difficult to determine Δf accurately. This ripple was caused for the most part by multiple echoes set up in our measurement system because of unavoidable reflections at the various transitions and connectors.

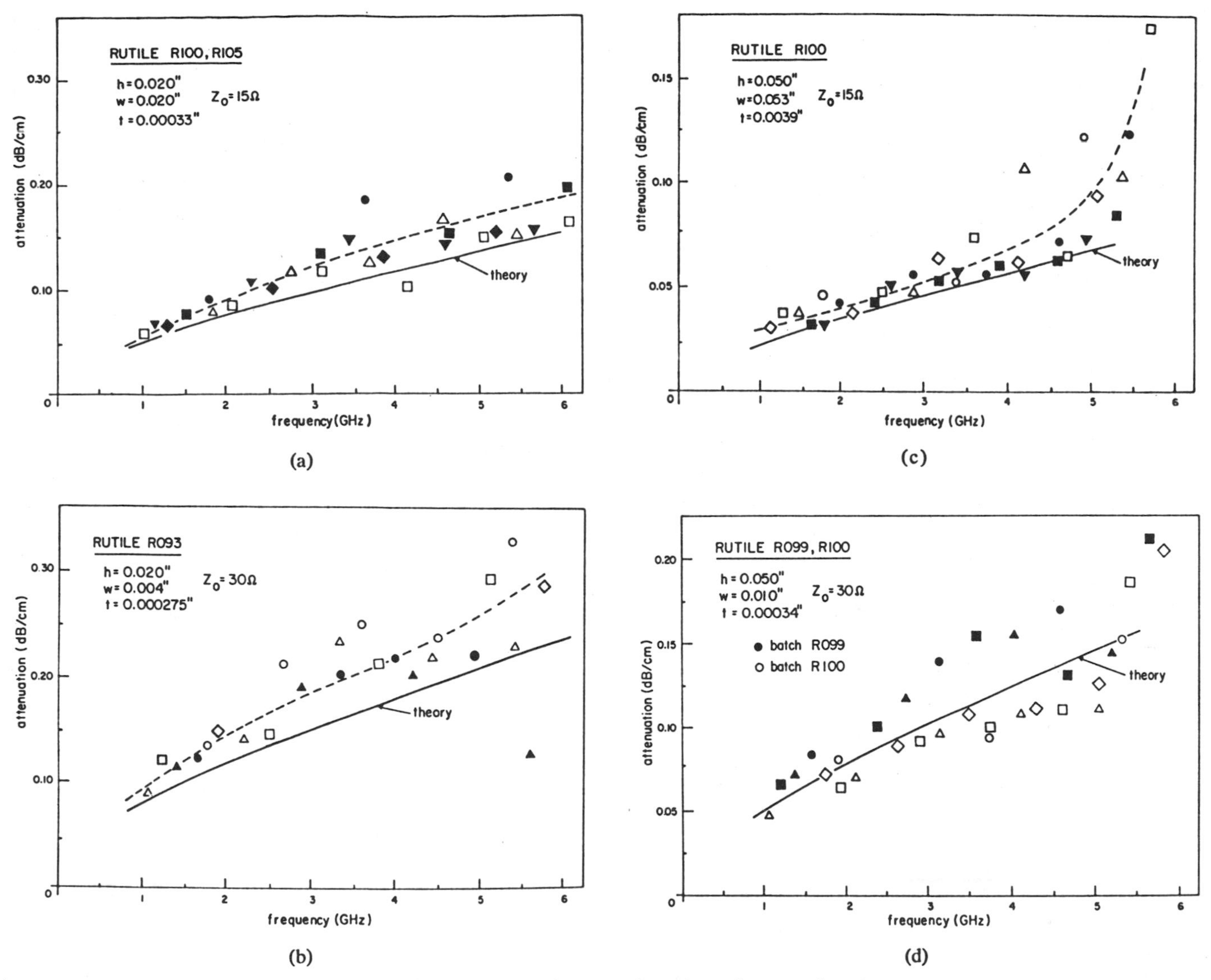

Fig. 9. Experimental attenuation data for microstrip on rutile substrates.

III. Conclusion and Summary

A theory of conductor attenuation in microstrip is derived for the practical case of a mixed dielectric system and a conductor strip of finite thickness. This theory, which is based on Wheeler's incremental inductance rule, predicts attenuations considerably lower than other theories now being used in industry.

Experimental data taken on rutile and alumina substrates for a wide range of geometrical parameter values shows good agreement with this theory. Possible explanation for the poor agreement observed in one case is given.

A technique for measuring low attenuation losses is described.

Appendix

Microstrip Attenuation Measurements

The attenuation constant of the microstrip has been measured by a resonance method commonly used to characterize waveguide.[16] It is particularly useful for lines of characteristic impedance $\neq 50$ ohms for which nonresonant transmission measurements are inhibited by mismatch loss. Furthermore, corrections can be made for any end effects.

A series of different lengths of a given microstrip line are inserted in a test jig as shown in Fig. 7. One end of each line terminates at the substrate edge and the other end terminates abruptly on the substrate. This latter end is found to behave as a low-loss open circuit. The center conductor of an ESCA coax-to-stripline transition is held against the substrate by a nylon screw, but not in contact with the stripline. Thus, the coupling is capacitive and can be varied by moving the substrate laterally on the base of the test jig. The loaded Q, Q_L, of the resultant resonator is then measured by a swept frequency technique.[17] A data reduction procedure has been derived for waveguide[16] and the analysis will not be repeated here. However, specific details inherent to microstrip evaluation are spelled out below.

In the measurement procedure, it is preferable to adjust for critical coupling. This is done to suppress spurious resonances produced by the mismatches of the various connectors in the measurement line. Another advantage is that critical coupling is easily monitored visually on an oscilloscope. For this coupling the unloaded Q, Q_0, is simply

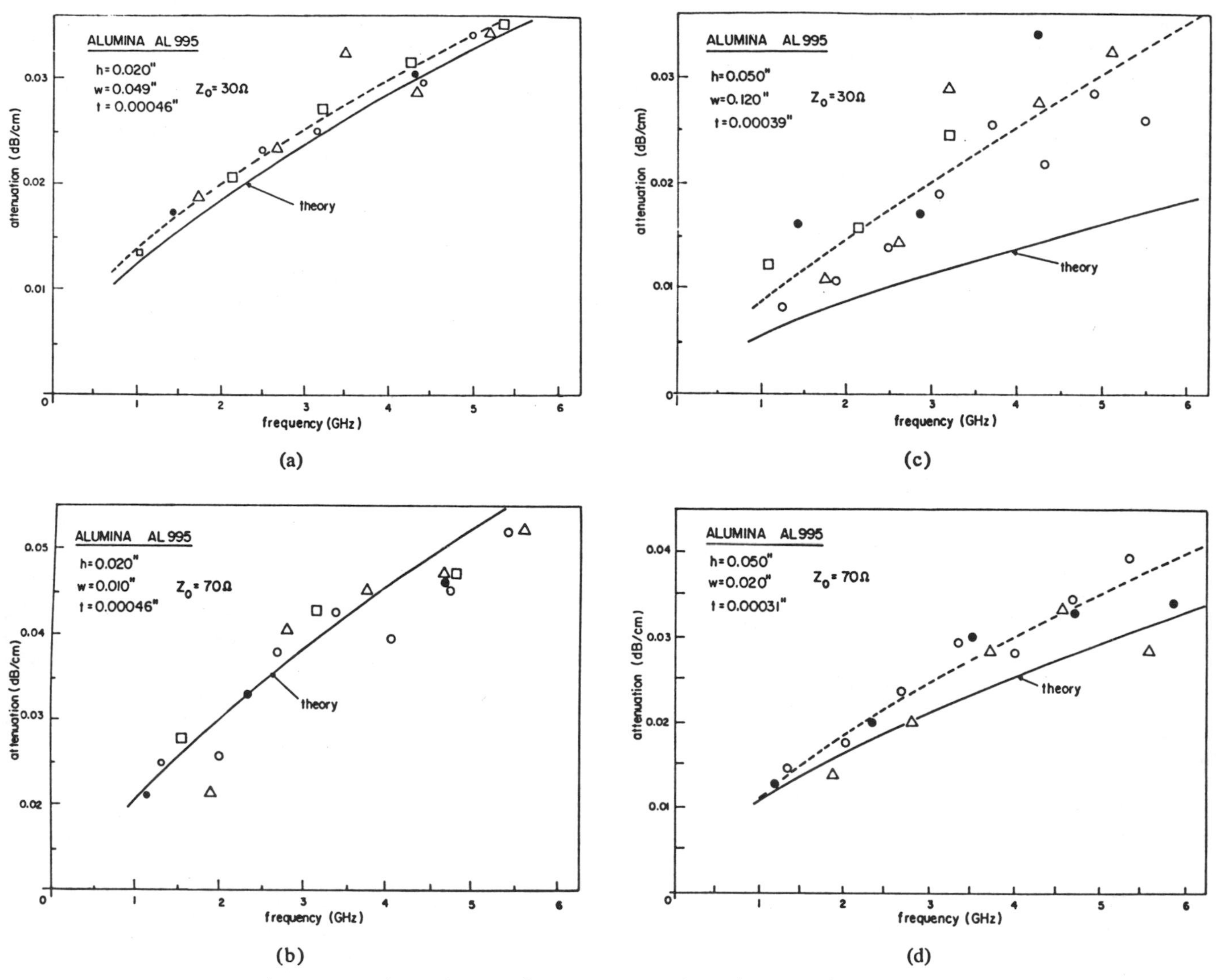

Fig. 10. Experimental attenuation data for the microstrip on alumina substrates.

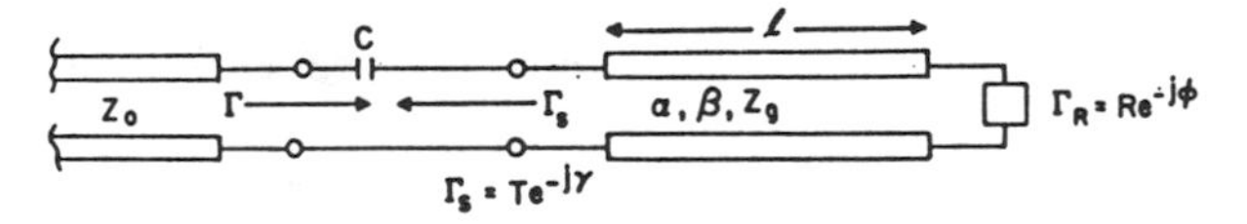

Fig. 11. Equivalent circuit of microstrip substrate in holder.

twice the loaded Q. Limitations of the method are presented in the following paragraphs.

An equivalent circuit of the structure of Fig. 7 is shown in Fig. 11. The microstrip line of physical length, l, is described with a characteristic impedance, Z_g, a phase constant, $\beta = 2\pi/\lambda_g$, and an attenuation constant α. For generality the open end is assumed to be an imperfect open circuit and is characterized by a reflection coefficient $(R)(e^{-j\phi})$ whose magnitude, R is real, positive and ≤ 1.

Critical coupling is achieved by adjustment of the coupling capacitance, C, and the test frequency, f, so as to satisfy (18) and (19) at some frequency, f_0.

$$Z_0 = \tfrac{1}{2} Z_g \sinh 2x[\sinh^2 x + \sin^2 y]^{-1} \tag{18}$$

$$(2\pi f_0 C)^{-1} = -Z_0 \sin 2y[\sinh 2x]^{-1} \tag{19}$$

where $2x = 2\alpha l - \ln R$ and $2y = 2\beta l + \phi$.

Equations (18) and (19) can be satisfied provided that the ratio $r = Z_g/Z_0$, where Z_g is the microstrip characteristic impedance and Z_0 is the source impedance, lies in the range

$$\coth x > r > \tanh x \tag{20}$$

determined by the loss in the resonator formed by the section of microstrip line under test.

The total electrical length, θ, includes the phase shift of the imperfect "open circuit," ϕ, and the phase, γ of Γ_s (see Fig. 11); i.e., $2\theta = 2\beta l + \gamma + \phi$. At resonance, $2\theta = 2m\pi$, where m is an integer. For a high Q (low loss) resonance, the frequency dependence of α and ϕ can be assumed negligible;

this is true of γ also provided

$$\tanh 2x \ll \frac{r}{r^2 + r + 1}. \tag{21}$$

For a small excursion in frequency, $\delta = f - f_0$, the phase can be described by $2\theta = 2m\pi + (4\pi l/v_g)\delta$, where $v_g(=d\omega/d\beta)$ denotes the group velocity. Thus the attenuation factor and the unloaded Q are interrelated by

$$a = (R^{-1})(e^{-2\alpha l}) = \sqrt{1 + \sin^2\left(\frac{l}{\lambda_g}\frac{v_p}{v_g}\frac{2\pi}{Q_0}\right)} + \sin\left(\frac{l}{\lambda_g}\frac{v_p}{v_g}\frac{2\pi}{Q_0}\right) \tag{22}$$

where $v_p(=\omega/\beta)$ denotes the phase velocity.

The attenuation factor in decibels per unit length is given by the expression

$$\alpha = \frac{4.34}{l}\left[\ln(a) + \ln R\right] \text{ dB/unit length.} \tag{23}$$

If

$$\frac{l}{\lambda_g}\frac{v_p}{v_g}\frac{2\pi}{Q_0} \ll 1,$$

as we find experimentally, and if dispersion[2] on the line is negligible ($v_p \cong v_g$), then (23) simplifies to

$$\alpha \cong 4.34\left\{\frac{\beta}{Q_0} + \frac{1}{l}\ln R\right\} \text{ dB/unit length.} \tag{24}$$

By measuring a number of lines of various lengths or a very long line within the limits of (20) one can eliminate R. A separate measurement of β[2] gives all the remaining information required to evaluate α. It must be stressed, however, that if $v_p \neq v_g$, (24) is not correct. This fact is of particular significance in the measurement of microstrip parameters on a ferrite substrate where the permeability displays a strong frequency dependence.

We have also taken measurements using a loosely coupled transmission resonator. The two methods agree within three percent. The reflection resonator method is preferred for several reasons. First, the transmission measurement requires either very loose coupling so that $Q_0 \approx Q_L$ or a tedious measurement of the coupling coefficient of each port if this is not the case. Second, because of the weak coupling the transmitted signal is attenuated severely and this makes measurements with a sweep generator very difficult because of the low power levels involved. Third, also because of weak coupling, the stripline exhibits high reflection coefficients at both ports, which combined with the mismatches produced by the various transitions, detector, etc., in the measurement circuit produce troublesome spurious resonances. Finally, the transmission measurement does not allow readily for corrections for losses in the transition.

Acknowledgment

The authors wish to express their appreciation to Dr. M. P. Lepie who prepared the dielectrics and the microstrip lines, and to A. Fine who obtained some of the microwave data.

References

[1] H. A. Wheeler, "Transmission-line properties of parallel strips separated by a dielectric sheet," *IEEE Trans. Microwave Theory and Techniques*, vol. MTT-13, pp. 172–185, March 1965.

[2] C. P. Hartwig, D. Massé, and R. A. Pucel, "Frequency dependent behavior of microstrip," presented at the G-MTT Internat'l Microwave Symp., Detroit, Mich., May 20–22, 1968.

[3] J. D. Welch and H. J. Pratt, "Losses in microstrip transmission systems for integrated microwave circuits," *NEREM Rec.*, vol. 8, pp. 100–101, 1966.

[4] F. Assadourian and E. Rimai, "Simplified theory of microstrip transmission systems," *Proc. IRE*, vol. 40, pp. 1651–1657, December 1952.

[5] J. M. C. Dukes, "An investigation into some fundamental properties of strip transmission lines with the aid of an electrolytic tank," *Proc. IEE* (*London*), vol. 103, pt. B, pp. 319–333, 1956.

[6] T. M. Hyltin, "Microstrip transmission on semiconductor dielectrics," *IEEE Trans. Microwave Theory and Techniques*, vol. MTT-13, pp. 777–781, November 1965.

[7] S. Ramo and J. R. Whinnery, *Fields and Waves in Modern Radio.* New York: Wiley, 1944, ch. 8.

[8] E. Gaál, "Current distribution on a strip line," *Acta Tech.* (Budapest), vol. 38, pp. 387–397, 1962.

[9] J. D. Cockcroft, "Skin effect in rectangular conductors at high frequencies," *Proc. Roy. Soc.* (London), vol. 122, pp. 533–542, 1929.

[10] M. Caulton, J. J. Hughes, and H. Sobol, "Measurements on the properties of microstrip transmission lines for microwave integrated circuits," *RCA Rev.*, vol. 27, pp. 377–391, 1966.

[11] G. D. Vendelin, "High-dielectric substrate for microwave hybrid integrated circuitry," *IEEE Trans. Microwave Theory and Techniques* (*Correspondence*), vol. MTT-15, pp. 750–752, December 1967.

[12] H. A. Wheeler, "Formulas for the skin effect," *Proc. IRE*, vol. 30, pp. 412–424, September 1942.

[13] *The Microwave Engineer's Handbook and Buyer's Guide.* Brookline, Mass.: Horizon House—Microwave, Inc., 1963, p. 62.

[14] R. E. Collin, *Field Theory of Guided Waves.* McGraw-Hill, New York: 1960, pp. 470–474.

[15] C. G. Shafer, "Higher mode of the microstrip transmission line," Cruft Lab., Harvard University, Cambridge, Mass., Tech. Rept. 258, November 25, 1957.

[16] M. Sucher and J. Fox, *Handbook of Microwave Measurements*, vol. I, 3rd ed. New York: Polytechnic Press 1963, pp. 359–363.

[17] *Ibid.*, vol. II, ch. 8, pp. 417–495.

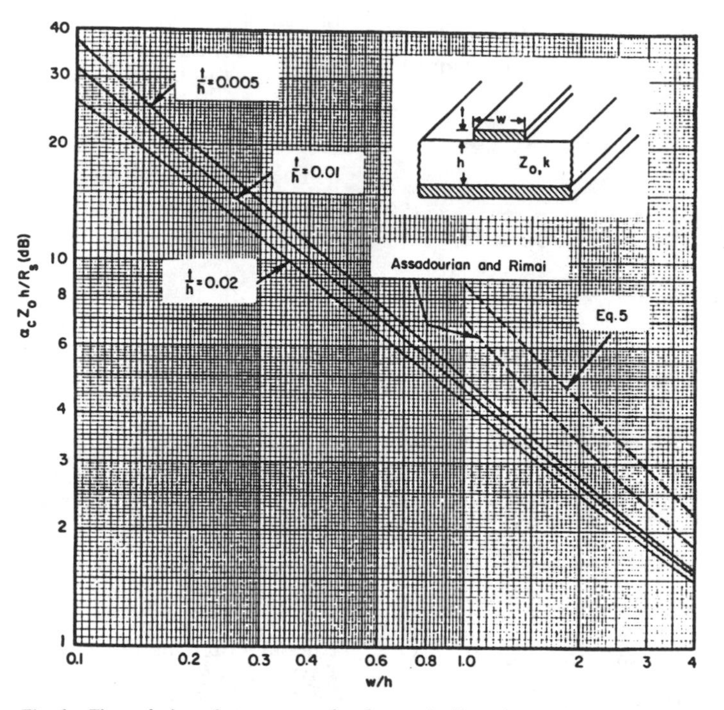

Fig. 6. Theoretical conductor attenuation factor of microstrip as a function of w/h.

Correction to "Losses in Microstrip"[1]

Equations (12a) and (12b) should read

$$\Delta w = \frac{t}{\pi}\left(\ln\frac{4\pi w}{t} + 1\right) \qquad w/h \leq 1/2\pi$$

$$(2t/h < w/h,\ 1/2\pi)$$

$$\Delta w = \frac{t}{\pi}\left(\ln\frac{2h}{t} + 1\right) \qquad w/h \geq 1/2\pi.$$

These corrections simplify (16a), (16b), and (16c) to read

$w/h \leq 1/2\pi$:

$$\frac{\alpha_c Z_0 h}{R_s} = \frac{8.68}{2\pi}\left[1 - \left(\frac{w'}{4h}\right)^2\right] \cdot \left[1 + \frac{h}{w'} + \frac{h}{\pi w'}\left(\ln\frac{4\pi w}{t} + \frac{t}{w}\right)\right], \qquad (16a)$$

$1/2\pi < w/h \leq 2$:

$$\frac{\alpha_c Z_0 h}{R_s} = \frac{8.68}{2\pi}\left[1 - \left(\frac{w'}{4h}\right)^2\right] \cdot \left[1 + \frac{h}{w'} + \frac{h}{\pi w'}\left(\ln\frac{2h}{t} - \frac{t}{h}\right)\right], \qquad (16b)$$

$2 \leq w/h$:

$$\frac{\alpha_c Z_0 h}{R_s} = \frac{8.68}{\left\{\frac{w'}{h} + \frac{2}{\pi}\ln\left[2\pi e\left(\frac{w'}{2h} + 0.94\right)\right]\right\}^2} \cdot \left[\frac{w'}{h} + \frac{w'/\pi h}{\frac{w'}{2h} + 0.94}\right] \cdot \left[1 + \frac{h}{w'} + \frac{h}{\pi w'}\left(\ln\frac{2h}{t} - \frac{t}{h}\right)\right]. \qquad (16c)$$

A corrected graph (Fig. 6) reflecting the changes in (16) is included. The corrected theoretical loss is slightly higher than what is predicted in our paper by approximately five to ten percent depending on the values of t/h and w/h. The corrected values are closer to our measured results.

We express our appreciation to S. Klug for bringing to our attention the error in (12) introduced by us in transcribing Wheeler's formulas from [1] of our paper.

On page 346 there is a typographical error. In the first column, second line below (16c), $\cdots\ w/h = \frac{1}{2}\pi \cdots$ should read $\cdots\ w/h = 1/2\pi$.

ROBERT A. PUCEL
DANIEL J. MASSÉ
CURTIS P. HARTWIG
Research Division
Raytheon Company
Waltham, Mass.

Manuscript received September 25, 1966.

[1] R. A. Pucel, D. J. Massé, and C. P. Hartwig, *IEEE Trans. Microwave Theory and Techniques*, vol. MTT-16, pp. 342–350, June 1968.

Reprinted from *IEEE Trans. Microwave Theory Tech.*, vol. MTT-16, no. 12, pp. 1064, December 1968.

Losses of Microstrip Lines

EDGAR J. DENLINGER, MEMBER, IEEE

Invited Paper

***Abstract*—This article summarizes state-of-the-art information on losses of single and coupled microstrip lines. Conductor loss, substrate loss (for pure dielectric or magnetic materials), and radiation loss are considered along with the effect of dispersion. Finally, a rough comparison is made between the losses of microstrip and that of several other types of lines used in microwave integrated circuits.**

Nomenclature

- a Height of ground plane above substrate surface in shielded microstrip.
- $\bar{E}_0$ Unperturbed electric field.
- $F(\epsilon_{eff})$ Radiation form factor.
- f Frequency.
- f_m Natural resonant frequency of magnetic substrate $=\gamma 4\pi M_s$.
- f_c Cutoff frequency of TE_1 surface wave.
- g_r Radiation conductance.
- h Substrate thickness.
- H_{IM} Separation between substrate surface and ground plane for inverted microstrip line.
- H_{TIM} Separation between strip and bottom of channel for trapped inverted microstrip.
- $\bar{H}_0$ Unperturbed magnetic field.
- $|H_t|$ Amplitude of magnetic field at conducting surfaces for the lossless case.
- $J_z(X)$ Current density distribution on strip.
- k_0 Free-space wavenumber.
- L Inductance.
- m Mutual resistive factor.
- $4\pi M$ Magnetization.
- $4\pi M_s$ Saturation magnetization.
- p Magnetic power factor.
- P_r Radiated power from microstrip discontinuity.
- P_t Total power incident on microstrip discontinuity.
- P_d Time-averaged power dissipated per-unit-length.
- P_F Time-averaged power flow along line.
- q Filling factor for the dielectric constant.
- Q_0 Circuit Q including conductor and substrate losses.
- Q_r Radiation Q.
- Q_t Total Q of resonator.
- R Series skin resistance of transmission line.
- R_s Surface skin resistivity of conductor (ohms/square).
- R_{sj} Surface skin resistivity of wall j.
- R_a Surface roughness.
- s Separation between edges of coupled pair of microstrip lines.
- s_{cp} Separation between edges of strip and coplanar ground plane of coplanar waveguide.
- s_{TIM} Separation between edge of strip and side of channel for trapped inverted microstrip.
- t Strip thickness.
- V_0 Velocity of light in free space.
- w Strip width.
- Z_0 Characteristic impedance.
- α_c Conductor attenuation constant.
- α_d Dielectric attenuation constant.
- α_{ST} Total substrate attenuation constant.
- α_t Total attenuation constant of line.
- α_d^e Dielectric attenuation constant for even mode.
- α_d^o Dielectric attenuation constant for odd mode.
- γ Gyromagnetic ratio $=2.8$ MHz/G.
- δ Skin depth of conductor.
- $\tan\delta_d$ Loss tangent of dielectric.
- $\tan\delta_m$ Magnetic loss tangent.
- ϵ Permittivity $=\epsilon_R\epsilon_0$.
- ϵ_R Relative dielectric constant of substrate.
- ϵ_{eff} Effective dielectric constant of line.
- ϵ_0 Permittivity of free space.
- ϵ_{eff}^e Effective dielectric constant of even mode.
- ϵ_{eff}^o Effective dielectric constant for odd mode.
- κ Off-diagonal term of permeability tensor.
- λ_0 Free-space wavelength.
- λ_g Guide wavelength along line.
- μ_0 Permeability of free space.
- μ_R Relative permeability of substrate.
- μ Diagonal term of permeability tensor.
- μ_{eff} Effective relative permeability of line.
- ρ Spacing between dipoles.
- σ Conductivity.
- ω Angular frequency.
- ω_m/ω Normalized saturation magnetization.

Manuscript received October 30, 1979; revised January 9, 1980.

The author is with the RCA Laboratories, David Sarnoff Research Center, Princeton, NJ 08540.

I. Introduction

MICROSTRIP transmission lines have had widespread use in microwave integrated circuits. The forms of single and coupled lines, shown in Fig. 1, are used for a great variety of functions. For performing accurate circuit design, it is necessary to have adequate knowledge of the phase velocity, impedance, and losses of the line. A great deal has been written on the first two

Reprinted from *IEEE Trans. Microwave Theory Tech.*, vol. MTT-28, no. 6, pp. 513–522, June 1980.

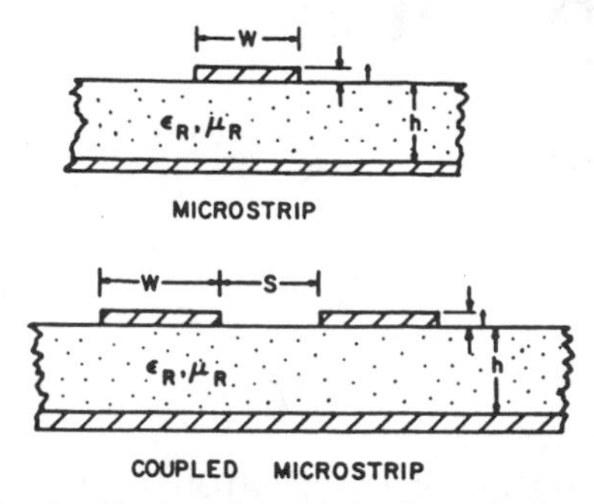

Fig. 1. Single and coupled microstrip lines.

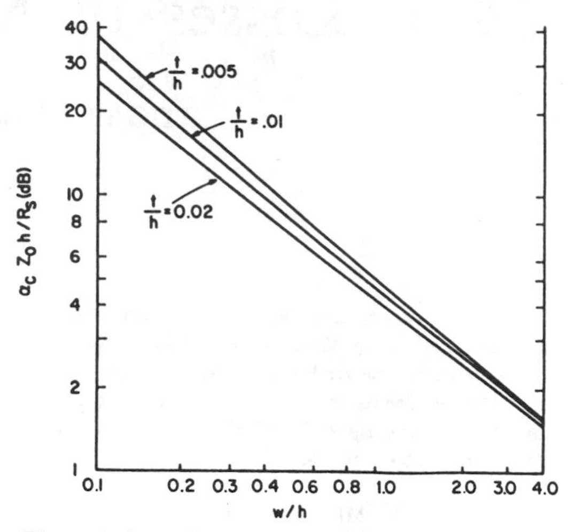

Fig. 2. Theoretical conductor attenuation factor of microstrip as a function of w/h (Pucel *et al.* [4]).

subjects, both of which are usually covered together because they are closely interrelated. However, microstrip losses have usually been treated separately in terms of either conductor and dielectric loss, magnetic loss, or radiation loss. The purpose of this paper is to bring together all that is available on this important subject and thus provide the circuit designer with an improved overall picture of the various losses that should be considered. For completeness, there is also a need to consider the dispersive behavior of microstrip in order to avoid coupling of energy to surface waves and higher order modes, and to accurately calculate losses (especially when using magnetic substrates as well as thick, high dielectric constant substrates which are highly dispersive). Finally, a rough comparison of losses in microstrip to those in other commonly used transmission lines such as coplanar line, slotline, inverted and trapped inverted microstrip, suspended substrate, and stripline-like microstrip will be made.

II. Conductor and Substrate Losses of Single Microstrip

The loss components of a single microstrip line (neglecting radiation) include dielectric loss, conductor loss, and for the case of a magnetic substrate, magnetic loss. For dielectric losses in a nonmagnetized mixed dielectric system, Welch and Pratt [1] followed the approach of Wheeler [2] in determining the filling factor q for the dielectric constant. They derived a corresponding filling factor for the loss tangent $\tan\delta_d$ and obtained the following results for the case when the upper dielectric is air (assumed lossless):

$$\alpha_d = 27.3 \frac{q\epsilon_R}{\epsilon_{\text{eff}}} \frac{\tan\delta_d}{\lambda_g} \text{dB/unit length} \tag{1}$$

where ϵ_R is the substrate dielectric constant and

$$\epsilon_{\text{eff}} = 1 + q(\epsilon_R - 1)$$

= microstrip's effective dielectric constant.

The filling factor for the loss tangent $q\epsilon_R/\sqrt{\epsilon_{\text{eff}}}$ is a function of ϵ_R and w/h and is an approximation based on the assumption that the air–dielectric interface is parallel to an electric field line. The correct equation for the filling factor derived by Schneider, Glance, and Bodtmann [3] is equal to the ratio r of stored electric field energy to the total field energy in microstrip

$$r = \frac{\epsilon_R}{\epsilon_{\text{eff}}} \frac{\partial \epsilon_{\text{eff}}}{\partial \epsilon_R}. \tag{2}$$

However, (1) is reasonably accurate since the dielectric–air interface is parallel to the electric field at both corners of the strip conductor where field intensities reach their maximum value.

Conductor loss was first analyzed by Pucel *et al.* [4] and more recently by Wheeler [5]. Both papers derived expressions for the series skin resistance in terms of the incremental inductance associated with the penetration of magnetic flux into the conducting surfaces. From Wheeler's incremental inductance rule [6], the series skin resistance is

$$R = \frac{1}{\mu_0} \sum_j R_{sj} \frac{\partial L}{\partial n_j} \tag{3}$$

where $\partial L/\partial n_j$ is the derivative of the inductance L with respect to the incremental recession of wall j; R_{sj} is the surface skin resistivity of wall j;

$$L = \mu_0 \epsilon_0 Z_0 \left(\frac{w + \Delta w(w,t)}{h}, \frac{t}{h}, \epsilon_R \right)$$

as derived by Wheeler [2]. The conductor attenuation constant is given by

$$\alpha_c = \frac{R}{2Z_0}. \tag{4}$$

Fig. 2 shows a plot from Pucel *et al.* of the conductor attenuation factor $\alpha_c Z_0 h/R_s$ as a function of w/h and t/h. Wheeler's recent theory [5] uses a more accurate correction for strip thickness than used in Pucel's calculations and also treated the case of square and circular conductor cross sections. Fig. 3 shows a plot of the magnetic power factor p normalized to the value δ/h (the value for a very wide strip) as a function of w/h and t/h. The two lower curves are for square and circular cross-sections. δ is the skin depth of the conductor. The conductor attenuation constant is obtained from the power factor

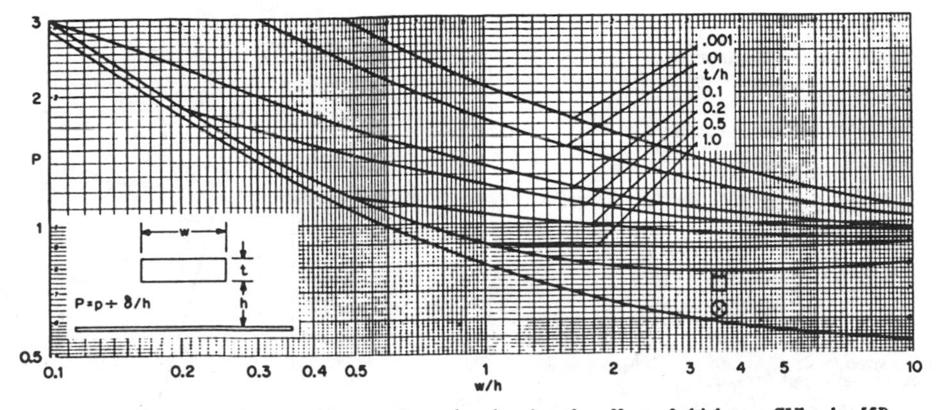

Fig. 3. The magnetic power factor of a strip, showing the effect of thickness (Wheeler [5]).

by

$$\alpha_c = \frac{27.3p}{\lambda_g} \text{ dB/unit length.} \quad (5)$$

For the case of magnetic substrates such as ferrites or garnets, the substrate's magnetic loss must also be included as well as the effect of the magnetic material on the conductor loss, since the characteristic impedance and thus the loss is a function of both ϵ_{eff} and the effective permeability μ_{eff}. Pucel and Massé [7] derived expressions for losses on magnetic substrates by first determining the functional dependence of μ_{eff} on w/h and the substrate's relative permeability μ_R. This was done by using the following duality relationship between dielectric and magnetic substrates and the known functional dependence of ϵ_{eff} on w/h and ϵ_R:

$$\mu_{eff}(w/h, \mu_R) = \frac{1}{\epsilon_{eff}(w/h, 1/\mu_R)}. \quad (6)$$

In short, the duality allows the conversions of $\epsilon_R \to 1/\mu_R$ and $\epsilon_{eff} \to 1/\mu_{eff}$ in the formulas for the pure dielectric case, which were already derived by Wheeler [2]. From Welch and Pratt's formula for the dielectric loss [1], an effective dielectric loss tangent can be defined by

$$\tan \delta_{d\,eff} = \left[\frac{1-\epsilon_{eff}^{-1}}{1-\epsilon_R^{-1}} \right] \tan \delta_d. \quad (7)$$

By the duality rule, an effective magnetic loss tangent can be written as

$$\tan \delta_{m\,eff} = \left[\frac{1-\mu_{eff}}{1-\mu_R} \right] \tan \delta_m. \quad (8)$$

Finally, the total magnetic substrate loss is the sum of dielectric and magnetic losses and is given by

$$\alpha_{ST} = \frac{27.3}{\lambda_g} (\tan \delta_{d\,eff} + \tan \delta_{m\,eff}) \text{ dB/unit length.} \quad (9)$$

What remains undetermined are the substrate's magnetic loss tangent and relative permeability. The former has been determined from experimental data for a number of commonly used magnetic substrate materials by Green *et al.* [8]. This parameter is a strong function of the material's normalized saturation magnetization $(\omega m/\omega)$. The relative permeability can be determined from a theoretical expression derived by Schloemann [9] for a demagnetized substrate and from an empirical expression based on experimental data by Green and Sandy [10] for a partially magnetized substrate. For the demagnetized case, Schloemann used a cylindrical model to obtain the following equation for μ_R:

$$\mu_{R\,dem} = \frac{1}{3} \left\{ 1 + 2\sqrt{1 - \left(\frac{\omega_m}{\omega} \right)^2} \right\}. \quad (10)$$

For the partially magnetized case (usually when the substrate is latched in its remanent state), the empirical formulas derived by Green and Sandy for the diagonal and off-diagonal terms of the substrate's permeability tensor are given by

$$\mu = \mu_{R\,dem} + (1 - \mu_{R\,dem}) \left[\frac{4\pi M}{4\pi M_s} \right]^{3/2} \quad (11)$$

$$\kappa = \gamma \frac{4\pi M}{\omega} \quad (12)$$

where $4\pi M$ is the material's magnetization, and $4\pi M_s$ is the material's saturation magnetization.

The analytical approximation for $\mu_{R\,mag}$ of the magnetized substrate in terms of w/h, μ, and κ was derived by Sandy [11] from a formula for the inductance per unit length. It is given by

$$\mu_R = \frac{\mu^2 - \kappa^2}{\mu} \frac{1}{1 - \frac{1}{7}\sqrt{h/w}\,(\kappa/\mu)^2 \ln[1 + \mu/(\mu^2 - \kappa^2)]}. \quad (13)$$

Now we have everything needed for calculating μ_{eff} from which we can determine $Z_0(\mu_{eff}, \epsilon_{eff})$ and, finally, the conductor and substrate losses. An example of the computed and measured loss data for microstrip on a garnet substrate is shown in Fig. 4. The large increase in loss near the material's natural resonant frequency f_m is due to the rapid increase in the magnetic loss tangent and the rapid decrease of Z_0 near this frequency. The latter phenomena causes the conductor loss to increase since $\alpha_c \propto 1/Z_0$.

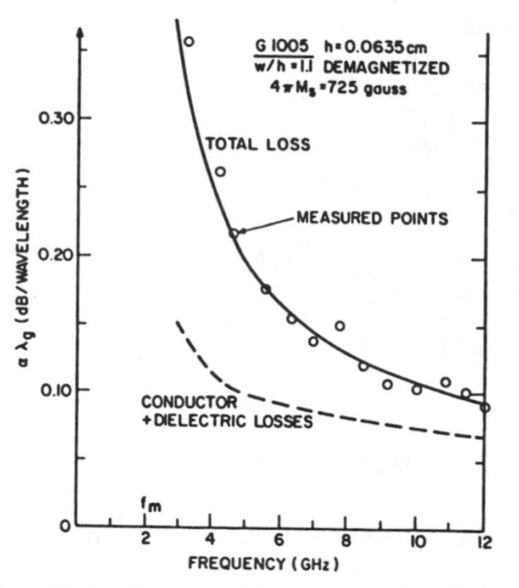

Fig. 4. Theoretical and measured loss per wavelength for microstrip on a demagnetized garnet substrate (Massé and Pucel [7]).

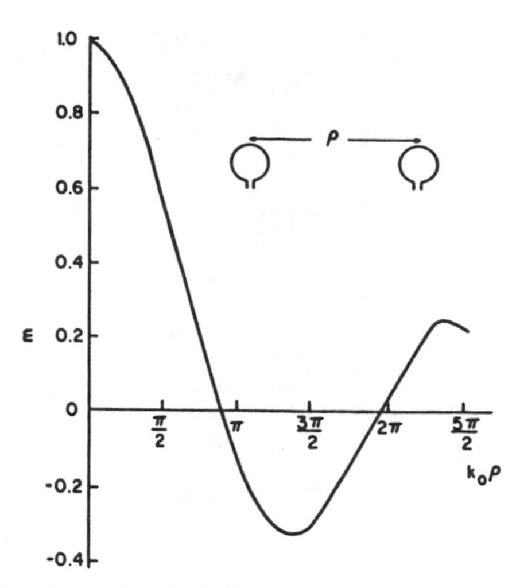

Fig. 5. Mutual coupling (resistive component) of two equatorially displaced magnetic dipoles, normalized to zero displacement (Lewin [13]).

III. Radiation Losses from Microstrip Discontinuities

Radiation loss from a number of different microstrip discontinuities such as open circuits, short circuits, and corners was first theoretically treated by Lewin [12] in 1960 with corrections and additions to this work recently published [13]. In his theory, the strip and polarization currents flowing from strip-to-ground plane are integrated with suitable phase and distance factors to evaluate the Hertz potential from which the fields may be found and the radiated power obtained. The radiated power from the discontinuities is expressed by $P_r = 60(k_0 h)^2 F(\epsilon_{eff})$, where k_0 is free-space wavenumber, and h is substrate thickness, and the form factor $F(\epsilon_{eff})$ is different for each of the discontinuities and can be found in references [12] and [13].

Table I shows the values of F for $\epsilon_{eff}=2.25$ and for large ϵ_{eff} [13]. Note that the two largest contributors to radiation among the various circuit discontinuities are the open circuit and the right-angle bend.

For many circuit applications there may be two discontinuities, such as open circuits spaced considerably less than a free-space wavelength apart. As shown in [13] and [14], the resultant radiation pattern is similar to that of a magnetic dipole. Fig. 5 shows a plot of the mutual resistive factor m as a function of the spacing ρ between two such dipoles. The form factor F for each of the open circuit discontinuities must then be multiplied by the term $(1+m)$ to obtain the radiated power. For an $n\lambda_g/2$ open-ended microstrip resonator, the value $k_0\rho$ is equal to $n\pi/\epsilon_{eff}$. Thus, for high dielectric constant substrates and low mode number, the mutual resistive factor can be close to one, which causes the radiated power to be nearly doubled when $n=1$.

A different theory on radiation by Van der Pauw [15] uses the fact that the power radiated by the microstrip discontinuity should be equal to the power necessary to maintain the current density on the strip at a stationary value. His results for an open-ended microstrip resonator agree to within 10 percent of Lewin's theoretical results. Other types of commonly used resonators analyzed by him were the circular resonator and the "hairpin" resonator. The former gives a radiation Q nearly equal to that of a stretched open-ended resonator, while the hairpin resonator exhibits particularly low radiation losses. The overall gain in quality factor Q_T may be small due to ohmic losses.

The influence of radiation losses based on Lewin's derivation on the overall Q of microstrip open-ended resonators for a variety of frequencies, characteristic impedances, substrate materials, and thicknesses, was treated by Belohoubek and Denlinger [16]. As shown in Fig. 6, radiation becomes a dominant factor for low impedance

TABLE I
Radiation Form Factors for Various Microstrip Discontinuities (Lewin [13])

VALUES OF F:

DISCONTINUITY	LARGE ϵ_{eff}	$\epsilon_{eff} = 2.25$
Open Circuit	$8/(3\,\epsilon_{eff})$	1.073
Short Circuit	$16/(15\,\epsilon_{eff}^2)$	0.246
Match	$2/(3\,\epsilon_{eff})$	0.330
90° Corner	$4/(3\,\epsilon_{eff})$	0.610
Impedance Change	$\frac{8}{3\epsilon_{eff}}\left(\frac{Z_2-Z_1}{Z_2+Z_1}\right)^2$	0.268 (3 to 1 change)
Side-Arm Divider	$\frac{4}{3\epsilon_{eff}}\left(\frac{Z_3}{Z_2+Z_3}\right)^2$	0.152 (3 dB case)
T-Junction	$2/(3\,\epsilon_{eff})$	0.349
Series Impedance	$\frac{8}{3\epsilon_{eff}}\left\vert\frac{Z}{Z+2Z_s}\right\vert^2$	0.119 $(Z=Z_s)$

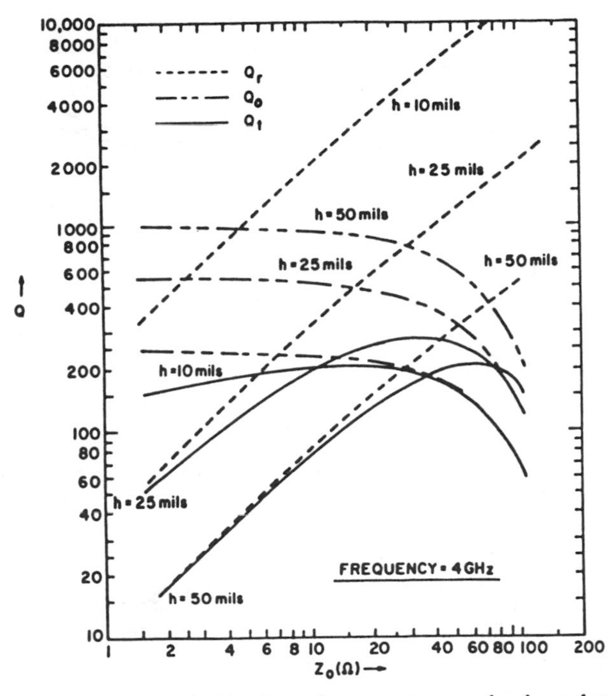

Fig. 6. Q factors for $\lambda_g/4$ microstrip resonator on alumina substrate for $f=4$ GHz (Belohoubek and Denlinger [16]).

lines and thick substrates. The graph in this figure is for a $\lambda_g/4$ resonator defined on alumina and shows the circuit quality factor Q_0, which includes the combined conductor and substrate losses, the radiation Q_r, and the overall Q_t of the resonator, which is given by

$$\frac{1}{Q_t}=\frac{1}{Q_0}+\frac{1}{Q_r}. \tag{14}$$

Similar graphs for alumina and teflon fiberglass substrates calculated for various frequencies (up to 8 GHz) illustrate that radiation becomes even more dominant for lower dielectric constant substrates and at higher frequencies. Although the energy may not truly be lost in fully enclosed circuits, the high radiation level will cause cross coupling between circuit elements. The radiation losses based on Lewin's derivation were compared with experimental results for $n\lambda_g/2$ open-ended resonators [16], [17]. The theoretical expressions for radiation Q_r and the fractional radiated power are given by [16]

$$Q_r=\frac{nZ_0}{480\pi(h/\lambda_0)^2(1+m)F(\epsilon_{\rm eff})} \tag{15}$$

$$\frac{P_r}{P_t}=\frac{Q_0}{Q_0+Q_r} \tag{16}$$

where the factor $(1+m)$ has been added to account for the mutual resistive effect between the two ends of the resonator [13]. In terms of measured results, P_r/P_t is given by

$$\frac{P_r}{P_t}=\frac{Q_0'-Q_t'}{Q_0'} \tag{17}$$

where Q_0' is the measured Q of the resonator when located in a waveguide below cutoff, and Q_t' is the value obtained

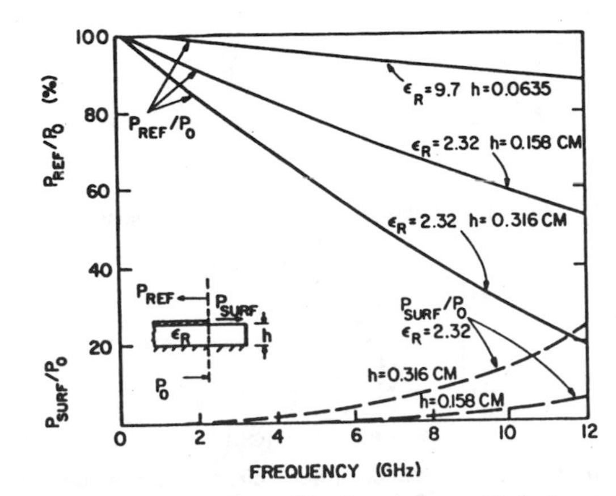

Fig. 7. Percentage of power reflected and transmitted at open end of wide microstrip lines against frequency for $w/h>3$ (Kompa [18]).

TABLE II
FRACTIONAL RADIATED POWER VERSUS NORMALIZED SUBSTRATE THICKNESS ($Z_0=50\Omega, \epsilon_{\rm eff}=2.25$)

					P_r/P_t (%)	
$1+m$	n	h/λ_0	Q_0	Q_t	Exper.	Theor.
0.66	2	.00537	340	300	12	9.5
0.66	2	.00609	345	298	13.6	12.0
1.038	3	.00807	360	291	19.0	20.7
1.038	3	.00931	370	286	22.7	26.4
0.66	2	.01069	540	366	32.0	39.6
1.038	3	.01603	605	194	68.0	63.5

TABLE III
FRACTIONAL RADIATED POWER VERSUS SUBSTRATE DIELECTRIC CONSTANT ($Z_0=50\,\Omega, h/\lambda_0\simeq 0.009$)

								P_r/P_t (%)	
$k_0\rho$	$1+m$	n	ϵ_R	ϵ_{eff}	$F(\epsilon_{eff})$	Q_0	Q_t	Exper.	Theor.
6.283	1.038	3	2.47	2.25	1.07	370	286	22.7	26.4
3.030	0.8868	2	6.0	4.3	0.5903	213	183	14.0	12.0
3.848	0.687	3	9.0	6.0	0.4293	273	263	4.0	5.7
1.987	1.362	2	16.0	10.0	0.2613	152	148	3.0	2.6

without shielding. A comparison between experimental and theoretical values of fractional radiated power is given in Tables II and III. Table II is for resonators with a fixed dielectric constant and varying h/λ_0 while Table III is for resonators having approximately the same h/λ_0 but varying ϵ_r.

Some other different theories on radiation from open-ended microstrip were reported by Kompa [18], Sobol [19], and Wood [20]. Kompa [18] has used the analysis due to Angulo and Chang [21] for open-ended dielectrically loaded parallel-plate waveguide to approximate the reflection coefficient of the end of a wide microstrip line. This theory is not applicable for narrow stripwidths but does allow one to separate the amount of power concentrated in the radiating field from that existing in the TM_0 surface mode. As shown in Fig. 7, both excitations increase with the substrate thickness and with lower dielectric constant; however, the surface wave power is considerably smaller than the radiation power.

Sobol [19] and Wood [20] have considered radiation

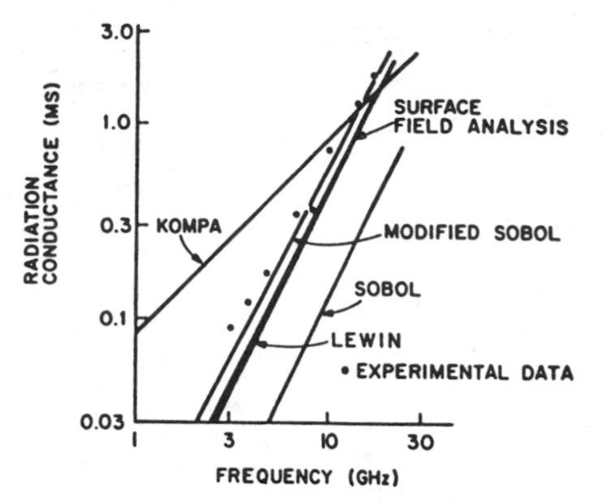

Fig. 8. Radiation conductance of open circuited 2.82-mm-wide microstrip line as function of frequency ($h=1.58$ mm, $\epsilon_R=2.32$; Wood *et al.* [20]).

from an open-circuited line as coming out of the aperture at the end. The radiating source used by Sobol [19] was a uniform electric field distribution confined to the physical strip width. Wood [20] tried to make allowance for the spreading of the source due to fringing fields at the strip edges by proposing to use an equivalent width rather than the physical strip width. This equivalent width is defined as the width of a parallel-plate line of separation h, filled with a medium of dielectric constant ϵ_{eff}, and having an impedance equal to that of the line

$$w_{\text{eq}}=\frac{120\pi h}{Z_0\sqrt{\epsilon_{\text{eff}}}}. \tag{18}$$

The resulting radiation conductance is given by

$$g_r=\frac{\sqrt{\epsilon_{\text{eff}}}}{240\pi^2}F\left(\sqrt{\epsilon_{\text{eff}}}\,\frac{2\pi}{\lambda_0}w_{\text{eq}}\right) \tag{19}$$

where

$$F(X)=X\,\text{Si}(X)-2\sin^2(X/2)-1+\sin(X)/X.$$

Wood also derived a conductance formula based on radiation from the electric-field distribution at the substrate surface

$$g_r=\frac{1}{120\pi^2}F\left(\frac{2\pi}{\lambda_0}w_{\text{eq}}\right) \tag{20}$$

where $F(X)$ is the same as shown above. Shown in Fig. 8 is Wood's comparison of various theoretical curves of radiation conductance versus frequency by Kompa, Lewin, Sobol, modified Sobol (using w_{eq}), and Wood's surface field analysis. Experimental data by Wood indicates best agreement with the modified Sobol analysis. Recently, James and Henderson [43] calculated that surface-wave generation becomes appreciable when $h/\lambda_0>0.09$ for $\epsilon_R\simeq2.3$ and $h/\lambda_0>0.03$ for $\epsilon_R\simeq10$.

IV. Other Microstrip Loss Considerations

The remaining contributors to microstrip losses are surface roughness, the thin films of high resistivity metal used for adhesion and metallurgical stability in a deposited metal system, and the thickness of the metal system. The effect of surface roughness on microwave losses was studied more than thirty years ago by Morgan [22], who calculated eddy current losses for triangular and

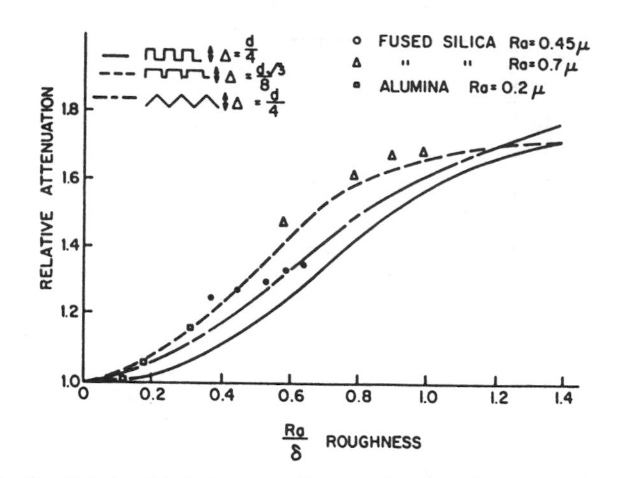

Fig. 9. Relationship between surface roughness and increase of loss in shielded open-ended line resonators on fused quartz (Van Heuven [23]).

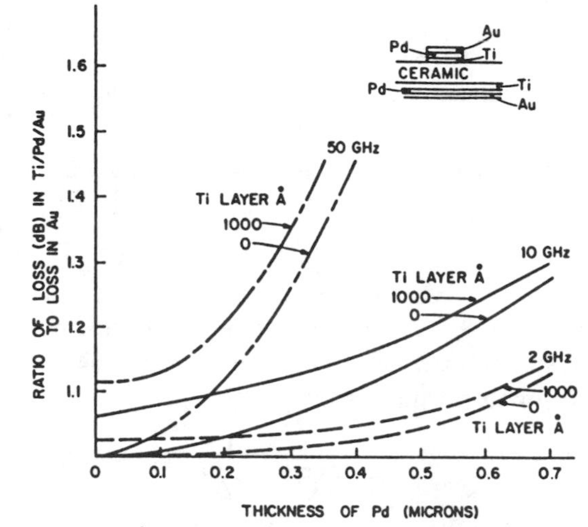

Fig. 10. Loss in Ti–Pd–Au microstrip conductor as a function of Ti and Pd Thickness (Sobol and Caulton [25]).

rectangular grooves perpendicular to the current flow. Data shown in Fig. 9 by Van Heuven [23] for both fuzed silica and alumina substrates show good agreement with Morgan's results. Here the increase in attenuation is plotted against the surface roughness R_a which is normalized to the skin depth δ. These measurements were made with shielded open-ended line resonators.

The losses due to the high resistivity metal films were calculated by Sobol ([24], [25]) for two-layered and three-layered metal systems. A typical adhesion metal, e.g., Cr, in a two-layered system produces only a negligible loss at frequencies well into the millimeter-wave region. However, the losses of the three-layered system should be seriously considered. For example, the Ti–Pd–Au system, which is attractive because of its metallurgical stability and compatibility with beam leads, has losses relative to that of a pure gold system as shown in Fig. 10. Typically, the thickness of the adhesion film Ti is a few hundred angstroms, and that of the buffer layer Pd is 2000–4000 Å. Thus, at X-band frequencies the losses may be 10 percent higher than those of the pure gold systems, and for millimeter waves the increase may be as high as 30–50 percent.

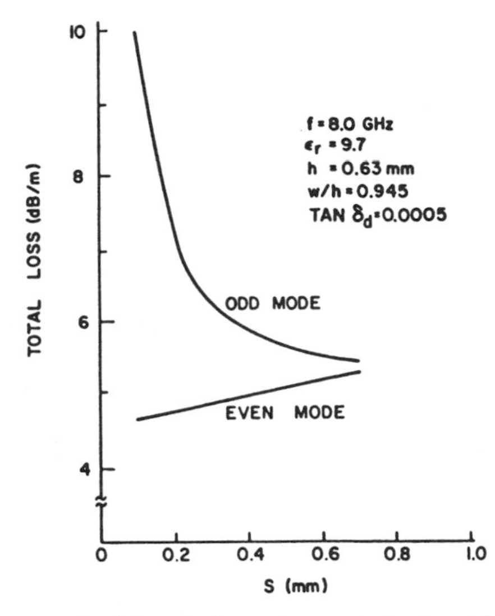

Fig. 11. Even- and odd-mode losses in coupled microstrip lines (Garg and Bahl [27]).

Conductor losses are also significantly affected by the ratio of metal thickness to the skin depth at the frequency of interest. It was shown in references [1] and [44] that the conductor loss reaches a minimum if the strip thickness is about three skin depths.

V. Losses for Coupled Microstrip

Conductor losses in coupled microstrip lines have been analyzed by Rao [26] and more recently by Garg and Bahl [27] using the incremental inductance rule of Wheeler [5]. The two solutions are similar and agree within 1 dB for the even- and odd-mode attenuation constants. Typical curves showing dependence of even- and odd-mode losses on the strip spacing s are presented in Fig. 11. Included in these losses are those due to dielectric loss which are given by [26]

$$\alpha_d^e = 27.3 \frac{\epsilon_R}{\sqrt{\epsilon_{\text{eff}}^e}} \frac{\epsilon_{\text{eff}}^e - 1}{\epsilon_R - 1} \frac{\tan\delta_d}{\lambda_0} \text{dB/unit length} \quad (21)$$

$$\alpha_d^o = 27.3 \frac{\epsilon_R}{\sqrt{\epsilon_{\text{eff}}^o}} \frac{\epsilon_{\text{eff}}^o - 1}{\epsilon_R - 1} \frac{\tan\delta_d}{\lambda_0} \text{dB/unit length.} \quad (22)$$

A second method of solution using the quasi-TEM propagation assumption was employed by Horton [28] and Spielman [29]. It involved finding the charge distributions on the conductors along with the even- and odd-mode effective dielectric constants and impedances; then, via a numerical integration, the even- and odd-mode attenuation constants could be calculated from a general formula given by

$$\alpha = \bar{P}_d/(2\bar{P}_F) \quad (23)$$

where $\bar{P}_d$ is time-averaged power dissipated per-unit-length; and $\bar{P}_F$ is time-averaged power flow along line. By relating the losses to the charge density distributions, the transmission-line geometry may contain any number of lossy conductors and inhomogenous dielectrics.

The final two methods by Jansen [30] and Mirshekar-Syahkal and Davies [31] use hybrid mode solutions for the fields and propagation constants to determine the losses for both single and coupled microstrip lines. Jansen [30] used the transmission-line formula

$$\alpha_t = \alpha_c + \alpha_d = 4.34 \frac{R'}{Z_0} + \frac{27.3}{\lambda_g} \tan\delta_d \text{ dB/unit length} \quad (24)$$

with the line resistance per-unit-length R' given by

$$R' = \left[\frac{R_s}{w_{\text{eff}}} + \frac{F_j R_s}{W_t} \right] F_{sR} \quad (25)$$

$$F_j = \frac{\int_0^1 |\bar{J}_Z(X)|^2 dx}{\left(\int_0^1 \bar{J}_Z(X) dX\right)^2}. \quad (26)$$

The factor F_j is the increase of strip loss due to the nonuniform current distribution compared with a uniform one. W_t is the correct strip width which takes into account the finite value of strip thickness while w_{eff} is the width of an equivalent parallel-plate waveguide of height h and with the frequency-dependent microstrip values of impedance and effective dielectric constant

$$w_{\text{eff}} = \frac{120\pi h}{\left(Z_0(f) \cdot \sqrt{\epsilon_{\text{eff}}(f)}\right)}. \quad (27)$$

Finally, R_s is the surface skin resistivity and F_{sR} is a factor which describes the additional loss due to substrate surface roughness.

In the solutions of microstrip losses by Mirshekar-Syahkal and Davies [31], perturbation theory is used to express the losses in terms of the unperturbed fields which were derived by the spectral domain method

$$\alpha_{\text{cond}} = \frac{R_s \int_c |H_t|^2 dl}{2R_e \int_s E_0 \times H_0^* \cdot \bar{a}_Z dS} \quad (28)$$

$$\alpha_{\text{diel}} = \frac{\omega\epsilon \tan\delta_d \int_{S_{\text{diel}}} |E_0|^2 dS}{2R_e \int_s E_0 \times H_0^* \cdot \bar{a}_Z dS}. \quad (29)$$

The results of the two hybrid mode methods described above show similar characteristics for the even and odd modes as those indicated in Fig. 11.

VI. Considerations of Dispersion and Higher Order Mode Propagation

Dispersion of microstrip becomes increasingly more important as the substrate's dielectric constant and thickness are made greater. In addition, when the substrate is magnetic, e.g., for a ferrite or garnet material, the microstrip line becomes highly dispersive near the material's natural resonant frequency $f_m = \gamma 4\pi M_s$. Hartwig *et al.* [32] described dispersion of microstrip as coupling of a fundamental TEM mode to TM_0 and TE_1 surface-wave modes. The TM_0 mode has a zero frequency cutoff while the TE_1 surface-wave mode starts propagating above the cutoff

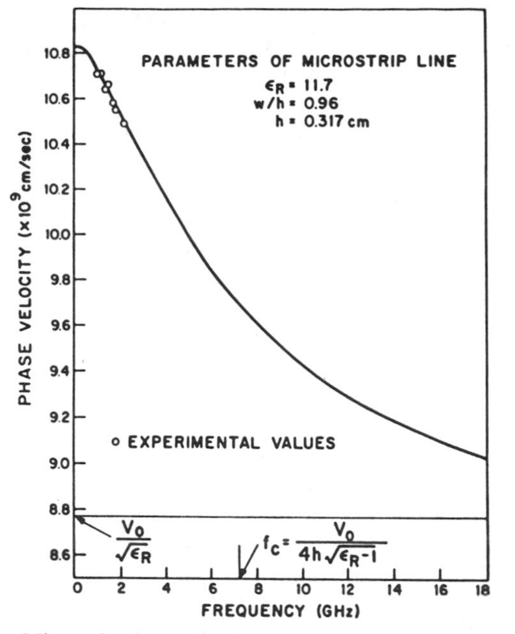

Fig. 12. Microstrip phase velocity versus frequency (Denlinger [33]).

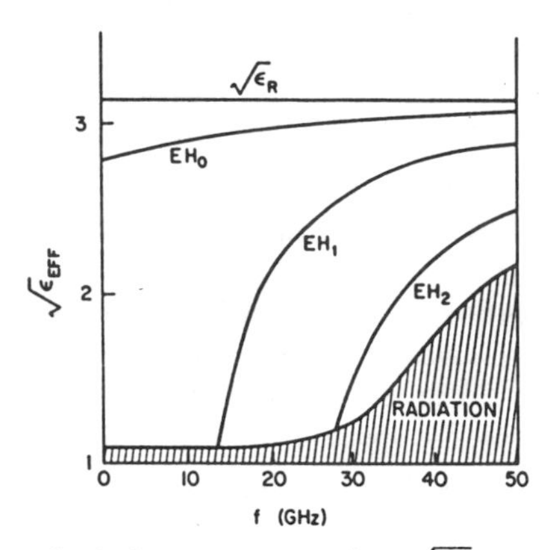

Fig. 13. Normalized phase constant ($\beta/k_0 = \sqrt{\epsilon_{eff}}$) versus frequency for some guided modes and region of radiation ($\epsilon_R = 9.8, w/h = 4.74, h = 0.635$ mm, $a/h = 5$; Ermert [35]).

frequency

$$f_c = \frac{V_0}{4h\sqrt{\epsilon_R - 1}}. \quad (30)$$

The resulting microstrip mode has a dispersive behavior that is illustrated in Fig. 12 and taken from Denlinger's hybrid mode solution [33]. There have been many other frequency-dependent solutions (e.g., [30], [31], [34], and [35]) involving spectral domain analysis, transverse mode-matching techniques, etc., which not only show the dispersion of the fundamental mode but also the existence of higher order modes such as illustrated in Fig. 13.

Dispersion of coupled microstrip lines has also been described by several authors [30], [31], [36], [37]. As shown in Fig. 14, the even mode is considerably more dispersive than the odd mode for typical strip spacings and widths. Work by Jansen [30] and Krage and Haddad [37] show that the characteristic impedance increases very slowly with frequency for both the even and odd modes and also

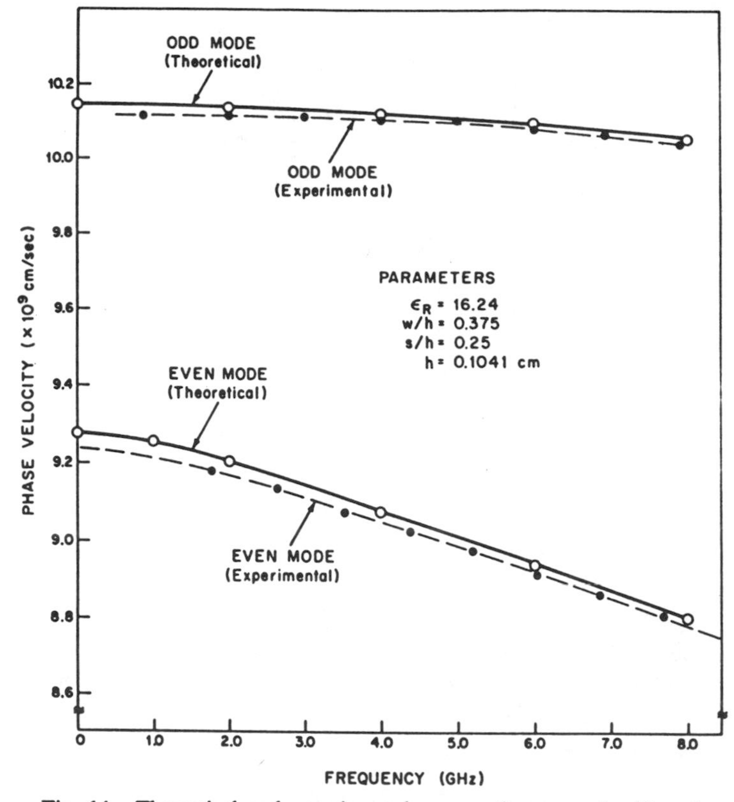

Fig. 14. Theoretical and experimental curves of even- and odd-mode phase velocity versus frequency for coupled microstrip lines (Denlinger [36]).

for single microstrip lines. Some other theories [33], [38], [39] claim that Z_0 decreases with frequency. The effect is so small that it is very hard to accurately measure and confirm the theoretical values.

VII. Comparison of Microstrip Losses to Other Commonly Used Quasi-TEM Lines

A comparison of losses in microstrip with those for other types of MIC lines is difficult since the variation of loss with impedance is different for the various lines and also the optimum substrate thicknesses may be different. In addition, the substrate's electrical characteristics (loss tangent, dielectric constant, etc.) has a significant impact on the relative losses of the various lines. Spielman's [29] computer-aided analysis of quasi-TEM lines, which was discussed previously, was used to compare losses of four types of 50-Ω transmission lines: microstrip, coplanar waveguide, trapped inverted microstrip and inverted microstrip. As shown in Fig. 15, the coplanar waveguide is considerably more lossy than microstrip whereas both the trapped inverted and inverted microstrip are considerably less lossy. The losses of suspended stripline is expected to be similar to the trapped inverted and inverted microstrip since all three types have strip widths two to three times greater than that for microstrip. By concentrating more of the field energies in air, a prescribed impedance level requires wider strips than for microstrip or coplanar lines, which results in lower conductor and dielectric losses. Another line, called stripline-like microstrip, where a top

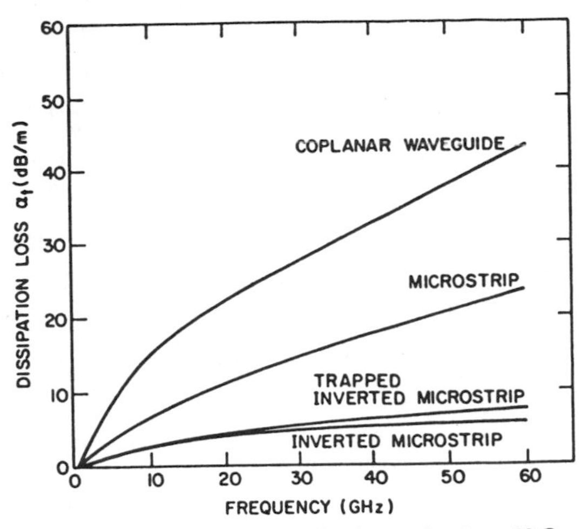

Fig. 15. Comparison of total dissipative losses for four 50-Ω transmission lines (Spielman [29]) (line parameters: (1) microstrip $-\epsilon_R = 10, w = 0.61$ mm, $h = 0.635$ mm, $t = 6.35$ μm; (2) coplanar waveguide $-\epsilon_R = 10.4, w = h = 0.635$ mm, $s_{cp} = 0.27$ mm, $t = 6.35$ μm; (3) trapped inverted microstrip $-\epsilon_R = 10, w = 1.25$ mm, $H_{TIM} = h = 0.508$ mm, $s_{TIM} = 1.27$ mm, $t = 6.35$ μm; (4) inverted microstrip $-\epsilon_R = 3.78, w = 1.524$ mm, $H_{IM} = 0.381$ mm, $h = 0.508$ mm, $t = 6.35$ μm).

ground plane is placed one substrate thickness above the strip, has also been shown to have slightly lower losses than microstrip, and about half the loss of stripline [40]. Due to having low radiation losses, stripline-like microstrip may be employed for circuits requiring high Q. Finally, losses for a non-TEM line called slotline has not been theoretically treated. Measurement of loss for substrates with $\epsilon_R = 16$ by Robinson and Allen [41] show about the same loss as for microstrip. However, these results do not agree with another experiment that showed unloaded-Q factors of slotline resonators to be about half that of microstrip resonators [42]. Radiation loss from slotline is expected to be quite high.

VIII. Conclusions

There is good understanding of conductor and substrate losses for dielectric and demagnetized ferrite-loaded microstrip. Losses for magnetized ferrite microstrip is less understood since it is still based on empirical formulas derived from experimental data. Dispersion of microstrip lines can have a significant effect on the calculation of losses. This is especially true for thick, high dielectric constant ceramic substrates as well as ferrite substrates. Theoretical radiation losses of open-ended microstrip agree fairly well with a limited amount of experimental data. However, no experimental data have been reported to confirm the theory for radiation from other types of microstrip discontinuities.

References

[1] J. D. Welch and H. J. Pratt, "Losses in microstrip transmission systems for integrated microwave circuits," *NEREM Rec.*, vol. 8, pp. 100–101, 1966.

[2] H. A. Wheeler, "Transmission-line properties of parallel strips separated by a dielectric sheet," *IEEE Trans. Microwave Theory Tech.*, vol. MTT-13, pp. 172–185, Mar. 1965.

[3] M. V. Schneider, B. Glance, and W. F. Bodtmann, "Microwave and millimeter wave hybrid integrated circuits for radio systems," *Bell Syst. Tech. J.*, vol. 48, no. 6, pp. 1703–1727, July/Aug. 1969.

[4] R. A. Pucel, D. J. Massé, and C. P. Hartwig, "Losses in microstrip," *IEEE Trans. Microwave Theory Tech.*, vol. MTT-16, pp. 342–350, June 1968.

[5] H. A. Wheeler, "Transmission-line properties of a strip on a dielectric sheet on a plane," *IEEE Trans. Microwave Theory Tech.*, vol. MTT-25, pp. 631–641, Aug. 1977.

[6] ——, "Formulas for the skin effect," *Proc. IRE*, vol. 30, pp. 412–424, Sept. 1942.

[7] R. A. Pucel and D. J. Massé, "Microstrip propagation on magnetic substrates—Part I: Design theory and Part II: Experiment," *IEEE Trans. Microwave Theory Tech.*, vol. MTT-20, pp. 304–313, May 1972.

[8] J. J. Green and F. Sandy, "A catalog of low power loss parameters and high power thresholds for partially magnetized ferrites," *IEEE Trans. Microwave Theory Tech.*, vol. MTT-22, pp. 645–651, June 1974.

[9] E. Schloemann, "Microwave behavior of partially magnetized ferrites," *J. Appl. Phys.*, vol. 41, p. 204, 1970.

[10] J. J. Green and F. Sandy, "Microwave characterizations of partially magnetized ferrites," *IEEE Trans. Microwave Theory Tech.*, vol. MTT-22, pp. 641–645, June 1974.

[11] F. Sandy, private communication to D. Massé and R. Pucel, Jan. 1971.

[12] L. Lewin, "Radiation from discontinuities in strip-line," *Proc. Inst. Elec. Eng.*, vol. 107C, pp. 163–170, 1960.

[13] ——, "Spurious radiation from microstrip," *Proc. Inst. Elec. Eng.*, vol. 125, pp. 633–642, July 1978.

[14] J. Watkins, "Radiation loss from open circuited dielectric resonators," *IEEE Trans. Microwave Theory Tech.*, vol. MTT-21, pp. 636–639, Oct. 1973.

[15] L. J. Van der Pauw, "The radiation of electromagnetic power by microstrip configurations," *IEEE Trans. Microwave Theory Tech.*, vol. MTT-25, pp. 719–725, Sept. 1977.

[16] E. Belohoubek and E. Denlinger, "Loss considerations for microstrip resonators," *IEEE Trans. Microwave Theory Tech.*, vol. MTT-23, pp. 522–526, June 1975.

[17] E. Denlinger, "Radiation from microstrip resonators," *IEEE Trans. Microwave Theory Tech.*, vol. MTT-17, pp. 235–236, Apr. 1969.

[18] G. Kompa, "Approximate calculation of radiation from open-ended wide microstrip lines," *Electron. Lett.*, vol. 12, pp. 222–224, Apr. 29, 1976.

[19] H. Sobol, "Radiation conductance of open-circuit microstrip," *IEEE Trans. Microwave Theory Tech.*, vol. MTT-19, pp. 885–887, Nov. 1971.

[20] C. Wood, P. S. Hall and J. R. James, "Radiation conductance of open-circuit low dielectric constant microstrip," *Electron. Lett.*, vol. 14, pp. 121–123, Feb. 16, 1978.

[21] C. M. Angulo and W. S. C. Chang, "The launching of surface waves by a parallel plate waveguide," *IRE Trans. Antennas Propagat.*, vol. AP-7, pp. 359–368, Oct. 1959.

[22] S. P. Morgan, "Effect of surface roughness on eddy current losses at microwave frequencies," *J. Appl. Phys.*, vol. 20, pp. 352–362, Apr. 1949.

[23] J. H. C. Van Heuven, "Conduction and radiation losses in microstrip," *IEEE Trans. Microwave Theory Tech.*, vol. MTT-22, pp. 841–844, Sept. 1974.

[24] H. Sobol, "Technology and design of hybrid microwave integrated circuits," *Solid State Technol.*, pp. 49–57, Feb. 1970.

[25] H. Sobol and M. Caulton, "Technology of microwave integrated circuits," in *Advances in Microwaves*. New York: Academic Press, 1974, vol. 8, pp. 12–64.

[26] B. R. Rao, "Effect of loss and frequency dispersion on the performance of microstrip directional couplers and coupled line filters," *IEEE Trans. Microwave Theory Tech.*, vol. MTT-22, pp. 747–750, July 1974.

[27] R. Garg and I. J. Bahl, "Characteristics of coupled microstriplines," *IEEE Trans. Microwave Theory Tech.*, vol. MTT-27, pp. 700–705, July 1979.

[28] R. Horton, "Loss calculations of coupled microstrip lines," *IEEE Trans. Microwave Theory Tech.*, vol. MTT-21, pp. 359–360, May 1973.

[29] B. E. Spielman, "Dissipation loss effects in isolated and coupled transmission lines," *IEEE Trans. Microwave Theory Tech.*, vol. MTT-25, pp. 648–655, Aug. 1977.

[30] R. H. Jansen, "High-speed computation of single and coupled microstrip parameters including dispersion, high-order modes, loss and finite strip thickness," *IEEE Trans. Microwave Theory Tech.*, vol. MTT-26, pp. 75–82, Feb. 1978.

[31] D. Mirshekar-Syahkal and J. B. Davies, "Accurate solution of microstrip and coplanar structures for dispersion and for dielectric and conductor losses," *IEEE Trans. Microwave Theory Tech.*, vol. MTT-27, pp. 694–699, July 1979.

[32] C. P. Hartwig, D. Massé and R. A. Pucel, "Frequency dependent behavior of microstrip," 1968 *Int. Symp. Dig.*, pp. 110–116.

[33] E. J. Denlinger, "A frequency dependent solution for microstrip transmission lines," *IEEE Trans. Microwave Theory Tech.*, vol. MTT-19, pp. 30-39, Jan. 1971.

[34] T. Itoh and R. Mittra, "Spectral-domain approach for calculating dispersion characteristics of microstrip lines," *IEEE Trans. Microwave Theory Tech.*, vol. MTT-21, pp. 496–498, July 1973.

[35] H. Ermert, "Guided modes and radiation characteristics of covered microstrip lines," *Arch. Electron. Übertragung.*, vol. 30, pp. 65–70, 1976.

[36] E. Denlinger, "Frequency dependence of a coupled pair of microstrip lines," *IEEE Trans. Microwave Theory Tech.*, vol. MTT-18, pp. 731–733, Oct. 1970.

[37] M. K. Krage and G. I. Haddad, "Frequency-dependent characteristics of microstrip transmission lines," *IEEE Trans. Microwave Theory Tech.*, vol. MTT-20, pp. 678–688, Oct. 1972.

[38] W. J. Getsinger, "Microstrip characteristic impedance," *IEEE Trans. Microwave Theory Tech.*, vol. MTT-27, p. 293, Apr. 1979.

[39] F. Arndt and G. U. Paul, "The reflection definition of the characteristic impedance of microstrips," *IEEE Trans. Microwave Theory Tech.*, vol. MTT-27, pp. 724–731, Aug. 1979.

[40] R. Garg, "Stripline-like microstrip configuration," *Microwave J.*, pp. 103–116, Apr. 1979.

[41] G. H. Robinson and J. L. Allen, "Slotline application to miniature ferrite devices," *IEEE Trans. Microwave Theory Tech.*, vol. MTT-17, pp. 1097–1101, Dec. 1969.

[42] G. P. Kurpis, "Coplanar and slotlines—Are they here to stay?" in *Proc. Int. Microelectric Symposium*, (Washington, DC), pp. 3B.6.1–3B.6.5, 1972.

[43] J. James and A. Henderson, "High-frequency behavior of microstrip open-circuit terminations," *IEEE J. Microwave, Optics, and Acoustics*, vol. 3, pp. 205–218, Sept. 1979.

[44] R. Horton, B. Easter, and A. Gopinath, "Variation of microstrip losses with thickness of strip," *Electron. Lett.*, vol. 17, no. 17, pp. 490–491, Aug. 26, 1971.

Dissipation Loss Effects in Isolated and Coupled Transmission Lines

BARRY E. SPIELMAN, MEMBER, IEEE

Abstract—**This paper describes a computer-aided analysis of dissipation losses in uniform isolated or coupled transmission lines for microwave and millimeter-wave integrated-circuit applications. The analysis employs a quasi-TEM model for isolated transmission lines and for the even- and odd-mode transmission lines associated with coupled-line structures. The conductor and dielectric losses are then related to equivalent charge density distributions, which are evaluated using a method-of-moments solution. The transmission lines treated by this analysis may contain any number of lossy conductors and inhomogeneous dielectrics, consisting of any number of different homogeneous dielectric regions. A development is provided to explicitly relate the four-port terminal-electrical performance of directional couplers to evaluated even- and odd-mode loss coefficients.**

Examples of evaluated losses are presented in graphical form for isolated lines of inverted microstrip and trapped inverted microstrip and edge-coupled microstrip with a dielectric overlay. The analysis accuracy has been confirmed using microstrip and coplanar waveguide configurations. A comparison is made of the total loss characteristics for microstrip, coplanar waveguide, inverted microstrip, and trapped inverted microstrip. Calculations are compared with measurements for the coupled-line structure. Accuracy of the solution and suggested refinements are discussed. Five computer programs are documented.

I. Introduction

THERE is considerable interest in investigating and exploiting new transmission lines for use in integrated circuits operating at higher microwave and millimeter-wave frequencies. This interest has been spurred by the success in effecting reductions in circuit cost, size, and weight through the application of microstrip at lower to intermediate microwave frequencies. Unfortunately, microstrip is discouragingly lossy and more difficult to fabricate at higher microwave and millimeter-wave frequencies. These considerations have prompted the search for transmission lines that are amenable to integrated-circuit fabrication methods (thin-film and photolithographic technology) and which have improved loss characteristics compared with microstrip.

To facilitate the investigation of transmission lines which offer potential for improvements over microstrip at the frequencies of interest, a flexible computer-aided analysis of transmission-line losses has been implemented. This analysis is suitable for application to a wide variety of transmission lines. This paper describes the implementation of that analysis as it applies to both isolated and coupled transmission lines, where losses due to both conductor and dielectric dissipation are taken into account. Various examples of loss evaluations using this formulation are presented for both isolated and coupled transmission lines of interest.

Manuscript received August 23, 1977; revised January 19, 1977.

The author is with the Microwave Techniques Branch, Electronics Technology Division, Naval Research Laboratory, Washington, DC 20375.

II. Formulation of Analysis for Evaluation of Losses

The approach used here for the analysis of conductor and dielectric loss characteristics of isolated or coupled uniform transmission lines is consistent with the quasi-TEM models described in [1] and [2]. For isolated transmission lines, the direction of propagation is taken to be along the z direction. Consistent with the quasi-TEM model and the transmission-line wave approach described in [3], the z dependence for voltage and current along the transmission line conductor and dielectric loss coefficients α_c and α_d, α is the attenuation constant due to conductor and dielectric losses and β is the phase constant. Following the development set forth in [3], the attenuation constant α is given by

$$\alpha = \bar{P}_d/(2\bar{P}_f) \tag{1}$$

where $\bar{P}_d$ is the time-averaged power dissipated per unit length and $\bar{P}_f$ is the time-averaged power flow along line.

In the following portion of this paper, explicit expressions are developed for use in evaluating isolated transmission-line conductor and dielectric loss coefficients α_c and α_d, respectively. The total coefficient α is obtained from these by summing α_c and α_d.

A. Isolated Transmission Lines

To obtain a useful expression for $\bar{P}_f$ in (1) a development, the details of which are found in [4], is summarized as follows. The complex flow P_f is represented in terms of a $+z$ traveling wave on a lossy transmission line. ReZ_0 and $|Z_0|$ are approximated [4] by $(Z_0)_{LL}$, the characteristic impedance of the same line without losses. By virtue of these considerations $\bar{P}_f$ can be expressed as

$$\bar{P}_f = |V_0|^2 vCe^{-2\alpha_z} \tag{2}$$

where v is the phase velocity, C is the electrostatic capacitance per unit length, and $|V_0|^2$ is the amplitude squared of the wave voltage at $z = 0$.

1) Conductor Losses: To obtain an expression which is useful for evaluating $\bar{P}_d$ in (1) for losses due to imperfect conductors, the approximation described in [5] is employed. $\bar{P}_d$ can be expressed approximately by

$$\bar{P}_{d,c} = \int |H_0|^2 R \, dl. \tag{3}$$

Here, $|H_0|^2$ is the amplitude squared of the magnetic field at conducting surfaces for the lossless case. R is the surface

Reprinted from *IEEE Trans. Microwave Theory Tech.*, vol. MTT-25, no. 8, pp. 648–656, August 1977.

resistance of the metals in the system. Here, the additional subscript "c" on $\bar{P}_{d,c}$ denotes power losses due to imperfect conductors. For good conductors, R is expressed [6] in terms of frequency f, free-space permeability μ_0, and dc conductivity σ. Using the quasi-TEM propagation assumption, $|H_0|^2$ is related to $|E_0|^2$, where $\vec{E}_0$ is the electric field in the lossless medium [4].

In the lossless TEM solution described in [1] and [2], the electric field in a medium with permittivity ε, the electric field at a point along the surface of a perfect conductor, is given by

$$E_0 = 2\pi q \tag{4}$$

where q is a flux-source distribution residing at the conductor surface and $2\pi\varepsilon q$ can be thought of as the equivalent charge-density distribution (sum of free and polarization charge densities). In [1], the flux-source distribution residing along the surface of the N_jth conductor, in a system having a total of N_c conductor surfaces, is approximated by a pulse expansion as

$$q^{N_j} = \sum_{i=1}^{(N_S)_j} q_i^{N_j} P^{N_j}(i), \qquad N_j = 1, \cdots, N_c. \tag{5}$$

This representation arises by subdividing the contour defined by the surface of the N_jth conductor into $(N_S)_j$ segments. Along the ith segment the free charge-density distribution is taken to be constant at the value $q_i^{N_j}$. $P(i)$ is a pulse function defined by

$$P^{N_j}(i) = \begin{cases} 1, & \text{on the } i\text{th section of } N_j \\ 0, & \text{on all other sections of } N_j. \end{cases} \tag{6}$$

Then, rewriting $\bar{P}_{d,c}$ in (3) and using (1), (2), (4)–(6), the loss coefficient due to conductor losses α_c can be written as

$$\alpha_c \approx \frac{20}{\ln 10} \cdot \frac{2\pi^2 \varepsilon_0 \varepsilon_{\text{eff}} \sqrt{\dfrac{\pi f \mu_0}{\sigma}}}{\mu_0 |V_0|^2 v C} \cdot \sum_{j=1}^{N_c} \sum_{i=1}^{(N_S)_j} (q_i^{N_j})^2 \Delta l_i^{N_j} \quad \text{dB/unit length} \tag{7}$$

where $\Delta l_i^{N_j}$ is the length of the ith segment on the N_jth conductor, ε_0 is the permittivity of free space, and ε_{eff} is given by

$$\varepsilon_{\text{eff}} = \frac{C}{C_0}. \tag{8}$$

Here, C_0 is the electrostatic capacitance of the transmission line under consideration, but with all dielectric materials fictitiously removed. It is the expression given in (7) which is embodied in the computer programs described in [4]. This expression has been used to provide the design information for conductor losses found in Section III.

2) *Dielectric Losses:* To obtain an expression which is useful for evaluating $\bar{P}_d$ in (1) for losses due to imperfect dielectrics, this quantity is initially written as

$$\bar{P}_{d,d} = \sum_{i=1}^{N_D} \int_{A_i} \omega \varepsilon_0 \varepsilon_i'' |E|^2 \, dS. \tag{9}$$

Here, the second subscript "d" on $\bar{P}_{d,d}$ denotes that the dissipation losses are due to imperfect dielectrics. N_D is the number of imperfect dielectric regions where the ith region has a complex permittivity given by $\varepsilon_i = \varepsilon_0(\varepsilon_i' - j\varepsilon_i'')$. A_i represents the area, in the transmission-line cross section, spanned by the ith simply connected homogeneous lossy dielectric region. Equation (9) can be rewritten as

$$\bar{P}_{d,d} = \sum_{i=1}^{N_D} \int_{A_i} \omega \tan \delta_i W_{ei} \, dS \tag{10}$$

where $\tan \delta_i$ is the loss tangent of the material in the ith region. The time-averaged energy stored in the electric field in the ith dielectric region is given by

$$\bar{W}_{ei} = \tfrac{1}{2} \int_{A_i} \varepsilon_0 \varepsilon_i' |E|^2 \, dS. \tag{11}$$

In [4] details of a development are presented which show that the time average of the total energy stored in the electric field per unit length of transmission line $\bar{W}_e$ is related to $\bar{W}_{ei}$ by

$$\bar{W}_{ei} = \varepsilon_i \frac{\partial \bar{W}_e}{\partial \varepsilon_i}. \tag{12}$$

While the development in [4] provides a general result for transmission structures with many conductors and dielectrics over the cross section, for isolated lines or coupled lines treated by an even- and odd-mode two-port interpretation $\bar{W}_e$ can be expressed as

$$\bar{W}_e = \tfrac{1}{2} C |V|^2 \tag{13}$$

where C is the electrostatic capacitance of the transmission line in the two-port configuration (isolated and even- or odd-mode line) and V is the voltage associated with a $+z$ traveling wave on the line.

To facilitate the evaluation of a dielectric loss coefficient in terms of readily computable parameters, equations (1), (2), (8), and (10)–(13) are combined to provide the following expression:

$$\alpha_d = \frac{20\pi f}{\ln 10 \, c \sqrt{\varepsilon_{\text{eff}}}} \sum_{i=1}^{N_D} \varepsilon_i' \tan \delta_i \frac{\partial \varepsilon_{\text{eff}}}{\partial \varepsilon_i'} \quad \text{dB/unit length.} \tag{14}$$

In (14), c is the speed of light in free space. To evaluate the partial derivative in (14) a "forward" difference quotient is employed, providing the computationally useful result

$$\alpha_d = \frac{20\pi f}{\ln 10 \, c \sqrt{\varepsilon_{\text{eff}}}} \sum_{i=1}^{N_D} \varepsilon_i' \tan \delta_i \left(\frac{\hat{\varepsilon}_{\text{eff}} - \varepsilon_{\text{eff}}}{\hat{\varepsilon}_i' - \varepsilon_i'} \right) \quad \text{dB/unit length.} \tag{15}$$

In (15) $\hat{\varepsilon}_{\text{eff}}$ is the value of ε_{eff} for the structure under consideration when the value of the permittivity for the ith homogeneous dielectric region is perturbed to a slightly different (higher) value $\hat{\varepsilon}_i'$. It is this expression which has been incorporated into the computer programs which are documented in [4]. For these programs the index "i" takes the value of "1" since there is only one lossy homogeneous-dielectric region in the overall inhomogeneous structures treated by the computer programs.

B. Coupled Transmission-Line Structures

For the purposes of this section the coupled transmission lines will be treated by an even- and odd-mode interpretation [7]. Although two transmission lines coupled over a given length truly represents a four-port structure, the even- and odd-modes associated with this structure are each two-port transmission lines which can be treated separately. To properly assess the effects of losses in such structures the problem is twofold. One problem is to determine the loss coefficients α_c and α_d for each of the even and odd modes. The second problem is to determine the effects of these coefficients on the four-port terminal-electrical performance of the entire structure. In the material to follow the former problem will be discussed first and will be followed by a treatment of the latter problem explicitly for four-port directional coupler loss effects.

1) Even- and Odd-Mode Loss Coefficients: Since the even- and odd-modes for a coupled line structure can each be depicted as two-port transmission lines whose length is that of the coupled-line region, the loss coefficients for conductor and dielectric losses can be determined using (7) and (15), respectively. The even- and odd-mode loss coefficients for losses due to imperfect conductors can be written as

$$\alpha_{ck} \approx \frac{20}{\ln 10} \cdot \frac{2\pi^2 \varepsilon_0 (\varepsilon_{\mathrm{eff}})_k}{\mu_0 |V_0|^2 v_k C_k} \sqrt{\frac{\pi f \mu_0}{\sigma}} \cdot \sum_{j=1}^{N_C} \sum_{i=1}^{(N_S)_j} (q_{ik}^{N_j})^2 \Delta l_{ik}^{N_j} \quad \text{dB/unit length} \tag{16}$$

where $k = e$ for the even mode and $k = o$ for the odd mode. The quantities v_k, C_k, q_{ik}, $(\varepsilon_{\mathrm{eff}})_k$, and Δl_{ik} are evaluated as described in [1]. The expression for even- and odd-mode loss coefficients for losses due to imperfect dielectrics can be expressed as

$$\alpha_{dk} \approx \frac{20\pi f}{\ln 10\, c\sqrt{(\varepsilon_{\mathrm{eff}})_k}} \sum_{i=1}^{N_D} \varepsilon_i' \tan \delta_i \cdot \left[\frac{(\hat{\varepsilon}_{\mathrm{eff}})_k - (\varepsilon_{\mathrm{eff}})_k}{\hat{\varepsilon}_i' - \varepsilon_i'} \right] \quad \text{dB/unit length} \tag{17}$$

where $k = e$ for the even mode and $k = o$ for the odd mode. In (17) $(\varepsilon_{\mathrm{eff}})_k$ represents the effective relative permittivity for the even- or odd-mode in the coupled-line configuration to be analyzed. This value is determined using the method described explicitly in [1]. $(\hat{\varepsilon}_{\mathrm{eff}})_k$ is evaluated the same way for the coupled-line structure where the relative permittivity of the ith dielectric region ε_i' is perturbed to the slightly higher value $\hat{\varepsilon}_i'$. Equations (16) and (17) have been used to evaluate the even- and odd-mode conductor and dielectric loss coefficients for the coupled-line structures described in Section III of this paper. The total loss coefficients α_e and α_o for the even- and odd-modes are obtained by adding the respective values of α_{ck} and α_{dk}.

2) Effects of Losses on Four-Port Terminal Performance: An understanding of the effects of losses on the four-port terminal performance of directional couplers is facilitated by considering the following. The problem at hand is one of determining the loss effects on measurable terminal voltages b_i ($i = 1,2,3,4$), leaving these ports, once the loss coefficients $(\alpha_{ce}, \alpha_{de})$ and $(\alpha_{co}, \alpha_{do})$ have been determined for the modal transmission lines, respectively. These loss coefficients are assumed to be known for this development, having been computed by the method described in the previous section.

The approach used here starts with a procedure similar to that described in [7]. The voltage b_i ($i = 1,2,3,4$) is written in terms of a sum or difference of Γ_{oe} and Γ_{oo} or T_{oe} and T_{oo} where (Γ_{oe}, T_{oe}) and (Γ_{oo}, T_{oo}) are pairs of reflection and transmission coefficients for the even- and odd-mode two ports, respectively. The reflection coefficients for the even or odd mode can be expressed in terms of the even- or odd-mode *ABCD* two-port network parameters [7]. The *ABCD* parameters for the lossy even- or odd-mode two-port are given by

$$A_i = D_i = \cosh \gamma_i l \tag{18}$$

$$B_i = Z_{0i} \sin \gamma_i l \tag{19}$$

$$C_i = \left(\frac{1}{Z_{0i}}\right) \sin \gamma_i l \tag{20}$$

where $i = e$ for the even mode and $i = o$ for the odd mode. In (18)–(20) *l represents the physical length of the coupled-line region;* Z_{0e} and Z_{0o} are the characteristic impedances of the even- and odd-mode transmission lines, respectively. γ_e and γ_o are the even- and odd-mode propagation constants, taken for this development to be given by

$$\gamma_e = \alpha_e + j\beta_e \tag{21}$$

$$\gamma_o = \alpha_o + j\beta_o \tag{22}$$

where α_e and α_o are the total (known) even- and odd-mode loss coefficients. β_e and β_o are the even- and odd-mode phase constants (known), respectively, which are determined by the lossless analysis described in detail in [1]. It is to be noted that the generality of this development permits β_e and β_o to be determined to allow for even and odd modes having different phase velocities.

Employing (18)–(20) the measurable terminal voltages of the four-port coupler under consideration can be expressed as follows:

$$b_2 = b_c = F_2 \left[\frac{\tanh \gamma_e l}{2 + F_1 \tanh \gamma_e l} + \frac{\tanh \gamma_o l}{2 + F_1 \tanh \gamma_o l} \right] \tag{23}$$

$$b_4 = b_t = \frac{\operatorname{sech} \gamma_e l}{2 + F_1 \tanh \gamma_e l} + \frac{\operatorname{sech} \gamma_o l}{2 + F_1 \tanh \gamma_o l} \tag{24}$$

$$b_3 = b_i = \frac{\operatorname{sech} \gamma_e l}{2 + F_1 \tanh \gamma_e l} - \frac{\operatorname{sech} \gamma_o l}{2 + F_1 \tanh \gamma_o l} \tag{25}$$

$$b_1 = b_r = F_2 \left[\frac{\tanh \gamma_e l}{2 + F_1 \tanh \gamma_e l} - \frac{\tanh \gamma_o l}{2 + F_1 \tanh \gamma_o l} \right] \tag{26}$$

where the subscripts c, t, i, and r denote coupled, transmitted, isolated, and reflected signals, respectively. In (23)–(26) F_1 and F_2 are quantities given by

$$F_1 \equiv \left(\frac{Z_{0e}}{Z_{0o}}\right)^{1/2} + \left(\frac{Z_{0o}}{Z_{0e}}\right)^{1/2} \tag{27}$$

$$F_2 \equiv \frac{1}{2}\left[\left(\frac{Z_{0e}}{Z_{0o}}\right)^{1/2} - \left(\frac{Z_{0o}}{Z_{0e}}\right)^{1/2}\right]. \tag{28}$$

For the special case where $\beta_e l = \beta_o l = \pi/2$, equations (21)–(26) simplify to become the following:

$$b_c = F_2 \left[\frac{1}{2\alpha_e l + F_1} + \frac{1}{2\alpha_o l + F_1} \right] \quad (29)$$

$$b_t = -j \left[\frac{1}{2\alpha_e l + F_1} + \frac{1}{2\alpha_o l + F_1} \right] \quad (30)$$

$$b_i = -j \left[\frac{1}{2\alpha_e l + F_1} - \frac{1}{2\alpha_o l + F_1} \right] \quad (31)$$

$$b_r = F_2 \left[\frac{1}{2\alpha_e l + F_1} - \frac{1}{2\alpha_o l + F_1} \right]. \quad (32)$$

It is to be noted that in (29)–(32) the even- and odd-mode loss coefficients appear explicitly. Also, it is easily seen from these relationships that, even for perfectly matched couplers, for α_e different from α_o the isolated and reflected signals are not zero. This effect is described quantitatively in Section IV. For couplers with other sources of impedance match and isolation degradation (e.g., different even- and odd-mode phase velocities) the additional degradation of isolation and reflected signal due to different modal loss characteristics will be superimposed on the other effects. The relationships developed in this section have been tested on experimental coupler models. The results are described in Section III.

III. Examples of Loss Evaluations and Design Information

In this section results are presented for specific isolated and coupled transmission-line structures. The accuracy of the analysis described in this paper has been confirmed [4] by comparing results calculated for microstrip and coplanar waveguide with well-accepted reference values. The results presented in the following sections for inverted microstrip and trapped inverted microstrip serve to provide useful currently unavailable design information for more complicated structures which offer potential for circuit applications. Following these results is a comparison of the loss characteristics for the four preceding transmission lines. The results for the edge-coupled microstrip structure with a dielectric overlay serve to confirm the utility of the coupled-line loss analysis described in Section II and to provide currently unavailable design information for this structure. This structure is presently the most viable approach for providing high-performance broad-band couplers, filters, and Schiffman phase-shift sections in a microstrip-compatible format [8].

A. Isolated Transmission Lines

1) Inverted Microstrip: The generic cross section of inverted microstrip is depicted in Fig. 1. Analyzed results for conductor and dielectric loss coefficients are shown in Fig. 2. The substrate dielectric constant chosen for this study corresponds to that of fused silica, which is deemed to be a suitable material for this transmission line at higher microwave and millimeter-wave frequencies. The characteristic impedance and phase velocities for the range of configurations treated in Fig. 2 are shown in Fig. 3 [4].

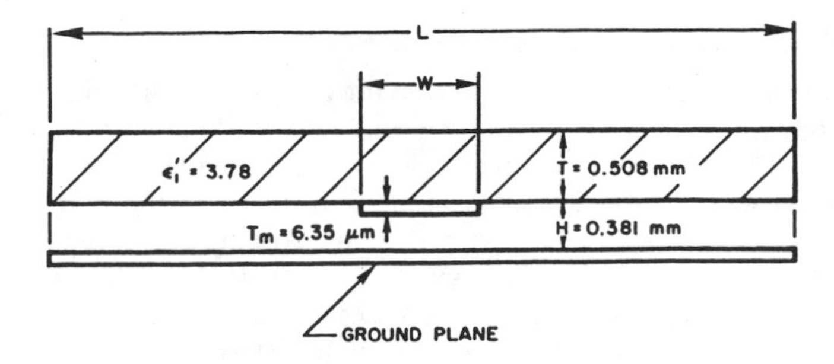

Fig. 1. Generic cross section of inverted microstrip.

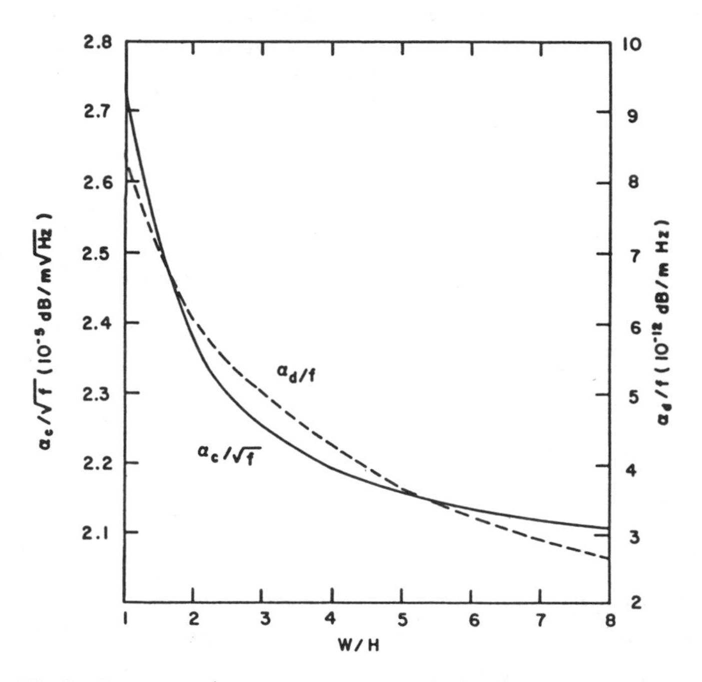

Fig. 2. Loss constants versus aspect ratio for inverted microstrip: $T = 20.0$ mils (0.508 mm), $\varepsilon'_1 = 3.78$, $\hat{\varepsilon}'_1 = 3.88$, $\tan \delta_1 = 2.0 \times 10^{-4}$, $T_m = 0.250$ mil (6.35 μm), $\sigma = 4.10 \times 10^7$ mho/m.

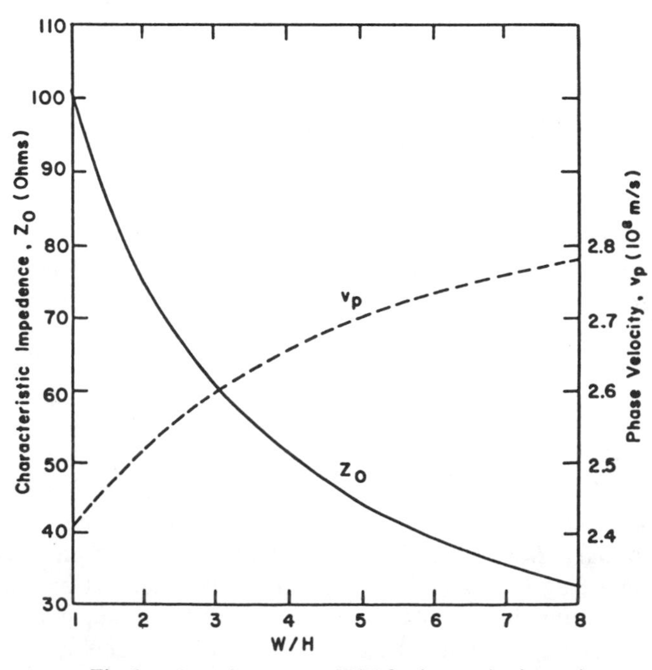

Fig. 3. Z_0 and v_p versus W/H for inverted microstrip.

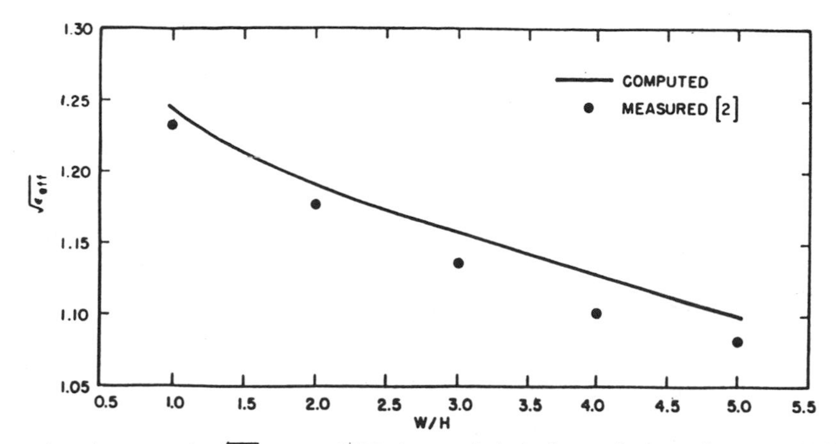

Fig. 4. Computed and measured $\sqrt{\varepsilon_{eff}}$ versus W/H characteristic for inverted microstrip: $H = 0.015$ in (0.381 mm), $T = 0.020$ in (0.508 mm), $\varepsilon_1' = 3.78$.

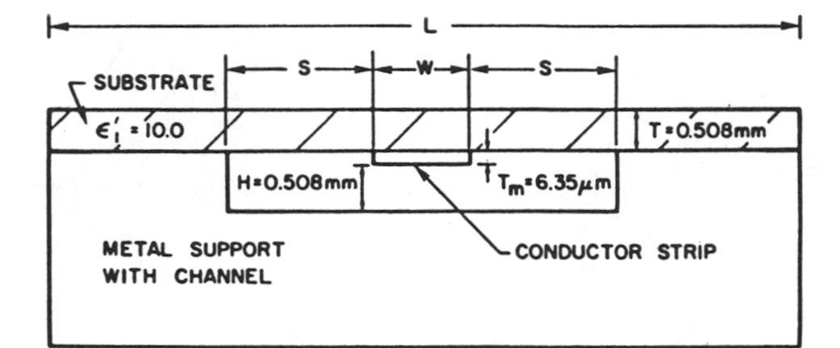

Fig. 5. Generic cross section for trapped inverted microstrip.

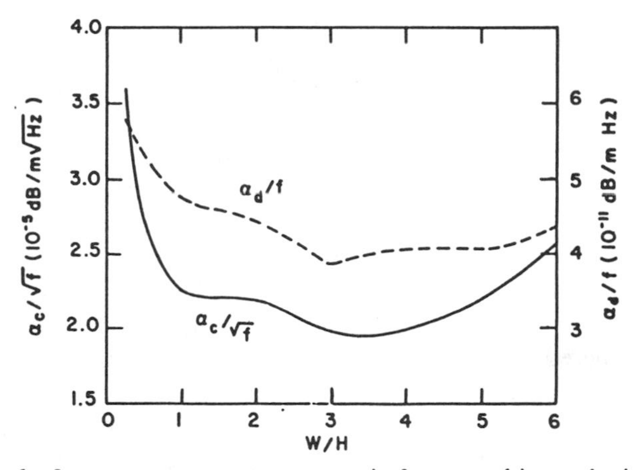

Fig. 6. Loss constants versus aspect ratio for trapped inverted microstrip: $S = 77.5$ mils (1.968 min), $T = 20.0$ mils (0.508 mm), $\varepsilon_1' = 10.0$, $\hat{\varepsilon}_1' = 10.1$, $\tan\delta_1 = 6.0 \times 10^{-4}$, $T_m = 0.250$ mil (6.35 μm), $\sigma = 4.10 \times 10^7$ mho/m.

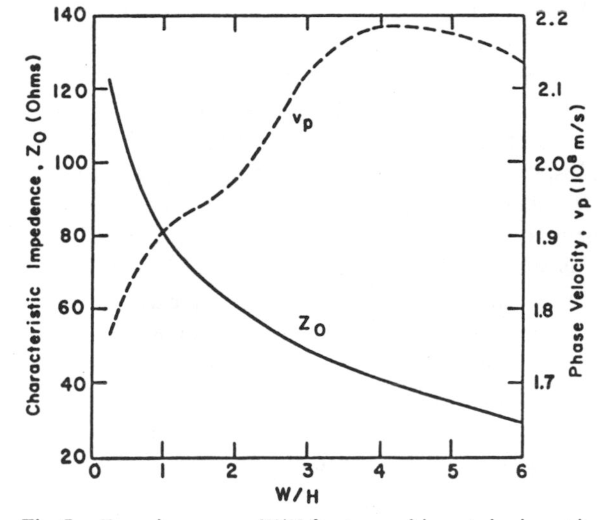

Fig. 7. Z_0 and v_p versus W/H for trapped inverted microstrip.

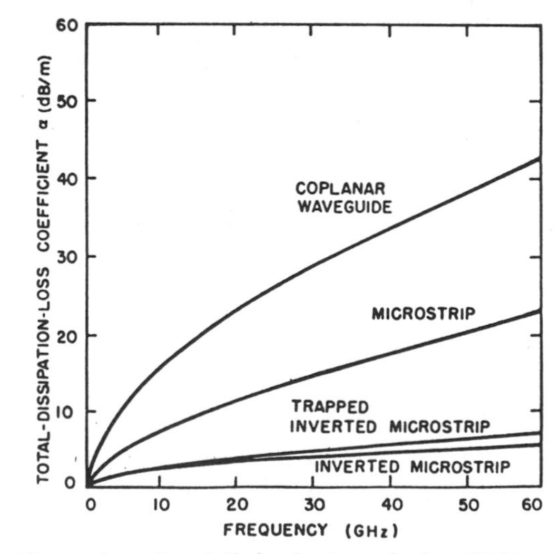

Fig. 8. Comparison of total dissipative losses for four 50-Ω transmission lines.

To demonstrate the correlation of the computed phase velocity compared to experimental data for an inverted microstrip, Fig. 4 shows computed values of $\sqrt{\varepsilon_{eff}}$ versus aspect ratio W/H as a solid curve. The measured points shown in this figure were obtained via time-domain reflectometer measurements [9]. The errors between measured and computed values are within 3 percent over this range of aspect ratios.

2) *Trapped Inverted Microstrip:* This section provides design information for trapped inverted microstrip, characterized by the generic cross-section configuration shown in Fig. 5. Calculated values of the conductor and dielectric loss coefficients are plotted in Fig. 6. The somewhat irregular appearance of these characteristics can be attributed to the effects of channel side-wall interaction with the conducting

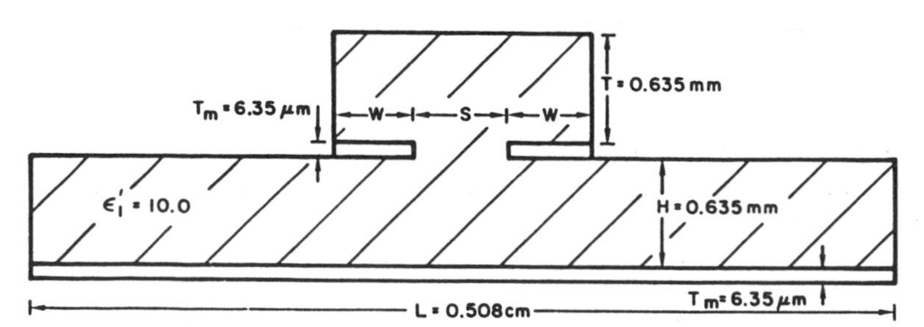

Fig. 9. Generic cross section of edge-coupled microstrip with a dielectric overlay.

strip, which is not explicitly accounted for by the "aspect ratio" definition as W/H. A similar situation is apparent in Fig. 7, where the characteristic impedance Z_0 and phase velocity v_p are shown [4].

The total dissipation loss characteristic for a nominally 50-Ω configuration of trapped inverted microstrip is plotted in Fig. 8 for comparison with those of microstrip, coplanar waveguide, and inverted microstrip.

3) Comparison of Loss Characteristics: This section presents a comparison of the loss characteristics for nominally 50 Ω, specific configurations of microstrip, coplanar waveguide, inverted microstrip, and trapped inverted microstrip transmission lines. The total loss coefficients (sum of the conductor and dielectric loss coefficients) for these lines are shown in Fig. 8 plotted versus frequency for the frequency range from 0 to 60 GHz. Effects due to higher order moding have been neglected, as well as those due to radiation losses. The curves in this figure for microstrip and coplanar waveguide correspond to configurations which are completely described in [4]. The cross-sectional configuration of microstrip chosen here represents a standard configuration of this transmission line as it has been used in many applications at frequencies up through about 12 GHz. The same is true for the selected coplanar waveguide configuration. The configuration for inverted microstrip was chosen to represent a version fabricated on a fused silica substrate, with gold being the predominant carrier of RF current in the metallization system. The trapped inverted microstrip configuration represents a model fabricated on a standard alumina substrate, metallized with a predominantly gold system. A discussion of the loss and fabrication aspects of this comparison is provided in Section IV.

B. Coupled Transmission Lines—Edge-Coupled Microstrip with a Dielectric Overlay

This section serves a twofold purpose. The first is to provide an illustrative example demonstrating the applicability of the coupled-line loss analysis developed in Sections II-B-1 and II-B-2. The second purpose is to provide a quantitative assessment of the losses encountered in quarter-wavelength long (at midband) sections of edge-coupled microstrip line with a dielectric overlay for various coupling levels.

The generic cross section of the edge-coupled microstrip with dielectric-overlay configuration is portrayed in Fig. 9.

Fig. 10. Photograph of a two-section coupler employing edge-coupled microstrip with a dielectric overlay. Substrate: alumina, 0.9 × 0.4 × 0.025 in (23 × 10 × 0.635 mm). Metallization: chromium-gold, thickness = 0.00025 in (6.35 μm).

Four configurations, differing in geometric parameters were analyzed and are described in Table I. The sections numbered 1 and 2 in this table were fabricated into a two-section asymmetric coupler of the type shown in the photograph in Fig. 10. The overlays were made of alumina pieces on each section and were attached to the substrate using Stycast Hi K epoxy ($K = 10.$). The two-section coupler, over the frequency band from 2 to 8.5 GHz, had a nominal coupling value of 6.7 dB. Using (29)–(32) and the computed values of loss coefficient shown in Table I, the calculated dissipation loss for the two coupled-line sections was found to be 0.2 dB, obtained for the coupler midband frequency of 5.4 GHz. Adding in measured losses due to connectors and theoretical lead-in line loss values, the total dissipation loss determined was 0.42 dB. This agreed well with the total measured coupler dissipation of 0.4 dB.

The sections numbered 3 and 4 in Table I were combined in the fabrication of a two-section asymmetric coupler. The fabrication details of this experimental model were the same as those described for the coupler in the previous paragraph. The experimental model had a nominal coupling characteristic of 20 dB over the 2–8.5-GHz frequency band. Using an

TABLE I
PARAMETERS FOR EDGE-COUPLED MICROSTRIP WITH DIELECTRIC OVERLAY SECTIONS

SECTION NO.	W (mils / mm)	S (mils / mm)	Z_o (ohms)	Z_{oe} (ohms)	Z_{oo} (ohms)	$\alpha_{ce}/\sqrt{f}$ (dB/m) x 10^{-5}	$\alpha_{co}/\sqrt{f}$ (dB/m) x10^{-5}	α_{de}/f (dB/m)x 10^{-10}	α_{do}/f (dB/m)x 10^{-10}
1	8.80 0.223	1.2 0.0305	50.0	115.	21.7	5.60	76.9	1.20	1.59
2	19.0 0.483	21.3 0.541	50.7	64.1	40.2	6.69	10.3	1.36	1.38
3	19.0 0.483	36.5 0.927	50.7	58.6	43.9	7.59	9.31	1.60	1.38
4	18.4 0.467	91.0 2.311	49.4	51.7	47.2	9.77	10.2	1.90	1.13

H = 25.0 mils (0.635 mm), Section Length = 0.200 in. (0.508 cm), T_m = 0.25 mil (6.35μm),

$\epsilon_1 = 10.0$, $\epsilon_1' = 10.1$

$\tan \delta_1 = 6.0 \times 10^{-4}$, $\sigma = 4.10 \times 10^7$ mho/m[10].

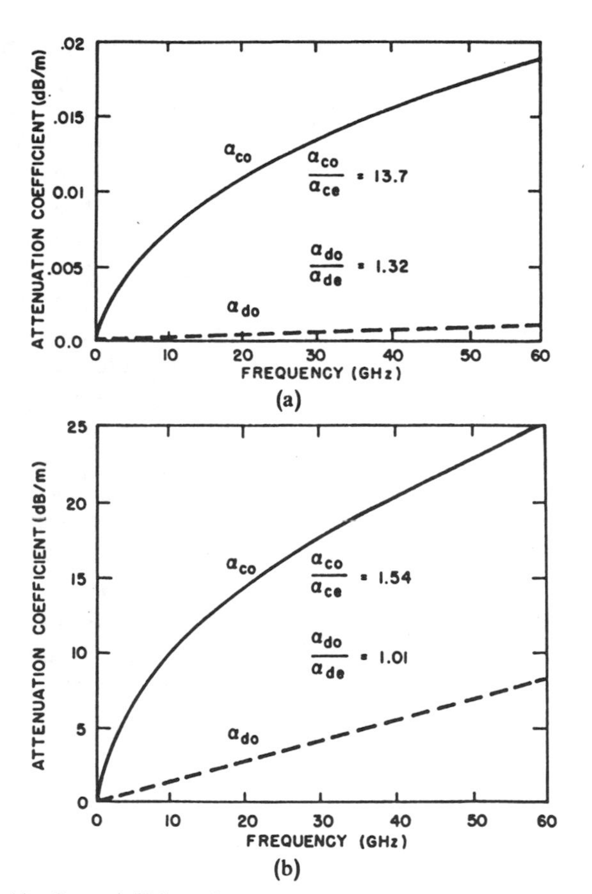

Fig. 11. Loss coefficients for sections of a 6.7-dB two-section coupler. (a) Tight section. (b) Loose section.

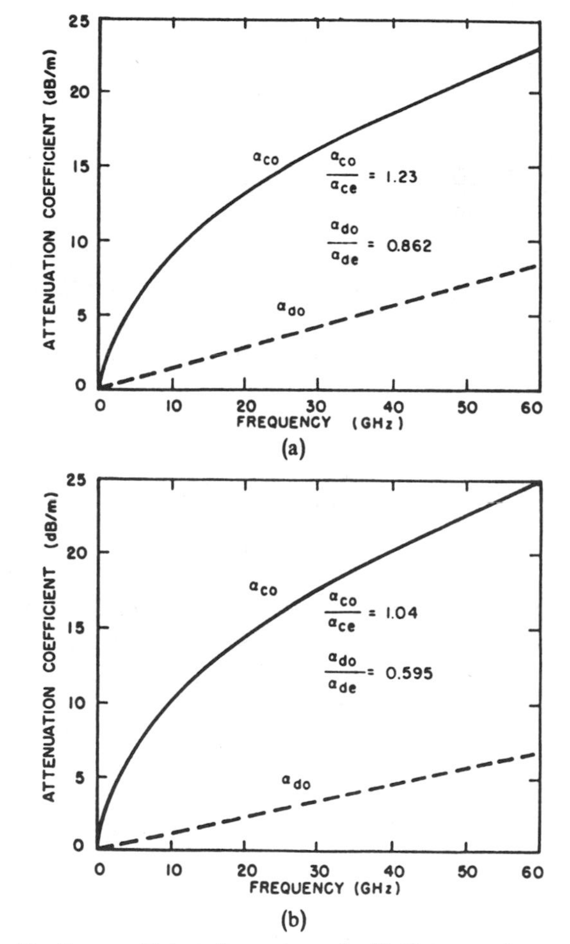

Fig. 12. Loss coefficients for sections of a 20-dB two-section coupler. (a) Tight section. (b) Loose section.

evaluation scheme similar to the one described in the previous paragraph the dissipation loss for the coupled line sections was determined to be 0.08 dB. After adding contributions for connector and lead-in line lengths the total dissipation loss was computed to be 0.38 dB. This agreed well with the measured midband (5.4 GHz) dissipation loss of 0.4 dB.

Another portrayal of the loss characteristics computed for these four coupled-line sections is shown in Figs. 11 and 12. Fig. 11 shows plots of α_{co} and α_{do} versus the frequency for the tight (section number 1) and loose (section number 2) coupling sections used in the 6.7-dB coupler. Also shown are the ratios α_{co}/α_{ce} and α_{do}/α_{de}. It is interesting to note the large disparity between the even- and odd-mode conductor losses for the tight section. Fig. 12 shows similar data for the tight (section number 3) and loose (section number 4) coupling sections used to make up the 20-dB coupler.

IV. Discussion

This paper details and illustrates analyses, amenable to computer programming, which can be used to evaluate dissipation losses in isolated or coupled microwave and millimeter-wave transmission lines. These lines can be composed of both imperfect conductors and piecewise-homogeneous dielectric materials. The analyses described here have been implemented in the form of digital computer programs [4] for treating: microstrip, coplanar waveguide, inverted microstrip, trapped inverted microstrip, and edge-coupled microstrip with dielectric overlay.

The programs for microstrip and coplanar waveguide were used to analyze a variety of configurations for these transmission lines. The computed values of conductor loss for microstrip on alumina have been compared [4] with the well-accepted values of Schneider [9] and were found to agree within better than 1.6 dB/m for frequencies up through 18 GHz. The computed values for dielectric losses in microstrip agree with Simpson *et al.* [11] and Schneider [12] and agreed to within 3 percent or better at frequencies up to 18 GHz. Computed total loss constants for microstrip have been compared with the results of Pucel *et al.* [13] and Gopinath *et al.* [14] and agree to within 0.01 dB/cm over the frequency ranges cited in these references. The sum of conductor and dielectric losses computed for coplanar waveguide agreed with the experimental values of McDade and Stockman [15] to within 0.1 dB/wavelength. The experimental determinations were made by two independent techniques.

Conductor and dielectric loss coefficients were also evaluated and presented here for inverted microstrip and trapped inverted microstrip. This information was presented in the form of useful design curves, providing loss values as functions of the aspect ratios for these lines. It was encouraging to note the agreement (better than 2 percent) of the computed phase velocities $\sqrt{\varepsilon_{eff}}$ for inverted microstrip when compared to measured data [9]. These calculations are made using the same equivalent charge densities that are used in the loss calculations.

A comparison of computed total loss coefficients for the four types of lines just discussed is shown in Fig. 8. It is interesting to note that the losses incurred in inverted and trapped inverted microstrip at frequencies as high as 60 GHz are comparable to those incurred in microstrip for frequencies in the 5–10-GHz frequency range. It is also seen that coplanar waveguide appears to be considerably more lossy than microstrip. The structures compared in this figure were selected to have characteristic impedance levels of nominally 50 Ω. Consistent with this characteristic, the conducting strip widths required for microstrip and coplanar waveguide were approximately two and one half to three times narrower than those required for inverted and trapped inverted microstrip. This feature enhances the attractiveness of the inverted and trapped inverted microstrip lines by virtue of the mitigation of fabrication difficulties. By concentrating more of the field energies in air (with correspondingly lower ε_{eff}), wider strips are possible for a prescribed impedance level (compared to the other two lines). These advantages should be realized even if the dimensions must be contracted to suppress higher order moding at higher frequencies. This should be investigated more extensively. Also, it should be noted that the transmission lines shown in Figs. 1, 5, and 9 may have to be shielded for some practical circuit applications.

The analyses for computing conductor and dielectric loss coefficients and for relating these parameters to terminal-electrical characteristics were applied to edge-coupled microstrip with a dielectric overlay. Computed results were compared to measured characteristics for two experimental models of two-section couplers, having nominal coupling values of 6.7 and 20 dB, respectively. The total dissipation values computed for these couplers agreed with experimental evaluations within 0.07 dB.

It is interesting to note that the difference between the even- and odd-mode loss coefficients for this structure gives rise to a finite isolation for an otherwise perfect coupled-line section. For the 6.7-dB coupler considered here, the tight and loose sections are limited to maximum isolation levels of 42 and 59 dB, respectively, due to this phenomenon. The tight and loose sections of the 20-dB coupler are similarly limited to 70- and 92-dB isolation levels, respectively.

There are several factors which can contribute to error in the analyses described here. One is the discrete representation of charge-density distributions employed in the quasi-TEM model employed. In one sense, this discretization can be viewed as a "surface roughness" which could lead to a high estimate of losses. Mathematical smoothing of the charge-density distributions might lead to improvement. An approach which could improve the accuracy is to solve the magnetostatic problem to evaluate the magnetic field at conductor surfaces for these transmission media. Certainly the difference quotient approximation used to arrive at (15) introduces error in the dielectric loss evaluations. Also, judicious choices must be made for the dc conductivity of the metal system and the loss tangents for dielectrics.

Acknowledgment

The author extends his sincere thanks to Ms. Rosemary Kelly for preparing the manuscript for this paper.

References

[1] B. E. Spielman, "Analysis of electrical characteristics of edge-coupled microstrip lines with a dielectric overlay," Naval Research Laboratory, Washington, DC, NRL Rep. 7810, Oct. 25, 1974.
[2] R. F. Harrington *et al.*, "Computation of Laplacian potentials by an equivalent source method," *Proc. Inst. Elec. Eng.*, vol. 116, no. 10, pp. 1715–20, Oct. 1969.
[3] R. F. Harrington, *Time-Harmonic Electromagnetic Fields*. New York: McGraw-Hill, 1961, pp. 61–67.
[4] B. E. Spielman, "Computer-aided analysis of dissipation losses in isolated and coupled transmission lines for microwave and millimeter wave integrated circuit applications," Naval Research Laboratory, Washington, DC, NRL Rep. 8009, July 1976.
[5] J. D. Jackson, *Classical Electrodynamics*. New York: Wiley, 1966, pp. 236–240.
[6] R. F. Harrington, *Time-Harmonic Electromagnetic Fields*. New York: McGraw-Hill, 1961, p. 50.
[7] Leo Young, Ed., *Advances in Microwaves*, Vol. 1. New York: Academic Press, 1966, pp. 115–209.

[8] B. Sheleg and B. E. Spielman, "Broad band directional couplers using microstrip with dielectric overlays," *IEEE Trans. Microwave Theory Tech.*, vol. MTT-22, pp. 1216–1220, Dec. 1974.
[9] M. V. Schneider, "Microstrip lines for microwave integrated circuits," *BSTJ*, vol. 48, no. 5, pp. 1421–1444, May–June 1969.
[10] C. D. Hodgman, Ed., *Handbook of Chemistry and Physics*. Cleveland, OH: Chemical Rubber Publishing Co., 1961, p. 2628.
[11] T. L. Simpson and B. Tseng, "Dielectric loss in microstrip lines," *IEEE Trans. Microwave Theory Tech.*, vol. MTT-24, pp. 106–108, Feb. 1976.
[12] M. V. Schneider, "Dielectric loss in integrated microwave circuits," *BSTJ*, vol. 48, pp. 2325–2332, Sept. 1969.
[13] R. A. Pucel, D. J. Masse, and C. P. Hartwig, "Losses in microstrip," *IEEE Trans. Microwave Theory Tech.*, vol. MTT-16, pp. 342–350, June 1968.
[14] A. Gopinath, R. Horton, and B. Easter, "Microstrip loss calculations," *Electron. Letts.*, vol. 6, no. 2, pp. 40–41, Jan. 22, 1970.
[15] J. McDade and D. Stockman, "Microwave integrated circuit techniques," AFAL/TEM, Wright Patterson AFB, OH, Tech. Rep. AFAL-TR-73-234, May 1973.

Part IV
Suspended Lines

AS the operating frequencies are increased to the millimeter wave band, the attenuation in a microstrip line is often too high to be tolerated. A possible remedy is the use of a suspended substrate configuration. In this arrangement, a conducting strip is placed on a low loss substrate which is suspended over the ground plane by a substantial air space. The thickness of the substrate is usually much smaller than the thickness of the air region under the substrate. Hence, the major portion of the electric field between the strip and the ground plane is in the lossless air region. To mechanically support the substrate, the structure is usually placed in a waveguide-type housing. The substrate is supported mechanically in small grooves created in the side walls along the propagating direction.

There are several useful features in this configuration. First, the attenuation due to the dielectric material is reduced. Second, it is possible to use both sides of the substrate. This feature is often used for fabrication of a broadside coupled stripline filter. On the other hand, there are some problems. For instance, the realizable range of the characteristic impedance is limited. A low impedance line is difficult to obtain. This is caused by the restriction placed on the size of the waveguide-type housing so as not to allow propagating higher order (waveguide) modes.

Since the suspended line automatically includes at least two layers of material media below the strip (such as the substrate and the air), the analysis method for the structure is often more generalized so that a number of layers with arbitrary permittivities exist both below and above the strip. Slot-type configurations in place of, or in addition to, the strips can be combined in such a generalized analysis (Jansen). Additionally, as mentioned above, the grooves on the side walls of the waveguide housing can affect the propagation characteristics of the suspended line mode (Yamashita, *et al.*; Vahldieck and Bornemann). The papers selected for this part deal with the problems addressed above.

Unified user-oriented computation of shielded, covered and open planar microwave and millimeter-wave transmission-line characteristics

R.H. Jansen

Indexing terms: *Striplines, Waveguide theory*

Abstract: A unified, rigorous and efficient hybrid-mode solution to the general planar 3-layer transmission-line problem with shielded, covered and open cross-sectional geometry is achieved by the spectral-domain method. Stress is put on the minimisation of the computational expense required to obtain the frequency-dependent design data of planar line structures with realistic geometrical dimensions. This implies the modal propagation constants, uniformly defined characteristic impedances and any field quantities. On the basis of user-oriented considerations a computer program has been developed which is applicable to fundamental, and higher, even and odd modes on single and coupled coplanar strips and slots. This computational approach includes most of the cases which today are of technical interest, e.g. stripline and microstrip, microstrip with dielectric overlay, suspended substrate strips and slots, slotline, coplanar line and grounded or isolated fin line. It combines analytical simplicity, rigorousness and high efficiency with the possibility of performing studies on numerical accuracy and convergence.

1 Introduction

The spectral domain hybrid mode approach to the planar transmission-line problem[1-25] can today be considered the most efficient rigorous approach in this field if it uses complete expansions and, particularly, if it includes the singular edge behaviour into its strip-current or slot-field representations. Recently, this technique has even been applied to arbitrary shielded coplanar transmission lines on a multilayer substrate, and is reported to result in short computing times.[20] With a proper choice of expansion functions, it leads to accurate numerical solutions with very low determinantal orders of the eigenvalue equations, not only for the phase constants but even for the more critical, field sensitive, characteristic impedances.[3,14,15,23-25] For designers who frequently employ a computer for the determination of reliable transmission-line data this is of great importance.

This is, at the same time, one of the crucial points in the spectral-domain approach. For practical applications of this technique low determinantal orders are mandatory since the computational expense required for the repeated generation of the coefficients of the final eigenvalue equations is high. These coefficients are improper integrals in the case of an open or covered line cross-section (as defined in Fig. 1, see e.g. Reference 6) and are infinite series in the shielded case.[9] Their convergence with respect to the upper integration or summation limit can be shown to be no more than linear if expansion functions are used which satisfy the edge condition. Otherwise, i.e. for non-singular basis functions, the coefficients converge more rapidly,[6,9,11] but this is paid for expensively by an increase of the determinantal order necessary to achieve the same accuracy. Some results concerning this point will be presented here.

Paper T291 M, received 25th September 1978

Dr. Jansen is with the Institut für Hochfrequenztechnik, Technical University of Aachen, Alte Maastrichter Strasse 25, D5100 Aachen, West Germany

Another user-oriented aspect in connection with the spectral-domain method is the proper choice of the type of cross-sectional geometry to be employed in the computations. Here, authors often argue that shielding is necessary to model the inevitable packaging in practice, and allows a study of its influence. This is certainly true for the conducting ground and cover planes shown in Fig. 1, particularly since these planes can be removed to infinity for a modelling of the open case without causing numerical problems or increasing computing times.[24] It applies in some way too for the lateral shielding, especially if the transmission lines considered are tightly encapsulated in a waveguide as, for example, in millimeter-wave applications.[21] However, for the modelling of realistic planar microwave-line geometries which can mostly be considered as laterally near-open structures, the shielded approach is not ideally suited.

The reason for this is an increase of the spatial spectral density of the electromagnetic-field representation directly proportional to the lateral shielding/line-width ratio. As a consequence, the number of terms to be taken into account in the series of coefficients also increases linearly.[12,21] For example, modelling an open structure with a shielding/line-width ratio of 30 and a determinantal order of 20 (Reference 21) requires the numerical generation of approximately 10^5 series contributions in each step of the eigenvalue search. Particularly if narrow strips or slots, tightly coupled lines, suspended substrate structures or slot configurations operating in the lower gigahertz region are analysed, this becomes a problem.[20,21] The unified approach to the covered, open and shielded geometries which will be described here provides flexibility and avoids these difficulties.

Beyond this, it is assumed here that the consideration of 3-layer structures with arbitrary dielectrics is sufficiently general for most practical applications. No restrictions, such as the symmetry condition used in Reference 21, are imposed concerning the positions of the upper and lower ground planes. The definition of characteristic impedance adopted here is uniformly based on the power transported along the line and an integral over the localised physical

Reprinted with permission from *Microwaves, Optics Acoust.*, vol. 3, no. 1, pp. 14–22, January 1979.

expansion quantities. The convergence of this design parameter in the spectral-domain method will be discussed. The numerical solutions show that first- or second-order basis functions do not provide accurate results for certain configurations. In addition, computed data are presented which throw some light on the question of numerical efficiency. The correctness of the computer solutions is proved by comparison with previous authors and several test measurements on standard alumina substrates.

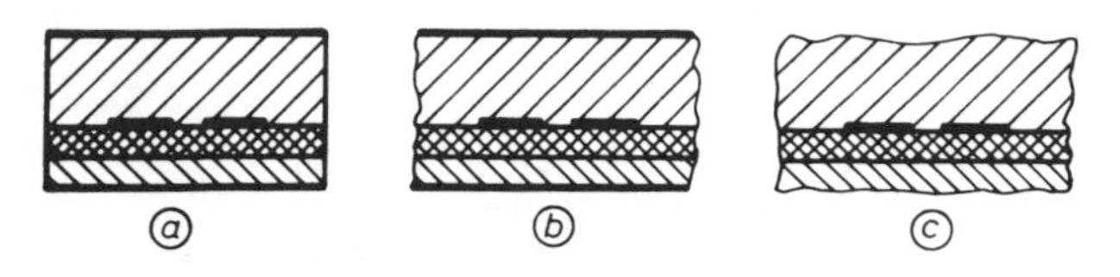

Fig. 1 *Cross-sectional geometry of planar 3-layer transmission line*

a Shielded type
b Covered type
c Open type

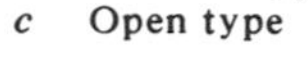

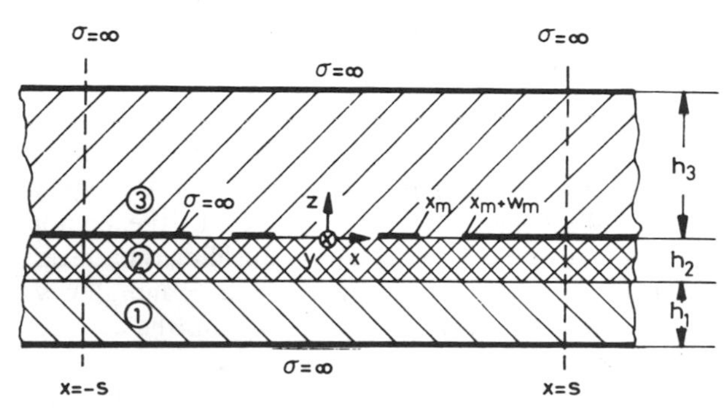

Fig. 2 *Representative planar transmission-line geometry*

2 Analysis and computational procedure

The setting up of the spectral domain field relations for planar transmission-line structures can, by now, be considered a well known analytical procedure. So, many steps may be omitted in the following mathematical derivation. This is based on the representative geometry of Fig. 2 which consists of three isotropic homogeneous layers of arbitrary thickness and relative permittivity. The metallic planes involved and the zero-thickness metallisation pattern are assumed to have infinite conductivity. For the formulation of the electromagnetic boundary-value problem vector potentials normal to the dielectric interfaces, i.e. LSE- and LSM- scalar potentials,[26,27] are employed. These are particularly suited to simplify the necessary algebraic manipulations from the beginning and can be written as

$$\phi_1 = +\sum_{n=1}^{\infty} W_n F_n \begin{Bmatrix} \sin \\ \cos \end{Bmatrix} (k_{xn}x) \frac{T_{n1} k_{zn2}}{k_{zn1}^2} \frac{\sin\{k_{zn1}(z+h_1+h_2)\}}{\sin(k_{zn1}h_1)\cos(k_{zn2}h_2)} \tag{1}$$

$$\psi_1 = +\sum_{n=1}^{\infty} W_n G_n \begin{Bmatrix} \cos \\ \sin \end{Bmatrix} (k_{xn}x) \frac{\cos\{k_{zn1}(z+h_1+h_2)\}}{\cos(k_{zn1}h_1)\cos(k_{zn2}h_2)}$$

$$\phi_2 = +\sum_{n=1}^{\infty} W_n F_n \begin{Bmatrix} \sin \\ \cos \end{Bmatrix} (k_{xn}x) \times \left(\frac{\sin\{k_{zn2}(z+h_2)\}}{\cos(k_{zn2}h_2)} + \frac{T_{n1}k_{zn2}}{k_{zn1}^2} \frac{\cos\{k_{zn2}(z+h_2)\}}{\cos(k_{zn2}h_2)} \right)$$

$$\psi_2 = +\sum_{n=1}^{\infty} W_n G_n \begin{Bmatrix} \cos \\ \sin \end{Bmatrix} (k_{xn}x) \times \left(\frac{\cos\{k_{zn2}(z+h_2)\}}{\cos(k_{zn2}h_2)} - \frac{T_{n1}\epsilon_2}{\epsilon_1 k_{zn2}} \frac{\sin\{k_{zn2}(z+h_2)\}}{\cos(k_{zn2}h_2)} \right)$$

$$\phi_3 = -\sum_{n=1}^{\infty} W_n F_n \begin{Bmatrix} \sin \\ \cos \end{Bmatrix} (k_{xn}x) \times \left\{ \left(\frac{T_{n1}}{k_{zn1}^2} + \frac{T_{n2}}{k_{zn2}^2} \right) k_{zn2} \right\} \frac{\sin\{k_{zn3}(z-h_3)\}}{\sin(k_{zn3}h_3)}$$

$$\psi_3 = -\sum_{n=1}^{\infty} W_n G_n \begin{Bmatrix} \cos \\ \sin \end{Bmatrix} (k_{xn}x) \times \left\{ \left(\frac{T_{n1}}{\epsilon_1} + \frac{T_{n2}}{\epsilon_2} \right) \frac{\epsilon_3}{T_{n3}} \right\} \frac{\cos\{k_{zn3}(z-h_3)\}}{\cos(k_{zn3}h_3)}$$

with

$$W_n = \pi \frac{w}{s} \qquad k_{xn} = \begin{Bmatrix} (n-0\cdot5) \\ (n-\ \ 1\ ) \end{Bmatrix} \frac{\pi}{s} \tag{1a}$$

and

$$W_n = \frac{\pi}{2} V_l, \qquad k_{xn} = \frac{\pi}{2w}(2k-1+X_l) \qquad n = \{(k-1)L+l\} \tag{1b}$$

respectively, and

$$k_{zn1} = \sqrt{k_0^2\epsilon_1 + \gamma^2 - k_{xn}^2} \qquad T_{n1} = k_{zn1}\tan(k_{zn1}) \quad \text{etc.} \tag{1c}$$

The notation used here, and in the following, is one in which the upper terms refer to even modes (magnetic wall in the plane of symmetry $x = 0$) whereas the lower ones describe odd modes (electric wall at $x = 0$). γ denotes the purely imaginary or real propagation constant and k_0 is the free-space wavenumber. The longitudinal dependence of $\exp(-\gamma y)$ and the harmonic time factor have already been extracted from the scalar potentials in eqn. 1. From these the hybrid-mode electromagnetic field is derived by superposition and in the standard fashion.[26,27] Note that the electric field associated with eqn. 1 satisfies all its continuity conditions at the two dielectric interfaces. Also, the continuity of the magnetic field tangential to the lower dielectric interface is warranted and the boundary conditions at the conducting ground and cover plane are incorporated. Therefore, only two sets of unknown spectral amplitudes are present in the above formulation, namely F_n and G_n. The boundary conditions on an eventual conducting lateral shielding at $x = s$ are satisfied by the choice of the discrete spectral variable k_{xn} given in eqn. 1*a*. For the laterally open case, eqn. 1*b* is valid with likewise discrete values of k_{xn} which, however, are now related to an Lth-order Gaussian integration algorithm[28] in intervals k of length π/w, with V_l as the weights and with the abscissas X_l. The quantity w is a normalisation width introduced for numerical reasons and not necessarily one of the strip or slot widths.

So, altogether, the formulation of eqn. 1 is a unified representation for guided waves on shielded, covered or open planar transmission lines. It naturally arises from the fact that for a real spectral variable k_x the finite Fourier transform[29] approaches the infinite case if the lateral shielding is removed far away from the line. Its advantage is a great flexibility in the choice of the spectral abscissas k_{xn} in the series of eqn. 1. As a consequence, the numerical efficiency of the shielded approach is preserved even for the limit of infinite shielding width.

In the following, it also turns out that great algebraic simplicity results from the use of eqn. 1. A surface-current density representing the jump of the magnetic field tangential to the metallisation in the upper dielectric interface at $z = 0$ is introduced. The electric field e at $z = 0$ is evaluated and, after suitable normalisation,[24] an analytical, compact relationship between the spectral components of e and the current density j can be set up as follows:

$$\begin{pmatrix} j_{xn} \\ j_{yn} \end{pmatrix} = (Y_n)\begin{pmatrix} e_{xn} \\ e_{yn} \end{pmatrix}, \quad \begin{pmatrix} e_{xn} \\ e_{yn} \end{pmatrix} = (Y_n)^{-1}\begin{pmatrix} j_{xn} \\ j_{yn} \end{pmatrix} = (Z_n)\begin{pmatrix} j_{xn} \\ j_{yn} \end{pmatrix} \tag{2}$$

$$(Y_n) = \begin{pmatrix} Y_{an} & Y_{bn} \\ Y_{bn} & Y_{cn} \end{pmatrix} = \begin{pmatrix} \text{sign}(\gamma^2)(k_{xn}^2 Y_{Gn} - \gamma^2 Y_{Fn}) & \mp |\gamma| k_{xn}(Y_{Gn} - Y_{Fn}) \\ \mp |\gamma| k_{xn}(Y_{Gn} - Y_{Fn}) & \gamma^2 Y_{Gn} - k_{xn}^2 Y_{Fn} \end{pmatrix}$$

$$Y_{Fn} = \frac{\left\{1 - \dfrac{T_{n1}T_{n2}}{k_{zn1}^2} + \left(\dfrac{T_{n1}}{k_{zn1}^2} + \dfrac{T_{n2}}{k_{zn2}^2}\right)\dfrac{k_{zn3}^2}{T_{n3}}\right\}}{\left\{(\gamma^2 k_{xn}^2)\left(\dfrac{T_{n1}}{k_{zn1}^2} + \dfrac{T_{n2}}{k_{zn2}^2}\right)\right\}}$$

$$Y_{Gn} = \frac{\left\{1 - \dfrac{\epsilon_2}{\epsilon_1}\dfrac{T_{n1}T_{n2}}{k_{zn2}^2} + \left(\dfrac{T_{n1}}{\epsilon_1} + \dfrac{T_{n2}}{\epsilon_2}\right)\dfrac{\epsilon_3}{T_{n3}}\right\}k_0^2}{\left\{(\gamma^2 - k_{xn}^2)\left(\dfrac{T_{n1}}{\epsilon_1} + \dfrac{T_{n2}}{\epsilon_2}\right)\right\}}$$

This relationship is real and symmetric for both propagation and attenuation modes. It is also valid for 2-layer structures like microstrip since the bottom layer can be eliminated simply by introducing $h_1 = 0$ ($T_{n1} = 0$) without causing numerical difficulties. In the limit of large values of h_1 and h_3 it should lead to the same results as those reported by Knorr and Kuchler[15] if the spectral variable of eqn. 1*b* is used in eqn. 2 and the expansion functions chosen for the Galerkin solution applied to eqn. 2 are the same. Extensive algebraic manipulations[15] are not necessary to achieve computational efficiency. The matrix of coefficients of the final Galerkin equations (see for example Reference 9) takes a very simple form, namely

$$(A_{ik}) = \sum_{n=1}^{\infty} \begin{pmatrix} Y_{an}e_{xn}^i e_{xn}^k & Y_{bn}e_{xn}^i e_{yn}^k \\ Y_{bn}e_{yn}^i e_{xn}^k & Y_{cn}e_{yn}^i e_{yn}^k \end{pmatrix} \qquad \begin{matrix} i = 1, \ldots, M \\ k = 1, \ldots, M \end{matrix} \tag{3}$$

for an array of slots and, with an interchange of Y to Z and e to j, for a strip configuration. Note that the elements of (A_{ik}) can be formulated as a symmetric product of diagonal matrices $(Y_{an}) \ldots (Z_{cn})$ with the frequency-independent column vectors $(e_{xn}^i) \ldots (j_{yn}^k)$ which is important to obtain numerical efficiency in broadband computations.[24] The spectral components $e_{xn}^i \ldots j_{yn}^k$ of the expansion functions used are generated easily in an automated fashion by the computer algorithm for arbitrary strip or slot patterns of reasonable complexity. At present up to four strips or slots (symmetrical pattern) can be handled by choosing a control parameter in the queue of the input data.

In order that this is feasible with low computing times a very careful choice of the set of expansion functions has to be made. Especially near modal,[5, 23-25] complete sets of basis functions satisfying the edge condition term by term have to be chosen for the prevailing general kind of problem which will be demonstrated in Section 3. Explicity, the choice made here is

$$\frac{\cos\{(i-1)\pi(x - x_m)/w_m\}}{[1 - \{2(x - x_m)/w_m - 1\}^2]^{1/2}}$$

$$\text{i.e. } C_{mn}\{B_{mn}(-i+1) \pm B_{mn}(i-1)\} \tag{4}$$

for e_{xn}^i, j_{yn}^i

$$\frac{\sin\{i\pi(x - x_m)/w_m\}}{[1 - \{2(x - x_m)/w_m - 1\}^2]^{1/2}}$$

$$\text{i.e. } S_{mn}\{B_{mn}(-i) \mp B_{mn}(i)\}$$

for e_{yn}^i, j_{xn}^i

$$C_{mn} = \begin{Bmatrix} \cos \\ \sin \end{Bmatrix}\left[\left(0{\cdot}5\frac{w_m}{w} + \frac{x_m}{w}\right)K_{xn}\right]$$

$$S_{mn} = \begin{Bmatrix} \sin \\ \cos \end{Bmatrix}\left[\left(0{\cdot}5\frac{w_m}{w} + \frac{x_m}{w}\right)K_{xn}\right]$$

$$B_{mn}(i) = B_0\left\{0{\cdot}5\left(\frac{w_m}{w}K_{xn} + i\pi\right)\right\}$$

which applies to strips of slots which are located between x_m and $x_m + w_m$. B_0 denotes the zero-order Bessel function of the first kind[28] and $K_{xn} = k_{xn}w$ is the normalised spectral variable. Depending on the actual value of i and the symmetry condition considered at $x = 0$ the trigonometric factors and the signs in the right half of eqn. 4 have to be properly interchanged. By insertion of the spectral components $e_{xn}^i \ldots j_{xn}^i$ of eqn. 4 into the series of coefficients in eqn. 3 the eigenvalue equation of the transmission-line problem is generated. The modal propagation constants γ are found as the roots of the determinant of $(A_{ik}(\gamma))$. Finally, a process of back-substitution yields the field quantities and the current-density distribution of the guided wave under consideration.

With the electromagnetic field information available, strip and slot characteristic impedances can be computed. However, owing to some arbitrariness in the definition of characteristic impedance for hybrid modes (see for example Reference 14), the criteria which form the basis of such a computation have first to be discussed. Here a dominant requirement is the usefulness of the quantity Z_L for design purposes. This means that the approximate solution of certain planar circuit problems, like for example that of a T-junction or a coupler,[15, 23] should be made possible by employing the value of Z_L in simple transmission-line formulas instead of solving a new boundary-value problem. So, this is equivalent to requiring the applicability of a network approach. Therefore, the physical quantities constituting the characteristic impedance should, as far as possible, be lumped, i.e. localised tightly compared to wavelength, like longitudinal strip current and transverse slot voltage. Transported power is also a relatively localised quantity since its lateral decay away from a planar line is much faster than that of the field itself. The use of strip-to-ground voltages or of the current parallel to a slot can be justified for quasi-TEM modes with strongly concentrated field distribution, e.g. for microstrip[23–25] or for fairly narrow slots.[21] Certainly it is not an optimal choice for wide slots, coupled slots and suspended substrate lines with large ground-plane separation.

Consequently a definition of characteristic impedance based on the power transported along the planar line promises maximum usefulness for design purposes, which is in agreement with previous authors.[15] In addition, it allows a unified computation for strips and slots. Note that the slot voltage U_x and the strip current I_y both result from a 1-dimensional integration over the unknowns e_x, j_y, respectively, by means of which the problem is formulated. So, finally, the definition adopted here for the mth slot or strip in a planar configuration is

$$Y_{Lm} = \frac{2P_m}{U_{xm}^2} = \frac{1}{U_{xm}^2}\,\mathrm{Re}\iint (e_m \times h^*)_y\,dx\,dz, \quad h = \sum_m h_m \tag{5a}$$

$$Z_{Lm} = \frac{2P_m}{I_{ym}^2} = \frac{1}{I_{ym}^2}\,\mathrm{Re}\iint (e \times h_m^*)_y\,dx\,dz, \quad e = \sum_m e_m \tag{5b}$$

In this definition eqn. $5b$ results from eqn. $5a$ by an interchange of dual quantities analogous to that performed in eqn. 3. By the utilisation of partial fields e_m, h_m each of which is associated only with the mth portion of the total of the expansion functions, the characteristic impedance (eqn. 5) is a generalisation which applies even if more than two slots or strips are present in a problem. It includes the special cases treated by Knorr and Kuchler.[15] In the limit of very loosely coupled structures it automatically approaches the single-line value.

For an easier numerical evaluation of line impedances, an explicit detailed expression of the power transported along a planar 3-layer configuration has been derived from the formulation of eqn. 1. This can be accomplished in a relatively compact form and will therefore by provided here. With the abbreviations

$$\left.\begin{aligned}
P_{Fn} &= \frac{w}{4} W_n \frac{\gamma}{j\omega\mu_0}(k_{xn}^2 - \gamma^2)\,|F_n k_{zn2}|^2 \\
P_{Gn} &= \frac{w}{4} W_n \frac{-\gamma^*}{j\omega\epsilon_0}(k_{xn}^2 - \gamma^2)\,|G_n|^2 \\
&\text{and} \\
P_{FGn} &= \frac{w}{2} W_n \frac{|\gamma|}{k_0}(k_{xn}^2 - \gamma^2)\,F_n^* G_n k_{zn2}^*
\end{aligned}\right\} \tag{6a}$$

and

$$C_{n1} = 1/\cos^2(k_{zn1}h_1) \qquad S_{n1} = 1/\sin^2(k_{zn1}h_1)$$

$$U_{n1} = T_{n1}/k_{zn1}^2 \quad \text{etc.}$$

in which P_{Fn} and P_{Gn} represent the LSE- and the LSM-field contributions, whereas P_{FGn} is the mixed-field term, the total transported power is

$$\begin{aligned}
P = \sum_{n=1}^{\infty} P_{Fn} \Bigg[& U_{n1}^2 \left(h_1 S_{n1} - \frac{1}{T_{n1}}\right) C_{n2} \\
& + U_{n1}^2 \left\{ U_{n2}\left(1 + 2\frac{U_{n2}}{U_{n1}} - \frac{1}{U_{n1}^2 k_{zn2}^2}\right) \right. \\
& \left. + h_2 C_{n2}\left(1 + \frac{1}{U_{n1}^2 k_{zn2}^2}\right)\right\} \\
& + \left(h_3 S_{n3} - \frac{1}{T_{n3}}\right)(U_{n1} + U_{n2})^2 \Bigg] \\
+ \sum_{n=1}^{\infty} P_{Gn} \Bigg[& \frac{1}{\epsilon_1}(U_{n1} + h_1 C_{n1}) C_{n2} \\
& + \frac{1}{\epsilon_2}\left\{ U_{n2}\left(1 - 2\frac{\epsilon_2}{\epsilon_1} T_{n1} U_{n2} - \frac{\epsilon_2^2}{\epsilon_1^2}\frac{T_{n1}^2}{k_{zn2}^2}\right) \right. \\
& \left. + h_2 C_{n2}\left(1 + \frac{\epsilon_2^2}{\epsilon_1^2}\frac{T_{n1}^2}{k_{zn2}^2}\right)\right\} \\
& + \frac{1}{\epsilon_3}(U_{n3} + h_3 C_{n3})\left\{\left(\frac{T_{n1}}{\epsilon_1} + \frac{T_{n2}}{\epsilon_2}\right)\frac{\epsilon_3}{T_{n3}}\right\}^2 \Bigg] \\
\pm \sum_{n=1}^{\infty} P_{FGn} \frac{k_{xn}}{|\gamma| k_0} \Bigg[& \frac{1}{\epsilon_1} U_{n1} C_{n2} \\
& + \frac{1}{\epsilon_2} U_{n2}\left\{1 - \frac{\epsilon_2}{\epsilon_1} T_{n1} U_{n1} - T_{n1} U_{n2}\left(\frac{\epsilon_2}{\epsilon_1} + \frac{k_{zn2}^2}{k_{zn1}^2}\right)\right\} \\
& + \frac{1}{\epsilon_3}(U_{n1} + U_{n2})\left(\frac{T_{n1}}{\epsilon_1} + \frac{T_{n2}}{\epsilon_2}\right)\frac{\epsilon_3}{T_{n3}} \Bigg].
\end{aligned} \tag{6b}$$

The partial transported-power terms P_m of eqn. 5 are computed in complete analogy, except that partial amplitudes F_{mn} and G_{mn} have to be used in eqn. $6a$. These are linearly related to the mth terms e_m, j_m of the solution in the same way as F_n, G_n, respectively, are to the total solution. In the case of symmetrically coupled lines $P_1 = P_2 = 0{\cdot}5P$ is valid, in equivalence to eqn. 5. So, independent of an interest in a general-case solution according to eqn. 5, the above power expressions are useful for single- and coupled-line computations.

3 Results

Without any exception, the numerical results presented in this paper have been obtained by means of a single computer program. The storage required by this program amounts to a total of 56000_8 words of memory on a Control Data Cyber 175 (computer centre of the Technical University of Aachen) if solutions up to the twelfth order ($M = 6$) and up to 600 series contributions ($N = 600$) per matrix coefficient shall be considered in a manner providing maximum computing speed. In principle, values of N

greater than 600 can also be handled, which is important when studying the limits of the widely shielded case; however there are no time-saving capabilities for the terms exceeding $N = 600$. This version is particularly suited for high-speed broadband computations. If storage requirements are a critical point, about 20000_8 words of memory can be saved by a slight reorganisation of this version of the program but, of course, at the cost of computer time. Since the numerical stability of the computations is excellent the wordlength of 60 bits at present (Cyber 175) can be reduced drastically and the implementation of the program on a comfortable desk-top computer seems feasible.

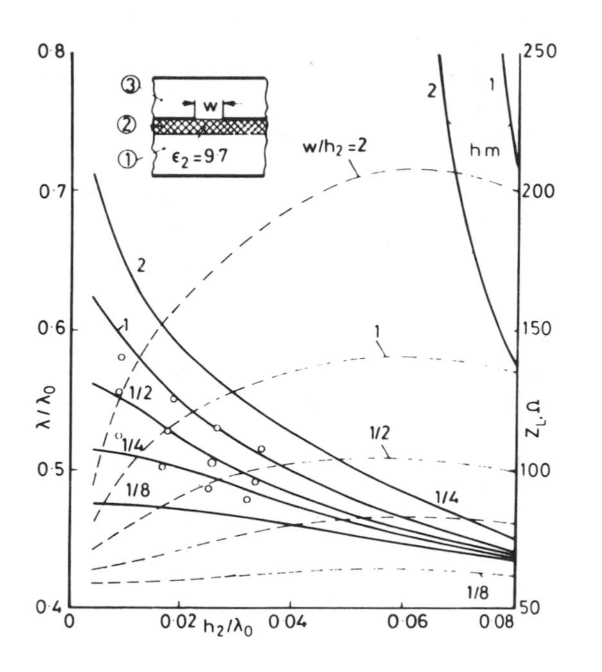

Fig. 3A *Covered slot-line characteristics*

$\epsilon_1 = \epsilon_3 = 1$
$h_1 = h_3 = 20\,h_2$
——— λ/λ_0
– – – Z_L Ω
o o o o test measurements
h.m. First higher mode

The correctness of the results reported here has been tested in several ways. For example, the papers by Cohn,[30] Mariani *et al.*,[31] Knorr *et al.*,[14, 15] Smith,[32] and Hofmann,[21] and recent computations performed by the author himself,[24, 25] have served as valuable sources of comparison. Reasonable agreement with all these sources has been achieved and some data taken from them have been incorporated into the Figures below. In addition, test measurements of the wavelengths of single and coupled slots on standard alumina substrates ($h_2 = 0{\cdot}64$ mm, $\epsilon_2 = 9{\cdot}7$) have been performed. These have been obtained by a modification of Deutsch's and Jung's method[33] to loosely coupled short-circuited resonating slots. Fig. 3A contains some of them and shows an average coincidence of about 1% with the computed characteristics. This method of measuring is not well suited for very narrow or very wide slots because, in the first case, the coupling probes have to be positioned extremely near the slots, whereas in the other case the Q-factors of the resonances are rather low. As to the impedance values computed in Fig. 3A these agree well with the results given in Fig. 2 of the paper by Mariani *et al.*[31] Nevertheless, in the present approach, the computed curves all run down to a normalised frequency of $h_2/\lambda_0 \approx 0{\cdot}004$ (2 GHz for $h_2 = 0{\cdot}64$ mm) which might be of interest for the application of narrow slotlines in the lower GHz-region, for example in mixers. Note, however, that in such low-frequency design applications the influence of the packaging must be considered. This is clearly visible in Fig. 3B even for a relatively wide slot of width $w/h_2 = 1$. Here, the influence of the lateral shielding on the slot wavelength is still noticeable for an aspect ratio of $s/w = 100$. The influence of s/w on the low-frequency characteristic impedance is of the same order of magnitude. Besides that, the approximate relative time factor t.f. depicted in Fig. 3B shows that at low-to-moderate frequencies the covered or open approaches are much faster than the shielded one if a wide shielding is used to model the open case (s/w chosen for less than 1% deviation from the equivalent covered solution).

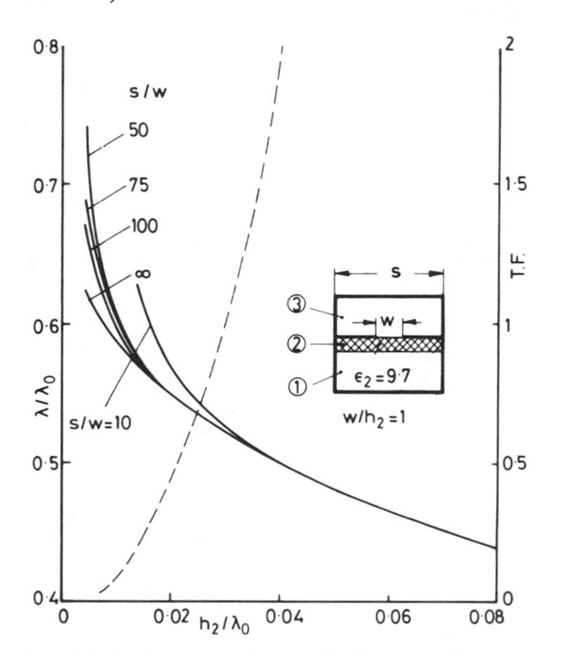

Fig. 3B *Influence of lateral shielding on the fundamental slot mode wavelength*

T.F. is factor relating the computing time of the covered ($s/w = \infty$) to that of the shielded approach
——— λ/λ_0
– – – t.f.

In the following, now, some general remarks have to be made concerning the numerical accuracy of the unified spectral-domain approach under discussion. For practical applications, usually, second-order single slot or strip solutions ($M = 1$, matrix order 2) and fourth-order solutions ($M = 2$, matrix order 4) for the symmetrically coupled versions provide sufficient accuracy, i.e. about 1% for both the phase constant and the characteristic impedance Z_L. Fig. 4 reveals the degree of inaccuracy which comes along with the use of a first-order only expansion as employed by Knorr and Kuchler[15] in the computation of coplanar strip characteristics (consider a factor of 2 in the definition of Z_L). Especially, for tightly coupled line structures the error introduced in this way cannot be tolerated. In such cases, at least one additional longitudinal expansion term is indespensable for a reasonable description of the true proximity distorted current or field distribution.

This interpretation, also, is supported by the convergence curves given in Fig. 5. These show indeed that a fourth-order edge-type solution is capable of producing an overall accuracy of 1% whereas a solution with $M = 1$

($K_{xN} = 10\pi M$) may still be far from the goal. In contrast to this, ordinary Fourier expansions, which have been studied in Fig. 5 by simply substituting Si-functions[28] for the zero-order Bessel terms in eqn. 4, exhibit relatively poor convergence properties. Particularly, the accuracy of the field-sensitive characteristic impedance is affected severely if the singular edge behaviour of the field is not incorporated explicitly into the numerical solution. In terms of the computational expense required to achieve the 1% margin, this means a roughly estimated increase of c.p. time by at least a factor of 25 compared to the edge-type solution. To illustrate this further, a relative, approximate measure of numerical expense, namely

$$\bar{M}:\bar{M}' = 40M + 2(2M+1)M^2 \quad \text{and}$$
$$\bar{M}'' = 20M^2 \qquad (7)$$

shall be given and discussed here. $\bar{M}'$ applies if a larger number of frequency points is considered and the frequency-independent structural information on the problem has been evaluated and stored away once, at the beginning of the computations. If the amount of storage necessary to take up the spectral contributions of the expansion functions cannot be spent the increased quantity $\bar{M}''$ applies. The total number of point operations or accesses which have to be performed during each step of the eigenvalue search is then obtained by multiplication of $\bar{M}$ with a factor ranging between 40 and 100. So, even for a very low determinantal order in the spectral-domain method, some thousands of point operations per step cannot be avoided, which puts an extreme stress on the proper choice of the expansion functions.

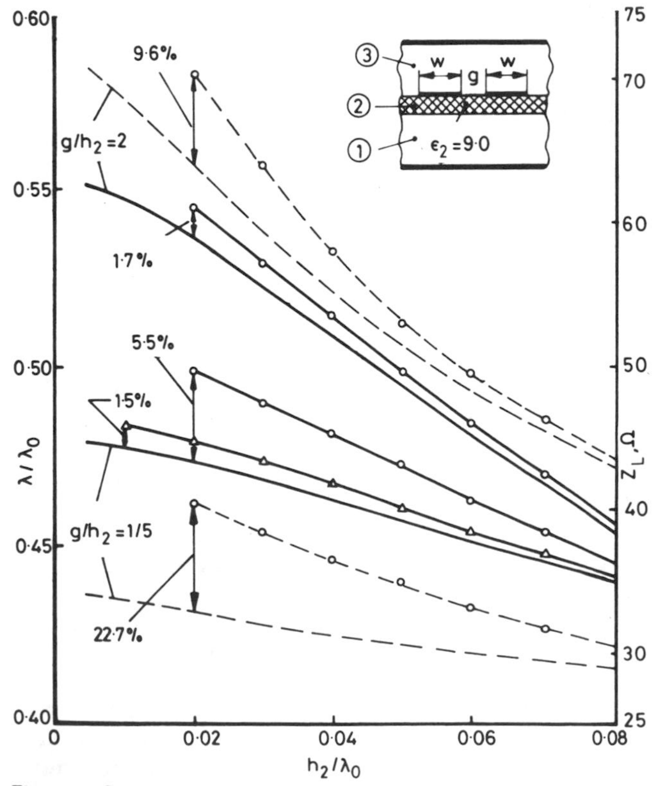

Fig. 4 *Coplanar-strip characteristics (fourth-order solution) in comparison with References 15 and 21*
△ Reference 21 (without edge terms)
○ Reference 15
——— λ/λ_0
– – – Z_L, Ω
$\epsilon_1 = \epsilon_3 = 1$
$h_1 = h_3 = 20\,h_2$
$w/h_2 = 1.5$

The situation is somewhat different for tightly shielded planar lines, as for example the fin lines of Fig. 6. Here, the aspect ratio and the selection of the variables in terms of which the problem is formulated determine the duration of a program run.[12, 13, 21] Except for small values of the line width w, the factor by which eqn. 7 has to be multiplied now is lower than for open problems. Besides this, the physical behaviour of the results depicted in Fig. 6 again confirms the correctness of the computations. As is expected, the corresponding electrical parameters of the isolated and the grounded fin line approach each other in the case of small slotwidths (consider the factor of 2 as a consequence of the definition in eqn. 5). That the line impedances calculated by Hofmann[21] on a voltage-per-current basis differ from the values resulting in this paper should not astonish very much. This is quite clear in view of the non-TEM character of the field of grounded fin lines.

In the next Figs. 7*a*–7*c* once more a laterally open, i.e. covered, configuration, is investigated. Specifically, the physical properties of the fundamental even mode (coplanar line quasi-TEM mode) and those of the odd mode on coupled slots are discussed. The Figures show that the phase velocity of the coplanar-type transmission-line mode depends only weakly on the width of the coupled slots. The same prevails for the low-frequency characteristic impedance of the odd mode. Moreover, due to its quasi-TEM character, the even mode exhibits only a few percent of frequency dispersion in its parameters and reacts fairly insensitively even to a rapid decrease of the ground-plane separation h_1. The results for both of the modes are again checked by some test measurements. Furthermore, the convergence speed of these two types of coupled-slot solutions is graphically represented in Fig. 7*c*. Here, the odd mode is similarly well behaved as for a single slot and could satisfactorily be described by a determinantal order of 2 ($M = 1$). On the other hand, the coplanar transmission-line mode poses the same accuracy problems as the coplanar strips in Fig. 5.

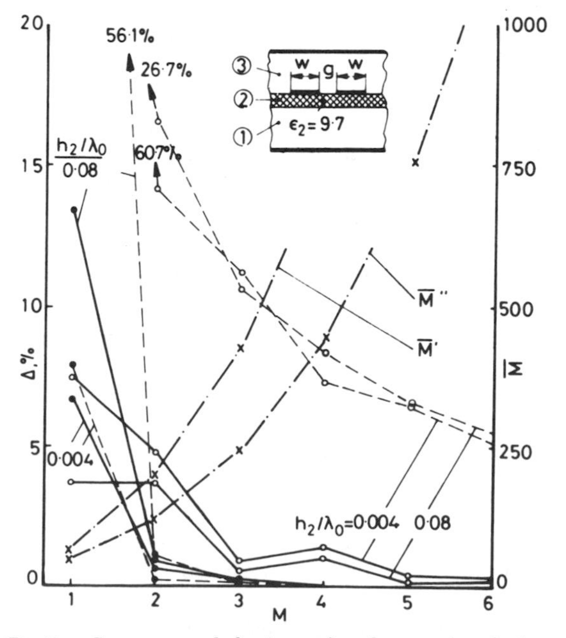

Fig. 5 *Convergence behaviour of coplanar strip solutions*
● with edge-type expansion functions
○ without edge-type expansion functions
$\bar{M}$ is a relative measure of numerical expense
——— $\Delta(\lambda/\lambda_0)$
– – – $\Delta(Z_L, \Omega)$
$w/h_2 = 1$
$g/h_2 = 1/8$

As a final interesting example of microwave and millimetre-wave transmission lines, the electrical properties of suspended substrate strips shall be analysed and interpreted. For the frequency-dependent characteristics of this class of lines merely a few specialised results[21] have been published to the present day. These may now serve for comparison and have, for this reason, been introduced into Fig. 8. Excellent agreement can be established there as far as the effective dielectric constant $\epsilon_{eff} = (\lambda_0/\lambda)^2$ of the various strips is concerned. Naturally, the high-frequency values of the line impedances Z_L do not coincide because the definitions applied are different. However, since the fundamental strip mode is a quasi-TEM wave this difference vanishes in the quasistatic limit.

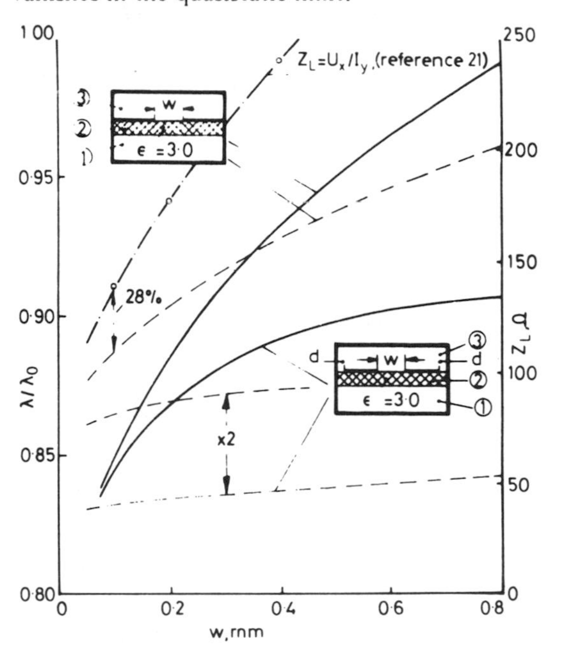

Fig. 6 *Characteristics of grounded and isolated fin lines in a WR 28 waveguide*

——— λ/λ_0
- - - - Z_L, Ω
$\epsilon_1 = \epsilon_3 = 1$
$h_1 = h_3 = 20\,h_2$; $h_2 = 0{\cdot}125$ mm
$h = 0{\cdot}125$ mm
$d = 0{\cdot}500$ mm
$f = 34$ GHz

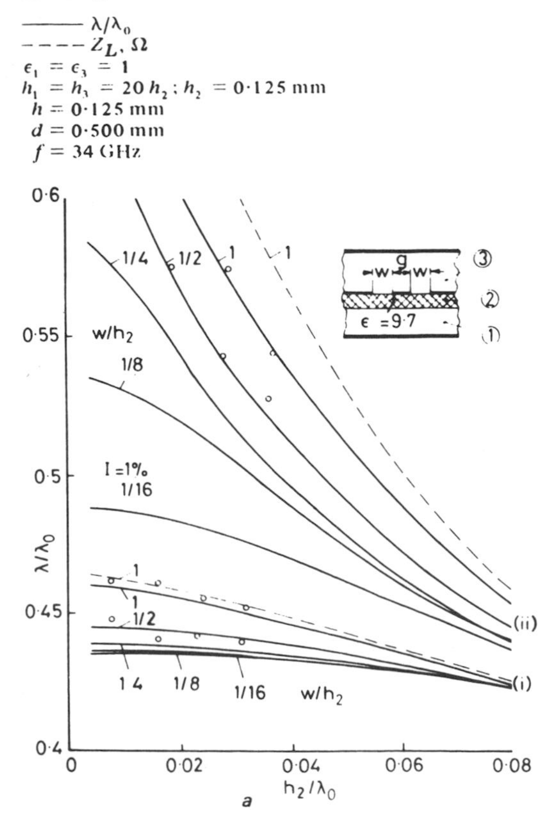

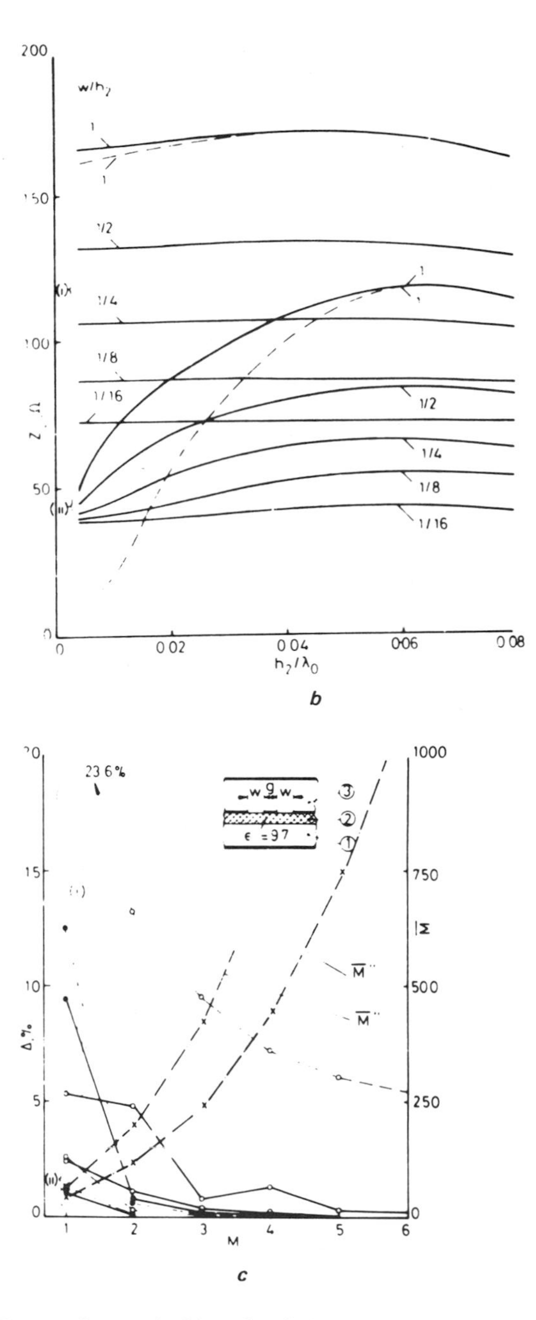

Fig. 7 *Even and odd mode solutions*

a Dispersion characteristics
——— $h_1 = 20\,h_2$
- - - $h_1 = h_2$
○ test measurements
$\epsilon_1 = \epsilon_3 = 1$
$g/h_2 = 1/2$
(i) even mode
(ii) Odd mode
$h_3 = 20\,h_2$

b Characteristic impedances for coupled slots ofdifferent widths
——— $h_1 = 20\,h_2$
- - - $h_1 = h_2$
$g/h_2 = 1/2$
$\epsilon_1 = \epsilon_3 = 1$
(i) Even mode
(ii) Odd mode
$h_3 = 20\,h_2$

c Convergence of solutions for coupled-slot configurations
● with edge-type expansion functions
○ without edge-type expansion functions
M is a measure of numerical expense
——— $\Delta(\lambda/\lambda_0)$
- - - $\Delta(Z_L, \Omega)$
$w/h_2 = 1$
$g/h_2 = 1/16$
$h_2/\lambda_0 = 0{\cdot}064$
(i) Even mode
(ii) Odd mode

Fig. 9 gives additional information on the behaviour of suspended strips. It features the line characteristics as a function of the normalised frequency h_2/λ_0 and the ground-plane separation h_1. Also, the influence of a lateral shielding is indicated. The parameter h_1 acts here as an effective means of designing transmission lines with an impedance level which is high compared to the microstrip case $h_1 = 0$. Therefore, shallow grooves in a metal plate attached at the substrate bottom ($h_1 = 0{\cdot}5 \ldots 2h_2$) can be utilised to create microstrip-compatible high-impedance lines without causing sensitivity problems.[25, 34] Similarly, the effect of the ground-plane separation on the even and odd mode characteristics of coupled suspended strips, as plotted in Fig. 10, can be employed with advantage. In this case the odd mode depends only modestly on the value of h_1 and the reason for this is physically obvious. So, the separation parameter can be adjusted for an equalisation of the even- and odd-mode phase velocities and high directivity couplers can be constructed[25] which seems to be practicable at least up to X-band frequencies. As a general rule for all of these suspended-strip geometries, flat packaging, i.e. especially a low ground-plane separation, reduces the sensitivity with respect to the lateral neighbourhood of metallic objects and thus enables higher circuit-integration densities.

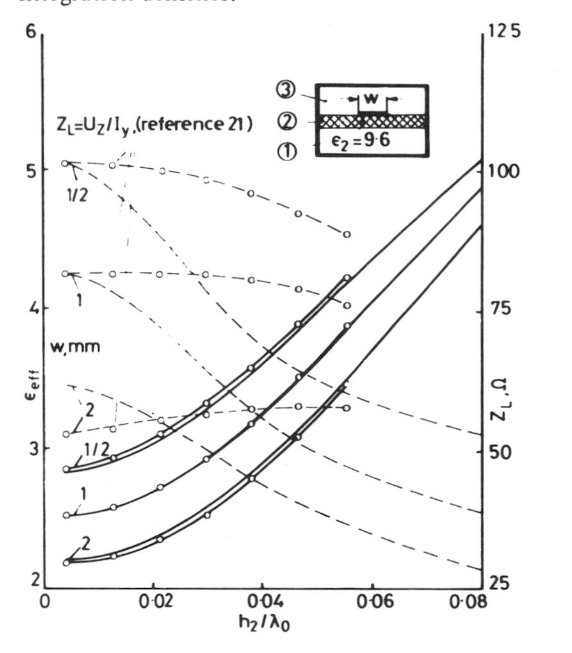

Fig. 8 *Electrical parameters of suspended substrate lines in a WR 28 waveguide compared to Reference 21*

o Reference 21
——— ϵ_{eff}
– – – Z_L, Ω
$h_2 = 0{\cdot}640$ mm

4 Conclusion

A unified, highly efficient approach to shielded, covered and open planar transmission-line configurations has been presented. The corresponding computer program is believed to be a valuable tool in the design of microwave and millimetre wave circuits. Its results throw some light on the problems of efficiency, accuracy and speed of convergence associated with the spectral-domain method. In addition, some of the physical aspects treated here may not have been published before. Due to space restrictions, the computed and measured results already obtained for structures with more than two strips or slots will be reserved for a later publication. It is intended, also, to implement the line losses into the computations.

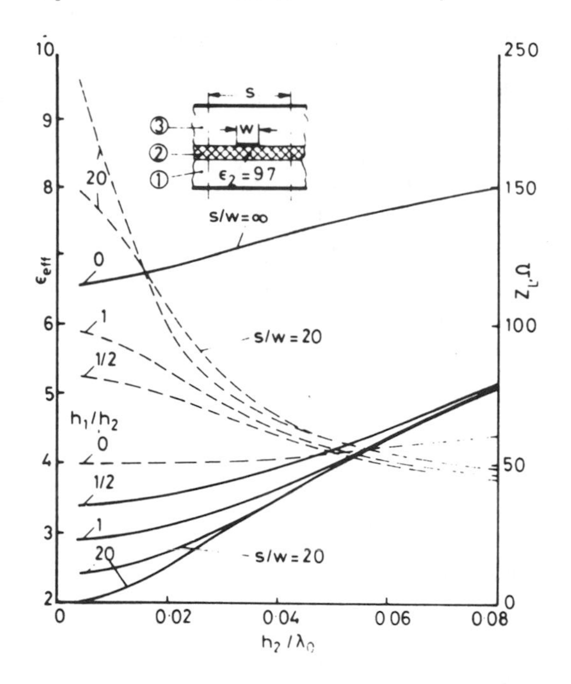

Fig. 9 *Characteristics of a laterally open suspended substrate line for different ground-plane separations h_1*

——— ϵ_{eff}
– – – Z_L, Ω
$w/h_2 = 1$
$h_3 = 20\,h_2$

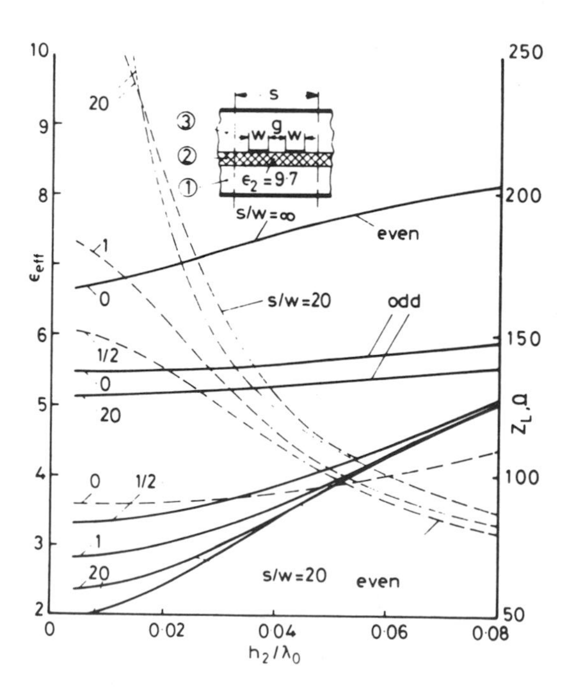

Fig. 10 *Even and odd mode parameters of coupled suspended substrate lines for different ground-plane separations h_1*

——— ϵ_{eff}
– – – Z_L, Ω
$w/h_2 = 1/2$
$g/h_2 = 1/4$
$\epsilon_1 = \epsilon_3 = 1$
$Z_L = 40, \cdots, 44\ \Omega$ for the odd mode
$h_3 = 20\,h_2$

5 References

1 DENLINGER, E.J.: 'A frequency dependent solution for microstrip transmission lines', *IEEE Trans.*, 1971, **MTT-19**, pp. 30–39

2 ITOH, T., and MITTRA, R.: 'Dispersion characteristics of slot lines', *Electron. Lett.*, 1971, 7, pp. 364–365

3 KRAGE, M.K., and HADDAD, G.I.: 'Frequency-dependent characteristics of microstrip transmission lines', *IEEE Trans.*, 1972, **MTT-20**, pp. 678–688

4 KOWALSKI, G., and PREGLA, R.: 'Dispersion characteristics of single and coupled microstrips', *Arch. Elektron. & Uebertragungstech.*, 1972, **26**, pp. 276–280

5 VAN DE CAPELLE, A.R., and LUYPAERT, P.J.: 'Fundamental- and higher-order modes in open microstrip lines', *Electron. Lett.*, 1973, 9, pp. 345–346

6 ITOH, T. and MITTRA, R.: 'Spectral-domain approach for calculating the dispersion characteristics of microstrip lines', *IEEE Trans.*, 1973, **MTT-21**, pp. 496–499

7 KOWALSKI, G., and PREGLA, R.: 'Dispersion characteristics of single and coupled microstrips with double-layer substrates', *Arch. Elektron. & Uebertragungstech.*, 1973, **27**, pp. 125–130

8 JANSEN, R.H.: 'A modified least-squares boundary residual method and its application to the problem of shielded microstrip dispersion', *ibid.*, 1974, **28**, pp. 275–277

9 ITOH, T., and MITTRA, R.: 'A technique for computing dispersion characteristics of shielded microstrip lines', *IEEE Trans.*, 1974, **MTT-22**, pp. 896–898

10 JANSEN, R.H.: 'Computer analysis of edge-coupled planar structures', *Electron. Lett.*, 1974, 10, pp. 520–522

11 JANSEN, R.H.: 'A moment method for covered microstrip dispersion', *Arch. Electron. & Uebertragungstech.*, 1975, **29**, pp. 17–20

12 JANSEN, R.H.: 'Computer analysis of shielded microstrip structures', *ibid.*, 1975, **29**, pp. 241–247

13 JANSEN, R.H.: 'Numerical computation of the eigenfrequencies and eigenfunctions of arbitrarily shaped microstrip structures'. Ph.D. thesis, Technical University of Aachen, 1975

14 KNORR, J.B., and TUFEKEIOGLU, A.: 'Spectral-domain calculation of microstrip characteristic impedance', *IEEE Trans.*, 1975, **MTT-23**, pp. 725–728

15 KNORR, J.B., and KUCHLER, K.D.: 'Analysis of coupled slots and coplanar strips on dielectric substrate', *ibid.*, 1975, **MTT-23**, pp. 541–548

16 YAMASHITA, E., and ATSUKI, K.: 'Analysis of microstrip-like transmission lines by nonuniform discretization of integral equations', *ibid.*, 1976, **MTT-24**, pp. 195–200

17 FARRAR, A., and ADAMS, A.T.: 'Computation of propagation constants for the fundamental and higher order modes in microstrip', *ibid.*, 1976, **MTT-24**, pp. 456–460

18 KITAZAWA, T., HAYASHI, Y., and SUZUKI, M.A.: 'A coplanar waveguide with thick metal coating', *ibid.*, 1976, **MTT-24**, pp. 604–608

19 SAMARDZIJA, N., and ITOH, T.: 'Double-layered slot line for millimeter-wave integrated circuits', *ibid.*, 1976, **MTT-24**, pp. 827–831

20 DAVIES, J.B. and MIRSHEKAR-SYAHKAL, D.: 'Spectral domain solution of arbitrary coplanar transimission line with multilayer substrate, *ibid.*, 1977, **MTT-25**, pp. 143–146

21 HOFMANN, H.: 'Dispersion of planar waveguides for millimeter-wave application', *Arch. Elektron. & Uebertragungstech.*, 1977, **31**, pp. 40–44

22 BORBURGH, J.: 'The behaviour of guided modes on the ferrite-filled microstrip line with the magnetization perpendicular to the ground plane', *ibid.*, 1977, **31**, pp. 73–77

23 JANSEN, R.H.: 'Fast accurate hybrid mode computation of nonsymmetrical coupled microstrip characteristics'. Presented at the 7th european microwave conference, Copenhagen, 1977

24 JANSEN, R.H.: 'High-speed computation of single and coupled microstrip parameters including dispersion, higher-order modes, loss and finite strip thickness', *IEEE Trans.*, 1978, **MTT-26**, pp. 75–82

25 JANSEN, R.H.: 'Microstrip lines with partially removed ground metallisation, theory and applications', *Arch. Elektron. & Uebertragungstech*, 1978, **32**, (to be published)

26 COLLIN, R.E.: 'Field theory of guided waves' (McGraw-Hill, New York, 1960)

27 HARRINGTON, R.F.: 'Time-harmonic electromagnetic fields' (McGraw-Hill, New York, 1961)

28 ABRAMOWITZ, M., and STEGUN, I.A.: 'Handbook of mathematical functions' (Dover Publications, New York, 1970)

29 SNEDDON, I.N.: 'The use of integral transform' (Tata McGraw-Hill, New Delhi, 1974)

30 COHN, S.B.: 'SLOT line on a dielectric substrate', *IEEE Trans.*, 1969, **MTT-17**, pp. 768–778

31 MARIANI, E.A. *et al.*: 'Slot line characteristics', *ibid.*, 1969, **MTT-17**, pp. 1091–1096

32 SMITH, J.I.: 'The even- and odd-mode capacitance parameters for coupled lines in suspended substrate', *ibid.*, 1971, **MTT-19**, pp. 424–431

33 DEUTSCH, J., and JUNG, H.J.: 'Messung der effektiven Dielektrizitätszahl von Mikrostrip-Leitungen im Frequenzbereich von 2 bis 12 GHz', *Machrichtentech. Z.*, 1970, **23**, pp. 620–624

34 DELFS, H.: 'Versatile construction of thin-film circuits for rf-applications', *NTG-Fachber.*, 1977, **60**, pp. 201–206

Effects of Side-Wall Grooves on Transmission Characteristics of Suspended Striplines

EIKICHI YAMASHITA, FELLOW, IEEE, BAI YI WANG, KAZUHIKO ATSUKI, AND KE REN LI

Abstract **—The use of suspended striplines is becoming an important transmission-line technique at millimeter wavelengths because of low attenuation, weak dispersion, and various merits in manufacturing processes. This paper estimates the effects of side-wall grooves of these lines on transmission characteristics within the TEM wave approximation.**

I. Introduction

SUSPENDED STRIPLINES (SSL's) have become very useful for millimeter-wave transmission recently (as discussed in the workshop on SSL filters in the 1984 International Microwave Symposium) because of their low attenuation, weak dispersion, and moderate wavelength reduction factor compared with microstrip lines, and various merits in manufacturing processes. The planar configuration also makes SSL's suitable for integration into millimeter-wave systems [1].

The transmission characteristics of SSL's as shown in Fig. 1 have already been analyzed both for thin-strip cases [2] and thick-strip cases [3]. However, effects of side-wall grooves, which have recently been employed in practical SSL's to support substrates mechanically, as shown in Fig. 2, have not been estimated theoretically, except a transverse resonance analysis [4] in which the cutoff frequency of the TE_{10} type mode has been calculated. In the case of finlines, the effects of grooves on hybrid-mode transmission have been estimated by using the mode-matching method [5]. The applications and technology of suspended striplines at millimeter wavelengths have been well described in other literature [6], [7].

This paper shows the method and results of the quantitative analysis of such effects, particularly on the characteristic impedance and wavelength reduction factor.

II. Method of Analysis

The present analysis of transmission characteristics of SSL's with side-wall grooves is carried out within the TEM-wave approximation since the weak dispersion of the dominant mode is expected because of the existence of two air regions. The past methods using Green's function [2], [3] cannot be applied to this structure because of its irregular shape.

Manuscript received March 19, 1985; revised June 25, 1985.

The authors are with the University of Electro-Communications, Chofu-shi, Tokyo, Japan 182.

The solution of Laplace's equation is first assumed for each of the three regions in Fig. 2. The strip conductor is assumed to be infinitely thin. Potential functions which satisfy boundary conditions at the surrounding conductor surface can be written in the form of Fourier series as

$$\phi_1(x,y) = \sum_{n=1,3,5}^{\infty} A_n \sinh\left(\frac{n\pi y}{a}\right)\cos\left(\frac{n\pi x}{a}\right) \quad \left(|x| \leqslant \frac{a}{2},\ 0 \leqslant y \leqslant h_1\right) \tag{1}$$

$$\phi_2(x,y) = \sum_{n=1,3,5}^{\infty} \left[B_n \sinh\left(\frac{n\pi y}{a+2d}\right) + C_n \cosh\left(\frac{n\pi y}{a+2d}\right)\right] \cdot \cos\left(\frac{n\pi x}{a+2d}\right) \quad \left(|x| \leqslant \frac{a}{2}+d,\ h_1 \leqslant y \leqslant h_1+h_2\right) \tag{2}$$

$$\phi_3(x,y) = \sum_{n=1,3,5}^{\infty} D_n \sinh\left[\frac{n\pi(b-y)}{a}\right]\cos\left(\frac{n\pi x}{a}\right) \quad \left(|x| \leqslant \frac{a}{2},\ h_1+h_2 \leqslant y \leqslant b\right). \tag{3}$$

When potential functions at $y = h_1 - 0$, $y = h_1 + 0$, $y = h_1 + h_2 - 0$, and $y = h_1 + h_2 + 0$ are given by $f_1(x)$, $f_2(x)$, $f_3(x)$, and $f_4(x)$, respectively, (1)–(3) lead to the Fourier series expansions of these functions. The Fourier coefficients A_n, B_n, C_n, and D_n are uniquely determined after knowing these functions. The variational method is used to find the form of these functions.

We use the fact that the total electric-field energy W of this structure per unit length is given by

$$W = \frac{1}{2}\sum_{i=1}^{3} \epsilon_i \iint_{S_i} \left[\left(\frac{\partial \phi_i}{\partial x}\right)^2 + \left(\frac{\partial \phi_i}{\partial y}\right)^2\right] dx\,dy \tag{4}$$

where S_i $(i = 1,2,3)$ denotes the cross-sectional area and ϵ_i $(i = 1,2,3)$ the dielectric constant of the region i. This energy can be minimized by varying the form of the above potential functions at $y = h_1$ and $y = h_1 + h_2$ as trial functions. The minimized energy W_{min} is related to the line capacitance C by

$$W_{min} = \frac{1}{2} C V^2 \tag{5}$$

Reprinted from *IEEE Trans. Microwave Theory Tech.*, vol. MTT-33, no. 12, pp. 1323–1328, December 1985.

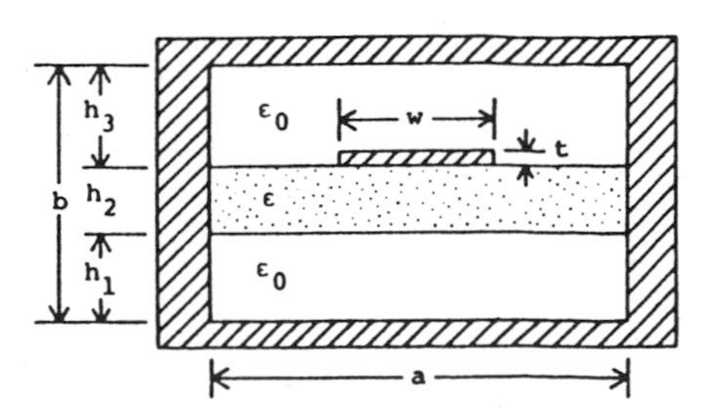

Fig. 1. Suspended striplines without side-wall grooves.

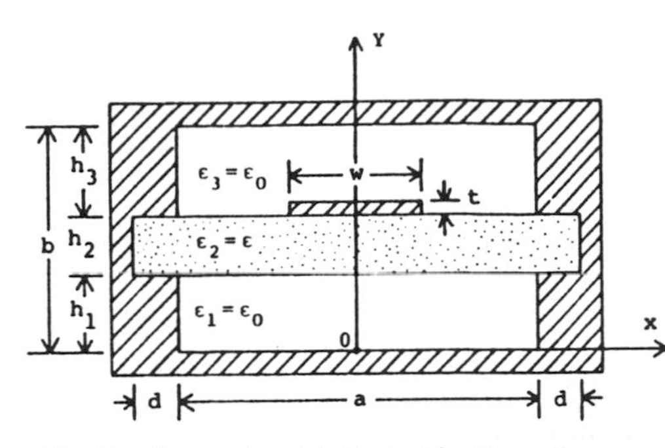

Fig. 2. Suspended striplines with side-wall grooves.

where V denotes the potential difference between the strip and wall conductor. Then, the characteristic impedance Z and the wavelength reduction factor λ/λ_0 within the TEM-wave approximation are given by

$$Z=\frac{1}{v_0\sqrt{CC_o}} \tag{6}$$

$$\frac{\lambda}{\lambda_0}=\sqrt{\frac{C_o}{C}} \tag{7}$$

where v_0 is the velocity of light in vacuum and C_0 is the line capacitance for the case $\epsilon_1=\epsilon_2=\epsilon_3=\epsilon_0$.

III. Trial Functions and Variational Formulation

Because of the symmetry of the structure, the potential functions have also even symmetry, and only one half of the structure has to be treated. The above four functions in the right half are mutually related as

$$f_1(x)=f_2(x)=f(x)\quad\left(0\leqslant x\leqslant\frac{a}{2},\ y=h_1-0\right) \tag{8}$$

$$f_2(x)=0\qquad\left(\frac{a}{2}\leqslant x\leqslant\frac{a}{2}+d,\ y=h_1+0\right) \tag{9}$$

$$f_3(x)=f_4(x)=g(x)\quad\left(0\leqslant x\leqslant\frac{a}{2},\ y=h_1+h_2-0\right) \tag{10}$$

$$f_3(x)=0\qquad\left(\frac{a}{2}\leqslant x\leqslant\frac{a}{2}+d,\ y=h_1+h_2+0\right). \tag{11}$$

Applying (8)–(11) to (1)–(3), we obtain the Fourier series whose coefficients are given by

$$A_n=\frac{4}{a\sinh\left(\frac{n\pi h_1}{a}\right)}\int_0^{a/2}f(x)\cos\left(\frac{n\pi x}{a}\right)dx \tag{12}$$

$$B_n=\frac{4}{(a+2d)\sinh\left(\frac{n\pi h_2}{a+2d}\right)}\left[\cosh\left(\frac{n\pi h_1}{a+2d}\right)\cdot\int_0^{a/2}g(x)\cos\left(\frac{n\pi x}{a+2d}\right)dx-\cosh\left[\frac{n\pi(h_1+h_2)}{a+2d}\right]\cdot\int_0^{a/2}f(x)\cos\left(\frac{n\pi x}{a+2d}\right)dx\right] \tag{13}$$

$$C_n=\frac{4}{(a+2d)\sinh\left(\frac{n\pi h_2}{a+2d}\right)}\left[-\sinh\left(\frac{n\pi h_1}{a+2d}\right)\cdot\int_0^{a/2}g(x)\cos\left(\frac{n\pi x}{a+2d}\right)dx+\sinh\left[\frac{n\pi(h_1+h_2)}{a+2d}\right]\cdot\int_0^{a/2}f(x)\cos\left(\frac{n\pi x}{a+2d}\right)dx\right] \tag{14}$$

$$D_n=\frac{4}{a\sinh\left(\frac{n\pi h_3}{a}\right)}\int_0^{a/2}g(x)\cos\left(\frac{n\pi x}{a}\right)dx. \tag{15}$$

Now, we choose the first-order spline function (or the polygonal line function) as trial functions to express $f(x)$ and $g(x)$ as given below, and as shown in Fig. 3, because the spline function has a simple form and is useful in approximating complicated curves

$$f(x)=\sum_{i=0}^{m_1}F_i(x) \tag{16}$$

$$g(x)=\sum_{j=0}^{m_2}G_j(x) \tag{17}$$

where

$$F_i(x)=\begin{cases}\dfrac{p_{i+1}-p_i}{a_{i+1}-a_i}x+\dfrac{p_ia_{i+1}-p_{i+1}a_i}{a_{i+1}-a_i} & (a_i\leqslant x\leqslant a_{i+1})\\ 0 & (\text{elsewhere})\end{cases} \tag{18}$$

$$G_j(x)=\begin{cases}\dfrac{q_{j+1}-q_j}{b_{j+1}-b_j}x+\dfrac{q_jb_{j+1}-q_{j+1}b_j}{b_{j+1}-b_j} & (b_j\leqslant x\leqslant b_{j+1})\\ 0 & (\text{elsewhere})\end{cases} \tag{19}$$

and the values of knot potentials of the trial function are denoted by p_i $(i=0,1,\cdots,m_1+1)$ for $f(x)$, and q_j $(j=0,1,\cdots,m_2+1)$ for $g(x)$ at the knot positions a_i $(i=0,1,\cdots,m_1+1)$ and b_j $(j=0,1,\cdots,m_2+1)$, respectively.

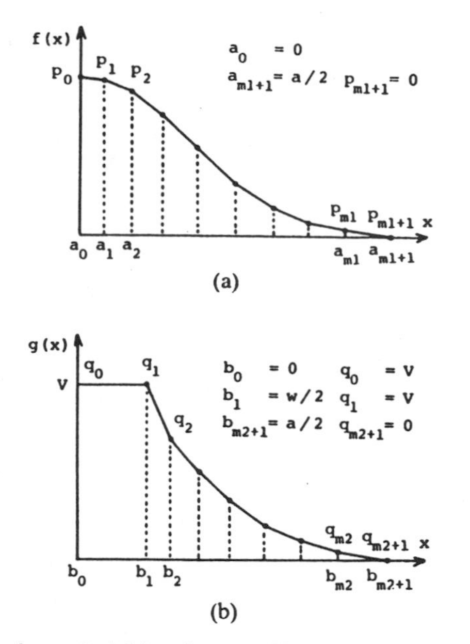

Fig. 3. The form of trial functions. (a) The first-order spline function as the trial function for $f(x)$. (b) The first-order spline function as the trial function for $g(x)$.

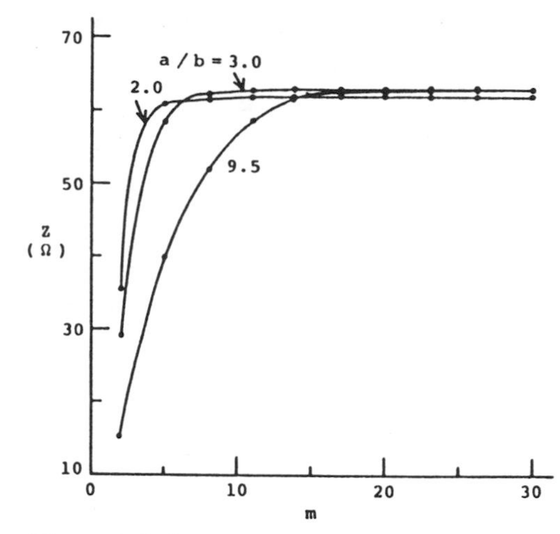

Fig. 4. The numerical convergence property of the characteristic impedance. $w/b=1.0$, $h_1/b=0.4$, $h_2/b=0.2$, $h_3/b=0.4$, $t=d=0$, $\epsilon^*=1.0$.

The total electric-field energy after the integration (4) is given by

$$W=\sum_{i=0}^{m_1}\sum_{k=0}^{m_1}\alpha_{ik}p_ip_k+\sum_{j=0}^{m_2}\sum_{k=0}^{m_2}\beta_{jk}q_jq_k-2\sum_{i=0}^{m_1}\sum_{j=0}^{m_2}\gamma_{ij}p_iq_j \quad (20)$$

where symbols α_{ik}, β_{jk}, and γ_{ij} are defined in the Appendix.

This energy expression includes a group of knot potentials (p_0, $p_1,\cdots,p_{m1}$, q_2, $q_3,\cdots$, and q_{m2}) as new variables and remaining knot potentials (p_{m1+1}, q_0, q_1, and q_{m2+1}) as constants. These new variables are adjusted to satisfy the following conditions to obtain the minimum of W:

$$\frac{\partial W}{\partial p_i}=0 \qquad (i=0,1,\cdots,m_1) \quad (21)$$

$$\frac{\partial W}{\partial q_j}=0 \qquad (j=2,3,\cdots,m_2). \quad (22)$$

By imposing these conditions on the energy expression (20), we obtain a set of linear, simultaneous, inhomogeneous equations as shown below which are the final equations to be solved on a computer

$$\sum_{k=0}^{m_1}\alpha_{ik}p_k-\sum_{j=2}^{m_2}\gamma_{ij}q_j=\sum_{j=0}^{1}\gamma_{ij}q_j \qquad (i=0,1,\cdots,m_1) \quad (23)$$

$$\sum_{i=0}^{m_1}\gamma_{ij}p_i-\sum_{k=2}^{m_2}\beta_{jk}q_k=\sum_{k=0}^{1}\beta_{jk}q_k \qquad (j=2,3,\cdots,m_2). \quad (24)$$

Since the knot potentials are known after solving these equations, the minimum value of the electric-field energy is found by substituting these values into (20).

V. Numerical Processing and Accuracy

The numbers of the knots of the trial function m_1 and m_2 play an important role in determining the required time for computation and accuracy. Since $g(x)$ changes sharply compared with $f(x)$, m_2 is necessarily larger than m_1. m_1 and m_2 can be minimized by taking the following measures.

1) The sizes of m_1 and m_2 are increased when those of $a/2b$ and $(a-w)/b$ are increased, respectively.

2) The knots of the spline function for $g(x)$ near the strip conductor are narrowly spaced to represent potential curves precisely. Fig. 4 shows the convergence property of the impedance values for increasing the sum of the two values $m=m_1+m_2$. The effect of the number of the Fourier series terms N is less important when N is larger than 100. Typical values of these numbers in the present analysis are: $m_1=7$, $m_2=13$, and $N=100$.

The accuracy of calculated transmission parameters can be examined by comparing numerical results for the case of $d=0$ by the present method with those by the previous method [2]. Fig. 5 indicates that both methods result in good agreement with the discrepancy by approximately 0.1 percent. This agreement is due to the fact that λ/λ_0 is expressed by $\sqrt{C_0/C}$, in which capacitance errors are mostly cancelled.

Fig. 6, on the other hand, indicates that there is a discrepancy of about 2 percent between the impedance values calculated by the two methods. This is naturally attributed to the expression of the characteristic impedance inversely proportional to $\sqrt{C_0C}$. However, because of the nature of the present and the previous variational method, it can be stated that the exact values of the characteristic impedance exist between these two curves. The time required for computing a set of impedance and the wave-

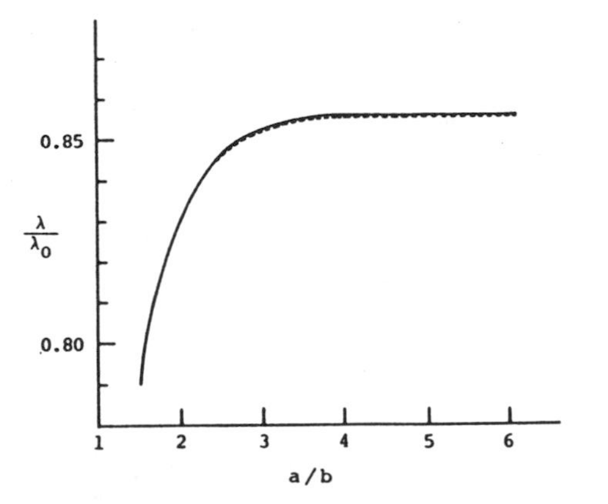

Fig. 5. A comparison of λ/λ_0 values by the present method (solid line) with those by the previous method (dashed line) for $d=0$. $w/b=1.0$, $h_1/b=0.4$, $h_2/b=0.2$, $h_3/b=0.4$, $t=d=0$, $\epsilon^*=3.78$.

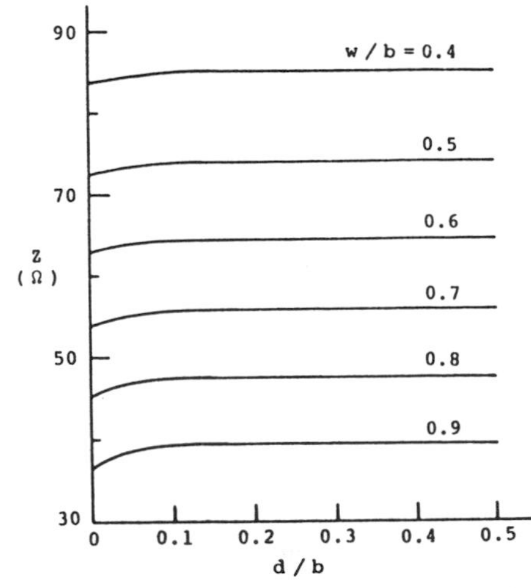

Fig. 7. Side-wall groove effects on the characteristic impedance Z. $a/b=1.0$, $h_1/b=0.4$, $h_2/b=0.2$, $h_3/b=0.4$, $t=0$, $\epsilon^*=2.22$.

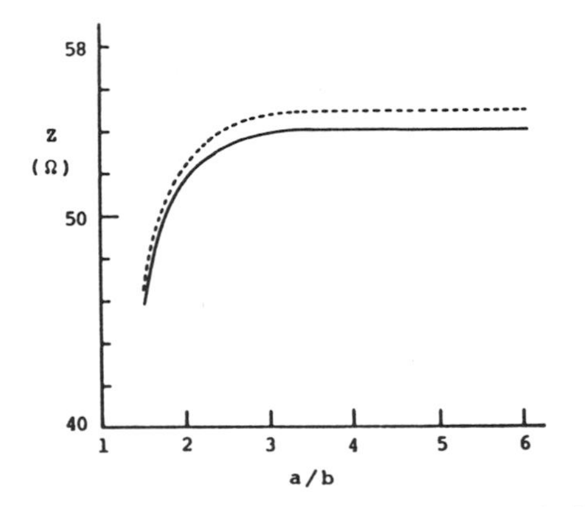

Fig. 6. A comparison of Z values by the present method (solid line) with those by the previous method (dashed line) for $d=0$. $w/b=1.0$, $h_1/b=0.4$, $h_2/b=0.2$, $h_3/b=0.4$, $t=d=0$, $\epsilon^*=3.78$.

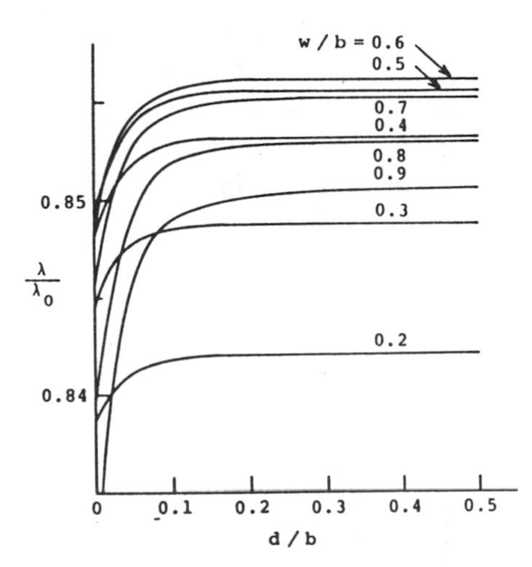

Fig. 8. Side-wall groove effects on the wavelength reduction factor λ/λ_0. $a/b=1.0$, $h_1/b=0.4$, $h_2/b=0.2$, $h_3/b=0.4$, $t=0$, $\epsilon^*=2.22$.

length reduction factor was about 1 s on a HITAC-M180 computer.

VI. Numerical Results

Since the common values of the characteristic impedance are 50 Ω and 75 Ω, the structural dimensions of the SSL's in this paper are also selected by considering these values. Relatively high walls are treated in order to see the strong effects of side-wall grooves. For Duroid (the dielectric constant $\epsilon^*=2.22$) used as substrates, the following dimensions are used:

$$a/b=1.0 \quad h_1/b=0.4 \quad h_2/b=0.2 \quad h_3/b=0.4$$
$$t=0.0 \quad w/b=0.2\sim 0.9 \quad d/b=0.0\sim 0.5.$$

The results of numerical calculations are shown in Figs. 7 and 8. When the depth of grooves is increased, the characteristic impedance and wavelength reduction factor are also increased rapidly but eventually no more effects are observed in the region of deep grooves. The flatness of these curves is important since it assures us the exactness of our method. Because groove effects are significant in the range of small d, these can not be neglected when precise

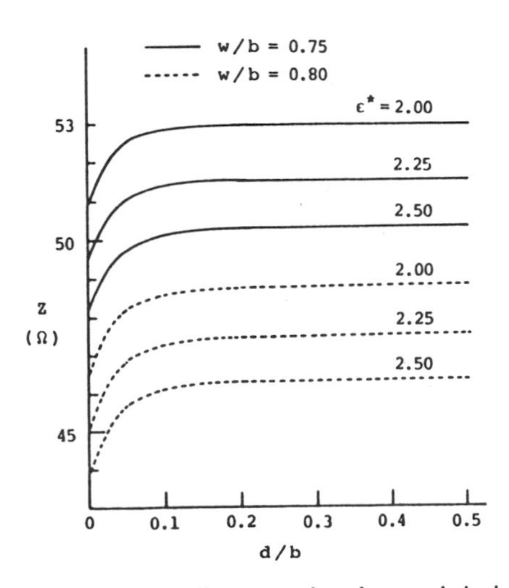

Fig. 9. Side-wall groove effects on the characteristic impedance for substrates of various dielectric constants. $a/b=1.0$, $h_1/b=0.4$, $h_2/b=0.2$, $h_3/b=0.4$, $t=0$.

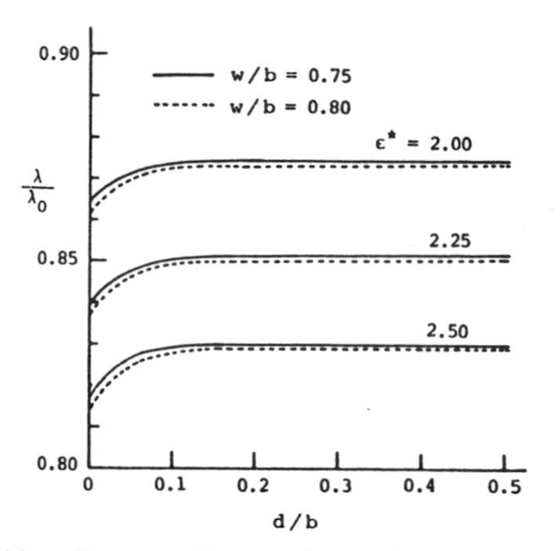

Fig. 10. Side-wall groove effects on the wavelength reduction factor for substrates of various dielectric constants. $a/b = 1.0$, $h_1/b = 0.4$, $h_2/b = 0.2$, $h_3/b = 0.4$, $t = 0$.

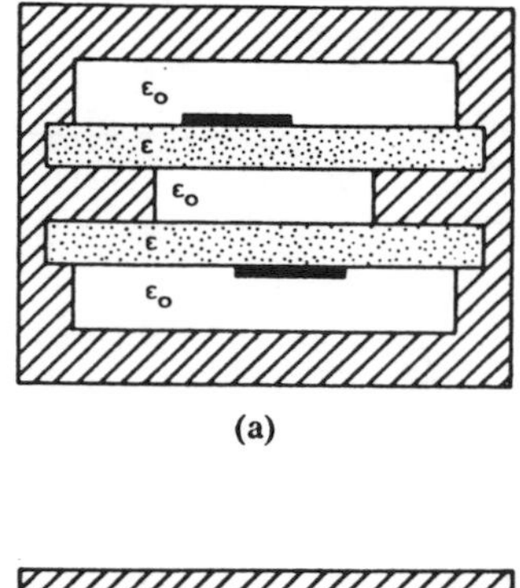

(a)

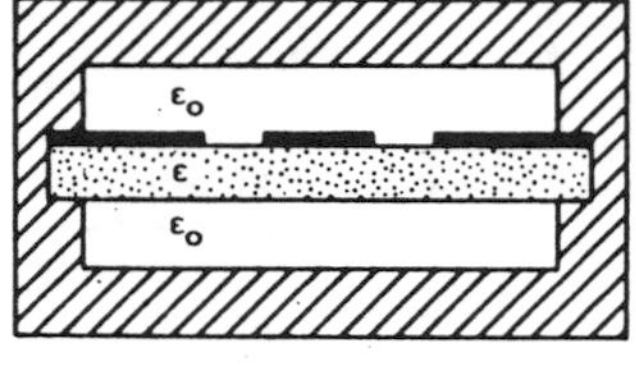

(b)

Fig. 11. Possible generalized structures for applying the present analysis method. (a) Coplanar SSL's. (b) SSL couplers.

filters are designed. In order to avoid these effects, very low side-walls ($b \ll a$) must be employed. However, it should be noted that the width a also determines the lowest cutoff frequency. Figs. 9 and 10 show the effects of the grooves on transmission characteristics with the dielectric constant of the substrate as parameters.

This method of analysis is simple and efficient in numerical computation, and can be applied to a variety of planar configurations as shown in Fig. 11.

Appendix

The symbols in the energy expression (20) are defined as follows:

$$\alpha_{ik} = \epsilon_1 \sum_{n=1,3,5}^{\infty} \frac{4}{n\pi \tanh\left(\frac{n\pi h_1}{a}\right)} K_n(a_i, a) K_n(a_k, a) + \epsilon_2 \sum_{n=1,3,5}^{\infty} \frac{4}{n\pi \tanh\left(\frac{n\pi h_2}{a+2d}\right)} \cdot K_n(a_i, a+2d) K_n(a_k, a+2d) \quad \text{(A1)}$$

$$\beta_{jk} = \epsilon_3 \sum_{n=1,3,5}^{\infty} \frac{4}{n\pi \tanh\left(\frac{n\pi h_3}{a}\right)} K_n(b_j, a) K_n(b_k, a) + \epsilon_2 \sum_{n=1,3,5}^{\infty} \frac{4}{n\pi \tanh\left(\frac{n\pi h_2}{a+2d}\right)} \cdot K_n(b_j, a+2d) K_n(b_k, a+2d) \quad \text{(A2)}$$

$$\gamma_{ij} = \epsilon_2 \sum_{n=1,3,5}^{\infty} \frac{4}{n\pi \sinh\left(\frac{n\pi h_2}{a+2d}\right)} \cdot K_n(a_i, a+2d) K_n(b_j, a+2d) \quad \text{(A3)}$$

where

$$K_n(x_i, A) = \cos\left(\frac{n\pi x_i}{A}\right) \cdot \left[\frac{2}{R_i}\left(\sin\frac{R_i}{2}\right)^2 + \frac{2}{R_{i+1}}\left(\sin\frac{R_{i+1}}{2}\right)^2\right] + \sin\left(\frac{n\pi x_i}{A}\right) \cdot \left[-\frac{1}{R_i}\sin R_i + \frac{1}{R_{i+1}}\sin R_{i+1}\right] \quad \text{(A4)}$$

$$R_i = \frac{n\pi(x_i - x_{i-1})}{A}. \quad \text{(A5)}$$

Acknowledgment

The authors thank Dr. Y. Suzuki for his helpful comments.

References

[1] M. H. Arain, "A 94 GHz suspended stripline circulator," in *1984 IEEE MTT-S Symp. Dig.*, pp. 78–79.
[2] E. Yamashita and K. Atsuki, "Strip lines with rectangular outer conductor and three dielectric layers," *IEEE Trans. Microwave Theory Tech.*, vol. MTT-18, pp. 238–242, May 1970.
[3] K. Atsuki and E. Yamashita, "Analytical method for transmission lines with thick-strip conductor, multi-dielectric layers, and shielding conductor," *IECE Japan*, vol. 53-B, pp. 322–328, June 1970.
[4] S. B. Cohn and G. D. Osterhues, "A more accurate model of the TE_{10} – type waveguide mode in suspended substrate," *IEEE Trans. Microwave Theory Tech.*, vol. MTT-30, pp. 293–294, Mar. 1982.
[5] A. Beyer, "Analysis of the characteristics of an earthed finline," *IEEE Trans. Microwave Theory Tech.*, vol. MTT-29, 676–680, July 1981.
[6] C. Chao, A. Contolatis, S. A. Jamison, and P. E. Bauhahn, "*Ka*-band monolithic GaAs balanced mixers," *IEEE Trans. Microwave Theory Tech.*, vol. MTT-31, pp. 11–15, Jan. 1983.
[7] R. S. Tahim, G. M. Hayashibara, and K. Chang, "Design and performance of *W*-band broad-band integrated circuit mixers," *IEEE Trans. Microwave Theory Tech.*, vol. MTT – 31, pp. 277–283, Mar. 1983.

A Modified Mode-Matching Technique and Its Application to a Class of Quasi-Planar Transmission Lines

RUEDIGER VAHLDIECK, MEMBER IEEE, AND JENS BORNEMANN

Abstract —A rigorous and versatile hybrid-mode analysis is presented to determine the normalized propagation constants in a class of quasi-planar transmission-line structures.

The method is accurate and covers the finite metallization thickness, mounting grooves, and an arbitrary number of dielectric subregions.

Utilizing a modified mode-matching technique, one can derive discontinuity and transmission-line matrices for each homogeneous subregion. Successively multiplying matrix equations of all subregions leads to the characteristic matrix system. This procedure makes it possible to create a modularized computer program which can be conveniently extended to a wide spectrum of conceivable configurations simply by inserting the matrix equations of additional subregions in the multiplication process. To demonstrate the efficiency of the proposed method, dispersion characteristics of dominant and next higher order hybrid modes in earthed and insulated finlines, suspended microstrips, and coupled striplines with tuning septa, are given as examples.

I. Introduction

E-PLANE MILLIMETER-WAVE integrated circuits are of considerable interest in systems requiring single-mode broad-band operation. Earthed and insulated finlines, coupled strip- and coplanar lines are the most common structures which have been succesfully applied, for example, in the design of broad-band directional couplers, taper sections, and filters. Their propagation characteristics have been obtained by various methods. An early paper by Meier [1] describes the dominant mode in an earthed finline as a variation of the corresponding mode in a ridged waveguide, but test measurements are still necessary to determine the equivalent dielectric constant of the configuration. In order to avoid these time-consuming and expensive measurements, several theoretical methods have been proposed to evaluate the dispersion characteristics of the dominant and sometimes even higher order modes in an earthed finline or more complicated configurations like coupled strip or coplanar lines.

The two-dimensional transmission-line matrix (TLM) has been applied by Shih and Hoefer [2] to determine the dominant and second-order mode cutoff frequencies in a unilateral, bilateral, and so-called insulated finline. More complicated structures have been analyzed by using the numerically very efficient spectral-domain method (SDM) (Itoh and Schmidt [3]–[5]), but only results with zero metallization thickness and neglected mounting grooves have been published.

More recently, two different mathematical treatments have been presented which take into account the finite metallization thickness (Kitazawa and Mittra [13]) and, additionally, the mounting grooves (Vahldieck [6] and Bornemann [7]). In [13], a hybrid-mode formulation has been proposed in which Green's functions are derived using conventional circuit theory. Results are given for unilateral and bilateral finlines.

Vahldieck [6] and Bornemann [7] extended a modified mode-matching technique which utilizes a transverse resonance relation [8], [14] to reduce the number of required eigenvalue equations. This method, which includes both the finite metallization thickness and the mounting grooves, has been applied successfully to a generalized finline configuration with more than one dielectric subregion and different fin thicknesses [6]. In [7], structures containing both a strip and a slot were treated, and results were presented for coupled slotlines and suspended microstrips. This already indicates that the method is quite general and can be extended very easily to more complicated combinations of strip- and slotlines.

For the numerical computation of the propagation constants, it is an advantage of this method that the order of the characteristic matrix equation is only $2NH-1$ (where NH is the number of orthogonal eigenfunctions considered in each subregion) and remains constant even for an increasing number of transverse discontinuities. This is in contrast, for example, to the well-known mode-matching procedure used by Siegl [11] and Beyer [12] in which the size of the characteristic matrix equation usually increases with the number of subregions. Moreover, in the latter method, all boundary conditions are satisfied before the tangential E- and H-fields at each interface are matched. This leads to an inflexible procedure and makes it normally necessary to create a new computer program when investigating different configurations with additional subregions.

However, the modified mode-matching technique combined with a transverse resonance relation avoids this disadvantage. Initially, the boundary conditions at the waveguide sidewalls ($x=0$ and $x=a$, Fig. 1(a)–(c)) are neglected. The configuration can then be regarded as a

Manuscript received February 5, 1985; revised May 31, 1985.

R. Vahldieck is with the Department of Electrical Engineering, University of Ottawa, 770 King Edward Ave., Ottawa, Ontario K1N 6N5, Canada.

J. Bornemann is with the Microwave Department, University of Bremen, D-2800 Bremen 33, Kufsteiner Str., NW1, West Germany.

Reprinted from *IEEE Trans. Microwave Theory Tech.*, vol. MTT-33, no. 10, pp. 916–928, October 1985.

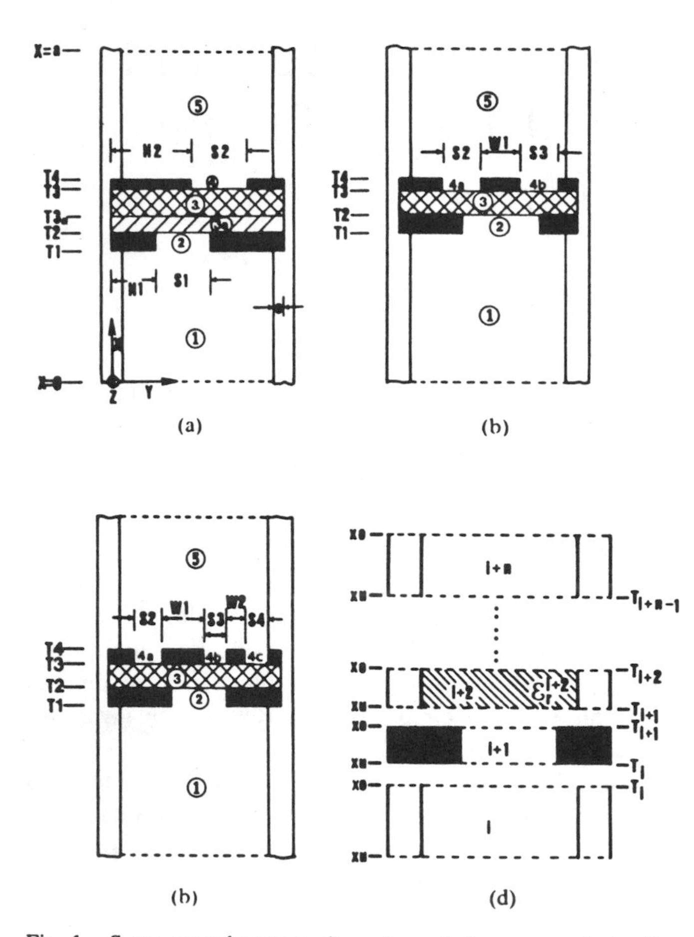

Fig. 1. Some general cross sections of quasi-planar transmission lines. (a) Earthed bilateral finline, (b) coupled slotline, (c) coupled suspended stripline, (d) subdivision of a cross section into an arbitrary number of homogeneous subregions.

parallel-plate line subdivided into homogeneous cross sections (Fig. 1(d)) in which partial wave amplitudes are defined, traveling in positive and negative x-directions with the still unknown propagation constant kx. After the elimination of the field coefficients in each subregion, a transverse transmission-line matrix results which relates the amplitudes at the lower ($x = xu$) and the upper boundary ($x = xo$, cf., Fig. 1(d)) of a subregion. Satisfying a modified field-matching condition at the common interfaces (e.g., T_i, Fig. 1(d)) transforms the partial wave amplitudes of cross-section i into those of cross-section $i+1$. Successively multiplying each transformation matrix with the appropriate transmission-line matrix of the corresponding subregion, a relation only between the partial wave amplitudes at the lower ($x = 0$) and upper ($x = a$) boundary of the waveguide is obtained. Finally, the inhomogeneous waveguide cross section can be regarded as a line resonator in which the resonance condition is satisfied by inserting the up-to-now neglected boundary conditions at the metallic sidewalls ($x = 0$ and $x = a$). This procedure reduces the size of the characteristic matrix equation to a quarter of the original size. Furthermore, it makes the method very flexible because an arbitrary number of subregions can be inserted easily in the matrix system simply by multiplying the additional transformation and transmission-line matrices with the previous.

The aim of this paper is to demonstrate the potential of this versatile and accurate treatment. As examples, earthed and insulated finlines will be analyzed, as well as coupled strip- and coupled slotlines. Finally, some detailed results on the effect of finite metallization thickness and mounting grooves on the dispersion characteristics of quasi-TEM, and next higher order hybrid modes will be presented.

II. Theory

Since the theoretical treatment has been described recently by Vahldieck [6] and Bornemann [7], only the principle steps will be explained. For further information, the reader is referred to [6] and [7].

It is well known that quasi-planar transmission lines support hybrid modes in which all six field components can occur. The transversely inhomogeneous configuration is divided into homogeneous subregions. In each of these subregions, the hybrid fields

$$\vec{E}^i = \nabla \times \nabla \times \vec{A}^i_e - j\omega\mu\nabla \times \vec{A}^i_m \tag{1}$$

$$\vec{H}^i = \nabla \times \nabla \times \vec{A}^i_m + j\omega\epsilon^i\nabla \times \vec{A}^i_e \tag{2}$$

can be expressed by a superposition of the axial z-components of two independent Hertzian vector potentials $\vec{A}_m$ and $\vec{A}_e$. The potential functions are a sum of suitable orthogonal eigenfunctions

$$\left.\begin{aligned} A^i_{mz} &= \sum_n^{\infty} f^i_{c(n)}(y)\cdot I^i_{m(n)}(x) \quad (3)\\ A^i_{ez} &= \sum_n^{\infty} f^i_{s(n)}(y)\cdot U^i_{e(n)}(x) \quad (4)\end{aligned}\right\}\cdot e^{-jkz\cdot z}$$

which satisfy the boundary condition at the metallic surface and obeys the scalar Helmholtz equation. Initially, however, we neglect the boundary conditions at the waveguide sidewalls ($x = 0$ and $x = a$, cf., Fig. 1) and regard each subregion as a part of a parallel-plate line in which partial wave amplitudes I^i_m and U^i_e are defined and travel in positive and negative x-directions. This procedure results in a somewhat higher theoretical effort, but is necessary in order to reduce the size of the characteristic matrix equation. Finally, it makes it easy to implement any desired number of subregions in the formulation. The functions f_c and f_s in (3) and (4) (given in the Appendix) implicitly satisfy the y-dependent boundary conditions in their respective subregions. I^i_m and U^i_e, as well as their derivatives

$$U^i_m = \frac{dI^i_m}{dx} \quad \text{and} \quad I^i_e = \frac{dU^i_e}{dx}$$

combined as vectors $\boldsymbol{U}^i$ and $\boldsymbol{I}^i$ at the upper boundary ($x = xo$) of each subregion, can be determined from the amplitudes at the lower boundary ($x = xu$), see Fig. 1(d). Therefore, a generalized transmission-line matrix $\boldsymbol{R}^i$ can be found which relates $\boldsymbol{U}^i$ and $\boldsymbol{I}^i$ at the two coordinates as follows:

$$\begin{bmatrix} \boldsymbol{U}^i \\ \boldsymbol{I}^i \end{bmatrix}_{x=xo} = \underbrace{\begin{bmatrix} \boldsymbol{R}^i_c & \boldsymbol{R}^i_s \\ \boldsymbol{R}^{i'}_s & \boldsymbol{R}^i_c \end{bmatrix}}_{\boldsymbol{R}^i} \cdot \begin{bmatrix} \boldsymbol{U}^i \\ \boldsymbol{I}^i \end{bmatrix}_{x=xo} \tag{5}$$

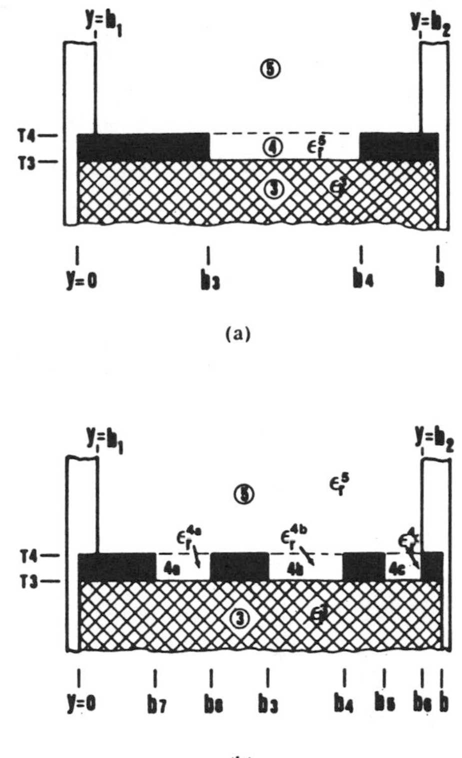

Fig. 2. Representative transition T4 and T3 for the configurations in Fig. 1. (a) Slot transition with only one slot. (b) Slot transition with three slots.

R^i_c, R^i_s, and R^i_c, are diagonal matrices containing sine and cosine functions. They are explained in the Appendix. Before a final relation between the amplitude vectors at the upper ($x = a$) and the lower ($x = 0$) sidewall of the waveguide can be given, it is necessary to modify the continuity equations at each transverse discontinuity, such that the amplitude vectors of subregion $i+1$, for example, are determined by those of region i.

The following mathematical derivation of this principle step is based on the representative transition $T4$ in Fig. 2. In the first case of Fig. 2(a), the continuity condition can be written as follows:

$$E^5_{y,z} = E^4_{y,z} \tag{6}$$

$$E^5_{y,z} = 0, \qquad b_1 \leqslant y < b_3, \quad b_4 < y \leqslant b_2 \tag{6a}$$

$$H^5_{y,z} = H^4_{y,z}, \qquad b_3 \leqslant y \leqslant b_4 \tag{7}$$

and in the second case of Fig. 2(b) in plane $T4$

$$E^5_{y,z} = \sum_\nu E^\nu_{y,z} \tag{8}$$

$$E^5_{y,z} = 0, \qquad b_1 \leqslant y < b_7, b_8 < y < b_3, b_4 < y < b_5 \tag{8a}$$

$$H^5_{y,z} = H^\nu_{y,z}, \qquad b_7 \leqslant y \leqslant b_8, b_3 \leqslant y \leqslant b_4, b_5 \leqslant y \leqslant b_2$$
$$(\nu \in 4a, 4b, 4c). \tag{9}$$

In order to determine the partial wave amplitudes of subregion $i = 5$ (Fig. 2(a)) from their corresponding amplitudes in the adjacent area $i = 4$, we first modify (6) and (7) such that the following expressions result:

$$E_y: \ \frac{\partial A^5_{mz}}{\partial x} = \frac{\partial A^4_{mz}}{\partial x} - K^4_\mu(\omega, kz)\frac{\partial A^4_{ez}}{\partial y} \tag{10}$$

$$E_z: \ A^5_{ez} = K^4(\omega, kz) A^4_{ez} \tag{11}$$

$$H_z: \ A^5_{mz} = K^4(\omega, kz) A^4_{mz} \tag{12}$$

$$H_y: \ \frac{\partial A^5_{ez}}{\partial x} = \frac{\epsilon^4_r}{\epsilon^5_r}\frac{\partial A^4_{ez}}{\partial x} + K^4_\epsilon(\omega, kz)\frac{\partial A^4_{mz}}{\partial y} \tag{13}$$

with

$$K^4(\omega, kz) = \frac{\epsilon^4_r - \beta^2}{\epsilon^5_r - \beta^2}, \qquad \beta = kz/ko$$

$$K^4_\mu(\omega, kz) = \frac{kz}{\omega\mu}(1 - K^4(\omega, kz))$$

$$K^4_\epsilon(\omega, kz) = \frac{kz}{\omega\epsilon^5}(1 - K^4(\omega, kz))$$

and similarly for (8) and (9)

$$E_y: \qquad \frac{\partial A^5_{mz}}{\partial x} = \sum_\nu \left[\frac{\partial A^\nu_{mz}}{\partial x} - K^\nu_\mu(\omega, kz)\frac{\partial A^\nu_{ez}}{\partial y}\right] \tag{14}$$

$$Ez: \qquad A^5_{ez} = \sum_\nu K^\nu(\omega, kz) A^\nu_{ez} \tag{15}$$

$$Hz: \ \bar{K}^\nu(\omega, kz) A^5_{mz} = A^\nu_{mz} \tag{16}$$

$$H_y: \ \frac{\epsilon^5_r}{\epsilon^\nu_r}\frac{\partial A^5_{ez}}{\partial x} + \bar{K}^\nu_\epsilon(\omega, kz)\frac{\partial A^5_{mz}}{\partial y} = \frac{\partial A^\nu_{ez}}{\partial x} \tag{17}$$

with

$$\bar{K}^\nu_\epsilon(\omega, kz) = \frac{kz}{\omega\epsilon^\nu}(1 - \bar{K}^\nu(\omega, kz))$$

$$\bar{K}^\nu(\omega, kz) = \frac{\epsilon^5_r - \beta^2}{\epsilon^\nu_r - \beta^2}$$

$$K^\nu_\mu(\omega, kz) = \frac{kz}{\omega\mu}(1 - K^\nu(\omega, kz)).$$

For the interface $T3$ (Fig. 2(a)), replace in (10)–(13) the index $i = 5$ with $i = 4$, and index $i = 4$ with $i = 3$. To obtain the interface expression for $T3$ in Fig. 2(b), replace in (14)–(17) $i = 5$ with $i = 3$. The discontinuities $T4$, $T3$ in Fig. 1(b) are then also included in (14)–(17). Moreover, the equation system (10)–(17) is very instructive. It shows that the coupling between TE and TM waves is automatically included, and that only in structures with homogeneously filled cross sections (e.g., ridged waveguide) or in inhomogeneously filled structures at cutoff ($kz = 0$) are these wave types decoupled since $K\mu$ (ω, kz) and $K\epsilon$ (ω, kz) in (10) and (13) vanish as well as $K\mu$ (ω, kz) and $K\epsilon$ (ω, kz) in (14) and (17).

Multiplication of the above equation system with appropriate orthogonal eigenfunctions [6], [7] separates the

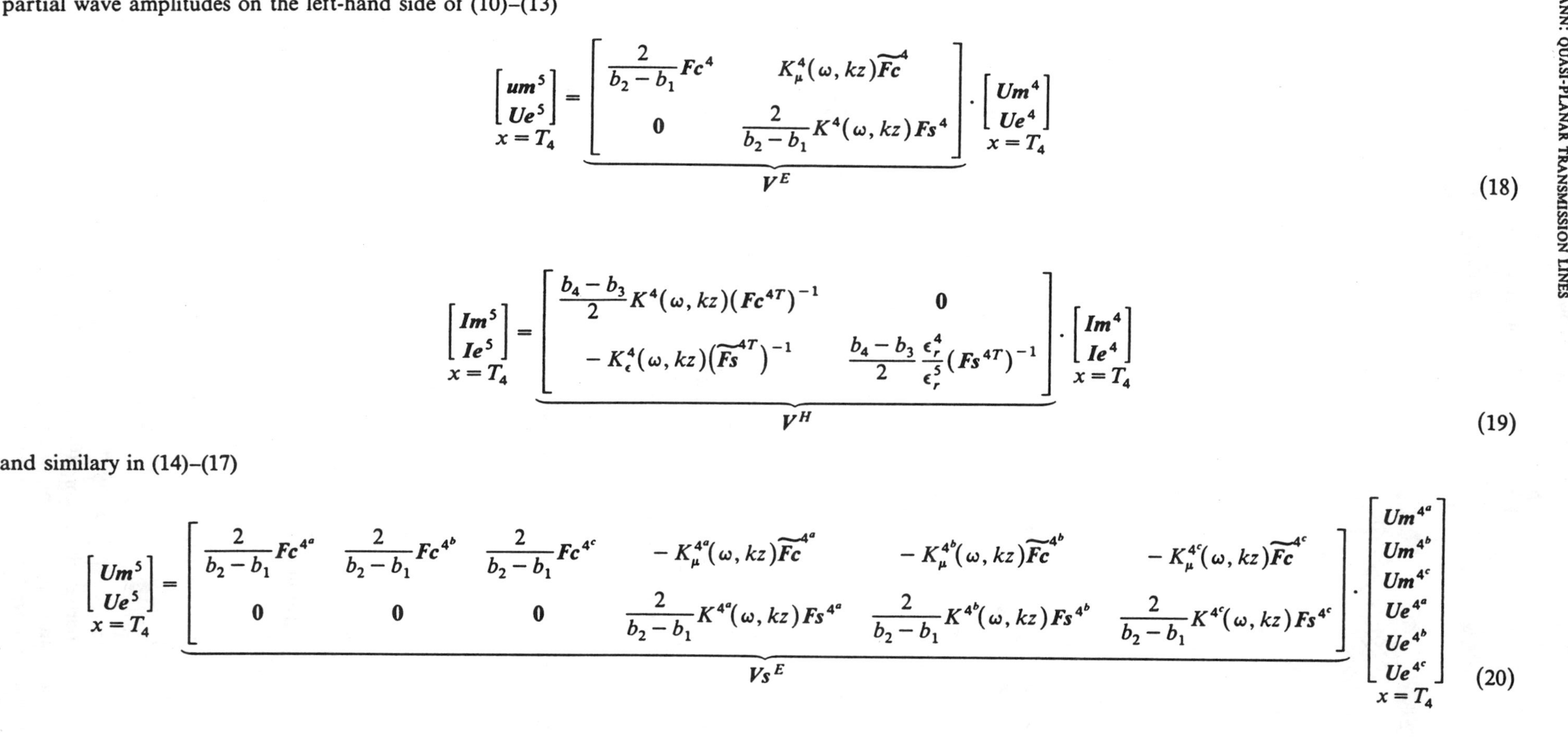

partial wave amplitudes on the left-hand side of (10)–(13)

$$\begin{bmatrix} um^5 \\ Ue^5 \end{bmatrix}_{x=T_4} = \underbrace{\begin{bmatrix} \frac{2}{b_2-b_1}\mathbf{Fc}^4 & K_\mu^4(\omega,kz)\widetilde{\mathbf{Fc}}^4 \\ \mathbf{0} & \frac{2}{b_2-b_1}K^4(\omega,kz)\mathbf{Fs}^4 \end{bmatrix}}_{V^E} \cdot \begin{bmatrix} Um^4 \\ Ue^4 \end{bmatrix}_{x=T_4} \tag{18}$$

$$\begin{bmatrix} Im^5 \\ Ie^5 \end{bmatrix}_{x=T_4} = \underbrace{\begin{bmatrix} \frac{b_4-b_3}{2}K^4(\omega,kz)(\mathbf{Fc}^{4T})^{-1} & \mathbf{0} \\ -K_\epsilon^4(\omega,kz)(\widetilde{\mathbf{Fs}}^{4T})^{-1} & \frac{b_4-b_3}{2}\frac{\epsilon_r^4}{\epsilon_r^5}(\mathbf{Fs}^{4T})^{-1} \end{bmatrix}}_{V^H} \cdot \begin{bmatrix} Im^4 \\ Ie^4 \end{bmatrix}_{x=T_4} \tag{19}$$

and similary in (14)–(17)

$$\begin{bmatrix} Um^5 \\ Ue^5 \end{bmatrix}_{x=T_4} = \underbrace{\begin{bmatrix} \frac{2}{b_2-b_1}\mathbf{Fc}^{4^a} & \frac{2}{b_2-b_1}\mathbf{Fc}^{4^b} & \frac{2}{b_2-b_1}\mathbf{Fc}^{4^c} & -K_\mu^{4^a}(\omega,kz)\widetilde{\mathbf{Fc}}^{4^a} & -K_\mu^{4^b}(\omega,kz)\widetilde{\mathbf{Fc}}^{4^b} & -K_\mu^{4^c}(\omega,kz)\widetilde{\mathbf{Fc}}^{4^c} \\ \mathbf{0} & \mathbf{0} & \mathbf{0} & \frac{2}{b_2-b_1}K^{4^a}(\omega,kz)\mathbf{Fs}^{4^a} & \frac{2}{b_2-b_1}K^{4^b}(\omega,kz)\mathbf{Fs}^{4^b} & \frac{2}{b_2-b_1}K^{4^c}(\omega,kz)\mathbf{Fs}^{4^c} \end{bmatrix}}_{Vs^E} \cdot \begin{bmatrix} Um^{4^a} \\ Um^{4^b} \\ Um^{4^c} \\ Ue^{4^a} \\ Ue^{4^b} \\ Ue^{4^c} \end{bmatrix}_{x=T_4} \tag{20}$$

$$\underbrace{\begin{bmatrix} \bar{K}^{4^a}(\omega,kz)\dfrac{2}{b_8-b_7}Fc^{4^aT} & 0 \\ \bar{K}^{4^b}(\omega,kz)\dfrac{2}{b_4-b_3}Fc^{4^bT} & 0 \\ \bar{K}^{4^c}(\omega,kz)\dfrac{2}{b_6-b_5}Fc^{4^cT} & 0 \\ \bar{K}_\epsilon^{4^a}(\omega,kz)\dfrac{2}{b_8-b_7}\widetilde{Fs}^{4^aT} & \dfrac{\epsilon_{r5}}{\epsilon_{r4^a}}\dfrac{2}{b_8-b_7}Fs^{4^aT} \\ \bar{K}_\epsilon^{4^b}(\omega,kz)\dfrac{2}{b_4-b_3}\widetilde{Fs}^{4^bT} & \dfrac{\epsilon_{r5}}{\epsilon_{r4^b}}\dfrac{2}{b_4-b_3}Fs^{4^bT} \\ \bar{K}_\epsilon^{4^c}(\omega,kz)\dfrac{2}{b_6-b_5}\widetilde{Fs}^{4^cT} & \dfrac{\epsilon_{r5}}{\epsilon_{r4^c}}\dfrac{2}{b_6-b_5}Fs^{4^cT} \end{bmatrix}}_{Vs^H}$$

$$\cdot\begin{bmatrix} Im^5 \\ Ie^5 \end{bmatrix}_{x=T_4} = \begin{bmatrix} Im^{4^a} \\ Im^{4^b} \\ Im^{4^c} \\ Ie^{4^a} \\ Ie^{4^b} \\ Ie^{4^c} \end{bmatrix}_{x=T_4}. \quad (21)$$

In the above equation system, F_c and F_s denote the coupling matrices. Combining vectors U_e and U_m into U as well as I_m and I_e into I finally yields the transformation matrix for the discontinuity $T4$ in Fig. 2(a)

$$\begin{bmatrix} U^5 \\ I^5 \end{bmatrix}_{x=T_4} = \underbrace{\begin{bmatrix} V^E & 0 \\ 0 & V^H \end{bmatrix}}_{V^4} \cdot \begin{bmatrix} U^4 \\ I^4 \end{bmatrix}_{x=T_4} \quad (22)$$

or more generally

$$\begin{bmatrix} U^{i+1} \\ I^{i+1} \end{bmatrix}_{x=T_i} = \underbrace{\begin{bmatrix} V^E & 0 \\ 0 & V^H \end{bmatrix}}_{V^i} \cdot \begin{bmatrix} U^i \\ I^i \end{bmatrix}_{x=T_i}.$$

Isolating the left-hand side amplitude vector I^5 in (21) by multiplying with the inverse of matrix Vs^H yields a similar transformation matrix at the interface T^4 in Fig. 2(b).

$$\begin{bmatrix} U^5 \\ I^5 \end{bmatrix}_{x=T_4} = \underbrace{\begin{bmatrix} Vs^E & 0 \\ 0 & (Vs^H)^{-1} \end{bmatrix}}_{Vs^4} \cdot \begin{bmatrix} U^{4a,b,c} \\ I^{4a,b,c} \end{bmatrix}_{x=T_4}. \quad (23)$$

Now, by successively multiplying the transmission-line matrix (5) of each subregion with the corresponding transformation matrix finally leads to the desired relation between amplitude vectors at the upper and lower waveguide sidewalls in the example of Fig. 1(c)

$$\begin{bmatrix} U^5 \\ I^5 \end{bmatrix}_{x=a} = \underbrace{\left[R^5 \cdot Vs^4 \cdot R^{4a,b,c} \cdot Vs^3 \cdot R^3 \cdot V^2 \cdot R^2 \cdot V^1 \cdot R^1\right]}_{G} \cdot \begin{bmatrix} U^1 \\ I^1 \end{bmatrix}_{x=0}. \quad (24)$$

Inserting $V^4 \cdot R^4 \cdot V^3$ instead of $Vs^4 \cdot R^{\nu} \cdot Vs^3$ in (24) yields the matrix equation for the structure in Fig. 1(a), and (24) then reads as follows:

$$\begin{bmatrix} U^5 \\ I^5 \end{bmatrix}_{x=a} = \underbrace{R^5 \prod_{i=1}^{5} V^i \cdot R^i}_{G} \begin{bmatrix} U^1 \\ I^1 \end{bmatrix}_{x=0}. \quad (25)$$

The partial wave amplitudes are directly proportional to the field components $E_{y,z} \sim U$ and $H_{z,y} \sim I$, respectively. Hence, it follows for the up-to-now neglected boundary conditions (the transverse resonance condition) at the waveguide sidewalls, that $U = 0$ (for $x = a$ and $x = 0$). Thus, the characteristic matrix equation is the upper-right quarter of the matrix product in both (25) and (24).

If NH is the number of summation terms in (3) and (4), G_{12} is of order $2NH - 1$ and remains constant even for an increasing number of discontinuities. The zeros of its determinant provide the desired propagation constants

$$\det(G_{12}(w, kz)) = 0. \quad (27)$$

It has thus been demonstrated that this method can handle generalized quasi-planar transmission-line configurations in which strips and slots are located arbitrarily in the waveguide; more than one dielectric subregion and different strip thicknesses in the same structure can be considered, as well as the finite depth of the mounting grooves.

In many practical cases, however, symmetry conditions can be included in the formulation process which make this general treatment as efficient as specialized procedures. If we consider, for example, a symmetrical bilateral finline with a symmetry plane at $x = T3$ (Fig. 1(a)), (25) can be written as follows:

$$\begin{bmatrix} U^3 \\ I^3 \end{bmatrix}_{x=T_3} = R^3 \prod_{i=4}^{3} V^i \cdot R^i \begin{bmatrix} U^1 \\ I^1 \end{bmatrix}_{x=0}. \quad (28)$$

Thus, only three subregions ($i = 1, 2, 3$) have to be included in the entire multiplication procedure, which makes the numerical analysis more efficient. For a magnetic wall at $x = T3$, the new boundary condition can now be satisfied by setting $I^3 = 0$, and the corresponding resonant condition changes the characteristic matrix equation from the upper-right quarter in (25) to the lower right in (28)

$$0 = G_{22} \cdot I^1. \quad (29)$$

For an electric wall at $x = T3$, however, the resonance condition is still the same as in the general case, and the characteristic matrix equation is again G_{12} in (28). Both submatriçes (G_{22} and G_{12}) can be solved independently and provide the propagation constants of all hybrid modes in symmetrical bilateral finlines. Other symmetry conditions may be implemented very easily so that solutions of a large spectrum of conceivable quasi-planar transmission-line configurations are available with only one general computer program.

For the numerical procedure, the infinite sum in (3) and (4) must be truncated after a certain number of terms. To

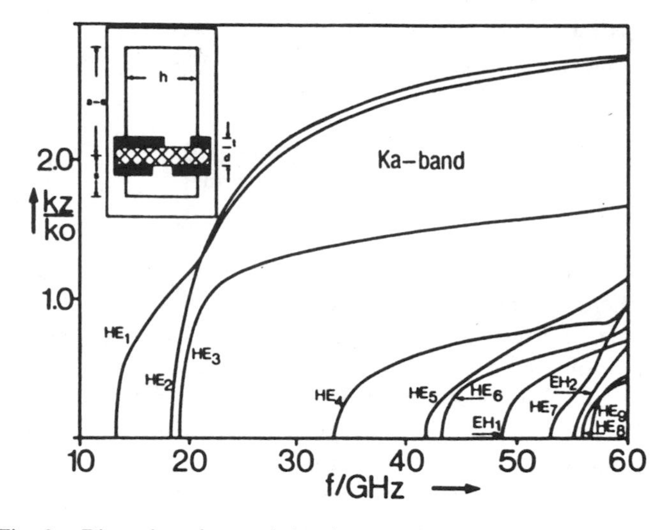

Fig. 3. Dispersion characteristic of an asymmetrical bilateral finline. $a = 7.112$ mm, $h = a/2$; $s = 2.2606$ mm, H1 = 1.201 mm, $s1 = 2.5$ mm, H2 = 2.156 mm, $s2 = 1.5$ mm, $e = 0.7$ mm; $d = 254.0$ μm, $\epsilon_r = 9.6$, $t = 70.0$ μm.

achieve a 0.5-percent accuracy in the propagation constant and to overcome a relative convergence effect which usually occurs between 11–15 terms, a minimum of 17 and a maximum of 25 summation terms have been found to be sufficient for most configurations investigated. The amount of CPU time required to evaluate the propagation constant at one frequency sample is between 1–3 min on a VAX 11/750, but only a few seconds on a main-frame computer like a Siemens 7880. A higher accuracy is attainable by increasing the number of summation terms, but at the expense of increased CPU time.

The behavior of some dispersion curves presented below has been verified by increasing the truncation index NH to 30 to exclude numerical inaccuracies. Furthermore, all dispersion curves are labeled at cutoff. That implies that their notation beyond the cutoff frequencies remains the same even when some curves cross each other.

III. Results

First of all, the potential of the present method is demonstrated by calculating quasi-planar configurations arbitrarily located in the waveguide mount. Fig. 3 shows the dispersion behavior of a bilateral finline with a slot offset, asymmetric location, finite metallization thickness, and mounting grooves. All higher order hybrid modes are excited by an incident H_{10} wave, and it is a remarkable result that the HE_2 mode crosses the HE_1 mode resulting in a higher propagation constant than the intrinsic dominant mode for frequencies beyond 20 GHz. This is due to two causes. Firstly, both modes are strongly affected by the mounting grooves; secondly, the insert is stepped away from the center of the waveguide (towards the higher field concentration of the H_{20} mode of the empty waveguide).

In an asymmetrically suspended microstrip combined with a bilateral finline, this curious dispersion behavior is even more pronounced (Fig. 4). It is obvious that there exists an interaction between the hybrid eigenmodes resulting, for example, in nontypical dispersion curve of the EH_0

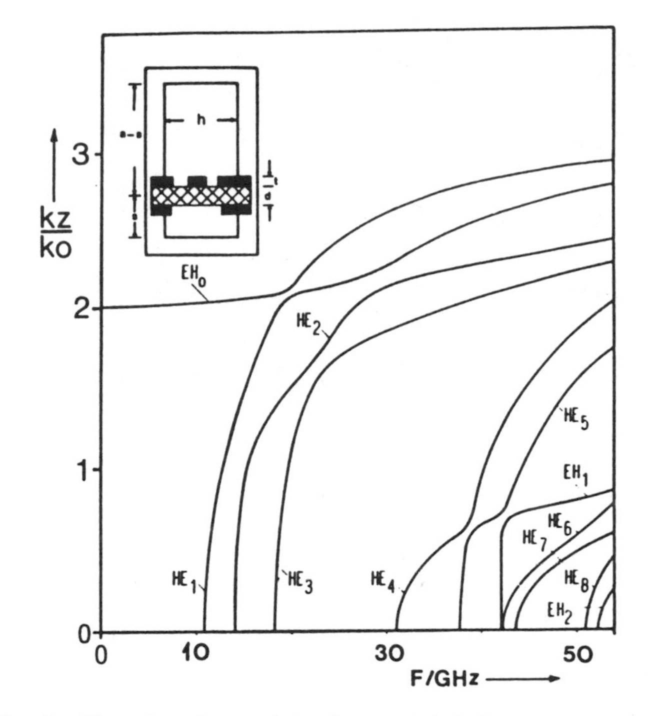

Fig. 4. Dispersion characteristic of a coupled slotline. $a = 7.112$ mm, $h = a/2$; $s = 1.778$ mm, $s1 = 2.2225$ mm, H1 = 1.1557 mm, $w1 = 0.592$ mm, $s2 = s3 = 0.592$ mm, H2 = 1.3038 mm; $d = 711.2$ μm, $\epsilon_r = 10.0$, $t = 71.0$ μm.

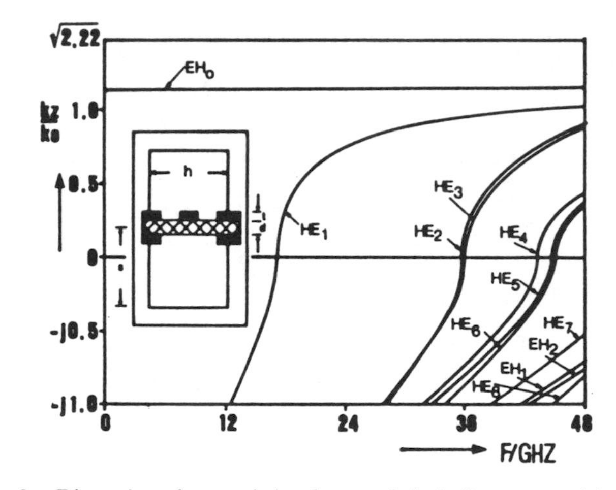

Fig. 5. Dispersion characteristic of a coupled slotline, centered in the waveguide. $a = 7.112$ mm, $h = a/2$; $s = 3.556$ mm, $s1 = 1.778$ mm, $w1 = 0.7112$ mm, $s2 = s3 = 0.7112$ mm; $d = 254.$ μm, $\epsilon_r = 2.22$, $t = 17.5$ μm.

and HE_1 mode. However, this is no longer so when a still asymmetric transmission line is centered in the waveguide (Fig. 5).

Fig. 6 shows the dispersion of both quasi-TEM and first higher order hybrid modes in a shielded coupled stripline combined with a tuning septum. All strips and slots are arbitrarily located on the substrate. As shown in [3] and [5], the different phase velocities of both quasi-TEM modes can be equalized by tuning the slot on the opposite side of the strips (Fig. 7) which is a prior condition for the design of broad-band contradirectional couplers with high directivity. We have compared our results on the influence of slot width on the quasi-TEM-propagation constant with data given by Schmidt [5]. Agreement is very good, as

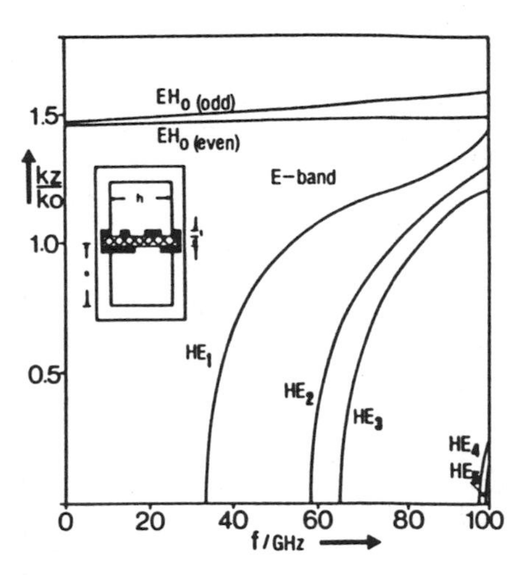

Fig. 6. Dispersion characteristic of a suspended coupled stripline centered in the waveguide but with arbitrarily located strips and slots. $a = 3.1$ mm, $h = a/2$; $s = a/2$, $s1 = 0.8$ mm, $s2 = 0.4$ mm, $s3 = 0.2$ mm, $s4 = 0.5$ mm, $w1 = 0.15$ mm, $w2 = 0.3$ mm, $H1 = 1.$ mm, $H2 = 0.5$ mm; $d = 110.$ μm, $\epsilon_r = 3.75$, $t = 10.$ μm.

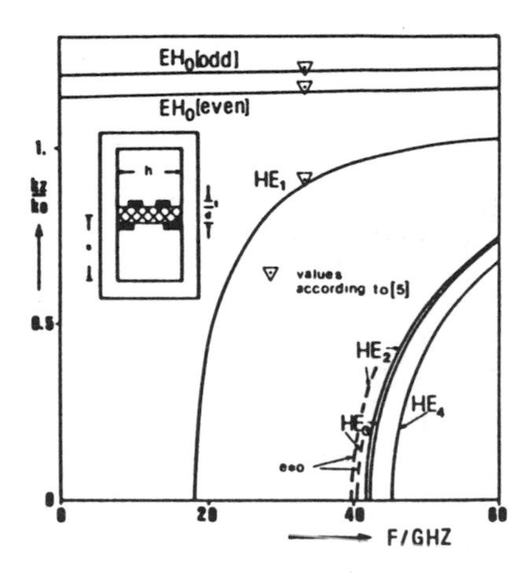

Fig. 7. Dispersion characteristic of a suspended coupled stripline compared with data given by [5]. $a = 7.112$ mm, $h = a/2$; $s = a/2$, $s1 = 0.2$ mm, $s2 = s4 = 1.378$ mm, $s3 = 0.4$ mm, $w1 = w2 = 0.2$ mm, $d = 254.$ μm, $\epsilon_r = 2.2$, $t = 5.$ μm; $e = 0$ (——), $e = 0.7$ mm (-----).

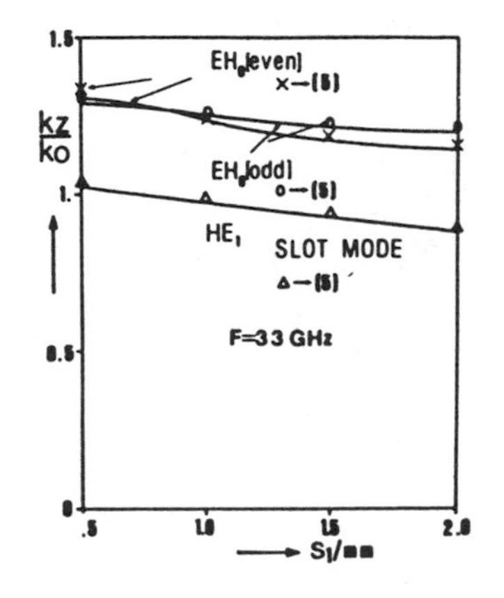

Fig. 8. Two coupled strips tuned by a slot on the opposite side of the dielectric. Dimensions identical to Fig. 7; $f = 33$ GHz; $e = 0$.

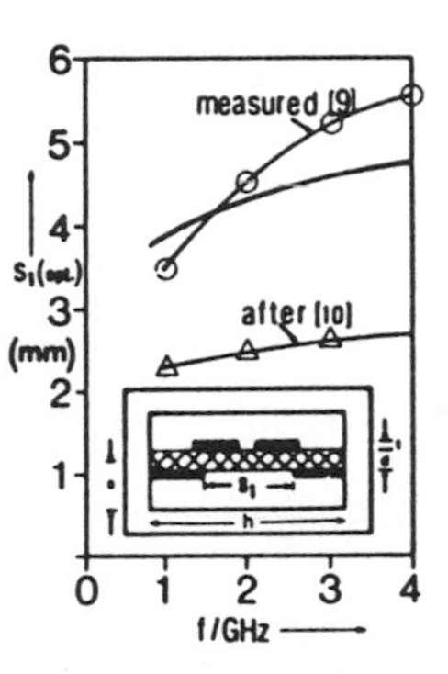

Fig. 9. Optimal slot width $S1$ (opt. in mm) of the tuning septum versus the frequency. $a = 7.27$ mm, $h = 25.$ mm; $s = a/2$, $w1 = w2 = 2.2$ mm, $s2 = s4 = 9.9$ mm, $s3 = 0.8$ mm; $d = 1.27$ mm, $\epsilon_r = 9.6$, $t = 17.5$ μm. measured [9], evaluated [10], —— this method.

shown in Fig. 8. Moreover, Fig. 7 shows the dispersion characteristics of the configuration investigated in Fig. 8. Differences in the propagation constant are less than 2 percent. The optimal slot width which equalizes both quasi-TEM modes was measured in [9] and is given in Fig. 9. While theoretical values published in [10] show only a poor agreement with these experimental results, our data agree relatively well.

Considering a symmetry plane with an electric or magnetic wall at discontinuity, T3 in Fig. 1(a) yields a so-called insulated finline [2], [6] (both fins are connected with the waveguide mount). For the same symmetry plane in Fig. 1(b), a single-side insulated finline is obtained in which only one fin is not connected with the waveguide mount, and Fig. 1(c) provides a dual-side insulated finline. The last two types of transmission lines play an important role when active components are used and one or both fins have to be insulated by a gasket to allow a dc voltage to be developed across the fins. The bias is introduced at the mounting grooves. To simplify the numerical procedure for determining the fundamental mode in these structures, the depth of the grooves is assumed to be a quarter wavelength at operating frequency and regarded as a choke section by which the RF continuity between the fins and the waveguide wall is achieved. Hence, the configuration has been investigated as an earthed finline. For practical application, however, these configurations support quasi-TEM modes, which can be seen in the following figures.

Fig. 10 presents the dispersion characteristics of a so-called insulated finline. A comparison with the configuration in Fig. 11, where only one fin is insulated, reveals the difference, which is essentially the occurence of the quasi-TEM mode (EH_0). In both cases, the HE_1 mode has almost the same cutoff frequency and is obviously not affected by the insulation of one fin. The insulation of both fins (Fig. 12) does not change the cutoff frequency of the HE_1 mode either, but a second quasi-TEM mode occurs as expected. Combining the single insulated with an earthed bilateral finline (Fig. 13) increases the propagation constant of EH_0 and decreases the cutoff frequencies of the higher order modes HE_2–HE_4. This tendency is also observed for the dual insulated finline in Fig. 14 and resembles somewhat the higher order mode behavior in an earthed bilateral finline [15]. Influences of the finite metallization thickness on the HE_1 and quasi-TEM modes are relatively small as can be seen from Fig. 15. A thicker metallization leads to decreasing quasi-TEM- and HE_1-mode propagation constants. This corresponds with the physical point of view that their electric fields are mainly concentrated in the air-filled region between the strips,

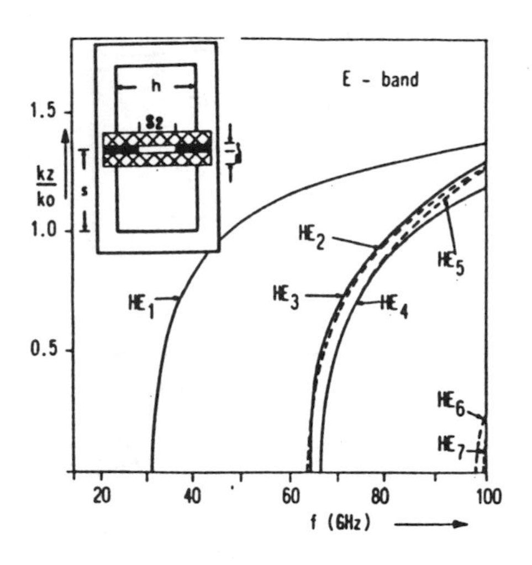

Fig. 10. Dispersion characteristic of a so-called insulated finline. $a = 3.1$ mm, $h = a/2$; $s = a/2$, $s2 = 0.6$ mm, $e = 0.5$ mm, $t = 5.$ μm, $d = 110.$ μm, $\epsilon_r = 3.75$. Symmetry plane at $s = a/2$: magnetic wall (——), electric wall (---------).

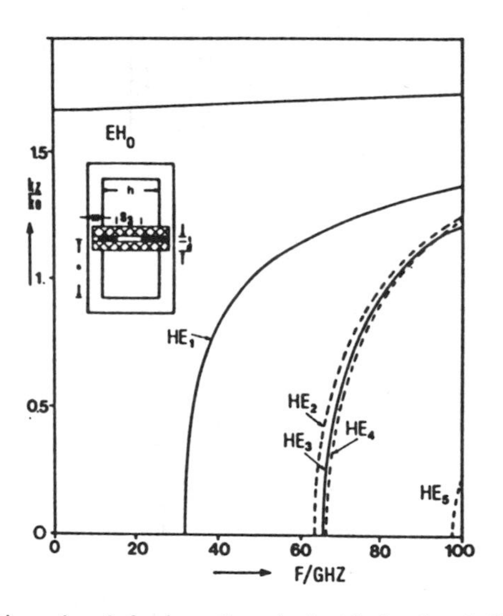

Fig. 11. Dispersion behavior of a single-side insulated finline. $a = 3.1$ mm, $h = a/2$; $s = a/2$, $e = 0.5$ mm, $c = s2 = 0.3$ mm (cf. Fig. 1(b)), $s3 = 0.6$ mm; $d = 110.$ μm, $\epsilon_r = 3.75$, $t = 5.$ μm. Symmetry plane at $s = a/2$: magnetic wall (——), electric wall (------).

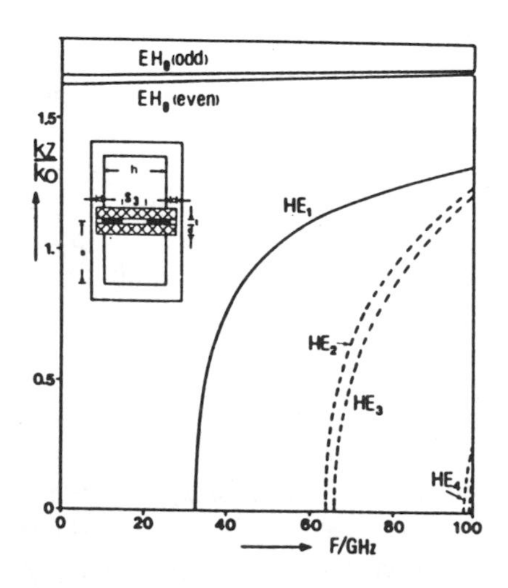

Fig. 12. Dispersion behavior of a dual-side insulated finline. $c = s2 = s4 = 0.3$ mm, all other dimensions identical to Fig. 11.

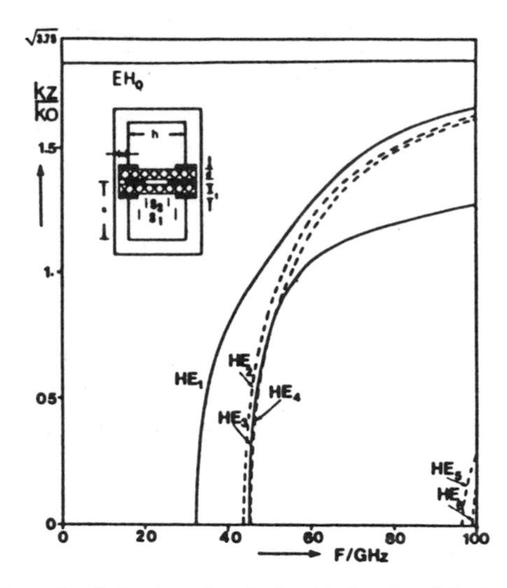

Fig. 13. Dispersion behavior of a single-side insulated finline combined with a bilateral finline. $s1 = 0.9$ mm, all other dimensions identical to Fig. 11.

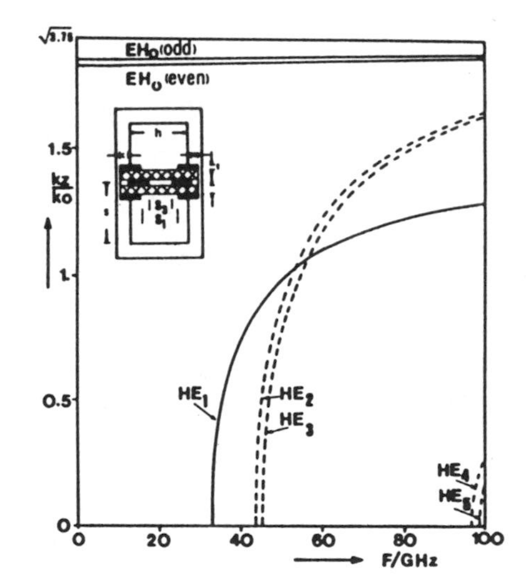

Fig. 14. Dispersion behavior of a dual-side insulated finline combined with a bilateral finline. Dimensions identical to Figs. 12 and 13.

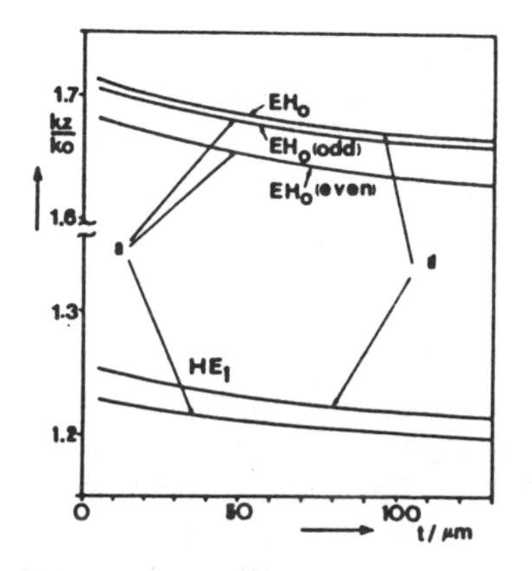

Fig. 15. Quasi-TEM and slot mode versus the metallization thickness, f-75 GHz. (a) Dual-side insulated finline (dimensions identical to Fig. 12. (b) Single-side insulated finline (dimensions identical to Fig. 11).

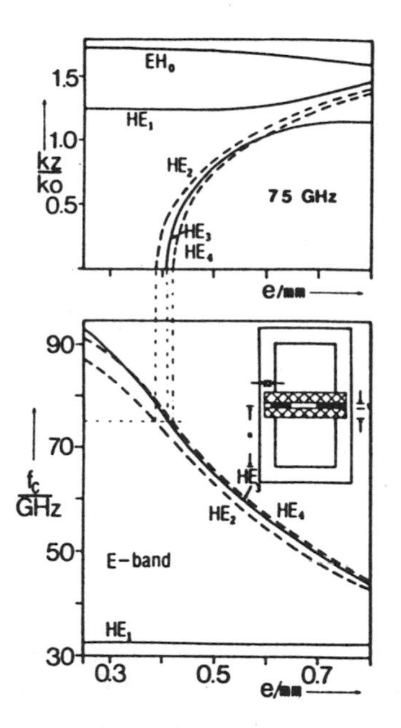

Fig. 16. Quasi-TEM and next higher order modes versus the groove depths e in a single-side insulated finline. f_c = cutoff frequency. Dimensions according to Fig. 11.

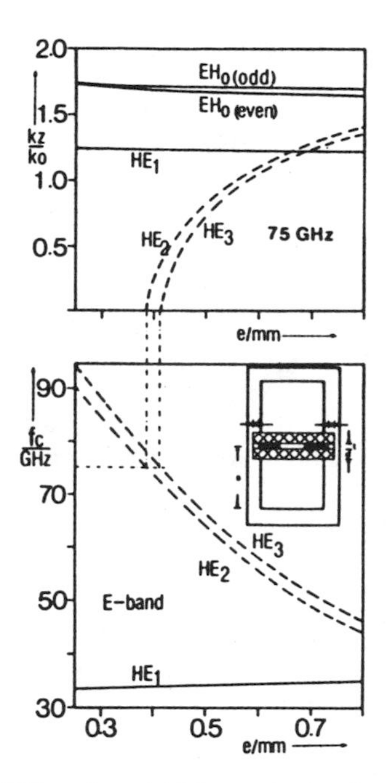

Fig. 17. Quasi-TEM and next higher order modes versus the groove depth e in a dual-side insulated finline. Dimensions according to Fig. 12.

resulting in a lower propagation constant when increasing the strip thickness.

Regarding the influence of the mounting grooves on the electrical characteristics of quasi-planar transmission lines, it was found out recently [6], [7] that higher order modes are strongly affected, whereas the HE_1 and the quasi-TEM modes are virtually insensitive to a wide range of changes in groove depth. In principle, this tendency is also observed for the single and dual insulated finline, and is shown in Figs. 16 and 17. For the single insulated configuration in Fig. 16, however, a mutual influence of the slot mode (HE_1) and the second higher order mode (HE_3) (which has the same field symmetry as the incident H_{10} mode) occurs, but was not observed for the dual-side insulated finline in Fig. 17. In that case (Fig. 17), both higher order modes (HE_2, HE_3) are obtained by considering an electric wall as the symmetry plane. It should be noted that, for operating frequencies beyond 75 GHz, the critical groove depth becomes shallower than given in Fig. 16, but can be increased a few percent by using a smaller substrate thickness.

IV. Conclusion

A versatile hybrid-mode analysis is presented taking into account the finite metallization thickness and mounting grooves in a class of various quasi-planar transmission lines. Using a modified mode-matching technique combined with a generalized transmission-line matrix exhibits the advantage of maintaining the size of the characteristic matrix equation constant even for an increasing number of discontinuities. Additionally, the method enables a user to create a modularized computer program which can easily be extended to a wide range of conceivable transmission-line configurations.

It is true that the method is not as fast as the spectral-domain technique; however, it includes the effect of finite metallization thickness and mounting grooves. Especially, the applications of quasi-planar transmission lines in the shorter millimeter-wave range have shown that the influence of the mounting grooves are more significant than the effect of finite metallization thickness. For example, a 0.7-mm-deep groove in a *Ka*-band waveguide mount causes a 2-GHz reduction of the cutoff frequency of the next higher order mode (Fig. 7), whereas the same groove depth reduces the next higher order mode in an *E*-band waveguide mount by at least 45 GHz (Fig. 16).

Appendix

Abbreviations for the partial wave amplitudes in (3) and (4) are

$$Im^i_{(n)}(x) = A^i_{(n)} e^{jkx^i_{(n)}x} + B^i_{(n)} e^{-jkx^i_{(n)}x}$$

$$Ue^i_{(n)}(x) = \frac{1}{jkx^i_{(n)}} \left\{ C^i_{(n)} e^{jkx^i_{(n)}x} - D^i_{(n)} e^{-jkx^i_{(n)}x} \right\}$$

$$Ie^i_{(n)}(x) = C^i_{(n)} e^{jkx^i_{(n)}x} + D^i_{(n)} e^{-jkx^i_{(n)}x}$$

$$Um^i_{(n)}(x) = jkx^i_{(n)} \left\{ A^i_{(n)} e^{jkx^i_{(n)}x} - B^i_{(n)} e^{-jkx^i_{(n)}x} \right\}.$$

The unknown coefficients $A^i_{(n)}$, $B^i_{(n)}$, $C^i_{(n)}$, and $D^i_{(n)}$ are eliminated by inserting the upper ($x = xo^i$) and the lower boundary ($x = xu^i$) values in the above equations. Thus, the generalized transmission-line matrix $\boldsymbol{R}^i$ for each subregion contains the diagonal matrices

$$Rc^i = \begin{bmatrix} \cos\left(kx^i_{(n)}xa^i\right) & 0 \\ 0 & \cos\left(kx^i_{(n)}xa^i\right) \end{bmatrix}$$

$$Rs^i = \begin{bmatrix} -kx_{(n)}\sin\left(kx^i_{(n)}xa^i\right) & 0 \\ 0 & \dfrac{1}{kx^i_{(n)}}\sin\left(kx^i_{(n)}xa^i\right) \end{bmatrix}$$

$$Rs^{i'} = \begin{bmatrix} \dfrac{1}{kx^i_{(n)}}\sin\left(kx^i_{(n)}xa^i\right) & 0 \\ 0 & -kx^i_{(n)}\sin\left(kx^i_{(n)}xa^i\right) \end{bmatrix} \quad xa = xo - xu$$

and the propagation constant

$$kx^i_{(n)} = \sqrt{ko^2\epsilon^i_r - \left(ky^i_{(n)}\right)^2 - kz^2}$$

with $ko^2 = \omega^2\mu_0\epsilon_0$.

The functions $fc^i_{(n)}(y)$ and $fs^i_{(n)}(y)$ in (3) and (4) are determined by the boundary condition in the y-direction

$$fc^i_{(n)}(y) = \frac{\cos\widetilde{ky}^i_{(n)}}{\sqrt{1+\delta_{0n}}} \text{ and } fs^i_{(n)}(y) = \sin\widetilde{ky}^i_{(n)},$$

$$\delta_{0n} = \text{Kronecker delta.}$$

For example, $\widetilde{ky}^4_{(n)}$ in Fig. 2(a) means

$$\widetilde{Ky}^4_{(n)} = ky^4_{(n)}\cdot(y - b3), \qquad ky^4_{(n)} = \frac{n\cdot\pi}{b4-b3}.$$

The coupling integrals Fc^ν and Fs^ν in (20) can be written as follows:

$$Fc^\nu_{(nk)} = \int_{yu^\nu}^{yo^\nu} fc^5_{(n)}(y) fc^\nu_{(k)}(y)\,dy, \qquad \nu \in 4a, 4b, 4c$$

with $yu^\nu = (b7, b3, b5)$ and $yo^\nu = (b8, b4, b6)$.

For Fs^ν, replace fc with fs. The abbreviations for $\widetilde{Fc}^\nu$ and $\widetilde{Fs}^{\nu T}$ in (20) and (21) are

$$\widetilde{Fc}^\nu = \frac{2}{b2-b1}\cdot Fc^\nu\cdot\text{Diag}\left(ky^\nu_{(n)}\right)$$

$$\widetilde{Fs}^{\nu T} = Fs^{\nu T}\cdot\text{Diag}\left(ky^\nu_{(n)}\right).$$

$Fs^{\nu T}$ means the transposed of matrix Fs^ν.

Acknowledgment

The authors wish to thank Dr. W. J. R. Hoefer, University of Ottawa, and Dr. F. Arndt, University of Bremen, for helpful discussions and suggestions.

References

[1] P. J. Meier, "Integrated fin-line millimeter components," *IEEE Trans. Microwave Theory Tech.*, vol. MTT-22, pp. 1209–1216, Dec. 1974.

[2] Y.-C. Shih and W. J. R. Hoefer, "Dominant and second-order mode cutoff frequencies in fin-lines calculated with a two-dimensional TLM program," *IEEE Trans. Microwave Theory Tech.*, vol. MTT-28, pp. 1443–1448, Dec. 1980.

[3] T. Itoh and A. S. Hebert, "A generalized spectral domain analysis for coupled suspended microstriplines with tuning septums," *IEEE Trans. Microwave Theory Tech.*, vol. MTT-26, pp. 820–826, Oct. 1978.

[4] L. P. Schmidt and T. Itoh, "Characteristics of unilateral fin-line structures with arbitrarily located slots," *IEEE Trans. Microwave Theory Tech.*, vol. MTT-29, pp. 352–355, Apr. 1981.

[5] L. P. Schmidt, "A comprehensive analysis of quasiplanar waveguides for millimeter-wave application," in *Proc. 11th Eur. Microwave Conf.* (Amsterdam), 1981, pp. 315–320.

[6] R. Vahldieck, "Accurate hybrid-mode analysis of various finline configurations including multilayered dielectrics, finite metallization thickness, and substrate holding grooves," *IEEE Trans. Microwave Theory Tech.*, vol. MTT-32, pp. 1454–1460, Nov. 1984.

[7] J. Bornemann, "Rigorous field theory analysis of quasiplanar waveguides," *Proc. Inst. Elec. Eng.*, vol. 132, pt. H, pp. 1–6, Feb. 1985.

[8] F. Arndt and G. U. Paul, "The reflection definition of the characteristic impedance of microstrips," *IEEE Microwave Theory Tech.*, vol. MTT-27, Aug. 1979.

[9] J. P. Villotte, M. Aubourg, and Y. Garault, "Modified suspended striplines for microwave integrated circuits," *Electron. Lett.*, vol. 14, no. 18, pp. 602–603, Aug. 31, 1978.

[10] W. Schuhmacher, "Wellentypen und Feldverteilung auf unsymmetrischen planaren Leitungen mit inhomogenem Dielektrikum," *Arch. Elec. Übertragung.*, vol. 34, pp. 445–453, 1980.

[11] J. Siegl, "Phasenkonstante und Wellenwiderstand einer Schlitzleitung mit rechteckigem Schirm und endlicher Metallisierungsdicke," *Frequenz*, vol. 31, pp. 216–220, July 1977.

[12] A. Beyer, "Analysis of the characteristics of an earthed fin-line," *IEEE Trans. Microwave Theory Tech.*, vol. MTT-29, pp. 676–680, July 1981.

[13] T. Kitazawa and R. Mittra, "Analysis of finline with finite metallization thickness," *IEEE Trans. Microwave Theory Tech.*, vol. MTT-32, pp. 1484–1487, Nov. 1984.

[14] H. Fritzsche, "Die frequenzabhaengigen Uebertragungseigenschaften gekoppelter Streifenleitungen im geschichteten Dielektrikum," *Nachrichtentech. Z.*, vol. 26, no. 1, pp. 1–8, 1973.

[15] R. Vahldieck and W. J. R. Hoefer, "The influence of metallization thickness and mounting grooves on the characteristics of finlines,"in *IEEE MTT-S Int. Microwave Symp. Dig.*, pp. 182-184.

Part V
Coupled Lines

A coupled pair of transmission lines has been extensively used in the development of microwave components such as directional couplers. Coupled line structures can also be made in a microstrip line form. When two identical strips are used in such a coupled pair, two orthogonal dominant modes are supported. They are typically called the even and odd modes. In the microstrip-type format, the phase constants of these two orthogonal modes are in general not equal, unlike the situations in the TEM transmission structure such as a coupled stripline. Once again, this phenomenon is caused by the non-TEM hybrid nature of the field configurations in the inhomogeneously filled structure under investigation. Therefore, in a coupled pair of microstrip lines, a total of four quantities are required to characterize the wave propagation. They are even and odd characteristic impedances (appropriately defined), and even and odd phase constants. These quantities must be computed by either a quasi-TEM method or a rigorous full-wave analysis. These methods are essentially identical to those treated in the previous parts except that they are extended for coupled line structures.

In many practical applications, it is desirable to equalize the phase constants of the even and odd modes so that the design methods developed for TEM transmission line structures can easily be adapted. Further, under such a condition of equal phase constants, the isolation of the directional couplers is theoretically infinite. Broadband design of such a configuration is of interest from a practical point of view. One of the papers (Su, Itoh, and Rivera) included in this part attempts to find an optimum structure.

Wave propagation of a coupled microstrip on an anisotropic substrate has been of considerable interest. This subject is included in this part (Kituzawa and Hayashi). Another topic of interest is a structure containing more than two strips or strips with nonsymmetric configurations. As the number of strips are increased, the number of orthogonal dominant modes also increases. Hence, the analysis and design of these structures are more involved. At the same time, flexibility of design is enhanced. In the case of nonsymmetric structures, the usual even and odd modes no longer exist. They are called the c and π modes. Several papers in this chapter deal with these interesting subjects (Tripathi; Kitazawa, Murahashi and Hayashi; Janiczak).

Parameters of Microstrip Transmission Lines and of Coupled Pairs of Microstrip Lines

THOMAS G. BRYANT, STUDENT MEMBER, IEEE, AND JERALD A. WEISS, SENIOR MEMBER, IEEE

Abstract—**A theoretical analysis is presented of microwave propagation on microstrip, with particular reference to the case of coupled pairs of microstrip lines. Data on this type of transmission line are needed for the design of directional couplers, filters, and other components in microwave integrated circuits. The inhomogeneous medium, consisting of the dielectric substrate and the vacuum above it, is treated in a rigorous manner through the use of a "dielectric Green's function" which expresses the discontinuity of the fields at the dielectric-vacuum interface. Results are presented in graphical form for substrate dielectric constants of 1, 9, and 16, and a range of values of width and spacing of the strips. Numerical tables for these and other cases are also available. The tables present capacitance, characteristic impedance, and velocity of propagation of the even and odd normal modes. The method lends itself to the treatment of other geometries which are of practical interest, such as "thick" strips, presence of an unsymmetrically located upper ground plane, etc.**

Manuscript received May 28, 1968; revised August 21, 1968. This work was performed under the sponsorship of the Array Radars Group at M.I.T. Lincoln Laboratory, in part through contracts with the University of Maine and Worcester Polytechnic Institute.

T. G. Bryant was with the University of Maine, Orono, Me. He is now with the Array Radars Group, M.I.T. Lincoln Laboratory, Lexington, Mass. (Operated with support from the U. S. Advanced Research Projects Agency.)

J. A. Weiss is with the Department of Physics, Worcester Polytechnic Institute, Worcester, Mass., and the Array Radars Group, M.I.T. Lincoln Laboratory, Lexington, Mass.

I. Introduction

FOR GUIDANCE in the design of integrated microwave circuit components, data are required on the parameters of symmetrical coupled pairs of microstrip transmission lines. The parameters needed to characterize this structure are the characteristic impedances and velocities of propagation of the two normal modes. In addition, for certain purposes such as investigation of spurious coupling, peak power capability, and gyromagnetic interaction (in the case of nonreciprocal substrate materials), information is also required on the RF field configuration.

The term "microstrip" is a nickname for a microwave circuit configuration which is constructed by printed-circuit techniques—modified where necessary to reduce loss, reflections, and spurious coupling, but retaining advantages in size, simplicity, reliability, and cost which such production techniques afford. Microstrip shares some of the trouble-

Reprinted from *IEEE Trans. Microwave Theory Tech.*, vol. MTT-16, no. 12, pp. 1021–1027, December 1968.

some properties of dispersive waveguide, in that conductor dimensions influence not only characteristic impedance, but also the velocity of propagation; also in that, on structures which support propagation in more than one mode, the velocities of the modes are in general unequal.

We present a solution of the microstrip problem in the "quasi-static" limit; i.e., for the frequency range in which propagation may be regarded as approximately TEM. Such a solution is valid in the range extending into the low gigahertz region, in the case of microstrip dimensions and substrate materials frequently used in integrated microwave circuit technology. For higher frequencies, the solution provides design guidance and can serve also as a basis for solution of the full propagation problem. The method developed for this work involves the determination of a "dielectric Green's function" which characterizes the effect on the field configuration due to the presence of the dielectric-vacuum interface. Such a function has potentially wide applicability to a variety of problems involving inhomogeneous dielectric media.

The physical construction of a coupled pair of microstrip lines is shown in Fig. 1. The configuration is conventionally specified by the parameters W/H and S/H, together with K, where W is the strip width, H is the substrate height, and S is the spacing between adjacent edges of the strips; K is the relative dielectric permittivity of the substrate material. The problem of a single microstrip may be identified with that of coupled strips in the limit $S/H \rightarrow \infty$. In the present report the strip thickness T is assumed to be negligibly small. This assumption is not intrinsic to the method, however, and application to the case of nonzero T/H will be presented in a future publication.

From the symmetry of the structure, the normal modes of propagation on a pair of parallel strips of equal dimensions have even and odd symmetry, respectively, with respect to reflection in a central bisecting plane. In the quasi-static limit, where propagation is approximately TEM, the characteristic impedances of the two modes can be determined from their respective dc capacitances and low-frequency velocities. The difference in impedances becomes large as the coupling between the strips is increased by reducing the spacing between them. The analogous problem of coupled strips in the case of balanced stripline has been treated by Cohn [1]. The microstrip problem presents additional difficulties due to the lower symmetry and the presence of the dielectric–vacuum boundary. Following the first thorough treatment by Wheeler [2], there have been calculations of capacitance, velocity, and impedance of single strips, and also some work on coupled pairs, using various approximate methods, by Cristal [3], Wolters and Clar [4], Policky and Stover [5], and no doubt by others. During the course of this study, we were apprised by a paper by Silvester [6] which treats the dielectric–vacuum boundary by means of a Green's function, similar in some respects to that employed in the present paper. A recent paper by Yamashita and Mittra [7] presents an analysis based on the variational principle.

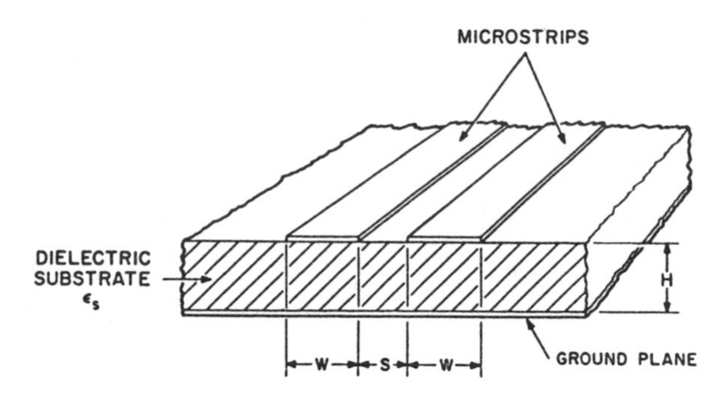

Fig. 1. A coupled pair of microstrip transmission lines.

II. The Dielectric Green's Function

We perform a rigorous solution of the electrostatic problem; namely, the determination of capacitance for a single strip, and for coupled strips, on a dielectric substrate. We then apply this solution to the propagation problem, incorporating the quasi-static approximation in which the longitudinal components of the fields are neglected.

Consider an arrangement of conducting boundaries, denoted collectively by S, and a harmonic function (potential) $\Phi_0(r)$ which satisfies given Dirichlet boundary conditions on S, and thus solves the electrostatic problem for the case of a vacuum or homogeneous dielectric medium. For the microstrip problem, S is simply the conducting ground plane, assumed to be infinite in extent. The potential is created by an elementary source, which we might take to be a uniformly charged line lying parallel to the plane at the position $\boldsymbol{r}_0$. A more suitable source is a narrow elementary strip, uniformly charged, since its potential is finite everywhere; the element actually used in this work is of that form.

Dielectric material, of relative permittivity K, is now introduced into this field, so as to occupy a region bounded by the surface T. In the following, we shall term "interface" that part of T which does not coincide with S. We seek the solution $\Phi(r)$ of the new electrostatic problem in the presence of the inhomogeneous dielectric medium.

Suppose we were to assume, as a first approximation, that the solution $\Phi_0(r)$ of the vacuum problem solves the dielectric problem also. This function is, indeed, the unique solution—not of the desired problem, but of a related one: namely, that in which, in addition to the charge distributions required to establish the specified values of potential on S, there is also a certain distribution of "free" charge $\sigma_0(\boldsymbol{r}_T)$ on the dielectric–vacuum interface. For, since Φ_0 and grad Φ_0 are continuous at T, the surface divergence of dielectric displacement $\boldsymbol{D}$ is

$$\epsilon_0(K - 1)\boldsymbol{n}_T \cdot \text{grad } \Phi_0 = \sigma_0(\boldsymbol{r}_T) \quad (1)$$

where $\boldsymbol{n}_T$ is the "outward" unit normal vector (directed from T into the vacuum).

Intuitively, we may imagine placing a compensating distribution of "bound" charge $\sigma(\boldsymbol{r}_T)$ on the interface so as to cancel $\sigma_0(\boldsymbol{r}_T)$. Let $\Phi(r)$ denote the new potential, including the effect of this additional charge. Such bound charge, corresponding to a surface divergence of electric field $\boldsymbol{E}$, creates a discontinuity in E such that

$$\epsilon_0 n_T \cdot [\text{grad } \Phi(r_{T-}) - \text{grad } \Phi(r_{T+})] = \sigma(r_T) \quad (2)$$

where r_{T+} and r_{T-} refer to the + (vacuum) and − (dielectric) sides of T, respectively.

Assuming that the correct bound charge distribution $\sigma(r_T)$ has been found, its effect is to annihilate the free charge distribution $\sigma_0(r_T)$:

$$\epsilon_0 n_T \cdot [K \text{ grad } \Phi(r_{T-}) - \text{grad } \Phi(r_{T+})] = 0. \quad (3)$$

To evaluate the contribution of $\sigma(r_T)$ to the potential, we may employ the (vacuum) Green's function $G(r|r')$ appropriate to the conducting boundary S. Let G satisfy

$$\nabla^2 G(r \mid r') = -4\pi\delta(r - r'), \quad (4)$$

with the boundary condition

$$G(r_S \mid r') = 0. \quad (5)$$

Then the total potential $\Phi(r|r_0)$ due to the source at r_0, including the contribution due to $\sigma(r_T)$, is

$$\Phi(r \mid r_0) = \Phi_0(r) + \frac{1}{4\pi\epsilon_0}\int_T \sigma(r'_T) G(r \mid r'_T) da'. \quad (6)$$

As will be shown, (6) is the dielectric Green's function which plays the role of kernel for the solution of the quasi-static microstrip problem. In terms of this potential, the boundary condition (3) becomes

$$\sigma_0(r_T) + \frac{1}{4\pi}\int_T \sigma(r'_T) n_T \cdot [K \text{ grad } G(r_{T-} \mid r'_T) - \text{grad } G(r_{T+} \mid r'_T)] da' = 0. \quad (7)$$

This constitutes an integral equation for the unknown bound charge distribution $\sigma(r_T)$. By making use of the properties of $G(r|r')$ we can put the relation (7) into a form suitable for computation.

We decompose the factor in brackets in (7) according to

$$\begin{aligned} K \text{ grad } G(r_{T-} \mid r'_T) &- \text{grad } G(r_{T+} \mid r'_T) \\ &= \tfrac{1}{2}(K+1)[\text{grad } G(r_{T-} \mid r'_T) - \text{grad } G(r_{T+} \mid r'_T)] \\ &\quad + \tfrac{1}{2}(K-1)[\text{grad } G(r_{T-} \mid r'_T) + \text{grad } G(r_{T+} \mid r'_T)]. \end{aligned} \quad (8)$$

In substituting in (7) according to (8), we note that the two terms on the right side of (8) have the following interpretations. The first term (the one with coefficient $K+1$) yields in effect an integral of the Green's-function field over the surface of a small region enclosing an element of T; with the aid of Gauss's theorem we replace it by a volume integral of $\nabla^2 G$ which, according to (4), becomes $-4\pi\delta(r-r')$. The second term (with coefficient $K-1$) represents an average of the Green's-function fields on the two sides of T. It is therefore continuous at T; in fact, it is the part of the field contributed by the "images" of the elementary source $\delta(r-r')$; or, put another way, it is the field due to the charge distributions induced on S by this source. We denote by $G_I(r|r')$ this "image part" of the Green's function.

With these interpretations, (7) becomes

$$\sigma_0(r_T) + \frac{1}{2}(K+1)\sigma(r_T) + \frac{1}{4\pi}(K-1)\int_T \sigma(r'_T) n_T \cdot \text{grad } G_I(r_T \mid r'_T) da' = 0. \quad (9)$$

Equation (9) is an inhomogeneous Fredholm's integral equation of the second kind [8] for the unknown bound charge distribution $\sigma(r_T)$, in terms of the known initial free charge distribution $\sigma_0(r_T)$, given by (1), and the image part of the Green's function for S, which is known, or in any case is derivable with more or less difficulty depending on the shape of the conducting boundary S. For the case of the flat microstrip ground plane, G_I takes a simple form. Of the various methods available for solution of such equations, one which we have used successfully in the present work is an iterative method based on a slight rearrangement of (9):

$$\sigma(r_T) = -\frac{2}{K+1}\sigma_0(r_T) - \frac{1}{2\pi}\frac{K-1}{K+1}\int_T \sigma(r'_T) n_T \cdot \text{grad } G_I(r_T \mid r'_T) da'. \quad (10)$$

As a first approximation, set the function $\sigma(r'_T)$ appearing under the integral sign equal to zero; (10) gives the second approximation as simply $2/(K+1)$ times the negative of the initial free charge σ_0. Using this in the integral, we obtain the third approximation of σ, and so on. The method is not guaranteed to converge under all conditions, but in any case we can directly determine the success of the process by evaluating the left side of (9) after each stage of iteration. For the microstrip problem contemplated here, convergence has been shown to take place without incident. (In experiments with other surfaces S, we have acquired some experience which enables us often to anticipate convergence difficulties.)

With σ known, the desired dielectric Green's function $\Phi(r|r_0)$ may now be generated according to (6). An illustration of the result is shown in Fig. 2. The graph refers to substrate $K=16$; it shows the potential due to a strip element of $W/H=0.025$ carrying unit charge. Also shown in the figure are the distributions of bound charge σ and of surface charge on the ground plane. The vacuum image Green's function G_I used for this calculation is obtained by elementary methods. The potential due to an element of width 2ξ located at height H is

$$\begin{aligned} G(x, y \mid 0, H) = &\frac{1}{2}\left(\frac{x}{\xi} - 1\right) \ln \frac{(x-\xi)^2 + (y-H)^2}{(x-\xi)^2 + (y+H)^2} \\ &- \frac{1}{2}\left(\frac{x}{\xi} + 1\right) \ln \frac{(x+\xi)^2 + (y-H)^2}{(x+\xi)^2 + (y+H)^2} \\ &+ \frac{y+H}{\xi}\left(\arctan \frac{x+\xi}{y+H} - \arctan \frac{x-\xi}{y+H}\right) \\ &- \frac{y-H}{\xi}\left(\arctan \frac{x+\xi}{y-H} - \arctan \frac{x-\xi}{y-H}\right). \end{aligned} \quad (11)$$

For use in (10), we require the component of field normal to T, due to the image part of (11):

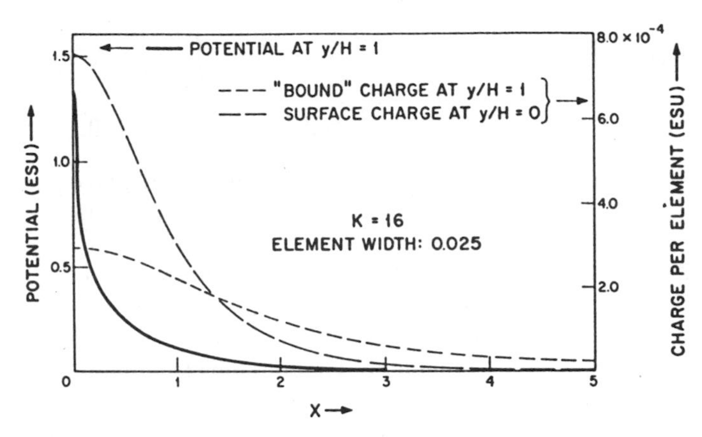

Fig. 2. The potential and charge distributions of the dielectric Green's function.

$$E_y(x, H) = \frac{1}{\xi}\left(\arc\tan\frac{x-\xi}{2H} - \arc\tan\frac{x+\xi}{2H}\right). \quad (12)$$

III. Calculation of Normal Mode Parameters

Consider a coupled-strip configuration composed of two strips, spacing S/H and equal widths W/H. One of these strips is at potential V; the other is at potential $\pm V$ for the even or odd mode, respectively. We divide the strips and the space between them into elements of width 2ξ (12); let $W/H=2\xi M$, $S/H=2\xi N$, and let $x_{j-i}=2\xi(j-i)$. The potentials on the strips are to be produced by the placement of charge λ_i on the ith element, $i=1, \cdots, M$, and $\pm\lambda_i$ on the symmetrically located element in the case of the even or odd mode, respectively. We determine the charges λ_i. Referring to the dielectric Green's function (6), let

$$\Phi_{ij} = \Phi(x_i, H \mid x_j, H) = \Phi(x_{j-i}, H \mid 0, H) \quad (13)$$

denote the potential at x_i due to the element at x_j carrying unit charge. Then the resultant potential at the center of the ith element is

$$\sum_{j=1}^{M} \lambda_j(\Phi_{ij} \pm \Phi_{i,2M+N+1-j}) = V. \quad (14)$$

This system is to be solved for the M values of charge λ_i. The solution yields the charge distribution (and therefore also the current distribution in the propagation problem) and the total charge

$$Q = \sum_{i=1}^{M} \lambda_i \quad (15)$$

from which the capacitance C_K per strip follows as $C_K = Q/V$. The method of calculation outlined above is obviously not limited to symmetrical pairs of strips, but may be applied to any strip configuration.

To determine the effective relative permittivity K_{eff}, the calculation must be performed for the vacuum $K=1$ as well as for the value of substrate K contemplated. The vacuum problem can, of course, be solved without iteration, since there is no bound charge in that case. Denoting the resulting capacitance per strip by C_1, we have

$$K_{eff} = C_K/C_1. \quad (16)$$

The velocity of propagation v is

$$v = \frac{c}{\sqrt{K_{eff}}} \quad (17)$$

where $c=2.997925\times10^8$ m·s^{-1} is the characteristic velocity of the vacuum. Within the quasi-static approximation, we may calculate the characteristic impedance per strip from

$$Z_0 = \frac{1}{vC_K} = \frac{1}{cC_1\sqrt{K_{eff}}}. \quad (18)$$

The wavelength ratio, which is occasionally convenient in design work, is

$$\frac{\lambda(K_{eff})}{\lambda(K)} = \sqrt{\frac{K}{K_{eff}}}. \quad (19)$$

IV. Data for Single and Coupled Strips

The method outlined above has been carried out for various values of K; in particular, for $K=9$, appropriate for an alumina substrate, and $K=16$, for the magnetic substrate material yttrium iron garnet. Fig. 3 shows the characteristic impedance of a single strip as a function of W/H for values of K from unity to 16; Fig. 4 shows the corresponding velocities.

The computational procedure contains its own means for determining accuracy, through the assessment of errors in the fulfillment of the boundary conditions, as will be discussed in Section VI. In addition, it is illuminating to compare these results with those which Wheeler [2] obtained analytically by means of a conformal transformation. The comparison is shown in Fig. 5, where the differences between our values of Z_0 and those of Wheeler are shown on a greatly expanded scale. Wheeler offers two approximations, appropriate for wide (Wheelw) and narrow (Wheeln) strips, respectively. Agreement between those two is best at about $W/H=1$. The figure shows that our results are intermediate between the two, the disagreement being well within one percent everywhere except for $W/H<0.2$. In that range, our values are higher. This is an error in our results, due to the fact that we have employed an elementary strip width of $W/H=0.025$, which means that for $W/H=0.2$ our representation of the strip is composed of only eight elements. With such coarse division of the strip, the steps in the charge distribution become a source of appreciable error, particularly at the edges where our discrete representation misses the sharp rise in concentration of charge. As the need arises for data on very narrow strips, this error can be made as small as desired by a straightforward modification of the program.

Results for coupled strips are shown graphically in Figs. 6–8, covering the ranges $W/H=0.2, \cdots, 2.0$, and $S/H=0.3$, 0.6, and 1.0 for values of substrate dielectric constant $K=1$, 9, and 16. Table I shows a sample page of tabular data. For given K and spacing parameter S/H, the table lists capacitance C_K, effective dielectric constant K_{eff} (16), char-

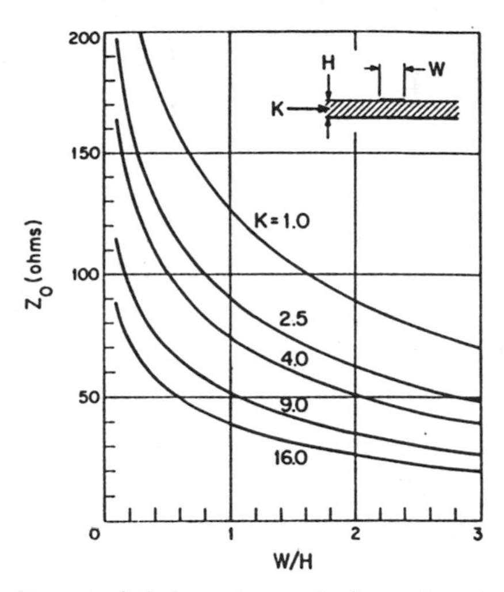

Fig. 3. Characteristic impedance of microstrip—single strip.

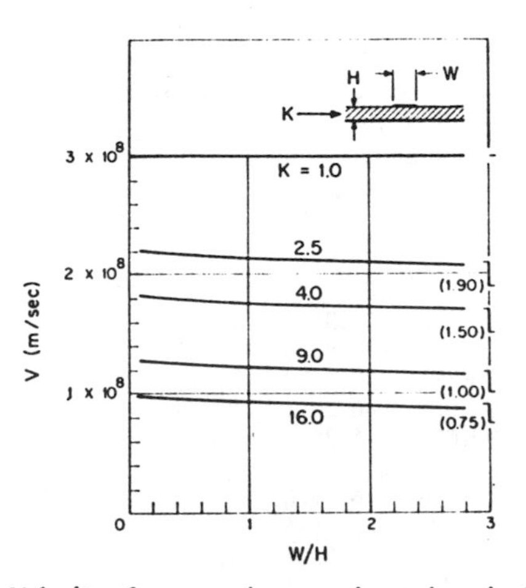

Fig. 4. Velocity of propagation on microstrip—single strip.

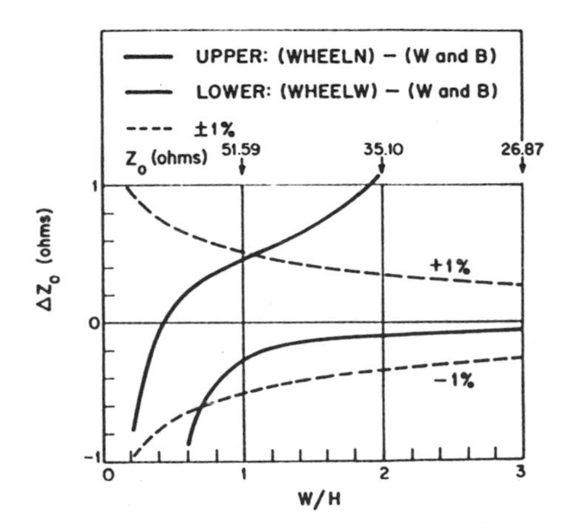

Fig. 5. Comparison of three values of characteristic impedance of microstrip.

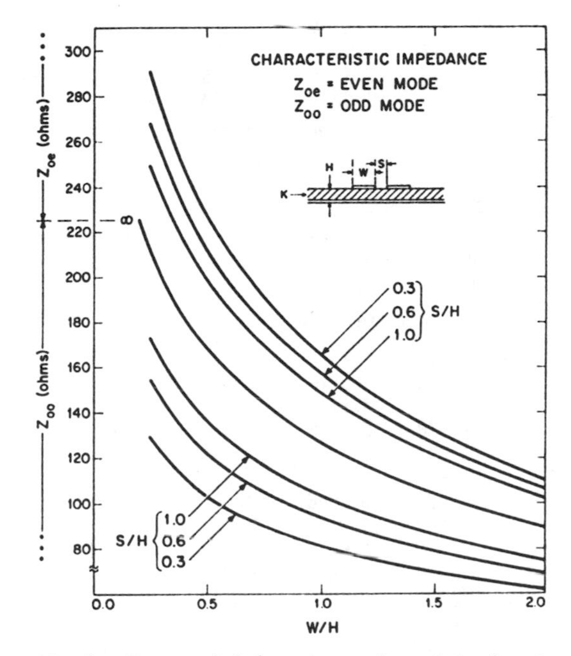

Fig. 6. Characteristic impedance of coupled pairs of microstrip transmission lines—$K=1$.

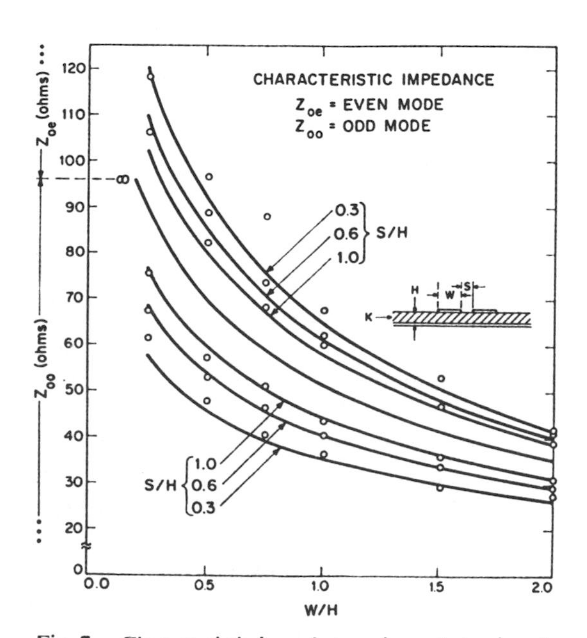

Fig. 7. Characteristic impedance of coupled pairs of microstrip transmission lines—$K=9.0$.

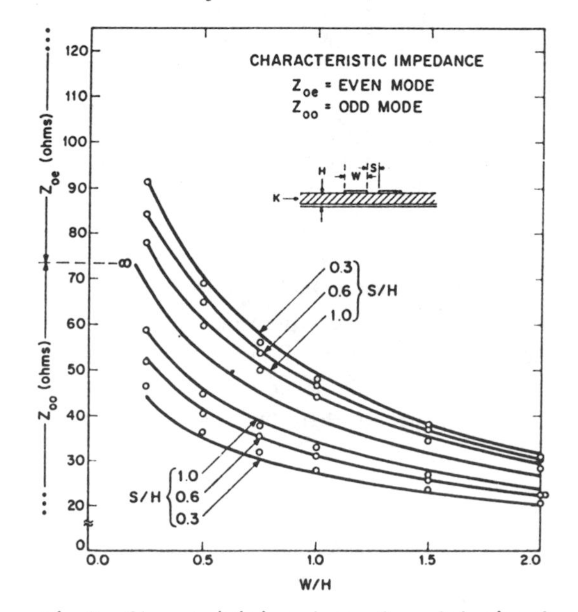

Fig. 8. Characteristic impedance of coupled pairs of microstrip transmission lines—$K=16.0$.

TABLE I
PARAMETERS OF MICROSTRIP
$S/H=0.2,\ K=10.0$

W/H	C_K (pF/m)	K_{eff}	Z_0 (ohms)	V (10^8 m/s)	$L(K_{eff})/L(K)$
Coupled Strips—Even Mode					
0.10	54.512	6.2506	152.984	1.1991	1.2648
0.20	66.463	6.4106	127.072	1.1841	1.2490
0.30	76.813	6.5429	111.079	1.1720	1.2363
0.40	86.558	6.6611	99.459	1.1616	1.2253
0.50	96.009	6.7695	90.395	1.1522	1.2154
0.60	105.298	6.8700	83.030	1.1438	1.2065
0.70	114.490	6.9635	76.882	1.1361	1.1984
0.80	123.621	7.0508	71.649	1.1290	1.1909
0.90	132.709	7.1326	67.128	1.1225	1.1841
1.00	141.767	7.2094	63.176	1.1165	1.1777
1.10	150.803	7.2815	59.687	1.1110	1.1719
1.20	159.822	7.3495	56.581	1.1058	1.1665
1.30	168.826	7.4135	53.796	1.1011	1.1614
1.40	177.819	7.4741	51.284	1.0966	1.1567
1.50	186.801	7.5313	49.004	1.0924	1.1523
1.60	195.775	7.5856	46.926	1.0885	1.1482
1.70	204.742	7.6372	45.023	1.0848	1.1443
1.80	213.701	7.6861	43.274	1.0814	1.1406
1.90	222.656	7.7328	41.659	1.0781	1.1372
2.00	231.604	7.7772	40.165	1.0750	1.1339
Coupled Strips—Odd Mode					
0.10	120.499	5.4993	64.916	1.2784	1.3485
0.20	150.072	5.5119	52.183	1.2769	1.3469
0.30	170.603	5.5275	45.968	1.2751	1.3450
0.40	186.988	5.5458	42.010	1.2730	1.3428
0.50	201.035	5.5666	39.148	1.2706	1.3403
0.60	213.624	5.5896	36.917	1.2680	1.3375
0.70	225.252	5.6146	35.089	1.2652	1.3346
0.80	236.220	5.6412	33.539	1.2622	1.3314
0.90	246.724	5.6693	32.191	1.2591	1.3281
1.00	256.892	5.6986	30.997	1.2558	1.3247
1.10	266.817	5.7289	29.923	1.2525	1.3212
1.20	276.560	5.7600	28.947	1.2491	1.3176
1.30	286.167	5.7917	28.052	1.2457	1.3140
1.40	295.671	5.8238	27.225	1.2423	1.3104
1.50	305.095	5.8563	26.458	1.2388	1.3067
1.60	314.457	5.8889	25.742	1.2354	1.3031
1.70	323.769	5.9217	25.071	1.2320	1.2995
1.80	333.042	5.9544	24.440	1.2286	1.2959
1.90	342.281	5.9870	23.845	1.2252	1.2924
2.00	351.494	6.0195	23.283	1.2219	1.2889

acteristic impedance Z_0, velocity of propagation v and the wavelength ratio (19). The table refers to the even and odd modes and covers a range of the width parameter W/H. Complete tables are available[1] covering the following ranges: $W/H=0.1(0.1)2.0$, single strip and even and odd modes for $S/H=0.2(0.2)1.0$, for substrate dielectric constant $K=9$, 10, 15, 16; also, a table for wide spacing: $S/H=1.0(0.5)4.0$, for $K=9.6$.

[1] Tables have been deposited as Document No. NAPS-00087 with ADI Auxiliary Publications Service, American Society for Information Sciences, c/o CCM Information Sciences, Inc., 22 West 34 Street, New York, N. Y. 10001. Microfiche copies are available at \$1.00 or 8½ by 11-inch photocopies at \$3.00. To order, include the document number and the advance payment.

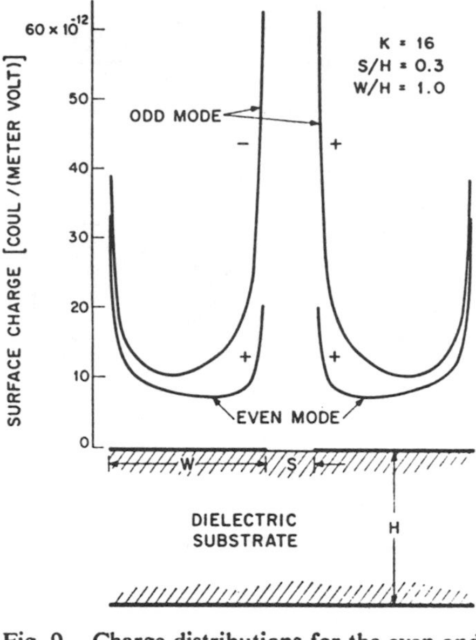

Fig. 9. Charge distributions for the even and odd modes on coupled strips.

The programs also furnish other useful information, including the determination of errors and the values of charge, potential, fields over the complete cross-section of the transmission line. An example of the charge distributions on the strips is shown in Fig. 9.

The data points in Figs. 7 and 8 are values measured by MacFarland [9].

V. PROGRAMS

The computations were performed at the computer facility at the University of Maine and on the CP/CMS system at M.I.T. Lincoln Laboratory. In the present form, the program is composed of two parts: the first computes the dielectric Green's function according to (10) for any specified K and stores these data. This computation has been used to generate data for a sequence of seven values of K in the range from 6 to 18. The dielectric Green's function for intermediate values of K is determined by interpolation. The second program uses these data to compute the single- and coupled-strip parameters as illustrated in Table I, according to (14).

VI. DETERMINATION OF ACCURACY

The calculation of the dielectric Green's function includes means for verification of its accuracy, through determination of errors in the fulfillment of the required boundary conditions. Using (9), we evaluate the surface divergence of electric displacement on the dielectric–vacuum interface. Typically, the residual free charge is less than one part in 10^4 or 10^5 relative to the bound charge at all points near the source, remaining roughly constant in magnitude—but therefore rising, relatively, to a few parts in 10^2 at distances about 10H from the source. We can assess the effect of this error by determining the total amount of "missing" bound charge. This estimate includes also the error resulting from the fact that the calculation was not normally carried to points beyond $\pm 10H$. As discussed below, this missing charge amounted to less than five percent of the required bound charge. If all the missing charge were lumped at the closest possible position, namely at $\pm 10H$, its contribution to the potential at points near the source would be less than one part in 10^4. This value is, of course, a great overestimate of the error, since in fact the missing charge would have been distributed over the entire infinite breadth of the interface.

All of the electric field originating on the source must, of course, terminate somewhere. The three locations of the terminating charge are 1) on the dielectric surface directly in contact with the source strip, 2) on the exposed area of the interface, and 3) on the ground plane. The accounting of these charges is as follows, for unit charge on the source: under the source, $(1-K)/(1+K)$—leaving net charge $2/(1+K)$ at the position of the strip; on the exposed interface, $(1-K)/K(1+K)$; on the ground plane, $-1/K$. We find that the calculated total of each of these agrees with requirements to within 1.6 percent at worst, and generally to within 0.7 percent or better, depending on the value of K. The greatest error is in the "missing" bound charge at the interface, as discussed above. It is noteworthy that the value of K for which the greatest deficiency is found to occur is in the vicinity of $K=2.5$, consistent with the fact that the maximum value of $(K-1)/K(K+1)$ occurs at $K=1+\sqrt{2}\cong 2.4$.

The conclusion from our various tests is that our calculation of the dielectric Green's function leads to errors in the microstrip parameters which are of the order of one part in 10^4 or less; completely adequate for all practical purposes.

As pointed out in Section IV, there is another type of error which enters in calculations involving values of W/H or S/H in the range less than 0.2. This is the consequence of the finite width of the discrete elementary strip, namely $W/H=0.025$ in the data reported here. As Fig. 5 shows, the effect is that the characteristic impedance is falsely high in that range: about one percent high at $W/H=0.2$ and becoming worse fast. As previously mentioned, this error can be easily corrected when the need arises.

Acknowledgment

We wish to acknowledge the contributions of Prof. F. Irons of the University of Maine and N. W. Cook of WPI. We are grateful to C. Blake and D. H. Temme of Lincoln Laboratory for their support and encouragement, and to H. T. MacFarland and J. D. Welch for the benefit of many discussions and other assistance. We also thank MacFarland for permission to quote his experimental results in advance of publication. Finally, we thank Mrs. J. Reid for her valuable instruction and assistance in the use of the CP/CMS time-sharing computer system.

References

[1] S. B. Cohn, "Shielded coupled-strip transmission line," *IRE Trans. Microwave Theory and Techniques*, vol. MTT-3, pp. 29–38, October 1955.
[2] H. A. Wheeler, "Transmission-line properties of parallel strips separated by dielectric sheet," *IEEE Trans. Microwave Theory and Techniques*, vol. MTT-13, pp. 172–185, March 1965.
[3] E. G. Cristal, Tech. Rept. USAEL Contract DA-28-043 AMC-02266 (E).
[4] K. C. Wolters and P. L. Clar, *Proc. 1967 G-MTT Internat'l Microwave Symp.*, paper V-2.
[5] G. Policky and H. L. Stover, "Parallel-coupled lines on microstrip," Texas Instruments, Inc., Rept. 03-67-61.
[6] P. Silvester, "TEM wave properties of microstrip transmission lines," *Proc. IEE* (London), vol. 115, pp. 43–48, January 1968.
[7] E. Yamashita and R. Mittra, "Variational method for the analysis of microstrip lines," *IEEE Trans. Microwave Theory and Techniques*, vol. MTT-16, pp. 251–256, April 1968.
[8] See, for example, L. P. Smith, *Mathematical Methods for Scientists and Engineers.* Englewood Cliffs, N. J.: Prentice-Hall, 1953.
[9] H. T. MacFarland, private communication.

Asymmetric Coupled Transmission Lines in an Inhomogeneous Medium

VIJAI K. TRIPATHI, MEMBER, IEEE

Abstract—**Terminal characteristic parameters for a uniform coupled-line four-port for the general case of an asymmetric, inhomogeneous system are derived in this paper. The parameters (impedance, admittance, etc.) are derived in terms of two independent modes that propagate in two uniformly coupled propagating systems. The four-port parameters derived are of the same form as those obtained for the symmetric case resulting in similar two-port equivalent circuits for various circuit configurations considered by Zysman and Johnson [1]. The results obtained should be quite useful in designing asymmetric coupled-line circuits in an inhomogeneous medium for various known applications.**

INTRODUCTION

UNIFORM coupled-line circuits are used for many applications including filters, couplers, and impedance matching networks. These circuits are usually designed by utilizing the impedance, admittance, chain, and other parameters characterizing the coupled-line four-port network. These parameters may be obtained in terms of the coupled-line impedances or admittances, and phase velocities for even and odd modes of excitation for the case of coupled TEM lines (homogeneous medium) [2], [3] or coupled identical lines in an inhomogeneous medium [1]. Recalling that even and odd modes of excitation correspond to the cases where the voltages and the currents on the two lines are equal in magnitude and are in phase for the even mode and out of phase for the odd mode, it is seen that such modes cannot propagate independently for the case of asymmetric coupled lines [5]. For asymmetric coupled-line cases these modes can be defined only for special cases [5]–[7] where the line parameters obey certain restrictive relationships.

In this paper the parameters of a general asymmetric asynchronous coupled-line four-port are obtained in terms of the line properties for two independent modes of excitations. These modes correspond to a linear combination of voltages and currents on the two lines which are related in magnitude and phase through terms involving line constants. The four-port circuit parameters are obtained by writing the solutions for voltages and currents on the two lines in terms of the two independent modes and deriving the relationships between port voltages and currents in a suitable form leading to impedance, admittance, chain, or any other parameters.

COUPLED-LINE ANALYSIS

The behavior of two coupled lines is described in general by the following set of equations:

$$-\frac{dv_1}{dx} = z_1 i_1 + z_m i_2 \tag{1a}$$

$$-\frac{dv_2}{dx} = z_2 i_2 + z_m i_1 \tag{1b}$$

$$-\frac{di_1}{dx} = y_1 v_1 + y_m v_2 \tag{2a}$$

$$-\frac{di_2}{dx} = y_2 v_2 + y_m v_1 \tag{2b}$$

where $z_j (j = 1,2)$ and $y_j (j = 1,2)$ are self-impedance and admittance per unit length of line j in the presence of line k ($k = 1,2; k \neq j$), z_m and y_m are mutual impedance and admittance per unit length, respectively, and an $e^{j\omega t}$ time variation has been assumed.

Differentiating (1a) and (1b) with respect to x and substituting (2a) and (2b), a system of equations for voltages on the uniformly coupled lines is obtained as

$$\frac{d^2 v_1}{dx^2} - a_1 v_1 - b_1 v_2 = 0 \tag{3a}$$

$$\frac{d^2 v_2}{dx^2} - a_2 v_2 - b_2 v_1 = 0 \tag{3b}$$

where

$$\begin{aligned} a_1 &= y_1 z_1 + y_m z_m \\ a_2 &= y_2 z_2 + y_m z_m \\ b_1 &= z_1 y_m + y_2 z_m \\ b_2 &= z_2 y_m + y_1 z_m. \end{aligned} \tag{4}$$

Since none of the coefficients in (3) varies with x, an x variation of the form $v(x) = v_0 e^{\gamma x}$ is assumed for the voltages. The solution of the resulting eigenvalue problem leads to the following four roots of γ:

$$\gamma_{1,2} = \pm\gamma_c$$

and

$$\gamma_{3,4} = \pm\gamma_\pi$$

where

$$\gamma_{c,\pi}^2 = \frac{a_1 + a_2}{2} \pm \tfrac{1}{2}[(a_1 - a_2)^2 + 4b_1 b_2]^{1/2}. \tag{5}$$

Manuscript received November 13, 1974; revised April 15, 1975.
The author is with the Department of Electrical and Computer Engineering, Oregon State University, Corvallis, Oreg. 97331.

Reprinted from *IEEE Trans. Microwave Theory Tech.*, vol. MTT-23, no. 9, pp. 734–739, September 1975.

For the case of lossless coupled systems these roots are the same as those obtained by Amemiya [8], Krage and Haddad [9], Marx [10], and others.

These values of γ_c and γ_π correspond to in phase and antiphase waves for a class of lossless lines. The relationship between the voltages on the two lines for each of these waves may be determined from (3) and (5) and is given as

$$\frac{v_2}{v_1} = \frac{\gamma^2 - a_1}{b_1} = \frac{b_2}{\gamma^2 - a_2} \tag{6}$$

or

$$R_c \triangleq \frac{v_2}{v_1} \quad \text{for} \quad \gamma = \pm\gamma_c$$

$$= \frac{1}{2b_1}\{(a_2 - a_1) + [(a_2 - a_1)^2 + 4b_1b_2]^{1/2}\} \tag{7}$$

and

$$R_\pi \triangleq \frac{v_2}{v_1} \quad \text{for} \quad \gamma = \pm\gamma_\pi$$

$$= \frac{1}{2b_1}\{(a_2 - a_1) - [(a_2 - a_1)^2 + 4b_1b_2]^{1/2}\}. \tag{8}$$

As seen from the expressions for R_c and R_π, v_2/v_1 is positive real for one mode, and negative real for the other mode for a large class of lossless coupled-line systems where $b_1b_2 > 0$. For the case of identical lines, $R_c = +1$ and $R_\pi = -1$ and the two modes correspond to the even and odd modes, respectively [4], and for homogeneous systems R_c and R_π correspond to lateral and diagonal excitations, respectively [11].

The general solutions for the voltages on the two lines in terms of all the four waves then are given by

$$v_1 = A_1e^{-\gamma_c x} + A_2e^{\gamma_c x} + A_3e^{-\gamma_\pi x} + A_4e^{\gamma_\pi x} \tag{9}$$

$$v_2 = A_1R_ce^{-\gamma_c x} + A_2R_ce^{\gamma_c x} + A_3R_\pi e^{-\gamma_\pi x} + A_4R_\pi e^{\gamma_\pi x}. \tag{10}$$

The corresponding currents for all four waves are determined by substituting the expressions for voltages (9) and (10) into (1a) and (1b) leading to

$$i_1 = A_1Y_{c1}e^{-\gamma_c x} - A_2Y_{c1}e^{\gamma_c x} + A_3Y_{\pi 1}e^{-\gamma_\pi x} - A_4Y_{\pi 1}e^{\gamma_\pi x} \tag{11}$$

$$i_2 = A_1R_cY_{c2}e^{-\gamma_c x} - A_2R_cY_{c2}e^{\gamma_c x} + A_3R_\pi Y_{\pi 2}e^{-\gamma_\pi x} - A_4R_\pi Y_{\pi 2}e^{\gamma_\pi x} \tag{12}$$

where Y_{c1}, Y_{c2}, $Y_{\pi 1}$, and $Y_{\pi 2}$ are the characteristic admittances of lines 1 and 2 for the two modes and are given by

$$Y_{c1} = \gamma_c \frac{z_2 - z_mR_c}{z_1z_2 - z_m^2} = \frac{1}{Z_{c1}} \tag{13}$$

$$Y_{c2} = \frac{\gamma_c}{R_c}\frac{z_1R_c - z_m}{z_1z_2 - z_m^2} = \frac{1}{Z_{c2}} \tag{14}$$

$$Y_{\pi 1} = \gamma_\pi \frac{z_2 - z_mR_\pi}{z_1z_2 - z_m^2} = \frac{1}{Z_{\pi 1}} \tag{15}$$

$$Y_{\pi 2} = \frac{\gamma_\pi}{R_\pi}\frac{z_1R_\pi - z_m}{z_1z_2 - z_m^2} = \frac{1}{Z_{\pi 2}}. \tag{16}$$

From these equations and (7) and (8) for R_c and R_π, respectively, it is seen that

$$\frac{Y_{c1}}{Y_{c2}} = \frac{Y_{\pi 1}}{Y_{\pi 2}} = -R_cR_\pi \tag{17}$$

and that the ratio of current amplitudes on the two lines are $i_2/i_1 = -1/R_\pi$ and $-1/R_c$ for the two modes $\gamma = \pm\gamma_c$ and $\gamma = \pm\gamma_\pi$, respectively.

Two independent modes can be excited on any two uniformly coupled systems. These modes correspond to a linear combination of voltages and currents which are related in magnitude and phase. The voltages and currents are related through $v_2/v_1 = R_c$ and R_π with $i_2/i_1 = -1/R_\pi$ and $-1/R_c$, respectively. This can be further illustrated from (1) and (2) by linearly combining the equations as $v_{c,\pi} = v_2 - R_{c,\pi}v_1$ and $i_{c,\pi} = i_2 + (1/R_{\pi,c})i_1$ resulting in uncoupled transmission-line equations for the two modes.

COUPLED-LINE FOUR-PORT

The impedance, admittance, or chain matrix for the coupled-line four-port as shown in Fig. 1 can now be obtained by solving for port current-voltage relationships from (9)–(12). For example, the impedance matrix for the four-port is found by solving for port voltages in terms of port currents. The port voltages are given as

$$\begin{bmatrix} V_1 \\ V_2 \\ V_3 \\ V_4 \end{bmatrix} = \begin{bmatrix} 1 & 1 & 1 & 1 \\ R_c & R_c & R_\pi & R_\pi \\ R_ce^{-\gamma_c l} & R_ce^{\gamma_c l} & R_\pi e^{-\gamma_\pi l} & R_\pi e^{\gamma_\pi l} \\ e^{-\gamma_c l} & e^{\gamma_c l} & e^{-\gamma_\pi l} & e^{\gamma_\pi l} \end{bmatrix} \cdot \begin{bmatrix} A_1 \\ A_2 \\ A_3 \\ A_4 \end{bmatrix}. \tag{18}$$

The port currents are given as

$$\begin{bmatrix} I_1 \\ I_2 \\ -I_3 \\ -I_4 \end{bmatrix} = \begin{bmatrix} Y_{c1} & -Y_{c1} & Y_{\pi 1} & -Y_{\pi 1} \\ R_cY_{c2} & -R_cY_{c2} & R_\pi Y_{\pi 2} & -R_\pi Y_{\pi 2} \\ R_cY_{c2}e^{-\gamma_c l} & -R_cY_{c2}e^{\gamma_c l} & R_\pi Y_{\pi 2}e^{-\gamma_\pi l} & -R_\pi Y_{\pi 2}e^{\gamma_\pi l} \\ Y_{c1}e^{-\gamma_c l} & -Y_{c1}e^{\gamma_c l} & Y_{\pi 1}e^{-\gamma_\pi l} & -Y_{\pi 1}e^{\gamma_\pi l} \end{bmatrix} \cdot \begin{bmatrix} A_1 \\ A_2 \\ A_3 \\ A_4 \end{bmatrix} \tag{19}$$

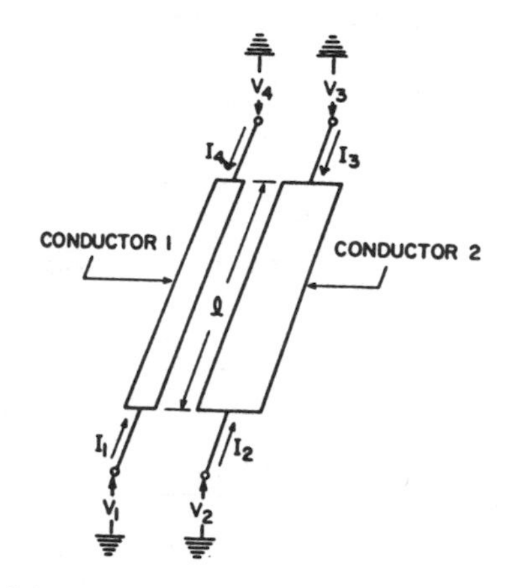

Fig. 1. Schematic of a uniform coupled-line four-port.

eliminating the amplitude coefficients A_1, A_2, A_3, and A_4 leads to four equations for V_1, V_2, V_3, and V_4 in terms of I_1, I_2, I_3, and I_4 of the form

$$[V] = [Z]\cdot[I]. \tag{20}$$

The elements of the 4×4 Z-matrix are given by

$$Z_{11} = Z_{44} = \frac{Z_{c1}\coth\gamma_c l}{(1 - R_c/R_\pi)} + \frac{Z_{\pi1}\coth\gamma_\pi l}{(1 - R_\pi/R_c)} \tag{21a}$$

$$Z_{12} = Z_{21} = Z_{34} = Z_{43} = \frac{Z_{c1}R_c\coth\gamma_c l}{(1 - R_c/R_\pi)} + \frac{Z_{\pi1}R_\pi\coth\gamma_\pi l}{(1 - R_\pi/R_c)}$$
$$= -\frac{Z_{c2}\coth\gamma_c l}{R_\pi(1 - R_c/R_\pi)} - \frac{Z_{\pi2}\coth\gamma_\pi l}{R_c(1 - R_\pi/R_c)} \tag{21b}$$

$$Z_{13} = Z_{31} = Z_{24} = Z_{42} = \frac{R_c Z_{c1}}{(1 - R_c/R_\pi)\sinh\gamma_c l} + \frac{R_\pi Z_{\pi1}}{(1 - R_\pi/R_c)\sinh\gamma_\pi l} \tag{21c}$$

$$Z_{14} = Z_{41} = \frac{Z_{c1}}{(1 - R_c/R_\pi)\sinh\gamma_c l} + \frac{Z_{\pi1}}{(1 - R_\pi/R_c)\sinh\gamma_\pi l} \tag{21d}$$

$$Z_{22} = Z_{33} = -\frac{R_c Z_{c2}\coth\gamma_c l}{R_\pi(1 - R_c/R_\pi)} - \frac{R_\pi Z_{\pi2}\coth\gamma_\pi l}{R_c(1 - R_\pi/R_c)}$$
$$= \frac{R_c^2 Z_{c1}\coth\gamma_c l}{(1 - R_c/R_\pi)} + \frac{R_\pi^2 Z_{\pi1}\coth\gamma_\pi l}{(1 - R_\pi/R_c)} \tag{21e}$$

$$Z_{23} = Z_{32} = \frac{R_c^2 Z_{c1}}{(1 - R_c/R_\pi)\sinh\gamma_c l} + \frac{R_\pi^2 Z_{\pi1}}{(1 - R_\pi/R_c)\sinh\gamma_\pi l}. \tag{21f}$$

The admittance parameters are found in a similar fashion and are given as

$$Y_{11} = Y_{44} = \frac{Y_{c1}\coth\gamma_c l}{(1 - R_c/R_\pi)} + \frac{Y_{\pi1}\coth\gamma_\pi l}{(1 - R_\pi/R_c)} \tag{22a}$$

$$Y_{12} = Y_{21} = Y_{34} = Y_{43} = -\frac{Y_{c1}\coth\gamma_c l}{R_\pi(1 - R_c/R_\pi)} - \frac{Y_{\pi1}\coth\gamma_\pi l}{R_c(1 - R_\pi/R_c)} \tag{22b}$$

$$Y_{13} = Y_{31} = Y_{24} = Y_{42} = \frac{Y_{c1}}{(R_\pi - R_c)\sinh\gamma_c l} + \frac{Y_{\pi1}}{(R_c - R_\pi)\sinh\gamma_\pi l} \tag{22c}$$

$$Y_{14} = Y_{41} = -\frac{Y_{c1}}{(1 - R_c/R_\pi)\sinh\gamma_c l} - \frac{Y_{\pi1}}{(1 - R_\pi/R_c)\sinh\gamma_\pi l} \tag{22d}$$

$$Y_{22} = Y_{33} = -\frac{R_c Y_{c2}\coth\gamma_c l}{R_\pi(1 - R_c/R_\pi)} - \frac{R_\pi Y_{\pi2}\coth\gamma_\pi l}{R_c(1 - R_\pi/R_c)} \tag{22e}$$

$$Y_{23} = Y_{32} = \frac{R_c Y_{c2}}{R_\pi(1 - R_c/R_\pi)\sinh\gamma_c l} + \frac{R_\pi Y_{\pi2}}{R_c(1 - R_\pi/R_c)\sinh\gamma_\pi l}. \tag{22f}$$

TWO-PORT CIRCUITS

The parameters (matrix elements) characterizing a general uniform coupled-line four-port obtained previously are of the same form as those for the case of symmetric four-port derived by Zysman and Johnson [1]. The resulting equivalent circuits may be obtained in a similar fashion as in [1]. For example, for an open-circuit interdigital section consisting of lossless lines as shown in Fig. 2, $I_2 = I_4 = 0$ and

$$\begin{bmatrix} V_1 \\ V_3 \end{bmatrix} = \begin{bmatrix} Z_{11} & Z_{13} \\ Z_{31} & Z_{33} \end{bmatrix} \begin{bmatrix} I_1 \\ I_3 \end{bmatrix}. \tag{23}$$

Substituting for Z_{11}, Z_{13}, Z_{31}, and Z_{33} yields

$$\begin{bmatrix} Z_{11} & Z_{13} \\ Z_{31} & Z_{33} \end{bmatrix} = -j\frac{Z_{c1}}{(1 - R_c/R_\pi)} \begin{bmatrix} \cot\theta_c & R_c\csc\theta_c \\ R_c\csc\theta_c & R_c^2\cot\theta_c \end{bmatrix} - j\frac{Z_{\pi1}}{(1 - R_\pi/R_c)} \begin{bmatrix} \cot\theta_\pi & R_\pi\csc\theta_\pi \\ R_\pi\csc\theta_\pi & R_\pi^2\cot\theta_\pi \end{bmatrix} \tag{24}$$

where $\theta_c = \beta_c l$ and $\theta_\pi = \beta_\pi l$.

This Z-matrix suggests an equivalent circuit as shown in Fig. 2 with its $ABCD$ parameters. The $ABCD$ parameters and the equivalent circuits for other configurations may be found in a similar manner. For the case of identical lines $R_c = -R_\pi = 1$ and the equivalent circuits and two-

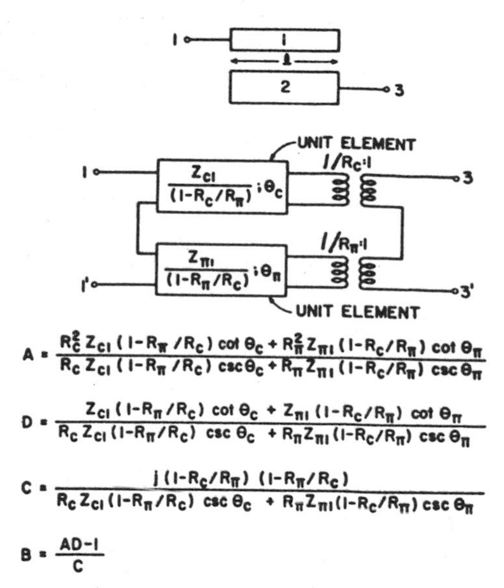

Fig. 2. Prototype open-circuited interdigital section.

port parameters are the same as those obtained by Zysman and Johnson [1].

SPECIAL CASES

The results obtained above are indeed a generalized case of known results for various coupled-line systems where even- and odd-mode analysis has been applied. For various cases studied involving coupled TEM or inhomogeneous lines, the equations are simplified leading to the respective known results.

Case 1—Symmetric Coupled Lines [1], [2]

For this case $y_1 = y_2 = y$; $z_1 - z_2 = z$. Then $R_c = 1$, and $R_\pi = -1$.

Expressing y's and z's in terms of line constants, i.e., self- and mutual inductances and capacitances, it is seen that

$$Z_{c2} = Z_{c1} = Z_{0e} \quad \text{the even-mode impedance}$$

and

$$Z_{\pi 2} = Z_{\pi 1} = Z_{0o} \quad \text{the odd-mode impedance}$$

with

$$\gamma_{c,\pi} = [(y \pm y_m)(z \pm z_m)]^{1/2} = \gamma_{e,o}[4] \tag{25}$$

and the resulting expressions for the coupled-line four-port parameters are the same as those in Zysman and Johnson [1] for an inhomogeneous medium and Jones and Bolljahn [2] for a homogeneous medium (for TEM case $y_m/y = -z_m/z$).

Case 2—Asymmetric Coupled Lines in a Homogeneous Medium [3], [6]

For lines with TEM waves

$$y_1 z_1 = y_2 z_2$$

and

$$\frac{y_m}{(y_1 y_2)^{1/2}} = -\frac{z_m}{(z_1 z_2)^{1/2}}. \tag{26}$$

Then

$$\gamma_c = \gamma_\pi = j\beta \tag{27}$$

$$R_c = -R_\pi = (Z_2/Z_1)^{1/2} \tag{28}$$

where $Z_1 = (z_1/y_1)^{1/2}$ and $Z_2 = (z_2/y_2)^{1/2}$.

The resulting expressions for the coupled-line four-port are the same as those in [3] and [6]. For example, the impedance parameters are [from (21a)–(21f)]

$$Z_{11} = Z_{44} = -j/2(Z_1/Z_2)^{1/2}(Z_c + Z_\pi) \cot\theta \tag{29a}$$

$$Z_{12} = Z_{21} = Z_{34} = Z_{43} = -j/2(Z_c - Z_\pi) \cot\theta \tag{29b}$$

$$Z_{13} = Z_{31} = Z_{24} = Z_{42} = -j/2(Z_c - Z_\pi) \csc\theta \tag{29c}$$

$$Z_{14} = Z_{41} = -j/2(Z_1/Z_2)^{1/2}(Z_c + Z_\pi) \csc\theta \tag{29d}$$

$$Z_{22} = Z_{33} = -j/2(Z_2/Z_1)^{1/2}(Z_c + Z_\pi) \cot\theta \tag{29e}$$

$$Z_{23} = Z_{32} = -j/2(Z_2/Z_1)^{1/2}(Z_c + Z_\pi) \csc\theta \tag{29f}$$

where

$$Z_{c,\pi} = (Z_1 Z_2)^{1/2}\left[\frac{1 \pm y_m/(y_1 y_2)^{1/2}}{1 \mp y_m/(y_1 y_2)^{1/2}}\right]^{1/2}. \tag{30}$$

Examination of Z_c and Z_π in terms of line constants reveals that the even- and odd-mode impedances of the two lines as defined by Z_{0e}^a, Z_{0o}^a for line 1 and Z_{0e}^b and Z_{0o}^b for line 2, respectively, [6] are given by

$$Z_{0e}^a + Z_{0o}^a = (Z_1/Z_2)^{1/2}(Z_c + Z_\pi) \tag{31a}$$

$$Z_{0e}^a - Z_{0o}^a = Z_{0e}^b - Z_{0o}^b = Z_c - Z_\pi \tag{31b}$$

and

$$Z_{0e}^b + Z_{0o}^b = (Z_2/Z_1)^{1/2}(Z_c + Z_\pi). \tag{31c}$$

Case 3—A Congruent Case [5]

If the line constants are such that

$$\frac{y_1 + y_m}{y_2 + y_m} = \frac{z_2 - z_m}{z_1 - z_m} \triangleq R_3 \tag{32}$$

which is approximately the case for tightly coupled lines, the even and odd modes can be redefined as in [5]. Substitution of (32) into expressions for R_c and R_π, (7) and (8), leads to

$$R_c = +1$$

and

$$R_\pi = -\frac{y_1 + y_m}{y_2 + y_m} = -R_3. \tag{33}$$

The corresponding ratio of currents on the two lines is

then given as

$$\frac{i_2}{i_1} = \frac{1}{R_3} \quad \text{for} \quad \gamma = \pm\gamma_c$$

and

$$\frac{i_2}{i_1} = -1 \quad \text{for} \quad \gamma = \pm\gamma_\pi. \tag{34}$$

Equations (33) and (34) correspond to the even- and odd-mode definitions for the coupled-line case where the condition given by (32) is satisfied [5]. Then the resulting matrix parameters are the same as those obtained by Speciale. These mode definitions have, of course, been experimentally verified for structures consisting of tightly coupled inhomogeneous lines.

CONCLUSIONS

It is shown that asymmetric, uniform coupled lines in an inhomogeneous medium, e.g., suspended substrate, microstrip lines, and others, may be analyzed in terms of the line properties for two independent modes of excitation. The mode characteristics, i.e., the propagation constants and the characteristic impedances, are derived in terms of the series impedances, the shunt admittances, and the mutual impedance and admittance per unit length of the lines. The 4×4 network matrices are then obtained in terms of these mode parameters. These circuit parameters characterizing the coupled-line four-port may be used to design various structures for all known applications including filters, couplers, and matching networks.

It should be noted that such structures can be treated utilizing the coupled-mode formulation [9]. However, the four-port circuit matrix is much easier and more convenient to use in formulating design procedures for various circuits particularly for the cases where multiple coupled-line sections are used. This paper has been primarily concerned with the study of inhomogeneous, asymmetric coupled lines. However, the formulation basically involves the evaluation of the properties of two linear uniformly coupled systems and coupled-line four-ports in terms of normal independent modes of the system and should provide a useful alternate tool for the study of many active and passive systems which have been studied using the coupled-mode theory.

REFERENCES

[1] G. I. Zysman and A. K. Johnson, "Coupled transmission line networks in an inhomogeneous dielectric medium," *IEEE Trans. Microwave Theory Tech.*, vol. MTT-17, pp. 753–759, Oct. 1969.

[2] E. M. T. Jones and J. T. Bolljahn, "Coupled-strip-transmission-line filters and directional couplers," *IRE Trans. Microwave Theory Tech.*, vol. MTT-4, pp. 75–81, Apr. 1956.

[3] H. Ozaki and J. Ishii, "Synthesis of a class of strip-line filters," *IRE Trans. Circuit Theory*, vol. CT-5, pp. 104–109, June 1958.

[4] E. G. Vlostovskiy, "Theory of coupled transmission lines," *Telecommun. Radio Eng.*, vol. 21, pp. 87–93, Apr. 1967.

[5] R. A. Speciale, "Fundamental even- and odd-mode waves for nonsymmetrical coupled lines in non-homogeneous media," in *1974 IEEE MTT Int. Microwave Symp. Digest Tech. Papers*, June 1974, pp. 156–158.

[6] E. G. Cristal, "Coupled-transmission-line directional couplers with coupled lines of unequal characteristic impedance," *IEEE Trans. Microwave Theory Tech.*, vol. MTT-14, pp. 337–346, July 1966.

[7] C. B. Sharpe, "An equivalence principle for nonuniform transmission line directional couplers," *IEEE Trans. Microwave Theory Tech.*, vol. MTT-15, pp. 398–405, July 1967.

[8] H. Amemiya, "Time domain analysis of multiple parallel transmission lines," *RCA Rev.*, vol. 28, pp. 241–276, June 1967.

[9] M. K. Krage and G. I. Haddad, "Characteristics of coupled microstrip transmission lines—I: Coupled-mode formulation of inhomogeneous lines," *IEEE Trans. Microwave Theory Tech.*, vol. MTT-18, pp. 217–222, Apr. 1970.

[10] K. D. Marx, "Propagation modes, equivalent circuits, and characteristic terminations for multiconductor transmission lines with inhomogeneous dielectrics," *IEEE Trans. Microwave Theory Tech.*, vol. MTT-21, pp. 450–457, July 1973.

[11] B. M. Oliver, "Directional electromagnetic couplers," *Proc. IRE*, vol. 42, pp. 1686–1692, Nov. 1954.

Design of an Overlay Directional Coupler by a Full-Wave Analysis

LUO SU, TATSUO ITOH, FELLOW, IEEE, AND JUAN RIVERA, MEMBER, IEEE

Abstract —**A full-wave analysis based on the spectral-domain method is applied to coupled overlay microstrips, coupled inverted microstrips, and coupled microstrips. Exclusive numerical data including frequency characteristics are included. A 10-dB overlay coupler was built according to the design theory, and experimental results are reported.**

Manuscript received April 4, 1983; revised August 14, 1983. This work was supported in part by a Grant from Texas Instruments, Equipment Group, and in part by the Army Research Office under Contract DAAG2981-K-0053.

L. Su is currently with the Department of Electrical Engineering, University of Texas at Austin, Austin, TX 78712, on leave from the Research Institute of Electronic Techniques of the Chinese Academy of Sciences, Guangzhou, China.

T. Itoh is with the Department of Electrical Engineering, University of Texas at Austin, Austin, TX.

J. Rivera is with TRW, Inc., Redondo Beach, CA 90278.

I. Introduction

It is known that when the even- and odd-mode propagation constants are identical, the isolation of a directional coupler is theoretically infinite. However, in an inhomogeneous structure such as microstrip, this condition is not always satisfied. A dielectric overlay is one way to improve the isolation of a microstrip coupler, by which the difference in even- and odd-phase velocities can be greatly reduced or even equalized [1]–[3]. To date, most of the

Reprinted from *IEEE Trans. Microwave Theory Tech.*, vol. MTT-31, no. 12, pp. 1017–1022, December 1983.

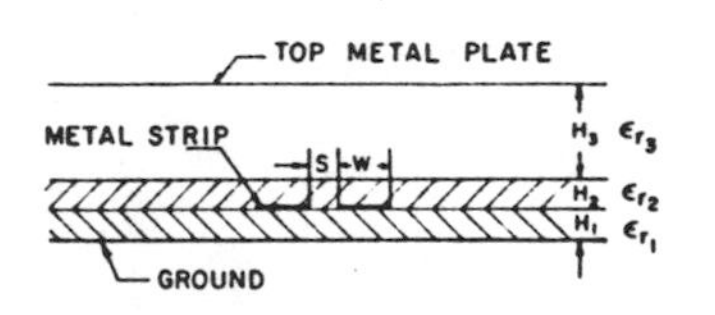

Fig. 1. Cross-sectional view of coupled microstrip lines with a dielectric overlay.

designs of overlay couplers are based on the quasi-TEM approximation. In this paper, the spectral-domain method [4] is used for a full-wave analysis with a view to designing an overlay coupler.

After the procedure for analyzing coupled line structures is introduced in Section II, we present in Section III numerical results for coupled microstrip lines, coupled overlay microstrip lines, and coupled inverted strip lines. In Section IV, directional couplers are designed and fabricated. These couplers are tested and their performance measured.

II. Analytical Process

Fig. 1 shows the cross section of a coupled overlay microstrip line structure. A full-wave analysis which includes the frequency dependent behavior of this structure is formulated based on the spectral-domain method [4]. Note that this structure is general enough to represent two other coupled line structures to be discussed in this paper. For instance, the coupled microstrip line is obtained by letting $\epsilon_{r2} = \epsilon_{r3} = 1$. The inverted microstrip line is realized by choosing $\epsilon_{r1} = \epsilon_{r3} = 1$. Therefore, once the formulation for Fig. 1 is done, we can generate data for microstrip and inverted strip structures in addition to overlay microstrip. In all of the calculations and experiments, we let $H_3 \to \infty$.

Since the spectral-domain method is now well known, we will not describe it in detail here. Basically, it solves the eigenvalue problem in the Fourier transform domain to obtain a pair of algebraic equations that relate the axial and transverse currents on the strips with the axial and tangential transverse electric fields at the interface containing strips. These solutions are subsequently transformed to a set of linear equations by Galerkin's procedure. This set is solved for the propagation constant β or the guide wavelength λ_g. Choice of the basis functions in Galerkin's procedure is important and, here, we used those proposed by Schmidt and Itoh [5] and Jansen [6]. They have correct edge singularities and can be analytically Fourier transformed to Bessel functions for use in the spectral-domain process. As a result of our convergence tests, we used three basis functions for calculations in this paper.

Once the propagation constant is available, we can calculate all the field coefficients in the cross section. From these quantities we can compute the characteristic impedance which is defined in this paper as [6]

$$Z_0 = 2P_{avg}/I_z^2 \tag{1}$$

where P_{avg} is the average power transmitted and I_z is the axial strip current.

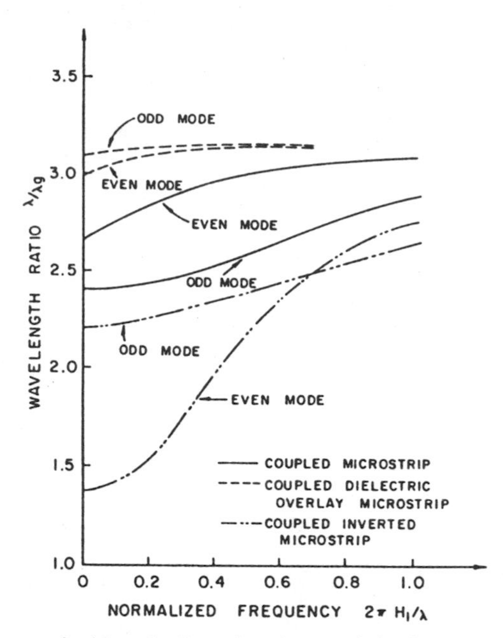

Fig. 2. Even- and odd-mode dispersion characteristics for $H_2/H_1 = 1.0$, $H_3 \to \infty$, $W/H_1 = 1.0$, $S/H_1 = 0.4$, $\epsilon_{r3} = 1.0$. For coupled microstrip, $\epsilon_{r1} = 10.0$, $\epsilon_{r2} = 1.0$. For coupled overlay microstrip, $\epsilon_{r1} = \epsilon_{r2} = 10.0$. For coupled inverted microstrip, $\epsilon_{r1} = 1.0$, $\epsilon_{r2} = 10.0$.

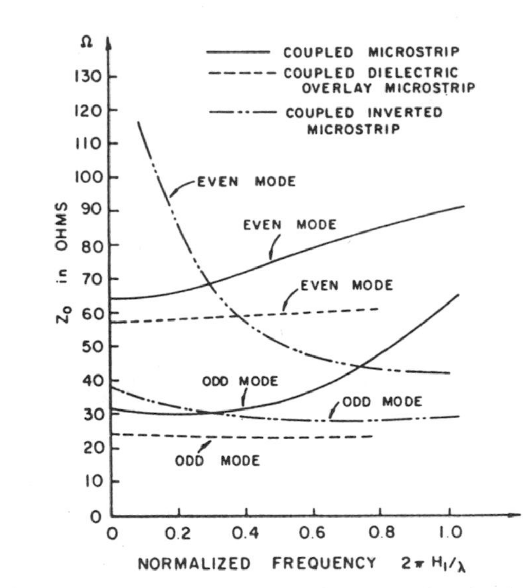

Fig. 3. Characteristic impedance versus frequency for the three structures. Parameters correspond to Fig. 2.

III. Numerical Results

Fig. 2 shows the dispersion characteristics for three types of coupled strip structures. From this figure, we find the following: the inverted configuration provides a frequency at which the even- and odd-mode phase velocities coincide; whereas, in the overlay construction, the difference in phase velocities becomes very small though they never become equal for this particular choice of structural parameters.

Fig. 3 shows the characteristic impedance of these lines. The impedances of the overlay structures are much less frequency dependent.

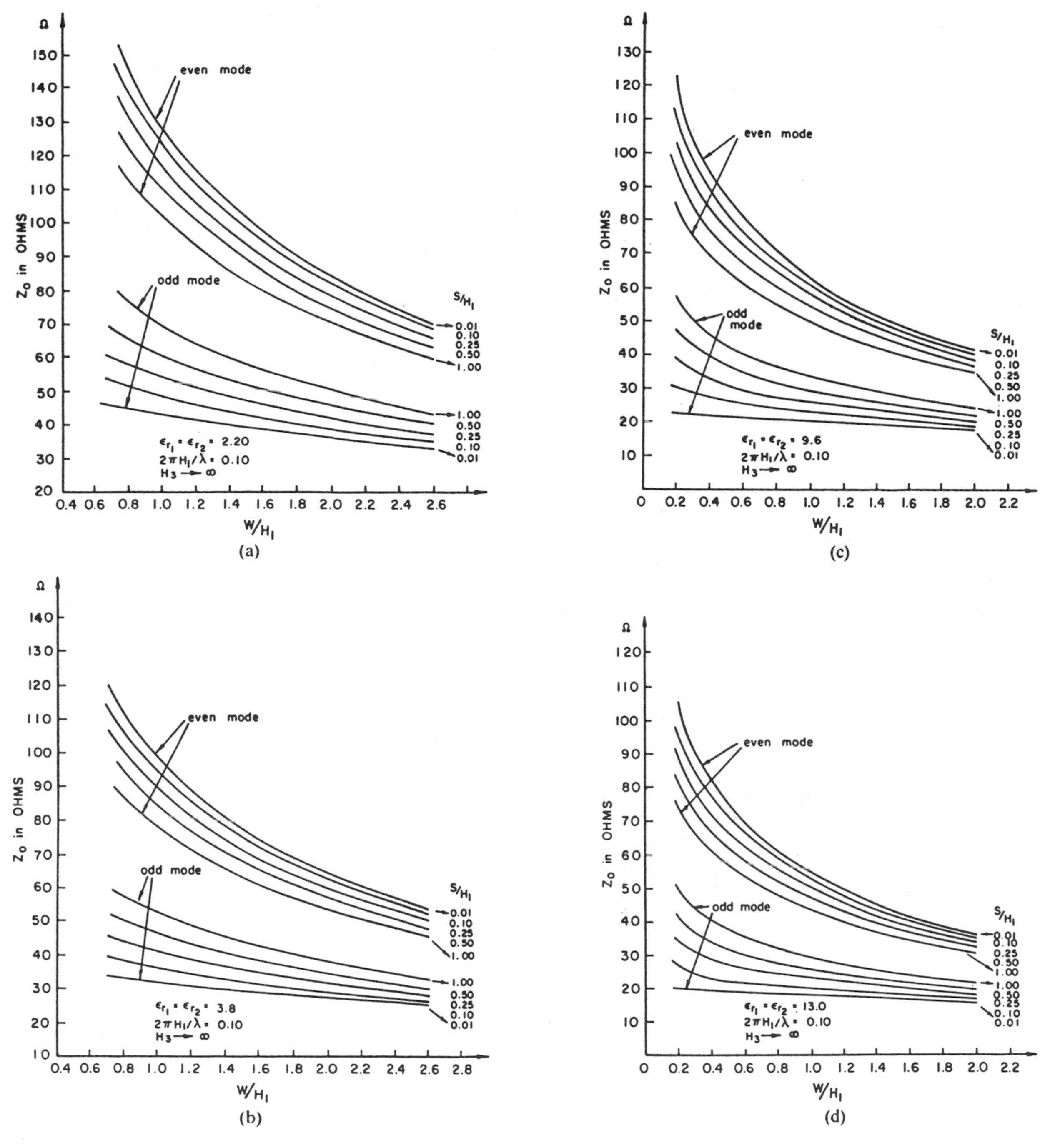

Fig. 4. Characteristic impedance versus the ratio of W to H_1 for the overlay coupled microstrip for a number of values of S/H_1. $H_1 = H_2$ and $H_3 \to \infty$. (a) $\epsilon_{r1} = \epsilon_{r2} = 2.20$. (b) $\epsilon_{r1} = \epsilon_{r2} = 3.8$. (c) $\epsilon_{r1} = \epsilon_{r2} = 9.6$. (d) $\epsilon_{r1} = \epsilon_{r2} = 13.0$.

From these figures, we find that a very narrow-band high-performance coupler may be constructed from the inverted configuration. However, when wide-band operation is desired, as in most practical applications, the overlay configuration is preferred. The frequency-dependent characteristics reported in Figs. 2 and 3 cannot be found from quasi-TEM approximations.

Fig. 4 presents the even- and odd-characteristic impedances of the overlay coupled microstrip line versus the normalized strip width for four commonly used substrates at a particular frequency. Fig. 5 shows the wavelength ratio versus the normalized strip width for four different substrates and five different strip spacing S/H_1. It is seen that there exists particular structures for which the even- and odd-mode phase velocities are equal. Even if the phase velocities are not equal, they are generally close to each other. Fig. 6 shows the characteristic impedance and wavelength ratio versus the normalized overlay thickness. Once

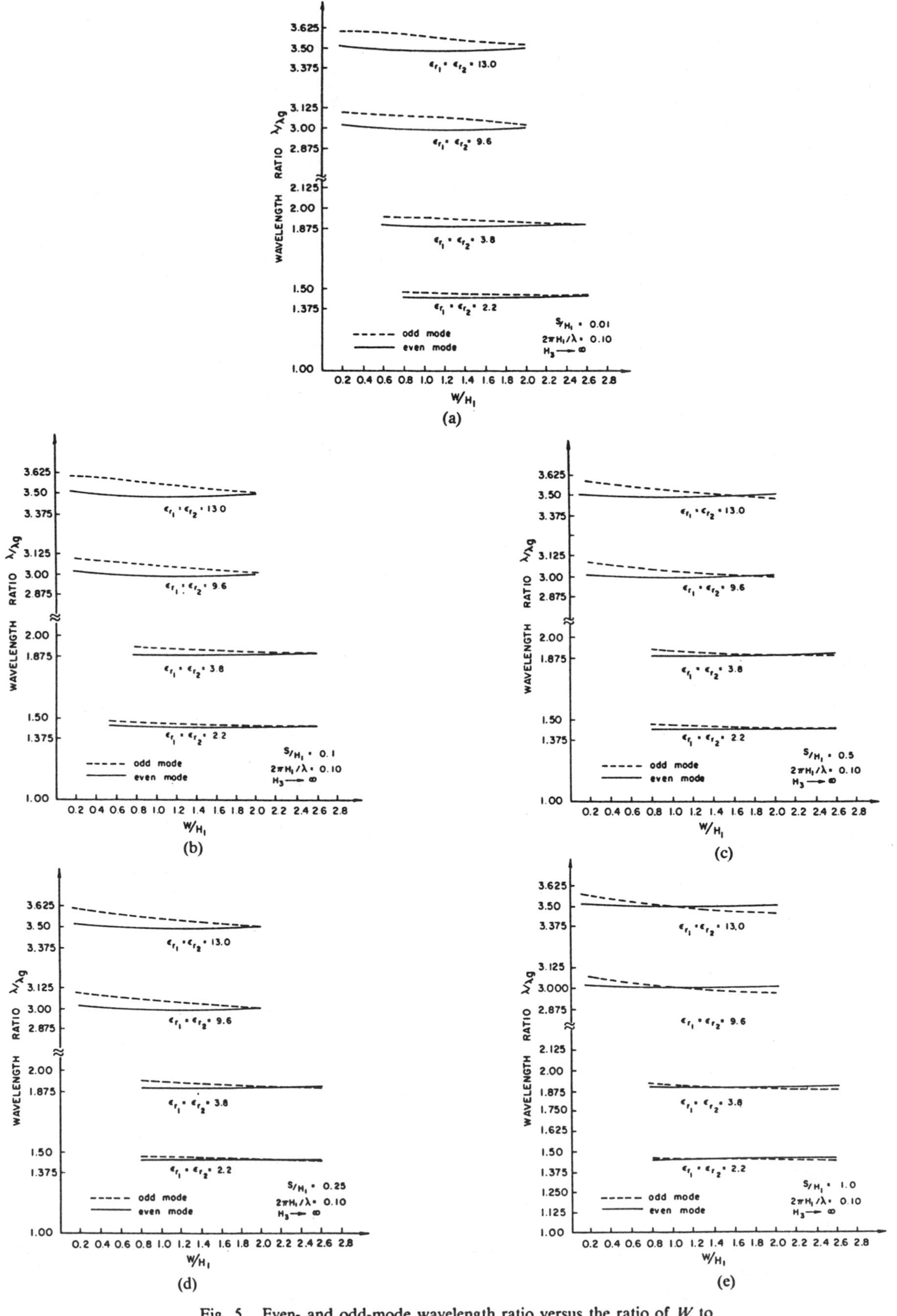

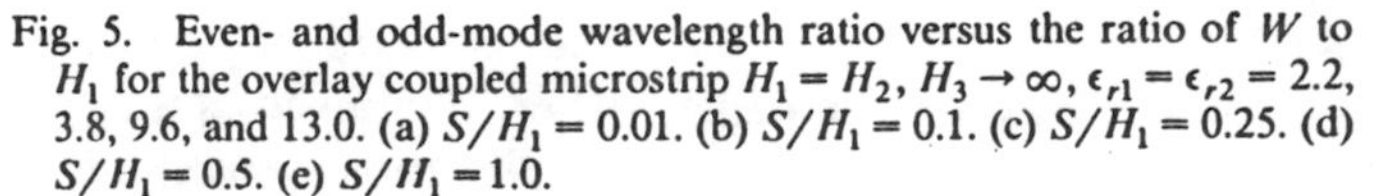
Fig. 5. Even- and odd-mode wavelength ratio versus the ratio of W to H_1 for the overlay coupled microstrip $H_1 = H_2$, $H_3 \rightarrow \infty$, $\epsilon_{r1} = \epsilon_{r2} = 2.2$, 3.8, 9.6, and 13.0. (a) $S/H_1 = 0.01$. (b) $S/H_1 = 0.1$. (c) $S/H_1 = 0.25$. (d) $S/H_1 = 0.5$. (e) $S/H_1 = 1.0$.

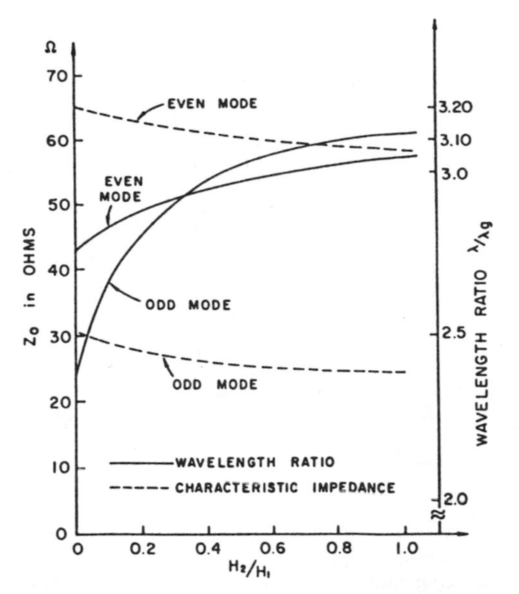

Fig. 6. Even- and odd-mode wavelength ratio and characteristic impedance versus H_2/H_1 for the overlay coupled microstrip. $\epsilon_{r1} = \epsilon_{r2} = 10.0$, $W/H_1 = 1.0$, $S/H_1 = 0.4$, $2\pi H_1/\lambda = 0.1$, $H_3 \to \infty$.

again, the even- and odd-mode phase velocities can be equalized.

IV. Design of Single-Section Directional Coupler

Once a computer program is available which calculates the even- and odd-mode characteristic impedance values Z_0^e and Z_0^o and the phase constants β_e and β_o of the coupled line, we make use of this program for designing a directional coupler.

In a hypothetical situation where $\beta_e = \beta_o$, we could proceed according to the standard approach [7]. If such were the case, the electrical length of the coupler is required to be $\beta L = \pi/2$. Assuming all the ports are terminated with Z_{0L}, the matching condition requires

$$Z_0^e Z_0^o = Z_{0L}^2. \tag{2}$$

The desired coupling coefficient K is related to Z_0^e and Z_0^o via

$$K = \frac{Z_0^e - Z_0^o}{Z_0^e + Z_0^o}. \tag{3}$$

These two equations result in specific values of Z_0^e and Z_0^o. We can find W and S that provide these values of impedance if all other structural parameters are fixed. We now calculate β and find L.

In the present case, $\beta_e \neq \beta_o$. We, therefore, initially assume that the $\beta L = \pi/2$ condition is satisfied and find W and S. The length L is then determined from β of the isolated line ($S \to \infty$) for the obtained value of W. This β is usually close to the average of β_e and β_o. Degradations of the coupler performance due to this choice of L are studied experimentally.

The design process may be summarized as follows.

i) Choose the center frequency, the substrate material,

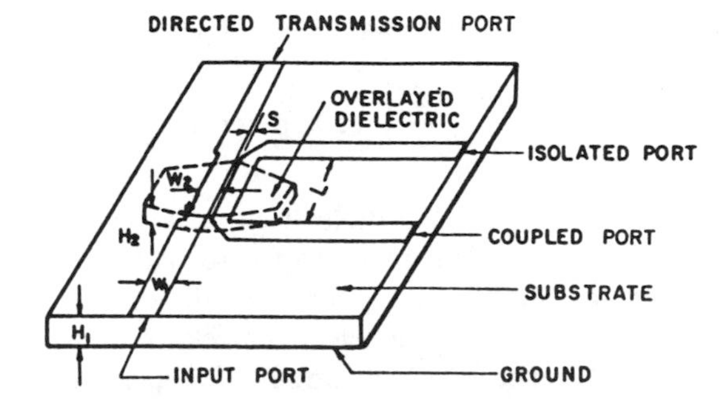

Fig. 7. Overlay directional coupler. $W_1 = 3.84$ mm, $W_2 = 3.20$ mm, $H_1 = H_2 = 1.42$ mm, $S = 0.4$ mm, $L = 20.5$ mm, $\epsilon_{r1} = \epsilon_{r2} = 2.48$.

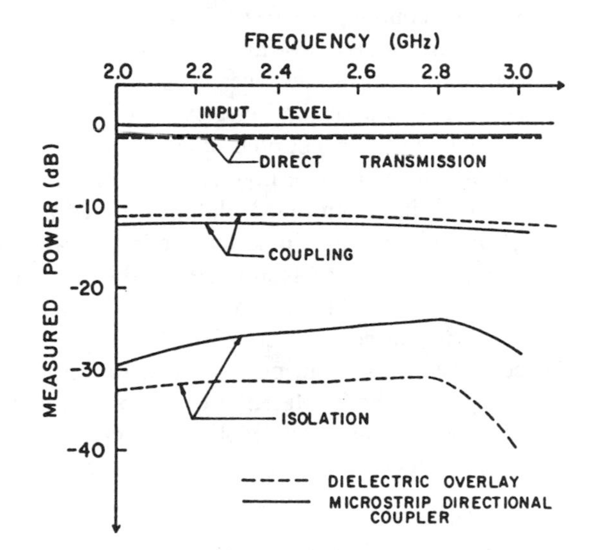

Fig. 8. Coupling and isolation of the directional coupler with and without a dielectric overlay.

and thickness. Other parameters such as the overlay material and thickness if applicable.

ii) Find Z_0^e and Z_0^o from (2) and (3).

iii) Obtain W and S values which result in Z_0^e and Z_0^o by using the computer program developed in Section II or the diagrams such as those in Fig. 4.

iv) If the coupling requirement cannot be satisfied or, though it is satisfactory, the value of S is too small for fabrication, we may choose a different substrate which has a larger thickness and/or higher dielectric constant and repeat the processes ii) and iii).

v) Finally, calculate β for an isolated line ($S \to \infty$) with the obtained value of W and find the coupler length L such that $\beta L = \pi/2$.

We have designed 10-dB, 6-dB, and 3-dB directional couplers. Since our fabrication facility is rather crude, only the 10-dB coupler was fabricated. Fig. 7 shows the structure, and Fig. 8 presents the measured results for the microstrip directional coupler and the dielectric overlay microstrip directional coupler. For these two structures, all of the dimensions are identical except for the presence or absence of the dielectric overlay. We can see that the dielectric overlay can significantly improve the isolation. This agrees with the works by Paolino [1]. A simple calculation shows that the relative propagation constant difference between the even mode and the odd mode

$$\left|\frac{\beta_e - \beta_o}{\beta_e}\right| \times 100 \text{ percent}$$

is only 0.54 percent in the dielectric overlay case and 7.35 percent in the coupled microstrip case. Additional measurements for an inverted microstrip directional coupler showed that at low frequencies, the relative propagation constant difference is large, about 12.7 percent, so the isolation is very poor.

In Fig. 7, the input and through arms of the coupler are in line with the coupling section. Experimental results showed that when these arms are perpendicular to the coupled section, the dielectric overlay did not improve the isolation. This is due to the bend existing between the transmission line and the coupled line, thus generating scattered waves. Dielectric overlays enhance the coupling of scattered waves, implying that isolation is deteriorating. In a configuration like Fig. 7 in which the input part is in line with the coupler, there is no measurable reflection caused by bends in the main line. From the network analysis [7] for any directional coupler, the magnitude of the reflected wave at the input port is equal to that of the wave appearing from the "isolated" port, implying that as the input VSWR increases, the isolation decreases. This shows that the structure shown in Fig. 7 seems to be practical.

V. Conclusion

In this paper, the spectral-domain approach was used for studying the coupled dielectric overlay microstrip and coupled inverted microstrip. The numerical computations were carried out with a CDC Dual Cyber 170/750 computer. The typical configuration time required for a structure at a given frequency was about 1.05 s. We designed and tested microstrip and overlay microstrip couplers. The frequency characteristics are presented.

References

[1] D. D. Paolino, "MIC overlay coupler design using spectral domain techniques," *IEEE Trans. Microwave Theory Tech.*, vol. MTT-26, pp. 646–649, Sept. 1978.

[2] B. Sheleg and B. E. Spielman, "Broad-band directional couplers using microstrip with dielectric overlays," *IEEE Trans. Microwave Theory Tech.*, vol. MTT-22, pp. 1216–1220, Dec. 1974.

[3] K. Atsuki and E. Yamashita, "Three methods for equalizing the even- and odd-mode phase velocity of coupled strip lines with an inhomogeneous medium," *Trans. IECE Japan*, vol. 55-B, no. 7, pp. 424–426, July 1972.

[4] T. Itoh, "Spectral domain immittance approach for dispersion characteristics of generalized printed transmission lines," *IEEE Trans. Microwave Theory Tech.*, vol. MTT-28, pp. 733–736, July 1980.

[5] L. Schmidt and T. Itoh, "Characteristics of a generalized fin-line for millimeter wave integrated circuits," *Int. J. Infrared and Millimeter Waves*, vol. 2, no. 3, pp. 427–436, May 1981.

[6] R. Jansen, "Unified user-oriented computation of shielded, covered and open planar microwave and millimeter-wave transmission-line characteristics," *Microwaves, Opt. Acoust.*, vol. 3, no. 1, pp. 14–22, Jan. 1979.

[7] R. Levy, "Directional couplers," *Advances in Microwaves*, vol. 1, L. Young, Ed. New York: Academic Press, 1966, pp. 115–206.

Propagation Characteristics of Striplines with Multilayered Anisotropic Media

TOSHIHIDE KITAZAWA AND YOSHIO HAYASHI

Abstract—Various types of striplines with anisotropic media are analyzed. The analytical approach used in this paper is based on the network analytical method of electromagnetic fields, and the formulation process is straightforward for complicated structures. Some numerical results are presented and comparison is made with the results available in the literature.

I. Introduction

The network analytical method of electromagnetic fields has been successfully applied to analyze the propagation characteristics of planar transmission lines. The hybrid-mode analysis of single and coupled slots was presented by employing this method [1], [2]. Recently, the dispersion characteristics of single microstrip on an anisotropic substrate have been obtained using this approach [3].

Single and coupled striplines on an anisotropic substrate have been analyzed by several investigators [3]–[7], but hybrid-mode analysis is available only for the single microstrip case [3], [7], [8].

The purpose of this paper is to outline a new approach which is an extension of the treatment used in [1]–[3] and is capable of giving the propagation characteristics of various types of striplines with anisotropic media inclusively. In what follows, the formulation process is illustrated using the general structure with multilayered uniaxially anisotropic media. Two methods of solution are presented. One is based on the quasi-static approximation and it derives the transformation from the case with anisotropic layers to the case with equivalent isotropic layers. The other is based on the hybrid-mode formulation and it gives the frequency dependent solutions. The numerical results will be presented for single and coupled microstrips, coupled suspended strips, and coupled strips with overlay.

II. The Network Analytical Method of Electromagnetic Fields

Fig. 1 shows the cross section of coupled strips having multilayered uniaxially anisotropic media, whose permittivity tensors are

$$\hat{\epsilon}_i = \begin{bmatrix} \epsilon_{i\perp} & 0 & 0 \\ 0 & \epsilon_{i\perp} & 0 \\ 0 & 0 & \epsilon_{i\parallel} \end{bmatrix}, \qquad i=1,2,3. \tag{1}$$

Manuscript received June 3, 1982; revised January 18, 1983.

T. Kitazawa is with the Department of Electrical Engineering, University of Illinois, Urbana, IL, on leave from the Kitami Institute of Technology, Kitami, Japan.

Y. Hayashi is with the Kitami Institute of Technology, Kitami, Japan.

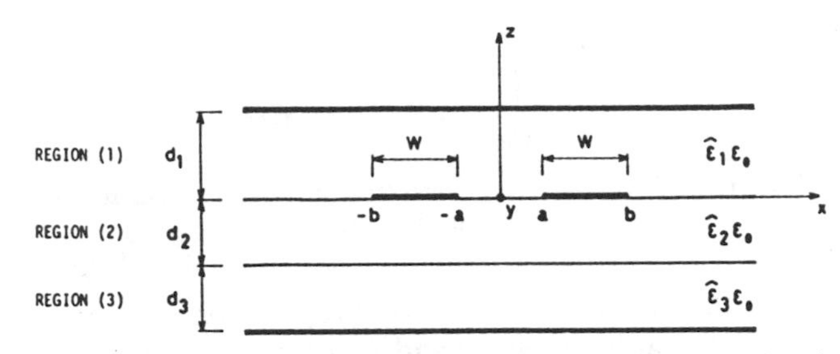

Fig. 1. General structure of coupled strips having multilayered anisotropic media.

As a first step we express the transverse fields in each region by the following Fourier integral:

$$\left.\begin{matrix} E_t^{(i)} \\ H_t^{(i)} \end{matrix}\right\} = \sum_{l=1}^{2} \int_{-\infty}^{\infty} \left\{ \begin{matrix} V_l^{(i)}(\alpha;z) f_l(\alpha;x) \\ I_l^{(i)}(\alpha;z) z_0 \times f_l(\alpha;x) \end{matrix} \right\} d\alpha\, e^{-j\beta_0 y}, \qquad i=1,2,3 \tag{2}$$

where

$$f_1 = \frac{j}{\sqrt{2\pi}} K_0 e^{-j\alpha x}, \qquad f_2 = f_1 \times z_0$$

$$K_0 = \frac{K}{K}$$

$$K = x_0\alpha + y_0\beta_0, \qquad K = |K| \tag{3}$$

where β_0 is the propagation constant in the y-direction, x_0, y_0, and z_0 are the x-, y-, and z-directed unit vectors, respectively, and $l=1$ and $l=2$ represent E waves ($H_z=0$) and H waves ($E_z=0$), respectively. Equation (2) shows that the field components are a superposition of inhomogeneous waves whose spatial variation is $\exp\{-j(\alpha x + \beta_0 y)\}$.

Substituting the above expression into Maxwell's field equation, we obtain the following transmission-line equation in each region:

$$-\frac{d}{dz} V_l^{(i)} = j\kappa_l^{(i)} z_l^{(i)} I_l^{(i)}$$

$$-\frac{d}{dz} I_l^{(i)} = j\kappa_l^{(i)} y_l^{(i)} V_l^{(i)} \tag{4}$$

where

$$\kappa_1^{(i)} = \sqrt{\epsilon_{i\perp}\mathcal{K}_0^2 - \frac{\epsilon_{i\perp}}{\epsilon_{i\parallel}}K^2} \qquad \kappa_2^{(i)} = \sqrt{\epsilon_{i\perp}\mathcal{K}_0^2 - K^2}$$

$$z_1^{(i)} = \frac{\kappa_1^{(i)}}{\omega\epsilon_0\epsilon_{i\perp}} \qquad z_2^{(i)} = \frac{\omega\mu_0}{\kappa_2^{(i)}}$$

$$y_l^{(i)} = \frac{1}{z_l^{(i)}}, \qquad (l=1,2) \qquad \mathcal{K}_0 = \omega\sqrt{\epsilon_0\mu_0}. \tag{5}$$

Reprinted from *IEEE Trans. Microwave Theory Tech.*, vol. MTT-31, no. 6, pp. 429–433, June 1983.

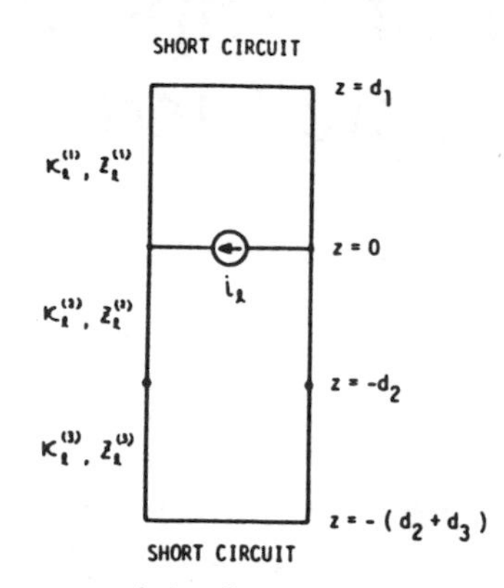

Fig. 2. Equivalent transmission-line circuits for transverse section of coupled strips.

Notice that $\kappa_1^{(i)}$ and $\kappa_2^{(i)}$ are the propagation constants in the z-direction for E waves and H waves, respectively, and $z_1^{(i)}$ and $z_2^{(i)}$ are the characteristic impedance for these waves.

The boundary conditions to be satisfied are expressed as follows:

$$V_l^{(1)}(d_1) = 0 \quad (6)$$

$$V_l^{(1)}(+0) = V_l^{(2)}(-0) \quad (7a)$$

$$I_l^{(1)}(+0) - I_l^{(2)}(-0) = i_l \quad (7b)$$

$$V_l^{(2)}(-d_2+0) = V_l^{(3)}(-d_2-0) \quad (8a)$$

$$I_l^{(2)}(-d_2+0) = I_l^{(3)}(-d_2-0) \quad (8b)$$

$$V_l^{(3)}(-d_2-d_3) = 0 \quad (9)$$

$$i_l = -\int_{-\infty}^{\infty} f_l^*(\alpha; x')\cdot i(x')\,dx' \quad (10)$$

where the asterisk signifies the complex conjugate functions, and $i(x')$ is the current density on the strip conductors at $z = 0$ and may be expressed as

$$i(x') = x_0 i_x(x') + y_0 i_y(x'). \quad (11)$$

Considering the transmission-line equation (4) together with the boundary conditions (6)–(9), we can obtain the equivalent circuits in the z-direction (Fig. 2). By conventional circuit theory, the mode voltages $V_l^{(i)}$ and currents $I_l^{(i)}$ in each region can be expressed in terms of i_l as

$$\begin{aligned} V_l^{(i)}(\alpha; z) &= Z_l^{(i)}(\alpha; z) i_l(\alpha) \\ I_l^{(i)}(\alpha; z) &= T_l^{(i)}(\alpha; z) i_l(\alpha). \end{aligned} \quad (12)$$

The electromagnetic fields in each region can be obtained by substituting (12) into (2).

III. Variational Expression for the Line Capacitance

In the quasi-static approximation, the characteristic impedance and the normalized propagation constant can be obtained from the line capacitance per unit length. We will derive a variational expression of the line capacitance of the general structure shown in Fig. 1.

The longitudinal component of the electric field in region (1) can be obtained from the transverse fields according to

$$E_z^{(1)} = \frac{1}{j\omega\epsilon_0\epsilon_{1\parallel}} \nabla\cdot\left(H_t^{(1)} \times z_0\right). \quad (13)$$

Substituting (2) and (12) into (13) and applying $\beta_0 \to 0$, $E_z^{(1)}$ can be obtained as

$$E_z^{(1)}(x,z) = \frac{1}{2\pi}\cdot\frac{1}{\omega\epsilon_0\epsilon_{1\parallel}} \cdot \iint_{-\infty}^{\infty} \alpha T_1^{(1)}(\alpha; z) i_x(x')\cdot e^{-j\alpha(x-x')}\,dx'\,d\alpha. \quad (14)$$

Performing the integration by parts, using the equation of continuity

$$-j\omega\sigma(x') = \frac{d}{dx'} i_x(x') \quad (15)$$

and applying the zero frequency approximation $\omega \to 0$ to (14), we get

$$E_z^{(1)}(x,z) = \frac{1}{2\pi\epsilon_0} \iint_{-\infty}^{\infty} F(\alpha) p_1 \frac{\cosh\{p_1(z-d_1)|\alpha|\}}{\sinh(p_1 d_1|\alpha|)} \cdot\sigma(x') e^{-j\alpha(x-x')}\,d\alpha\,dx' \quad (16)$$

where

$$F(\alpha) = \frac{1}{\epsilon_{1e}\coth(p_1 d_1|\alpha|) + \epsilon_{2e} L} \quad (17)$$

$$L = \frac{1 + \frac{\epsilon_{2e}}{\epsilon_{3e}} \tanh(p_2 d_2|\alpha|)\tanh(p_3 d_3|\alpha|)}{\tanh(p_2 d_2|\alpha|) + \frac{\epsilon_{2e}}{\epsilon_{3e}}\tanh(p_3 d_3|\alpha|)} \quad (18)$$

$$p_i = \sqrt{\frac{\epsilon_{i\perp}}{\epsilon_{i\parallel}}}, \quad \epsilon_{ie} = \sqrt{\epsilon_{i\parallel}\epsilon_{i\perp}} \quad (19)$$

and $\sigma(x')$ is the charge distribution on the strip conductors. The potential distribution at $z = 0$ becomes

$$V(x) = \int_0^{d_1} E_z(x,z)\,dz = \int_a^b \int_0^{\infty} G(\alpha; x|x')\sigma(x')\,d\alpha\,dx' \quad (20)$$

where

$$\begin{aligned} G(\alpha; x|x') &= \frac{2}{\pi\epsilon_0}\cdot\frac{F(\alpha)}{|\alpha|}\cos\alpha x \cos\alpha x' \text{ (for even modes)} \\ &= \frac{2}{\pi\epsilon_0}\cdot\frac{F(\alpha)}{|\alpha|}\sin\alpha x \sin\alpha x' \text{ (for odd modes).} \end{aligned} \quad (21)$$

On the strip conductor $a < x < b$, $V(x)$ is equal to a constant V_0, that is, the potential difference between the strip and the ground conductors

$$V(x) = V_0 = \int_a^b \int_0^{\infty} G(\alpha; x|x')\sigma(x')\,d\alpha\,dx', \quad a < x < b. \quad (22)$$

From (22), the variational expression for the line capacitance can be obtained [9]

$$\frac{1}{C}=\frac{V_0}{Q}$$
$$=\frac{\iint_a^b\int_0^\infty \sigma(x)G(\alpha;x|x')\sigma(x')\,d\alpha\,dx'\,dx}{\left\{\int_a^b \sigma(x)\,dx\right\}^2} \tag{23}$$

where Q is the total charge on the strip conductor $a<x<b$

$$Q=\int_a^b \sigma(x)\,dx. \tag{24}$$

Equation (23), together with (21) and (17)–(19), suggests that, in the quasi-static approximation, coupled strips with multilayered uniaxially anisotropic media can be transformed into the case with effective isotropic layers, of which the effective thickness and the relative permittivity are $\sqrt{\epsilon_{i\perp}/\epsilon_{i\parallel}}\cdot d_i$ and $\sqrt{\epsilon_{i\perp}\epsilon_{i\parallel}}$, respectively.

IV. Hybrid-Mode Analysis

The analytical method for the frequency-dependent characteristics of coupled strips shown in Fig. 1 is explained here. This method is analogous to those used in [1]–[3] and will be outlined briefly.

The transverse electric fields, which were obtained in the integral representation in Section II, must be zero on the strip conductors at $z=0$. This gives the integral equation on the current density $\boldsymbol{i}(x)$ and the propagation constant in the y-direction β_0. The unknown current densities $i_x(x)$ and $i_y(x)$ are expanded in terms of known sets of basis functions as follows:

$$i_x(x)=\sum_{k=1}^{N_x} a_{xk}f_{xk}(x)$$
$$i_y(x)=\sum_{k=1}^{N_y} a_{yk}f_{yk}(x) \tag{25}$$

where a_{xk} and a_{yk} are unknown coefficients. Substituting (25) into the integral equation and applying Galerkin's procedure, we obtain a set of simultaneous equations on the unknown a_{xk} and a_{yk}. The propagation constant can be obtained by searching the nontrivial solution.

The definition for the characteristic impedance is not uniquely specified due to the propagation of the hybrid mode. The definition chosen here is

$$Z_0=\frac{P_{\text{ave}}}{I_0^2} \tag{26}$$

where I_0 is the total current on one strip conductor, and P_{ave} is the average power flow along the y-direction.

V. Basis Functions

The line capacitance is calculated by applying the Ritz procedure to the variational expression (23). In this procedure, we express the unknown charge distribution $\sigma(x)$ as

$$\sigma(x)=f_0(x)+\sum_{k=1}^{N} A_k f_k(x) \tag{27}$$

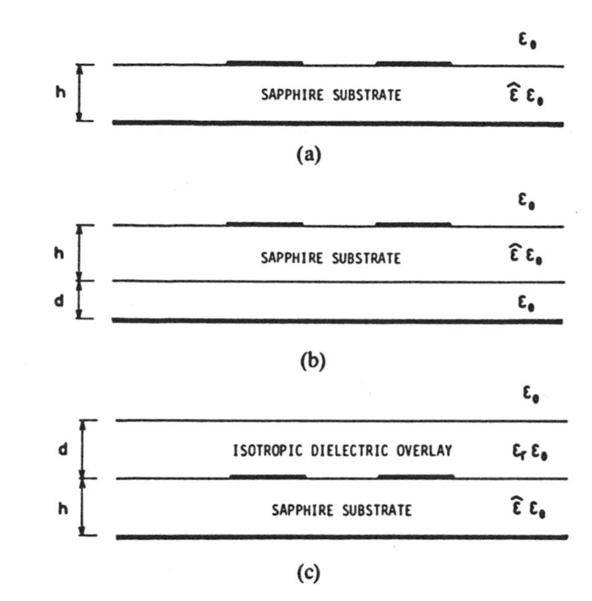

Fig. 3. (a) Coupled microstrips. (b) Coupled suspended strips. (c) Coupled strips with overlay.

where A_k are variational parameters which are determined so that the best approximation is obtained.

In the numerical computations, the choice of the basis functions, $f_k(x)$ in (27) and $f_{xk}(x)$ and $f_{yk}(x)$ in (25), is important. It is desirable that the edge effect should be properly accounted for, and that the approximation to the true value should be systematically improved by increasing the number of basis functions. Taking these requirements into account, we adopt the following families of functions for basis functions:

$$f_{xk}(x)=U_k\left\{\frac{2(x-S)}{W}\right\}$$
$$\left.\begin{matrix} f_{k-1}(x)\\ f_{yk}(x)\end{matrix}\right\}=\frac{T_{k-1}\left\{\frac{2(x-S)}{W}\right\}}{\sqrt{1-\left\{\frac{2(x-S)}{W}\right\}^2}}$$
$$S=(a+b)/2,\quad W=b-a \tag{28}$$

where $T_k(y)$ and $U_k(y)$ are Chebyshev's polynomials of the first and second kind, respectively. By the use of these basis functions, the fast convergence to the exact values is obtained. Preliminary computations show that $N=2$ in (27) and $N_x=N_y=2$ in (25) are sufficient for any case.

VI. Numerical Results

Numerical computations were carried out for single and coupled microstrips (Fig. 3(a)), coupled suspended strips (Fig. 3(b)), and coupled strips with overlay (Fig. 3(c)). In the open microstrip configurations of Fig. 3, the boundary condition (6) or (9) for Fig. 1 should be replaced by the radiation condition. However, the resulting equations thus obtained are the same as those for Fig. 1 in which $d_1\to\infty$ or $d_3\to\infty$. These calculations were performed using the same computer program with very little modification.

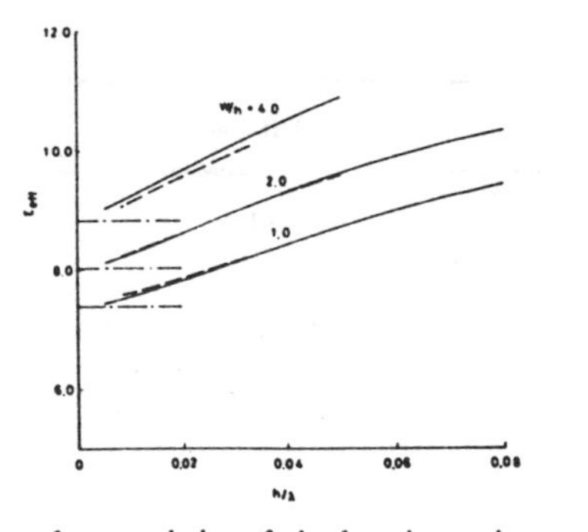

Fig. 4. Dispersion characteristics of single microstrip on sapphire. ($\epsilon_{\perp} = 9.4$, $\epsilon_{\parallel} = 11.6$; —hybrid-mode; —-—quasi static; ——— El-Sherbiny's [4].)

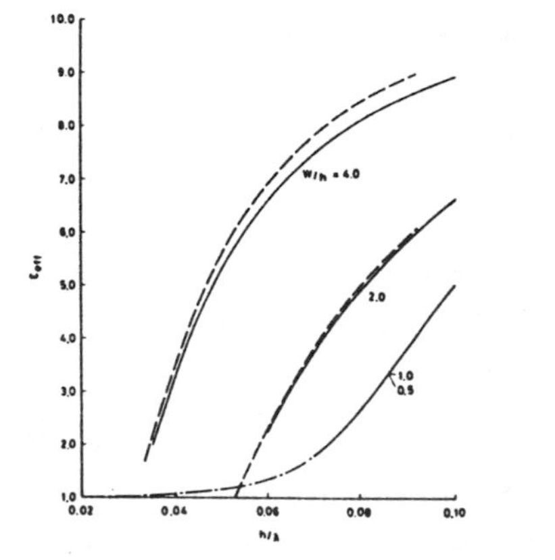

Fig. 5. Dispersion characteristics of the first higher order mode of single microstrip on sapphire. (—this theory; —-—TM_0 mode of a sapphire-coated conductor; ———El-Sherbiny's [4].)

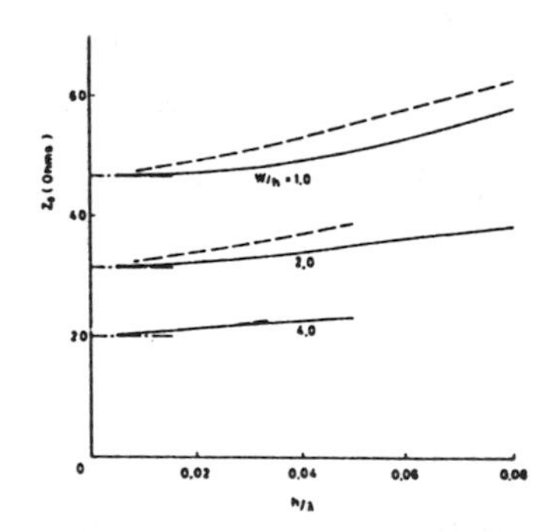

Fig. 6. Characteristic impedance of single microstrip on sapphire.

Fig. 4 shows the dispersion characteristics, the frequency dependence of the effective dielectric constant $\epsilon_{\text{eff}} = \beta_0^2/\omega^2\epsilon_0\mu_0$, of single microstrip on sapphire substrates, where ϵ_{eff} for the dominant mode is reported and compared with the results of El-Sherbiny [8]. The agreement is quite good, although some disagreement appears for wide strips.

Fig. 5 shows the dispersion characteristics of the first higher order mode, which are also compared with those from [8]. Fig. 5 also presents the dispersion characteristics of the TM_0 surface wave of the sapphire coated conductor which results when $W = 0$. When the strip is not so wide compared with the substrate, the dispersion characteristics

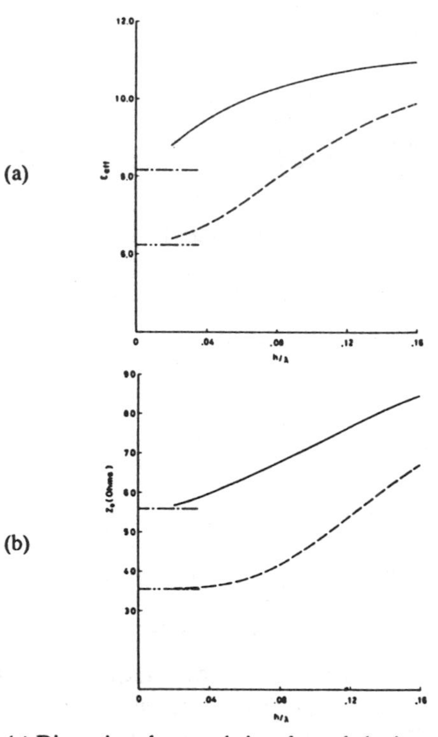

Fig. 7. (a) Dispersion characteristics of coupled microstrips on sapphire. (b) Characteristic impedance of coupled microstrips on sapphire. ($W/h = 1$, $a/h = 0.25$; —even mode (hybrid-mode); ———odd mode (hybrid-mode); —-—even mode (quasi-static); —--—odd mode (quasi-static).)

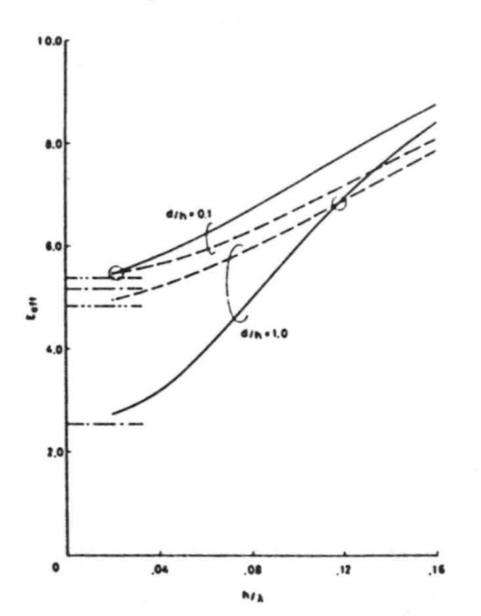

Fig. 8. Dispersion characteristics of coupled suspended strips. ($\epsilon_{\perp} = 9.4$, $\epsilon_{\parallel} = 11.6$, $W/h = 1$, $a/h = 0.25$; —even mode (hybrid-mode); ———odd mode (hybrid-mode); —-— even mode (quasi-static); —--— odd mode (quasi-static).)

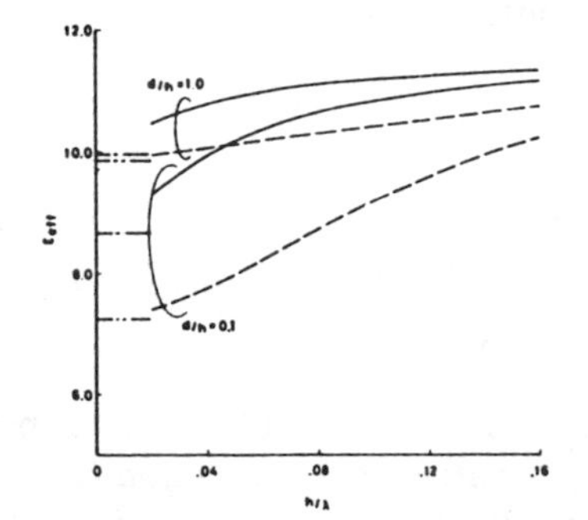

Fig. 9. Dispersion characteristics of coupled strips with overlay. ($\epsilon_{\perp} = 9.4$, $\epsilon_{\parallel} = 11.6$, $\epsilon_r = 9.6$, $W/h = 1$, $a/h = 0.25$; —even mode (hybrid-mode); ———odd mode (hybrid mode); —-— even mode (quasi-static); —--—odd mode (quasi-static).)

of the first higher order mode are indistinguishable from those of the TM_0 surface wave.

The frequency dependence of the characteristic impedance of single microstrip is shown in Fig. 6. Comparison of the results by this method and those from [8] shows that both results converge to the quasi-static values calculated from (23), but that some discrepancies appear at high frequencies. For single microstrip, the characteristic impedance is defined as

$$Z_0 = \frac{2P_{ave}}{I_0^2} \tag{29}$$

instead of (26) in our calculations, whereas it is defined as the ratio of the voltage at the center of the strip to the total longitudinal current in [8].

The dispersion characteristics of coupled microstrips, coupled suspended strips, and coupled strips with a dielectric overlay are depicted in Figs. 7, 8, and 9, respectively. It should be noted that the dispersion characteristics of the even mode of coupled suspended strips is more sensitive than that of the odd mode to the variation in d/h, therefore the frequency at which both modes have the equal phase velocity varies largely.

VII. Conclusions

Various types of striplines with anisotropic media have been analyzed using the same approach, which is based on the network analytical method of electromagnetic fields. In this analytical approach, the derivation of Green's functions is based on the conventional circuit theory, therefore the formulation for the complicated structures is straightforward.

Computations have been carried out by employing the efficient method based on the Ritz and Galerkin procedure to calculate the propagation characteristics of single and coupled microstrips, coupled suspended strips, and coupled strips with overlay. Numerical results of single microstrip were compared with other available data.

References

[1] T. Kitazawa, Y. Hayashi, and M. Suzuki, "Analysis of the dispersion characteristic of slot line with thick metal coating," *IEEE Trans. Microwave Theory Tech.*, vol. MTT-28, pp. 387–392, Apr. 1980.

[2] T. Kitazawa and Y. Hayashi, "Coupled slots on an anisotropic sapphire substrate," *IEEE Trans. Microwave Theory Tech.*, vol. MTT-29, pp. 1035–1040, Oct. 1981.

[3] Y. Hayashi and T. Kitazawa, "Analysis of microstrip transmission line on a sapphire substrate," *J. Inst. Electron. Commun. Eng. Jap.*, vol. 62-B, pp. 596–602, June 1979.

[4] N. G. Alexopoulos and C. M. Krowne, "Characteristics of single and coupled microstrips on anisotropic substrates," *IEEE Trans. Microwave Theory Tech.*, vol. MTT-26, pp. 387–393, June 1978.

[5] M. Kobayashi and R. Terakado, "Method for equalizing phase velocities of coupled microstrip lines by using anisotropic substrate," *IEEE Trans. Microwave Theory Tech.*, vol. MTT-28, pp. 719–722, July 1980.

[6] M. Horno, "Quasistatic characteristics of microstrip on arbitrary anisotropic substrates," *Proc. IEEE*, vol. 68, pp. 1033–1034, Aug. 1980.

[7] F. J. K. Lange, "Analysis of shielded strip- and slot-lines on a ferrite substrate transversely magnetized in the plane of the substrate," *Arch. Elek. Ubertragung.*, vol. 36, pp. 95–100, Mar. 1982.

[8] A-M. A. El-Sherbiny, "Hybrid mode analysis of microstrip lines on anisotropic substrates," *IEEE Trans. Microwave Theory Tech.*, vol. MTT-29, pp. 1261–1265, Dec. 1981.

[9] R. E. Collin, *Field Theory of Guided Waves.* New York: McGraw-Hill, 1960, p. 162.

DISPERSION CHARACTERISTICS OF UNSYMMETRICAL BROADSIDE-COUPLED STRIPLINES WITH ANISOTROPIC SUBSTRATE

Indexing terms: Microwave circuits and systems, Striplines, Dispersion

Dispersion characteristics of the unsymmetrical broadside-coupled striplines of unequal width are presented for the first time. The frequency-dependent hybrid-mode solutions converge to the quasistatic values at lower frequencies for various structural parameters of the broadside-coupled suspended striplines (BCSSs) and broadside-coupled inverted striplines (BCISs) with the anisotropic substrates.

The broadside-coupled striplines (BCSs) have been investigated extensively, which include the quasistatic[1–3] and hybrid-mode analysis[4] for the case with the isotropic and/or anisotropic substrates. However, most of them deal with the symmetrical structure, i.e. the strips of equal width placed symmetrically. Recently, a variational technique was introduced to treat the unsymmetrical BCS of unequal width.* It provides accurate numerical results for the case with uniaxially anisotropic substrates, but it is based on the quasistatic approximation. This letter presents the frequency-dependent characteristics of the unsymmetrical BCS of unequal width to extend the application of this transmission line in higher-frequency ranges.

Fig. 1 shows the cross-section of the unsymmetrical BCS of unequal width with the uniaxially anisotropic substrates whose permittivities are given by

$$\bar{\bar{\varepsilon}}_i = (\hat{x}\varepsilon_{i,\perp}\hat{x} + \hat{y}\varepsilon_{i,\parallel}\hat{y} + \hat{z}\varepsilon_{i,\perp}\hat{z})\varepsilon_0 \tag{1}$$

The frequency-dependent characteristics can be obtained by extending the procedure for the case with the strips of equal width.[4] This method is based on the hybrid-mode formulation, and no approximations for simplification are used in the formulation procedure. Galerkin's procedure[4–7] is employed for the numerical computations, and the basis functions, which are used to represent the unknown current density components, are similar to those used in Reference 6, and they

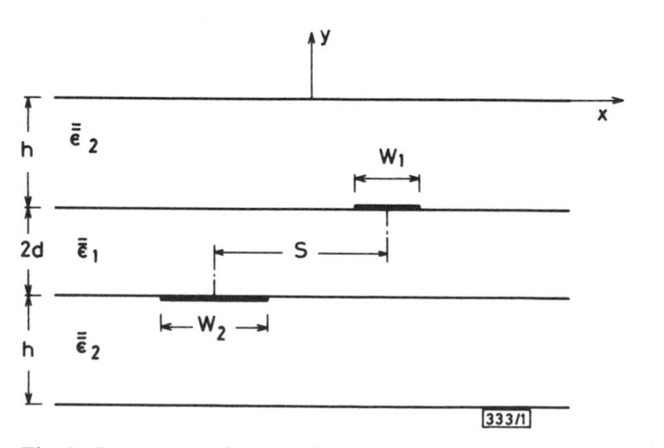

Fig. 1 *Cross-sectional view of unsymmetrical broadside-coupled striplines (BCSs) of unequal width*

* KITAZAWA, T., and HAYASHI, Y.: 'Analysis of unsymmetrical broadside-coupled striplines with anisotropic substrates', *IEEE Trans.*, **MTT** (to be published)

incorporate the edge effect. However, the coupling between the strip conductors for this structure is tighter than that for the planar structure (e.g. coupled microstrips[5,6] and coupled

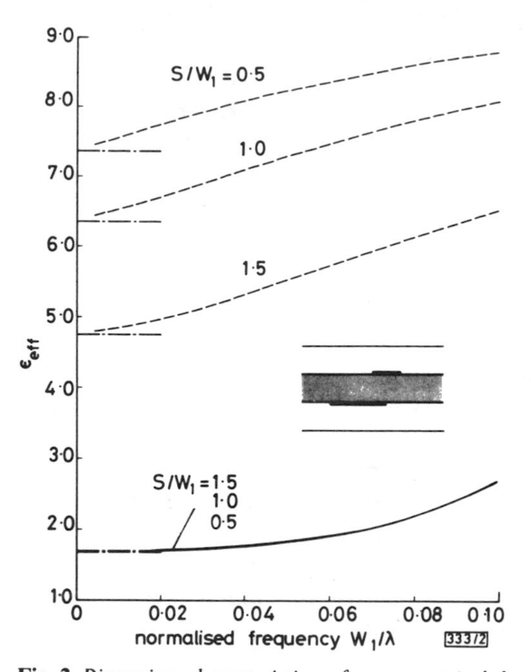

Fig. 2 *Dispersion characteristics of unsymmetrical broadside-coupled suspended striplines (BCSSs) with S/W_1 as parameter*

λ is free-space wavelength

$\varepsilon_{1,\parallel} = 10{\cdot}2$, $\varepsilon_{1,\perp} = 13{\cdot}0$, $\varepsilon_{2,\parallel} = \varepsilon_{2,\perp} = 1{\cdot}0$, $d/W_1 = 0{\cdot}2$, $h/W_1 = 0{\cdot}8$, $W_2/W_1 = 2{\cdot}0$

——— c-modes
– – – – π-modes
— - — quasistatic*

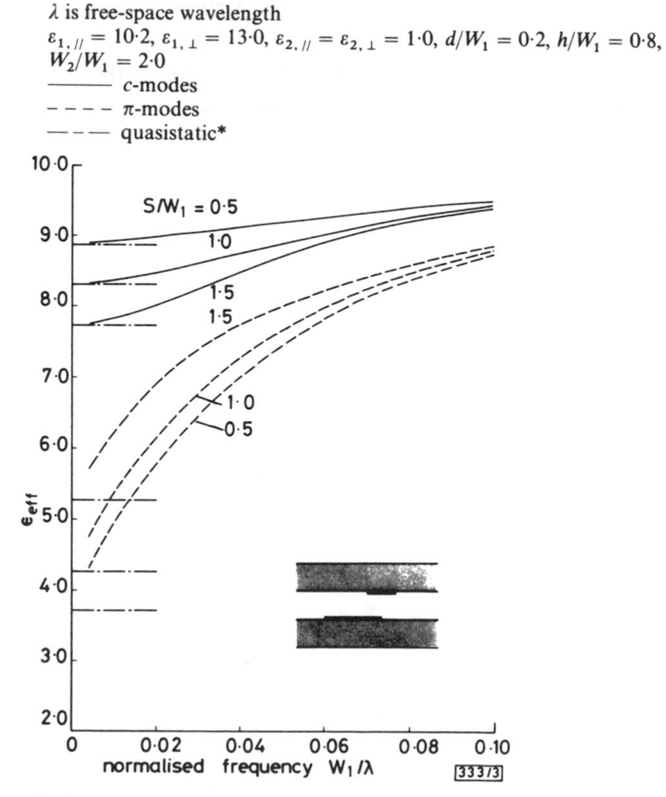

Fig. 3 *Dispersion characteristics of unsymmetrical broadside-coupled inverted striplines (BCISs) with S/W_1 as parameter*

$\varepsilon_{1,\parallel} = \varepsilon_{1,\perp} = 1{\cdot}0$, $\varepsilon_{2,\parallel} = 10{\cdot}2$, $\varepsilon_{2,\perp} = 13{\cdot}0$, $d/W_1 = 0{\cdot}2$, $h/W_1 = 0{\cdot}8$, $W_2/W_1 = 2{\cdot}0$

——— c-modes
– – – – π-modes
— - — quasistatic

Reprinted with permission from *Electronics Letters,* vol. 22, no. 5, pp. 281–283, February 1986.

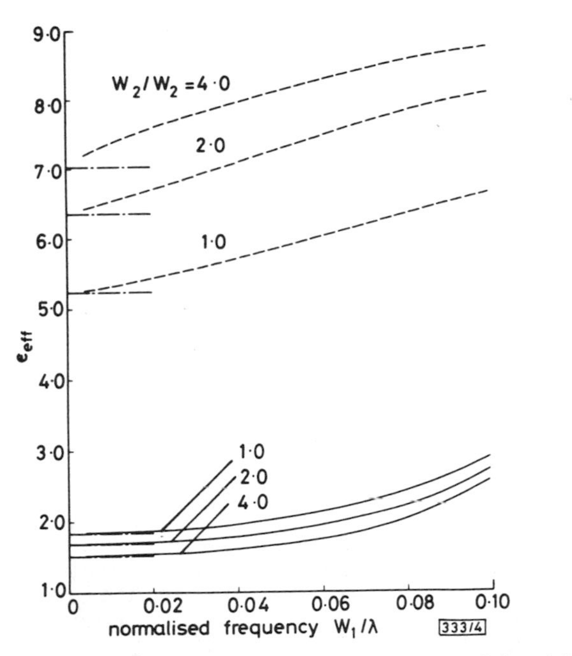

Fig. 4 *Dispersion characteristics of unsymmetrical broadside-coupled suspended striplines (BCSSs) with W_2/W_1 as parameter*

$\varepsilon_{1,//} = 10{\cdot}2$, $\varepsilon_{1,\perp} = 13{\cdot}0$, $\varepsilon_{2,//} = \varepsilon_{2,\perp} = 1{\cdot}0$, $d/W_1 = 0{\cdot}2$, $h/W_1 = 0{\cdot}8$, $S/W_1 = 1{\cdot}0$
——— c-modes
– – – – π-modes
——— quasistatic*

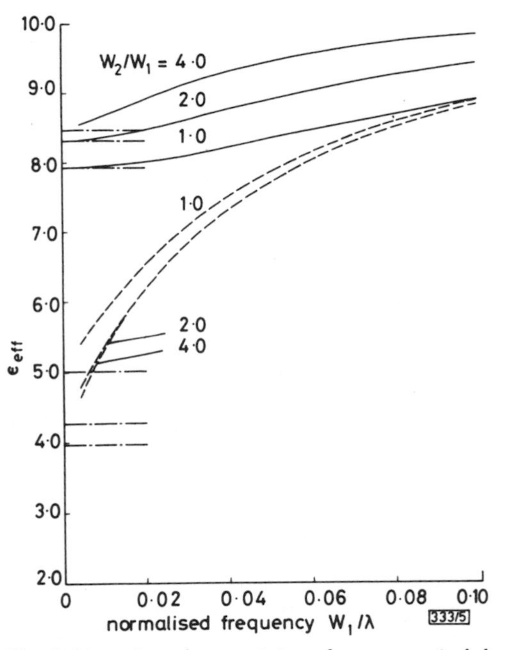

Fig. 5 *Dispersion characteristics of unsymmetrical broadside-coupled inverted striplines (BCISs) with W_2/W_1 as parameter*

$\varepsilon_{1,//} = \varepsilon_{1,\perp} = 1{\cdot}0$, $\varepsilon_{2,//} = 10{\cdot}2$, $\varepsilon_{2,\perp} = 13{\cdot}0$, $d/W_1 = 0{\cdot}2$, $h/W_1 = 0{\cdot}8$, $S/W_1 = 1{\cdot}0$
——— c-modes
– – – – π-modes
——— quasistatic

slots[7]), and more basis functions are required to obtain accurate results. Two basis functions were sufficient for most planar structures,[5,6] whereas four basis functions are used in the following calculations. First, a special case with the strips of equal width was considered, and the numerical results obtained using the present method were in good agreement with those from Reference 4, showing the validity of the method. The frequency dependence of the effective dielectric constants is plotted for the unsymmetrical broadside-coupled suspended striplines (BCSSs) and broadside-coupled inverted striplines (BCISs) in Figs. 2 and 3 with S/W_1 as parameter and in Figs. 4 and 5 with W_2/W_1 as parameter, respectively. The frequency-dependent hybrid-mode solutions converge to the quasistatic values* at lower frequencies for all cases, while the dispersion effect cannot be neglected in higher-frequency ranges. It should be noted that the dispersion characteristics of the c-mode of BCSSs are not affected by the spacing of the strips S/W_1, and that the dispersion curves of c-modes cannot be distinguished in a wide frequency band (Fig. 2). Note also that the phase velocities of two fundamental modes are equalised at some frequency for BCISs with strips of equal width (Fig. 5).

In conclusion, the dispersion characteristics of the unsymmetrical broadside-coupled striplines of unequal width are presented for the first time. The results are shown for the case with the anisotropic substrates, and the accuracy is verified by comparison with the quasistatic values.

T. KITAZAWA
A. MURANISHI
Y. HAYASHI

Department of Electronic Engineering
Kitami Institute of Technology
Kitami 090, Japan

21st January 1986

References

1 COHN, S. B.: 'Characteristic impedances of broadside-coupled strip transmission lines', *IRE Trans.*, 1960, **MTT-8**, pp. 633–637
2 BAHL, I. J., and BHARTIA, P.: 'Characteristics of inhomogeneous broadside-coupled striplines', *IEEE Trans.*, 1980, **MTT-28**, pp. 529–535
3 D'ASSUNÇÃO, A. G., GIAROLA, A. J., and ROGERS, D. A.: 'Characteristics of broadside-coupled microstrip lines with iso/anisotropic substrates', *Electron. Lett.*, 1981, **17**, pp. 264–265
4 KITAZAWA, T., HAYASHI, Y., FUJITA, K., and MUKAIHARA, H.: 'Analysis of broadside-coupled strip lines with anisotropic substrate', *Trans. IECE Jpn.*, 1983, **66-B**, pp. 1139–1146
5 KITAZAWA, T., and HAYASHI, Y.: 'Propagation characteristics of strip lines with multilayered anisotropic media', *IEEE Trans.*, 1983, **MTT-31**, pp. 429–433
6 KITAZAWA, T., and MITTRA, R.: 'Analysis of asymmetric coupled striplines', *ibid.*, 1985, **MTT-33**, pp. 643–646
7 KITAZAWA, T., and HAYASHI, Y.: 'Coupled slots on an anisotropic sapphire substrate', *ibid.*, 1981, **MTT-29**, pp. 1035–1040

MULTICONDUCTOR PLANAR TRANSMISSION-LINE STRUCTURES FOR HIGH-DIRECTIVITY COUPLER APPLICATIONS

Bogdan J. Janiczak

Microwave Division of Telecommunication Institute
Technical University of Gdańsk, Majakowski St.11/12
80-952 Gdańsk-Wrzeszcz, Poland

ABSTRACT

The spectral domain technique is used to analyze multiconductor printed transmission-line structures containing layered dielectric substrate. Starting from general analytical formulation a useful modification of the procedure is suggested for accurate description of multiconductor interdigitated microstrip circuit with equally dimensioned conductors. Frequency dependent numerical results for some new microstrip and coplanar structures with more than two coupled lines are shown and discussed with respect to high-directivity multiconductor coupler applications.

INTRODUCTION

In recent years, an increasing interest has been observed in the field of accurate description of the behaviour of multiconductor printed transmission-line systems (1-5). This problem takes important place not only in microwave engineering but also in high-speed computer units using microstrip or coplanar strip-lines as a transmission paths for logic information signal (5). For microwave circuitry the knowledge of the transmission behaviour of various multiconductor printed lines is needed to indicate theoretical limitations in the designing of topology of MIC and MMIC layouts /with respect to the prediction of the possibility of the greatest density packaging/ as well as to determine transmission property of novel composed waveguiding structures.

The effect of coupling between nonadjacent lines and propagation characteristics of various multimode systems of edge-coupled conductors with single-layer substrate have already been reported in literature (1-5). The transmission behaviour of layered lines, however, is generally not exactly explained to date. Recently, it was shown that frequency characteristics of three-line microstrip directional coupler can be remarkably improved for both interdigital (6) and three-mode (7) operation. In this paper, therefore, some new alternative microstrip and coplanar structures filled with combined double-layer dielectric substrate are considered for possible application in high-directivity multiconductor coupler design.

In the calculations, the spectral domain technique /SDT/ is used to provide phase constant characteristics of various multiconductor-line structures. This method, comparing with other realisations, has not necessarily lead to unacceptable increase of analytical complexity in case of multimode systems even with multilayered substrates and always results in the well-known set of coupled algebraic equations easily and efficiently solved by numerical techniques.

REMARKS ON ANALYTICAL PROCEDURE

Comparing various realisations of the SDT one can conclude that the most important and sensitive stage of this method lies in the proper choice of the analytical description of unknown field-quantities at the nonuniform boundary-condition cross-sectional interfaces interrelated by the spectral domain Green's function components involving the associated hybrid-mode boundary-value problem. The latter, in case of layered dielectric media, can be directly and simply obtained by inspection of equivalent transverse transmission-line circuits for TE- and TM-waves (8). Although, from the theoretical point of view we have always two possible descriptions of the transmission-line, in terms of impedance or admittance matrices of spectral Green's function components, in practice the manner of analytical description is determined by variables defined over finite intervals in space-domain (4),(9) to avoid unnecessary computational effort.

In order to analyze general nonsymmetrical structures the unknown field-quantities are expanded in double-closed series-type representation in terms of assu-

Reprinted from *IEEE Microwave Symposium,* pp. 215–218, 1985.

med set of known expansion functions(1), (9). This procedure, however, may be too complicated in many practical cases when the symmetry with respect to vertical mid plane exist or the knowledge of only some chosen modes of the structure is required as the resulting dimensions of a uniform system of homogeneous equations to solve are increased linearly with both the number of lines and of expansion functions used in the calculations. In this case , the method can be significantly improved allowing more efficient numerical calculations of various multiconductor printed line structures like, for example, interdigital microstrip circuit with two fundamental modes corresponding to magnetic wall in the plane of symmetry in structures with odd number of strips and to magnetic and electric wall in these with even (10), respectively. Assuming equally dimensioned strip and spacing slot widths relatively simple for numerical implementation functional factors appearing in the spectral representation of the expansion functions are obtained:

$$F_{q\text{-}e} = \begin{cases} \sum_{n=1}^{\frac{N}{2}} \cos\left[(2n-1)\alpha \frac{s+w}{2}\right], & N\text{- even} \\ 1 + 2\sum_{n=1}^{\frac{N-1}{2}} \cos\left[n\alpha(s+w)\right], & N\text{- odd} \end{cases}$$

$$F_{q\text{-}o} = \begin{cases} \sum_{n=1}^{\frac{N}{2}} (-1)^{\frac{N}{2}+n} \sin\left[(2n-1)\alpha \frac{s+w}{2}\right], & N\text{-even} \\ (-1)^{\frac{N-1}{2}} + 2\sum_{n=1}^{\frac{N-1}{2}} (-1)^{\frac{N-1}{2}-n} \cos\left[n\alpha(s+w)\right], & N\text{-odd} \end{cases}$$

with subscripts q-e and q-o corresponding to quasi-even and quasi-odd mode, respectively, α - being the Fourier transform variable and N the number of strip conductors. Thus, arbitrary system of N equally dimensioned interdigital microstrip lines can be exactly analyzed even with matrices 2×2 if expansion functions are correctly chosen.

NUMERICAL RESULTS

In this section exemplary obtained results of normalized to the free-space wave number phase constants of various microstrip and coplanar structures with double layer dielectric substrate are shown and briefly discussed. These structures, due to the possibility of the compensation of phase velocities of propagating fundamental modes are especially intended for high-directivity coupler application. The presentation, therefore, is oriented on phase-velocity equalization effect and the methods to achieve this. In numerical calculation the series of sine and cosine functions with Maxwell term have been used as the expansion functions which are proved to be very efficient in many similar field-theory problems.

In Figs. 1, 2 and 3 characteristics of three-line microstrip and coplanar structures resulting from general analytical formulation are shown versus relative dielectric permittivity of lower substrate layer. The first structure - coupled coplanar strip-lines /C-CPS/, can be treated as two edge-coupled slot lines with finite ground planes and designed as typical four port device. The remaining two structures illustrated in Figs. 2 and 3 may be used both as four- and six-port devices. Microstrip version should provide better frequency behaviour due to the smaller differences between permittivities required for the equalization of any two modal phase velocities.

Fig. 4 presents characteristics of interdigital microstrip structure which are obtained in terms of suggested modification of SDT. The plotted curves show that single layer network does not provide the decrease of phase velocity ratio with increasing N. The use of layered substrate results in the equalization of two modal phase velocities. This technique is especially suited for the interdigital structure as the use of crossovers does not allow the efficient use of an overlay techniques commonly employed in conventional two-line microstrip couplers.

REFERENCES

(1) B.Janiczak,"Behaviour of guided modes in systems of parallelly located transmission lines on dielectric substrates", Electron. Lett., 19,1983, pp.778-779

(2) H.Lee,V.K.Tripathi:"New perspectives on the Green's function for quasi-TEM planar structures",IEEE MTT-S Digest,1983,pp.571-573

(3) S.B.Worm,R.Pregla:"Hybrid mode analysis of of arbitrarily shaped planar microwave structures by the method of lines, Vol.MTT-32, pp.191-196

(4) H.Lee,V.K.Tripathi:"Generalized spectral domain analysis of planar structures having semi-infinite ground planes",IEEE MTT-S Digest,1984,pp.327-329

(5) C.Chan,R.Mittra:"Spectral iterative technique for analyzing multiconductor microstrip lines",ibid.,pp.463-465

(6) K.Shibata,H.Yanagisawa,Y.Ishihata:"Three-line microstrip coupler with dielectric overlay",Electron. Lett.,19,1983,pp.911-912

(7) B.Janiczak:"Accurate hybrid-mode approach for computing modes in three-line coupled microstrip structure in overlaid configuration",ibid.,20,1984,pp.825-826

(8) T.Itoh:"Spectral domain immitance approach...", Vol.MTT-28,1981,pp.352-355

(9) B.Janiczak:"On the analysis of multimodes...", Mikrowellen Magazin,Vol.10,1984,pp.157-165

(10) Y.Tajima,S.Kamihashi:"Multiconductor couplers", Vol.MTT-26,1978,pp.795-801

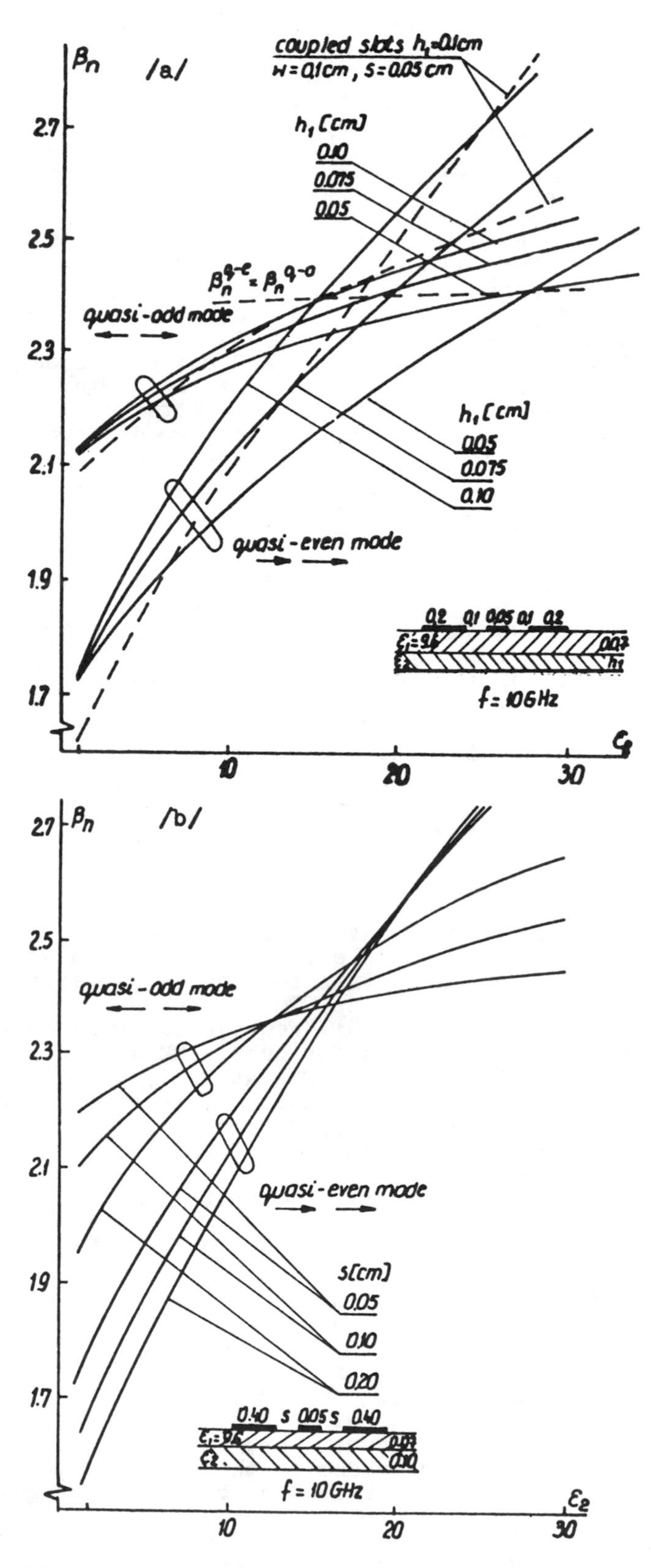

Fig.1 - Characteristics of two modal phase constants of double-layer C-CPS versus the permittivity of lower substrate layer with the thickness of lower layer /a/ and spacing slot width /b/ as a parameter

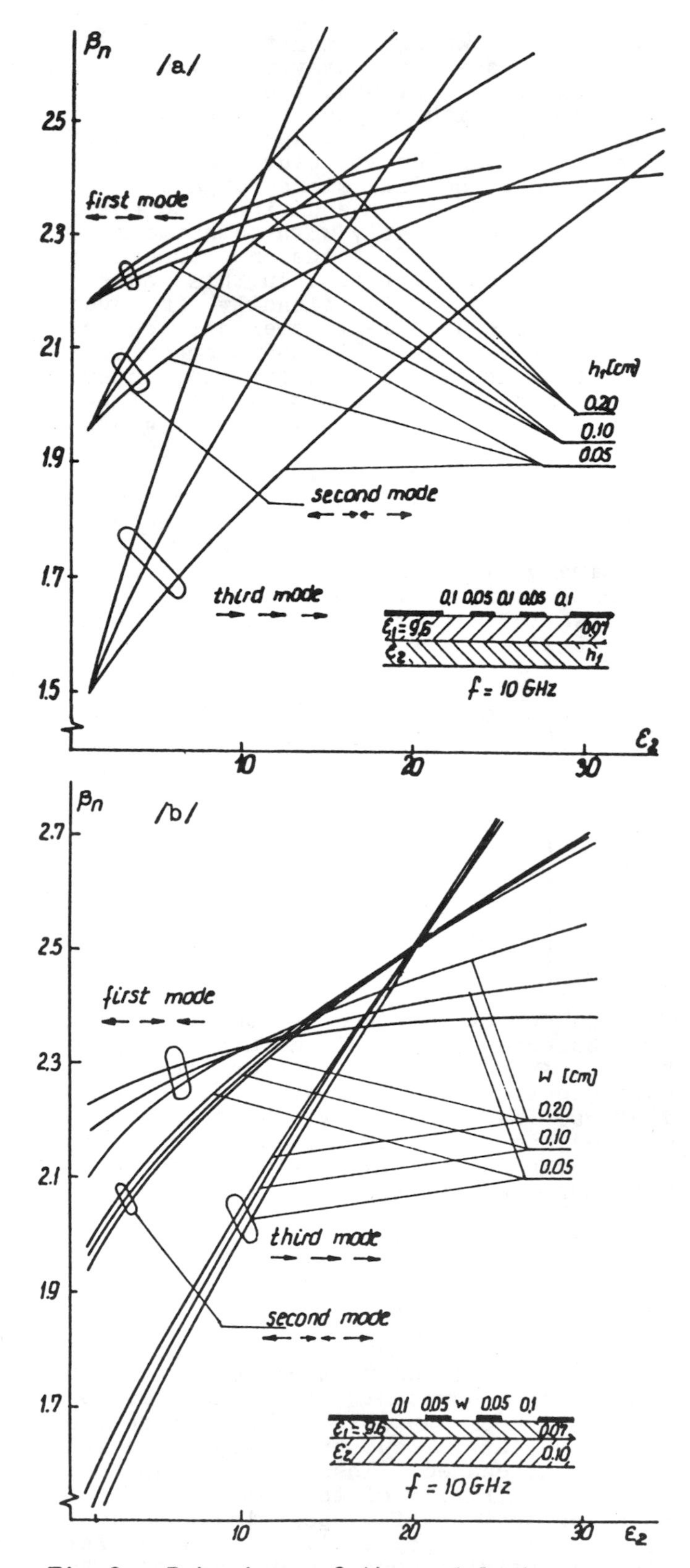

Fig.2 - Behaviour of the modal phase constants of three coupled slot lines with double-layer substrate versus permittivity of lower layer with the thickness of the lower layer /a/ and center slot width /b/ as a parameter

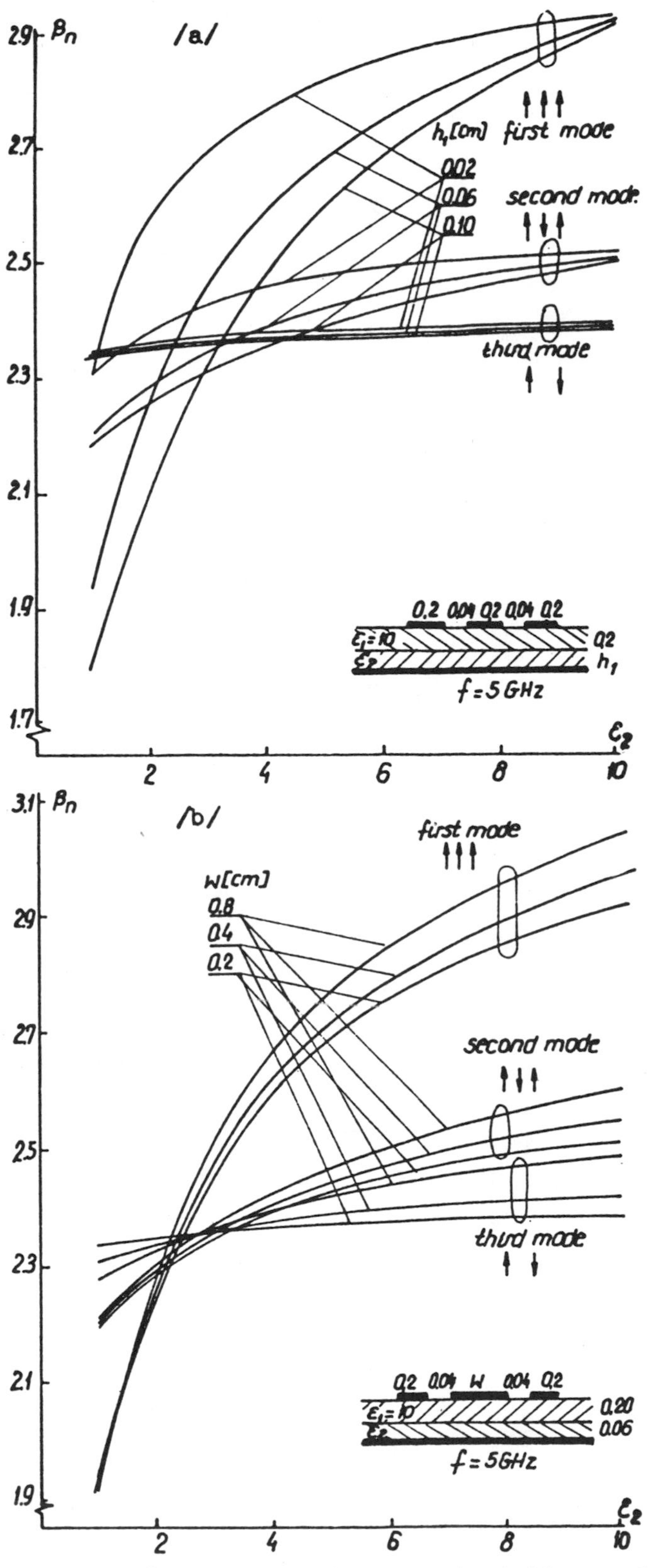

Fig.3 - Phase constant characteristics of three-line coupled microstrip structure in double-layer configuration versus permittivity of lower layer with the thickness of the lower layer /a/ and center strip width /b/ as a parameter

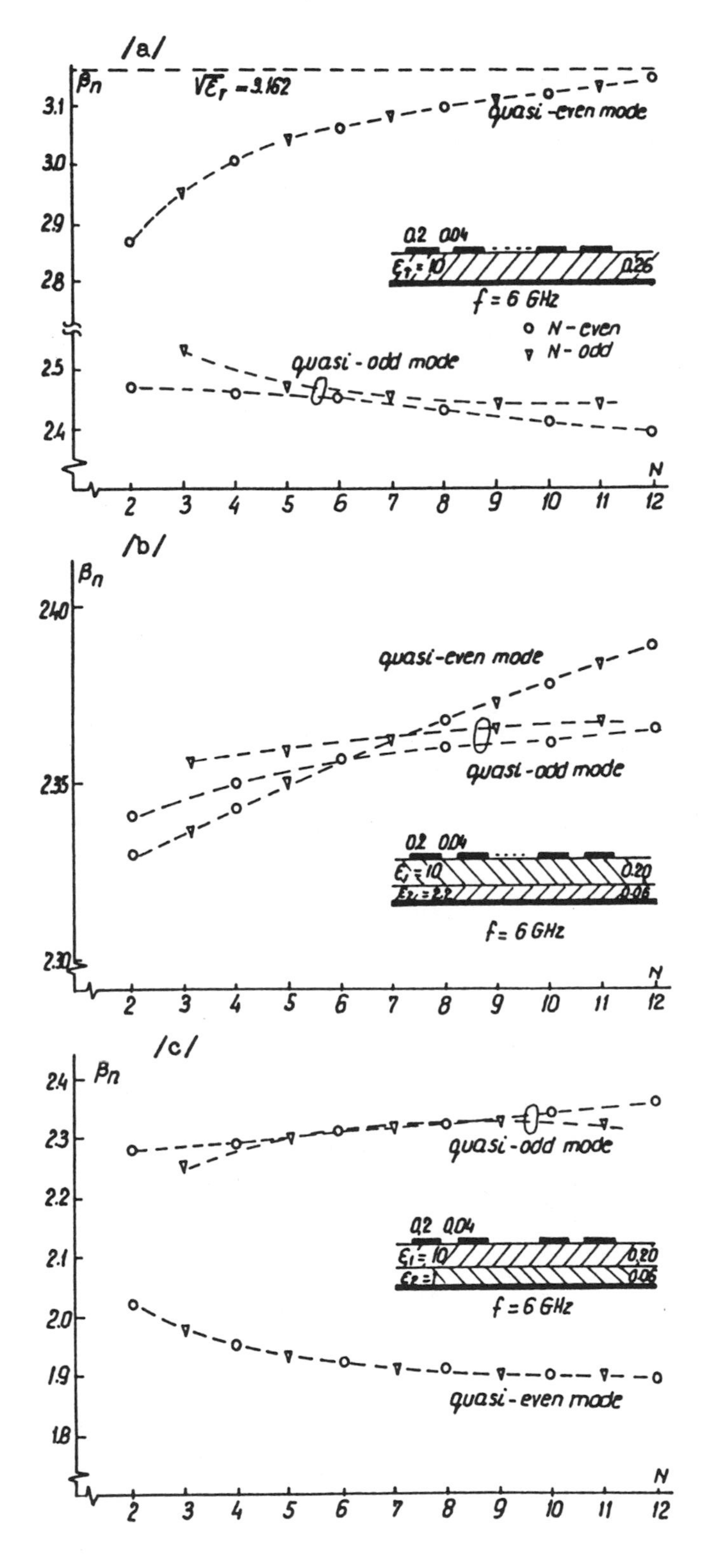

Fig.4 - Characteristics of the quasi-even and quasi-odd mode of interdigital microstrip structure versus the number of coupled strip conductors for single /a/, double /b/ and suspended /c/ substrate

Part VI
Model of Microstrip Lines

IT has already been mentioned that the characteristics of the microstrip line cannot be obtained in analytical forms. Although numerical solutions of rigorous analytical methods are available, there are a number of instances in which such a practice is undesirable. In spite of recent advances in numerical algorithms and in computer hardware, numerical evaluations of these quantities require a considerable amount of time. In the area of computer aided design (CAD), the need for a quick evaluation of microstrip line characteristics is particularly acute, because the characteristic quantities have to be evaluated repeatedly for an assumed set of structural parameters until the desired quantities are obtained. To accelerate the design process, some forms of closed-form expressions are needed in the CAD or, for that matter, in any form of design process. It may not be scientific, but use of such a formula is often an engineering necessity.

The papers in this part address the modeling aspects of the microstrip line analysis. Several papers present empirical closed-form expressions for the characteristic impedance and the propagation constant of microstrip lines. Many formulas published need to be used with caution in terms of expected accuracy and of the applicable range of parameters. A number of efforts exist which try to derive closed-form expressions with the widest possible range of applicability. One paper aiming for such a goal is included in this part (Kirschning and Jansen). It is generally advisable to check the accuracy of particular closed-form expressions by comparing the data from these formulas with numerical or experimental results, unless accuracy has previously been confirmed.

Microstrip Dispersion Model

WILLIAM J. GETSINGER

Abstract—The assumption that the quasi-TEM mode on microstrip is primarily a single longitudinal-section electric (LSE) mode leads to a transmission line model whose dispersion behavior can be analyzed and related to that of microstrip. Appropriate approximations yield simple, closed-form expressions that allow slide-rule prediction of microstrip dispersion.

Nomenclature

a, a', b, b', s, w	Mechanical dimensions of conventional microstrip and the LSE mode model (Fig. 2).
c	Speed of light in free space = 11.8 in/ns.
C'	Capacitance per unit length of microstrip line at zero frequency.
D	Width of the zero-frequency parallel-plate microstrip equivalent structure.
f	Frequency.
f_i	Frequency of inflection of the dispersion curve.
f_p	Parameter of the dispersion function.
G	Empirical parameter used to simplify the microstrip dispersion function.
k_o	Free-space wavenumber.
L'	Inductance per unit length of microstrip line at zero frequency.
Z_f	Microstrip characteristic impedance at frequency f.
Z_0	Microstrip characteristic impedance at zero frequency.
γ	Propagation constant along the microstrip line.
γ_a	Transverse propagation constant in the air-filled part of the microstrip model.
γ_s	Transverse propagation constant in the dielectric-filled part of the microstrip model.
ϵ_e	Microstrip effective dielectric constant (a function of frequency).
ϵ_{ei}	Microstrip effective dielectric constant at the inflection point.
ϵ_{e0}	Microstrip effective dielectric constant at zero frequency.
ϵ_o	Permittivity of free space = 8.85×10^{-12} F/m.
ϵ_s	Substrate relative dielectric constant.
η_o	Impedance of free space = 376.7 Ω.
μ_o	Permeability of free space = 31.92 nH/in, or $4\pi\times10^{-7}$ H/m.
ω	Radian frequency.

Manuscript received April 13, 1972; revised July 10, 1972. This paper is based upon work performed at COMSAT Laboratories under Corporate sponsorship.

The author is with COMSAT Laboratories, Clarksburg, Md. 20734.

Reprinted from *IEEE Trans. Microwave Theory Tech.*, vol. MTT-21, no. 1, pp. 34–39, January 1973.

Introduction

PROPAGATION on microstrip is usually handled as though the line were filled with dielectric and carried a TEM mode. This is an adequate representation except that the effective dielectric constant changes slowly with frequency, making microstrip dispersive [1].

Both analytical [2] and empirical [3], [4] attempts to describe microstrip dispersion have been published. (A good bibliography is given in [4].) The analytical techniques have been nearly exact, but have required numerical solution on large electronic computers. Thus these techniques have been too ponderous for practical engineering application. The empirical techniques, on the other hand, have had limited ranges of applicability and inadequate theoretical foundations for confidence in application.

With the intention of achieving analytical simplicity, this paper considers microstrip propagation as a single longitudinal-section electric (LSE) [5] mode. Physical reasoning indicates that this might be a practical approximation for investigating dispersion on microstrip. However, the structure of microstrip precludes analysis by direct means. Thus a structure (the model) has been conceived that resembles microstrip in all but shape, but whose LSE-mode propagation can be analyzed directly. It is assumed that the propagation characteristics (dispersion) of the model can be applied to microstrip by appropriate adjustment of parameters.

Since it does not follow from theory that the dispersion functions of the two structures must be the same, as it does for differently shaped, homogeneously filled waveguides, the validity of the model must be tested by its agreement with measured dispersion of actual microstrip.

It turns out that the model yields a simple closed-form algebraic expression that closely describes measured dispersion in microstrip. It is found that only one parameter in addition to those available from static analyses of microstrip, such as the MSTRIP program [6], is necessary to describe microstrip dispersion.

For convenience, the results of this paper are illustrated in Fig. 1. The symbols are defined in the Nomenclature list. The dispersion relationships shown in Fig. 1 have been found to agree with a theoretical prediction [2] based on coupled integral equations, with published [1] measurements of a 20-Ω microstrip line on a rutile ($\epsilon_s = 104$) substrate and with measurements on 0.025- and 0.050-in alumina ($\epsilon_s \approx 10$) substrates.

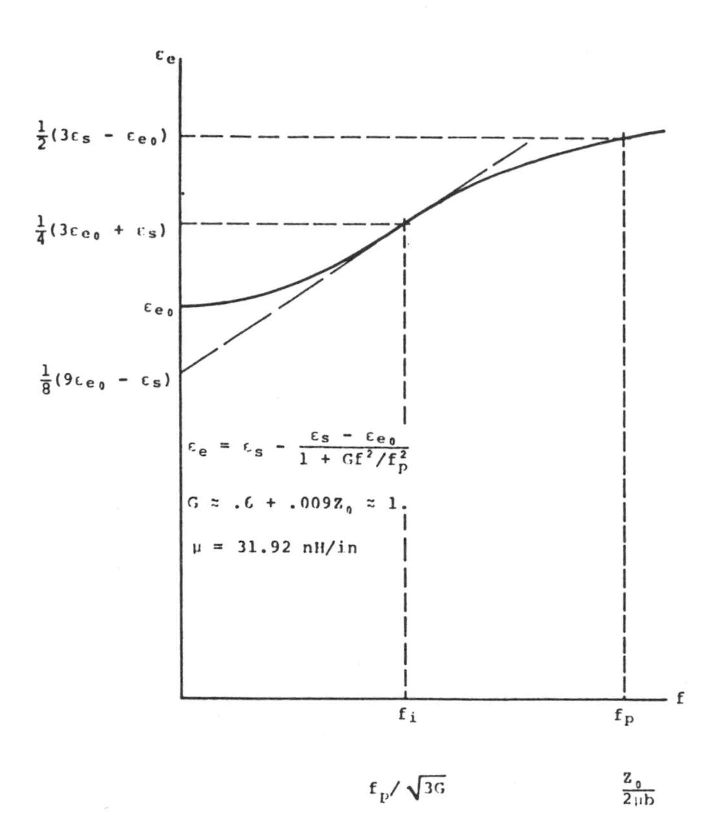

Fig. 1. Microstrip dispersion relationships.

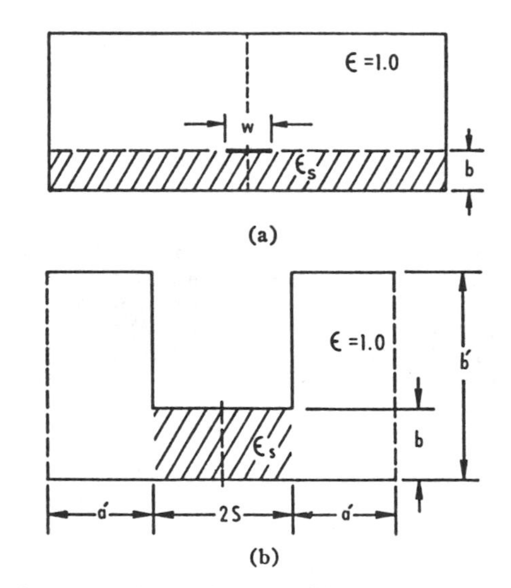

Fig. 2. (a) Conventional microstrip. (b) LSE model for microstrip.

The Analytical Model

A conventional microstrip structure is shown in Fig. 2(a). The fields are concentrated around the edges of the strip and in the dielectric beneath the center strip. Near the strip edges, the magnetic field is predominantly normal to and the electric field predominantly tangential to the air–dielectric interface. This is characteristic of the LSE mode [5], [7]. The structure of Fig. 2(a) is intractable to direct analysis on this basis, but its boundaries can be distorted to result in a model, shown in Fig. 2(b), that can be analyzed.

The electric field lines emanating from the lower surface of the center strip of the microstrip in Fig. 2(a) pass only through the substrate dielectric, as do the electric field lines emanating from the center portion of the model of Fig. 2(b). The electric fields emanating from the upper surface of the center strip of the microstrip occupy a much larger space, which is mostly filled with air. This space is approximated by the large, air-filled end sections of the model. The magnetic wall (indicated by a dashed line) above the center strip of Fig. 2(a) is split and the upper wall of the center strip is unfolded at the edge, stretched out, and bent to form the end-section boundaries of Fig. 2(b). Thus the model consists of one parallel-plate transmission line, which has a dielectric constant ϵ_s, width $2s$, and height b, connected without junction effect to other parallel-plate transmission lines that have a dielectric constant of one, width a', and height b'.

The heuristic assumption made is that because the two regions, air filled and dielectric filled, of the model and the microstrip are grossly similar, the two structures will have the same dispersion behavior for the same mode of propagation. It is clear that junction capacitance could be included at the steps of the model to make it more realistic, or more

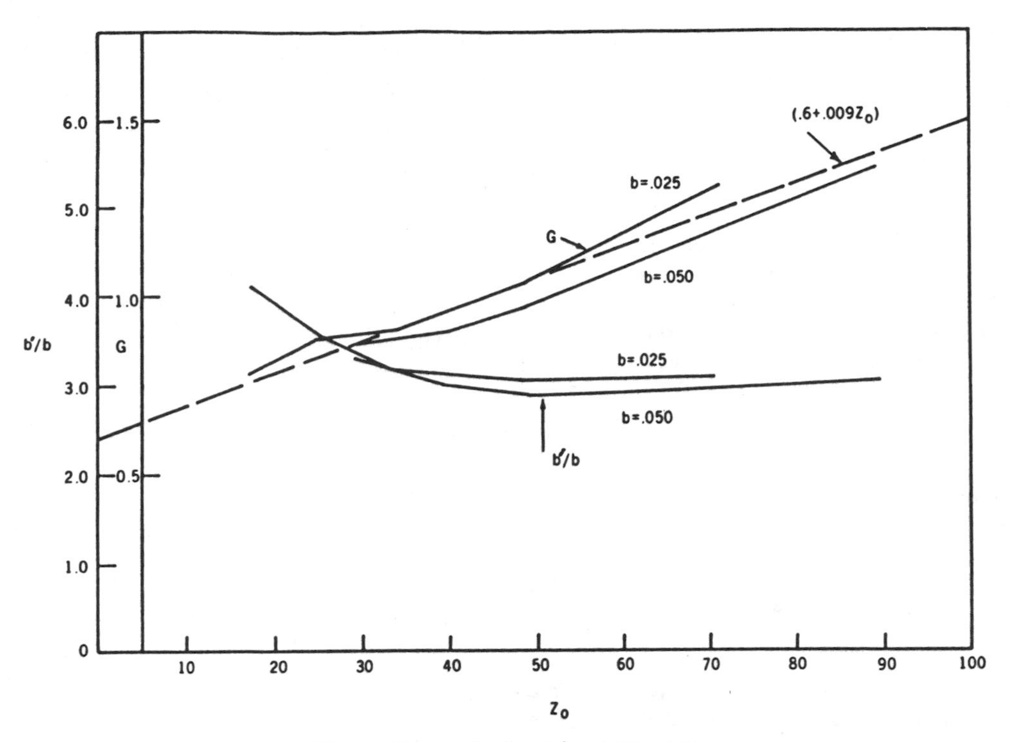

Fig. 3. Microstrip dispersion parameters.

like microstrip, but this would greatly complicate the analysis and has not been found necessary for practical results.

The model is, after all, only an intuitive aid in setting up the simplest mathematics that lead to a useful dispersion relation; it need not be physically realizable.

The analysis proceeds by forcing the model to have the same electrical characteristics at zero frequency as the microstrip. (These characteristics can be found from well-known and widely available computer programs, such as MSTRIP [6].) Next, a transverse resonance analysis of the model relates the propagation constant (or effective dielectric constant) to frequency. A closed-form approximation of this function is then found and compared with measured microstrip dispersion to determine the unknown parameter b'/b [see Fig. 2(b)]. Finally, the results show that b'/b and related parameters are nearly constant or linear with characteristic impedance. Hence, it is possible to derive simple formulas that can be used to predict the dispersion of microstrip transmission lines.

Zero-Frequency Relationships

A static analysis such as the MSTRIP program [6] is employed to yield the effective relative dielectric constant ϵ_{e0} and the characteristic impedance Z_0 for given w/b, ϵ_s, and possibly other dimensional parameters, such as strip thickness or proximity of an upper ground plane. [See Fig. 2(a).]

The inductance L' and capacitance C' per unit length of the microstrip can be written as

$$\frac{L'}{\mu_o} = \frac{Z_0}{\eta_o}\sqrt{\epsilon_{e0}} \tag{1a}$$

$$\frac{C'}{\epsilon_o} = \frac{\eta_o\sqrt{\epsilon_{e0}}}{Z_0} \tag{1b}$$

respectively, where

$$\eta_o = \sqrt{\frac{\mu_o}{\epsilon_o}} = 376.7\ \Omega. \tag{2}$$

The subscript o indicates free-space values of the constitutive parameters, while the subscript 0 indicates zero-frequency values of the characteristic impedance and effective dielectric constant.

Inductance and capacitance per unit length for the LSE model [Fig. 2(b)] at zero frequency can be written as

$$\frac{L'}{\mu_o} = \frac{1}{2[(a'/b') + (s/b)]} \tag{3a}$$

$$\frac{C'}{\epsilon_o} = 2\left(\frac{a'}{b'} + \epsilon_s\frac{s}{b}\right) \tag{3b}$$

respectively. Equating equivalent parameters yields

$$\frac{a'}{b'} = \frac{\eta_o}{2Z_0\sqrt{\epsilon_{e0}}}\cdot\frac{\epsilon_s - \epsilon_{e0}}{\epsilon_s - 1} \tag{4a}$$

$$\frac{s}{b} = \frac{\eta_o}{2Z_0\sqrt{\epsilon_{e0}}}\cdot\frac{\epsilon_{e0} - 1}{\epsilon_s - 1}. \tag{4b}$$

Transverse Resonance Solution

The sum of the admittances on the left and right of either air–dielectric interface of Fig. 2(b) must equal to zero according to the transverse resonance [5], [7] technique. The propagation constants are related by

$$\gamma_a^2 + \gamma^2 + k_o^2 = 0 \tag{5}$$

in the air-filled section and by

$$\gamma_s^2 + \gamma^2 + \epsilon_s k_o^2 = 0 \tag{6}$$

in the dielectric-filled section. In (5), (6), γ is the propagation constant along the transmission line and applies to both air-

and dielectric-filled sections, while γ_a is the constant in the transverse direction in the air-filled section and γ_s is the constant in the transverse direction in the dielectric-filled section. Finally,

$$k_o = \omega/c \tag{7}$$

is the free-space wave number.

The vertical dashed lines of Fig. 2(b) indicate magnetic walls or open-circuit boundaries. The characteristic admittances in the two sections are proportional to their propagation constants and inversely proportional to their heights. Thus the sum of the admittances at the interface is

$$\frac{\gamma_a}{b'} \tanh \gamma_a a' + \frac{\gamma_s}{b} \tanh \gamma_s s = 0. \tag{8}$$

The following approximation is used to solve the preceding transcendental equation:

$$\tanh x \approx \frac{1}{(1/x) + (x/3)}. \tag{9}$$

Equation (9) is in error by about 1.5 percent at $x=1$ rad. As an example of the range of applicability, $\gamma_s s=1.0$ for a 25-Ω line on a 0.05-in alumina substrate at about 10 GHz. The use of higher impedances and thinner substrates raises the frequency at which an error of this magnitude occurs.

Substituting (9) into (8) yields

$$\frac{b'/a'}{\gamma_a^2} + \frac{b/s}{\gamma_s^2} = -\frac{a'b' + sb}{3} \tag{10}$$

after some manipulation.

The longitudinal propagation constant can be expressed in terms of the effective dielectric constant; i.e.,

$$\gamma^2 = -k_o^2 \epsilon_e. \tag{11}$$

Substituting (11) into (5) and (6) results in

$$\gamma_a^2 = k_o^2(\epsilon_e - 1) \tag{12}$$

$$\gamma_s^2 = -k_o^2(\epsilon_s - \epsilon_e). \tag{13}$$

Substituting (12) and (13) into (10) yields

$$\frac{b/s}{\epsilon_s - \epsilon_e} - \frac{b'/a'}{\epsilon_e - 1} = \frac{a'b' + sb}{3} k_o^2 \tag{14}$$

which is the basic dispersion relationship.

The unknown parameters a' and b' can be reduced to a single unknown by assuming that a is the solution of (4a) when b' is given the value of b, which is known. That is,

$$a' = a\left(\frac{b'}{b}\right) \tag{15}$$

where b'/b is the new unknown parameter. When (15) is substituted into (14), the basic dispersion relationship becomes

$$\frac{1/s}{\epsilon_s - \epsilon_e} - \frac{1/a}{\epsilon_e - 1} = \frac{a(b'/b)^2 + s}{3} k_o^2. \tag{16}$$

When (16) is solved for $\epsilon_s - \epsilon_e$ as the dependent variable, a quadratic results. Its solution is

$$\epsilon_s - \epsilon_e = \frac{B}{2}\left\{1 - \sqrt{1 - \frac{4(\epsilon_s - 1)/s}{B^2[\{[a(b'/b)^2+s]/3\}k_o^2]}}\right\} \tag{17}$$

where

$$B = (\epsilon_s - 1) + \frac{(a+s)/as}{\{[a(b'/b)^2+s]/3\}k_o^2} \tag{18}$$

and the negative root has been selected because it is physically meaningful.

Equation (17) can be simplified by observing that the second term under the radical is considerably less than one for practical cases and then by using the usual square-root approximation. (For a 25-Ω line on a 0.05-in alumina substrate at 12.5 GHz, the error is about 5 percent.) After a small amount of algebra, the result is

$$\epsilon_e = \epsilon_s - \frac{[(\epsilon_s - 1)a]/(a+s)}{1 + k_o^2(as/3)(\epsilon_s - 1)\{[a(b'/b)^2+s]/(a+s)\}}. \tag{19}$$

Substituting (4) and (15) into (19) makes it possible to express (19) in terms of known quantities, except for the parameter b'/b; i.e.,

$$\epsilon_e = \epsilon_s - \frac{\epsilon_s - \epsilon_{e0}}{1 + G(f^2/f_p^2)} \tag{20}$$

where

$$f_p = \frac{Z_0}{2\mu_o b} \tag{21}$$

and

$$G = \frac{\pi^2}{12} \frac{[(\epsilon_{e0} - 1) + (b'/b)^2(\epsilon_s - \epsilon_{e0})](\epsilon_{e0} - 1)(\epsilon_s - \epsilon_{e0})}{\epsilon_{e0}(\epsilon_s - 1)^2}. \tag{22}$$

Nonmagnetic substrates are assumed; therefore, $\mu_o = 31.9186$ nH/in. Equation (20) is the final analytical expression for dispersion of microstrip. Investigation of experimental results shows that G approximates unity.

Evaluation of Parameters

Dispersion curves for microstrip lines on alumina substrates 0.025 and 0.050 in thick were measured. The microwave measurements were made on ring resonators [8], and the 1-MHz points were determined from the MSTRIP program by using the value of the substrate dielectric constant ϵ_s found from capacitance measurements of each fully metallized substrate. These data were used to calculate the multiplier of f^2 in (20) that forced a fit at 10 GHz for each microstrip line. Then, values of b'/b were calculated using (21) and (22). The results are shown in Fig. 3, which indicates that $b'/b \simeq 3$ for characteristic impedances above about 35 Ω. The experimentally determined values of G are also plotted in Fig. 3.

Equations (20) and (21) clearly demonstrate the nature of the dependence of the effective dielectric constant on the substrate thickness b and microstrip characteristic impedence Z_0.

In many engineering applications of microstrip, dispersion can be treated as a correction factor to the zero-frequency effective dielectric constant ϵ_{e0}; thus only approximate values are required. In such situations, it is sufficient to assume that $G=1.0$ in (20). For greater accuracy, the curves in Fig. 3, or an equivalent based on other careful measurements, can be used. A linear approximation of curves of Fig. 3 is

$$G = 0.6 + 0.009\, Z_0. \tag{23}$$

The Inflection Point

Study of the dispersion function (20) can provide some general information about typical dispersive behavior. Equation (20) shows that the effective dielectric constant goes from a value of ϵ_{e0} at zero frequency to a value of ϵ_s at infinite frequency, in agreement with theory [2], and that the slope of ϵ_e with respect to frequency is zero at both extremes. The frequency of the maximum slope between these two points is called the inflection point. The requirement that the second derivative of ϵ_e with respect to frequency must equal zero gives the inflection frequency

$$f_i = \frac{f_p}{\sqrt{3G}}. \tag{24}$$

Using (24) in (20) yields the value of the effective dielectric constant ϵ_{ei} at the inflection frequency:

$$\epsilon_{ei} = \tfrac{1}{4}(\epsilon_s + 3\epsilon_{e0}) \tag{25}$$

and the slope with respect to frequency at ϵ_{ei}:

$$\left.\frac{d\epsilon_e}{df}\right|_{f=f_i} = \tfrac{3}{8}(\epsilon_s - \epsilon_{e0}). \tag{26}$$

A graphical construction that makes it possible to draw a straight line tangent to the inflection point of the dispersion curve has been shown in Fig. 1.

It can be observed that the dispersion and the inflection relationships agree closely with measured data. Equations (24) and (25) do in fact predict the frequency and effective dielectric constant values at which measured dispersion curves have maximum slope, and that slope is in very good agreement with (26) for an ideal dispersion function. This detailed agreement between theory and experiment supports the validity of the LSE model of microstrip propagation.

Comparison with Measurements

The theory developed in this paper will first be compared with the theoretical prediction of Zysman and Varon [2] for a microstrip line having the following characteristics: $\epsilon_s=9.7$, $\epsilon_{e0}=6.50$, $Z_0=50$, and $b=0.05$ in. The unknown parameter G will be found from the inflection point formulas and the graphical data of [2].

Using (25) to calculate the inflection point gives $\epsilon_{ei}=7.3$; [2, fig. 5] then gives $f_i=9$ GHz. Equation (24) predicts

$$\frac{G}{f_p^2} = \frac{1}{3f_i^2} = 0.00412. \tag{27}$$

Using values given above in (21) yields

$$f_p = \frac{Z_0}{2\mu b} = \frac{50}{2 \times 31.92 \times 0.05} = 15.66 \text{ GHz}. \tag{28}$$

Substituting (28) into (27) yields

$$G = 0.00412 \times 15.66^2 = 1.01. \tag{29}$$

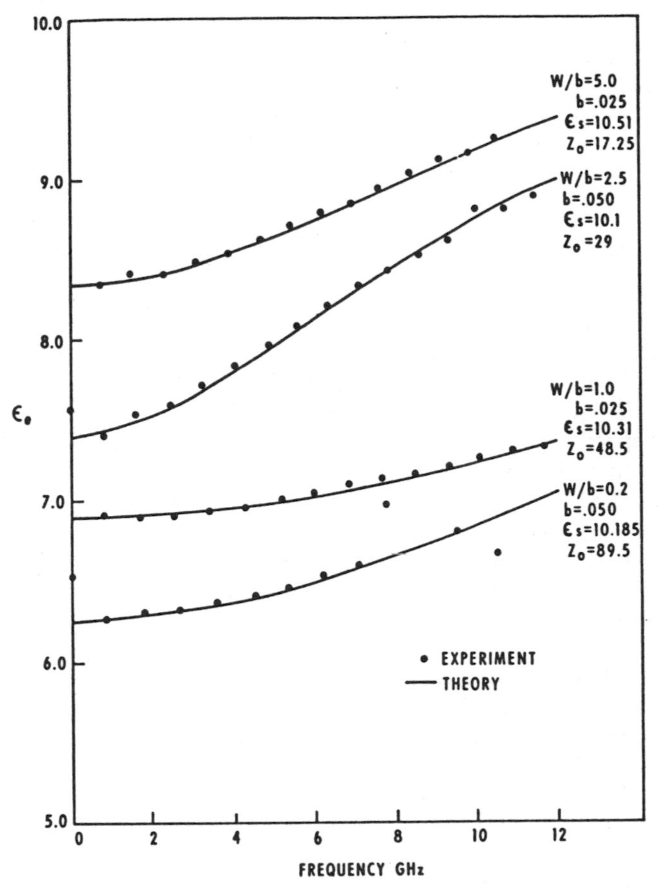

Fig. 4. Comparison of dispersion function and measurements.

This value can be compared with the approximate value $G=1.0$ or the curve-fit formula (23), which gives $G=1.05$.

The next problem is to determine the model parameters of the dispersion measurements on rutile reported by Hartwig *et al.* [1]. The given parameters are $\epsilon_s=104$, $\epsilon_{e0}=62$ (calculated from capacitance measurement), $b=0.05$, and $Z_0=20$ Ω. However, since the dispersion curve must have a zero slope at zero frequency, an extrapolation of the measured points on [1, fig. 3] clearly yields a greater value of ϵ_{e0} than 62. A value of 63.5 will be used in these calculations.

Following the same procedure used in the preceding example gives $\epsilon_{ei}=73.6$ and $f_i \approx 4.4$, so that $G/f_p^2=0.0172$. Equation (21) gives $f_p=6.22$ GHz; thus $G=6.22^2 \times 0.0172 = 0.667$. The approximate formula (23) gives $G=0.78$. Either value of G gives a calculated curve that is within the scatter of the measured points. The first value of G, based on inflection point formulas, seems to average out a little better, however. This is reasonable because it is based in part on the data it characterizes.

The last comparison of theory and experiment is shown in Fig. 4. The solid-line curves were calculated from the dispersion function (20) by using (23) to set the value of G in each case. The round points are values of effective dielectric constant measured at COMSAT Laboratories. The circuits used were ring resonators [8] on commercial 0.025- and 0.05-in alumina substrates. In each case, the value of ϵ_{e0} was found by extrapolating the curve of microwave measurements to zero frequency. Then, curves generated by the MSTRIP program [6] were used to determine a value of ϵ_s for the appropriate width-to-height ratio of the line. The shapes and values of the experimental and theoretical curves are found to be in good agreement.

Limitations and Applications

The basic hypothesis of this paper is that the dispersion function (20) describes the propagation characteristics of any microstrip-like transmission line. So far, measurements on microstrip have supported this point of view.

A more general theoretical investigation than given in this paper would be necessary to explore the fundamental limitations on applying the dispersion function. Some points can be considered, however.

The dispersion relation applies only to the fundamental LSE mode. It probably holds closely only for thin ($b < \lambda/4$ in ϵ_s) substrates and strips that are not very wide ($w < \lambda/3$ in ϵ_s) to insure that the LSE mode is dominant, but these restrictions seldom arise in microstrip applications. Also, the dispersion relation takes on the correct value at infinite frequency, and so there is no clearly defined upper-frequency limit at which it no longer applies. The practical upper-frequency limit of microstrip, where every junction and discontinuity radiate strongly via surface wave modes [1], probably occurs before the dispersion function becomes unreliable.

Since the dispersion function appears to have general applicability to all structures having the same types of boundaries as microstrip and propagating an LSE mode, it would be expected to hold for microstrip with or without an enclosure, for the even and odd modes of the parallel-coupled microstrip, and possibly for other quasi-TEM structures, such as inhomogeneously loaded coaxial line. It would, of course, be necessary to have appropriate values for ϵ_s, ϵ_{e0}, Z_0, and G for each structure.

Acknowledgment

The author wishes to thank Dr. W. J. English for his technical discussions and T. J. Lynch for his careful measurements.

References

[1] C. Hartwig, D. Massé, and R. Pucel, "Frequency dependent behavior of microstrip," in *1968 G-MTT Symp. Dig.*, pp. 110–116.
[2] G. Zysman and D. Varon, "Wave propagation in microstrip transmission lines," in *1969 G-MTT Symp. Dig.*, pp. 3–9.
[3] O. Jain, V. Makios, and W. Chudobiak, "Coupled-mode model of dispersion in microstrip," *Electron. Lett.*, vol. 7, pp. 405–407, July 15, 1971.
[4] M. V. Schneider, "Microstrip dispersion," *Proc. IEEE* (*Special Issue on Computers in Design*) (Lett.), vol. 60, pp. 144–146, Jan. 1972.
[5] R. Collin, *Field Theory of Guided Waves.* New York: McGraw-Hill, 1960, p. 224.
[6] T. G. Bryant and J. A. Weiss, "MSTRIP (parameters of microstrip)," *IEEE Trans. Microwave Theory Tech.*, vol. MTT-19, pp. 418–419, Apr. 1971.
[7] C. Montgomery, R. Dicke, and E. Purcell, *Principles of Microwave Circuits* (M.I.T. Radiation Laboratory Series), vol. 8. New York: McGraw-Hill, 1948.
[8] P. Troughton, "Measurement techniques in microstrip," *Electron. Lett.*, vol. 5, pp. 25–26, Jan. 23, 1969.

An Approximate Dispersion Formula of Microstrip Lines for Computer-Aided Design of Microwave Integrated Circuits

EIKICHI YAMASHITA, MEMBER, IEEE, KAZUHIKO ATSUKI, AND TOMIO UEDA

***Abstract*—A simple approximate formula for the computer-aided design of the dispersion property of microstrip lines is reported in this paper which well fits the calculated curves based on a rigorous analysis.**

I. Introduction

Microstrip lines have been widely used as basic components of microwave integrated circuits. The TEM approximation [1] is employed, in many cases, for analyzing the propagation modes of these lines. Two features of the TEM mode are that the wave velocity is independent of frequencies and the longitudinal components of electromagnetic fields are neglected.

When the wavelength becomes comparable to cross-sectional dimensions of the microstrip line, the wave velocity is no longer independent of frequencies. This phenomenon is known as the dispersion property and should be considered in practice. An exact analysis of this property is possible by using Maxwell's equations. The numerical results of an analysis based on the integral equation method [2] agree well with those based on the mode matching method [3] in a wide frequency range. This is an indication of accuracy in both methods.

On the other hand, a simple approximate formula to express the dispersion property is needed for the purpose of desk calculations or the computer aided design of microwave integrated circuits. Most of rigorous calculation methods including the above two are not appropriate in such applications because they usually require a complicated computer programming. Though a few approximate formulas have been reported based on some physical considerations and experimental data [4]–[7], these empirical formulas have had limited ranges of applicability and inadequate theoretical foundation for confidence in application.

Our approximate dispersion formula to be derived here has a simple form and is based on a rigorous theory in contrast to other approximate formulas.

II. Computation of Propagation Constant

Our computation of the propagation constant of microstrip lines is based on the integral equation method which has been reported in [2]. Fig. 1 shows the transmission line structure treated with this method. The microstrip line is

Manuscript received April 22, 1979; revised September 21, 1979.
The authors are with the University of Electro-Communications, Chofu-shi, Tokyo, Japan 182.

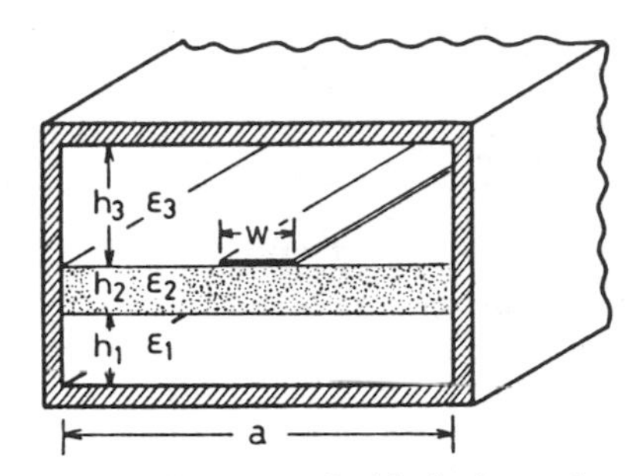

Fig. 1. Transmission lines treated with the integral equation method.

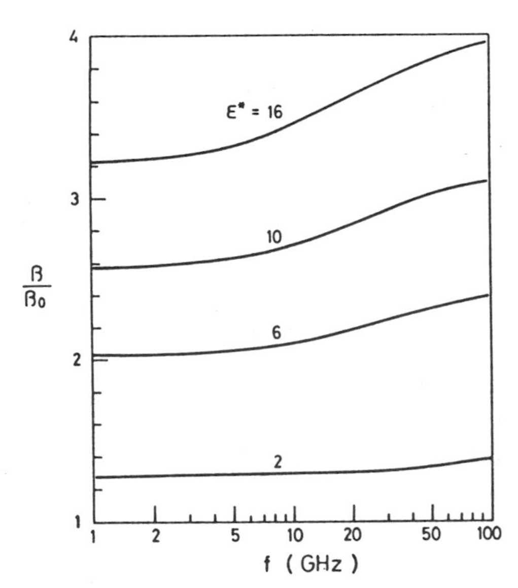

Fig. 2. Computed propagation constants for $h = 1$ mm, $W/h = 1$.

only a special case of Fig. 1. For instance, a good approximation of such structures may be obtained by setting parameters as $h_1 = 0$, $h_2 = h$, $h_3 = 1000h$, $a = W + 8h$, $\epsilon_2 = \epsilon^*\epsilon_0$, $\epsilon_3 = \epsilon_0$. Fig. 2 shows the evaluated propagation constant of the dominant mode, β, by setting ϵ^* as a parameter, where β_0 is the propagation constant in vacuum. The dispersion or the dependence of β/β_0 on frequencies is observed quite well here. Fig. 3 shows the same quantity by setting W/h as a parameter.

III. Derivation of Approximate Dispersion Formula

Let us find a simple formula to fit these curves with least errors in a wide range of frequencies, dielectric constant, and physical dimensions. The propagation con-

Reprinted from *IEEE Trans. Microwave Theory Tech.*, vol. MTT-27, no. 12, pp. 1036–1038, December 1979.

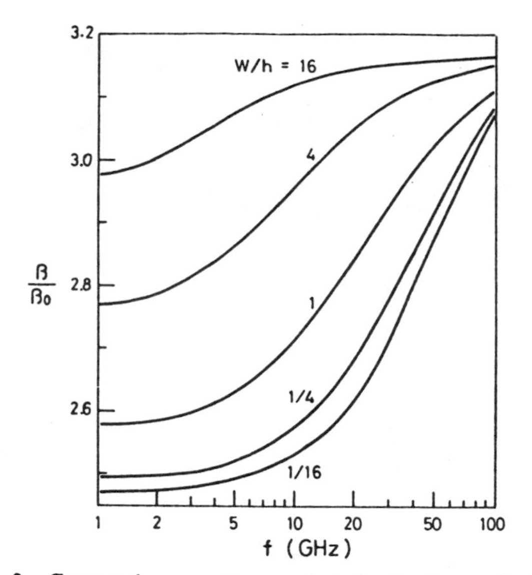

Fig. 3. Computed propagation constants for $h = 1$ mm, $\epsilon^* = 10$.

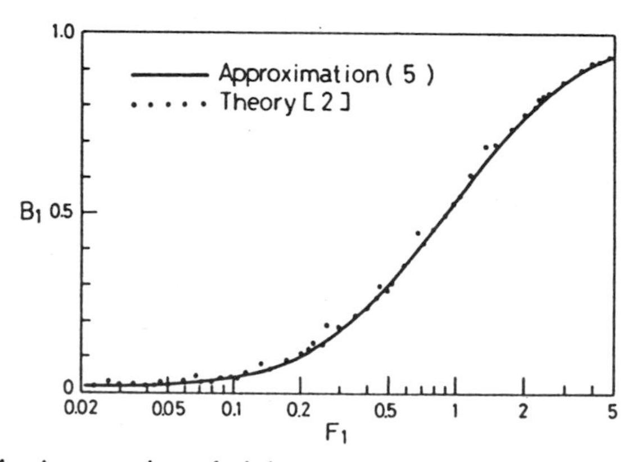

Fig. 4. An approximate logistic curve to fit the theoretical values of propagation constants in Fig. 2.

stant of the dominant mode has the following two extremes: $\beta \to \beta_{TEM} (f \to 0)$ and $\beta \to \beta_0 \sqrt{\epsilon^*}$ $(f \to \infty)$. β_{TEM} is the propagation constant obtained with the TEM wave approximation. At first, we define the normalized propagation constant B as

$$B = \frac{\beta/\beta_0 - \beta_{TEM}/\beta_0}{\sqrt{\epsilon^*} - \beta_{TEM}/\beta_0} \tag{1}$$

so as to confine all curves between the two extremes, $B \to 0 (f \to 0)$ and $B \to 1 (f \to \infty)$. Frequencies are also normalized as

$$F_1 = \frac{f}{f_{TE1}} \tag{2}$$

to collect all curves as only one curve. f_{TE1} is the cutoff frequency of the surface-wave TE_1 mode given by

$$f_{TE1} = \frac{v_0}{4h\sqrt{\epsilon^* - 1}} \tag{3}$$

where v_0 is the velocity of light in vacuum. Fig. 4 shows the result of the normalization drawn on a log scale.

We found a resemblance between the logistic curve given by

$$y(x) = \frac{1}{1 + e^{-bx}}, \qquad -\infty < x < \infty \tag{4}$$

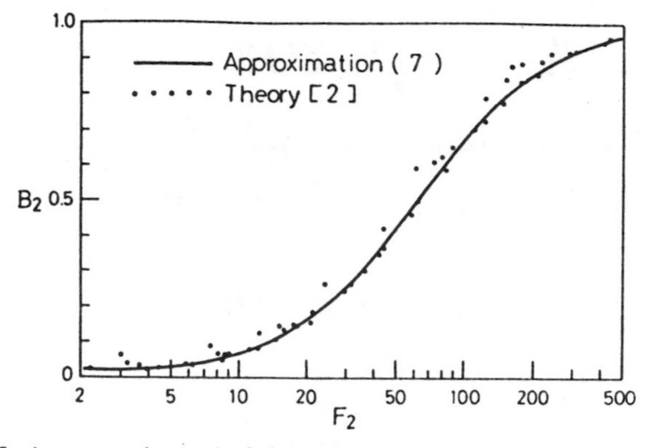

Fig. 5. An approximate logistic curve to fit the theoretical values of propagation constants in Fig. 3.

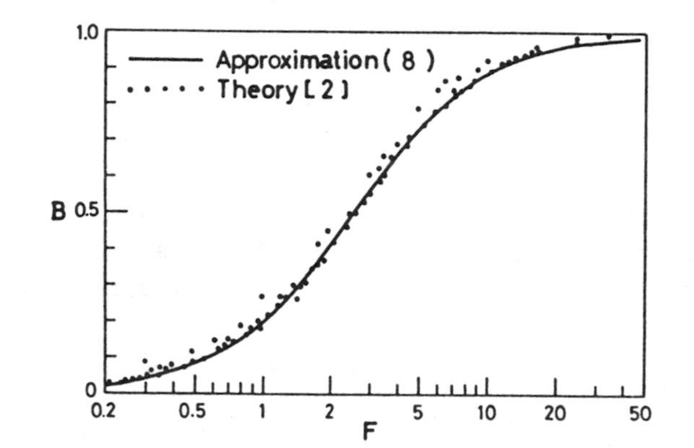

Fig. 6. An approximate logistic curve to fit the theoretical values of propagation constants in Figs. 2 and 3.

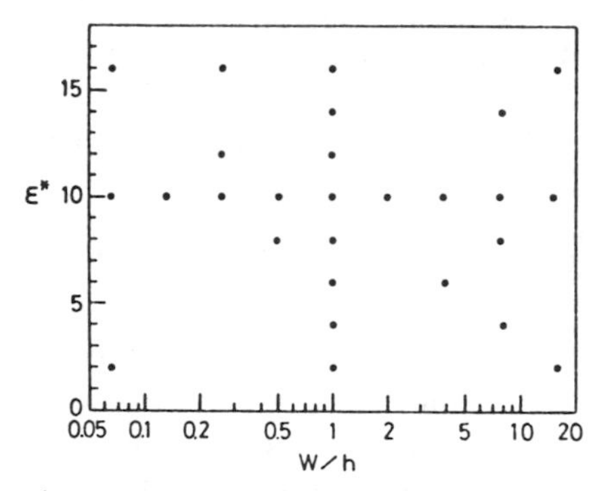

Fig. 7. Sample parameters used to estimate errors of the proposed approximate formula (10).

and the one in Fig. 4. Considering the logarithmic scale in Fig. 4, we rewrite (4) in a form

$$B_1 = \frac{1}{1 + a_1 F_1^{-b_1}}. \tag{5}$$

Selected values for the parameters, a_1 and b_1, to fit the curves in Fig. 4 are $a_1 = 0.785$ and $b_1 = 1.5$.

We also normalized frequencies in Fig. 3 as

$$F_2 = f\left[0.5 + \left\{1 + 2\log\left(1 + \frac{W}{h}\right)\right\}^2\right] \tag{6}$$

so that all curves shrink to almost one curve. The result of the normalization is shown in Fig. 5. Again, this curve can be expressed by a logistic curve

$$B_2 = \frac{1}{1 + a_2 F_2^{-b_2}}. \tag{7}$$

Selected parameters to fit the curve in Fig. 5 are $a_2=0.785\times10^{-1.5}$ and $b_2=1.5$.

The above two curves, (5) and (7), are combined together and expressed by

$$B=(1/1+aF^{-b}) \tag{8}$$

where the normalized frequency is given by

$$F=(F_1F_2/f) \tag{9}$$

and selected values of parameters are $a=4$ and $b=1.5$.

As a conclusion, we obtained an approximate dispersion formula:

$$\frac{\beta}{\beta_0}=\frac{\sqrt{\epsilon^*}-\dfrac{\beta_{TEM}}{\beta_0}}{1+4F^{-1.5}}+\frac{\beta_{TEM}}{\beta_0} \tag{10}$$

where

$$F=\frac{4h\sqrt{\epsilon^*-1}}{\lambda_0}[0.5+\{1+2\log(1+\frac{W}{h})\}^2]. \tag{11}$$

The estimated values of the propagation constants by the approximate formula (10) and the theoretical values given in Figs. 2 and 3 are compared as a solid line and dots in Fig. 6. The differences between these values were found to be within 1 percent for sampled parameters shown as dots in Fig. 7 in a wide frequency range. The above formula assumes the knowledge of β_{TEM} but a simple formula of β_{TEM} can be easily found in literatures such as Wheeler [8].

References

[1] E. Yamashita *et al.*, *IEEE Trans. Microwave Theory Tech.*, vol. MTT-16, p. 251, Apr. 1968.
[2] E. Yamashita *et al.*, *IEEE Trans. Microwave Theory Tech.*, vol. MTT-24, p. 195, Apr. 1976.
[3] G. Kowalski *et al.*, *Arch. Elek. Übertragung*, vol. 107, p. 163, Apr. 1971.
[4] O. P. Jain *et al.*, *Electron. Lett.*, vol. 7, p. 405, July 1971.
[5] W. J. Getsinger, *IEEE Trans. Microwave Theory Tech.*, vol. MTT-21, p. 34, Jan. 1973.
[6] M. V. Schneider, *Proc. IEEE*, vol. 60, p. 144, Jan. 1972.
[7] H. J. Carlin, *IEEE Trans. Microwave Theory Tech.*, vol. MTT-21, p. 589, Sept. 1973.
[8] H. A. Wheeler, *IEEE Trans. Microwave Theory Tech.*, vol. MTT-13, p. 172, Mar. 1965.

ACCURATE MODELS FOR MICROSTRIP COMPUTER-AIDED DESIGN

E.Hammerstad and Ø.Jensen
ELAB
7034 Trondheim-NTH, Norway

Summary

Very accurate and simple equations are presented for both single and coupled microstrip lines' electrical parameters, i.e. impedances, effective dielectric constants, and attenuation including the effect of anisotropy in the substrate. For the single microstrip the effects of dispersion and non-zero strip thickness are also included.

Introduction

In microstrip design it is often observed that the physical circuit performance differs significantly from that theoretically calculated. This is due to factors such as attenuation, dispersion, and discontinuity effects, which again are functions of physical parameters and the actual circuit lay-out. The conclusion which may be drawn from this observation, is that optimum microstrip design should be based directly upon physical dimensions. This approach requires accurate models and for computer-aided design the models must also be in an easily calculable form.

The models presented in this paper represent partly completely new models, but also revisions and expansions of some which have been given earlier [1,2]. It has been a goal to obtain an accuracy which gives errors at least less than those caused by physical tolerances. The line models are based on theoretical data [1,2], while the discontinuity models also are based on experimental data.

Single microstrip

The model of the single microstrip line is based upon an equation for the impedance of microstrip in an homogeneous medium, Z_{01}, and an equation for the microstrip effective dielectric constant, ε_e.

$$Z_{01}(u)=\frac{\eta_o}{2\pi}\ln\left[\frac{f(u)}{u}+\sqrt{1+\left(\frac{2}{u}\right)^2}\right] \quad (1)$$

$$f(u)=6+(2\pi-6)\exp\left[-\left(\frac{30.666}{u}\right)^{0.7528}\right] \quad (2)$$

Here η_o is wave impedance of the medium (376.73Ω in vacuum) and u is the strip width normalized with respect to substrate height (w/h). The accuracy of this model is better then 0.01% for $u \leq 1$ and 0.03% for $u \leq 1000$.

$$\varepsilon_e(u,\varepsilon_r)=\frac{\varepsilon_r+1}{2}+\frac{\varepsilon_r-1}{2}\left(1+\frac{10}{u}\right)^{-a(u)\,b(\varepsilon_r)} \quad (3)$$

$$a(u)=1+\frac{1}{49}\ln\frac{u^4+(u/52)^2}{u^4+0.432}+\frac{1}{18.7}\ln\left[1+\left(\frac{u}{18.1}\right)^3\right] \quad (4)$$

$$b(\varepsilon_r)=0.564\left(\frac{\varepsilon_r-0.9}{\varepsilon_r+3}\right)^{0.053} \quad (5)$$

The accuracy of this model is better than 0.2% at least for $\varepsilon_r \leq 128$ and $0.01 \leq u \leq 100$.

Compared to earlier equations, these give much better accuracy, in the order of 0.1% for both impedance and wavelength. They are also complete with respect to range of strip width, while earlier ones where given in two sets.

Strip thickness correction

In correcting the above results for non-zero strip thickness, a method described by Wheeler[3] is used. However, some modifications in his equations have been made, which give better accuracy for narrow strips and for substrates with low dielectric constant.
For homogeneous media the correction is

$$\Delta u_1=\frac{t}{\pi}\ln\left(1+\frac{4\exp(1)}{t\coth^2\sqrt{6.517\,u}}\right) \quad (6)$$

where t is the normalized strip thickness. For mixed media the correction is

$$\Delta u_r=\frac{1}{2}\left(1+1/\cosh\sqrt{\varepsilon_r-1}\right)\Delta u_1 \quad (7)$$

By defining corrected strip widths, $u_1=u+\Delta u_1$ and $u_r=u+\Delta u_r$, the effect of strip thickness may be included in the above equations:

$$Z_0(u,t,\varepsilon_r)=Z_{01}(u_r)/\sqrt{\varepsilon_e(u_r,\varepsilon_r)} \quad (8)$$

$$\varepsilon_{eff}(u,t,\varepsilon_r)=\varepsilon_e(u_r,\varepsilon_r)\cdot[Z_{01}(u_1)/Z_{01}(u_r)]^2 \quad (9)$$

Dispersion

As microstrip propagation is not purely TEM, both impedance and effective dielectric constant vary with frequency. Getsinger[4] has propsed the following model for dispersion in effective dielectric constant

$$\varepsilon_{eff}=\varepsilon_r-\frac{\varepsilon_r-\varepsilon_{eff}(0)}{1+G(f/f_p)^2} \quad (10)$$

where $f_p=Z_0/(2\mu_0 h)$ may be regarded as an approximation to the first TE-mode cut-off frequency, while G is a factor which is empirically determined. Getsinger gave an expression for G which fitted experimental data for alumina substrates, but this did not give correct results for other substrates. The following expression for G has been shown to give very good results for all types of substrates now in use:

$$G=\frac{\pi^2}{12}\,\frac{\varepsilon_r-1}{\varepsilon_{eff}(0)}\sqrt{\frac{2\pi Z_0}{\eta_0}} \quad (11)$$

There is today not general agreement on a model for microstrip impedance dispersion. Based on a parallel-plate model and using the theory of dielectrics, the following equation was found.

$$Z_o(f)=Z_o(o)\sqrt{\frac{\varepsilon_{eff}(o)}{\varepsilon_{eff}(f)}}\cdot\frac{\varepsilon_{eff}(f)-1}{\varepsilon_{eff}(o)-1} \quad (12)$$

Reprinted from *IEEE Microwave Symposium*, pp. 407–409, 1980.

While we would accept theoretical objections to this model on the grounds that microstrip impedance is rather arbitrarily defined, it may be pointed out that it agrees very well with the calculations of Krage & Haddad [5] and that in the limit $f \to \infty$ it is agreement with three of the definitions discussed by Bianco & al. [6]. It may also be noted that the model predicts a rather small but positive increase in impedance with frequency and that this would seem to fit experimental observations.

Coupled microstrips

The equations given below represent the first generally valid model of coupled microstrips with an acceptable accuracy. The model is based upon perturbations of the homogeneous microstrip impedance equation and the equations for effective dielectric constant. The equations have been validated against theoretically calculated data in the range $0.1 \le u \le 10$ and $g \ge 0.01$ (g is the normalized gap width (s/h)), a range which should cover that used in practise.

The homogeneous mode impedances are

$$Z_{01m}(u,g) = Z_{01}(u)/[1 - Z_{01}(u)\phi_m(u,g)/\eta_0] \quad (13)$$

where the index m is set to e for the even mode and o for the odd. The effective dielectric constants are

$$\varepsilon_{em}(u,g,\varepsilon_r) = \frac{\varepsilon_r+1}{2} + \frac{\varepsilon_r-1}{2} F_m(u,g,\varepsilon_r) \quad (14)$$

$$F_e(u,g,\varepsilon_r) = \left[1 + \frac{10}{\mu(u,g)}\right]^{-a(\mu)b(\varepsilon_r)} \quad (15)$$

$$F_o(u,g,\varepsilon_r) = f_o(u,g,\varepsilon_r)\left(1 + \frac{10}{u}\right)^{-a(u)b(\varepsilon_r)} \quad (16)$$

The modifying equations are as follows

$$\phi_e(u,g) = \varphi(u)/\{\psi(g)[\alpha(g)u^{m(g)} + [1-\alpha(g)]u^{-m(g)}]\} \quad (17)$$

$$\varphi(u) = 0.8645u^{0.172} \quad (18)$$

$$\psi(g) = 1 + \frac{g}{1.45} + \frac{g^{2.09}}{3.95} \quad (19)$$

$$\alpha(g) = 0.5\exp(-g) \quad (20)$$

$$m(g) = 0.2175 + \left[4.113 + \left(\frac{20.36}{g}\right)^6\right]^{-0.251} + \frac{1}{323}\ln\frac{g^{10}}{1 + \left(\frac{g}{13.8}\right)^{10}} \quad (21)$$

$$\phi_o(u,g) = \phi_e(u,g) - \frac{\theta(g)}{\psi(g)}\exp[\beta(g)u^{-n(g)}\ln u] \quad (22)$$

$$\theta(g) = 1.729 + 1.175\ln\left(1 + \frac{0.627}{g + 0.327g^{2.17}}\right) \quad (23)$$

$$\beta(g) = 0.2306 + \frac{1}{301.8}\ln\frac{g^{10}}{1 + \left(\frac{g}{3.73}\right)^{10}} + \frac{1}{5.3}\ln(1 + 0.646g^{1.175}) \quad (24)$$

$$n(g) = \left\{\frac{1}{17.7} + \exp\left[-6.424 - 0.76\ln g - \left(\frac{g}{0.23}\right)^5\right]\right\} \cdot \ln\frac{10 + 68.3g^2}{1 + 32.5g^{3.093}} \quad (25)$$

$$\mu(u,g) = g\exp(-g) + u \cdot \frac{20+g^2}{10+g^2} \quad (26)$$

$$f_o(u,g,\varepsilon_r) = f_{o1}(g,\varepsilon_r) \cdot \exp[p(g)\ln u + q(g) \cdot \sin(\pi\frac{\ln u}{\ln 10})] \quad (27)$$

$$p(g) = \exp(-0.745g^{0.295})/\cosh(g^{0.68}) \quad (28)$$

$$q(g) = \exp(-1.366 - g) \quad (29)$$

$$f_{o1}(g,\varepsilon_r) = 1 - \exp\left\{-0.179g^{0.15} - \frac{0.328g^{r(g,\varepsilon_r)}}{\ln[\exp(1) + (g/7)^{2.8}]}\right\} \quad (30)$$

$$r(g,\varepsilon_r) = 1 + 0.15\left[1 - \frac{\exp(1 - (\varepsilon_r - 1)^2/8.2)}{1 + g^{-6}}\right] \quad (31)$$

When the development of the above equations was done (1) and (2) were not available. Instead a simpler equation for homogeneous microstrip impedance with a maximum error of 0.3% for $u > 0.06$ was used:

$$Z_{01}(u) = \eta_o/(u + 1.98u^{0.172}) \quad (32)$$

The errors in the even and odd mode impedances where then found to be less than 0.8% and less than 0.3% for the wavelengths. The above model does not include the effect of non-zero strip thickness or assymetry. It is a goal at ELAB to include these factors, although the parameter range probably will have to be restricted to the strictly practically useful in regard of the complexity of the above equations.

Dispersion is also not included presently. Getsinger[7] has proposed modifications to his single strip dispersion model, but unfortunately it is easily shown that the results are asymptotically wrong for extreme values of gap width. However, it is possible to modify the Getsinger dispersion model so that the asymptotic values are satisfied for coupled lines, but the details remain to be worked out.

Attenuation

The strip capacitive quality factor is

$$Q_d = \frac{(1-q) + q\varepsilon_r}{(1-q)/Q_A + q\varepsilon_r/Q_s} \quad (33)$$

where q is the mixed dielectric filling fraction, Q_A is the dielectric quality factor of the upper half space medium, and Q_s is the substrate quality factor.

The strip inductive quality factor is approximated by[8]:

$$Q_c = \frac{\pi Z_{01}h f}{R_s c} \cdot \frac{u}{K} \quad (34)$$

where R_s is the skin resistance and K is the current distribution factor. Surface roughness will increase the skin resistance[1].

$$R_s(\Delta) = R_s(0)\left[1 + \frac{2}{\pi}\,\mathrm{Arc\,tan}\,1.4\left(\frac{\Delta}{\delta}\right)^2\right] \quad (35)$$

where δ is the skin depth and Δ is the rms surface roughness.

For the single microstrip current distribution factor we have found

$$K = \exp(-1.2(\frac{Z_{01}(u)}{\eta_0})^{0.7}). \quad (36)$$

to be a very good approximation to the current distribution factor due to Pucel et al. [9] provided that the strip thickness exceeds three skin depths.

The mictrostrip quasi-TEM mode quality factor, Q_o, is now given by

$$\frac{1}{Q_o} = \frac{1}{Q_d} + \frac{1}{Q_c} \quad (37)$$

and the attenuation factor becomes

$$\alpha[\frac{dB}{m}] = \frac{20\Pi}{\ln 10} \frac{c}{Q_o f\sqrt{\varepsilon_{eff}}} \quad (38)$$

The above loss equations are also valid for coupled microstrips [8], provided that the dielectric filling factor, homogenious impedance, and current distribution factor of the actual mode are used. Presently, no equations are available for the odd and even mode current distribution factors. Except for very tight coupling, however, the following approximation gives good results

$$K_e = K_o = \exp[-1.2(\frac{Z_{ole}+Z_{olo}}{2\eta_o})^{0.7}] \quad (39)$$

Anistropy

From recent work [10,11] it may be shown that the above models are also valid with anisotropic substrates by defining an isotropyized substrate where

$$h_{eq} = h\sqrt{\varepsilon_x/\varepsilon_y} \quad (40)$$

$$\varepsilon_{eq} = \sqrt{\varepsilon_x \varepsilon_y} \quad (41)$$

This method requires that one of the substrate's principal axes is parallel to the substrate (x-axis) and the other normal to the substrate (y-axis). The strip dimensions remain unchanged, but are normalized with respect to h_{eq}. The procedure for utilizing the above models then proceed through calculation of the mode capacitances, C_m,

$$C_m = 1/(v_{eqm}\, Z_{eqm}) \quad (42)$$

Here v_{eqm} is the mode phase velocity and Z_{eqm} the mode impedance on the isotropyized substrate. Calculating the homogeneous mode capacitance, C_{ml}, the impedance and effective dielectric constant on the anistropic substrate are then

$$Z_{0m} = 1/(c\sqrt{C_m\, C_{ml}}) \quad (43)$$

$$\varepsilon_{effm} = C_m/C_{ml} \quad (44)$$

The inductive quality factor of the anisotropic substrate, may be directly calculated with the above equations, while the substrate quality factor has to be modified:

$$Q_s = 2(\frac{1}{Q_{sx}} + \frac{1}{Q_{sy}})^{-1} \quad (45)$$

Discontinuities

Due to the illnes, modelling of microstrip discontinuities could unfortunately not be finished to meet the digest deadline. The final paper presented at the symposium will give models for the microstrip open end, the gap, compensated right-angle bend, and the T-junction.

Acknowledgements

This work has been supported by the European Space Agency under contract no. 3745/78.

References

1. E.O.Hammerstad & F.Bekkadal:"Microstrip Handbook" ELAB-report, STF44 A74169, Feb. 1975, Trondheim.

2. E.O.Hammerstad:"Equations for Microstrip Circuit Design". Conference Proceedings, 5th. Eu.M.C., Sept. 75, Hamburg.

3. H.A.Weeler:"Transmission-Line Properties of a Strip on a Dielectric Sheet on a Plane". IEEE-trans. Vol. MTT-25, No. 8, Aug. 1977, p 631.

4. W.J.Getsinger:"Microstrip Dispersion Model" IEEE-trans., Vol. MTT-21, No.1, Jan. 1973, p. 34.

5. M.K.Krage & G.I.Haddad:"Frequency-Dependent Characteristics of Microstrip Transmission Lines" IEEE-trans, Vol. MTT-20, No. 10, Oct. 1972, P. 678.

6. B.Bianco et al.:"Some Considerations About the Frequency Dependence of the Characteristic Impedance of Uniform Microstrip". IEEE-trans, Vol. MTT-26, No. 3, March 1978, p. 182.

7. W.J.Getsinger: "Dispersion of Parallell-Coupled Microstrip" IEEE-trans. Vol. MTT-21, No. 3, March 1973, p. 144.

8. M.Mæsel:"A Theoretical and Experimental Investigation of Coupled Microstrip Lines". ELAB-report TE-168, April 1971, Trondheim.

9. R.A.Pucel et al.:"Losses in Microstrip". IEEE-trans., Vol. MTT-16, No. 6, June 1968, p. 342.

10. Ø.Jensen:"Single and Coupled Microstrip Lines on Anistropic Substrates" ELAB Project Memo No. 3/79, Project no.441408.04 June 1979, Trondheim.

11. M.Kobayashi & R.Terakado:"Accuratly Approximate Formula of Effective Filling Fraction for Microstrip Line with Isotropic Substrate and Its Application to the Case with Anistropic Substrate". IEEE-trans.,Vol.MTT-27,No 9,Sept.1979, p.776.

Accurate Wide-Range Design Equations for the Frequency-Dependent Characteristic of Parallel Coupled Microstrip Lines

MANFRED KIRSCHNING AND ROLF H. JANSEN, MEMBER, IEEE

Abstract —**In this paper, closed-form expressions are presented which model the frequency-dependent even- and odd-mode characteristics of parallel coupled microstrip lines with hitherto unattained accuracy and range of validity. They include the effective dielectric constants, the characteristic impedances using the power–current formulation, as well as the open-end equivalent lengths for the two fundamental modes on coupled microstrip. The formulas are accurate into the millimeter-wave region. They are based on an extensive set of accurate numerical data which were generated by a rigorous spectral-domain hybrid-mode approach and are believed to represent a substantial improvement compared to the state-of-the-art and with respect to the computer-aided design of coupled microstrip filters, directional couplers, and related components.**

I. Introduction

THE STATIC NUMERICAL solution of Bryant and Weiss in 1968 [1] provided one of the first reliable and accurate sources of information on coupled microstrip transmission-line characteristics. Their MSTRIP computer program [2] has been used by numerous authors as a reference, and has been validated by comparison with many other sources. Several years after the appearance of this computer program, the first frequency-dependent spectral domain analyses of coupled microstrip lines became available [3]–[7]. Today, the algorithms used to perform such computations have been developed to a higher maturity level, and efficient program packages exist which are suited for industrial application [8]–[13]. Using these, the frequency-dependent characteristics of coupled microstrip lines can be calculated to practically any required degree of accuracy.

Parallel to the mentioned rigorous computational efforts, a larger number of contributions have dealt with the description of coupled microstrip design data in the form of closed analytical expressions. A selection of the more representative papers on this topic is given in the references of this paper [14]–[25]. Since reviews of the state-of-the-art of coupled microstrip design formulas, up to about the end of 1979, can be found in three recent books on computer-aided microstrip circuit design [26]–[28], only the latest developments have to be recapitulated here. The accumulation of analytical approaches to the problem of describing coupled microstrips within the last few years can partially be explained as a consequence of microwave technology improvements, in so far as circuits and substrates of decreased tolerances justify and inspire descriptions of increased accuracy. These descriptions should preferrably be available as closed-form analytical models which are a requirement resulting from the growing application of computer-aided design tools in the microwave industry. Numerical algorithms, like those listed in [1]–[13], are accurate and are reliable sources of design information, but are too time-consuming for direct use in circuit optimization routines. Finally, the recent trend toward analytical modeling efforts on the whole has been accelerated by the rapid development of monolithic microwave integrated circuits during the last time period.

Today, it appears that the most accurate and generally valid static model of coupled microstrips has been given by Hammerstad and Jensen in 1980 [24]. The goal of these authors was to obtain results with errors at least less than those caused by physical tolerances. Recently March [25] verified the accuracy specifications of [24] through detailed comparison with the MSTRIP computer program [2] and the results reported by Jansen [10]. Also, comparison with the spectral-domain hybrid-mode program used in this paper [11], [13] confirms that the static equations presented by Hammerstad and Jensen generally have maximum errors less than 1 percent, except for some limiting situations. In addition, the computer program used here as a reference has been validated itself extensively by single and coupled microstrip dispersion measurements since its generation in 1978. Therefore, Hammerstad and Jensen's static equations can be judged to be superior to those of Garg and Bahl [22], who report a 3-percent (characteristic impedances) and 4-percent (effective dielectric constants) accuracy for their static semiempirical coupled microstrip expressions. Furthermore, the formulas of [24] are valid in an extended range of geometrical parameters and are essentially constructed to incorporate the correct asymptotic behavior with respect to these.

Nevertheless, there still exists no accurate frequency-

Manuscript received April 26, 1983; revised August 4, 1983.

M. Kirschning was with the University of Duisburg, Department of Electrical Engineering, FB9/ATE, Bismarckstr. 81, D-4100 Duisburg 1, West Germany. He is now with Honeywell GmbH, P.O. Box 1109, D-6457, Maintal, West Germany.

R. H. Jansen is with the University of Duisburg, Department of Electrical Engineering, FB9/ATE, Bismarckstr. 81, D-4100 Duisburg 1, West Germany.

Reprinted from *IEEE Trans. Microwave Theory Tech.*, vol. MTT-32, no. 1, pp. 83–90, January 1984.

dependent model for parallel coupled microstrip lines [24], [27]. Getsinger's dispersion model [15] can be shown to be asymptotically wrong for extreme values of gap width [24], and some of its limitations were reported already in 1973 [16]. This model gives relatively good results for alumina substrates [22], [25] for which it has been adjusted. However, it becomes inaccurate when a wide range of geometrical and substrate parameters is considered. In our experience, this limitation not only exists with respect to the modal effective dielectric constants, but also for the even- and odd-mode characteristic impedances of coupled microstrip lines. Furthermore, the ability of even a slightly modified version of Getsinger's dispersion model to describe the frequency dependence of coupled microstrip characteristic impedances with fair accuracy has again been demonstrated only for alumina [22], [25], and with respect to the voltage–current definition of impedance used in [10]. Theoretical and experimental work performed very recently indicates that the power–current formulation of characteristic impedance with its smaller frequency dependence should be preferred for microstrip computer-aided design [29]–[31]. Therefore, a thorough modification of Getsinger's coupled microstrip expressions as invoked in recent publications [24], [27] is performed here to provide accurate wide-range frequency-dependent design formulas. In addition, further refinement of part of the static equations of Hammerstad and Jensen [24] is performed in order to achieve better error margins as a start for wide-range modeling of dispersion. This paper sets forth a modeling approach previously applied to single microstrip lines [31]–[33]. For completeness, it includes formulas for the even- and odd-mode open-end equivalent lengths of the coupled microstrip section. The necessary reference data for these come from a three-dimensional hybrid-mode approach developed by Jansen [34], [35], which was meanwhile verified by Hornsby's results [36]. Therefore, the whole set of equations presented here is based upon reliable, validated, and cross-referenced hybrid-mode data.

II. Dispersive Coupled Microstrip Model

The analytical expressions which follow describe the effective dielectric constants, the power–current characteristic impedances, and the equivalent open-end lengths of coupled microstrip lines. The named quantities are functions of the substrate dielectric constant ϵ_r, the coupled microstrip cross-sectional geometry as depicted in Fig. 1, and are functions of frequency, except for the modal open-end length Δl_e and Δl_o. The latter are relatively small quantities and do not vary with frequency to a considerable degree in the usual range of applications up to about 18 GHz [34], [35].

The expressions given here have all been derived by successive computer matching to converged numerical results stemming from a rigorous spectral-domain hybrid-mode approach [11], [13], and [34]. The accuracies specified for them are with respect to the numerical data basis which, typically, consisted of several thousand test values.

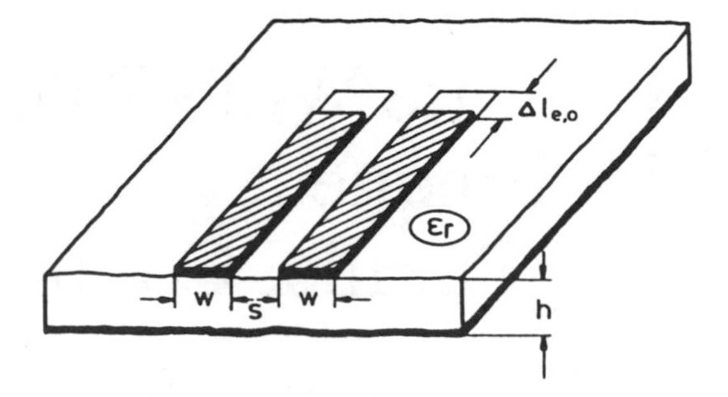

Fig. 1. Pictorial representation of coupled microstrip cross-sectional geometry and even- and odd-mode open-end equivalent lengths.

The range of validity to which the given accuracies apply is

$$0.1 \leqslant u \leqslant 10 \qquad 0.1 \leqslant g \leqslant 10 \qquad 1 \leqslant \epsilon_r \leqslant 18 \tag{1}$$

where $u = w/h$ denotes normalized strip width and $g = s/h$ is the normalized line spacing. This covers the typical range of technically meaningful parameters. Since the correct asymptotic behavior, with respect to the physical parameters, has been incorporated into the equations described here, the equations can even be applied beyond the limits of range (1), although with reduced accuracy. This is also true with respect to the frequency dependence, which is introduced here using the normalized quantity f_n, where

$$f_n = (\mathrm{f/GHz}) \cdot (\mathrm{h/mm}) \tag{2}$$

and the upper limit of which will be given separately for each set of expressions. The influence of a conducting cover plate or of lateral shielding walls is not taken into account in the derived expressions, i.e., the shielding is assumed to be far enough and of negligible effect on the line parameters. This is in agreement with usual microstrip circuit design practice. If corrections for small cover height appear to become necessary in a design, the correction formulas given by March [25] can be applied. Also, the finite thickness of the strip conductors is not accounted for in the formulas listed below. If conclusion of its effect is desired, for example, in the case of very small widths w or small spacings s, it is proposed to proceed in the way outlined in previous publications [10], [22], and [25]. The width correction introduced by Hammerstad and Jensen [24] is not recommended, since this exaggerates the effect of finite thickness on the characteristics; probably, there is a misprint in the named reference. Starting with the modeling process, an accuracy check of the static model of Hammerstad and Jensen [24] with respect to the reference basis used here [11], [13] shows that the static even-mode effective dielectric constant of [24], denoted by $\epsilon_{\mathrm{eff}_e}(0)$, is in error by not more than 0.7 percent for the range of applicability (1), while the corresponding odd-mode value, $\epsilon_{\mathrm{eff}_o}(0)$, is in error by about 4 percent in the case of low dielectric constants and large widths and spacings. For this reason, only the static value of $\epsilon_{\mathrm{eff}_e}(0)$ has been adopted here from [24] and is rewritten as

$$\epsilon_{\mathrm{eff}_e}(0) = 0.5(\epsilon_r + 1) + 0.5(\epsilon_r - 1) \cdot (1 + 10/v)^{-a_e(v) \cdot b_e(\epsilon_r)}$$

$$v = u(20+g^2)/(10+g^2)+g\cdot\exp(-g)$$
$$a_e(v) = 1+\ln\left(\left(v^4+(v/52)^2\right)/\left(v^4+0.432\right)\right)/49 + \ln\left(1+(v/18.1)^3\right)/18.7$$
$$b_e(\epsilon_r) = 0.564((\epsilon_r-0.9)/(\epsilon_r+3))^{0.053}. \quad (3)$$

In contrast to this, the static odd-mode parameter $\epsilon_{\text{eff}_o}(0)$ has been remodeled for an improved accuracy of 0.5 percent over the range (1) and is given by

$$\epsilon_{\text{eff}_o}(0) = (0.5(\epsilon_r+1)+a_o(u,\epsilon_r)+\epsilon_{\text{eff}}(0))\cdot\exp(-c_o\cdot g^{d_o})+\epsilon_{\text{eff}}(0)$$
$$a_o(u,\epsilon_r) = 0.7287(\epsilon_{\text{eff}}(0)-0.5(\epsilon_r+1))\cdot(1-\exp(-0.179u))$$
$$b_o(\epsilon_r) = 0.747\epsilon_r/(0.15+\epsilon_r)$$
$$c_o = b_o(\epsilon_r)-(b_o(\epsilon_r)-0.207)\cdot\exp(-0.414u)$$
$$d_o = 0.593+0.694\cdot\exp(-0.562u) \quad (4)$$

where the quantity $\epsilon_{\text{eff}}(0)$ without additional subscript refers to zero-thickness single microstrip [24] of width w. Frequency dispersion is introduced for both modes in the same form as has been done previously for single microstrip [33], namely by

$$\epsilon_{\text{eff}_{e,o}}(f_n) = \epsilon_r-(\epsilon_r-\epsilon_{\text{eff}_{e,o}}(0))/(1+F_{e,o}(f_n)). \quad (5)$$

This is a generalization of Getsinger's dispersion relation [15] and incorporates a more complicated form of frequency dependence in the terms denoted by $F_e(f_n)$ and $F_o(f_n)$. For the even mode on a coupled microstrip, this results in

$$F_e(f_n) = P_1P_2((P_3P_4+0.1844P_7)\cdot f_n)^{1.5763} \quad (6)$$

with

$$P_1 = 0.27488+\left(0.6315+0.525/(1+0.0157f_n)^{20}\right)\cdot u - 0.065683\cdot\exp(-8.7513u)$$
$$P_2 = 0.33622\cdot(1-\exp(-0.03442\epsilon_r))$$
$$P_3 = 0.0363\cdot\exp(-4.6u)\cdot\left(1-\exp\left(-(f_n/38.7)^{4.97}\right)\right)$$
$$P_4 = 1+2.751\cdot\left(1-\exp\left(-(\epsilon_r/15.916)^8\right)\right)$$
$$P_5 = 0.334\cdot\exp\left(-3.3(\epsilon_r/15)^3\right)+0.746$$
$$P_6 = P_5\cdot\exp\left(-(f_n/18)^{0.368}\right)$$

and

$$P_7 = 1+4.069P_6g^{0.479}\exp(-1.347g^{0.595}-0.17g^{2.5}).$$

For the odd-mode effective dielectric constant, the effect of dispersion is described by

$$F_o(f_n) = P_1P_2\cdot((P_3P_4+0.1844)\cdot f_n\cdot P_{15})^{1.5763} \quad (7)$$

with

$$P_8 = 0.7168(1+1.076/(1+0.0576(\epsilon_r-1)))$$
$$P_9 = P_8-0.7913\cdot\left(1-\exp\left(-(f_n/20)^{1.424}\right)\right)\cdot\arctan\left(2.481(\epsilon_r/8)^{0.946}\right)$$
$$P_{10} = 0.242\cdot(\epsilon_r-1)^{0.55}$$
$$P_{11} = 0.6366\cdot(\exp(-0.3401f_n)-1)\cdot\arctan\left(1.263(u/3)^{1.629}\right)$$
$$P_{12} = P_9+(1-P_9)/(1+1.183u^{1.376})$$
$$P_{13} = 1.695\cdot P_{10}/(0.414+1.605P_{10})$$
$$P_{14} = 0.8928+0.1072\cdot\left(1-\exp\left(-0.42(f_n/20)^{3.215}\right)\right)$$
$$P_{15} = \text{abs}\left(1-0.8928(1+P_{11})P_{12}\cdot\exp(-P_{13}\cdot g^{1.092})/P_{14}\right).$$

The upper frequency limit of these expressions, in conjunction with the range of applicability (1), is $f_n = 25$, i.e., 25 GHz for substrates of 1-mm thickness, and even higher for thinner substrates. Within this limit, the maximum error involved is not greater than 1.4 percent. Additional tests have shown that this error does not exceed if, for example, dielectric constants near $\epsilon_r = 12.9$ and a normalized frequency of $f_n = 30$ are considered. Note, however, that these equations describe only undisturbed coupled microstrip lines, and other limitations might become effective in a circuit.

For the coupled microstrip characteristic impedances, again the attempt was made to use Hammerstad and Jensen's static formulas [24] as a starting point. However, detailed test computations revealed that the error in these, as compared to the hybrid-mode computer program employed here [11], [13], increases to about 1.5 percent for special parameter combinations near the limits of (1). Therefore, their usage would present problems, in so far as they would not provide sufficient error margin for the later inclusion of frequency dependence. So, for the static values of the even- and odd-mode characteristic impedances of coupled microstrip lines further improved expressions have been derived. Specifically, for the even mode, the static characteristic impedance is

$$Z_{L_e}(0) = Z_L(0)\cdot(\epsilon_{\text{eff}}(0)/\epsilon_{\text{eff}_e}(0))^{0.5}\cdot 1/\left(1-(Z_L(0)/377\Omega)\cdot(\epsilon_{\text{eff}}(0))^{0.5}\cdot Q_4\right) \quad (8)$$

with

$$Q_1 = 0.8695\cdot u^{0.194}$$
$$Q_2 = 1+0.7519g+0.189\cdot g^{2.31}$$
$$Q_3 = 0.1975+\left(16.6+(8.4/g)^6\right)^{-0.387} + \ln\left(g^{10}/\left(1+(g/3.4)^{10}\right)\right)/241$$

$$Q_4 = (2Q_1/Q_2) \cdot \left(\exp(-g)\cdot u^{Q_3} + (2-\exp(-g))\cdot u^{-Q_3}\right)^{-1}.$$

The quantities without the subscript e in the main expression are again those for a zero-thickness single microstrip [24] of width w. Similarly, the odd-mode impedance is written

$$Z_{L_o}(0) = Z_L(0)\cdot\left(\epsilon_{\text{eff}}(0)/\epsilon_{\text{eff}_o}(0)\right)^{0.5} \cdot 1/\left(1-(Z_L(0)/377\Omega)\cdot(\epsilon_{\text{eff}}(0))^{0.5}\cdot Q_{10}\right) \quad (9)$$

with

$$Q_5 = 1.794 + 1.14\cdot\ln\left(1+0.638/(g+0.517g^{2.43})\right)$$
$$Q_6 = 0.2305 + \ln\left(g^{10}/\left(1+(g/5.8)^{10}\right)\right)/281.3 + \ln(1+0.598g^{1.154})/5.1$$
$$Q_7 = (10+190g^2)/(1+82.3g^3)$$
$$Q_8 = \exp\left(-6.5-0.95\ln(g)-(g/0.15)^5\right)$$
$$Q_9 = \ln(Q_7)\cdot(Q_8+1/16.5)$$
$$Q_{10} = Q_2^{-1}\cdot\left(Q_2Q_4 - Q_5\cdot\exp(\ln(u)\cdot Q_6\cdot u^{-Q_9})\right).$$

The accuracy of these new static expressions is better than 0.6 percent for both modes in the range of validity (1). Considering the frequency dependence of the characteristic impedances, the power–current formulation prevails by analogy to the treatment of a single microstrip [31]. Impedance dispersion, as resulting from numerical hybrid-mode computations [11], [13], is included for the even mode in the form

$$Z_{L_e}(f_n) = Z_{L_e}(0)\cdot\left(0.9408(\epsilon_{\text{eff}}(f_n))^{C_e} - 0.9603\right)^{Q_o} \cdot 1/\left((0.9408-d_e)(\epsilon_{\text{eff}}(f_n))^{C_e} - 0.9603\right)^{Q_o} \quad (10)$$

with

$$C_e = 1 + 1.275\left(1-\exp\left(-0.004625p_e\epsilon_r^{1.674}\cdot(f_n/18.365)^{2.745}\right)\right) - Q_{12} + Q_{16} - Q_{17} + Q_{18} + Q_{20}$$
$$d_e = 5.086q_e\cdot\left(r_e/(0.3838+0.386q_e)\right)\cdot\left(\exp(-22.2u^{1.92})/(1+1.2992r_e)\right)\cdot\left((\epsilon_r-1)^6/\left(1+10(\epsilon_r-1)^6\right)\right)$$
$$p_e = 4.766\cdot\exp(-3.228\cdot u^{0.641})$$
$$q_e = 0.016 + (0.0514\epsilon_r\cdot Q_{21})^{4.524}$$
$$r_e = (f_n/28.843)^{12}$$

and

$$Q_{11} = 0.893\cdot\left(1-0.3/(1+0.7(\epsilon_r-1))\right)$$
$$Q_{12} = 2.121\left((f_n/20)^{4.91}/\left(1+Q_{11}\cdot(f_n/20)^{4.91}\right)\right)\cdot\exp(-2.87g)\cdot g^{0.902}$$
$$Q_{13} = 1 + 0.038(\epsilon_r/8)^{5.1}$$
$$Q_{14} = 1 + 1.203(\epsilon_r/15)^4/\left(1+(\epsilon_r/15)^4\right)$$
$$Q_{15} = 1.887\cdot\exp(-1.5g^{0.84})\cdot g^{Q_{14}}\cdot\left(1+0.41(f_n/15)^3\cdot u^{2/Q_{13}}/(0.125+u^{1.626/Q_{13}})\right)^{-1}$$
$$Q_{16} = \left(1+9/\left(1+0.403(\epsilon_r-1)^2\right)\right)\cdot Q_{15}$$
$$Q_{17} = 0.394\cdot\left(1-\exp\left(-1.47(u/7)^{0.672}\right)\right)\cdot\left(1-\exp\left(-4.25(f_n/20)^{1.87}\right)\right)$$
$$Q_{18} = 0.61\cdot\left(1-\exp\left(-2.13(u/8)^{1.593}\right)\right)/(1+6.544g^{4.17})$$
$$Q_{19} = 0.21g^4\left((1+0.18g^{4.9})\cdot(1+0.1u^2)\left(1+(f_n/24)^3\right)\right)^{-1}$$
$$Q_{20} = \left(0.09+1/\left(1+0.1(\epsilon_r-1)^{2.7}\right)\right)\cdot Q_{19}$$
$$Q_{21} = \text{abs}\left(1-42.54g^{0.133}\cdot\exp(-0.812g)\cdot u^{2.5}/(1+0.033u^{2.5})\right).$$

In the above equations, $\epsilon_{\text{eff}}(f_n)$ denotes the single microstrip effective dielectric constant as described in [33] as a function of frequency. The auxiliary quantity Q_o also refers to single microstrips. It is the exponential term which appears in the description of single-line impedance dispersion in [31, eq. (5)] and is denoted by R_{17} there. Since it consists of a chain of several expressions, it is not repeated explicitly here. Instead, it is recommended that the previously given single-line expressions [31]–[33] and the equations presented here be used for implementation on a desktop computer as a whole. For the odd-mode on coupled microstrip lines, the frequency-dependent characteristic impedance is modeled by

$$Z_{L_o}(f_n) = Z_L(f_n) + \left(Z_{L_o}(0)\cdot\left(\epsilon_{\text{eff}_o}(f_n)/\epsilon_{\text{eff}_o}(0)\right)^{Q_{22}} - Z_L(f_n)Q_{23}\right)\cdot\left(1+Q_{24}+(0.46g)^{2.2}\cdot Q_{25}\right)^{-1}$$

with

$$Q_{22} = 0.925(f_n/Q_{26})^{1.536}/\left(1+0.3(f_n/30)^{1.536}\right)$$
$$Q_{23} = 1 + 0.005f_n\cdot Q_{27}\cdot\left(\left(1+0.812(f_n/15)^{1.9}\right)\cdot(1+0.025u^2)\right)^{-1}$$
$$Q_{24} = 2.506Q_{28}\cdot u^{0.894}\cdot\left((1+1.3u)f_n/99.25\right)^{4.29}\cdot(3.575+u^{0.894})^{-1}$$
$$Q_{25} = \left(0.3f_n^2/(10+f_n^2)\right)\cdot\left(1+2.333(\epsilon_r-1)^2/\left(5+(\epsilon_r-1)^2\right)\right)$$
$$Q_{26} = 30 - 22.2\left(((\epsilon_r-1)/13)^{12}/\left(1+3((\epsilon_r-1)/13)^{12}\right)\right) - Q_{29}$$

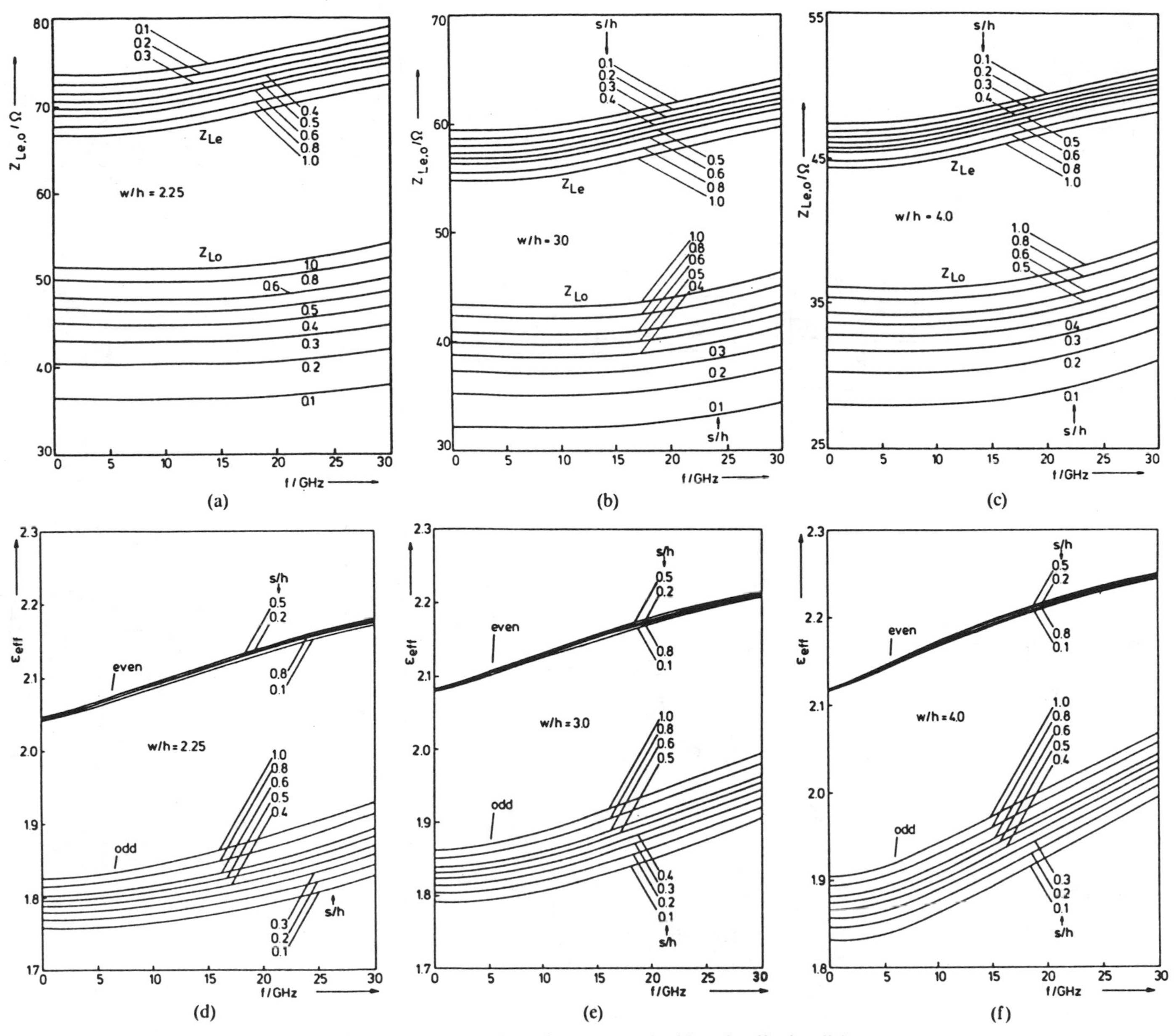

Fig. 2. The frequency-dependent even- and odd-mode effective dielectric constants and characteristic impedances of coupled microstrip lines on a plastic substate (RT-Duroid 5870, $\epsilon_r = 2.35$, $h = 0.79$ mm). (a), (d) $w/h = 2.25$. (b), (e) $w/h = 3.0$. (c), (f) $w/h = 4.00$.

$$Q_{27} = 0.4g^{0.84} \cdot \left(1 + 2.5(\epsilon_r - 1)^{1.5} / \left(5 + (\epsilon_r - 1)^{1.5}\right)\right)$$

$$Q_{28} = 0.149(\epsilon_r - 1)^3 / \left(94.5 + 0.038(\epsilon_r - 1)^3\right)$$

$$Q_{29} = 15.16 / \left(1 + 0.196(\epsilon_r - 1)^2\right).$$

The quantity $Z_L(f_n)$ is the frequency-dependent power–current characteristic impedance formulation of a single microstrip with width w [31]. The range of applicability (1) applies again, and the impedance equations (10) and (11) are valid up to $f_n = 20$, with a maximum error smaller than 2.5 percent. If the specified upper value of the substrate dielectric constant is reduced from 18 to 12.9, the expressions can be used up to $f_n = 25$ without a decrease in accuracy. This would correspond to about 40 GHz for a 25-mil-thick substrate. Modeling accuracy is typically better than 1.5 percent if usage is restricted to $\epsilon_r \leqslant 12.9$ and $f_n \leqslant 15$. As outlined for the effective dielectric constants, additional limitations may have to be regarded in an actual microstrip circuit.

Pictorial representation of (5)–(7), and (10), (11) is given in Figs. 2(a)–(f) and 3(a)–(f) for two widely used, commercially available substrates, namely RT-Duroid 5870 ($\epsilon_r = 2.35$, $h = 0.79$ mm) and alumina ($\epsilon_r = 9.70$, $h = 0.64$ mm). These are included as an immediate design aid and as a reference for the installation of the formulas on a computer. The line widths in Figs. 2 and 3 were chosen so that the equivalent single microstrip impedances, i.e., those for very loose coupling, are grouped around 50 Ω. With the

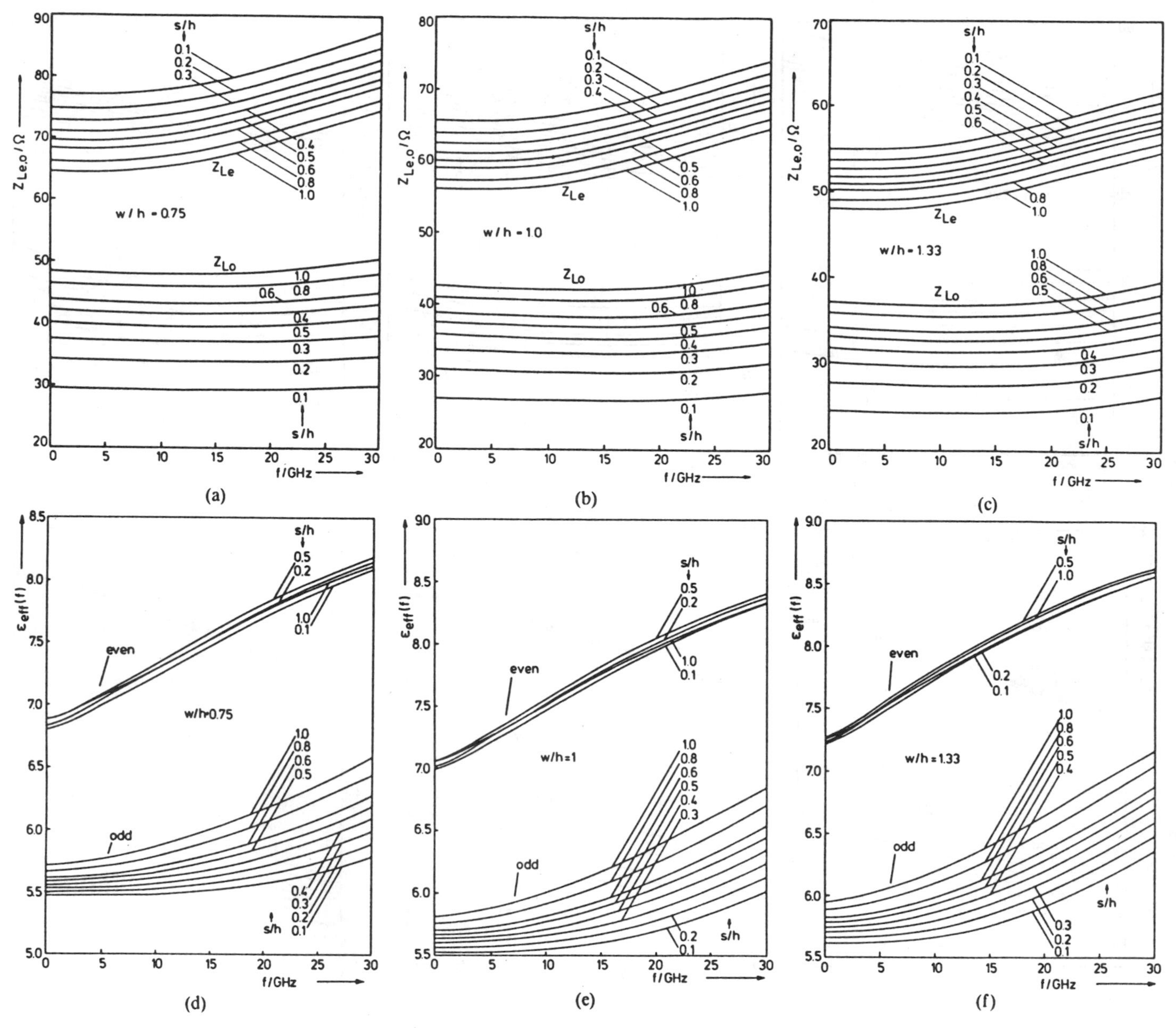

Fig. 3. The frequency dependent even- and odd-mode effective dielectric constants and characteristic impedances of coupled microstrip lines on a ceramic substrate (Alumina, $\epsilon_r = 9.70$, $h = 0.64$ mm). (a), (d) $w/h = 0.75$. (b), (e) $w/h = 1.0$. (c), (f) $w/h = 1.33$.

normalized line spacing $g = s/h$ as a parameter, these curves allow interpolation to obtain frequency-dependent even- and odd-mode impedance values for design.

Analytical expressions for the even- and odd-mode open-end equivalent lengths of the coupled microstrip section [37] are also reproduced here for completeness. The meaning of these quantities is illustrated in Fig. 1 and is analogous to the single microstrip open-end effect [32]. Their application for improved filter and coupler design has been outlined in a separate paper [37]. As for the single microstrip end effect, the inclusion of frequency dependence is not necessary for most design applications up to about 18 GHz [32], [34], and [35], and often even beyond that. In detail, the modal coupled microstrip equivalent end-effect lengths are described by

$$\Delta l_e = \left(\Delta l(2u, \epsilon_r) - \Delta l(u, \epsilon_r) + 0.0198 \cdot h \cdot g^{R_1}\right) \cdot \exp\left(-0.328 g^{2.244}\right) + \Delta l(u, \epsilon_r) \quad (12)$$

with

$$R_1 = 1.187 \cdot \left(1 - \exp\left(-0.069 u^{2.1}\right)\right)$$

and

$$\Delta l_o = \left(\Delta l(u, \epsilon_r) - h \cdot R_3\right) \cdot \left(1 - \exp\left(-R_4\right)\right) + h \cdot R_3 \quad (13)$$

with

$$R_2 = 0.343 \cdot u^{0.6187} + \left(0.45\epsilon_r/(1+\epsilon_r)\right) \cdot u^{(1.357 + 1.65/(1 + 0.7\epsilon_r))}$$

$$R_3 = 0.2974 \cdot (1 - \exp(-R_2))$$

$$R_4 = (0.271 + 0.0281\epsilon_r) \cdot g^{(1.167\epsilon_r/(0.66+\epsilon_r))} + (1.025\epsilon_r/(0.687+\epsilon_r)) \cdot g^{(0.958\epsilon_r/(0.706+\epsilon_r))}.$$

The end-effect quantities $\Delta l(u, \epsilon_r)$ and $\Delta l(2u, \epsilon_r)$, without subscript, represent single-line values for widths w and $2w$, respectively. The range of applicability is again defined by (1). The associated accuracy is 5 percent compared to the numerical hybrid-mode data basis [34], [35] employed for a fixed frequency of 4 GHz (actually, there is a slight increase with frequency). This should be accurate enough for most microstrip design purposes, since the length corrections (12), (13) themselves seldom contribute to the electrical length of a coupled microstrip section by more than 10–15 percent. It is observed that the even-mode equivalent end-effect length Δl_e decreases asymptotically to that of the corresponding single line Δl, if the spacing s/h is increased. Oppositely, the odd-mode length correction Δl_o approaches the single-line value Δl from below.

Coupled microstrip loss has not been considered here. The physical parameters required for such a computation, like surface roughness, dielectric loss tangent, and sheet resistivity, are not always known with good accuracy in practice. Therefore, approximate calculations, as found in several publications on the topic [10], [22], [24]–[28], seem adequate for many microstrip design cases. Also, sensitivity data for coupled microstrip lines has not been presented here, but is available in the technical literature [26], [38]–[40], or can be computed by the designer from the closed-form expressions provided here.

III. Conclusion

Novel frequency-dependent analytical expressions have been reported for the even- and odd-mode characteristics of parallel coupled microstrip lines. They are given in a form such that they can be implemented easily on a desktop computer or a programmable pocket calculator. In terms of their accuracy, which is specified for a very wide range of validity up to millimeter-wave frequencies, the described closed-form equations represent a considerable improvement to the foundations of coupled microstrip filter and coupler design. The equations are primarily set up for use in computer-aided microstrip circuit optimization routines, and are fully compatible with the needs and trends of modern computer-aided microwave integrated circuit design.

References

[1] T. G. Bryant, and J. A. Weiss, "Parameters of microstrip transmission lines and of coupled pairs of microstrip lines," *IEEE Trans. Microwave Theory Tech.*, vol. MTT-16, pp. 1021–1027, 1968.

[2] ——, "MSTRIP (parameters of microstrip). Comp. Prog. Des.," *IEEE Trans. Microwave Theory Tech.*, vol. MTT-19, pp. 418–419, 1971.

[3] G. Kowalski and R. Pregla, "Dispersion characteristics of single and coupled microstrips," *Arch. Elek. Übertragung (AEÜ)*, vol. 26, pp. 276–280, 1972.

[4] M. K. Krage and G. I. Haddad, "Frequency-dependent characteristics of microstrip transmission lines," *IEEE Trans. Microwave Theory Tech.*, vol. MTT-20, pp. 678–688, 1972.

[5] R. Pregla and G. Kowalski, "Simple formulas for the determination of the characteristic constants of microstrips," *Arch. Elek. Übertragung (AEÜ)*, vol. 28, pp. 339–340, 1974.

[6] R. H. Jansen, "Computer analysis of edge-coupled planar structures," *Electron. Lett.*, vol. 10, pp. 520–522, 1974.

[7] J. B. Knorr and K. D. Kuchler, "Analysis of coupled slots and coplanar strips on dielectric substrate," *IEEE Trans. Microwave Theory Tech.*, vol. MTT-23, pp. 541–548, 1975.

[8] J. B. Davies and D. Mirshekar-Syahkal, "Spectral domain solution of arbitrary coplanar transmission line with multilayer substrate," *IEEE Trans. Microwave Theory Tech.*, vol. MTT-25, pp. 143–146, 1977.

[9] R. H. Jansen, "Fast accurate hybrid mode computation of nonsymmetrical coupled microstrip characteristics," in *Proc. 7th European Microwave Conf.*, (Copenhagen, Denmark), 1977, pp. 135–139.

[10] ——, "High-speed computation of single and coupled microstrip parameters including dispersion, higher-order modes, loss and finite strip thickness," *IEEE Trans. Microwave Theory Tech.*, vol. MTT-26, pp. 75–82, 1978.

[11] ——, "Unified user-oriented computation of shielded, covered and open planar microwave and millimetre-wave transmission line characteristics," *Proc. Inst. Elec. Eng.*, vol. MOA-3, pp. 14–22, 1979.

[12] D. Mirshekar-Syahkal and J. B. Davies, "Accurate solution of microstrip and coplanar structures for dispersion and for dielectric and conductor losses," *IEEE Trans. Microwave Theory Tech.*, vol. MTT-27, pp. 694–699, 1979.

[13] R. H. Jansen, *Microstrip Computer Programs—Short Descriptions.* Aachen, W. Germany: Verlag H. Wolff, 1981.

[14] A. Schwarzmann, "Microstrip plus equations adds up to fast designs," *Electronics*, pp. 109–112, Oct. 2, 1967.

[15] W. J. Getsinger, "Dispersion of parallel-coupled microstrip," *IEEE Trans. Microwave Theory Tech.*, vol. MTT-21, pp. 144–145, 1973.

[16] B. Easter and K. C. Gupta, "More accurate model of the coupled microstrip line section," *Proc. Inst. Elec. Eng.*, vol. MOA-3, pp. 99–103, 1973.

[17] H. J. Carlin and P. P. Civalleri, "A coupled-line model for dispersion in parallel-coupled microstrips," *IEEE Trans. Microwave Theory Tech.*, vol. MTT-23, pp. 444–446, 1975.

[18] S. Akhtarzad *et al.*, "The design of coupled microstrip lines," *IEEE Trans. Microwave Theory Tech.*, vol. MTT-23, pp. 486–492, 1975.

[19] S. D. Shamasundara and N. Singh, "Design of coupled microstrip lines," *IEEE Trans. Microwave Theory Tech.*, vol. MTT-25, pp. 232–233, 1977.

[20] J. Zehntner, "Analysis of coupled microstrip lines," *Microwave J.*, pp. 82–83, May 1978.

[21] A. E. Ros, "Design charts for inhomogeneous coupled microstrip lines," *IEEE Trans. Microwave Theory Tech.*, vol. MTT-26, pp. 394–400, 1978.

[22] R. Garg and I. J. Bahl, "Characteristics of coupled microstrip lines," *IEEE Trans. Microwave Theory Tech.*, vol. MTT-27, pp. 700–705, 1979.

[23] J. H. Hinton, "On design of coupled microstrip lines," *IEEE Trans. Microwave Theory Tech.*, vol. MTT-28, p. 272, 1980.

[24] E. Hammerstad and O. Jensen, "Accurate models for microstrip computer-aided design," in *IEEE MTT-S Int. Microwave Symp. Dig.* (Washington, DC), 1980, pp. 407–409.

[25] S. March, "Microstrip packaging: Watch the last step," *Microwaves*, pp. 83–84, 87–88, 90, and 92–93, Dec. 1981.

[26] K. C. Gupta *et al.*, *Microstrip Lines and Slotlines.* Dedham, MA: Artech House, 1979, pp. 303–358.

[27] T. C. Edwards, *Foundations for Microstrip Circuit Design.* Chichester: Wiley, 1981, pp. 129–158, 221–230.

[28] K. C. Gupta *et al.*, *Computer-Aided Design of Microwave Circuits.* Dedham MA: Artech House, 1981, pp. 76–85, 120–126.

[29] R. H. Jansen and N. H. L. Koster, "New aspects concerning the definition of microstrip characteristic impedance as a function of frequency," in *IEEE MTT-S Int. Microwave Symp. Dig.* (Dallas, TX), 1982, pp. 305–307.

[30] W. J. Getsinger, "Measurement of the characteristic impedance of microstrip over a wide frequency range," in *IEEE MTT-S Int. Microwave Symp. Dig.* (Dallas, TX), 1982, pp. 342–344 and assoc. manuscript.

[31] R. H. Jansen and M. Kirschning, "Arguments and an accurate model for the power-current formulation of microstrip characteris-

tic impedance," *Arch. Elek. Übertragung (AEÜ)*, vol. 37, pp. 108–112, 1983.

[32] M. Kirschning *et al.*, "Accurate model for open end effect of microstrip lines," *Electron. Lett.*, vol. 17, pp. 123–125, 1981.

[33] M. Kirschning and R. H. Jansen, "Accurate model for effective dielectric constant of microstrip with validity up to millimetre-wave frequencies," *Electron. Lett.*, vol. 18, pp. 272–273, 1982.

[34] R. H. Jansen, "Hybrid mode analysis of end effects of planar microwave and millimeterwave transmission lines," *Proc. Inst. Elec. Eng.*, Pt. H, vol. 128, pp. 77–86, 1981.

[35] R. H. Jansen and N. H. L. Koster, "Accurate results on the end effect of single and coupled microstrip lines for use in microwave circuit design," *Arch. Elek. Übertragung (AEÜ)*, vol. 34, pp. 453–459, 1980.

[36] J. S. Hornsby, "Full-wave analysis of microstrip resonator and open-circuit end effect," *Proc. Inst. Elec. Eng.*, Pt. H, vol. 129, pp. 338–341, 1982.

[37] M. Kirschning *et al.*, "Coupled microstrip parallel gap model for improved filter and coupler design," *Electron. Lett.*, vol. 19, pp. 377–379, 1983.

[38] R. J. Roberts, "Effect of tolerances on the performance of microstrip parallel-coupled bandpass filters," *Electron. Lett.*, vol. 7, pp. 255–257, 1971.

[39] B. Ramo Rao, "Effect of loss and frequency dispersion on the performance of microstrip directional couplers and coupled filters," *IEEE Trans. Microwave Theory Tech.*, vol. MTT-22, pp. 747–750, 1974.

[40] C. Gupta, "Design of parallel coupled line filter with discontinuity compensation in microstrip," *Microwave J.*, vol. 22, pp. 39–40, 42–43, 57, 1979.

Part VII
End Effects and Discontinuity of Microstrip Lines

IN actual microstrip line circuits, a number of discontinuities appear. These discontinuities occur at the junctions between an active device and a passive transmission line and at the junctions formed by passive components only. Some of the structurally simpler discontinuities of the latter type include an open-ended microstrip line, a series gap in a microstrip line, a step discontinuity between the two microstrips with different strip widths, a T-junction, a bend, and a cross junction. Though structurally simple, these configurations belong to the class of problems which do not permit easy electromagnetic solutions. One of the reasons for difficulty comes from the fact that the discontinuity analysis seeks information on "local" phenomena. For instance, in the evaluation of the edge capacitance of an open ended microstrip line, this quantity is often obtained by taking one-half of the difference between the total capacitance of a finite, but reasonably long, microstrip line section and the capacitance per unit length multiplied by the strip length. The latter two quantities are of the same order. Hence, to obtain an accurate value of the edge capacitance, evaluations of the two latter quantities must be super-accurate. Although better methods to alleviate this difficulty have been developed (Gopinath and Gupta), the numerical labor is still much higher than that required for a uniform line analysis. Nevertheless, basic information on these discontinuities is needed before all but the simplest circuit structures are designed.

The papers in this part can be classified into three groups. The first two papers are concerned with the discontinuity inductances and capacitances evaluated based on the quasi-static approximation. Under this low frequency approximation, the electric field and the magnetic field are treated separately. Hence, instead of the dynamic wave equation for E and H, a static equation for E only or H only (or their respective potentials) is solved. Accuracy of the results by these methods becomes questionable as the operating frequency is increased. The second group of two papers incorporates wave scattering phenomena at the discontinuity in an approximate manner. The approximation introduces a hypothetical closed structure for which analytical formulations are more tractable. In spite of a rather heuristic approximation, the results appear to be reasonably accurate up to considerably high frequencies. The last group of papers treat the electromagnetic scattering problems at the discontinuities more rigorously. A particularly noteworthy trend in recent years is the significant efforts spent on the analysis of the radiation and leakage from microstrip discontinuities (last three papers). This trend is due to several factors. One of them is the use of increasingly high frequencies for millimeter-wave integrated circuits in which many components are located in close proximity of each other. Another reason is the use of high permittivity substrates such as GaAs.

It should be noted that the effort represented by the third group of papers is still in its infancy and many more dedicated efforts are needed.

Microstrip Discontinuity Inductances

BRIAN M. NEALE AND A. GOPINATH, MEMBER, IEEE

Abstract—**A method of calculation for microstrip discontinuity inductance parameters is described which overcomes various problems encountered in previous methods. A computer program has been written to implement this method, and typical calculated data is presented showing good agreement with experimental measurements.**

I. Introduction

ALTHOUGH several papers have provided a range of results for the capacitive component of microstrip discontinuities [1]–[8], there is still little data available in the literature for the inductive components. Early calculations for these parameters [9], [10] were apparently in error and not confirmed by experimental measurement. The magnetic wall model [11] modified from the method used for triplate lines suffers from a lack of theoretical justification. The inductive components of some discontinuity geometries have been calculated from an approximation-to-quasi-static theory based on the skin effect [12], [13]. The results obtained agree with experimentation only over a restricted range of discontinuities, probably because current continuity through the discontinuity is not always maintained. A second quasi-static formulation [14] sets the normal component of the magnetic field penetrating the strip to zero, which is the normal time-varying boundary condition when the skin effect is fully established. However, these authors used the moment method to obtain the solution, and, as may be seen from their paper, this technique suffers the disadvantage of requiring a large amount of computer store. Additionally, the method cannot be implemented easily for a range of discontinuity geometries.

The present paper follows the formulation set forth in [14] but uses the finite-element technique with polynomial approximations. Thus the computer store requirements are greatly reduced, and, in principle, any microstrip geometry may be studied.

In the following sections we discuss the formulation and method of solution and include some results on T junctions and right-angle bands with experimental measurements for comparison.

Manuscript received October 26, 1977; revised January 16, 1978. This work is supported by the U.K. Ministry of Defence, Procurement Executive under the CVD Directorate. B. M. Neale is supported by an assistantship under this Contract.

The authors are with the School of Electronic Engineering Science, University College of North Wales, Bangor, Gwynedd, U.K.

II. Method of Solution

The method adopted is to calculate the current distribution over the discontinuity structure, subject to appropriate governing and boundary conditions, and from this to obtain the inductance by an analysis of the stored energy in the structure. The quasi-static approximation is assumed neglecting radiation and retardation and implying continuity of current. It is also assumed that the skin effect is completely established. Thus the frequency dependence of the inductance components is not predicted in this work. The governing condition used here is that the normal component of the magnetic flux on the strip is zero which is the normal high-frequency boundary condition.

The discontinuity is described in terms of rectangular elements (as this enables most discontinuities to be described while retaining ease of integration) and semi-infinite lines, which are assumed to exist from $\pm\infty$ to reference planes within the discontinuity. For example, in Fig. 1 both semi-infinite lines terminate at the common plane ss' with no junction region between them. Secondary reference planes are defined at suitable distances from the discontinuity (for example, pp', qq' in Fig. 1) at which it is assumed that the perturbations due to the discontinuity are negligible and thus beyond which infinite line conditions may be assumed. Continuity of current is maintained through the structure.

The infinite line current distributions are calculated subject to the above governing condition; these currents are assumed to exist up to the planes of termination of the lines. A further current distribution is then calculated over the discontinuity region which redistributes the infinite line currents so that the total current at any point satisfies both governing and boundary conditions. The latter are that perturbations from the infinite line solution are negligible at the secondary reference planes, that current is confined to the conducting strip, and that the continuity of current is maintained across interelement boundaries.

Let us define a current potential $\overline{W}$ such that

$$\bar{J}=\nabla X \overline{W}$$

where $\bar{J}$ is the current density. We assume the conductor to be a lamina having no variation of current through its thickness. The current density distribution is then characterized by only two components, J_x and J_y. Therefore, it is sufficient for $\overline{W}$ to have a single z-directed component W_z, and hence

Reprinted from *IEEE Trans. Microwave Theory Tech.*, vol. MTT-26, no. 10, pp. 827–831, October 1978.

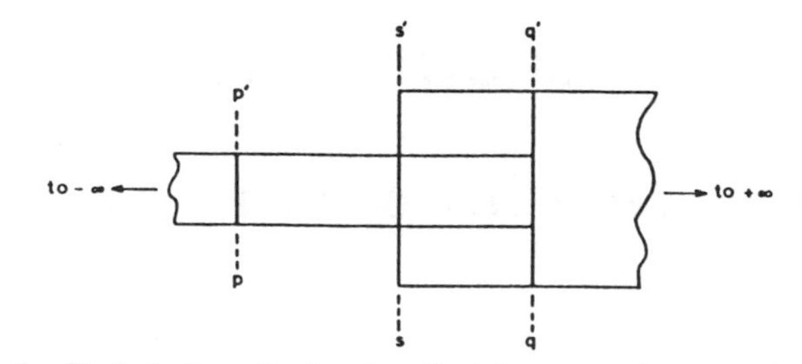

Fig. 1. Typical discontinuity described in terms of rectangular elements.

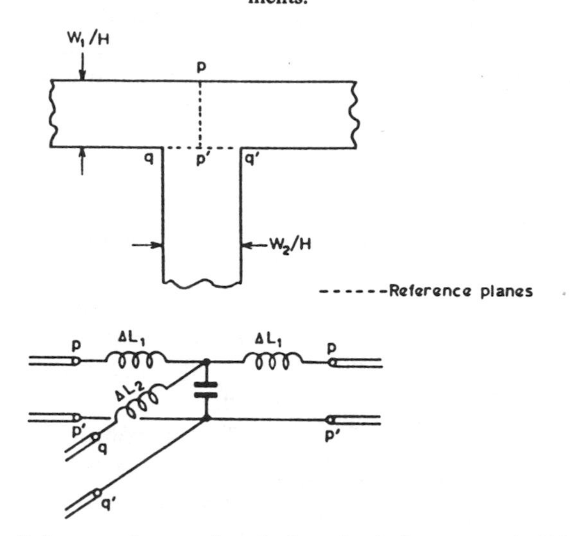

Fig. 2. Reference planes and equivalent circuit for symmetric T junction.

$$J_x=\frac{\partial}{\partial y}W_z$$

and

$$J_y=-\frac{\partial}{\partial x}W_z.$$

In addition,

$$\nabla\cdot\nabla X\overline{W}=0$$

hence

$$\nabla\cdot\bar{J}=0.$$

The current so defined is nondivergent and hence continuous.

Now, W_z is described within each element in terms of a trial function set $f_i\ (x,y)$ defined solely within that element, such that

$$W_z(x,y)=\sum_i a_i f_i(x,y)$$

where a_i is the set of unknown coefficients. The governing equation is set up in matrix form by application of the Galerkin method in terms of these unknown coefficients. Boundary conditions are applied by means of the generalized matrix inverse technique (see the Appendix), which provides a matrix equation relating the original coefficients to a subset of independent coefficients so that any arbitrary choice for these new coefficients will result in a solution satisfying the impressed boundary conditions. The resulting matrix equation is solved to give the independent coefficients, which are then used to evaluate the set a_i as shown in the Appendix.

The discontinuity inductance is calculated from the relationships

$$I=\int\bar{J}\,dS$$

and

$$I^2L=\int\bar{A}\cdot\bar{J}\,ds$$

where $\bar{A}$ is the magnetic vector potential. The advantage of describing the current distribution in terms of infinite line and excess (redistributing) components is found in these latter calculations. The discontinuity inductance would otherwise be found by subtracting the infinite line inductance between the secondary reference planes from the total inductance. As the discontinuity inductance is generally much smaller than these two quantities, large numerical errors may arise easily from taking the difference of two nearly equal numbers. Instead, the excess current is itself a measure of the excess inductance, and this enables a more accurate result to be calculated.

In performing the various integrations, both in the governing equation and in the final inductance calculation, mathematical singularities are encountered. These are all handled by coordinate transformation methods analogous to those used by Silvester and Benedek [16].

All numerical integrations are carried out by means of the Gaussian quadrature [17].

III. Results and Discussion

A computer program has been written to implement the method described above. This program will handle all

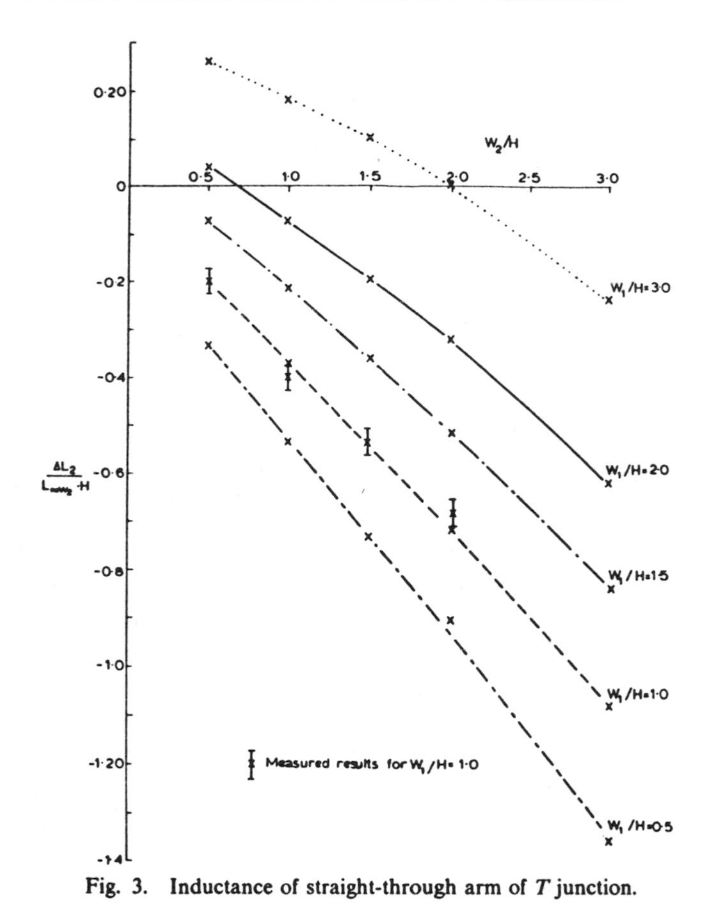

Fig. 3. Inductance of straight-through arm of T junction.

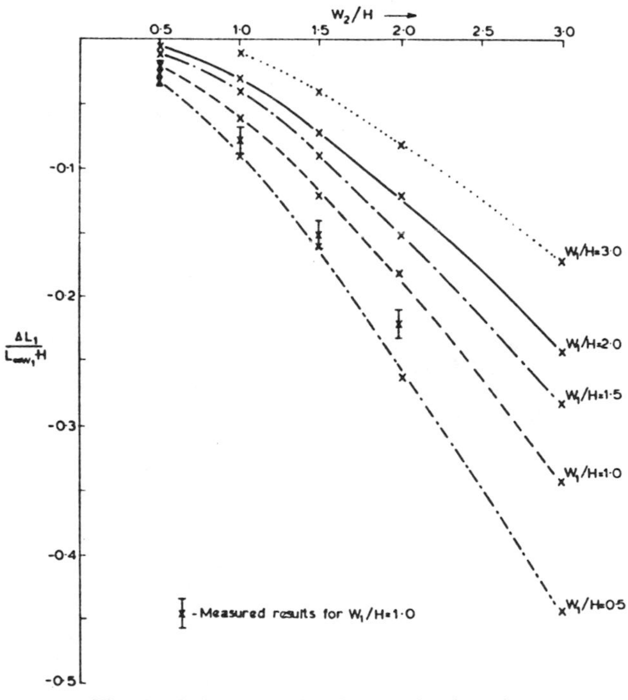

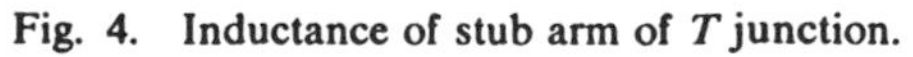

Fig. 4. Inductance of stub arm of T junction.

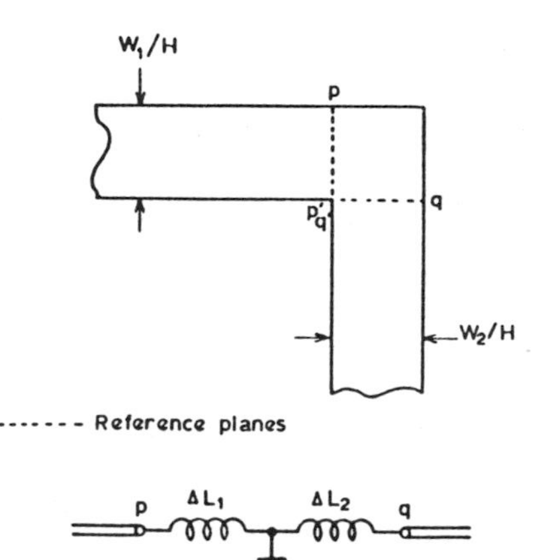

Fig. 5. Reference planes and equivalent circuit for right-angle corner.

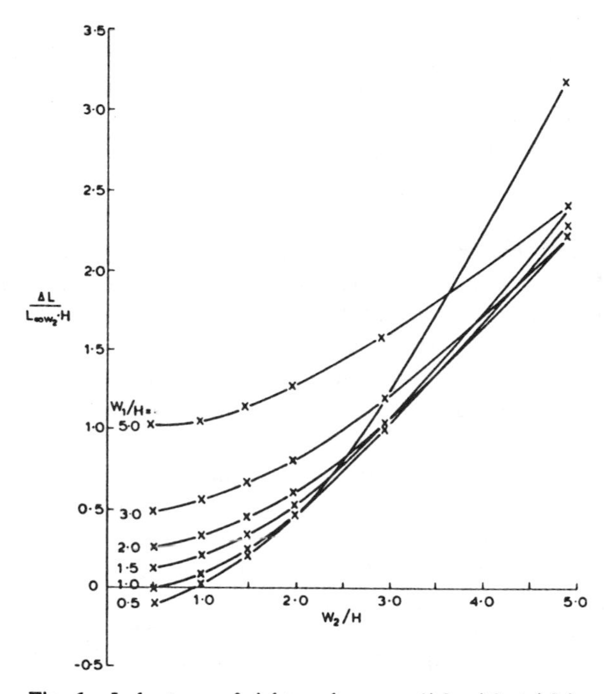

Fig. 6. Inductance of right-angle corner ($\Delta L = \Delta L_1 + \Delta L_2$).

discontinuities having rectangular geometries and involving changes in line widths of up to about 1:10. Typical data generated by this program on the T junction and right-angle bend discontinuities are given here. This information, together with the chosen reference planes and equivalent circuits assumed, is set forth in Figs. 2–6. All results are plotted normalized-to-infinite-line inductance per unit length (L_∞) and substrate height (h). A curve of L_∞ against w/h is given in Fig. 7 for convenience. In the case of the T junction, measured results are given for comparison [6]. Measured and calculated values generally agree to better than an equivalent line length of 15 μm on a 660-μm alumina substrate. This is about the estimated error in the measurements. It should be noted that these results are independent of substrate material provided that it is nonmagnetic.

The program will handle any discontinuity which may be described in terms of rectangular elements. In the case of an n-port junction, the discontinuity may be considered two ports at a time, leaving the others open circuit.

The trial function set used to date has been the bivariate monomial set $1, x, y, x^2, xy, y^2, \cdots$ taken to the fourth order. This set has been chosen because it is fast

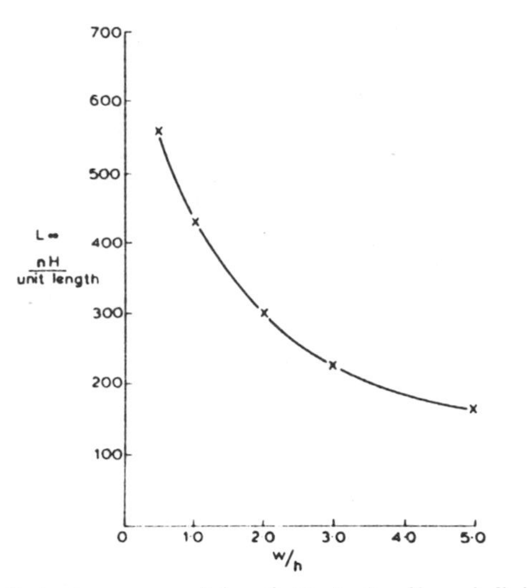

Fig. 7. Inductance per unit length (L_∞) of uniform infinite line.

and easy to evaluate at a given point. Its disadvantage is that with high-order approximations it may give poorly conditioned matrices, but this has not been found to be a problem in the present work.

IV. Conclusions

A finite-element method of calculating quasi-static inductive components of microstrip discontinuities has been outlined. A computer program to perform these calculations has been written, and the results obtained show good agreement with measurements where available. Since the formulation is based on quasi-static assumptions, the data is valid up to some maximum frequency. Experimental observations suggest that this is about 6–8 GHz on 660-μm alumina substrates (this frequency scales inversely with substrate thickness). Above this frequency, these results form a starting point for designs in the absence of other reliable data. When this data is used in conjunction with the capacitance data available elsewhere, a range of discontinuities is accurately characterized. Inductance results on other discontinuities are currently being collated and will be published in due course.

V. Appendix
Satisfying the Boundary Conditions

The solution of operator equations such as the governing equation described in this paper usually is required to satisfy additionally a set of boundary condition equations. In general, we have

$$[\bar{L}][\bar{a}]=[\bar{g}] \quad \text{(A1)}$$

and

$$[B][\bar{a}]=[\bar{b}] \quad \text{(A2)}$$

representing the governing equation (commonly set up by some weighted residual technique, for instance) and boundary condition equations, respectively. $[\bar{L}]$ is a square matrix and $[B]$ rectangular. Application of the generalized matrix inversion technique [18] to (A2) allows us to find further rectangular matrices $[N]$ and $[\bar{k}]$ such that

$$[\bar{a}]=[N][\bar{d}]+[\bar{k}] \quad \text{(A3)}$$

where $[N]$ and $[\bar{k}]$ may be determined from (A2). The vector $[\bar{d}]$ represents a new set of basis functions, independent of the boundary conditions. The original set $[\bar{a}]$ is expressed in terms of $[\bar{d}]$ such that any $[\bar{d}]$ together with a constant vector $[\bar{k}]$, which derives from the right-hand side of (A2), will give a vector $[\bar{a}]$ which satisfies the boundary conditions as originally set up in (A2). The rectangular matrix $[N]$ may be considered as a transformation matrix mapping the n-dimensional function space $[\bar{a}]$ into an $(n-m)$-dimensional function space $[\bar{d}]$ where m is the number of independent boundary conditions specified by $[B]$. Substitution of (A3) into (A1) together with premultiplication of both sides by $[N]$ leads to a system which may be solved for $[\bar{d}]$ and, hence, $[\bar{a}]$ obtained by back substitution in (A3). The resulting solution will, in general, satisfy the boundary conditions exactly and give an approximation to the operator equation. This technique has been found to be both powerful and flexible in the solution of operator equations such as that described in the body of this paper.

The technique has been implemented in the present case using a modified version of the IBM Scientific Subroutine Library package MFGR. Better behaved subroutines for such implementations have also been discussed by Peters and Wilkinson [19].

Acknowledgment

Our thanks are due to B. Easter and Prof. I. M. Stephenson for advice and guidance.

References

[1] A. Farrar and A. T. Adams, "Computation of lumped microstrip capacitances by matrix method—Rectangular sections and end effects," *IEEE Trans. Microwave Theory Tech.*, vol. MTT-19, pp. 495–496, May 1971.

[2] ——, "Matrix method for microstrip three-dimensional problems," *IEEE Trans. Microwave Theory Tech.*, vol. MTT-20, pp. 497–504, Aug. 1972.

[3] D. S. James and S. H. Tse, "Microstrip end effects," *Electron. Lett.*, vol. 8, pp. 46–47, 1972.

[4] I. Wolff, "Static capacitances of rectangular and circular microstrip disc capacitors," *Arch. Elek. Ubertragung.*, vol. 27, pp. 44–47, 1973.

[5] Y. Rahmat-Samii, T. Oh, and R. Mittra, "A spectral domain analysis for solving microstrip discontinuity problems," *IEEE Trans. Microwave Theory Tech.*, vol. MTT-22, pp. 372–375, Apr. 1971.

[6] B. Easter, "The equivalent circuit of some microstrip discontinuities," *IEEE Trans. Microwave Theory Tech.*, vol. MTT-23, pp. 655–660, Aug. 1975.

[7] C. Gupta and A. Gopinath, "Equivalent circuit capacitance of microstrip step change in width," *IEEE Trans. Microwave Theory Tech.*, vol. MTT-25, pp. 819–822, Oct. 1977.

[8] C. Gupta, B. Easter, and A. Gopinath, "Some results on the end effects of microstrip lines," to be published.

[9] R. Horton, "Electrical characterization of a right-angled bend in microstrip line," *IEEE Trans. Microwave Theory Tech.* vol. MTT-21, pp. 427–429, June 1973.

[10] ——, "Electrical representation of an abrupt impedance step in microstrip line," *IEEE Trans. Microwave Theory Tech.*, vol. MTT-21, pp. 562–564, Aug. 1973.

[11] I. Wolff, G. Kompa, and R. Mehran, "Calculation method for microstrip discontinuities and T junction," *Electron. Lett.*, vol. 8, pp. 177–79, 1972.
[12] A. Gopinath and P. Silvester, "Calculation of inductance of finite length strips and its variation with frequency," *IEEE Trans. Microwave Theory Tech.*, vol. MTT-21, pp. 380–386, June, 1973.
[13] A. Gopinath and B. Easter, "Moment method of calculating discontinuity inductance of microstrip right-angled bends," *IEEE Trans. Microwave Theory Tech.*, vol. MTT-22, pp. 880–883, Oct. 1974.
[14] A. F. Thomson and A. Gopinath, "Calculation of microstrip discontinuity inductances," *IEEE Trans. Microwave Theory Tech.*, vol. MTT-23, pp. 648–655, Aug. 1975.
[15] P. Silvester, "TEM-wave properties of microstrip transmission lines," *Proc. Inst. Elec. Eng.*, vol. 115, pp. 43–48, Jan. 1968.
[16] P. Silvester and P. Benedek, "Capacitance of parallel rectangular plates separated by a dielectric sheet," *IEEE Trans. Microwave Theory Tech.*, vol. MTT-20, pp. 504–510, Aug. 1972.
[17] Stroud and Secrest, *Gaussian Quadrature Formulas.* Englewood Cliffs, NJ: Prentice Hall, 1966.
[18] Z. Csendes, A. Gopinath and P. Silvester, "Generalised matrix inverse techniques for local approximations of operator equations," in *Mathematics of Finite Elements and Its Applications*, J. Whiteman, Ed. New York: Academic Press.
[19] G. Peters and J. H. Wilkinson, "The least-squares problem and pseudo-inverses", *Comput. J.*, vol. 13, pp. 308–316, Aug. 1970.

Capacitance Parameters of Discontinuities in Microstriplines

A. GOPINATH, MEMBER, IEEE, AND CHANDRA GUPTA, STUDENT MEMBER, IEEE

Abstract—Capacitance components of microstrip equivalent circuit discontinuities of gaps and the _T_ junctions have been calculated using the quasi-static formulation. Comparison of the calculated results with experimental measurements where available show good agreement. The range of data extends that data previously published, and new results on parallel gaps also are included.

I. Introduction

MICROSTRIP circuits comprise lengths of line and a variety of discontinuities, some examples of which are open circuits, gaps (series and parallel types), T junctions, cross junctions, corners, and step changes in width. The implementation of a particular design in microstrip is a cut-and-try process involving several trials before a satisfactory layout is obtained. This is largely due to the inadequate data on discontinuities. The present paper goes some way towards filling this gap by providing a range of data on the capacitance components of the equivalent circuits of some widely used discontinuities.

Manuscript received October 26, 1977; revised January 17, 1978. This work is supported by the U.K. Ministry of Defense, Procurement Executive under CVD Directorate. C. Gupta is supported by a U.C.N.W. scholarship.

The authors are with the School of Electronic Engineering Science, University College of North Wales, Bangor LL57 1UT, North Wales, U.K.

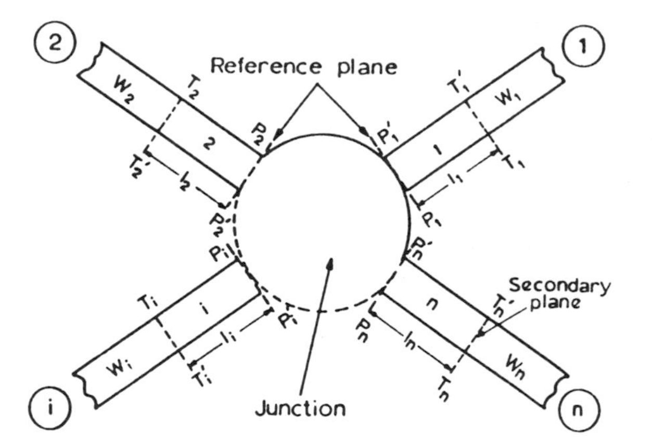

Fig. 1. General microstripline discontinuity.

Several methods of calculating the capacitances associated with discontinuities exist, ranging from quasistatic calculations using the moment method [1], the spectral domain technique [2], and the variational method [3] to the magnetic wall model [4]. The present set of data was generated using the quasi-static approach and the concept of excess charge due to Silvester and Benedek [5]–[7]. The advantage with this method is that the results

Reprinted from _IEEE Trans. Microwave Theory Tech._, vol. MTT-26, no. 10, pp. 831–836, October 1978.

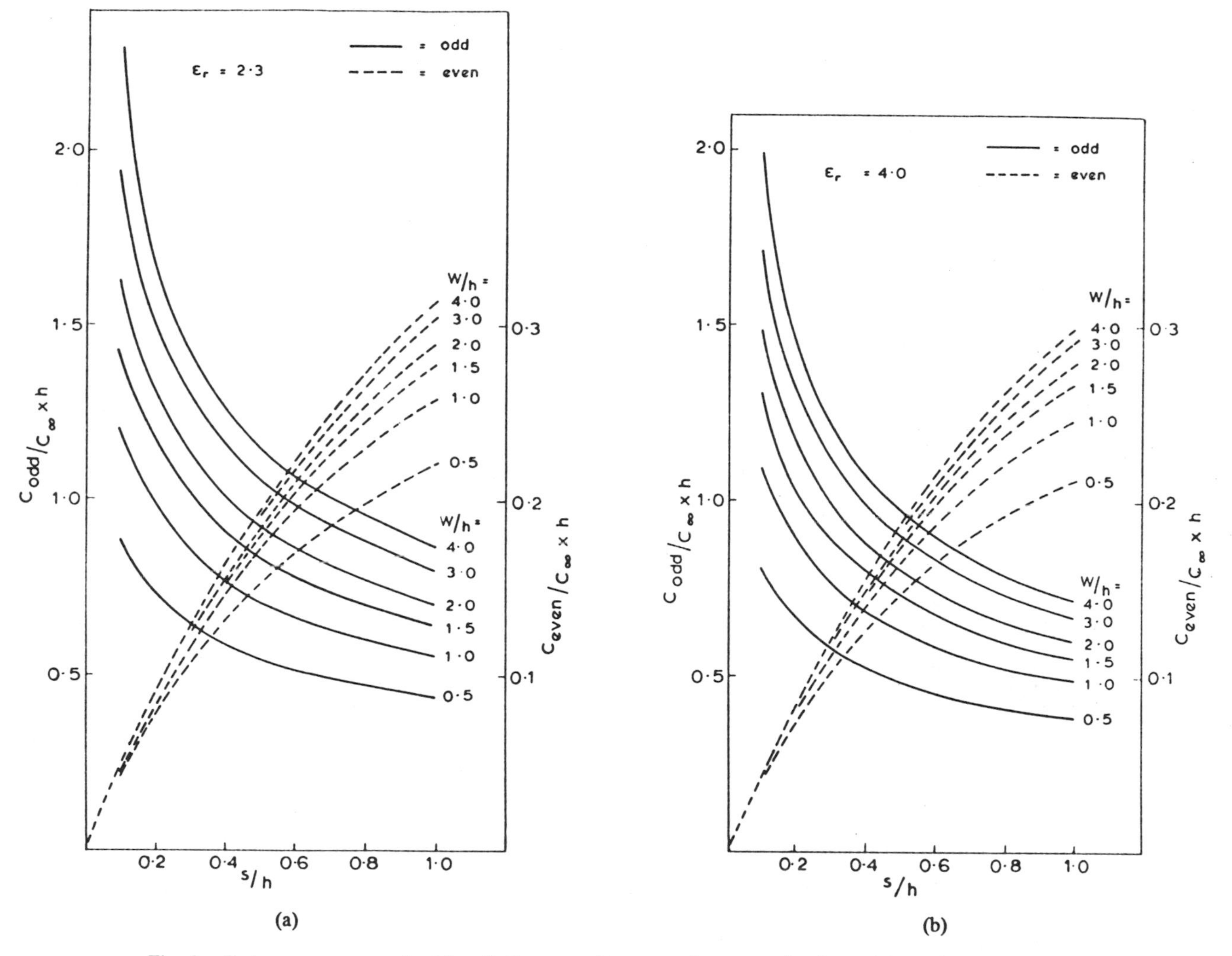

Fig. 2. Series gap even- and odd-excitation capacitances against normalized gap. (a) $\epsilon_r=2.3$, (b) $\epsilon_r=4.0$.

in our experience agree with experimental data. However, these quasi-static calculations do not predict the frequency dependence of this data. The alternative magnetic wall model, while predicting the frequency dependence, does not separate the discontinuity parasitics from the impedance change in the predicted S parameters.

For discontinuities where the capacitance effect is dominant, as in the series gap [8], the quasi-static data is valid up to about 17 GHz on 0.66-mm thick alumina substrates, but investigations into the validity of the results at frequencies beyond 17 GHz were not performed. In more complex junctions, for example, in the T junction, these results are valid up to a few GHz on 0.66-mm thick alumina. These quasi-static results provide a starting point for first designs. A companion paper [9] deals with the inductance component of T-junction equivalent circuits. Together, these papers fully characterize the T junctions.

Since details of the method have been discussed in a previous paper [10], only a brief outline is provided in Section II. Results on gaps and the T junction are given with comments on their agreement with experiment in Section III.

II. Formulation and Method of Solution

A microstrip discontinuity, in general, comprises a junction of semi-infinite lines together with a junction area, if any, as shown in Fig. 1. The discontinuity is represented by an equivalent circuit, defined between reference planes at which the semi-infinite lines terminate.

The method of solution of the discontinuity capacitance starts by assuming that the infinite line charge distributions extend from infinity to the respective reference plane P_iP_i' in Fig. 1. These charge distributions are estimated for the infinite lines for some constant potential on the strip, using the concept of partial images [11]. An additional charge is added in the junction region and near the junction on the infinite lines to maintain this same potential through the structure. This excess charge is assumed to exist over well-defined areas in the junction region, and in the present approach these are discretised into rectangular elements $P_iP_i'T_iT_i'$. It is assumed that beyond these areas and the secondary plane T_iT_i' the charge distributions on the semi-infinite lines remain unchanged from the corresponding infinite line ones. The excess charge is expressed over each element by expansion in a trial func-

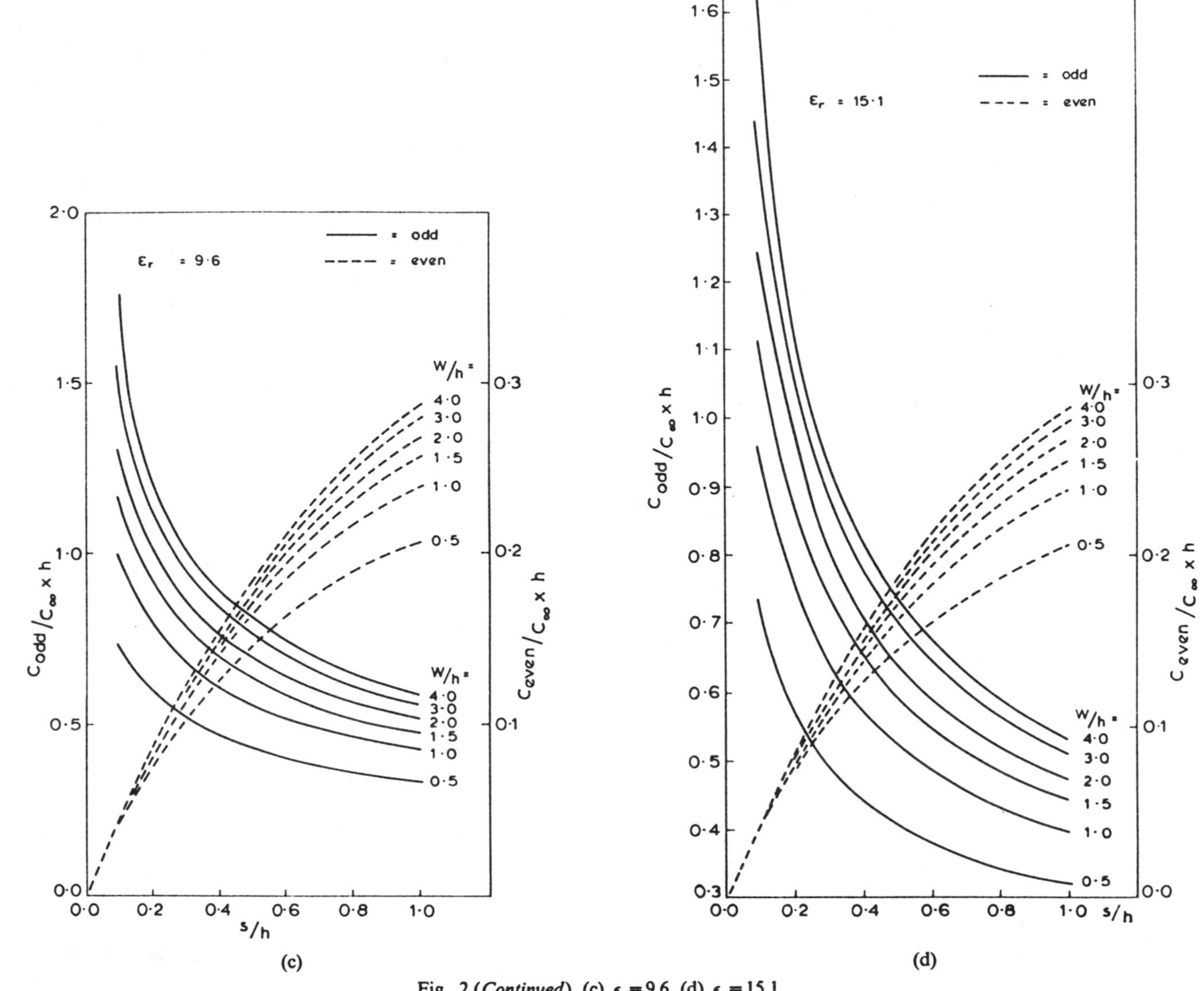

Fig. 2 (*Continued*). (c) $\epsilon_r = 9.6$, (d) $\epsilon_r = 15.1$.

tion set with unknown coefficients. The coefficients are determined by the requirement that the potential over the entire structure remains constant at the specified value. The equation to be solved takes the form

$$\sum_{i=1}^{n} \int_{P_iP_i'T_iT_i'} G_{ki}\sigma_{e_i}\, dS + \sum_{j=1}^{m} \int_{w_i} G_{\pm\infty/2}\, \sigma_{\infty_j}\, dl$$

$$+ \int_{\substack{\text{Junction}\\ \text{area}}} G_k J \sigma_{eJ}\, dS = \phi, \qquad \text{for } k = 1, \cdots, n \tag{1}$$

where σ_{ei} is the unknown excess charge distribution in the ith element, $\sigma_{\infty j}$ is the known infinite line charge distribution in the jth line, G_{ki} is the Green's function with the observation point in element k due to the excess charge in ith element [10], and $G_{\pm\infty/2}$ is the Green's function for the semi-infinite lines [10].

Now σ_{ei} is approximated in a set of interpolation polynomials valid only over the ith rectangle.

$$\sigma_{ei} = \sum_{j=1}^{m} a_j^i f_j^i(x,y). \tag{2}$$

The $f_j^i(x,y)$ are of the form

$$f_j^i(x,y) = \sum_{\substack{k=1\\ k_1 \neq j_1}}^{n_1} \frac{\left(x - x_{k_1}^i\right)}{\left(x_{j_1}^i - x_{k_1}^i\right)} \sum_{\substack{k_2=1\\ k_2 \neq j_2}}^{n_2} \frac{\left(y - y_{k_2}^i\right)}{\left(y_{j_2}^i - y_{k_2}^i\right)}$$

where n_1 and n_2 are the number of nodes along the x and y axis, respectively; $(x_{k_1}^i, y_{k_2}^i)$ are the equispaced nodes in the ith rectangle.

A matrix equation is obtained by taking the inner product of (1) with the same expansion function set term by term, and this equation is solved for the a_j^i. The discontinuity capacitance is now given by

$$C = \frac{\left(\sum_{i=1}^{n} \int \sigma_{ei}\, dS\right)^2}{\sum_{i=1}^{n} \int \phi \sigma_{ei}\, dS}. \tag{3}$$

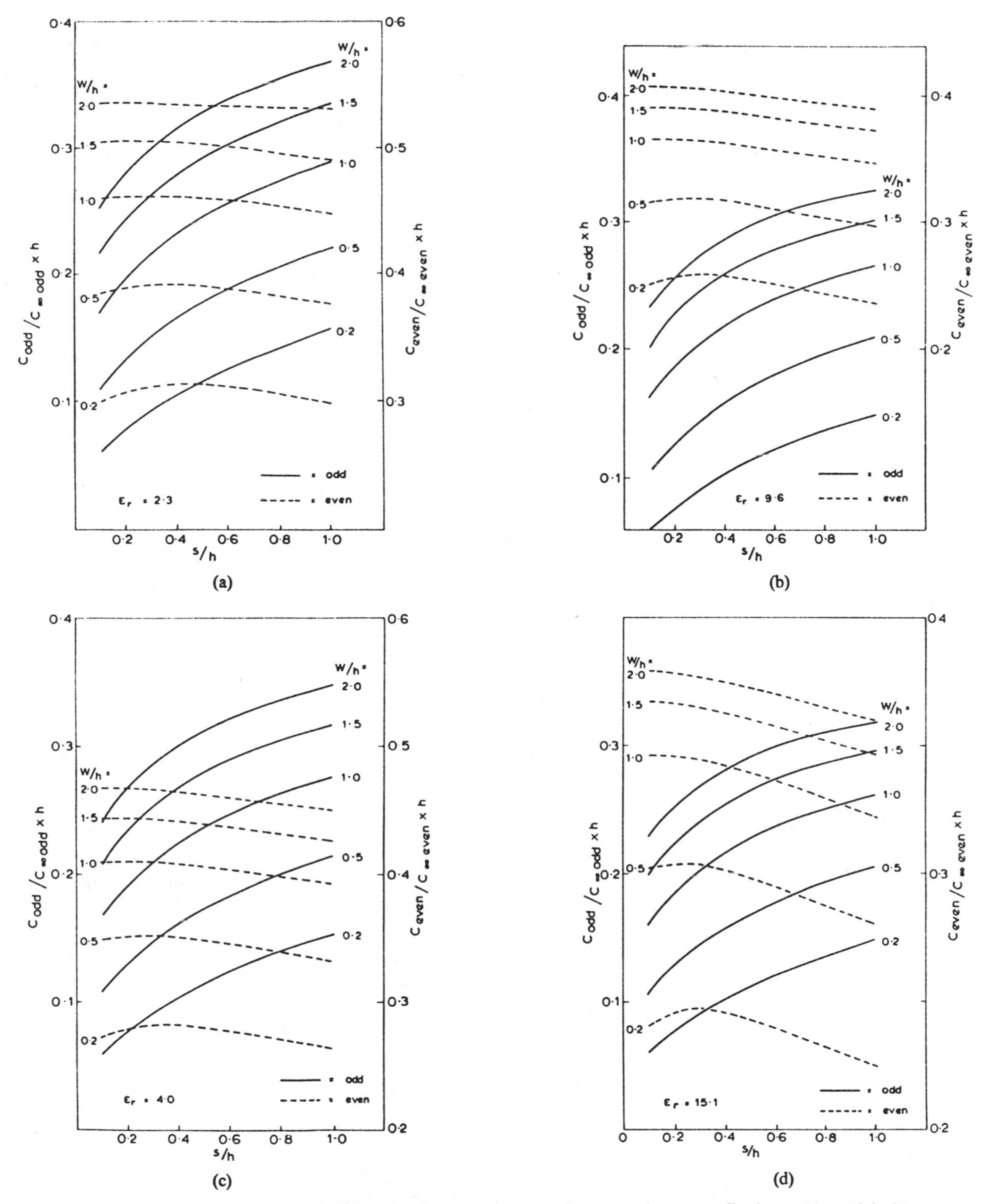

Fig. 3. Parallel gap even and odd mode of propagation capacitances against normalized gap. (a) $\epsilon_r = 2.3$, (b) $\epsilon_r = 4.0$, (c) $\epsilon_r = 9.6$, (d) $\epsilon_r = 15.1$.

The advantage of using the above interpolation polynomials in preference to the rather limited set of bivariate monomials used by other authors [5]–[7] is that these are better behaved and yield results which show closer agreement with experiment.

A computer program which is capable of examining all discontinuities with rectangular geometry has been written, and results generated for this paper are discussed below.

III. Results

In this paper, results on series gaps, parallel gaps, and T junctions are presented in Figs. 2–4.

These have been obtained for commonly used substrate dielectric constants, $\epsilon_r = 2.3$, 4.0, 9.6, 15.1, and the interpolation for the intermediate values of permittivities can be readily performed. Several runs were carried out to determine the degree of polynomial for consistent results

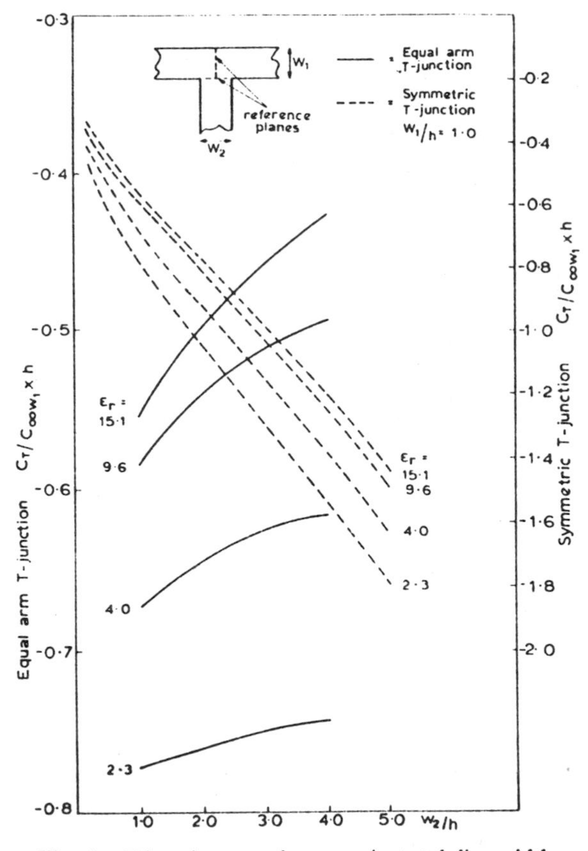

Fig. 4. T-junction capacitance against stub-line width.

in the interest of optimum time and accuracy, and these were the eleventh degree for the infinite line cases and the sixth degree in x and y for the excess charge distributions. The number of Gaussian-quadrature points in the integration were also determined in a similar manner. The length of the excess charge rectangles on the semi-infinite lines away from the junctions were also varied until consistent results were obtained in each test case. In all these cases care was taken to ensure that the results converged. Table I provides the equivalent circuits of the discontinuity geometries considered in this paper, and this is to be used with the above results.

The calculations on some discontinuities for $\epsilon_r = 9.6$ are compared with published data [1], [2], [5]–[7] and experiment [8], [12], [13] in Table II. The results from our program in most cases are somewhat higher than the previous results, show good agreement with experiment, and provide the basis for the validity of the data. As mentioned earlier, the results of the series gaps show excellent agreement [6] with the experiments over a wide frequency range. In the case of the T, good agreement exists with low-frequency measurements.

The low-frequency measurements were performed at 1 kHz using a GR-1616 bridge capable of measuring down to 10^{-15} F, again employing substitution techniques as used in the resonant measurements to preserve accuracy. The T results, in common with previous ones [5]–[7], show poorer agreement with the extrapolated values of Easter [9]. Note that in some cases the comparison is with

TABLE I
SOME DISCONTINUITY EQUIVALENT CIRCUITS

Type of discontinuity	Equivalent Circuit of discontinuity	Parameter evaluated
SERIES - GAP (P, T, P', T', s/h, w/h)	P, T, P', T', C_{12}, C_1, C_1	$C_{ODD} = 2C_{12} + C_1$; $C_{EVEN} = C_1$
PARALLEL - GAP (P, P', w/h, s/h, w/h)	P, P', P, P', C_{12}, C_1, C_1	$C_{ODD} = 2C_{12} + C_1$; $C_{EVEN} = C_1$
T - JUNCTION (w_1/h, w_2/h)	L_1, L_1, L_2, C_T	C_T

TABLE II
COMPARISON OF RESULTS

Reference	$C_{oc}/C_\infty \times h$, w/h = 0.5	1.0	2.0	$(C_{12}+C_1)/C_\infty \times h$, s/h = 0.3, w/h = 0.5	1.0	2.0	$C_1/C_\infty \times h$, w/h = 1.0	2.0	$C_{ODD}/C_{\infty ODD} \times h$ (w/h = 0.7, s/h = 0.6)	$C_{even}/C_{\infty even} \times h$	$C_T/C_\infty \times h$ (w/h = 1.0)	$C_*/C_\infty \times h$ (w/h = 1.0)
Present Program	0.260	0.324	0.378	0.295	0.377	0.466	0.893	1.93	0.208	0.331	-0.584	-0.200
[5]	0.268	0.311	0.364	0.307	0.409	0.466	0.906	1.916	-	-	-0.614	-0.190
[1]	0.244	0.270	0.321		0.321	-	-	-	-	-	-	-
[2]	0.228	0.259	0.310	-	0.335	-	-	-	-	-	-	-
[6] *1,2) resonators	-	-	-	0.296* ±.015	0.378* ±.015	0.455* ±.015	-	-	-	-	-	-
[10]	0.258 ±.015	0.318 ±.015	0.378 ±.015	0.303 ±.015	0.363 ±.015	0.417 ±.015	0.930 ±.03	2.05 ±.03	-	-	-0.425±0.03	0.364±0.040
[11]	-	-	-	-	-	-	-	-	0.233 ±.02	0.337 ±.02	-	-
GR1616 Measurements	-	-	-	-	-	-	-	-	-	-	-0.59±.04	-0.200±0.01

measurements on a $\epsilon_r = 9.8$ substrate which are expected to be only very slightly different from $\epsilon_r = 9.6$ results.

These curves have been normalized to the product of the infinite line capacitance and the substrate height. Information on the infinite line and coupled line capacitances may be obtained from standard computer programs [14] or from the authors.

ACKNOWLEDGMENT

Our thanks are due to Prof. I. M. Stephenson and B. Easter for advice and suggestions.

REFERENCES

[1] A. Farrar and A. T. Adams, "Matric method for microstrip three dimensional problems," *IEEE Trans. Microwave Theory Tech.*, vol. MTT-20, pp. 497–503, Aug. 1972.

[2] Y. Rahmat-Samii, T. Itoh, and R. Mittra, "A spectral domain analysis for solving microstrip discontinuity problems," *IEEE Trans. Microwave Theory Tech.*, vol. MTT-22, pp. 372–378, Apr. 1974.

[3] M. Meada, "An analysis of gap in microstrip transmission lines," *IEEE Trans. Microwave Theory Tech.*, vol. MTT-20, pp. 390–396, June 1972.

[4] W. Menzel and I. Wolff, "A method for calculating the frequency-dependent properties of microstrip discontinuities," *IEEE Trans. Microwave Theory Tech.*, vol. MTT-25, pp. 107–112, Feb. 1977.

[5] P. Silvester and P. Benedek, "Equivalent capacitances of microstrip open cicuits," *IEEE Trans. Microwave Theory Tech.*, vol.

MTT-20, pp. 390–395, June 1972.

[6] ——, "Equivalent capacitances for microstrip gaps and steps," *IEEE Trans. Microwave Theory Tech.*, vol. MTT-20, pp. 729–733, Nov. 1972.

[7] ——, "Microstrip discontinuity capacitances for right-angle-bends, *T* Junctions and Crossings," *IEEE Trans. Microwave Theory Tech.*, vol. MTT-21, pp. 341–346, May 1973.

[8] C. Gupta, B. Easter, and A. Gopinath, "Some results on the end effects of microstrip lines," in *IEEE Trans. Microwave Theory Tech.*, to be published.

[9] B. M. Neale and A. Gopinath, "Microstrip discontinuity inductances," *IEEE Trans. Microwave Theory Tech.*, this issue, pp. 827–831.

[10] C. Gupta and A. Gopinath, "Equivalent circuit capacitance of microstrip step change in width," *IEEE Trans. Microwave Theory Tech.*, vol. MTT-25, pp. 819–822, Oct. 1977.

[11] P. Silvester, "TEM wave properties of microstrip transmission lines," *Proc. Inst. Elec. Eng.* (*GB*), vol. 115, pp. 43–48, Jan. 1968.

[12] B. Easter, "The equivalent circuit of some microstrip discontinuities," *IEEE Trans. Microwave Theory Tech.*, vol. MTT-23, pp. 655–660, Aug. 1975.

[13] C. Gupta, "Characterization of the discontinuities present in the design of broadband wilkinson power divider (hybrid *T*) on microstrip," B.Sc. (Hons.) dissertation 1975, University College of North Wales, p. 22.

[14] T. G. Bryant and J. A. Weiss, "Parameters of microstrip transmission lines and of coupled pairs of microstrip lines," *IEEE Trans. Microwave Theory Tech.*, vol. 16, pp. 1021–1027, Dec. 1968.

A Method for Calculating the Frequency-Dependent Properties of Microstrip Discontinuities

WOLFGANG MENZEL AND INGO WOLFF

Abstract—A method is described for calculating the dynamical (frequency-dependent) properties of various microstrip discontinuities such as unsymmetrical crossings, T junctions, right-angle bends, impedance steps, and filter elements. The method is applied to an unsymmetrical T junction with three different linewidths. Using a waveguide model with frequency-dependent parameters, a field matching method proposed by Kühn is employed to compute the scattering matrix of the structures. The elements of the scattering matrix calculated in this way differ from those derived from static methods by a higher frequency dependence, especially for frequencies near the cutoff frequencies of the higher order modes on the microstrip lines. The theoretical results are compared with measurements, and theory and experiment are found to correspond closely.

I. Introduction

MICROSTRIP discontinuities such as crossings, T junctions, bends, and impedance steps are elements of many complex microstrip circuits like filters, power dividers, ring couplers, and impedance transformers. Therefore, knowledge of the exact reflection and transmission properties in dependence on the frequency is of great importance. Various approaches have been made to calculate equivalent circuits for those discontinuities. Oliner [1] used Babinet's principle to describe stripline discontinuities, Silvester and Benedek [2]–[4], and Stouten [5] calculated the capacitances of microstrip discontinuities, and Gopinath and Silvester [6], Gopinath and Easter [7], and Thomson and Gopinath [8] computed the inductive elements of the equivalent circuits. All the methods described in these papers are based on static approximations, and therefore are valid with sufficient accuracy only for low frequencies.

A method is presented in this paper for calculating the transmission properties of the discontinuities, taking into account the frequency-dependent energy stored in higher order cutoff modes of the microstrip line. A waveguide model for the microstrip line, described in [9], [10], and a field matching technique proposed by Kühn [11] are used. This method has the advantage that complex microstrip circuits containing discontinuities can be calculated in a way similar to that shown by Rozzi and Mecklenbräuker [12] for waveguide circuits recently. Earlier papers, which used the waveguide model, only described less complex structures, as for example symmetrical T junctions [17].

As comparisons of the theoretical results and the measurements show, the dependence of the scattering matrix of microstrip discontinuities on the frequency is well approximated in a wide frequency range by the theory given in this paper.

Manuscript received January 29, 1976; revised May 2, 1976.
The authors are with the Department of Electrical Engineering, University of Duisburg, Duisburg, Germany.

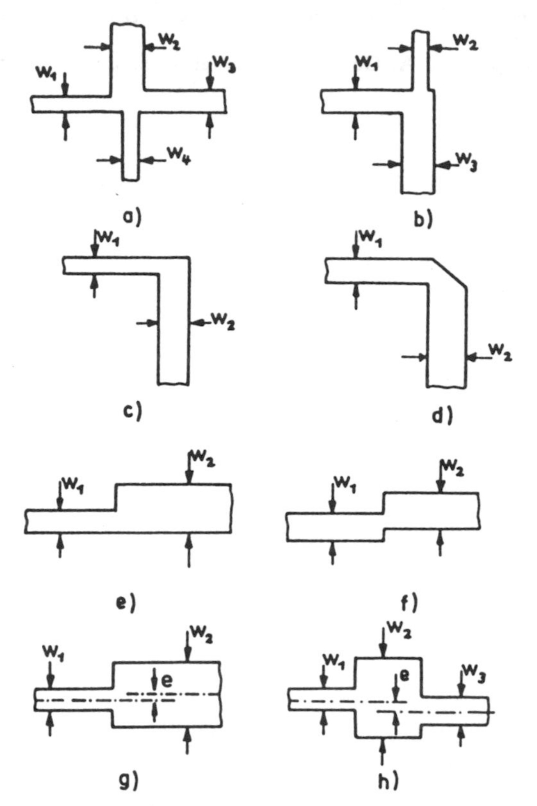

Fig. 1. Microstrip discontinuities which can be calculated by the theory described in this paper.

II. The Formulation of the Field Problem

It is the aim of this paper, to develop a field theoretical method for calculating the transmission properties of microstrip discontinuities, some of which are shown in Fig. 1.

For all discontinuities it is allowed that lines of different widths are connected to each other. In contrast to most of the theories described in previous papers, the method can be applied to unsymmetrical discontinuities. It is to be explained by the example of an unsymmetrical microstrip T junction (Fig. 3).

The dynamical properties of the microstrip lines are described using a waveguide model (Fig. 2). It consists of a parallel plate waveguide of width w_{eff} and height h with

Reprinted from *IEEE Trans. Microwave Theory Tech.*, vol. MTT-25, no. 2, pp. 107–112, February 1977.

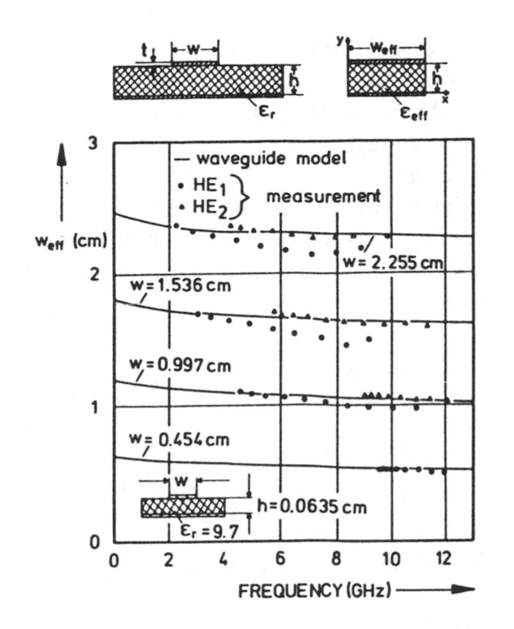

Fig. 2. Waveguide model of a microstrip line and effective width of the waveguide in dependence on the frequency. After [10].

plates of infinite conductivity at the top and bottom and with magnetic side walls. It is filled with a dielectric medium of the dielectric constant ε_{eff}. The effective width as well as the effective dielectric constant are frequency dependent. The height h of the waveguide model is equal to the height of the microstrip substrate material; ε_{eff} is the frequency-dependent effective dielectric constant as it can be computed for the microstrip line (e.g., [13]). The width of the waveguide model to a first approximation can be assumed to be equal to the frequency-independent effective width given by Wheeler [14]. As further investigations show [10], the effective width must also be frequency dependent. This is due to the physical fact that for higher frequencies the electromagnetic field is increasingly concentrated in the dielectric medium. The frequency-dependent effective width can be calculated from the characteristic impedance of the quasi-TEM mode (e.g., [15]), if the frequency-dependent effective dielectric constant is known. In Fig. 2 theoretical and experimental results for w_{eff} in dependence on the frequency are shown [10].

As in the case of the effective dielectric constant [16], a simple formula can be found to describe the frequency dependence of w_{eff} with sufficient accuracy in the relevant frequency range [17]

$$w_{eff}(f) = w + \frac{w_{eff}(0) - w}{1 + f/f_c} \tag{1}$$

where w is the width of the microstrip line, $w_{eff}(0)$ is the static value of the effective width according to Wheeler [14], and $f_c = c_0/(2w\varepsilon_r^{1/2})$, where ε_r is the dielectric constant of the substrate material.

It is assumed that the height h of the microstrip line and the waveguide model is so small that the fields of the waveguide model are independent of the y coordinate (Fig. 2) in the relevant frequency range. At the top and bottom the tangential electric field strength must vanish. So only a

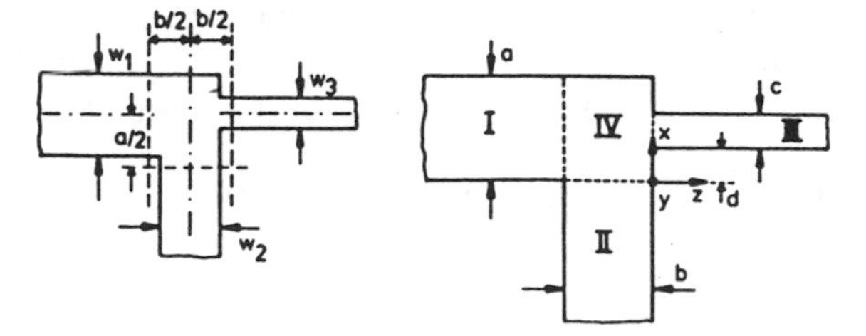

Fig. 3. Unsymmetrical microstrip T junction and introduced coordinate system.

TEM mode and TE_{n0} modes with E_y, H_x, and H_z components exist. The transversal electromagnetic field can be described using a scalar potential

$$\psi_{p0} = \sqrt{\frac{\varepsilon_p}{w_{eff}h}} \frac{\sin\left(\frac{p\pi}{w_{eff}} x\right)}{\frac{p\pi}{w_{eff}}}, \qquad \varepsilon_p = \begin{cases} 1 \text{ for } p = 0 \\ 2 \text{ for } p \neq 0 \end{cases} \tag{2}$$

by the equations

$$E_t = \sum_{p=0}^{\infty} (A_p e^{-\gamma z} + B_p e^{+\gamma z})(a_z \times \nabla_t \psi_{p0})$$

$$H_t = -\sum_{p=0}^{\infty} Y_{p0}(A_p e^{-\gamma z} - B_p e^{+\gamma z}) \nabla_t \psi_{p0}. \tag{3}$$

A_p and B_p are the amplitude coefficients, $\gamma = [(p\pi/w_{eff})^2 - \omega^2 \varepsilon_{eff}\varepsilon_0\mu_0]^{1/2}$ is the propagation constant, a_z is the unit vector of the z coordinate, ∇_t is the transversal Nabla operator, and $Y_{p0} = 1/Z_{p0} = \gamma/(j\omega\mu_0)$ is the complex characteristic wave admittance. Using normalized wave amplitudes and introducing the potential function, the transversal components of the electric and magnetic fields are given by

$$E_y = \sum_{p=0}^{\infty} (a_p e^{-\gamma z} + b_p e^{+\gamma z}) \sqrt{Z_{p0}} \sqrt{\frac{\varepsilon_p}{w_{eff}h}} \cos\left(\frac{p\pi}{w_{eff}} x\right)$$

$$H_x = -\sum_{p=0}^{\infty} (a_p e^{-\gamma z} - b_p e^{+\gamma z}) \sqrt{Y_{p0}} \sqrt{\frac{\varepsilon_p}{w_{eff}h}} \cos\left(\frac{p\pi}{w_{eff}} x\right) \tag{4}$$

where $a_p = Z_{p0}^{1/2} \cdot A_p$ and $b_p = Z_{p0}^{1/2} \cdot B_p$.

In Fig. 3 an unsymmetrical microstrip T junction and the equivalent waveguide circuit is shown. a, b, and c are the effective widths of the three microstrip lines. The waveguide T junction is divided into four regions. The regions I, II, and III are filled with a dielectric medium of effective dielectric constants ε_{eff1}, ε_{eff2}, and ε_{eff3}, corresponding to the three microstrip lines. In region IV an equivalent dielectric constant as defined in [18] for a microstrip disk capacitor is introduced, taking into account the electric stray field only at those sides of the region where no microstrip line is connected. The reference planes in the waveguide model are chosen to be the interfaces between region IV and the regions I, II, and III. Accordingly, the reference planes of the microstrip T junction are defined as shown in Fig. 3.

A complete solution for the electromagnetic field in the regions I, II, and III can be given by analogy with (4) taking into consideration the change of the coordinates:

$$E_y^{\mathrm{I}} = \sum_{p=0}^{\infty} \sqrt{Z_p^{\mathrm{I}}}(a_p^{\mathrm{I}} e^{-\gamma(z+b)} + b_p^{\mathrm{I}} e^{+\gamma(z+b)}) \cdot \sqrt{\frac{\varepsilon_p}{ah}} \cos\left(\frac{p\pi}{a} x\right)$$

$$H_x^{\mathrm{I}} = -\sum_{p=0}^{\infty} \sqrt{Y_p^{\mathrm{I}}}(a_p^{\mathrm{I}} e^{-\gamma(z+b)} - b_p^{\mathrm{I}} e^{+\gamma(z+b)}) \cdot \sqrt{\frac{\varepsilon_p}{ah}} \cos\left(\frac{p\pi}{a} x\right) \tag{5a}$$

$$E_y^{\mathrm{II}} = \sum_{k=0}^{\infty} \sqrt{Z_k^{\mathrm{II}}}(a_k^{\mathrm{II}} e^{-\gamma x} - b_k^{\mathrm{II}} e^{+\gamma x}) \sqrt{\frac{\varepsilon_k}{bh}} \cos\left(\frac{k\pi}{b} z\right)$$

$$H_z^{\mathrm{II}} = \sum_{k=0}^{\infty} \sqrt{Y_k^{\mathrm{II}}}(a_k^{\mathrm{II}} e^{-\gamma x} - b_k^{\mathrm{II}} e^{+\gamma x}) \sqrt{\frac{\varepsilon_k}{bh}} \cos\left(\frac{k\pi}{b} z\right) \tag{5b}$$

$$E_y^{\mathrm{III}} = \sum_{m=0}^{\infty} \sqrt{Z_m^{\mathrm{III}}}(a_m^{\mathrm{III}} e^{+\gamma z} + b_m^{\mathrm{III}} e^{-\gamma z}) \cdot \sqrt{\frac{\varepsilon_m}{ch}} \cos\left(\frac{m\pi}{c}(x-d)\right)$$

$$H_x^{\mathrm{III}} = \sum_{m=0}^{\infty} \sqrt{Y_m^{\mathrm{III}}}(a_m^{\mathrm{III}} e^{+\gamma z} - b_m^{\mathrm{III}} e^{-\gamma z}) \cdot \sqrt{\frac{\varepsilon_m}{ch}} \cos\left(\frac{m\pi}{c}(x-d)\right). \tag{5c}$$

Following Kühn [11], the field in region IV is found by superimposing three standing wave solutions:

$$E_y^{\mathrm{IV}a} = \sum_{p=0}^{\infty} \sqrt{\bar{Z}_p^{\mathrm{I}}}\, c_p^{\mathrm{IV}a} \cos(\beta^{\mathrm{I}} z) \cos\left(\frac{p\pi}{a} x\right) \sqrt{\frac{\varepsilon_p}{ah}}$$

$$H_x^{\mathrm{IV}a} = j \sum_{p=0}^{\infty} \sqrt{\bar{Y}_p^{\mathrm{I}}}\, c_p^{\mathrm{IV}a} \sin(\beta^{\mathrm{I}} z) \cos\left(\frac{p\pi}{a} x\right) \sqrt{\frac{\varepsilon_p}{ah}} \tag{6a}$$

$$E_y^{\mathrm{IV}b} = \sum_{k=0}^{\infty} \sqrt{\bar{Z}_k^{\mathrm{II}}}\, c_k^{\mathrm{IV}b} \cos(\beta^{\mathrm{II}}(x-a)) \cos\left(\frac{k\pi}{b} z\right) \sqrt{\frac{\varepsilon_k}{bh}}$$

$$H_z^{\mathrm{IV}b} = -j \sum_{k=0}^{\infty} \sqrt{\bar{Y}_k^{\mathrm{II}}}\, c_k^{\mathrm{IV}b} \sin(\beta^{\mathrm{II}}(x-a)) \cos\left(\frac{k\pi}{b} z\right) \cdot \sqrt{\frac{\varepsilon_k}{bh}} \tag{6b}$$

$$E_y^{\mathrm{IV}c} = \sum_{m=0}^{\infty} \sqrt{\bar{Z}_m^{\mathrm{I}}}\, c_m^{\mathrm{IV}c} \cos(\beta^{\mathrm{I}}(z+b)) \cos\left(\frac{m\pi}{a} x\right) \sqrt{\frac{\varepsilon_m}{ah}}$$

$$H_x^{\mathrm{IV}c} = j \sum_{m=0}^{\infty} \sqrt{\bar{Y}_m^{\mathrm{I}}}\, c_m^{\mathrm{IV}c} \sin(\beta^{\mathrm{I}}(z+b)) \cos\left(\frac{m\pi}{a} x\right) \sqrt{\frac{\varepsilon_m}{ah}}. \tag{6c}$$

β^{ν} ($\nu = \mathrm{I,II}$) are the phase constants of region I or II, respectively, calculated with the equivalent dielectric constant of region IV instead of ε_{eff1} or ε_{eff2}. In the same way $\bar{Z}_\nu^{\mu}$ is calculated from Z_ν^{μ}.

Matching the magnetic field strength of regions I, II, and III to that of region IV, in each case only one term of the superimposed field after (6) must be taken into account, because of the boundary conditions of the magnetic walls. The connections between the field amplitudes of regions I and IV, and II and IV, can be found simply by comparison of the coefficients

$$c_p^{\mathrm{IV}a} = -j \frac{a_p^{\mathrm{I}} - b_p^{\mathrm{I}}}{\sin(\beta_p^{\mathrm{I}} b)} \sqrt{\frac{\bar{Z}_p^{\mathrm{I}}}{Z_p^{\mathrm{I}}}} \tag{7}$$

and

$$c_k^{\mathrm{IV}b} = -j \frac{a_k^{\mathrm{II}} - b_k^{\mathrm{II}}}{\sin(\beta_k^{\mathrm{II}} a)} \sqrt{\frac{\bar{Z}_k^{\mathrm{II}}}{Z_k^{\mathrm{II}}}} \tag{8}$$

whereas the coefficients of regions III and IV must be determined by a normal mode matching procedure at the interface III–IV:

$$c_M^{\mathrm{IV}c} = -j \sum_{m=0}^{\infty} \sqrt{\frac{\bar{Z}_M^{\mathrm{I}}}{Z_m^{\mathrm{III}}}} \frac{a_m^{\mathrm{III}} - b_m^{\mathrm{III}}}{\sin(\beta_M^{\mathrm{I}} b)} \cdot K_{[m,M]}^{(1)} \tag{9}$$

with

$$K_{[m,M]}^{(1)} = \frac{\sqrt{\varepsilon_m \varepsilon_M}}{\sqrt{ac \cdot h}} \iint_{z=0} \cos\left(\frac{m\pi}{c} x\right) \cos\left(\frac{M\pi}{a} x\right) dx\, dy.$$

If the electric field strength of regions I, II, and III is to be matched to the electric field of region IV, the tangential components of all three standing wave solutions [see (6)] have to be taken into account. This will be shown by the example of the field matching between regions I and IV. The tangential electric field of region IV in the plane $z = -b$ [Fig. 3(b)] is given by

$$E_y^{\mathrm{IV}}|_{z=-b} = E_y^{\mathrm{IV}a}|_{z=-b} + E_y^{\mathrm{IV}b}|_{z=-b} + E_y^{\mathrm{IV}c}|_{z=-b}. \tag{10}$$

Matching this field to that of region I leads to an equation which connects the amplitude coefficients of regions I and IV:

$$\sqrt{Z_P^{\mathrm{I}}}(a_P^{\mathrm{I}} + b_P^{\mathrm{I}}) = \sqrt{\bar{Z}_P^{\mathrm{I}}}\, c^{\mathrm{IV}a} \cos(\beta_P^{\mathrm{I}} b) + \sqrt{\bar{Z}_P^{\mathrm{I}}}\, c_P^{\mathrm{IV}c} + \sum_{k=0}^{\infty} \sqrt{\bar{Z}_k^{\mathrm{II}}}\, c_k^{\mathrm{IV}b} K_{[k,P]}^{(2)} \tag{11}$$

with

$$K_{[k,P]}^{(2)} = (-1)^P \iint_{z=-b} \sqrt{\frac{\varepsilon_k \varepsilon_P}{ba}} \frac{1}{h} \cdot \cos(\beta_k^{\mathrm{II}}(x-a)) \cos\left(\frac{P\pi}{a} x\right) dx\, dy.$$

In the same way the electric fields of regions II and III are matched to that of region IV.

From (7)–(9) the coefficients $c_\nu^{\mathrm{IV}a}$, $c_\nu^{\mathrm{IV}b}$, and $c_\nu^{\mathrm{IV}c}$ can be eliminated and introduced into the matching conditions of the electric field strength. In this way a set of $P + K + M + 3$ equations results, which connects the amplitudes $a_\nu^{\mathrm{I}}, a_\mu^{\mathrm{II}}, a_\eta^{\mathrm{III}}$ ($\nu,\mu,\eta = 0,1,2,\cdots$) of the incident waves to the coefficients $b_\nu^{\mathrm{I}}, b_\mu^{\mathrm{II}}, b_\eta^{\mathrm{III}}$ of the reflected or transmitted waves.

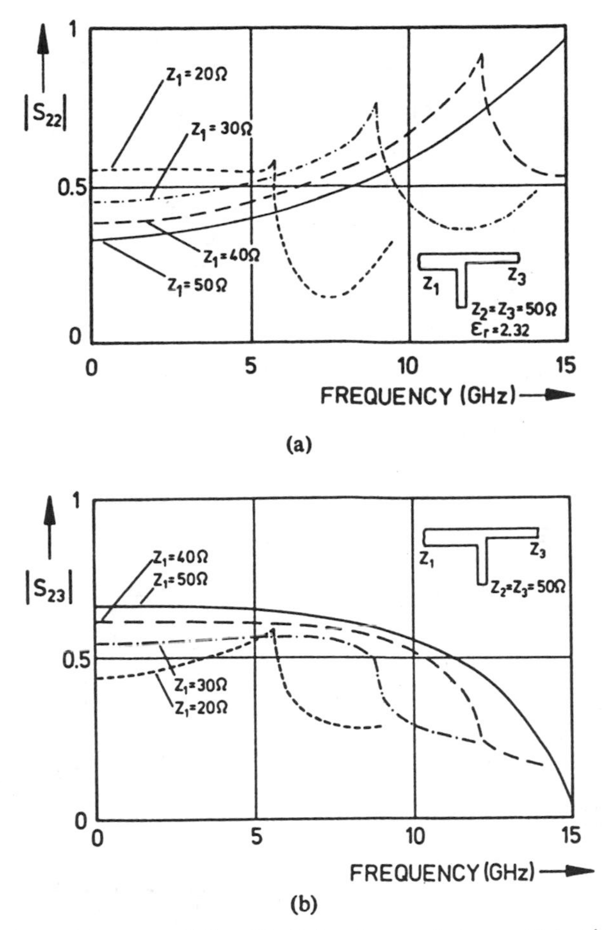

Fig. 4. Reflection coefficient $|S_{22}|$ (a) and transmission coefficient $|S_{23}|$ (b) of an unsymmetrical T junction with $Z_2 = Z_3 = 50\ \Omega$ and different values of Z_1. Substrate material Polyguide: $\varepsilon_r = 2.32$, $h = 0.158$ cm.

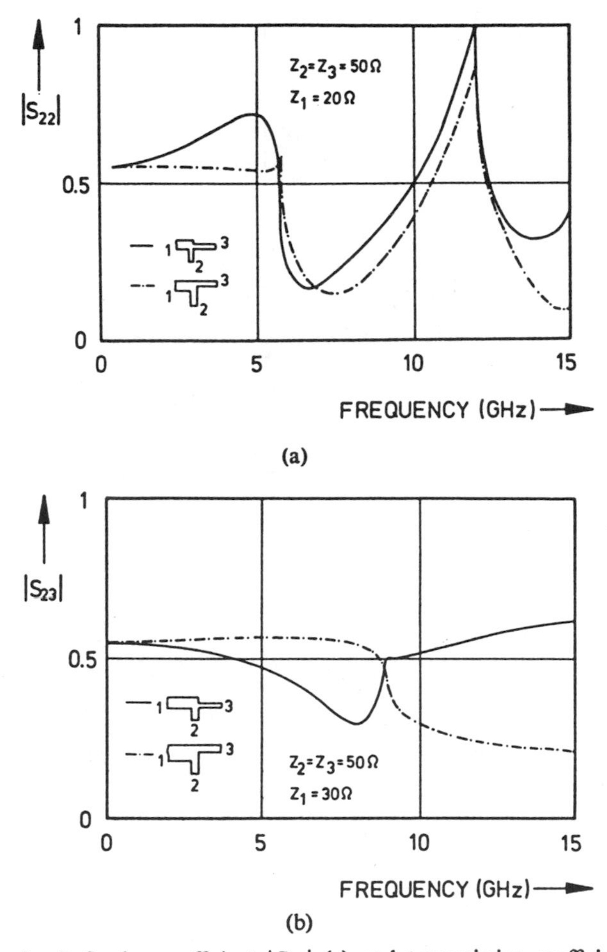

Fig. 5. Reflection coefficient $|S_{22}|$ (a) and transmission coefficient $|S_{23}|$ (b) for two T junctions of equal characteristic impedances but different geometrical structure. Substrate material Polyguide: $\varepsilon_r = 2.32$, $h = 0.158$ cm.

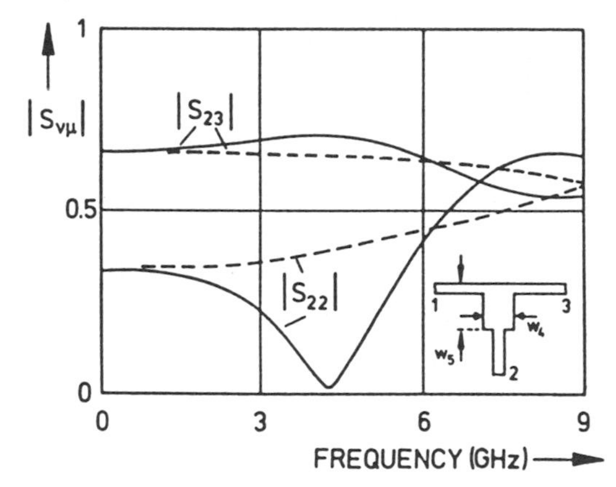

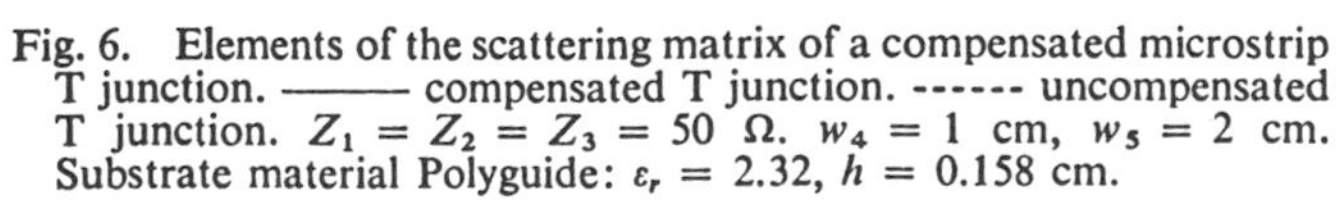
Fig. 6. Elements of the scattering matrix of a compensated microstrip T junction. ——— compensated T junction. ------ uncompensated T junction. $Z_1 = Z_2 = Z_3 = 50\ \Omega$. $w_4 = 1$ cm, $w_5 = 2$ cm. Substrate material Polyguide: $\varepsilon_r = 2.32$, $h = 0.158$ cm.

P,K,M are the numbers of the highest order modes which are taken into account if the equations are evaluated numerically. If only an incident TEM mode at one port of the T junction is considered, the scattering parameters can easily be computed from the amplitudes of the incident and reflected or transmitted waves.

III. Numerical Results

The resulting equations have been evaluated numerically for different unsymmetrical microstrip T junctions on Polyguide substrate material. No relative convergence problems occur, and the results are of sufficient accuracy (error < 0.5 percent) even if only five higher order modes are taken into account in each line and the connecting field region.

The results shown in Figs. 4–6 have been computed with eight higher order modes. In this case a computing time (central processing time on a CD Cyber 72/76) of 50 ms is required for the calculation of the scattering matrix at one frequency. Fig. 4 shows the reflection coefficients $|S_{22}|$ and the transmission coefficients $|S_{23}|$ of T junctions with the characteristic impedances $Z_2 = Z_3 = 50\ \Omega$ and different values of Z_1 on Polyguide material. At low frequencies ($f < 2$ GHz) the coefficients can be calculated from the static characteristic impedances. For higher frequencies the coefficients become frequency dependent. The reflection coefficients increase with increasing frequency, until they reach a maximum at the cutoff frequency of the first higher order mode. The transmission coefficients decrease with increasing frequency. If $Z_1 = Z_3$, the maximum value of $|S_{22}|$ becomes 1, whereas $|S_{23}|$ decreases to zero. For different impedances of line 1 and line 2 the maximum value of $|S_{22}|$ decreases with increasing difference between the impedances. For frequencies higher than the cutoff frequency of the first higher order mode (normally this frequency range is not of great interest for practical use),

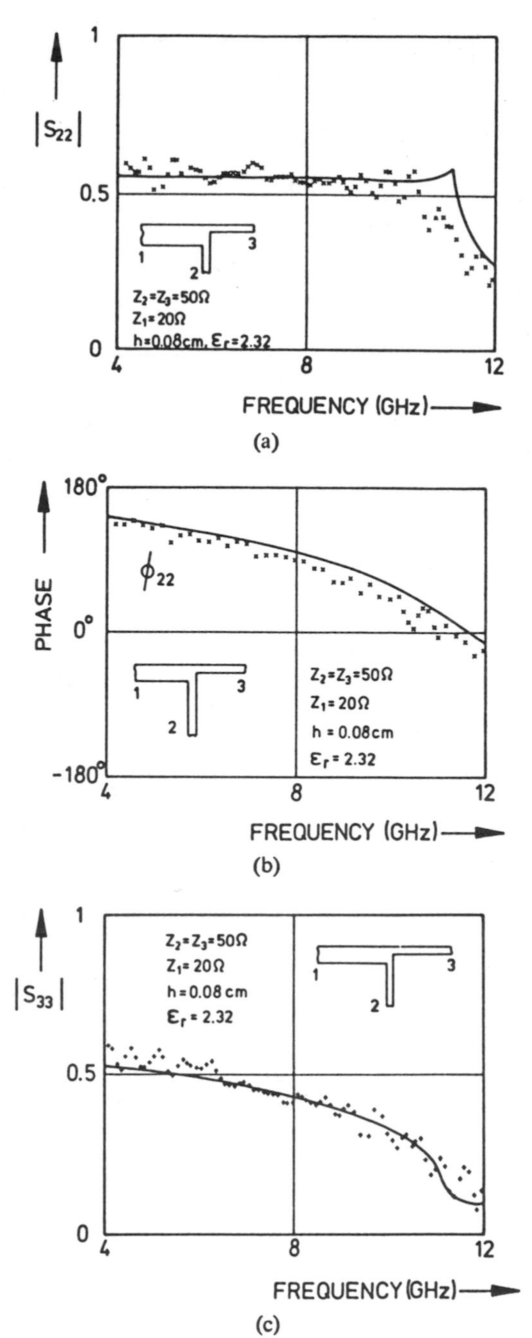

the reflection coefficients strongly decrease and again have a second maximum at the next cutoff frequency. For a 0.158-cm-thick Polyguide substrate material, which has a small dielectric constant, the variation of the reflection coefficient $|S_{22}|$ from 0 to 10 GHz is about 100 percent. The frequency dependence of the scattering matrix becomes smaller with decreasing values of height h and increasing dielectric constant ε_r, if the same frequency range is considered.

The calculation method described can also be used to study the influence of the geometrical structure on the transmission properties of discontinuities. By way of an example, Fig. 5 shows the reflection coefficients $|S_{22}|$ and the transmission coefficients $|S_{23}|$ for two unsymmetrical T junctions with equal characteristic impedances but different geometrical structures. It can clearly be recognized that a stronger frequency dependence arises if the discontinuity contains additional edges.

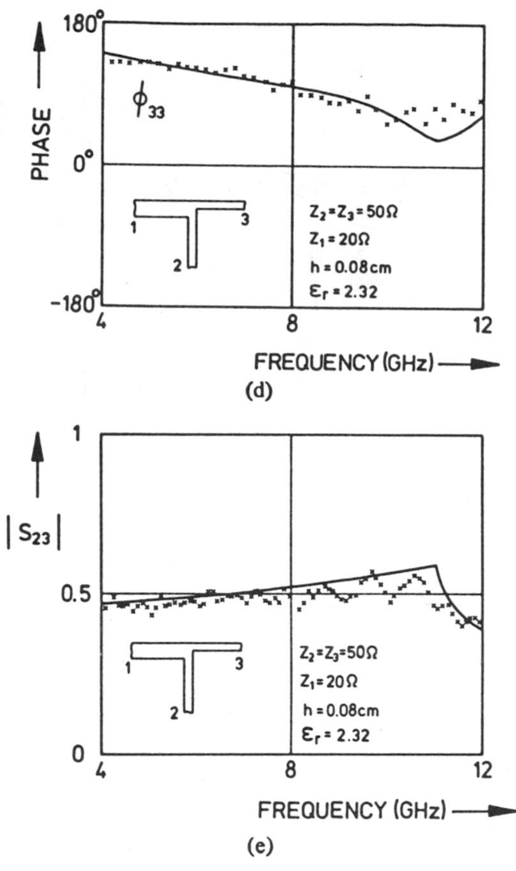

Fig. 7. Measured and calculated scattering parameters of an unsymmetrical T junction on Polyguide material.

This result can be used, for example, to design compensated microstrip power dividers. As is shown in Fig. 6, the reflection coefficient $|S_{22}|$ of a T junction can be made zero at one frequency, if the widths w_1 and w_2 of the first and second microstrip line are smaller than the widths of the connecting field region. This T junction can be calculated in a similar manner to that described previously. As Fig. 6 shows, the reflection coefficient $|S_{22}|$ over a frequency range of 6 GHz is smaller than that of the uncompensated T junction with the same characteristic impedances.

IV. Experimental Results

Measurements have been performed with T junctions on Polyguide material ($\varepsilon_r = 2.32$, $h = 0.0794$ cm) and have been compared to theoretical results. The measurements were carried out on an HP network analyzer connected to an automatic data acquisition system. An attempt was made to eliminate the influence of the microstrip–coax transitions on the measurements by determining the reflection and transmission coefficients of the transitions and correcting the measured scattering parameters. Because of difficulties with the terminations, deviations larger than the inaccuracies of the network analyzer occur. The influence of the line losses on the results have been taken into consideration, whereas radiation losses, which occur especially for substrate materials of small dielectric constant and large height, could not be taken into account.

Figs. 7 and 8 show the measured and computed scattering parameters of two T junctions with 20–50–50 Ω and 50–

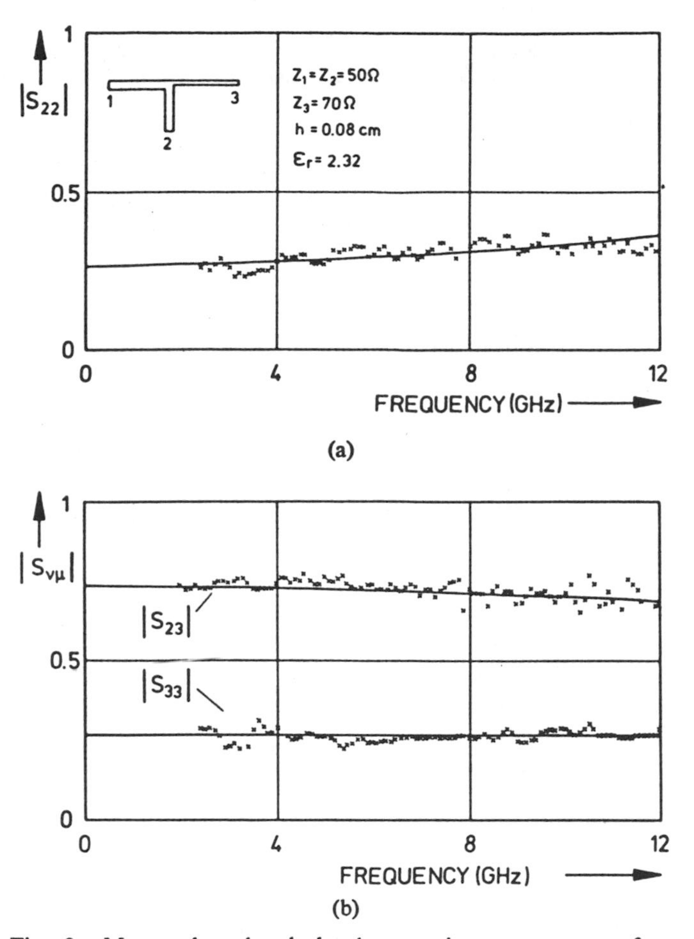

Fig. 8. Measured and calculated scattering parameters of an unsymmetrical T junction on Polyguide material.

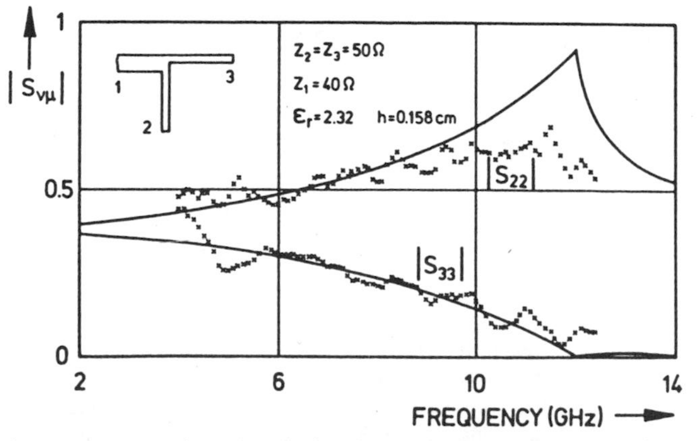

Fig. 9. Measured and calculated scattering parameters of an unsymmetrical T junction on thick Polyguide material.

50–70 Ω impedances as examples. All the measured absolute values and phases, in the light of the remarks made previously, correspond closely to theoretical results. Larger deviations between theory and experiment occur for thicker substrates, especially near the cutoff frequencies, for at these frequencies the radiated power becomes large. Additional difficulties occur with the measurements of those discontinuities, for the termination of lines of larger width is much more complicated. For example, Fig. 9 shows the calculated and measured results for a T junction on Polyguide material of height $h = 0.158$ cm. At low frequencies the termination of the 40-Ω line is very difficult, giving rise to the deviations between theory and experiment at 4 GHz. For frequencies near the cutoff frequency ($f \approx 12$ GHz) the measured reflection coefficient is smaller than the calculated one because of the radiated power.

V. Conclusions

A method is presented which can be used to calculate the dynamical properties of many microstrip discontinuities. As far as the authors know, this is the first method described in the literature for calculating the frequency-dependent properties of unsymmetrical microstrip discontinuities. In the case of normally used frequency ranges and microstrip lines, the method leads to results which are strongly supported by measurements. The computer programs do not need much computer storage and are relatively fast, so they can also be used in computer-aided design methods. Further developments of the existing programs may lead to the possibility of studying compensation methods for power dividers or similar circuits.

References

[1] A. A. Oliner, "Equivalent circuits for discontinuities in balanced strip transmission line," *IEEE Trans. Microwave Theory Tech.* (*Special Issue on Symp. on Microwave Strip Circuits*), vol. MTT-3, pp. 134–143, Nov. 1955.

[2] P. Silvester and P. Benedek, "Equivalent capacitances of microstrip open circuits," *IEEE Trans. Microwave Theory Tech.*, vol. MTT-20, pp. 511–516, Aug. 1972.

[3] P. Benedek and P. Silvester, "Equivalent capacitances of microstrip gaps and steps," *IEEE Trans. Microwave Theory Tech.*, vol. MTT-20, pp. 729–733, Nov. 1972.

[4] P. Silvester and P. Benedek, "Microstrip discontinuity capacitances for right-angle bends, T-junctions and crossings," *IEEE Trans. Microwave Theory Tech.*, vol. MTT-21, pp. 341–346, May 1973.

[5] P. Stouten, "Equivalent capacitances of T-junctions," *Electron. Lett.*, vol. 9, pp. 552–553, Nov. 1973.

[6] A. Gopinath and P. Silvester, "Calculation of inductance of finite-length strips and its variation with frequency," *IEEE Trans. Microwave Theory Tech.*, vol. MTT-21, pp. 380–386, June 1973.

[7] A. Gopinath and B. Easter, "Moment method of calculating discontinuity inductance of microstrip right angled bends," *IEEE Trans. Microwave Theory Tech.* (Short Papers), vol. MTT-22, pp. 880–883, Oct. 1974.

[8] A. Thompson and A. Gopinath, "Calculation of microstrip discontinuity inductance," *IEEE Trans. Microwave Theory Tech.*, vol. MTT-23, pp. 648–655, Aug. 1975.

[9] I. Wolff, G. Kompa, and R. Mehran, "Calculation method for microstrip discontinuities and T-junctions," *Electron. Lett.*, vol. 8, p. 177, Apr. 1972.

[10] G. Kompa and R. Mehran, "Planar waveguide model for calculating microstrip components," *Electron. Lett.*, vol. 11, pp. 459–460, Sept. 1975.

[11] E. Kühn, "A mode-matching method for solving field problems in waveguide and resonator circuits," *Arch. Elek. Übertragung*, vol. 27, pp. 511–513, Dec. 1973.

[12] T. E. Rozzi and W. F. G. Mecklenbräuker, "Wide-band network modeling of interacting inductive irises and steps," *IEEE Trans. Microwave Theory Tech.*, vol. MTT-23, pp. 235–245, Feb. 1975.

[13] R. Jansen, "A moment method for covered microstrip dispersion," *Arch. Elek. Übertragung*, vol. 29, pp. 17–20, 1975.

[14] H. A. Wheeler, "Transmission-line properties of parallel wide strips by a conformal-mapping approximation," *IEEE Trans. Microwave Theory Tech.*, vol. MTT-12, pp. 280–289, 1964.

[15] H. J. Schmitt and K. H. Sarges, "Wave propagation in microstrip," *Nachrichtentech. Z.*, vol. 24, pp. 260–264, 1971.

[16] M. V. Schneider, "Microstrip dispersion," *Proc. Inst. Elect. Electron. Engrs.*, vol. 60, pp. 144–146, 1972.

[17] R. Mehran, "The frequency-dependent scattering matrix of microstrip right-angle bends, T-junctions and crossings," *Arch. Elek. Übertragung*, vol. 29, pp. 454–460, Nov. 1975.

[18] I. Wolff, "Statische Kapazitäten von rechteckigen und kreisförmigen Mikrostrip-Scheibenkondensatoren," *Arch. Elek. Übertragung*, vol. 27, pp. 44–47, Jan. 1973.

Generalized Scattering Matrix Method for Analysis of Cascaded and Offset Microstrip Step Discontinuities

TAK SUM CHU AND TATSUO ITOH, FELLOW, IEEE

***Abstract*—Detailed algorithms are presented for characterizations of cascaded microstrip step discontinuities, symmetric stubs, and offset step. The analysis is based on the generalized scattering matrix techniques after the equivalent waveguide model is introduced for the microstrip line.**

I. INTRODUCTION

THE EQUIVALENT waveguide model has been advantageously used for characterizing a number of discontinuities appearing in microstrip circuits [1]–[3]. Although the radiation and surface-wave excitation are neglected, these characterizations provide useful and accurate information in many practical applications. Formulations for these discontinuity problems are typically done by the mode-matching technique. However, detailed formulation algorithms are not readily available.

Recently, the present authors carried out an assessment for a number of different formulations for a microstrip step discontinuity within the framework of the waveguide model. The most economical and yet most accurate formulation was suggested [4]. The step discontinuity has also been analyzed by the modified residue calculus technique (MRCT) [5].

The present paper extends the analysis of the step discontinuity to a cascaded step discontinuity and an offset step discontinuity. The symmetric stub can be treated as the cascaded step discontinuity. The offset discontinuity will be treated as the limiting case of a cascaded discontinuity. In each case, the individual step discontinuity is characterized by either the MRCT or the mode-matching method. The analysis results in a generalized scattering matrix for each step. The analysis of the cascaded step is undertaken by invoking the generalized scattering matrix technique in which the generalized scattering matrices of two step junctions are combined [6], [7]. The ultimate result is the generalized scattering matrix of the cascaded junction as a whole.

It should be noted that the waveguide model is presumed to be an acceptable model for the present analysis. The results are compared with the experimental data as well as those reported in the literature.

Manuscript received August 2, 1985; revised October 1, 1985. This work was supported in part by the U.S. Army Research Office, under Contract DAAG29-84-K-0076.

The authors are with the Electrical Engineering Research Laboratory, University of Texas, Austin, TX 78712.

IEEE Log Number 8406468.

II. FORMULATION OF THE PROBLEM

A. Cascaded Step Discontinuity

The algorithm for the analysis will be best illustrated by means of the cascaded step discontinuity. Specific changes required for the offset discontinuity will be explained later. The first step of the analysis is to replace the microstrip circuit under study with its equivalent waveguide model [8]. The top and bottom are electric walls and the sidewalls are magnetic walls. The height h in Fig. 1 remains unchanged. The effective dielectric constant ϵ_1 and the effective width $\overline{W}_1$ of region I with the microstrip width W_1 can be found from the dominant mode phase constant β_1 and the characteristic impedance Z_{01} of the microstrip line

$$\epsilon_1 = (\beta_1/k_0)^2 \tag{1}$$

$$Z_{01} = \left[120\pi/\sqrt{\epsilon_1}\right](h/\overline{W}_1). \tag{2}$$

β_1 and Z_{01} must be calculated from the structural parameters by a standard full-wave analysis [9], [10] or a curve-fit formula [11]. Other regions may be modeled in a similar manner.

The next step is to characterize all the discontinuities involved in the waveguide model of the microstrip circuit under study. This characterization is done in terms of the generalized scattering matrix [6], [7]. This matrix is closely related to the scattering matrix used in the microwave network theory, but differes in that the dominant as well as higher order modes are included. Therefore, the generalized scattering matrix will be, in general, of infinite order. Consider, for instance, the TE_{p0} excitation with unit amplitude from the left to junction 1 in Fig. 2. If the amplitude of the nth mode of the reflected wave to the left is A_n, the (n, p) entry of the scattering matrix $S^{11}(n, p)$ is A_n. Similarly, if the amplitude of the mth mode of the wave transmitted to the right in B_m, $S^{21}(m, p)$ is B_m. Other matrix elements can be derived similarly. Hence, the generalized scattering matrix S_1 of the junction 1 can be written in terms of four submatrices of infinite order

$$S_1 = \begin{bmatrix} S^{11} & S^{12} \\ S^{21} & S^{22} \end{bmatrix}. \tag{3}$$

The corresponding matrix of the junction 2 is

$$S_2 = \begin{bmatrix} S^{33} & S^{34} \\ S^{43} & S^{44} \end{bmatrix}. \tag{4}$$

Reprinted from *IEEE Trans. Microwave Theory Tech.*, vol. MTT-34, no. 2, pp. 280–284, February 1986.

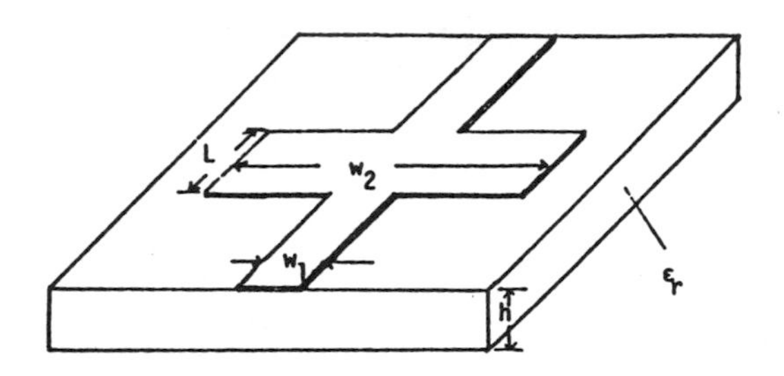

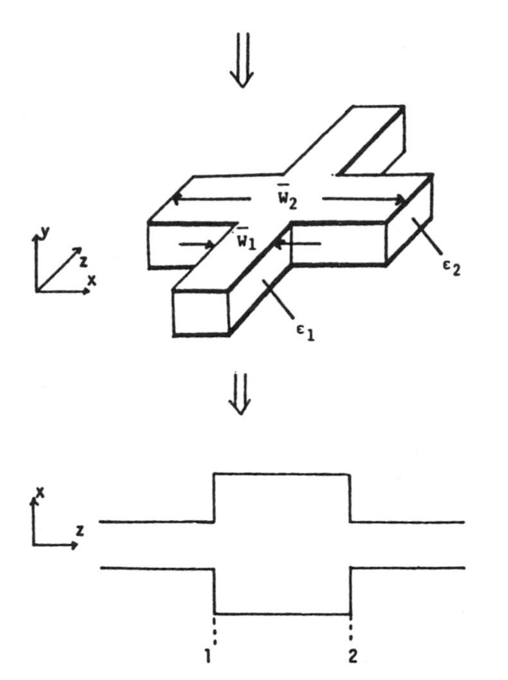

Fig. 1. Cascaded step discontinuities, equivalent waveguide model, and top view.

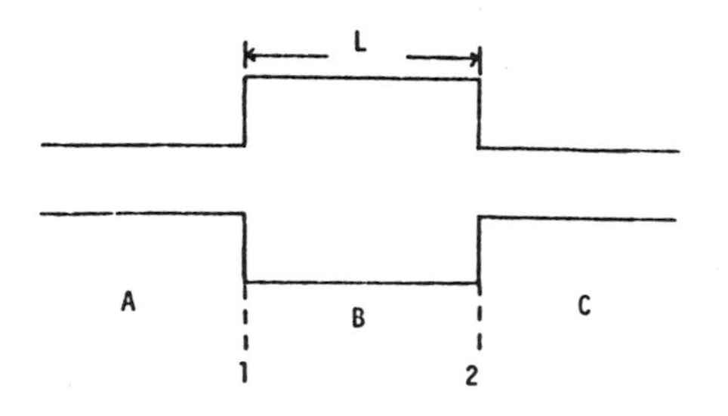

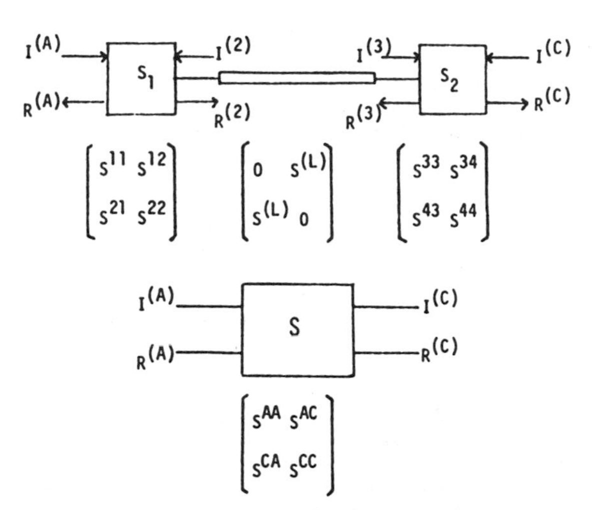

Fig. 2. Derivation of S parameters for the cascaded step discontinuity structures.

All of the elements of the generalized scattering matrix can be obtained by solving the electromagnetic problem of the junction scattering by means of a standard technique such as the mode-matching technique [4] or a modified residue calculus technique [5]. Since the details of these methods for an isolated step discontinuity are reported in [4] and [5], they are not repeated here. We presume all of these quantities are now available.

The remaining step is to combine these generalized scattering matrices of the cascaded junctions and to derive the composite matrix.

$$S=\begin{bmatrix} S^{AA} & S^{AC} \\ S^{CA} & S^{CC} \end{bmatrix}. \tag{5}$$

To this end, we introduce the transmission matrix $S^{(L)}$ of the waveguide between junctions 1 and 2. The wave travels a distance L so that each mode is multiplied by $\exp(-\gamma L)$

$$S^{(L)}=\begin{bmatrix} e^{-\gamma_1 L} & & & \\ & e^{-\gamma_2 L} & & 0 \\ & & \ddots & \\ 0 & & & e^{-\gamma_n L} \end{bmatrix} \tag{6}$$

where γ_n is the propagation constant of the nth mode of Region B. Hence $\gamma_1 = j\beta_2 L$, where β_2 is the dominant-mode phase constant of Region B. Our algebraic process to derive S is detailed in the Appendix. The results are

$$S^{AA}=S^{11}+S^{12}S^{(L)}U_2S^{33}S^{(L)}S^{21} \tag{7a}$$

$$S^{AC}=S^{12}S^{(L)}U_2S^{34} \tag{7b}$$

$$S^{CA}=S^{43}S^{(L)}U_1S^{21} \tag{7c}$$

$$S^{CC}=S^{44}+S^{43}S^{(L)}U_1S^{22}S^{(L)}S^{34} \tag{7d}$$

where

$$U_1=(I-S^{22}S^{(L)}S^{33}S^{(L)})^{-1} \tag{8a}$$

$$U_2=(I-S^{33}S^{(L)}S^{22}S^{(L)})^{-1} \tag{8b}$$

and I is the unit matrix. The above matrices are formally of infinite size. However, in practice, these matrices must be truncated to a finite size. It is found that excellent convergence is obtained when 3×3 or even 2×2 submatrices S^{11}, etc., are used.

It should be noted here that the use of generalized scattering matrices is increasingly more important as the distance between two junctions is smaller. Therefore, the present technique can be used for analysis of the symmetric stub from the knowledge of the generalized scattering matrix of a single step discontinuity.

B. Offset Step Discontinuity

Next, the technique described above will be applied to an offset step discontinuity shown in Fig. 3. This offset discontinuity occurs in a microstrip circuit either intentionally or unintentionally. As we will see shortly, a small amount of offset Δ significantly affects the scattering characteristic of the discontinuity.

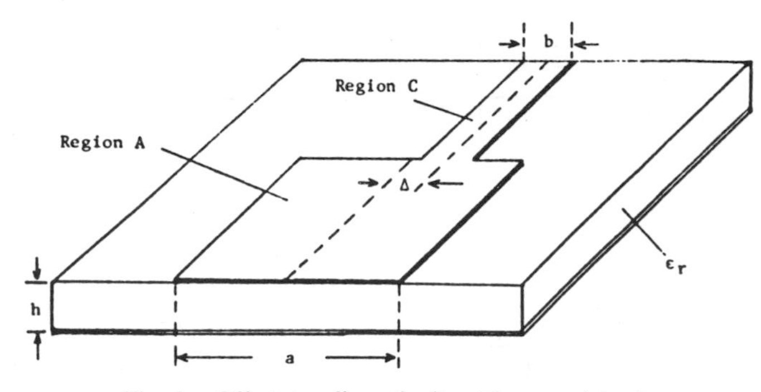

Fig. 3. Offset step discontinuity with eccentricity Δ.

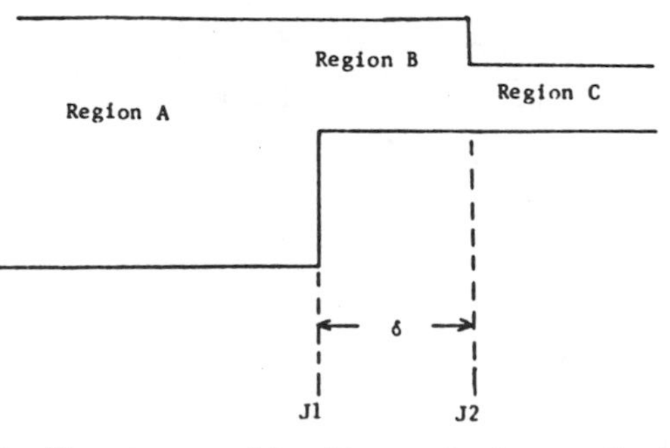

Fig. 5. Auxiliary structure of the offset step for the generalized scattering matrix technique.

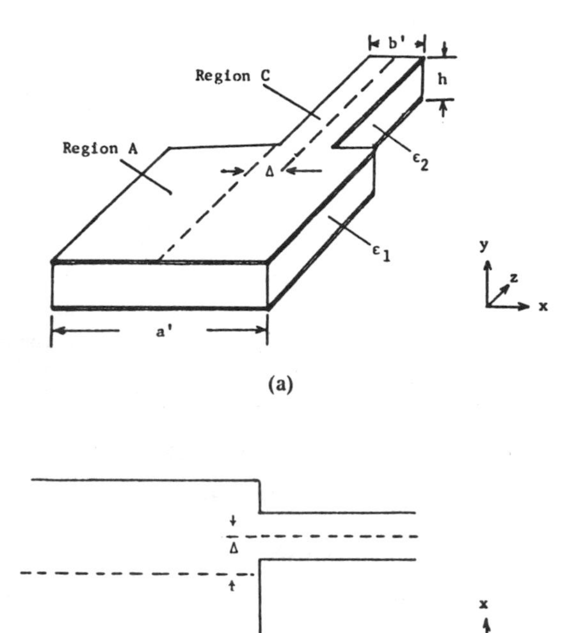

Fig. 4. Equivalent waveguide model of the offset step: (a) perspective view and (b) top view.

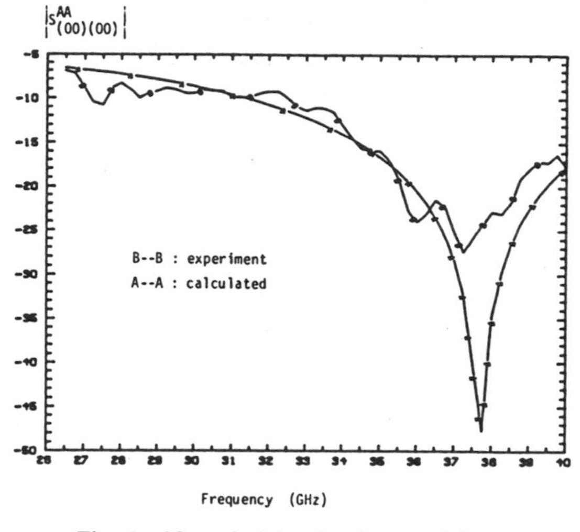

Fig. 6. Numerical data for the cascaded step.

Once again, the first step is to derive the equivalent waveguide model depicted in Fig. 4. Although a direct formulation of Fig. 4 is possible by way of the mode-matching technique, we will take an alternative approach in which the formulations and the generalized scattering matrices of the symmetric step discontinuities are advantageously used. To this end, an auxiliary structure in Fig. 5 is introduced. Notice that the original offset step discontinuity structure can be recovered by letting δ in Fig. 5 to zero after all the formulations are carried out. Also, the individual discontinuities $J1$ and $J2$ in Fig. 5 are one-half of the symmetric step discontinuities. Hence, all of the previous results for the symmetric step discontinuities excited by the even-mode can be directly used. In fact, in [4] and [5], only one-half of the structure has been used for analysis.

Once the scattering matrices of $J1$ and $J2$ are available, the scattering matrix of the composite discontinuity can be derived from (7) and (8) except that $S^{(L)} = I$ when $\delta \to 0$. This completes the formulation for the offset discontinuity.

III. Results and Discussions

Fig. 6 shows typical results of the amplitude of the dominant mode (quasi-TEM) reflection coefficient from a cascaded step discontinuity. The results are found to agree very well with the experimental data taken at Hughes Torrance Research Center for the microstrip circuit on a Duroid substrate.

For efficient calculation, it is desirable to be able to truncate the matrices at as small a size as possible with accurate results. In the case of the cascaded discontinuity, $S^{(L)}$ contains a convergent factor since all of the higher order modes have real values of γ_n. In the case of an offset discontinuity, such exponentially decaying factors disappear because the length $\delta \to 0$, and hence $S^{(L)} = I$. To test the convergence of the solution, the dominant-mode transmission coefficient calculated using generalized scattering matrices of sizes 2×2, 3×3, and 5×5 are compared. Physical parameters are chosen to be the same as those studied by Kompa [12] so as to permit a comparison of results. Fig. 7 shows the results of this convergence study. It is seen that even the 2×2 matrix gives reasonably accurate results. To establish the validity of the results, they are compared with those calculated by Kompa [12]. This comparison is shown in Fig. 8(a) and (b). Finally, the dominant-mode transmission coefficient for various eccentricities are calculated. It is evident in Fig. 9 that the

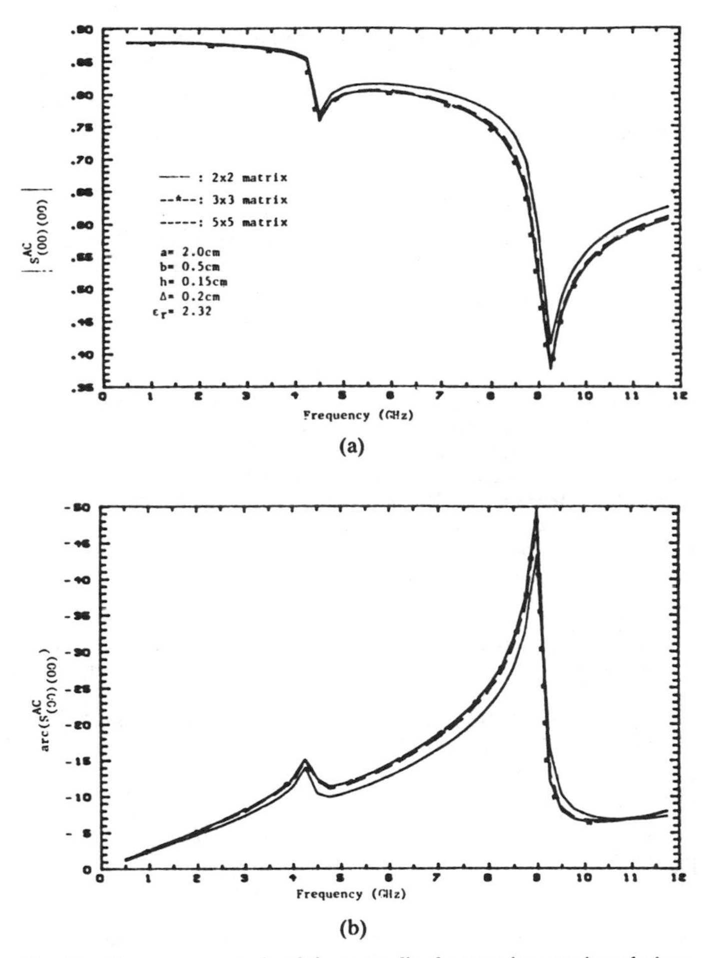

Fig. 7. Convergence study of the generalized scattering matrix technique for an offset step: (a) magnitude and (b) phase.

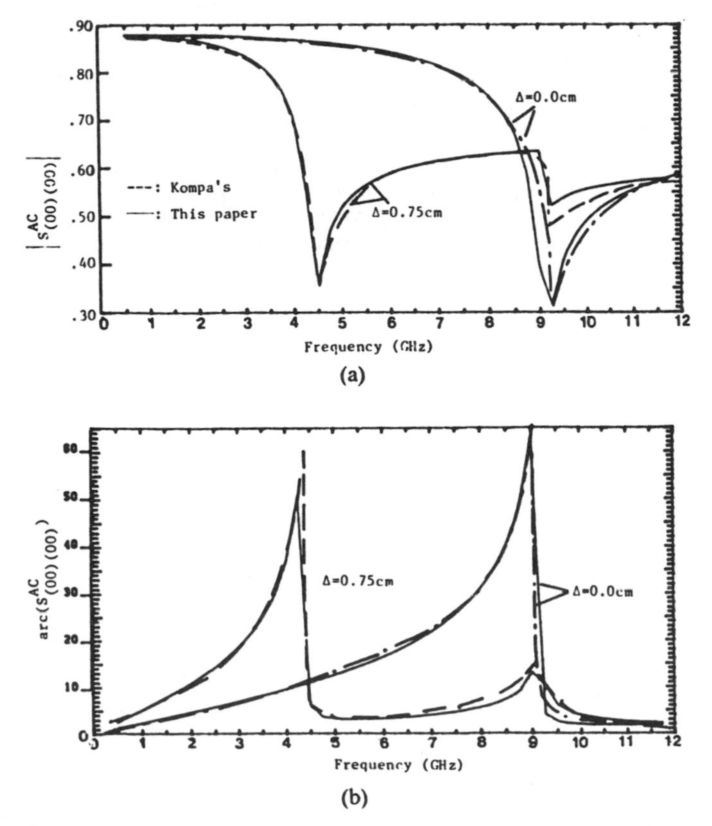

Fig. 8. Comparison with Kompa's results: (a) magnitude and (b) phase.

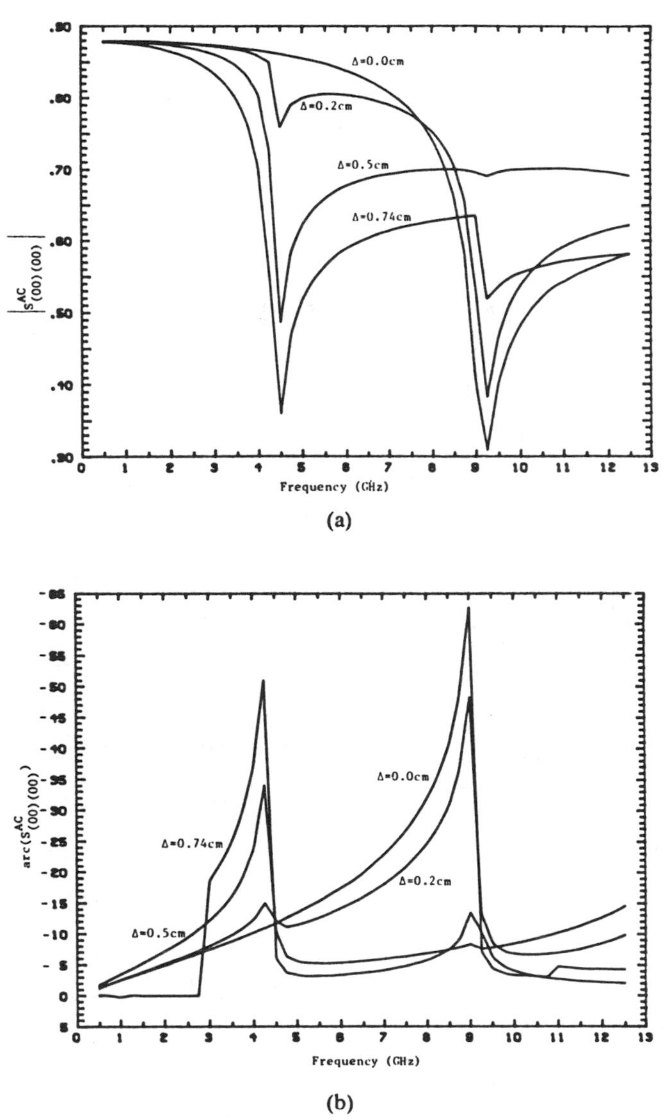

Fig. 9. Effect of eccentricity: (a) magnitude and (b) phase.

effect of the first odd-mode cutoff is exhibited as soon as the eccentricity is nonzero. Also, for $\Delta = 0.5$ cm, the effect of the second- (even-) mode cutoff is quite small due to the fact that the center of the smaller waveguide is located at the second-mode null and, hence, little coupling via this mode exists.

IV. Conclusions

The generalized scattering matrix technique has been applied to the problems of a cascaded step discontinuity and an offset step discontinuity. The waveguide model has been presumed to be applicable for analysis. Individual discontinuities are characterized first and the results are used for the description of the composite discontinuity via the generalized scattering matrix technique. In spite of the two-step process, the overall effort in numerical processing is quite efficient. In actual numerical software, the programs for isolated discontinuities can be used as subroutines for the composite structures.

APPENDIX
DERIVATION OF SCATTERING MATRIX FOR THE CASCADED STEP DISCONTINUITIES

The following matrix equations can be derived from Fig. 2:

$$\begin{bmatrix} R^{(A)} \\ R^{(2)} \end{bmatrix} = \begin{bmatrix} S^{11} & S^{12} \\ S^{21} & S^{22} \end{bmatrix} \begin{bmatrix} I^{(A)} \\ I^{(2)} \end{bmatrix} \quad \text{(A1)}$$

$$\begin{bmatrix} I^{(2)} \\ I^{(3)} \end{bmatrix} = \begin{bmatrix} 0 & S^{(L)} \\ S^{(L)} & 0 \end{bmatrix} \begin{bmatrix} R^{(2)} \\ R^{(3)} \end{bmatrix} \quad \text{(A2)}$$

$$\begin{bmatrix} R^{(3)} \\ R^{(C)} \end{bmatrix} = \begin{bmatrix} S^{33} & S^{34} \\ S^{43} & S^{44} \end{bmatrix} \begin{bmatrix} I^{(3)} \\ I^{(C)} \end{bmatrix}. \quad \text{(A3)}$$

Next, (A2) is substituted into (A1) to get

$$R^{(A)} = S^{11}I^{(A)} + S^{12}S^{(L)}R^{(3)} \quad \text{(A4)}$$

$$R^{(2)} = S^{21}I^{(A)} + S^{22}S^{(L)}R^{(3)}. \quad \text{(A5)}$$

Next, (A2) is substituted into (A3) to get

$$R^{(3)} = S^{33}S^{(L)}R^{(2)} + S^{34}I^{(C)} \quad \text{(A6)}$$

$$R^{(C)} = S^{43}S^{(L)}R^{(2)} + S^{44}I^{(C)}. \quad \text{(A7)}$$

Equations (A5) and (A6) are used to isolate $R^{(2)}$ and $R^{(3)}$

$$R^{(2)} = U_1S^{21}I^{(A)} + U_1S^{22}S^{(L)}S^{34}I^{(C)} \quad \text{(A8)}$$

where

$$U_1 = (I - S^{22}S^{(L)}S^{33}S^{(L)})^{-1}$$

and

$$R^{(3)} = U_2S^{33}S^{(L)}S^{21}I^{(A)} + U_2S^{34}I^{(C)} \quad \text{(A9)}$$

where

$$U_2 = (I - S^{33}S^{(L)}S^{22}S^{(L)})^{-1}.$$

Finally, (A8) and (A9) are substituted into (A4) and (A7) to get

$$R^{(A)} = S^{11}I^{(A)} + S^{12}S^{(L)}U_2S^{33}S^{(L)}S^{21}I^{(A)} + S^{12}S^{(L)}U_2S^{34}I^{(C)}$$

$$R^{(C)} = S^{43}S^{(L)}U_1S^{21}I^{(A)} + S^{43}S^{(L)}U_1S^{22}S^{(L)}S^{34}I^{(C)} + S^{44}I^{(C)}.$$

S^{AA}, S^{AC}, S^{CA}, and S^{CC} can be identified easily from the above equations.

ACKNOWLEDGMENT

The authors thank Dr. Y. C. Shih of Hughes Torrance Research Center for providing experimental data and for technical discussions.

REFERENCES

[1] I. Wolff, G. Kompa, and R. Mehran, "Calculation method for microstrip discontinuities and T-junctions," *Electron. Lett.*, vol. 8, pp. 177–179, Apr. 1972.

[2] G. Kompa, "S-matrix computation of microstrip discontinuities with a planar waveguide model," *Arch. Elek. Übertragung.*, vol. 30, pp. 58–64, Feb. 1976.

[3] W. Menzel and I. Wolff, "A method for calculating the frequency-dependent properties of microstrip discontinuities," *IEEE Trans. Microwave Theory Tech.*, vol. MTT-25, pp. 107–112, Feb. 1977.

[4] T. S. Chu, T. Itoh, and Y-C. Shih, "Comparative study of mode-matching formulations for microstrip discontinuity problems," *IEEE Trans. Microwave Theory Tech.*, vol. MTT-33, pp. 1018–1023, Oct. 1985.

[5] T. S. Chu and T. Itoh, "Analysis of microstrip step discontinuity by the modified residue calculus technique," *IEEE Trans. Microwave Theory Tech.*, vol. MTT-33, pp. 1024–1028, Oct. 1985.

[6] J. Pace and R. Mittra, "Generalized scattering matrix analysis of waveguide discontinuity problems," in *Quasi-Optics XIV*. New York: Polytechnic Institute of Brooklyn Press, 1964, pp. 172–194.

[7] Y.-C. Shih, T. Itoh, and L. Q. Bui, "Computer-aided design of millimeter-wave E-plane filters," *IEEE Trans. Microwave Theory Tech.*, vol. MTT-31, pp. 135–142, Feb. 1983.

[8] I. Wolff and N. Knoppik, "Rectangular and circular microstrip disk capacitors and resonators," *IEEE Trans. Microwave Theory Tech.*, vol. MTT-22, pp. 857–864, Oct. 1974.

[9] T. Itoh, "Spectral domain immittance approach for dispersion characteristics of generalized printed transmission lines," *IEEE Trans. Microwave Theory Tech.*, vol. MTT-28, pp. 733–736, July 1980.

[10] R. Jansen and M. Kirschning, "Arguments and an accurate model for the power-current formulation of microstrip characteristic impedance," *Arch. Elek. Übertragung.*, vol. 37, pp. 108–112, Mar. 1983.

[11] E. Hammerstad and O. Jensen, "Accurate models for microstrip computer-aided design," in *IEEE MTT-S Int. Symp. Dig.* (Washington, DC), 1980, pp. 407–409.

[12] G. Kompa, "Frequency-dependent behavior of microstrip offset junction," *Electron. Lett.*, vol. 11, no. 22, pp. 172–194, Oct. 1975.

Hybrid mode analysis of end effects of planar microwave and millimetrewave transmission lines

Prof. R.H. Jansen, Dr.Ing., C.Eng., M.I.E.E.E. M.V.D.E.(NTG)

Indexing terms: *Waveguides, End effects*

Abstract: The phase shift associated with the reflection of a guided wave at the abrupt open end of planar strip and slot waveguides is determined numerically by solving the relevant frequency-dependent hybrid-mode boundary-value problem. In addition, nonideally short-circuited ends are considered, i.e. those where the planar guiding structure terminates in a metallisation layer. The method of analysis employed is a Galerkin approach in conjunction with a spectral-domain Green's function interpolation technique and especially suited expansion functions. Its main features are described, and numerical results are presented for most of the technically important types of planar waveguides (microstrip, suspended substrate, slot/fin and coplanar lines). The results are compared with published measured data and quasistatic computations, as far as these are available. The frequency-dependent behaviour of the different end effects is discussed. In an Appendix, it is shown how the numerical approach of this paper can be extended to the analysis of planar n-ports.

1 Introduction

One of the basic presuppositions to a successful design of planar distributed microwave and millimetre-wave circuits is the accurate knowledge of the frequency-dependent characteristics of the involved planar waveguides. This is a consequence of the fact that these circuits consist of components, filters, matching networks etc. which are assembled from elementary strip and/or slot transmission line structures. As an efficient aid to their design fairly general and efficient computer algorithms have been developed by now, for the determination of the propagation coefficients and characteristic impedances of the involved planar transmission lines, see for example References 1 & 2. However, except for the specialised case of the microstrip open end, few or no theoretical results have been published for the end effects of planar guiding structures. Nevertheless, the open or short-circuited end of a strip or slot configuration constitutes an important type of discontinuity which exists in nearly every planar circuit. In a certain sense, the end effect, i.e. the associated phase shift, can be considered as one of the parameters characterising a transmission line. Therefore, its quantitative knowledge is important for an improved design of MIC and MMW circuits.

This paper describes the rigorous hybrid-mode computation of the end effects caused by the abrupt end configurations of Fig. 1*a*. The transmission line cross-sections considered are of the microstrip (two dielectric layers) and the suspended substrate, and slot and coplanar type (three dielectric layers) (see Fig. 1*b*), respectively. A quantitative characterisation of the different end effects is given in terms of the equivalent line length Δl which defines the position of a hypothetical ideal short or open circuit with respect to the abrupt physical end. Since typical values of Δl are in the range of 2%–10% of a line wavelength, its accurate determination requires a solution of the associated electromagnetic hybrid-mode boundary-value problem which is more accurate than that of the line or resonator problem by at least one order of magnitude. This high accuracy is achieved on the base of an efficient three-dimensional generalisation of a user-oriented approach which has recently been applied to planar MIC transmission lines [2]. Methods similar to that used in this paper have in the past been employed for the analysis of microstrip resonators [3–6] and, as an accessory result, microstrip open-end data have been published [4]. However, whereas these original techniques are suited for the evaluation of end effects in principle, it will become obvious here that without some essential modifications they are subject to severe restrictions concerning their accuracy, efficiency and practicability for end effect computations. In this paper a spectral-domain Green's function interpolation technique is used in conjunction with abrupt end expansion functions and a Fourier series summation scheme which minimises computer time and storage requirements. As a consequence, realistically large microwave shielding cases can be simulated with reasonable computational expense. Also, the interaction of the planar waveguide end with its RF housing at high frequencies can be studied. The results given refer to microstrip, suspended substrate, slot fin and coplanar lines, respectively. Because of the special interest the microstrip open end has found in the technical literature, this case is treated in detail in a separate paper [7].

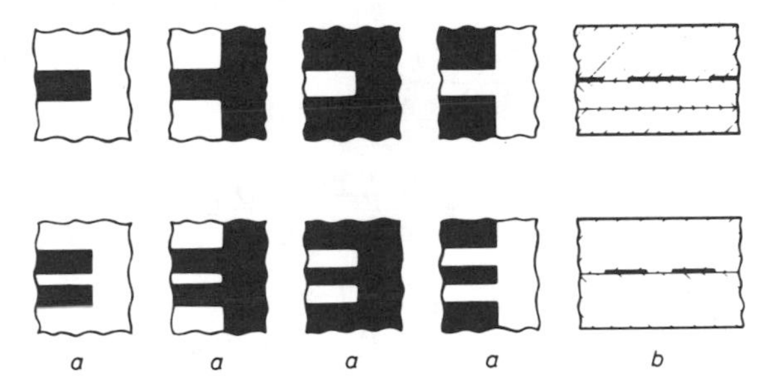

Fig. 1 *Open and short-end planar metallisation patterns discussed in this paper*

a Associated representative cross-sectional geometries
b Metallisation hatched

2 Outline of the method of analysis

In complete analogy to the microstrip case [3–6] the basically three-dimensional electromagnetic-field problem of general planar ciruit structures can be reduced to two dimensions. This means a drastic alteration to the analysis of planar configurations and is because these can usually be treated with good accuracy under the assumption of negligible metallisation thickness. In the present approach, this advantage is utilised and the abrupt end problem is formulated in terms of the field and surface current density distribution existing in the plane of the circuit metallisation pattern which is assumed to

Paper 1262H, first received 4th November 1980 and in revised form 6th January 1981
Prof. Jansen is with the University of Duisburg, FB9, Department of Electrical & Electronic Engineering, Bismarek-Strasse 81, D-4100 Duisburg 1, West Germany

Reprinted with permission from *IEE Proceedings,* vol. 128, Pt. H, no. 2, pp. 77-86, April 1981.

be lossless. The dielectric constants of the different layers involved are arbitrary real scalars. They refer to a lossless, piecewise homogeneous and isotropic medium of arbitrary layer thicknesses. All the structures considered are thought to be enclosed in an ideally conducting box of rectangular cross-section and sufficiently large dimensions to model the RF housing of the physical configuration.

With reference to the representative geometry of Fig. 2*a*, a solution to Maxwell's equations is most easily obtained by superposition of LSE and LSM scalar potentials $\phi(x,y,z)$ and $\psi(x,y,z)$ [8, 9, 2], i.e. z-directed vector potentials. This takes into account the hybrid-mode character of the fields and leads to a very compact Fourier (spectral) domain representation of $\phi(x,y,z)$ and $\psi(x,y,z)$. Specifically, this results in the two-dimensional spectral domain coefficients:

$$\phi^{(1)}_{mn}(z) = f_{mn}\cdot\frac{\cos\{k^{(1)}_{zmn}(z+h_1+h_2)\}}{\cos(k^{(1)}_{zmn}h_1)\cos(k^{(2)}_{zmn}h_2)} \tag{1}$$

$$\psi^{(1)}_{mn}(z) = g_{mn}\cdot\frac{\sin\{k^{(1)}_{zmn}(z+h_1+h_2)\}}{\sin(k^{(1)}_{zmn}h_1)\cos(k^{(2)}_{zmn}h_2)}\cdot\frac{T^{(1)}_{mn}k^{(2)}_{zmn}}{k^{(1)2}_{zmn}}$$

$$\phi^{(2)}_{mn}(z) = f_{mn}\cdot\left(\frac{\cos\{k^{(2)}_{zmn}(z+h_2)\}}{\cos(k^{(2)}_{zmn}h_2)} - \frac{\epsilon_2 T^{(1)}_{mn}}{\epsilon_1 k^{(2)}_{zmn}}\cdot\frac{\sin\{k^{(2)}_{zmn}(z+h_2)\}}{\cos(k^{(2)}_{zmn}h_2)}\right)$$

$$\psi^{(2)}_{mn}(z) = g_{mn}\cdot\left(\frac{\cos\{k^{(2)}_{zmn}(z+h_2)\}}{\cos(k^{(2)}_{zmn}h_2)}\cdot\frac{T^{(1)}_{mn}k^{(2)}_{zmn}}{k^{(1)2}_{zmn}} + \frac{\sin\{k^{(2)}_{zmn}(z+h_2)\}}{\cos(k^{(2)}_{zmn}h_2)}\right)$$

$$\phi^{(3)}_{mn}(z) = f_{mn}\cdot(-1)\cdot\frac{\cos\{k^{(3)}_{zmn}(z-h_3)\}}{\cos(k^{(3)}_{zmn}h_3)}\cdot\left(\frac{T^{(1)}_{mn}}{\epsilon_1}+\frac{T^{(2)}_{mn}}{\epsilon_2}\right)\frac{\epsilon_3}{T^{(3)}_{mn}}$$

$$\psi^{(3)}_{mn}(z) = g_{mn}\cdot(-1)\cdot\frac{\sin\{k^{(3)}_{zmn}(z-h_3)\}}{\sin(k^{(3)}_{zmn}h_3)}\cdot\left(\frac{T^{(1)}_{mn}}{k^{(1)2}_{zmn}}+\frac{T^{(2)}_{mn}}{k^{(2)2}_{zmn}}\right)k^{(2)}_{zmn}$$

with

$$k^{(\nu)2}_{zmn} = k_0^2\epsilon_\nu - k^2_{xm} - k^2_{yn},$$

$$T^{(\nu)}_{mn} = k^{(\nu)}_{zmn}\cdot\tan(k^{(\nu)}_{zmn}\cdot h_\nu);\quad \nu = 1,2,3,$$

and

$$k_{xm} = (m-\delta)\frac{2\pi}{a},\quad k_{yn} = (n-\delta)\frac{2\pi}{l_b+l_a},$$

$$\delta = 0.5 \text{ and } 1.0, \text{ respectively},\quad m,n = 1,\ldots,\infty.$$

The associated doubly infinite Fourier series expansions with respect to the x- and y-direction describe the electromagnetic field rigorously and completely. They satisfy all boundary and continuity conditions, except the continuity of the tangential magnetic field in the plane of the substrate $z=0$ and the vanishing of the electric field tangential to the metallisation pattern existing there. The even and odd portions of the field, which refer to different combinations of the values of δ in the separation constants k_{xm} and k_{yn}, are simply superimposed in this stage of the analysis. Note that by means of eqn. 1, the determination of the three-dimensional field distribution for an arbitrary metallisation in the plane $z=0$ has already been reduced to the computation of the two sets of spectral coefficients f_{mn} and g_{mn}. An even more compact formulation could be achieved by the utilisation of a two-dimensional orthonormalised base of functions for the expansion of the field [5, 6]. However, since separability plays an important role in the numerical part of the computation this is not used here.

The next step in the analysis is the derivation of the field and surface current density components implicitly contained in the potential representation eqn. 1. Since this is a conventional procedure it will not be described here. As a result, a pair of spectral-domain equations is set up which links the transverse electric field $e=(e_x, e_y)$ and the current distribution $i=(i_x, i_y)$ in the plane $z=0$ on top of the substrate. This spectral relationship is of the form

$$\begin{pmatrix}e_{xmn}\\ e_{ymn}\end{pmatrix} = (z_{mn})\begin{pmatrix}i_{xmn}\\ i_{ymn}\end{pmatrix},\quad \begin{pmatrix}i_{xmn}\\ i_{ymn}\end{pmatrix} = (z_{mn})^{-1}\begin{pmatrix}e_{xmn}\\ e_{ymn}\end{pmatrix} \tag{2}$$

$$(z_{mn}),(z_{mn})^{-1} = \begin{pmatrix}k^2_{xm}X_{mn}+k^2_{yn}Y_{mn} & \pm k_{xm}k_{yn}(X_{mn}-Y_{mn})\\ \pm k_{xm}k_{yn}(X_{mn}-Y_{mn}) & k^2_{yn}X_{mn}+k^2_{xm}Y_{mn}\end{pmatrix}$$

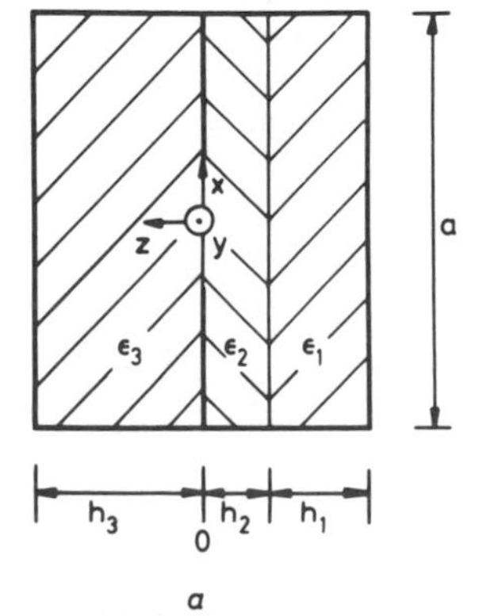

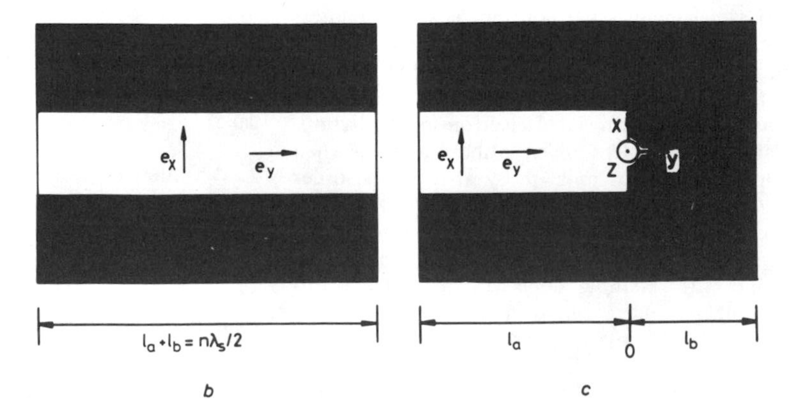

Fig. 2 *Representative slot structure*

a Cross-section
b Slot transmission line resonator
c Abrupt shorted end configuration

with

$$X_{mn} = X(k_{\rho mn}^2), \quad Y_{mn} = Y(k_{\rho mn}^2),$$
$$k_{\rho mn}^2 = k_{xm}^2 + k_{yn}^2,$$

i.e. it is formulated in a compact manner and in terms of transcendental functions $X(k_{\rho mn}^2)$, $Y(k_{\rho mn}^2)$ which exclusively depend on the radial (cylindrical) spectral domain variable $k_{\rho mn}$. Furthermore, this relationship retains its basic mathematical structure given in eqn. 2 if inverted to express (i_x, i_y) by (e_x, e_y), as is indicated in eqn. 2. For both, strip and slot circuits, this mirrors the interrelation of the Fourier representation (eqn. 1) to an equivalent radial wave expansion. Indeed, the functions X and Y can easily be shown to have the properties of radial waveguide eigenfunctions [9] in the covered planar layered medium under discussion and that of radial surface wave functions if the cover is removed to infinity, respectively. In the original domain (x, y), eqns. 2 represent a pair of coupled integral equations mapping the two-dimensional vector quantities $e(x, y)$ and $i(x, y)$ into each other. Therefore, (Z_{mn}) and $(Z_{mn})^{-1}$ can be interpreted as the spectral domain dyadic Green's function of the prevailing class of boundary value problems. Because of the symmetry of (Z_{mn}) and its inverse, the numerical solution of the integral equation equivalent to eqn. 2 is most favourably achieved by Galerkin's method in the spectral domain which is in agreement with the results obtained for planar waveguides [10, 11]. However, the additional spatial dimension being present in the end-effect problem introduces difficulties that may put into question the realisability of such a solution on even a powerful modern digital computer. Consider, for example, the computational expense necessary for a rough spatial resolution of the field at the end of a 100 μm wide shorted slot which is positioned somewhere in a circuit area of one square inch. For this problem approximately a number of terms of $m = 1, \ldots, 1000$, $n = 1, \ldots, 1000$ in eqn. 2, i.e. a total of $m \cdot n = 10^6$ Fourier coefficients, have to be taken into account.

The numerical solution presented here starts from the original domain analogue of the spectral relationship, eqn. 2. In the symbolical operator form the mathematical problem posed in this way can be written as

$$\begin{pmatrix} e_x \\ e_y \end{pmatrix} = L_\infty(f, l_a) \begin{pmatrix} i_x \\ i_y \end{pmatrix}$$

and

$$\begin{pmatrix} i_x \\ i_y \end{pmatrix} = L_\infty^{-1}(f, l_a) \begin{pmatrix} e_x \\ e_y \end{pmatrix}, \quad \text{respectively} \tag{3}$$

$$e_x = e_y = 0 \text{ on } F_c$$

and

$$i_x = i_y = 0 \text{ on } \bar{F}_c, \quad \text{respectively}$$

where L_∞ denotes the strip problem integral operator and L_∞^{-1} is its dual counterpart relevant for slot configurations. F_c is the metallised substrate surface circuit area and $\bar{F}_c$ the associated nonmetallised part. The linear operators L_∞, L_∞^{-1} depend on the frequency of operation f, the geometry and the dielectric constants, and are nonlinear functions of the parameter l_a (see Fig. 2c and Eqn. 1), which defines the length of the short or open circuited planar transmission line stub under consideration. For a given operating frequency f there exist one or more resonant lengths l_a for which a nonvanishing surface current density i (slot electric field e) is mapped into a complementary electric field e (surface current density i) which vanishes on F_c (and $\bar{F}_c$). The required end effect phase information is directly deduced from the value of l_a which, in principle, is a numerical version of the Weissfloch or tangent method [8]. The advantage of proceeding in this way is that results obtained for the quantity l_a exhibit a stationary character with respect of the expansion functions to be selected for $i(x, y)$ or $e(x, y)$. In addition, with the special choice of expansion functions made here, all boundary conditions applying to the end-effect problem can be satisfied *a priori*, except for those in the immediate vicinity of the abrupt planar waveguide ends. The extension of this approach to planar 2-, 3- and 4-ports is a straight forward matter (see Appendix). It is currently being performed and provides a means for the rigorous frequency-dependent characterisation of planar n-ports. Its disadvantage is, that for a systematical variation of the stub length l_a, until an accurate solution is found, the operators L_∞ and L_∞^{-1}, respectively, have to be generated approximately five to eight times even if a good start estimate is available. However, compared to a deterministic source-type solution which has also been performed by the author for the purpose of comparison [12], it is an advantage that it is not necessary to specify suitable sources. In addition, the number of expansion functions can be held especially small since, with the absence of sources, no boundary conditions in the surrounding of sources have to be satisfied.

For a more detailed description of the numerical solution, consider eqns. 2 and 3 together with the representative sketches of Figs. 2 and 3. If the operators L_∞ or L_∞^{-1} are applied to a homogeneous planar transmission line extending between $y = -l_a$ and $y = l_b$ and the subscript of the separation constant k_{yn} is restricted to $n = 1$, they automatically reduce to operators with a one-dimensional space dependency. The associated solutions l_a then provide the guide wavelength λ_g (phase coefficient β) for the fundamental or any higher-order mode since $l_a + l_b$ must equal a multiple of half wavelengths. At the same time by means of eqn. 3 the field and surface current density of the planar wavefuide mode (and its characteristic impedance) i.e. its spatial or spectral transverse distribution, can be computed (see for example Fig. 3a). This is a one-dimensional problem and is solved by the same computer routine as that for the end effect with, however, a computational expense negligible compared to that of the end-effect problem. Then, according to Fig. 3c, a truncated version of the longitudinal pattern (Fig. 3b) in conjunction with the transverse distribution (Fig. 3a) is introduced into the set of expansion functions describing the abrupt end at $y = 0$, namely

$$\left.\begin{matrix} j_x(x,y) \\ e_x(x,y) \end{matrix}\right\} = a_{xo} \cdot \begin{Bmatrix} j_{xo}^T(x,y) \\ e_{xo}^T(x,y) \end{Bmatrix} + \sum_{\mu=1}^{M} \sum_{\nu=1}^{N} \alpha_{x\mu\nu} f_{x\mu}(x) g_{x\nu}(y) \tag{4}$$

$$\left.\begin{matrix} j_y(x,y) \\ e_y(x,y) \end{matrix}\right\} = a_{yo} \cdot \underbrace{\begin{Bmatrix} j_{yo}^T(x,y) \\ e_{yo}^T(x,y) \end{Bmatrix}}_{-l_a \leqslant y \leqslant 0} + \underbrace{\sum_{\mu=1}^{M} \sum_{\nu=1}^{N} \alpha_{y\mu\nu} f_{y\mu}(x) g_{y\nu}(y)}_{-\Delta y \leqslant y \leqslant 0}$$

where the superscript T denotes the truncation at $y = 0$. This kind of expansion ensures that, only in a certain region $-\Delta y \leqslant y \leqslant 0$, boundary conditions remain to be satisfied.

Physically, this is equivalent to an excitation of the abrupt end with a planar waveguide mode of known spatial distribution. The end-effect disturbance region Δy has to be chosen not too small so that the perturbation of the strip current or slot field caused by the abrupt end can be expected to have decayed at the left end of Δy. Estimates of Δy can be deduced from the knowledge of the substrate thickness and the line width. The funcitons $g_{x\nu}(y)$ and $g_{y\nu}(y)$ in eqn. 4 are chosen in such a way that they are suited to satisfy the edge condition and form a complete set in $-\Delta y \leqslant y \leqslant 0$. They contain a common y-dependent factor and with

$$g_{x\nu}(y), g_{y\nu}(y) \sim \left(1 + \frac{y}{\Delta y}\right)^3, \quad -\Delta y \leqslant y \leqslant 0 \qquad (5)$$

the total expansion eqn. 4 can be made two times continuously differentiable in its region of definition, $-l_a \leqslant y \leqslant 0$. This latter criterion is applied in agreement with related earlier results, for example [11], in order to avoid spurious, i.e. nonphysical solutions. The numerical end effect result is achieved by expanding the surface current density i, and the electric field e in the plane of the substrate $z = 0$ according to eqn. 4, and testing the complementary quantity e and i, respectively, with the same functions which is Galerkin's method [13]. The evaluation of the scalar products (matrix coefficients A_{ik}) representing the discrete system of eqns. 3 is performed either in the spatial domain or in the spectral domain making use of the relationship 3. Computationally, this is completely analogous because of the validity of Parseval's theorem for sine- and cosine-transforms [14] and

As has been indicated before, the crucial point concerning the realisability of a spectral-domain approach to planar two-dimensional structures is the generation of the elements (eqn. 6) which is extremely laborious. Owing to the use of a precalculated excitation term in eqn. 4, in the present approach the number of functions necessary to describe the end effect accurately is held very small. However, the possibility of choosing the values of MM and NN up to $MM = 1000$ $NN = 1000$ must be taken into account if an acceptable spatial resolution is to be achieved within realistic microwave housing dimensions. Especially, this prevails for low frequencies and narrow planar transmission line widths. For high frequencies, average widths, and in particular for millimeter-wave circuits, MM and NN of the order of 100 may be typical. Nevertheless, even this can be managed only by a high speed digital computer if the summations are performed straightforwardly according to eqn. 6.

For this reason, the approach so far outlined, is modified. First, the order in which the summations are performed is interchanged in eqn. 6 so that m is the inner summation variable. As a consequence of formulating the problem in terms of the stublength l_a the transverse distributions $f_\mu(x), f_\xi(x)$ and likewise the spectral coefficients $f_{\mu m}, f_{\xi m}$ remain fixed during the search for the resonant values of l_a. So, these coefficients have to be computed only once at the beginning, and are stored into a suitable array for the summation process. Note, that because of the frequency dependence of the excitation functions appearing in eqn. 4 this would not be possible if the solution were in terms of the operating frequency.

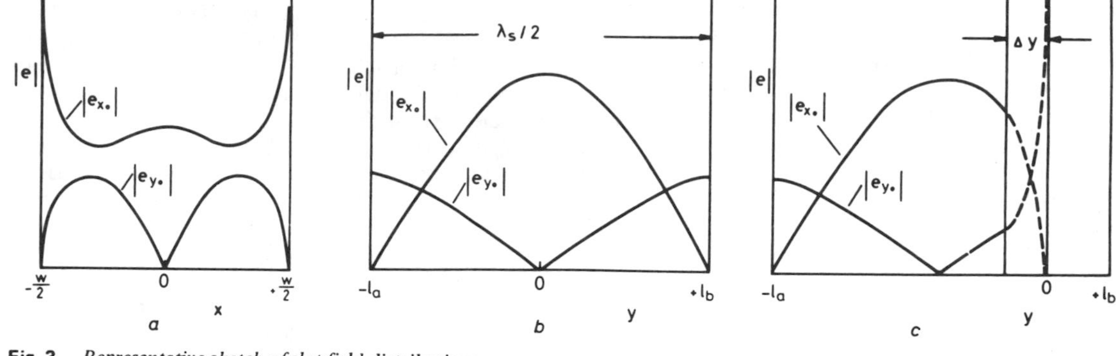

Fig. 3 *Representative sketch of slot field distributions*
a x-dependence
b y-dependence for ideal slot resonator
c Abrupt shorted end

the interchangeability of the integrations and summations contained in L_∞, L_∞^{-1}. The result is a set of infinite double series of the type

$$A_{ik}(f, l_a) = \sum_{m=1}^{\infty} \sum_{n=1}^{\infty} z_{mn}(f, l_a) f_{\mu m} g_{\nu n} f_{\xi m} g_{on} \qquad (6)$$

where

$$i = (\mu - 1)N + \nu, \; k = (\xi - 1)N + 0; \; \mu, \xi = 1, \ldots, M$$

and the subscripts x, y of the expansion and testing functions have been omitted for simplicity. For practical computations and with reasonably fast decrease of the Fourier coefficients $f_{\mu m}, \ldots, g_{on}$ of $f_\mu(x), \ldots, g_o(y)$ the upper summation limits of m and n in eqn. 6 are truncated, say at $m = MM$, $n = NN$. The resonant stub lengths l_a containing the required end-effect phase information are computed as the zeroes of the determinant of the linear system of equations which is symbolically represented by the element $A_{ik}(f, l_a)$.

As a second drastic improvement, use is made of the dominantly radial dependency of the spectral-domain Green's function, the structure of which is described by eqn. 2. Instead of generating the transcendental functions X_{mn} and Y_{mn} laboriously again and again, a radial, i.e. one-dimensional, fast spectral domain interpolant is employed. This is chosen to be of the three point Lagrangian type, or even linearly, if certain numerical precautions are made initially. As a result, the evaluation of eqn. 6 can be speeded up enormously and is now in the form

$$A_{ik}(f, l_a) = \sum_{n=1}^{NN} g_{\nu n} g_{on} \alpha_{\mu\xi}(n)$$

with

$$a_{\mu\xi}(n) = \sum_{m=1}^{MM} f_{\mu m} f_{\xi m} \bar{Z}_{mn},$$

$$\bar{Z}_{mn} = Z_{mn}\{\bar{X}(k_{\rho mn}^2), \bar{Y}(k_{\rho mn}^2)\} \qquad (7)$$

where the bar denotes the interpolation with respect to the variable k_ρ. Only with these modifications, high spatial resolutions can be achieved on a substrate surface of realistic dimensions and with moderate time and storage requirements. Nevertheless, some care is necessary to prevent numerical difficulties associated with the summation scheme eqn. 7. The presence of radial wave poles $k_\rho = k_{\rho p}$ (surface wave poles) in the continuous-spectral-domain Green's functions analogue $Z(k_\rho^2)$ and $Z^{-1}(k_\rho^2)$, respectively in eqn. 2 have to be taken into account properly. These poles can be swept by the discrete radial variable $k_{\rho mn}$, since this depends on the stublength l_a and varies continuously with the variation of l_a. In this way, poles are introduced into the eigenfunction of the prevailing planar one-port problem (also, of course, into that of n-ports). Fortunately, they can be eliminated analytically and this has been done in the present approach. So, the final eigenfunction to be investigated here is a smooth function and is free of poles, which speeds up the determination of the resonant stublength l_a considerably. In addition to this step, the numerical condition of the matrix with elements A_{ik} (f, l_a), i.e. the discrete version of the operator eqn. 3, is optimised here. Otherwise, the roundoff errors introduced by the interpolation (eqn. 7) may become noticeable in the high number of summations to be performed. This conditioning step is achieved by a proper choice of the upper summation limits MM and NN in conjunction with the specific choice of the expansion functions and cannot be discussed in detail here. It is found that with a ratio of MM/NN equal to the ratio of the sidewalls of the conducting RF-enclosure and MM, NN chosen large enough the numerical results exhibit an excellent stability. Visually, this means equal spatial resolutions with respect to the x- and the y-direction. Similar but not directly related rules for the solution of electromagnetic boundary value problems have been elaborated by Klein and Mittra [15].

A summarising inspection of the validity of the method described reveals the following: The approach takes into account the hybrid mode character and the frequency dependence of Maxwell's equations and, therefore, is a rigorous one. Its sole restriction comes from the presupposition introduced into the expansion 4, namely, that there exists a transmission-line region $-l_a \leqslant y \leqslant -\Delta y$ (see Fig. 3) in which the total field is most predominently that of a standing planar-transmission-line wave. Nevertheless, even this restriction can be removed by a proper choice of the reference plane $y = -\Delta y$. If necessary, the choice $l_a = \Delta y$ has to be made in order to ensure the completeness of the expansion 4. This is, however, accompanied by an increase of the number $2 \cdot M \cdot N$ of terms required to characterise the strip current or slot field distribution with sufficient accuracy. In practical computations, it turns out that only in the

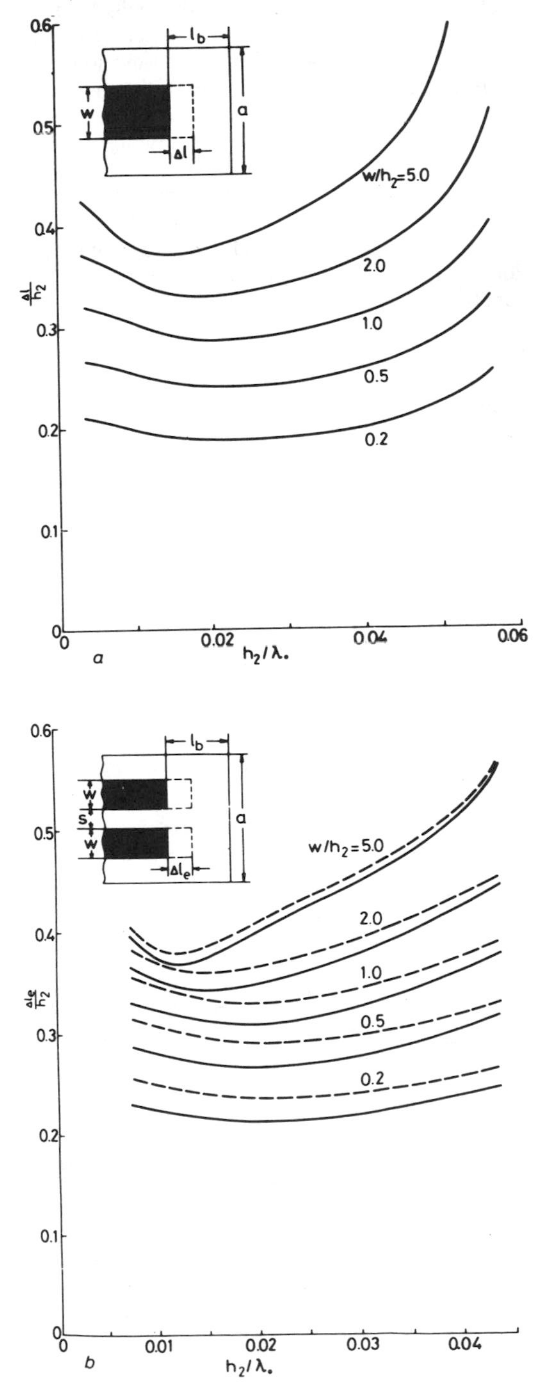

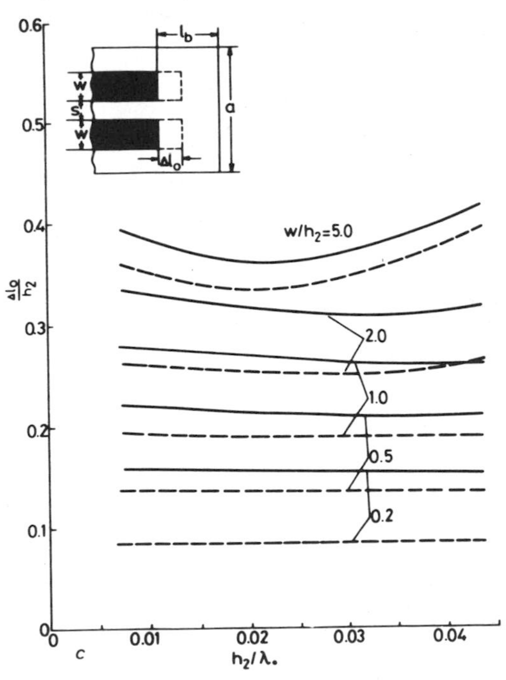

Fig. 4 *End effects of open-ended single and coupled microstrip lines*

a Single microstrip
b Coupled microstrip in even mode
c Coupled microstrip in odd mode
$h_2 = 0.635$ mm, $\epsilon_r = 9.7$
$h_1/h_2 = 0$, $h_3/h_2 = 10$, $a/h_2 = w/h_2 + 10$ and $a/h_2 = 2w/h_2 + s/h_2 + 10$, respectively.
$l_b/h_2 = 5$
---- $s/h_2 = 0.2$
—— $s/h_2 = 1.0$

neighbourhood of box resonances of the RF-housing or of higher-order-mode resonances the dominance of the excitation term is invalidated. These cases can be recognised easily from the behaviour of the amplitude coefficients a_{xo} and a_{yo} in eqn. 4, which weight the longitudinal and transverse excitation function towards each other. If the region of disturbance Δy has been chosen large enough and the representation 4 is able to satisfy the strip or slot boundary conditions the equality

$$|a_{xo}| = |a_{yo}| \quad \text{for } l_a = l_{areson}. \tag{8}$$

must emerge from the numerical solution automatically. The components of the vector excitation $j_o^T = (j_{xo}^T, j_{yo}^T)$ and $e_o^T = (e_{xo}^T, e_{yo}^T)$, respectively, are then related in the same way as on a homogeneous ideally shorted transmission line. A solution which invalidates the criterion 8 indicates the existence of higher-order modes or strong surface wave (radial wave) coupling in the circuit area under consideration. In any case, such operating conditions have to be avoided in planar circuit design. Therefore, the special form of the expansion 4 is just what one needs for design purposes and provides useful results with a minimum of computer time and storage.

3 Results

The numerical results presented in this paper have been generated on a Control Data CYBER 76 high-speed computer. They are given in normalised form, i.e. with respect to the substrate thickness h_2, so that maximum usefulness is gained for design purposes. The cross-sectional geometry prevailing for all of the following graphs Fig. 4 to Fig. 8 is that of Fig. 2*a* with vanishing layer thickness h_1 in the case of microstrip configurations. The parameter h_3 appearing in the figure captions denotes the cover height, i.e. the height above substrate of the conducting enclosure. The meaning of the variables describing the relevant planar transmission line geometry, the reference plane extension due to the end effect and the RF-shielding dimensions is obvious from a sketch included in each of the Figures. The quantity λ_0 denotes the free space wavelength.

The control variables prevailing for the computations have been chosen in such a way that the numerical accuracy achieved is about 2%, or better, with reference to a converged solution. For example, this means that the number of expansion functions used to describe the end-effect current density or electric-field disturbance has been chosen accordingly. This results in a typical matrix order of as low as 10 for the discrete version of the operator eqn. 3 and is due to the special choice of expansion functions made with eqn. 4.

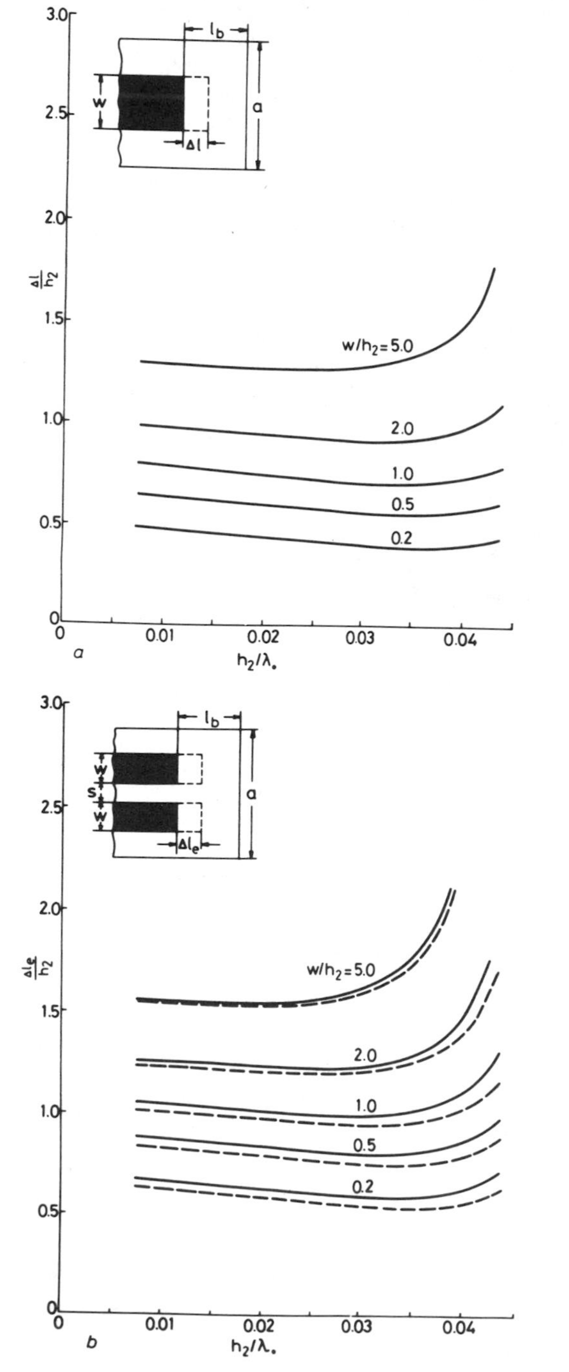

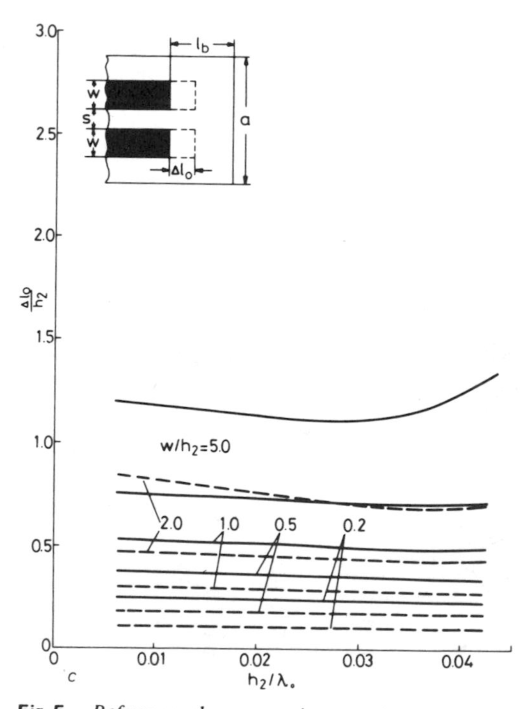

Fig. 5 *Reference plane extension associated with the open end of single and coupled suspended substrate lines*

a Single strip
b Coupled strips in even mode
c Coupled strips in odd mode
$h_2 = 0.635$ mm, $\epsilon_r = 9.7$
$h_1/h_2 = 1$, $h_3/h_2 = 10$
$a/h_2 = w/h_2 + 10$ and $a/h_2 = 2w/h_2 + s/h_2 + 10$, respectively
$l_b/h_2 = 5$
---- $s/h_2 = 0.2$
—— $s/h_2 = 1.0$

For a first proof of the validity of the numerical approach presented here, single and coupled microstrip open-end data are given in Figs. 4a, b and c. Though the microstrip case has been dealt with in detail in a separate paper [7] it will be discussed here briefly for the purpose of completeness. The reference plane extension Δl applying to the single microstrip open end and its frequency dependence is depicted in Fig. 4*a*. It turns out that the values of Δl on the standard alumina substrate considered are of the order of 20%–40% of the substrate thickness. The quasistatic limits of Δl are in good agreement with the data published by Silvester and Benedek [16] and Hammerstad and Bekkadal's microstrip handbook [17], and as is explicitly shown in Reference 7. The frequency dependence of the equivalent line length Δl is such that it first decreases from its static value by some per cent. In accordance with Itoh's interpretation [4] this is thought to be due to the lumped inductance associated with the strip current density disturbance at the open end. Then, with further increasing frequency, the magnitude of Δl rises again, which can be explained as a consequence of interaction of the open strip end with the field in the volume of the shielding box [7]. The height h_3 above substrate of the cover screen has only a negligible effect as long as it exceeds some times the substrate thickness h_2 and is not large enough to make the existence of vertical box resonances possible. Therefore, for practical circuit design, the cover height h_3 should not be chosen too large, as is discussed in detail in Reference 7. However, numerical or convergence problems do not arise for large screen separations in the method described since the value of h_3 enters the computations in an analytical fashion. So, the numerical technique of this paper is capable of supplying accurate physical results independent of the height of the shielding box prevailing in a given practical situation.

Nevertheless, it should be pointed out that this does not imply that radiation effects can be modeled for large screen separations h_3. The frequency-dependent results given here do not involve radiation in the sense of the computations as presented, for example, by James and Henderson [18]. In the present paper, radiation and surface wave excitation by an abrupt planar transmission-line end is considered only in so far as, by backscattering from the shielding walls, it is part of the total electromagnetic field. As a consequence, i.e. because of interaction with the shielding box, and particularly in the vicinity of volume resonances, there may result values of the equivalent line length Δl, different from those computed for the unshielded case [18]. Only for the quasistatic limit and for configurations (not necessarily at low frequencies) in which interaction of the transmission-line end and the shielding box is weak, good agreement between end-effect results of the shielded and unshielded type can be expected (see References 7 and 18 for comparison).

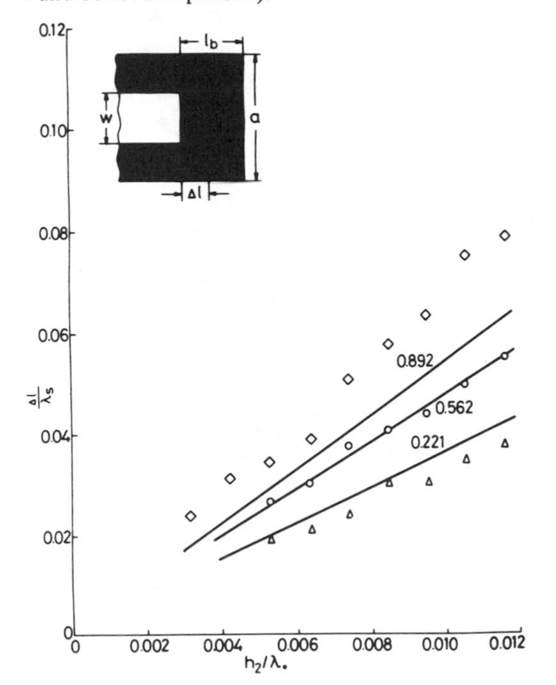

Fig. 7 *Shorted slot end-effect data in comparison with the measured results of Knorr and Saenz (Single slot)*

$h_2 = 0.121$ in, $\epsilon_r = 12$
$h_1/h_2 = 5$, $h_3/h_2 = 5$,
$a/h_2 = w/h_2 + 15$, $l_b/h_2 = 3$
◇◇◇◇ $w/h_2 = 0.892$, Reference 20
○○○○ $w/h_2 = 0.562$, Reference 20
△△△△ $w/h_2 = 0.221$, Reference 20
——— this paper.

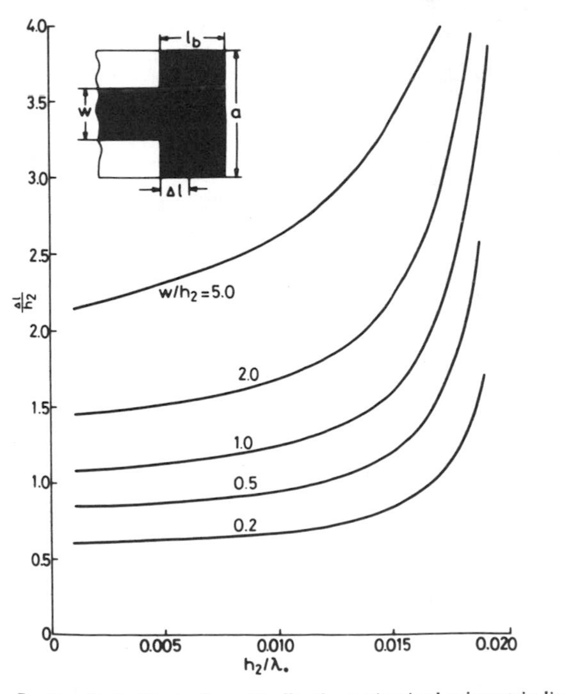

Fig. 6 *End effect of nonideally short circuited microstrip lines as a function of strip width and frequency (single microstrip)*

$h_2 = 0.635$ mm, $\epsilon_r = 9.7$
$h_1/h_2 = 0$, $h_3/h_2 = 10$,
$a/h_2 = w/h_2 + 10$, $l_b/h_2 = 5$

For the end effect of open-ended coupled microstrip lines in the even and odd mode the argument is completely analogous and the associated values of Δl against frequency are given in Figs. 4*b* and 4*c*. The even mode value of Δl behaves roughly like that of a single strip of double width $2w$ if coupling between the strips is not too loose. If the gap width s between the strips increases, the single-strip case should be approached, i.e. Δl must become smaller as can be seen in Fig. 4*b*. In contrast to this, the odd-mode reference plane extension Δl grows up with increasing gap width s (see Fig. 4*c*). Apparently, for the odd-mode case the behaviour of the end effect is dominated by the slot s formed between the strips if coupling is tight. So, as a consequence of the odd-mode field distribution the single-strip limit of Δl for $s \to \infty$ is approached from the other side. In addition, the influence of the spacing s on the end effect is much more noticeable than for the even mode. For both modes the quasistatic values of Δl have been compared with the static-capacitance data of Gopinath and Gupta [19] and have been found to be in good agreement [7].

A short inspection of the end-effect data referring to

suspended substrate lines in Figs. 5*a*, *b* and *c* immediately reveals that for this type of transmission line the average value of Δl is considerably larger than for microstrip lines on the same substrate. Here, typical magnitudes of Δl are 50%–100% of the substrate thickness if the material is alumina. The frequency behaviour of the end effect is similar to that of microstrip open ends. This should be expected since the field distribution of suspended substrate lines is not very different from that of microstrip lines, however, with a considerable fraction of a electromagnetic field energy extending into the air regions around the guiding strip and strips, respectively. As in the microstrip case, a gap *s* between coupled strips has only a little influence on the reference plane extension Δl applying to the even mode. Furthermore, the value of Δl associated with the odd mode exhibits a tendency to increase with the gap width *s* which is completely analagous to the behaviour of coupled microstrip lines.

A situation completely different from those discussed before, is featured in Fig. 6. There, a nonideally short circuited single microstrip is considered together with the equivalent line length Δl describing this configuration quantitatively. The microstrip sketched in Fig. 6 terminates vertically on a metallised plane which extends towards the conducting walls of the shielding box. In this way, a short circuited connection to the substrate bottom metallisation is realised at least for relatively low operating frequencies. For practical design purposes, particularly on plastic substrates, it is proposed to use through-metallisations left and right of the microstrip end to get short circuits with quantifiable electrical properties. The resulting value of Δl is considerably larger than for the microstrip open end, namely 50%–200% of the substrate thickness for the standard alumina material under discussion. From the physical situation, it should be clear and has been confirmed by computations that there is a reltively strong influence of the shielding width *a* on the equivalent line length Δl. Likewise, the frequency dependence of Δl is strong and its value increases steadily as if it were approaching a pole. The associated frequency is easily detected to be the cutoff frequency of the dielectric-filled rectangular wave guide which is formed by the top and bottom metallisation of the substrate together with the lateral shielding walls and is excited by the microstrip end. However, regarding the numerical quantity of Δl, it should be kept in mind that for such a case of strong field interaction the expansion 4 may have been applied in an insufficient manner. Here, the author would like to refer to his discussion of the validity of the method described.

As a first set of results obtained for slot lines, a comparison has been performed with the measured data published by Knorr and Saenz [20] (see Fig. 7). The agreement between

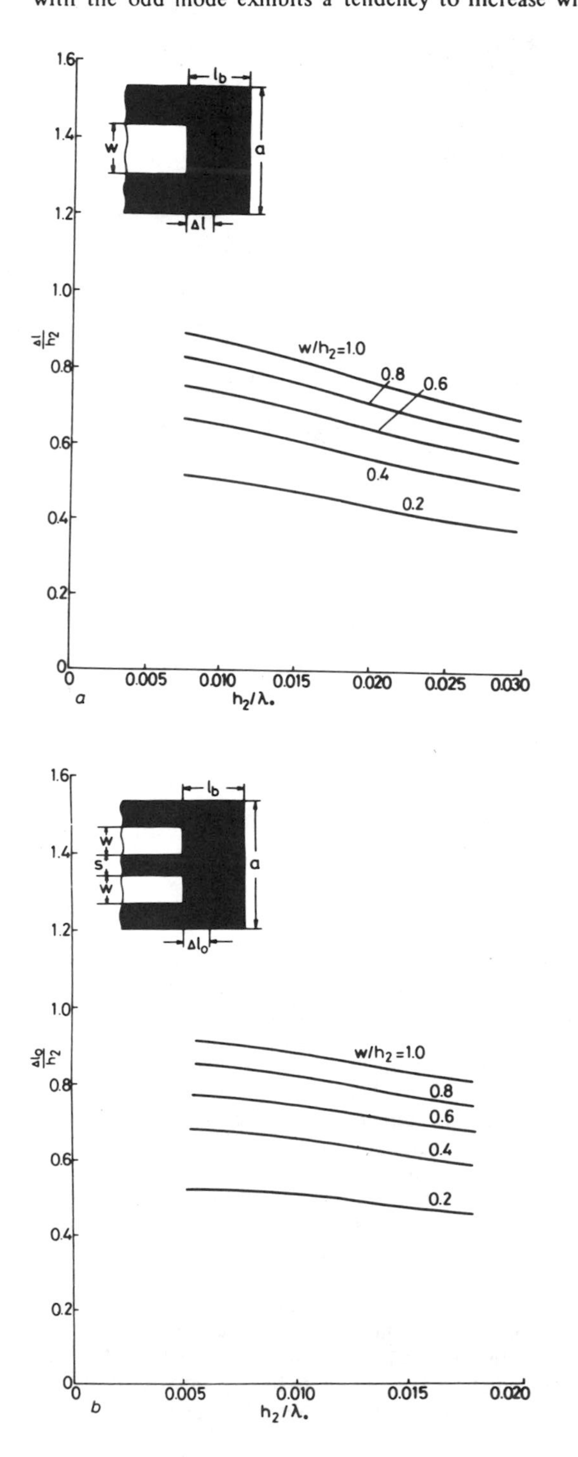

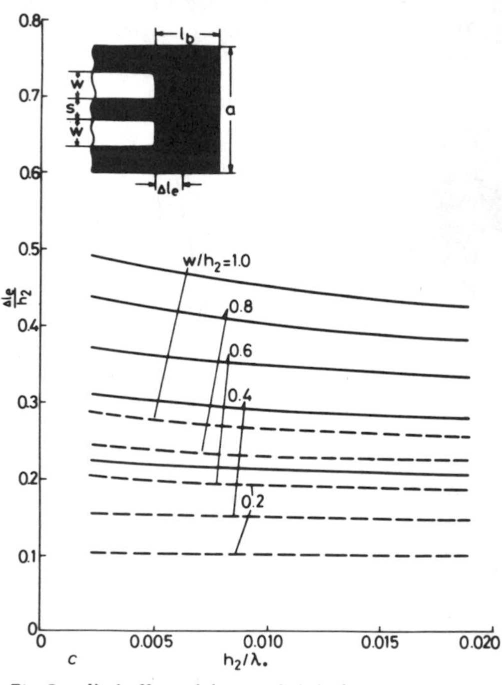

Fig. 8 *End effect of short-ended single and coupled slot lines*

a Single slot
b Coupled slots in odd mode
c Coupled slots in even mode (coplanar line)
$h_2 = 0.635$ mm, $\epsilon_r = 9.7$
$h_1/h_2 = 10$, $h_3/h_2 = 10$
$a/h_2 = w/h_2 + 20$ and $a/h_2 = 2w/h_2 + s/h_2 + 20$, respectively
l_b/h_2
– – – – $s/h_2 = 0.2$
——— $s/h_2 = 1.0$

these data and the computed end-effect reference-plane extensions Δl of this paper is satisfactory if one considers that the details of the measuring set up used in Reference 20 are not available to the author. For example, the substrate dimensions, the cover height above the substrate, the elevation of the substrate above the ground plate are all parameters which influence the computed data more or less sensitively. Furthermore, the coupling probes used by Knorr and Saenz for their measurements are suspected to have had some influence. In the author's experience, such probes can introduce measurement errors, especially, if slots are wide, i.e. the field is only loosely guided by the slot, or, on the other hand, if slots are particularly narrow. In this latter case, the electric field is concentrated well within the slots, but, for measureable coupling, the probes have to be in near contact with the slot. So, there are a lot of potential sources of error which could be responsible for the slight discrepancies in Fig. 7.

A more systematic set of results for the shorted end of single and coupled slot lines is depicted in Figs. 8*a*, *b* and *c*. The single slot results of Fig. 8*a* show equivalent line lengths Δl which are in the order of magnitude of 50%–90% of the alumina substrate thickness. The values of Δl exhibit a decrease with increasing operating frequency. They become smaller rapidly if the slot width w tends to zero which is understandable from the slot field distribution.

Coupled slots in the odd mode, and the associated end-effect data, are shown in Fig. 8*b*, and can be seen to behave in a very similar way to single slots. In contrast to previous coupled line figures of this paper, only one single value of slot distance $s/h_2 = 0.2$ has been introduced into Fig. 8*b*, since the set of curves which would have resulted from $s/h_2 = 1.0$ could hardly have been distinguished on the scale of this graph. This is equivalent to the fact that only for very tight slot coupling the effect of s on the end-effect quantity Δl becomes noticeable. The situation is different, however, for coupled slots in the even mode which implies quasi-TEM coplanar line operation. Here, the relevant end-effect values of Δl are smaller, namely 20%–50% of the substrate thickness, and exhibit a considerable dependency with respect to the width s of the coplanar line centre strip. The equivalent line length Δl decreases slightly with frequency and increases in a pronounced way with the centre strip width s. To explain this it should be kept in mind that the parameter s strongly affects the coplanar line characteristics as a whole, particularly the line chacteristic impedance.

4 Conclusion

A numerical solution to the frequency-dependent hybrid mode boundary value problem of planar transmission line end effects has been presented. The method applied is a rigorous, efficient and unified one, i.e. it is not restricted to a specialised planar waveguide geometry. For some technically important guiding structures end effect data have been generated numerically as an aid to improved microwave and millimetre-wave circuit design. The computational expense and the storage requirements involved are moderate since for the open- or short-end problem the method employed results in low matrix orders, typically eight to ten. Beyond this, the numerical approach of this paper can easily be extended to planar n-ports as is shown in the Appendix. So, the end-effect results presented here are considered to be just the first step towards the frequency dependent characterisation of more complex planar configurations. The next steps are at present being done in the author's group with the aim of improving microwave CAD capabilities.

5 Acknowledgment

Thanks are due to N.H.L. Koster for useful discussions on the problem and for his support in the preparation of the numerical results.

6 References

1 DAVIES, J.B., and MIRSHEKAR-SYAHKAL, D.: 'Spectral domain solution of arbitrary coplanar transmission line with multilayer substrate', *IEEE Trans.*, 1977, **MTT-25**, pp. 143–146

2 JANSEN, R.H.: 'Unified user-oriented computation of shielded, covered and open planar microwave and millimeter-wave transmission-line characteristics', *IEE Proc. H, Microwave, Opt. & ant.*, 1979, **3**, pp. 14–22

3 JANSEN, R.H.: 'Shielded rectangular microstrip disc resonators', *Electron. Lett.*, 1974, **10**, pp. 299–300

4 ITOH, T.: 'Analysis of microstrip resonators', *IEEE Trans.*, 1974, **MTT-22**, pp. 946–951

5 JANSEN, R.H.: 'Computer analysis of shielded microstrip structures,' (in Germany) Electron. & Commun. (AEÜ), 1975, **29**, pp. 241–247

6 JANSEN, R.H.: 'High-order finite element polynomials in the computer analysis of arbitrary shaped microstrip resonators', *ibid.*, 1976, **30**, pp. 71–79

7 JANSEN, R.H., and KOSTER, N.H.L.: 'Accurate results on the end-effect of single and coupled microstrip lines for use in microwave circuit design', *ibid.*, 1980, **34**, pp. 453–459

8 COLLIN, R.E.: 'Field theory of guided waves' (McGraw-Hill, 1960)

9 HARRINGTON, R.F.: 'Time-harmonic electromagnetic fields' (McGraw-Hill, 1961)

10 ITOH, T., and MITTRA, R.: 'A technique for computing dispersion characteristics of shilded microstrip lines', *IEEE Trans.*, 1974, **MTT-22**, pp. 896–898

11 JANSEN, R.H.: 'High-speed computation of single and coupled microstrip parameters including dispersion, high-order modes, loss and finite strip thickness', *ibid.*, 1978, **MTT-26**, pp. 75–82

12 JANSEN, R.H.: 'SFPMIC-Source formulation approach to planar microwave integrated circuits'. Unpublished results and computer program, University of Duisburg, W. Germany, 1979/80

13 HARRINGTON, R.F.: 'Field computation by moment methods' (Macmillan, 1968)

14 SNEDDON, I.N.: 'The use of integral transforms' (Tata McGraw-Hill, New Delhi, India, 1974)

15 KLEIN, Ch., and MITTRA, R.: 'Stability of matrix equations arising in electromagnetics', *IEEE Trans.*, 1973, **AP-21**, pp. 902–905

16 SILVESTER, P., and BENEDEK, P.: 'Equivalent capacitances of microstrip open circuits', *ibid.*, 1972, **MTT-20**, pp. 511–516

17 HAMMERSTAD, E., and BEKKADAL, F.: 'Microstrip handbook', ELAB Rep. STF44A74169, University of Trondheim, Norway, 1975

18 JAMES, J.R., and HENDERSON, A.: 'High-frequency behaviour of microstrip open-circuit terminations', *IEE Proc. H. Microwaves Opt. & ant.*, 1979, **3**, pp. 205–218

19 GOPINATH, A., and GUPTA, Ch.: 'Capacitance parameters of discontinuities in microstrip lines', *IEEE Trans.*, 1978, **MTT-26**, pp. 831–836

20 KNORR, J.B., and SAENZ, J.: 'End effect in a shorted slot', *ibid.*, 1973, **MTT-21**, pp. 579–580

21 ROS, A., *et al.*: 'Analysis of some three-dimensional microstrip discontinuities in terms of frequency', 8th European MW Conference, 1978, pp. 111–115

22 ROS, A., *et al.*: 'Application of an accelerated TLM-method to microwave systems', 10th European MW Conference, 1980, pp. 382–388

7 Appendix

The extension of the numerical method of this paper to the rigorous frequency-dependent characterisation of planar n-ports is a straightforward matter. Essentially, it is achieved by a generalisation of the Weissfloch or tangent method [8] in conjunction with the efficient spectral domain approach described. A first but not systematic step into the direction followed here has recently been done by Ros *et al.* [21], who combined the Weissfloch method with John's TLM-technique for the analysis of microstrip discontinuities. However, this approach results in excessively large orders of the involved matrices [21, 22], since it is a truly three-dimensional one and is thus not very practicable.

The concept presented here contains a computational scheme by which the complex power based scattering parameters of a planar microwave n-port are directly derived from a set of n real number resonator field solutions, i.e. without the need for specifying equivalent circuits. The numerical computation of the resonator fields for a given frequency is performed by exactly the same technique as has been applied to the end-effect problem. Again, expansions of the type in eqn. 4 can be used, however, with preformulated transmission line terms (subscript 0 in eqn. 4) for each of the n ports. Typically, this results in low matrix orders. Moreover, this concept does not require a detailed knowledge of the field in the direct vicinity of a junction or discontinuity due to the somewhat stationary character of the associated resonator solution as has been evolved in numerical practice. Nevertheless, it provides accurate phase and magnitude S-parameter data by purely real number field computations. It is also applicable to mixed-type planar structures, like for example microstrip-slot transitions.

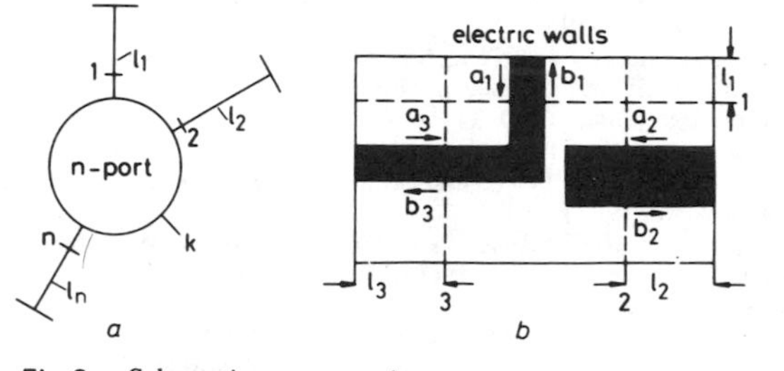

Fig. 9 *Schematic representations*

a Planar n-port and associated physical strip
b Slot n-port configurations

For a detailed description of the concept, consider Figs. 9*a* and 9*b*. According to Fig. 9*a*, a number of n ideally short-circuited microstrip, slot or coplanar stubs of lengths $l_1, \ldots, l_n$ are thought to be attached to the reference planes $1, \ldots, n$ of a lossless reciprocal n-port. The reference planes are chosen close to the junction, but, far enough away to allow the assumption that current density or field disturbances have decayed, see for example Fig. 9*b*. For a fixed combination of stub lengths, one of these, say l_k, is chosen in such a way, that the lossless n-port resonates at the predescribed operating frequency. This is computer simulated for $k = 1, \ldots, n$ and results in n sets of wave amplitudes which are used to assemble the complex scattering matrix of the n-port. Mathematically, this scheme is described in terms of the vectors of complex wave amplitudes (a) and (b) in the following way:

$$(b) = (S)(a) \quad \text{and} \quad (a) = (D)(b), \quad D_{ii} = -e^{-j2\beta_i l_i}$$

$$\{(s) - (D)^{-1}\}(a) = 0 \rightarrow \det\{(S) - (D)^{-1}_{ok}\} = 0$$

$$\begin{array}{lll} (l)_{o1} \rightarrow (a)_{o1}, & (b)_{o1} = (D)^{-1}_{o1}(a)_{o1}, & (b)_{o1} = (S)(a)_{o1} \\ \vdots & \vdots & \vdots \\ (l)_{on} \rightarrow (a)_{on}, & (b)_{on} = (D)^{-1}_{on}(a)_{on}, & (b)_{on} = (S)(a)_{on} \end{array} \tag{9}$$

$$\begin{gathered} \{(b)_{o1}, \ldots, (b)_{on}\} = (S) \cdot \{(a)_{o1}, \ldots, (a)_{on}\}, \\ (B)_o = (S) \cdot (A)_o \rightarrow (S) = (B)_o (A)_o^{-1} \end{gathered} \tag{10}$$

Owing to the special circuit configuration depicted in Fig. 9a the wave amplitudes (a) and (b) are interrelated in two ways, namely by the n-port scattering matrix (S) and the diagonal matrix (D). For a certain set of stub lengths $(l)_{ok}$ the configuration resonates, i.e. the determinant of (S) minus the inverse of $[D\{(l)_{ok}\}] = (D)_{ok}$ must vanish. This fictitious experiment is repeated n times and leads to n sets of stub lengths $(l)_{o1}, \ldots, (l)_{on}$ n eigenvectors of associated wave amplitudes $(a)_{o1}, \ldots, (a)_{on}$ and n vectors of reflected waves $(b)_{o1}, \ldots, (b)_{on}$. In each of these experiments, the relationship symbolised by the scattering matrix (S) is preserved. So, the results can be assembled as in eqn. 10 and the matrix (S) can be computed according to eqn. 10. All the information necessary to perform these steps is extracted from a number of planar resonator field solutions and the precomputations made to generate the planar transmission line terms contained in the expansion 4.

The concept described is especially practicable for a number of ports $n = 1, \ldots, 4$ since these cases can be dealt with using a rectangular shielding box. This involves the majority of technically important cases. In addition, it has the advantage that the number of necessary numerical resonator analyses becomes not too high. Compared to a source formulation type of solution it is advantageous that boundary conditions which would have to be satisfied in the vicinity of sources are eliminated. By this, the number of expansion functions can be held particularly small. The concept elaborated here is presently developed further to become a useful tool in the computer-aided design of planar microwave and millimetre-wave circuits.

THE NATURE OF THE LEAKAGE FROM HIGHER MODES ON MICROSTRIP LINE

A. A. Oliner and K. S. Lee

Polytechnic University
Brooklyn, New York 11201

ABSTRACT

Some confusion in the literature is clarified regarding the properties of microstrip line higher modes in the neighborhood of cutoff. It is shown that those modes become leaky in that range, and that the leakage occurs in two forms, a surface wave and a space wave. Numerical values obtained from an accurate analysis are presented that illustrate the nature of the leakage for microstrip lines with either open or covered tops.

1. Introduction and Summary

During the late 1970's, a paper presented by H. Ermert at the European Microwave Conference stimulated instant controversy. That paper and a subsequent publication [1] presented a thorough mode-matching analysis of modes on microstrip line, treating numerically the dominant mode and the first two higher modes. A principal conclusion was that a "radiation" region exists close to the cutoff of those modes. Because the description of this region, made in that talk and in published papers [1,2], was incomplete and therefore unclear to many, confusion persisted and certain practical consequences remained hidden.

Also in this general period, a paper by W. Menzel [3] presented a new traveling-wave antenna on microstrip line fed in its first higher mode and operated near to the cutoff of that mode. Menzel assumed that the propagation wavenumber of the first higher mode was real in the very region where Ermert said no such solutions exist; since his guided wave, with a real wavenumber, was fast in that frequency range, Menzel presumed that it should radiate. His approximate analysis and his physical reasoning were therefore also incomplete, but his proposed antenna was valid and his measurements demonstrated reasonably successful performance.

The first feature of the present paper involves the clarification of the confusion or contradictions implicit in the paragraphs above. These apparent contradictions are resolved when it is realized that leaky modes are present in this "radiation" region, and particularly so if the region can be characterized by only a single leaky mode. Not all leaky modes are physically significant, and more than one leaky mode may be present at the same time; each case must be examined separately for the physical significance of the role of leaky waves in any given "radiation" region. We have conducted such an examination, making use of the steepest descent plane, and we found that Ermert's "radiation" region is characterized in a highly convergent manner by essentially a single leaky mode.

Once we recognize the relevance of leaky modes to the "radiation" region of microstrip line higher modes, the application to leaky wave antennas becomes evident. In particular, it is clear that Menzel's antenna is a leaky wave antenna in principle, even though he did not recognize this fact and did not discuss the antenna's design or behavior in those terms. A leaky wave analysis explains quantitatively the performance features and the limitations of his antenna, and it also tells us how to improve the antenna performance in a controlled way. We will not be discussing any application to antennas here, however, but we will confine our attention to other interesting features of these modes, such as the nature of the leakage, and the need to modify our concepts of "cutoff" for higher modes on microstrip line.

It is shown that the power in any of these leaky modes is contained in onlv a surface wave in part of the frequency range and in a combination of a space wave and a surface wave in the remainder of the range. Simple conditions define the relevant portions of the range.

When the microstrip line is open above as well as on the sides, the space wave corresponds to radiation in a specified pattern. When the microstrip line is open on the sides but has a top cover, the "space wave" consists of a number of non-surface-wave modes that propagate away in the outside waveguide, composed of the dielectric layer between parallel plates. If the plate spacing is made larger (relative to wavelength), the number of propagating modes will increase. From a mode-matching analysis, we have found that, when the surface wave and the "space wave" are both present, the portion of power in the surface wave decreases substantially as the plate spacing is increased. Results as a function of top cover spacing will be

Reprinted from *IEEE MTT-S Digest*, pp. 57-60, 1986.

presented.

We have derived an accurate transverse resonance formulation for the propagation characteristics of the higher modes, both in the purely bound range (real wavenumbers) and the "radiation" range (complex wavenumbers). Using this solution, we present numerical comparisons with special cases in the literature, showing the differences between the values for an open top and a covered top. In this derivation, we employed a rigorous (Wiener-Hopf) solution derived by D. C. Chang and E. F. Kuester [4] for the reflection from one side of that microstrip line. We made a parametric dependence study of the leakage and phase constants of the first three higher modes in the "radiation" range; we found that the leakage rate α grows rapidly as the mode approaches "cutoff," as expected, but that the phase constant β behaved unexpectedly. After approaching zero, it slowly increased again and continued to increase as the frequency was lowered further, requiring us to alter our earlier understandings of the nature of "cutoff" for higher microstrip line modes in open regions.

2. The "Radiation" Region and Leaky Modes

One of the figures presented by Ermert [1,2] is reproduced here, with modifications, as Fig. 1. His curves are the solid ones shown, for the lowest mode and the first two higher modes of microstrip line. All of his wavenumber values are real, meaning that the modes are purely bound in those ranges. He states, however, that in the region shown lined no real solutions exist, and he called this region the "radiation region." We have added the dashed lines appearing in this region in Fig. 1, which corresponds to complex solutions, and where, of course, only the real part is plotted. Physically, these complex solutions signify that this mode has become leaky in this region.

Ermert selects a spectral description for the modes of microstrip, and in his second paper [2] he rejects any inclusion of leaky modes since they are nonspectral (true). He then concludes that these leaky modes are "no longer of importance" in his analysis (false). His rejection of leaky modes not only caused much initial confusion, but it prevents one from understanding certain practical consequences. Not all leaky modes are physically significant, but we have shown by employing the steepest descent plane that for this problem the continuous spectrum in Ermert's radiation region is characterized in a highly convergent manner by essentially a single leaky mode. The physical importance of leaky modes despite their nonspectral nature is quite an old story, but it must be shown in each case that a particular leaky mode is physically valid; in this case, we have shown that it is, in agreement with obvious physical intuition.

3. The Two Forms of Leakage

It is shown next that leakage can occur in two forms: a surface wave and a space wave. Furthermore, the onset of leakage for each form is given by simple conditions.

A top view of the strip and the dielectric region around it is shown in Fig. 2. With this figure, we examine the case of leakage away from the strip in the form of a surface wave on the dielectric layer outside of the strip region. When there is leakage into the surface wave, the modal field propagates axially (in the z direction) with phase constant β, and the surface wave propagates away (on both sides) at some angle with phase constant k_s, as shown in Fig. 2. The surface wave wavenumber k_s has components k_z and k_x in the z and x directions, respectively, where k_z must be equal to β, since all field constituents are part of the same leaky modal field. We may therefore write:

$$k_x^2 = k_s^2 - \beta^2 \tag{1}$$

For actual leakage, k_x must be real, so that the condition for leakage is $k_x^2 > 0$. (When there is no leakage, i.e., the mode is purely bound, the modal field decays transversely and k_x is imaginary.) Applying this condition to (1), we find that, for leakage,

$$\beta < k_s \tag{2}$$

Relation (2) defines the lined region in Fig. 1; the upper boundary of that region is actually the dispersion curve for the surface wave, of wavenumber k_s, that can be supported by the dielectric layer on a ground plane, if the microstrip line is open above, or by the dielectric layer between parallel plates, if there is a metal top cover. At the onset of the surface wave, it emerges essentially parallel to the strip axis, consistent with the condition $\beta = k_s$.

As β (and therefore the frequency) is decreased below the value k_s, power leaks away in the form of a surface wave, as discussed above. As β is decreased further, power is then also leaked away in another form, the space wave. If the microstrip line is open above, this space wave actually corresponds to radiation at some angle, the value of this angle changing with the frequency. At the onset of this space wave, the wave emerges essentially parallel to the strip axis, so that $\beta = k_o$ then, where $k_o = (2\pi/\lambda_o)$ is the free space wavenumber. This boundary corresponds to the horizontal line $\beta/k_o = 1$ in Fig. 1. For values of $\beta/k_o<1$, or

$$\beta < k_o \tag{3}$$

power will leak into a space wave in addition to the surface wave.

What happens when the microstrip line has a top cover, of height h? If $h<\lambda_o/2$, approximately, such that only the surface wave can propagate in the dielectric-loaded parallel plate region, then all the other modes are below cutoff, and power can leak away only in surface wave form. If the plate spacing is increased, then some of the non-surface-wave modes are above cutoff, and these modes can also carry away power. The "space wave" then corresponds to the sum of those modes.

At what value of β do these "space wave" modes begin to contribute to the leakage? The value depends on the height h of the top cover. Space does not permit us to present the appropriate relations, but, as examples, we find that for h/λ_o = 5, 2, and 1, β/k_o = 0.995, 0.97, and 0.87, respectively.

4. Numerical Values for the Leakage and Phase Constants

We have derived an accurate expression for the propagation characteristics of the first higher microstrip mode, both in its purely bound range (real wavenumbers) and in its leakage range (complex wavenumbers). Our analysis involves a horizontal transverse resonance in which we employed as a key constituent a solution for the strip sides given by Chang and Kuester [4].

We compared our numerical values for open tops with special cases appearing in the literature for covered tops, to determine where they differ. First, we compared our results in the real wavenumber range with those of Ermert [1]; the curves will be presented in the talk. Then, in the complex wavenumber range, the only available numbers for a special case (a top cover with small height) are given by J. Boukamp and R. H. Jansen as part of a larger paper [5]. Our comparison with their numbers is presented in Fig. 3. It is interesting to note that entirely different theoretical approaches were used in these three cases; Ermert employed a horizontal mode-matching procedure, and Boukamp and Jansen used a vertical spectral domain approach.

In Fig. 3, we plot the leakage constant α/k_o and the phase constant β/k_o as a function of frequency. The solid line curves represents our solution, which is valid for an open microstrip line, and the dashed curves are the numbers presented by Boukamp and Jansen for a line with a top cover of small height. The dimensional parameters, which are identical for the two cases except for the top cover, are given in the figure caption.

It is seen that in the plots in Fig. 3(a) for the leakage constant the two curves are roughly parallel to each other, with the covered case leaking more strongly, even though the leakage for each begins at about the same frequency. We should also note that the forms of leakage are different for each. For the solid line, the leakage is mostly in the form of a space wave, whereas the dashed line, for the covered top case, has all the leakage in surface wave form. Another important feature evident from Fig. 3(a) is that the values of α/k_o are quite high, meaning that a substantial amount of power can leak per unit length if the frequency is made somewhat lower than the one corresponding to the onset of leakage. For example, for a frequency of 12.8 GHz for this case, for the structure with a top cover, the value of α/k_o is about 0.20, corresponding to a leakage attenuation rate of about 4.6 dB/cm! For the open structure, the leakage rate is lower but still high, being about 2.4 dB/cm.

5. References

1. H. Ermert, "Guided Modes and Radiation Characteristics of Covered Microstrip Lines," A.E.U., Band 30, pp. 65-70, February 1976.

2. H. Ermert, "Guiding and Radiation Characteristics of Planar Waveguides," IEE Microwave, Optics and Acoustics, Vol. 3, pp. 59-62, March 1979.

3. W. Menzel, "A New Travelling-Wave Antenna in Microstrip," A.E.U., Band 33, pp. 137-140, April 1979.

4. D. C. Chang and E. F. Kuester, "Total and Partial Reflection from the End of a Parallel-Plate Waveguide with an Extended Dielectric Loading," Radio Science, Vol. 16, pp. 1-13, January-February 1981.

5. J. Boukamp and R. H. Jansen, "Spectral Domain Investigation of Surface Wave Excitation and Radiation by Microstrip Lines and Microstrip Disk Resonators," Proc. European Microwave Conference, Nurnberg, Germany, September 5-8, 1983.

Acknowledgment

This research has been supported in part by the Air Force Rome Air Development Center, Hanscom Air Force Base, under Contract No. F19628-84-K-0025, and in part by the Joint Services Electronics Program, under Contract No. F49620-85-C-0078.

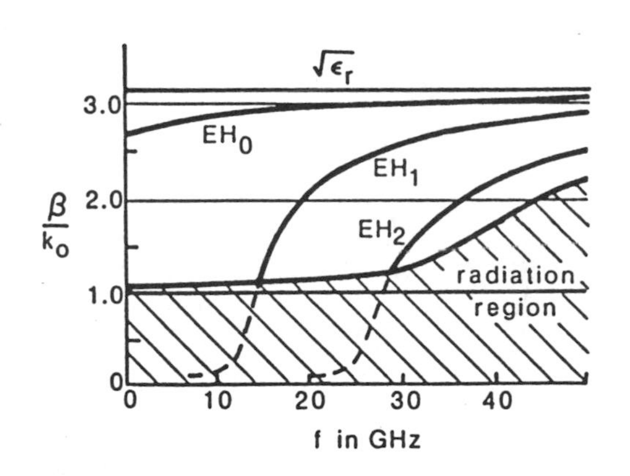

Fig. 1 Dispersion curves for the lowest mode and the first two higher modes in microstrip line with a top cover. The normalized phase constant β/k_o is plotted against frequency. The solid lines (given by Ermert [1,2]) represent real wavenumbers, whereas the dashed lines correspond to the real parts of the leaky mode (complex) solutions in the "radiation region." The microstrip line dimensions are: strip width = 3.00 mm; dielectric layer thickness = 0.635 mm, ϵ_r = 9.80, and the height of the top cover is five times the dielectric layer thickness.

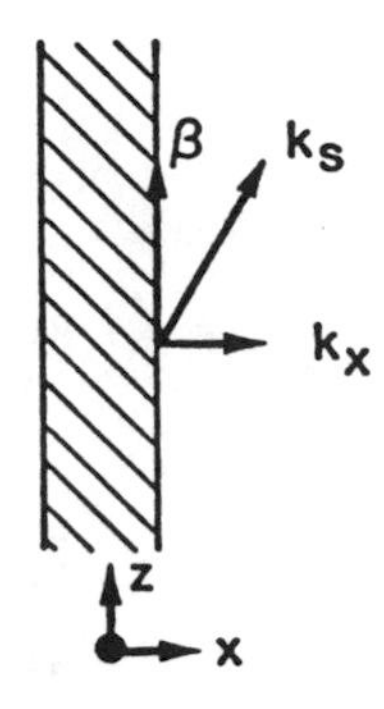

Fig. 2 Top view of the strip of microstrip line and the dielectric region around it. Wavenumbers β and k_s correspond, respectively, to the phase constant of the leaky mode guided by the strip and the wavenumber of the surface wave that propagates away at some angle during the leakage process.

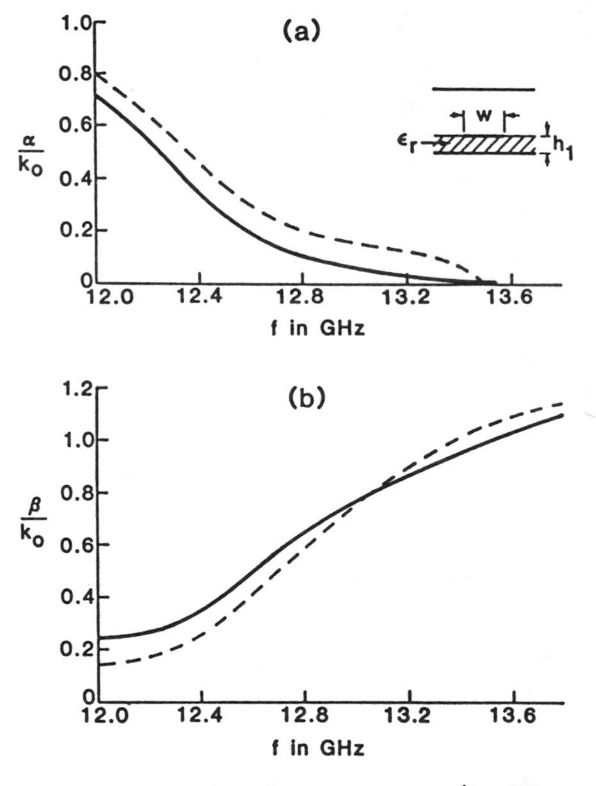

Fig. 3 Variation of leakage constant α/k_o (figure (a)) and phase constant β/k_o (figure (b)) with frequency for the first higher microstrip mode in its leakage range. The solid lines in both (a) and (b) represent our solution for an open microstrip line, and the dashed lines are the numbers presented by Boukamp and Jansen [5] for a line with a top cover of small height. The microstrip line dimensions are those given in reference [5]: dielectric layer thickness h_1, ε_r = 9.7, and, for the covered case, the height of the top cover = 10 h_1.

Frequency-Dependent Characteristics of Microstrip Discontinuities in Millimeter-Wave Integrated Circuits

PISTI B. KATEHI AND NICÓLAOS G. ALEXOPOULOS, SENIOR MEMBER, IEEE

Abstract —**A theoretical approach for the representation of microstrip discontinuities by equivalent circuits with frequency-dependent parameters is presented. The model accounts accurately for the substrate presence and associated surface-wave effects, strip finite thickness, and radiation losses. The method can also be applied for the solution of microstrip components in the millimeter frequency range.**

I. Introduction

THE LITERATURE on the theory of microstrip lines and microstrip discontinuities is extensive but, almost without exception, the published methods do not account for radiation and discontinuity dispersion effects. Microstrip discontinuity modeling was initially carried out either by quasi-static methods [1]–[12] or by an equivalent waveguide model [13]–[22]. The former approach gives a rough estimate of the discontinuity parameters valid at low frequencies, while the latter gives some information about dispersion effects at higher frequencies. However, the applicability of the latter model is also of limited value since it does not account for losses due to radiation and surface-wave excitation at the microstrip discontinuity under investigation. Therefore, it is reasonable to assume that the data obtained with this model are accurate only at the lower frequency range, i.e., before radiation losses become significant.

In this paper, three types of microstrip discontinuities are presented by equivalent circuits with frequency-dependent parameters (see Fig. 1). The implemented method accounts accurately for all the physical effects involved including surface-wave excitation [23], [24]. The model developed in this paper also accounts for conductor thickness and it assumes that the transmission-line and resonator widths are much smaller than the wavelength. The latter assumption insures that the error incurred by neglecting the transverse vector component of the current distribution on each conducting strip is a second-order effect. For each type of discontinuity, the method of moments is applied to determine the current distribution in the longitudinal direction, while the longitudinal current dependence in the transverse direction is chosen to satisfy the edge condition at the effective width location [28]. Upon determining the current distribution, transmission-line theory is invoked to evaluate the elements of the admittance matrix for the open-end and the gap discontinuities (Fig. 1(a) and (b)). The same method is also applied for evaluation of the resonant frequency of the coupled microstrip resonator (Fig. 1(c)). The equivalent circuits for the first two discontinuities are evaluated and compared with the results obtained by a quasi-static method based on the concept of excess length and equivalent capacitance. Although the quasi-static model does not include the discon-

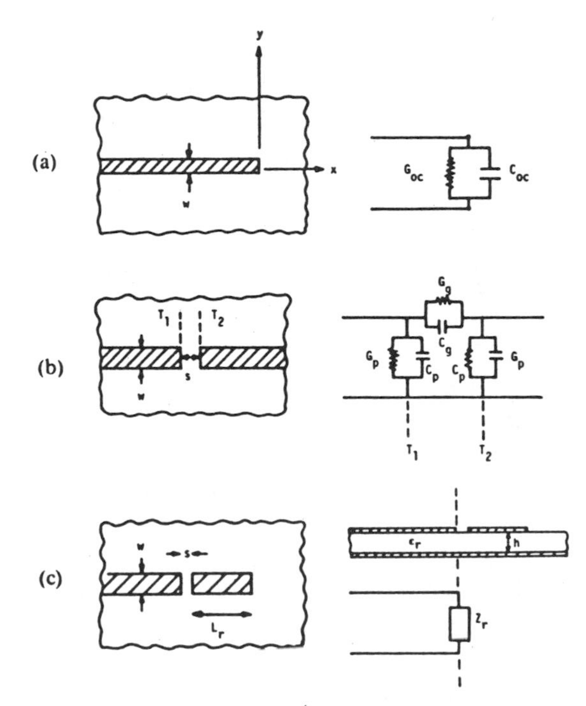

Fig. 1. Microstrip discontinuities and equivalent circuits. (a) Open-circuit microstrip line. (b) Microstrip gap. (c) Coupled microstrip resonator.

Manuscript received March 13, 1985; revised May 31, 1985. This research was supported in part by the U.S. Army under Research Contract DAAG 29-84-K-0067, and in part by the Northrop Corp. under Research Grant 84-110-1006.

P. B. Katehi is with the Department of Electrical Engineering and Computer Science, University of Michigan, Ann Arbor, Michigan 48109.

N. G. Alexopoulos is with the Electrical Engineering Department, University of California, Los Angeles, CA 90024.

Reprinted from *IEEE Trans. Microwave Theory Tech.*, vol. MTT-33, no. 10, pp. 1029–1035, October 1985.

tinuity's radiation conductance in the equivalent circuits, it yields results which at low frequencies are in good agreement with previously published data [5]. However, a comparison of the quasi-statically obtained results with those of the dynamic model developed in this paper shows the inadequacy of the quasi-static approach.

II. Analytical Formulation

A. Current Distribution Evaluation

The current distribution on the transmission-line sections for the discontinuities considered here radiates an electric field given by Pocklington's integral equation

$$\vec{E}(\vec{r}) = \iint_S \bar{\bar{G}}(\vec{r}/\vec{r}')\cdot\vec{J}(\vec{r}')\,ds' \tag{1}$$

where $\vec{E}(\vec{r})$ is the total electric field at the point $\vec{r} = (r,\theta,\phi)$, $\bar{\bar{G}}(\vec{r}/\vec{r}')$ is the dyadic Green's function and $\vec{J}(\vec{r}')$ is the unknown current distribution at the point $\vec{r}' = (r',\theta' = \pi/2,\phi')$. The dyadic Green's function is given by the expression

$$\bar{\bar{G}}(\vec{r}/\vec{r}') = \int_0^\infty \left[k_0^2 \bar{\bar{I}} + \bar{\nabla}\bar{\nabla}\right]\cdot\left(J_0(\lambda|\vec{r}-\vec{r}'|)\bar{\bar{F}}(\lambda)\right) d\lambda \tag{2}$$

with $\bar{\bar{I}}$ the unit dyadic, $k_0 = 2\pi/\lambda_0$, and $\bar{\bar{F}}(\lambda)$ a known dyadic function of the form

$$\bar{\bar{F}}(\lambda) = \frac{\bar{\bar{A}}(\lambda,\epsilon_r,h)}{f_1(\lambda,\epsilon_r,h)f_2(\lambda,\epsilon_r,h)}. \tag{3}$$

In (3), ϵ_r is the relative dielectric constant of the substrate, h is the substrate thickness, and f_1, f_2 are analytic functions of their variables [29], [30].

The integrand in (2) has poles whenever either one of the functions $f_1(\lambda,\epsilon_r,h)$, $f_2(\lambda,\epsilon_r,h)$ becomes zero. The contribution from these poles gives the field propagating in the substrate in the form of TE or TM surface waves [30]. Particularly, the zeros of $f_1(\lambda,\epsilon_r,h)$ correspond to TE surface waves, while the zeros of $f_2(\lambda,\epsilon_r,h)$ to TM surface waves.

Since the widths of the microstrip sections are fractions of the wavelength in the dielectric, it can be assumed that the currents are unidirectional and parallel to the x-axis. Therefore, the current vector in (1) may be written in the form

$$\vec{J}(\vec{r}') = \hat{x}f(x')g(y') \tag{4}$$

where $f(x')$ is an unknown function of x' and $g(y')$ is assumed to be of the form

$$g(y') = \frac{2}{w_e\pi}\left\{1-\left(\frac{2y'}{w_e}\right)^2\right\}^{-1/2}. \tag{5}$$

In (5), w_e is the effective strip width given by $w_e = w + 2\delta$, where δ is the excess half-width and it accounts for fringing effects due to conductor thickness. Formulas for effective width exist in the literature [25], [26] and they have been adopted in this formulation [28].

In order to solve (1) for the current density $\bar{J}$, the method of moments is employed. Each section of the microstrip is divided into a number of segments, and the current is written as a finite sum

$$\vec{J}(\vec{r}') = \hat{x}g(y')\sum_{n=1}^{N} I_n f_n(x') \tag{6}$$

where N is the total number of segments considered and the expansion functions $f_n(x')$ have been chosen to be piecewise sinusoidal functions given by

$$f_n(x') = \begin{cases} \dfrac{\sin[k_0(x'-x_{n-1})]}{\sin(k_0 l_x)}, & x_{n-1} \le x' \le x_n \\ \dfrac{\sin[k_0(x_{n+1}-x')]}{\sin(k_0 l_x)}, & x_n \le x' \le x_{n+1} \\ 0, & \text{otherwise} \end{cases} \tag{7}$$

with l_x being the length of each subsection.

If the electric field is projected along the axis $y = 0$, $z = 0$ using as weighting functions the basis functions (Galerkin's method), (1) will reduce to a matrix equation of the form

$$\underset{N\times N}{[Z_{mn}]}\ \underset{(N\times 1)}{[I_n]} = \underset{(N\times 1)}{[V_m]} \tag{8}$$

where $[I_n]$ is the vector of unknown coefficients and $[V_m]$ is the excitation vector which depends on the impressed feed model. $[Z_{mn}]$ is the impedance matrix with elements given by

$$Z_{mn} = \delta(y)\delta(z)\int_{-w/2}^{w/2} \frac{dy'}{\left[1-\left(\frac{2y'}{w_e}\right)^2\right]^{1/2}} \cdot \int_C dx \int_C dx' \left\{k_0^2 F_{xx} + \frac{\partial^2}{\partial x^2}(F_{xx}-F_{zx})\right\} f_m(x)f_n(x') \tag{9}$$

where

$$F_{xx} = 2\left(\frac{j\omega\mu_0}{4\pi k_0^2}\right)\int_0^\infty \left(\frac{\sinh(uh)}{f_1(\lambda,\epsilon_r,h)}\right) J_0(\lambda\rho)e^{-u_0 t}\,d\lambda \tag{10}$$

and

$$F_{zx} = 2\left(\frac{j\omega\mu_0}{4\pi k_0^2}\right)(\epsilon_r - 1)\int_0^\infty \left(\frac{u_0\cosh(uh)}{f_1(\lambda,\epsilon_r,h)}\right)\left(\frac{\sinh(uh)}{f_2(\lambda,\epsilon_r,h)}\right) \cdot J_0(\lambda\rho)e^{-u_0 t}\,d\lambda. \tag{11}$$

In (9) and (10)

$$u_0 = \left[\lambda^2 - k_0^2\right]^{1/2} \quad u = \left[\lambda^2 - \epsilon_r k_0^2\right]^{1/2} \tag{12}$$

$$f_1(\lambda,\epsilon_r,h) = u_0\sinh(uh) + u\cosh(uh) \tag{13}$$

$$f_2(\lambda,\epsilon_r,h) = \epsilon_r u_0\cosh(uh) + u\sinh(uh) \tag{14}$$

and

$$\rho = \left[(x-x')^2 + (y-y')^2\right]^{1/2}. \tag{15}$$

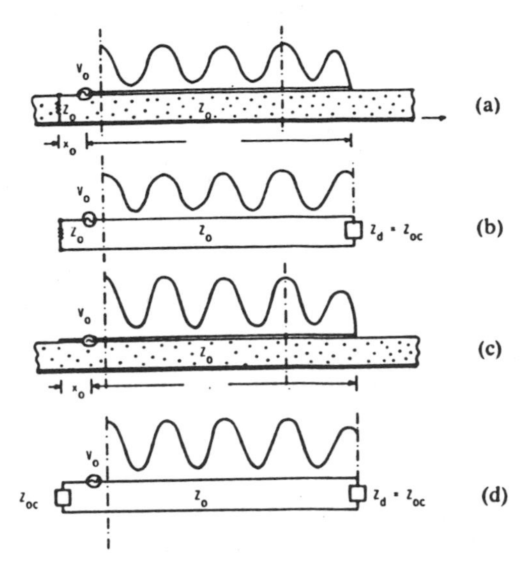

Fig. 2. Modeling of the excitation mechanism for a microstrip transmission line.

Using this form for the evaluation of the elements of the impedance matrix, one can solve the matrix equation shown in (8) to find the unknown coefficients for the current.

B. Excitation Mechanism—Equivalent Impedance

One difficulty always encountered in this type of problem is the implementation of a practical excitation mechanism which can be included in the mathematical modeling. In most applications, the microstrip line is kept as close to the ground as possible and is excited by a coaxial line of the same characteristic impedance. As a result, a unimodal field is excited under the transmission line, reflection at the excitation end is minimized, and the current distribution on the line beyond an appropriate reference plane forms standing waves of a transverse electromagnetic (TEM)-like mode. Thus, this microstrip line can be approximated by an ideal transmission line of the same characteristic impedance Z_0 terminated to an unknown equivalent impedance Z_d (see Fig. 2). It can be shown (see Appendix A) that the reflection coefficient does not change in amplitude and phase if the coaxial line is substituted with a voltage gap generator and the line is left open at the excitation end (see Fig. 2(c)). This excitation mechanism is adopted for the application of moments method in the solution of Pocklington's integral equation and results in an excitation vector $[V_m]=[\delta_{im}]$ with

$$\delta_{im}=\begin{cases}1 & \text{at the position of the gap generator } (x_i=x_m)\\ 0 & \text{everywhere else } (x_i\neq x_m)\end{cases}.$$

The equivalent ideal transmission line for this type of excitation is shown in Fig. 2(d) where $Z_d=Z_{oc}$ for the case of an open-end microstrip discontinuity. The quasi-TEM mode considered has a wavelength λ_g equal to the dominant spatial frequency of the amplitude of the current which is derived by the method of moments.

If the origin of the x coordinate is taken at the position of Z_d, then the equivalent impedance, normalized with respect to the characteristic impedance, is given by [28]

$$Z_d=\frac{1+\Gamma(0)}{1-\Gamma(0)} \tag{16}$$

where

$$\Gamma(0)=-\frac{SWR-1}{SWR+1}e^{j\beta|x_{\max}|} \tag{17}$$

and $x_{\max}$ is the position of a maximum.

From (16) and (17), one can see that the accurate determination of equivalent circuits for different discontinuities depends on the accuracy of evaluating the characteristic impedance Z_0 and guided wavelength $\lambda_g(\beta=2\pi/\lambda_g)$. The characteristic impedance of the transmission line of Fig. 2(d) is given by (see Appendix B)

$$Z_0=\frac{1}{|I_{\max}|}\left|\frac{Z_{oc}}{1-Z_{oc}}\right|\cdot\left|e^{j\beta|x_{\max}|}+\frac{1-Z_{oc}}{1+Z_{oc}}e^{-j\beta|x_{\max}|}\right| \tag{18}$$

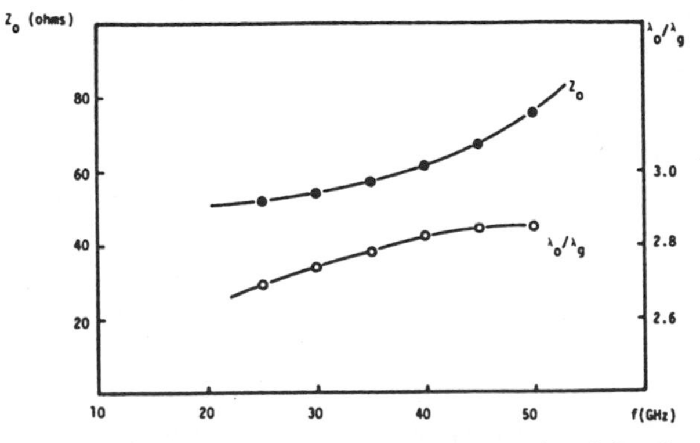

Fig. 3. Characteristic impedance Z_0 and guide wavelength λ_g/λ_0 of a microstrip line as a function of frequency ($\epsilon_r=9.6, w/h=1, h=0.6$ mm).

where $x_{\max}$ is the position of a maximum and $|I_{\max}|$ is the maximum amplitude of the current. Z_{oc} is the normalized equivalent impedance of an open-circuited microstrip line of length $l=(n/2)\lambda_g(n\geq 8)$ and the gap generator is placed at a position $\lambda_g/4$ from one end. The characteristic impedance and guided wavelength are shown as functions of frequency in Fig. 3. These results are in excellent agreement with already existing data [25].

III. Numerical Results

A. Microstrip Open-Circuit Discontinuity

The equivalent circuit from an open-end discontinuity, as shown in Fig. 1, consists of a capacitance C_{oc} in parallel with a conductance G_{oc}, which is proportional to radiation losses. Using the method presented previously, values for the normalized capacitance C_{oc}/w in pF/m and for the conductance G_{oc} (in mmhos) are plotted as functions of frequency for a microstrip line with $w/h=1$ on a 0.6-mm Alumina substrate (see Fig. 4). From these data, one can conclude that the radiation conductance increases with frequency while the capacitance decreases.

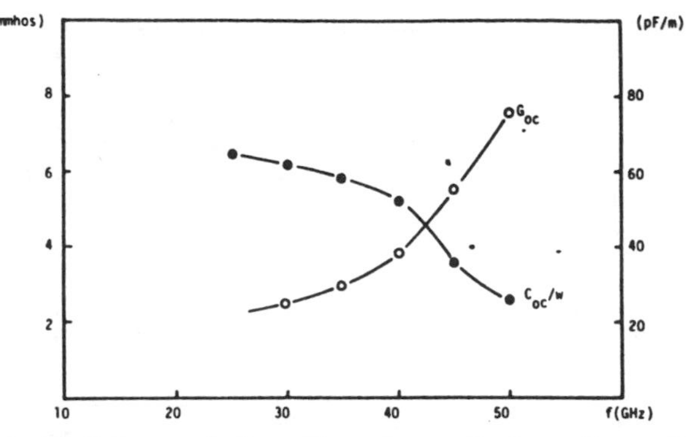

Fig. 4. Radiation conductance G_{oc} and normalized capacitance C_{oc}/w of an open-circuited microstrip line as functions of frequency ($\epsilon_r = 9.6, w/h = 1, h = 0.6$ mm).

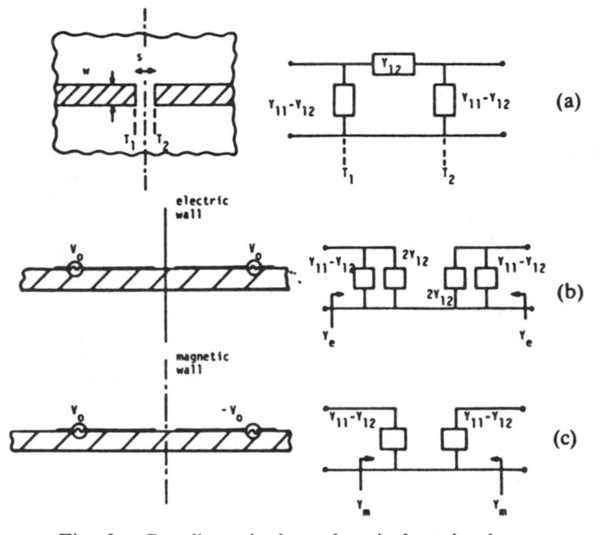

Fig. 5. Gap discontinuity and equivalent circuits.

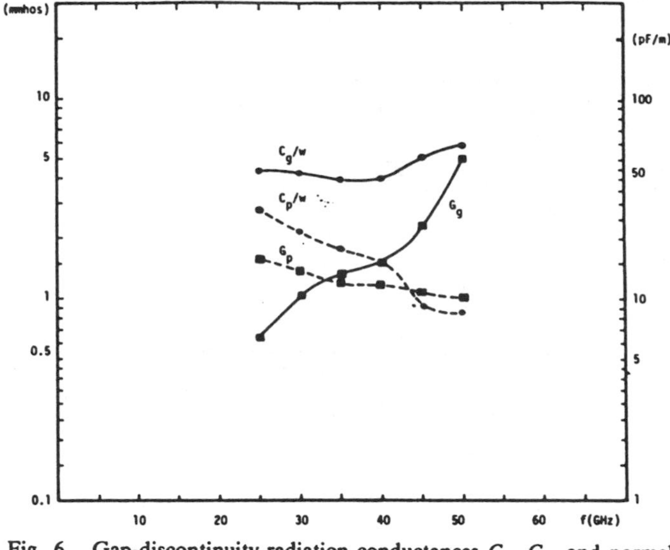

Fig. 6. Gap-discontinuity radiation conductances G_g, G_p, and normalized capacitances C_g/w, C_p/w as functions of frequency ($\epsilon_r = 9.6, w/h = 1, h = 0.6$ mm, $s/h = 0.3762$).

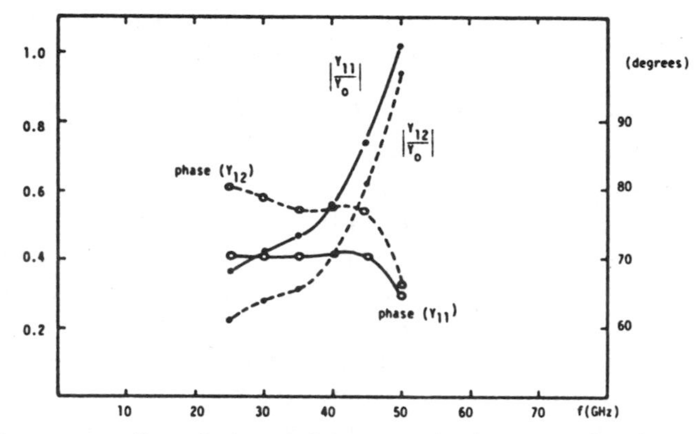

Fig. 7. Gap-discontinuity admittance matrix elements as functions of frequency ($\epsilon_r = 9.6, w/h = 1, h = 0.6$ mm, $s/h = 0.3762$).

B. Microstrip Gap Discontinuity

The equivalent circuit for a microstrip gap is shown in Fig. 1. For the evaluation of G_p, C_p, G_g, and C_g, both sections of the microstrip are excited by gap generators which are either in phase (see Fig. 5(a)) or out of phase (Fig. 5(c)). The former case is equivalent to the presence of a magnetic wall in the middle of the gap, while the latter to an electric wall at the same position. The equivalent circuits for these two excitations are shown in Fig. 5(b) and (c) and give

$$Y_e = G_e + j\omega C_e = Y_{11} + Y_{12} \tag{19}$$

and

$$Y_m = G_m + j\omega C_m = Y_{11} - Y_{12} \tag{20}$$

where Y_e, Y_m are the equivalent normalized admittances for the case of the electric and magnetic wall, respectively. From (19) and (20), the conductances and capacitances of a p-type equivalent circuit are given by

$$G_p = G_m \tag{21}$$

$$C_p = C_m \tag{22}$$

$$G_g = \tfrac{1}{2}(G_e - G_m) \tag{23}$$

and

$$C_g = \tfrac{1}{2}(C_e - C_m) \tag{24}$$

Values of G_p, C_p, G_g, and C_g are plotted as functions of frequency for a microstrip line with $w/h = 1$ on a 0.6-mm Alumina substrate (Fig. 6). From the values for the gap and open-end conductances (Figs. 4 and 6), one can see that as the frequency increases, the radiation losses become higher and, therefore, the inter-circuit coupling through space and surface waves becomes a dominant factor in the design of printed circuits. From (19)–(24), the elements of the admittance matrix of the discontinuity considered as a two port can be found in amplitude and phase as functions of frequency (see Fig. 7).

C. Excess Length and Equivalent Capacitances

Another method of deriving equivalent circuits is based on the evaluation of excess length. The method developed here results in values for the excess length in such a way so as to take into account dispersion and radiation losses. The dominant reason for loss of accuracy at high frequencies by this method is the way the equivalent capacitance is

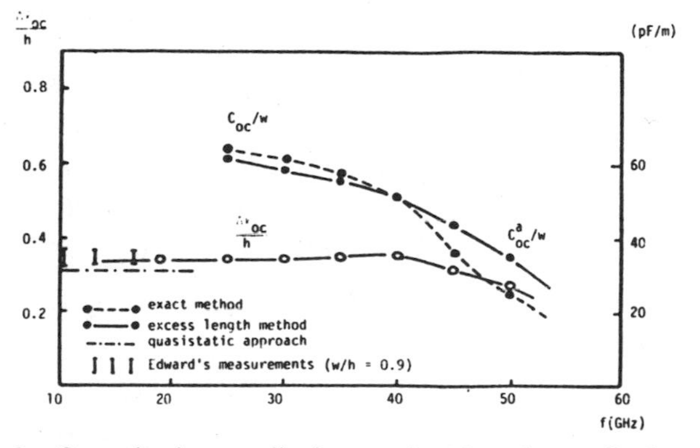

Fig. 8. Open-circuit normalized excess length and normalized capacitances as functions of frequency ($\epsilon_r = 9.6, w/h = 1, h = 0.6$ mm).

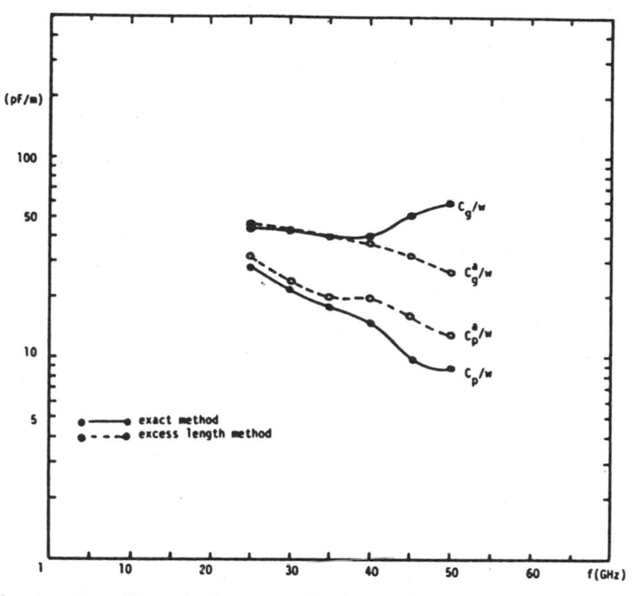

Fig. 9. Gap-discontinuity normalized capacitances C_p/w, C_g/w as functions of frequency ($\epsilon_r = 9.6, w/h = 1, h = 0.6$ mm, $s/h = 0.3762$).

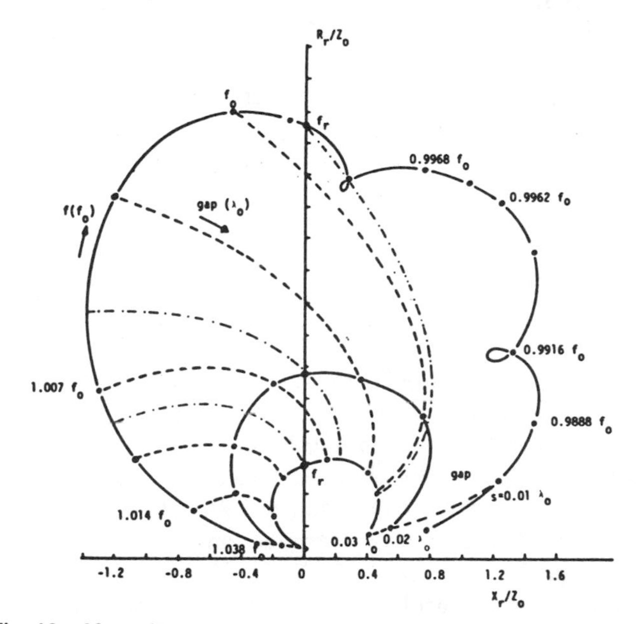

Fig. 10. Normalized equivalent impedance Z_r as a function of frequency f and gap s ($\epsilon_r = 9.9, w/h = 1, h = 0.05\lambda_0, L_r = 0.032\lambda_0$).

evaluated. The excess length Δl_d is measured from the standing waves of the amplitude of the computed current

$$\Delta l_d = \frac{\lambda_g}{4} - d_{\max} \tag{25}$$

where λ_g is the guided wavelength and $d_{\max}$ is the position of the first maximum of the current amplitude from the open end. The discontinuity capacitance C^a_d is evaluated as a function of Δl_d from the following relation [25]:

$$\frac{C^a_d}{w} = \frac{\Delta l_d}{h}\sqrt{\epsilon_{\text{eff}}}\,\frac{1}{Z_0}\,\frac{h}{w} \tag{26}$$

where ϵ_{eff} is the effective dielectric constant and Z_0 the characteristic impedance. Equation (26) gives quasi-static values for the capacitance and, therefore, is accurate for low frequencies only.

In Fig. 8, the open-end normalized equivalent capacitance C^a_{oc}/w in pF/m is plotted as a function of frequency and is compared to the values derived with the exact method. As shown, the values of the two capacitances agree at the lower part of the frequency range and they shift away as the frequency becomes higher. In addition, the capacitances C_p, C_g of the p-type equivalent circuit derived with the two methods are compared and they show a big discrepancy at high frequencies where (25) becomes much less accurate (Fig. 9).

D. Coupled Microstrip Resonator

For this discontinuity, the transmission-line model is used for the evaluation of the normalized equivalent impedance Z_r as a function of frequency (see Fig. 1(c)). The normalized impedance Z_r is measured at $\lambda_g/4$ from the open end. In Fig. 10, the real and imaginary parts of Z_r are plotted as functions of frequency for a microstrip line with $w/h = 1$, $t = 0.0001\,\lambda_0$ on a $0.05\lambda_0$-thick Alumina substrate ($\epsilon_r = 9.9$), and for a gap $s = 0.01\,\lambda_0$. The lengths have been measured in terms of the free space wavelength λ_0 at a specified frequency f_0. The resonator length L_r varies between $0.031\,\lambda_0$ and $0.033\,\lambda_0$. Fig. 10 implies that there exists a particular length L_r for which the VSWR on the transmission line at the resonant frequency f_r becomes unity. Fig. 11(a) indicates that the resonant frequency decreases as the resonator length becomes larger. It is also very interesting to see how the resonant frequency changes as a function of the length of the gap. Fig. 11(b) shows that the resonant frequency increases as the gap becomes larger, while the resonator length was kept constant and equal to $0.032\,\lambda_0$.

IV. Conclusion

The representation of microstrip discontinuities by equivalent circuits or admittance matrices has been treated by an effective method. The current distribution is computed by the method of moments in the longitudinal direction of the microstrip discontinuities, while in the transverse direction it is chosen as the Maxwell distribution, thus satisfying the edge condition. Upon determining

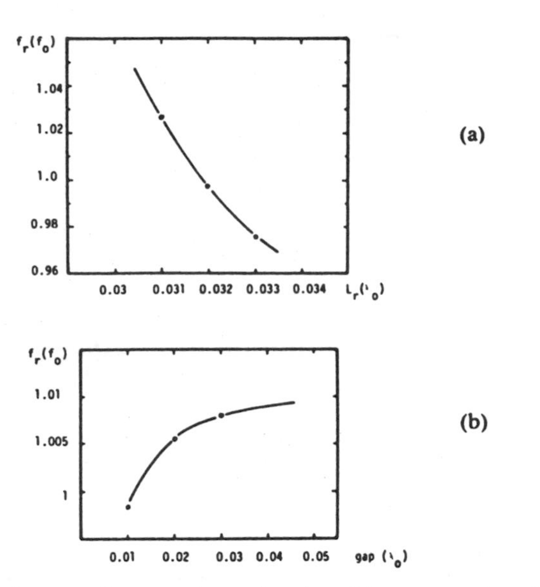

Fig. 11. Resonant frequency f_r as a function of the resonator length and the gap.

the current, a transmission-line model is used for the representation of the unimodal field excited under the microstrip line.

Therefore, the normalized admittance matrix is evaluated in terms of equivalent admittances. This method takes into account dispersion and radiation losses, provides us with equivalent circuits which are an accurate representation of the discontinuities under consideration, and does not have any frequency limitations. Furthermore, this method was compared with results obtained through a quasi-static as well as a waveguide model and was found superior. The accuracy of the method depends on the accuracy of evaluating λ_g, x_{max}, and VSWR. This implies that if more subsections are considered in the method of moments, the error will become smaller. In the present derivations, the estimated error is up to two percent.

Appendix A

The transmission line of Fig. 2(d) is considered and it is assumed that the voltage generator is at the position $x = 0$ and that $Z_d \neq Z_{oc}$. The coupled differential equations for the voltage and current in this transmission line are of the form

$$-\frac{dI(x)}{dx} = j\beta Y_0(x) \tag{1A}$$

and

$$-\frac{dV(x)}{dx} = j\beta Z_0 I(x) + V_0\delta(x). \tag{2A}$$

From (1A) and (2A), the differential equation for the current is given by

$$\frac{d^2I(x)}{dx^2} + \beta^2 I(x) = j\beta Y_0 V_0 \delta(x). \tag{3A}$$

From (1A) to (3A), one can conclude that the voltage and the current in this transmission line are of the form

$$I(x<) = B_1 e^{-j\beta x} + B_2 e^{j\beta x} \qquad (x \le 0) \tag{4A}$$

$$I(x>) = A_1 e^{-j\beta x} + A_2 e^{j\beta x} \qquad (x \ge 0) \tag{5A}$$

$$V(x<) = Z_0\left(A_1 e^{-j\beta x} - A_2 e^{j\beta x}\right) \qquad (x \le 0) \tag{6A}$$

and

$$V(x>) = Z_0\left(B_1 e^{-j\beta x} - B_2 e^{j\beta x}\right) \qquad (x \ge 0). \tag{7A}$$

The boundary conditions for this problem are

$$\frac{V(l)}{I(l)} = Z_d \tag{8A}$$

$$\frac{V(-x_0)}{I(-x_0)} = Z_{oc} \tag{9A}$$

$$I(x<) = I(x>) \text{ at } x = 0 \tag{10A}$$

and

$$\left.\frac{dI(x>)}{dx}\right|_{x=0} - \left.\frac{dI(x<)}{dx}\right|_{x=0} = j\beta Y_0 V_0. \tag{11A}$$

From (4A) through (11A), one can find that

$$I(x>) = -\frac{1+Z_d}{Z_0}\frac{\cos(\beta x_0) - jZ_{oc}\sin(\beta x_0)}{(Z_{oc}-Z_d)\cos[\beta(l-x_0)] + j(Z_{oc}Z_d - 1)\sin[\beta(l-x_0)]} \cdot \left\{e^{-j\beta(x-l)} + \frac{1-Z_{oc}}{1+Z_{oc}}e^{+j\beta(x-l)}\right\}. \tag{12A}$$

From (5A) and (12A), the current reflection coefficient is given by

$$\rho_i = \frac{A_2}{A_1} = \frac{1-Z_{oc}}{1+Z_{oc}}e^{-2j\beta l}. \tag{13A}$$

If the case is considered where the left end of the microstrip line is matched (see Fig. 2(b)), then $B_1 = 0$ and

$$I^m(x>) = -\frac{1}{Z_0}\left\{e^{-j\beta x} + \frac{1-Z_{oc}}{1+Z_{oc}}e^{j\beta x}e^{-2j\beta l}\right\} \tag{14A}$$

with

$$\rho_i^m = \frac{1-Z_{oc}}{1-Z_{oc}}e^{-2j\beta l}. \tag{15A}$$

From (13A) and (15A), one can conclude that the amplitude and phase of the reflection coefficient do not change when the line is matched at the generator end.

Appendix B

If x_0, l are equal to $\lambda_g/4$ and $\eta(\lambda_g/2)$, then

$$l - x_0 = \frac{2n-1}{4}\lambda_g \tag{B1}$$

and

$$I(x>) = \frac{1+Z_d}{2Z_0}\frac{Z_{oc}}{Z_{oc}Z_d - 1}\left\{e^{-j\beta x} + \frac{1-Z_{oc}}{1+Z_{oc}}e^{j\beta x}\right\}. \tag{B2}$$

At $x = l - x_{max}$ for $Z_d = Z_{oc}$, the absolute value of the current is given by

$$|I_{max}| = \frac{1}{Z_0}\left|\frac{Z_{oc}}{Z_{oc}-1}\right|\left|e^{+j\beta|x_{max}|} + \frac{1-Z_{oc}}{1+Z_{oc}}e^{-j\beta|x_{max}|}\right| \tag{B3}$$

where x_{max} is the position of a maximum measured from the equivalent impedance Z_d. Equation (17) is derived from (B3).

References

[1] A. Farrar and A. T. Adams, "Computation of lumped microstrip capacitances by matrix methods: Rectangular sections and end effects," *IEEE Trans. Microwave Theory Tech.*, vol. MTT-19, pp. 495–497, 1971.

[2] M. Maeda, "Analysis of gap in microstrip transmission lines," *IEEE Trans. Microwave Theory Tech.*, vol. MTT-20, 1972.

[3] T. Itoh, R. Mittra, and R. D. Ward, "A method of solving discontinuity problems in microstrip lines," in 1972 *IEEE-GMTT Int. Microwave Symp. Dig.*

[4] P. Silvester and P. Benedek, "Equivalent capacitance of microstrip open circuits," *IEEE Trans. Microwave Theory Tech.* vol. MTT-20, 1972.

[5] P. Benedek and P. Silvester,, "Equivalent capacitance for microstrip gaps and steps," *IEEE Trans. Microwave Theory Tech.*, vol. MTT-20, 1972.

[6] P. Silvester and P. Benedek, "Microstrip discontinuity capacitances for right-angle bends, T-junctions and crossings," *IEEE Trans. Microwave Theory Tech.*, vol. MTT-21, 1973.

[7] R. Horton, "The electrical characterization of a right-angled bend in microstrip line," *IEEE Trans. Microwave Theory Tech.*, vol. MTT-21, 1973.

[8] R. Horton, "Equivalent representation of an abrupt impedance step in microstrip line," *IEEE Trans. Microwave Theory Tech.*, vol. MTT-21, 1973.

[9] A. F. Thompson and A. Gopinath, "Calculation of microstrip discontinuity inductances," *IEEE Trans. Microwave Theory Tech.*, vol. MTT-23, 1975.

[10] A. Gopinath *et al.*, "Equivalent circuit parameters of microstrip step change in width and cross junctions," *IEEE Trans. Microwave Theory Tech.*, vol. MTT-24, 1976.

[11] R. Garg, and I. J. Bahl, "Microstrip discontinuities," *Int. J. Electron.* vol. 45, July 1978.

[12] C. Gupta and A. Gopinath, "Equivalent circuit capacitance of microstrip step change in width," *IEEE Trans. Microwave Theory Tech.*, vol. MTT-25, 1977.

[13] T. Itoh, "Analysis of microstrip resonators," *IEEE Trans. Microwave Theory Tech.*, vol. MTT-22, 1974.

[14] I. Wolf, G. Kompa, and R. Mehran, "Calculation method for microstrip discontinuities and T-junctions," *Electron. Lett.*, vol. 8, 1972.

[15] G. Kompa and R. Mehran, "Planar waveguide model for calculating microstrip components," *Electron. Lett.*, vol. 11, 1975.

[16] G. Kompa, "*S*-matrix computation of microstrip discontinuities with a planar waveguide model," *Arch. Elek. Übertragung.*, vol. 30, 1976.

[17] R. Mehran, "The frequency-dependent scattering matrix of microstrip right-angle bends, T-junctions and crossings," *Arch. Elek. Übertragung.*, vol. 29, 1975.

[18] W. Menzel and I. Wolf, "A method for calculating the frequency dependent properties of microstrip discontinuities," *IEEE Trans. Microwave Theory Tech.*, vol. MTT-25, 1977.

[19] I. W. Stephenson and B. Easter, "Resonant techniques for establishing the equivalent circuits of small discontinuities in microstep," *Electron. Lett.*, vol. 7, pp. 582–584, 1971.

[20] B. Easter, "The equivalent circuits of some microstrip discontinuities," *IEEE Trans. Microwave Theory Tech.*, vol. MTT-23, pp. 655–660, 1975.

[21] J. R. James and A. Henderson, "High frequency behavior of microstrip open-circuit terminations," *Proc. Inst. Elec. Eng., Microwaves, Optics and Acoustics*, vol. 3, no. 4, pp. 205–218, Sept. 1979.

[22] R. Mehran, "Computer-aided design of microstrip filters considering dispersion loss and discontinuity effects," *IEEE Trans. Microwave Theory Tech.*, vol. MTT-27, pp. 239–245, Mar. 1979.

[23] P. B. Katehi and N. G. Alexopoulos, "On the effect of substrate thickness and permittivity on printed circuit dipole properties," *IEEE Trans. Antennas Propagat.*, vol. AP-31, Jan. 1983.

[24] N. G. Alexopoulos, P. B. Katehi, and D. B. Rutledge, "Substrate optimization for integrated circuit antennas," *IEEE Trans. Microwave Theory Tech.*, vol. MTT-31, July 1983.

[25] K. G. Gupta, R. Garg, and I. J. Bahl, *Microstrip Lines and Slotlines.* Dedham, MA: Artech House, 1979.

[26] T. C. Edwards, *Foundations for Microstrip Circuit Design.* New York: Wiley, 1981.

[27] P. B. Katehi and N. G. Alexopoulos, "On the theory of printed circuit antennas for millimeter waves," in Conf. Dig., *Sixth Int. Conf. Infrared and Millimeter Waves*, Dec. 1981, pp. F.2.9–F.2.10.

[28] P. B. Katehi and N. G. Alexopoulos, "On the modeling of electromagnetically coupled microstrip antennas—The printed strip dipole," *IEEE Trans. Antennas Propagat.*, vol. AP-32, Nov. 1984.

[29] P. B. Katehi, "A generalized solution to a class of printed circuit antennas," Ph.D. dissertation, Univ. of California, Los Angeles, 1984.

[30] N. G. Alexopoulos, D. R. Jackson, and P. B. Katehi, "Criteria for nearly omnidirectional radiation patterns for printed antennas," *IEEE Trans. Antennas Propagat.*, vol. AP-33, Feb. 1985.

[31] P. B. Katehi and N. G. Alexopoulos, "Real axis integration of Sommerfeld integrals with applications to printed circuit antennas," *J. Math. Phys.*, vol. 24(3), Mar. 1983.

Full-Wave Analysis of Microstrip Open-End and Gap Discontinuities

ROBERT W. JACKSON, MEMBER, IEEE, AND DAVID M. POZAR, MEMBER, IEEE

Abstract—A solution is presented for the characteristics of microstrip open-end and gap discontinuities on an infinite dielectric substrate. The exact Green's function of the grounded dielectric slab is used in a moment method procedure, so surface waves as well as space-wave radiation are included. The electric currents on the line are expanded in terms of longitudinal subsectional piecewise sinusoidal modes near the discontinuity, with entire domain traveling-wave modes used to represent incident, reflected, and, for the gap, transmitted waves away from the discontinuity. Results are given for the end admittance of an open-ended line, and the end conductance is compared with measurements. Results are also given for the reflection coefficient magnitude and surface-wave power generation of an open-ended line on substrates with various dielectric constants. Loss to surface and space waves is calculated for a representative gap discontinuity.

I. Introduction

THIS PAPER DESCRIBES a "full-wave" solution of the open-end and symmetric gap discontinuities in microstrip line. The solution is rigorous in that space-wave radiation and surface-wave generation from discontinuities is explicitly included through the use of the exact Green's function for a grounded dielectric slab. A moment method procedure is used whereby the electric surface current density on the microstrip line is expanded in terms of four different types of expansion modes: one mode represents a traveling wave incident on the discontinuity, another mode represents a traveling wave reflected from the discontinuity, a third represents a traveling wave transmitted through the discontinuity (gap case only), and a number of subsectional (piecewise sinusoidal) modes are used in the vicinity of the discontinuity to model the nonuniform current in that region. The result is a physically meaningful solution in terms of incident, reflected, and transmitted-wave amplitudes, with only a small number of unknown coefficients to solve for (typically four to five for the open-end case and twice that for the gap case). For the open-end case, the complex reflection coefficient can then be determined, as well as an "end admittance," referred to the end of the microstrip line. For the gap, scattering parameters can be determined. In addition, the amount of real power delivered to radiation and surface waves can be calculated. It is assumed that only the fundamental microstrip mode is propagating on the line away from the open end, although higher order mode fields are accounted for in the vicinity of the discontinuity.

Manuscript received January 11, 1985; revised June 3, 1985. This work was supported in part by the Rome Air Development Center, Hanscom Air Force Base, MA, under U.S. Air Force Contracts F19628-84-K-0022 and F49620-82-C-0035.

The authors are with the Department of Electrical and Computer Engineering, University of Massachusetts, Amherst, MA 01003.

Much of the previous work on the open-circuited microstrip line has used quasi-static approximations, with the results of Hammerstad and Bekkadal [1], [2] being widely referenced. Jansen [3] has calculated length extensions for an enclosed microstrip using a spectral-domain method. Lewin [4] used an assumed current distribution to calculate the radiated power from an open line. James and Henderson [5], [6] developed an improved analysis using a variational technique, including surface-wave effects, and compared their results favorably with measurements of the end conductance of an open-ended line on a thin, low dielectric constant substrate. Compared with the present solution, the results of James and Henderson appear to be quite good for such substrates, and their relatively simple expressions are an advantage computationally.

Likewise, the gap has also been analyzed by predominately quasistatic methods [7]–[10], the results of which are used extensively in computer-aided design routines. Fully electromagnetic solutions have been calculated by Jansen and Koster [11] using a spectral-domain method, but their gap is surrounded on four sides by perfect conductors. None of the aforementioned approaches include surface waves and radiation losses.

There were two motivations for the present work. First, the increasing interest in monolithic and millimeter-wave integrated circuits requires rigorous analyses to characterize such microstrip discontinuities on electrically thick, high dielectric constant substrates (such as GaAs, with $\epsilon_r = 12.8$). Quantities such as radiation and surface waves are more important with such substrates than with thin, low dielectric constant substrates. Second, the present work is an ancillary result from the solution to the problem of a microstrip patch antenna on an electrically thick substrate fed by a microstrip line. This problem is also of interest in terms of MMIC design, and may be addressed in the future.

Section II presents the theory of the solution, which is based on the moment method/Green's function solutions for printed dipole and microstrip patch antennas [12], [13]. The propagation constant for the fundamental mode of an infinite microstrip line is also developed in terms of a "full-wave" solution in this section, and the opportunity is taken to dispel a few myths about propagation on microstrip lines. Section III presents results for terminal conductance and the Δl length extension for $\epsilon_r = 2.32$ and 12.8 substrates, and is compared with measurements and calculations from [1] and [6]. Reflection coefficient magnitudes

Reprinted from *IEEE Trans. Microwave Theory Tech.*, vol. MTT-33, no. 10, pp. 1036–1042, October 1985.

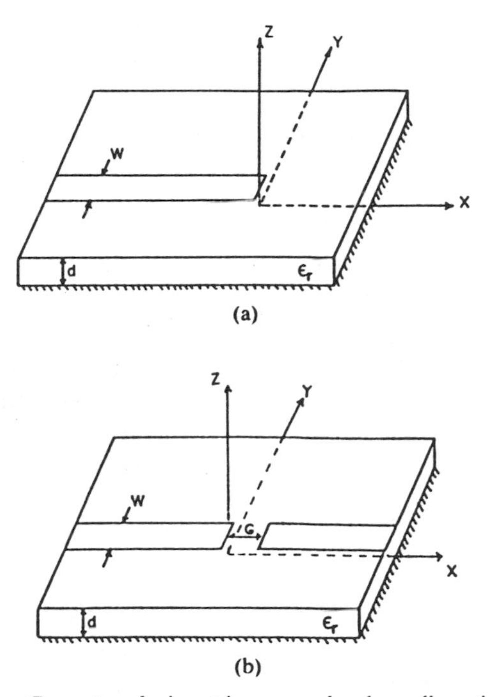

Fig. 1. Geometry of microstrip open-end and gap discontinuities.

and a radiation efficiency e, defined as the ratio of radiated power to radiated plus surface-wave power [13], are plotted versus substrate thickness for $\epsilon_r = 2.55$ and $\epsilon_r = 12.8$. The fraction of incident power launched into surface waves is seen to increase sharply with increasing substrate thickness and/or dielectric constant. Results are also presented for radiation loss at a representative gap discontinuity on $\epsilon_r = 12.8$.

II. Theory

Fig. 1 shows the geometry of the open-end and gap discontinuities in width W. The substrate is assumed infinitely wide in the x- and y-directions, and of thickness d, and relative permittivity ϵ_r. Only $\hat{x}$-directed electric surface currents are assumed to flow on the microstrip line, which, as was found in [14] and other references, is a good approximation when thin lines (with respect to wavelength) are used on substrates of any thickness.

A. Green's Function for the Grounded Dielectric Slab

The canonical building block for the present solution is the plane-wave spectral representation of the grounded dielectric slab Green's function, representing the $\hat{x}$-directed electric field at (x, y, d) due to an $\hat{x}$-directed infinitesimal dipole of unit strength at (x_0, y_0, d). This field can be written as [12]

$$E_{xx}(x, y|x_0, y_0) = -\iint_{-\infty}^{\infty} Q(k_x, k_y) e^{jk_x(x-x_0)} \cdot e^{jk_y(y-y_0)}\, dk_x\, dk_y \quad (1)$$

where

$$Q(k_x, k_y) = \frac{jZ_0}{4\pi^2 k_0} \frac{(\epsilon_r k_0^2 - k_x^2) k_2 \cos k_1 d + jk_1 (k_0^2 - k_x^2) \sin k_1 d}{T_e T_m} \sin k_1 d \quad (2)$$

$$\begin{aligned} T_e &= k_1 \cos k_1 d + jk_2 \sin k_1 d \\ T_m &= \epsilon_r k_2 \cos k_1 d + jk_1 \sin k_1 d \\ k_1^2 &= \epsilon_r k_0^2 - \beta^2, \qquad \operatorname{Im} k_1 < 0 \\ k_2^2 &= k_0^2 - \beta^2, \qquad \operatorname{Im} k_2 < 0 \\ \beta^2 &= k_x^2 + k_y^2 \\ k_0 &= \omega\sqrt{\mu_0 \epsilon_0} = 2\pi/\lambda_0 \\ Z_0 &= \sqrt{\mu_0/\epsilon_0}. \end{aligned} \quad (3)$$

As discussed in [12], the zeros of the T_e, T_m functions constitute surface-wave poles. During the integration in (1), which is done numerically, special care must be given to these pole contributions, and a method for doing this is presented in [12]. The integration in (1) is further facilitated by a conversion to polar coordinates, as described in [12].

B. Propagation Constant of an Infinite Microstrip line

The solution for the open-circuited line requires the propagation constant of an infinitely long microstrip line. It is assumed that the electrical thickness of the substrate is such that only the fundamental microstrip mode propagates. A quasi-static value [2] could be used with reasonable results, but the more rigorous "full-wave" solution involves only a small fraction of the total effort for the open-circuit problem, and so the propagation constant was computed in this manner. The method is very similar to [14].

Consider an infinitely long microstrip line of width W with a traveling-wave current of the form $e^{-jk_e x_0}$, where k_e is the effective propagation constant to be determined. Substituting this current into (1) and integrating over x_0, y_0 yields the electric field at (x, y, d) due to this line source

$$E_{xx}^l = -2\pi \iint_{-\infty}^{\infty} Q(k_x, k_y) \delta(k_x + k_e) \cdot e^{jk_x x} e^{jk_y y} F_y(k_y)\, dk_x\, dk_y \quad (4)$$

where F_y is the Fourier transform of the distribution of current in the y-direction, which is, for now, assumed uniform. Thus

$$F_y(k_y) = \frac{2\sin(k_y W/2)}{k_y}. \quad (5)$$

Now, the above electric field must vanish at all points on the microstrip line, since it is assumed to be a perfect conductor. This boundary condition is enforced across the width of the strip by integrating on y over the width. After carrying out the k_x integration, the following characteristic equation for k_e results:

$$\int_{-\infty}^{\infty} Q(k_e, k_y) F_y^2(k_y)\, dk_y = 0. \quad (6)$$

This equation can be solved relatively quickly for k_e using a simple search technique, such as the interval halving method. The characteristic impedance of the uniform line (used later) can also be derived from this solution by computing the voltage between the strip and the ground plane [12]. In the interest of brevity, this derivation is not presented.

Two points of interest regarding propagation on uniform microstrip lines can be inferred from the above solution. First, there exists in the literature (for example, [15] and [16]) the idea that surface-wave modes can be excited by the fundamental mode of the uniform microstrip line. This is false, as can be seen by noting that the fundamental propagation constant k_e (as determined numerically) is always greater than any surface-wave pole β_{sw}. Thus, the integration path of (6) never crosses a surface-wave pole, with the result that no surface-wave power is generated by the uniform line. Discontinuities in the line can, of course, excite surface waves, as can higher order propagating modes.

Second, it is sometimes stated that a uniform microstrip line does not radiate *any* power into space waves. Again, this is false, as a stationary phase evaluation of the field above the substrate due to a uniform line will show far-zone radiated power is generated. As a practical matter, however, this loss to radiation is much less than either conductor loss or dielectric loss.

C. Current Expansion Modes

The method of solution for both the open-circuited microstrip line and the gap basically involves expanding the electric surface current density on the line and formulating an integral equation which can be solved by the method of moments for the unknown expansion currents. The choice of basis functions affects the computational efficiency quite significantly, so a judicious choice is important. We first describe in detail the basis functions for the open end and then describe the modifications needed to compute the gap.

In this formulation, only $\hat{x}$-directed currents are assumed, which should be adequate for lines that are not too wide [14]. Sinusoids, several cycles in length, are used to represent incident and reflected traveling waves of the fundamental microstrip mode, and subsectional (piecewise sinusoidal) modes are used near the open end, to represent currents that do not conform to the fundamental mode. This approach thus differs from a recent solution to the microstrip dipole antenna proximity fed by a microstrip line [17], where subsectional expansion modes were used. We also note that, in contrast to [5], the exciting wave is not assumed to be TEM.

Thus, define an incident electric current of unit amplitude as

$$I^{\text{inc}} = e^{-jk_e x} \tag{7}$$

and a reflected current as

$$I^{\text{ref}} = -Re^{jk_e x} \tag{8}$$

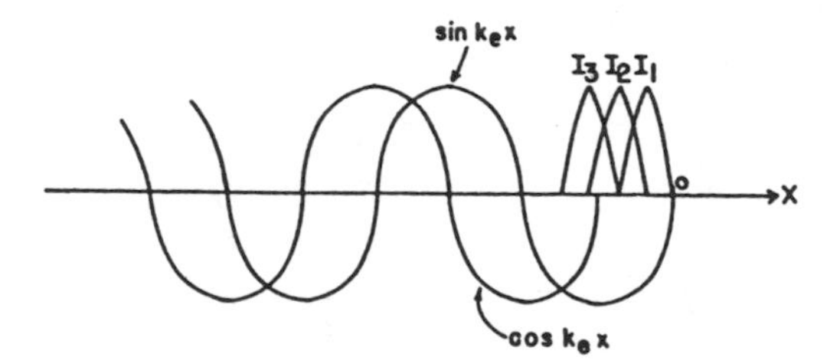

Fig. 2. Layout of expansion modes on the open-ended microstrip line.

where R is the reflection coefficient referenced to the end of the line ($x = 0$). Now, because of the method of numerically integrating (1) [12], it is useful to deal with only real expansion modes; thus, a simple transformation from exponential form to sine and cosine form is made

$$I^{\text{inc}} + I^{\text{ref}} = (1-R)\cos k_e x - j(1+R)\sin k_e x, \qquad x<0. \tag{9}$$

At $x = 0$, the total electric current must be zero. The sine term of the above current satisfies this condition, but not the cosine term, so the cosine term is truncated at $x = -\pi/2k_e$. Also, both terms are truncated after several cycles. The incident and reflected current components can then be written as

$$I^{\text{inc}} + I^{\text{ref}} = (1-R)f_s(k_e x + \pi/2) - j(1+R)f_s(k_e x) \tag{10}$$

where

$$f_s(u) = \begin{cases} \sin u, & 0 > u > -m\pi \\ 0, & \text{otherwise} \end{cases}.$$

It has been found that choosing the length of the sinusoids to be an integer number of half wavelengths speeds the convergence of the integrals shown in the next section. (Physically, this means that no end charges exist on the lines, as would be the case if the end currents were nonzero.) Typically, the solutions are insensitive to sinusoid length for lengths greater than three or four wavelengths.

Piecewise sinusoidal (PWS) modes are defined starting at the end and working left. These modes can be defined as

$$I_n f_n(x,y) = I_n \frac{\sin k_e(h - |x - x_n|)}{\sin k_e h},$$

$$\text{for } |x - x_h| < h,\ |y| < W/2 \tag{11}$$

where I_n is the unknown expansion coefficient, h is the half-length of the mode, and x_n is the terminal location, which is chosen as $x_n = -nh$, for $n = 1,2,3,\cdots$. The current is assumed uniform across the strip width. Fig. 2 shows how the various modes are arranged on the microstrip line. Typically convergence is achieved with three or four PWS modes.

D. Integral Equation/Moment Method Solution

An integral equation for the discontinuity is written by enforcing the boundary condition that the total $\hat{x}$ electric field due to all the currents on the line must be zero on the line. Equation (1) then yields

$$\int_{x_0}\int_{y_0}\left[I^{\mathrm{inc}}+I^{\mathrm{ref}}+\sum_{n=1}^{N}I_nf_n\right]E_{xx}\,dx_0\,dy_0=0,$$

$$\text{for } x<-\infty,\ |y|<W/2 \quad (12)$$

where N is the total number of PWS modes. This equation is enforced by multiplying by $N+1$ weighting or test functions (since there are $N+1$ unknowns), taken here as PWS modes as defined in (11) for $n=1$ to $N+1$, and integrating over x and y. Impedance matrix elements can then be defined as

$$Z_{mn}=\iint_{-\infty}^{\infty}Q(k_x,k_y)F_y^2(k_y)F_{xm}(k_x)F_{xn}^*(k_x)\,dk_x\,dk_y \quad (13)$$

$$Z_{mc}=\iint_{-\infty}^{\infty}Q(k_x,k_y)F_y^2(k_y)F_{xm}(k_x)F_{xc}^*(k_x)\,dk_x\,dk_y \quad (14)$$

$$Z_{ms}=\iint_{-\infty}^{\infty}Q(k_x,k_y)F_y^2(k_y)F_{xm}(k_x)F_{xs}^*(k_x)\,dk_x\,dk_y \quad (15)$$

where F_y is the Fourier transform defined in (5), and F_{xn}, F_{xc}, F_{xs} are Fourier transforms of the mode currents

$$F_{xn}=\int_{x_n-h}^{x_n+h}f_n(x)e^{jk_xx}\,dx \quad (16)$$

$$F_{xs}=\int_{-m\pi/k}^{0}\sin k_exe^{jk_xx}\,dx \quad (17)$$

$$F_{xc}=\exp\left[-jk_x\pi/(2k_e)\right]F_{xs}. \quad (18)$$

This results in a matrix equation for the unknown coefficients $R, I_1, I_2,\cdots, I_N$. For example, with two PWS modes, $N=2$, and a 3×3 matrix equation results

$$\begin{bmatrix} Z_{11} & Z_{12} & -(Z_{1c}+jZ_{1s}) \\ Z_{21} & Z_{22} & -(Z_{2c}+jZ_{2s}) \\ Z_{31} & Z_{32} & -(Z_{3c}+jZ_{3s}) \end{bmatrix}\cdot\begin{bmatrix} I_1 \\ I_2 \\ R \end{bmatrix}=\begin{bmatrix} -Z_{1c}+jZ_{1s} \\ -Z_{2c}+jZ_{2s} \\ -Z_{3c}+jZ_{3s} \end{bmatrix}. \quad (19)$$

Note that the above testing procedure only enforces (12) near the open end, where the testing modes are located. Farther away from the end (but still much greater than $-m\pi/k_e$), (12) is automatically satisfied since then the line looks locally as if it were infinitely long, and (6) implies that the E_x field is near zero. In other words, Z_{nc} and Z_{ns} quickly approach zero as n increases.

E. Surface-Wave Power

With the above formulation, the reflection coefficient of the open-circuited microstrip line can be found. Since the termination is not an *ideal* electrical open circuit, the reflection coefficient magnitude is always less than unity, implying that some incident power is lost to radiation of space and surface waves. In addition, an end admittance Y can be defined using the characteristic impedance Z_0 of the line. Thus

$$Y=\frac{1-R}{Z_0(1+R)}. \quad (20)$$

In order to quantify the separation of the total power loss into space-wave radiation and surface-wave excitation, an efficiency is defined as in [13]

$$e=\frac{P_{\mathrm{rad}}}{P_{\mathrm{rad}}+P_{\mathrm{sw}}} \quad (21)$$

where P_{rad} is the power lost to space waves and P_{sw} is the power lost to surface waves. This efficiency was originally defined for printed antennas [13] and is used here in the interest of consistency. The powers in (21) are found as follows:

$$P_{\mathrm{rad}}+P_{\mathrm{sw}}=\mathrm{Re}\left\{\sum_{i=1}^{N+2}\sum_{j=1}^{N+2}I_iZ_{ij}I_j^*\right\} \quad (22)$$

$$P_{\mathrm{sw}}=\mathrm{Re}\left\{\sum_{i=1}^{N+2}\sum_{j=1}^{N+2}I_iZ_{ij}^{sw}I_j^*\right\} \quad (23)$$

where

$$I_i=\begin{cases} I_i, & \text{for } 1\leqslant i\leqslant N \\ (1-R), & \text{for } i=N+1 \\ -j(1+R), & \text{for } i=N+2 \end{cases} \quad (24)$$

and Z_{ij} is an impedance matrix element with indices i, j for $i, j\leqslant N$, $i, j=c$ for $i, j=N+1$, and $i, j=s$ for $i, j=N+2$. The Z_{ij}^{sw} elements represent only the surface-wave contribution of the Z_{ij} elements, as computed from the residues of the surface-wave poles [12].

F. Gap Formulation

It is not difficult to modify the formulation for the open end in order to compute gap-discontinuity parameters. The configuration is shown in Fig. 1(b) with a gap G between the input and output microstrip lines. To analyze this discontinuity, three entire domain modes are used to represent incident, reflected, and transmitted currents. In addition to I^{inc} and I^{ref}, we therefore add

$$I^{\mathrm{tr}}=Te^{-jk_e(x-G)} \quad (25)$$

which is then modified in a manner similar to (9) and (10) to eliminate current discontinuities and to impose a finite length. This results in

$$I^{\mathrm{tr}}=T\left[-g\left(k_e[x-G]-\frac{\pi}{2}\right)-jg\left(k_e[x-G]\right)\right] \quad (26)$$

where

$$g(u)=\begin{cases}\sin u, & 0<u<m\pi \\ 0, & \text{otherwise}\end{cases}$$

which is then placed, with I^{inc} and I^{ref}, in (12) along with additional piecewise sinusoidal modes (eq. (11)). Piecewise modes will now exist at $x_n=-nh$ and at $x_n=nh+G$ for

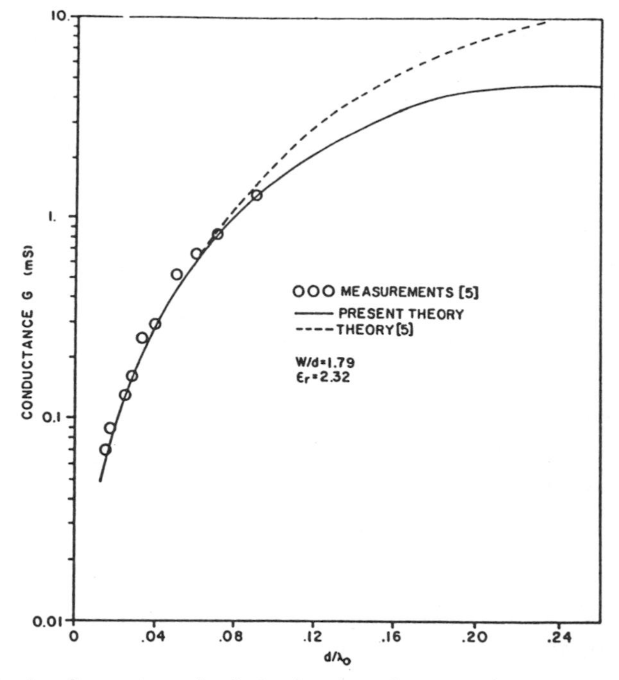

Fig. 3. Comparison of calculated end conductance of an open-circuit microstrip line compared with measurements and calculations of [5].

$n = 1, 2, \cdots, N$. This gives a total of $2N$ piecewise modes. Equation (12) has thus been modified such that R, T, and I_n are $2N+2$ unknowns for which one can solve once (12) has been tested with $2N+2$ piecewise testing functions. The remaining formulation is analogous to (13)–(19). We note that in computing the impedance matrix elements there are several redundancies due to reciprocity and due to the physical symmetry of the gap configuration. Making use of these redundancies considerably reduces computation time.

III. Results

Fig. 3 shows the terminal conductance of an open-circuited microstrip line as computed by this theory and compared with the measurements of [5] and calculations of [5] and [6]. The agreement of both theories and the measured data is good for substrate thickness up to about $0.1\lambda_0$, while the theories depart slightly above this value. In contrast to this theory, James and Henderson assume a TEM parallel-plate mode as an excitation. For thicker, higher dielectric constant substrates, their assumption is questionable and the two theories may diverge more readily. Note the trend that, as the substrate thickness increases, the termination looks less like an ideal open circuit.

Fig. 4(a) and (b) shows the reflection coefficient magnitude and efficiency e versus substrate thickness for $\epsilon_r = 2.55$, and various microstrip line widths. The efficiency e is practically independent of widths. Observe from Fig. 4(a) that the reflection coefficient magnitude drops well below unity for substrate thicknesses greater than a few hundreths of a wavelength, and is smaller for wider strips, as would be expected. The efficiency data of Fig. 4(b) shows that very little of the total power loss is caused by surface-wave

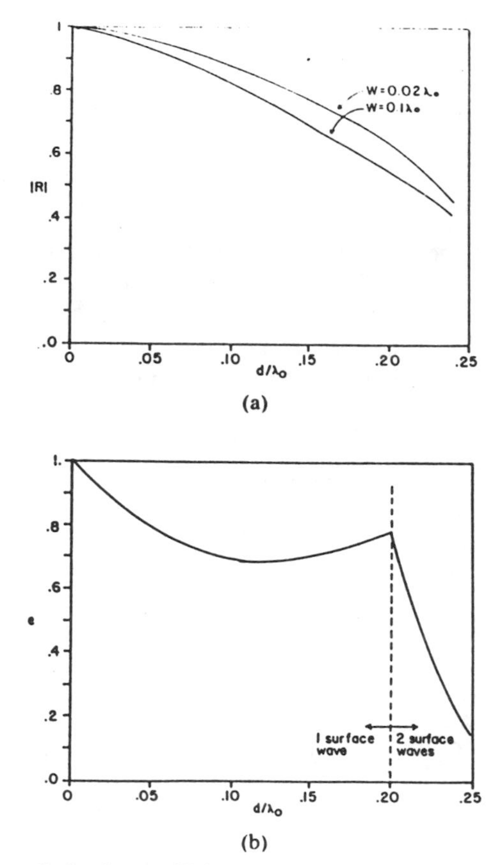

Fig. 4. (a) Reflection coefficient magnitude and (b) efficiency for an open-circuit microstrip line on an $\epsilon_r = 2.55$ substrate.

excitation when the substrate is thin, but that the surface-wave power increases (e decreases) for thicker substrates, with a cusp in the data at the cutoff point of the TE_1 surface-wave mode. This curve is very similar to that obtained for printed antennas [12], [18]. Fig. 5(a) and (b) shows corresponding data for a substrate with $\epsilon_r = 12.8$. It can be seen that the reflection coefficient magnitude drops off more rapidly with increased permittivity, and that significantly more power is launched into surface waves, for a given substrate thickness.

The data given in Figs. 3 and 4 allow one to determine the amount of power loss and the amount of surface-wave power generation of an open-circuit line. For example, assume that $1w$ of power is incident on an open microstrip line of width $0.1\lambda_0$ on a GaAs ($\epsilon_r = 12.8$) substrate $0.04\lambda_0$ thick. Then, from Fig. 5(a) and (b), $|R| = 0.96$ and $e = 0.53$, so there is $0.922w$ reflected on the line, $0.0416w$ delivered to space-wave radiation, and $0.0368w$ delivered to surface waves.

For MIC design, an open-circuit microstrip line is often modeled as having a reflection coefficient with unit magnitude and a phase accounted for by a length extension $\Delta l/d$. When radiation loss occurs, a conductance in parallel with a length extension is necessary, yielding

$$YZ_0 = GZ_0 + j\tan(k_e \Delta l) \tag{27}$$

where Y is given in (20). Fig. 6 gives normalized end conductance for common microstrip parameters on a Gal-

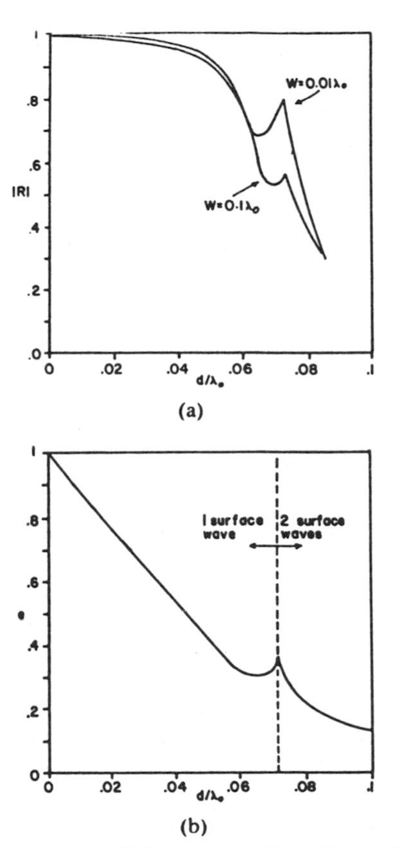

Fig. 5. (a) Reflection coefficient magnitude and (b) efficiency for an open-circuit microstrip line on an $\epsilon_r = 12.8$ substrate.

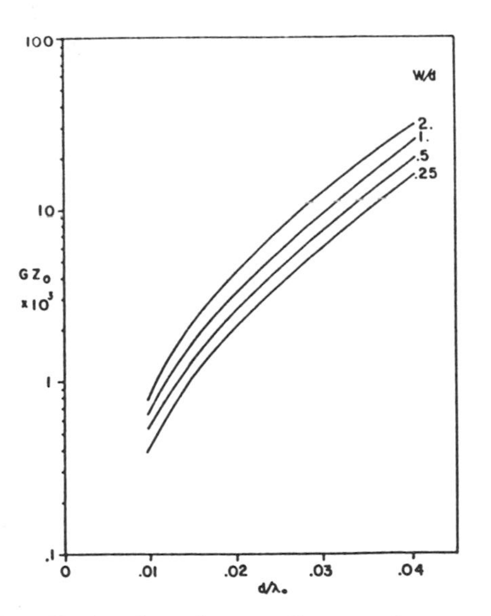

Fig. 6. Normalized end conductance for several common microstrip parameters on $\epsilon_r = 12.8$.

lium Arsenide substrate. For lossless cases, the length extension has been computed by quasi-static analysis [1], [2], as well as other methods [3], [5]. In contrast to the end conductance and reflection coefficient magnitude, we have found the length extension to be much more sensitive to the number of expansion modes, current distribution across the microstrip line, and truncation of the integrations in (13)–(15), especially for thin substrates. Instead of a uniform current distribution with respect to y, we found that a current distributed to enforce the edge condition

$$I^{\text{inc}}, I^{\text{ref}} \propto \frac{1}{\sqrt{1-(2y/w)^2}} \tag{28}$$

gave better agreement with quasistatic results in the thin substrate limit. Fig. 7 presents calculated results for the length extension using this theory and the quasi-static result [2]. Good agreement occurs for narrow lines but not for lines greater than two substrate thicknesses. Based on this result and on results with substrates having smaller dielectric constants, we conclude that the simple transverse current distribution assumed in (28) will give a reasonable length extension result for widths less than an eighth of a wavelength in the dielectric. For precise length extension calculations, or for wider microstrip lines, a more complicated transverse current distribution is necessary. In this paper, our calculation of length extension is primarily to help validate the less sensitive conductance calculations.

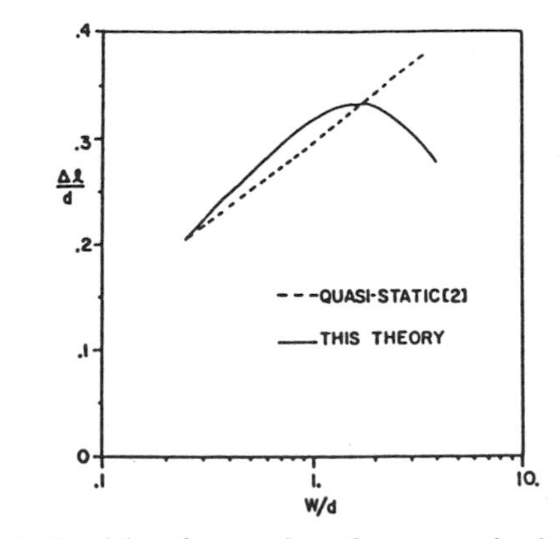

Fig. 7. Calculated length extension of an open-circuit microstrip line compared with quasi-state theory [2] on a substrate with $\epsilon_r = 12.8$, $d = 0.02\lambda_0$.

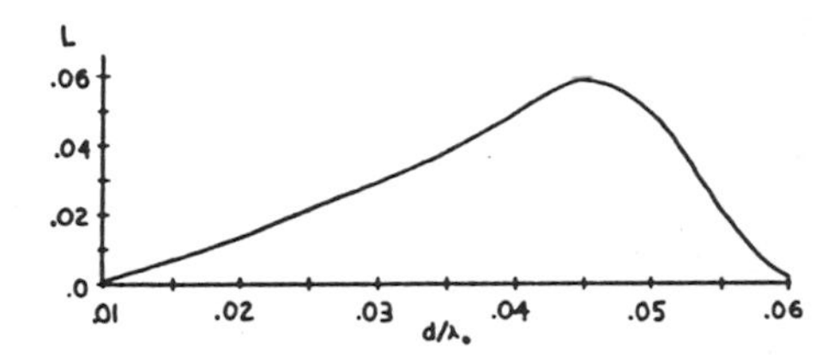

Fig. 8. Loss for a microstrip gap with $\epsilon_r = 12.8$, $G = 0.2d$, $w = 2.5d$.

For the gap, we have calculated total power loss due to the combination of surface-wave and space-wave radiation by computing

$$L = 1 - |R|^2 - |T|^2.$$

In Fig. 8, we plot this quantity versus substrate electrical thickness for a representative configuration on Gallium Arsenide. Loss increases, peaks, and then decreases as frequency increases. The decrease occurs due to the fact that a gap, intuitively considered as a series capacitance, looks like a series short at high frequencies and thus like less of a radiation-producing discontinuity. This loss data is insensitive to the distribution of current in the transverse direction. We have calculated circuit models for a gap using this method and found models which roughly agree with quasi-static models but which are somewhat sensitive to transverse current distribution. For very accurate calculation of gap and fringing capacitances, a higher degree of modal approximation is desirable. Investigations of this type are underway.

IV. Conclusion

A full-wave analysis has been presented for the problems of microstrip open-end and gap discontinuities. For the open end, the reflection coefficient, radiated power, and surface-wave power have been calculated and compared with previous calculations and measured data, when available. Plots of end conductance and length extension have been presented for a high dielectric substrate. Loss at a gap discontinuity has also been calculated. This type of analysis should aid in the design of microwave integrated circuits, particularly for higher frequencies and high dielectric constant substrates. Similar analysis can be used to characterize more complicated microstrip discontinuities.

References

[1] E. O. Hammerstad and F. Bekkadal, "Microstrip handbook," ELAB Rep. STF44A74169, University of Trondheim, 1975.

[2] K. C. Gupta, R. Garg, and I. J. Bahl, *Microstrip Lines and Slotlines*. Dedham, MA: Artech House, 1979.

[3] R. Jansen, "Hybrid mode analysis of end effects of planar microwave and millimeterwave transmission lines," *Proc. Inst. Elec. Eng.*, vol. 128, pt. H. pp. 77–86, Apr. 1978.

[4] L. Lewin, "Radiation from discontinuities in stripline," *Proc. Inst. Elec. Eng.*, vol. 107C, pp. 163–170, Feb. 1960.

[5] J. R. James and A. Henderson, "High frequency behavior of microstrip open-circuit terminations," *IEE J. Microwave Opt. Acoust.*, vol. 3, pp. 205–211, Sept. 1979.

[6] J. R. James, P. S. Hall, and C. Wood, *Microstrip Antenna Theory and Design*. London: Peter Peregrinus, 1981.

[7] M. Maeda, "An analysis of gaps in microstrip transmission lines," *IEEE Trans. Microwave Theory Tech.*, vol. MTT-20, pp. 390–396, June 1972.

[8] P. Benedek and P. Silvester, "Equivalent capacitances for microstrip gaps and steps," *IEEE Trans. Microwave Theory Tech.*, vol. MTT-20, pp. 729–733, Nov. 1972.

[9] Y. Rahmat-Samii, T. Itoh, and R. Mittra, "A spectral domain analysis for solving microstrip discontinuity problems," *IEEE Trans. Microwave Theory Tech.*, vol. MTT-24, pp. 372–378, Apr. 1984.

[10] A. Gopinath and C. Gupta, "Capacitance parameters of discontinuities in microstrip lines," *IEEE Trans. Microwave Theory Tech.*, vol. MTT-26, pp. 831–836, Oct. 1978.

[11] R. Jansen and N. Koster, "A unified CAD basis for the frequency dependent characterization of strip slot and coplanar MIC components," in *Proc. 11th Eur. Microwave Conf.* (Amsterdam), 1981, pp. 682–687.

[12] D. M. Pozar, "Input impedance and mutual coupling of rectangular microstrip antennas," *IEEE Trans. Antennas Propagat.*, vol. AP-30, pp. 1191–1196, Nov. 1982.

[13] D. M. Pozar, "Considerations for millimeter wave printed antennas," *IEEE Trans. Antennas Propagat.*, vol. AP-31, pp. 740–747, Sept. 1983.

[14] T. Itoh and R. Mittra, "Spectral-domain approach for calculating the dispersion characteristics of microstrip lines," *IEEE Trans. Microwave Theory Tech.*, vol. MTT-21, pp. 496–499, July 1973.

[15] C. P. Hartwig, D. Masse, and R. A. Pucel, "Frequency dependent behavior of microstrip," in *1968 G-MTT Int. Symp. Dig.*, pp. 11–116.

[16] I. J. Bahl and D. K. Trivedi, "A designer's guide to microstrip line," *Microwaves*, pp. 174–182, May 1977.

[17] P. B. Katehi and N. G. Alexopoulos, "On the modeling of electromagnetically coupled microstrip antennas—The printed strip dipole," *IEEE Trans. Microwave Theory Tech.*, this issue, pp. 1029–1035.

[18] P. B. Katehi and N. G. Alexopoulos, "On the effect of substrate thickness and permittivity on printed circuit dipole properties," *IEEE Trans. Antennas Propagat.*, vol. AP-31, pp. 34–39, Jan. 1983.

Part VIII
Coplanar Waveguide

THE coplanar waveguide invented by C. P. Wen is a surface oriented planar transmission line made of three pieces of conductors. A center strip is placed on the surface of a dielectric substrate with two ground planes running adjacent and parallel to the strip on the same surface. In its original form, there is no conductor plane on the other side of the substrate. The dominant mode is once again hybrid but is quasi-TEM. This is a balanced mode, meaning that the electric field in the transverse cross section is symmetrically oriented from the center strip to the two ground planes. Since there are gaps on both sides of the center strip, the structure may alternatively be thought of as two coupled slot lines. Although the nature of the slot line is a topic in the next part, it is worthwhile to mention here that a coupled slot line mode can be excited in a coplanar waveguide as an unbalanced mode. For example, when a discontinuity is encountered or when a deviation from a symmetric construction exists, this coupled slot mode can be excited. To prevent excitation of this mode, equal RF potentials of the two ground planes are often ensured by air bridges between two ground planes. Because of the availability of three separate conductors, the coplanar waveguide is thought to be convenient for mounting three terminal devices such as MESFET's. One of the problems of the coplanar waveguide is heat removal from an active device. It is difficult to provide an efficient heat sink without perturbing the electromagnetic field in the coplanar waveguide. An additional ground plane on the other side of the three conductors is sometimes used.

The papers included in this part cover a number of aspects of the coplanar waveguide. In addition to the quasi-TEM analysis in the original paper by Wen, several papers treat more rigorous full-wave analysis methods for the propagation characteristics in the coplanar waveguide and its modifications. Attenuation characteristics are also important and are treated by two papers (Gopinath; Jackson). For a number of years, interest in the coplanar waveguide has been less than interest in the microstrip line. However, with the advent of monolithic integrated circuits, the last few years have seen renewed interest in the coplanar waveguide. One of the papers in this part discusses an interesting comparison between the microstrip line and the coplanar waveguide (Jackson). The last paper in this part presents one of the empirical closed-form expressions for the coplanar waveguide.

Coplanar Waveguide: A Surface Strip Transmission Line Suitable for Nonreciprocal Gyromagnetic Device Applications

CHENG P. WEN, MEMBER, IEEE

Abstract—A coplanar waveguide consists of a strip of thin metallic film on the surface of a dielectric slab with two ground electrodes running adjacent and parallel to the strip. This novel transmission line readily lends itself to nonreciprocal magnetic device applications because of the built-in circularly polarized magnetic vector at the air–dielectric boundary between the conductors. Practical applications of the coplanar waveguide have been experimentally demonstrated by measurements on resonant isolators and differential phase shifters fabricated on low-loss dielectric substrates with high dielectric constants. Calculations have been made for the characteristic impedance, phase velocity, and upper bound of attenuation of a transmission line whose electrodes are all on one side of a dielectric substrate. These calculations are in good agreement with preliminary experimental results. The coplanar configuration of the transmission system not only permits easy shunt connection of external elements in hybrid integrated circuits, but also adapts well to the fabrication of monolithic integrated systems. Low-loss dielectric substrates with high dielectric constants may be employed to reduce the longitudinal dimension of the integrated circuits because the characteristic impedance of the coplanar waveguide is relatively independent of the substrate thickness; this may be of vital importance for low-frequency integrated microwave systems.

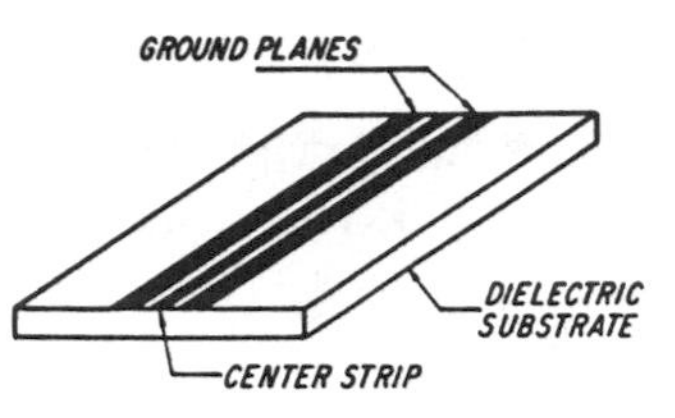

Fig. 1. Coplanar waveguide (CPW), a surface strip transmission line.

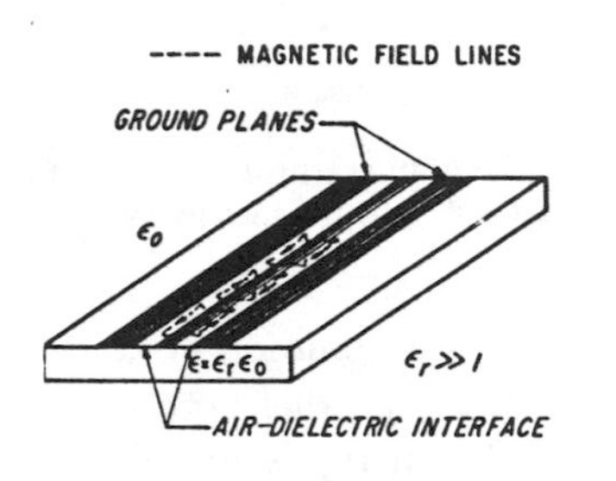

Fig. 2. RF magnetic field configuration in a CPW.

MICROWAVE integrated circuits offer system engineers prospects of small, batch processed modules for radar and communication systems. In the past, radar and communication systems included a variety of nonreciprocal magnetic devices. These devices require circularly polarized RF magnetic fields for their operation, and present microstrip and strip lines do not provide such fields. In addition, the ground plane of these lines, which is located on the opposite side of a dielectric substrate, is not easily accessible for shunt connections necessary for many active microwave devices. Direct dependence of the characteristic impedance on the thickness of the substrate makes it difficult to use a low-loss, high-dielectric-constant material, such as a temperature-compensated ceramic. This is a definite drawback for low-frequency operation where size consideration dominates. All these disadvantages may either be overcome or alleviated by a novel integrated-circuit transmission-line configuration in which all conducting elements, including the ground planes, are on the same side of a dielectric substrate. This is called the coplanar waveguide, a surface-strip transmission line.

A coplanar waveguide (CPW) consists of a strip of thin metallic film deposited on the surface of a dielectric slab with two ground electrodes running adjacent and parallel to the strip on the same surface, as shown in Fig. 1. There is no low-frequency cutoff because of the quasi-TEM mode of propagation. However, the RF electric field between the center conducting strip and the ground electrodes tangential to the air–dielectric boundary produces a discontinuity in displacement current density at the interface, giving rise to an axial, as well as transverse, component of RF magnetic field. These components provide the elliptical polarization needed for nonreciprocal gyromagnetic microwave device applications [1]. If the relative dielectric constant ϵ_r of the substrate is very large compared to unity, the magnetic field at the interface is nearly circularly polarized with the plane of polarization perpendicular to the surface of the substrate, as shown in Fig. 2. Such a transmission line readily lends itself to integrated-circuit fabrication techniques and nonreciprocal gyromagnetic device applications because of the built-in circularly polarized magnetic vector which is easily accessible at the surface of the substrate. The coplanar configuration of all conducting elements permits easy connection of external shunt elements such as active devices as well as the fabrication of series and shunt capacitances. It is also ideal for connecting various elements in monolithic microwave integrated circuits built on semiconducting substrates or ferromagnetic semiconductors. All ground planes may be connected together through a metallic capsule as shown in

Manuscript received March 28, 1969; revised July 9, 1969. This paper was presented at the International Microwave Symposium, Dallas, Tex., May 5–7, 1969.

The author is with RCA Laboratories, David Sarnoff Research Center, Princeton, N. J. 08540.

Reprinted from *IEEE Trans. Microwave Theory Tech.*, vol. MTT-17, no. 12, pp. 1087–1090, December 1969.

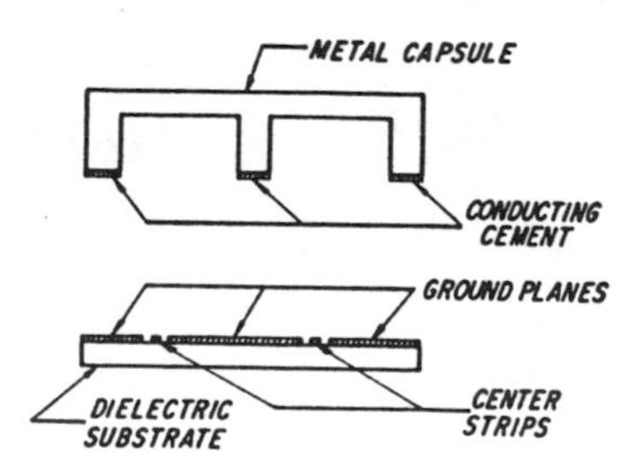

Fig. 3. Metallic capsule for CPW ground connections.

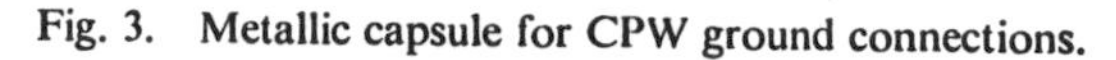

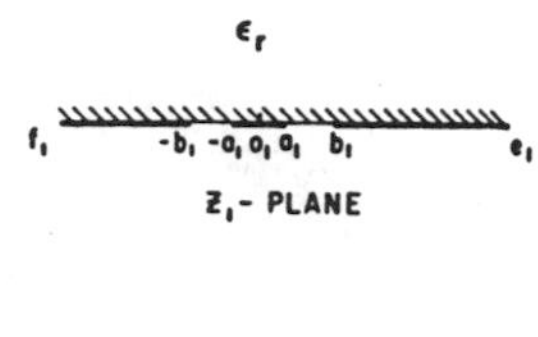

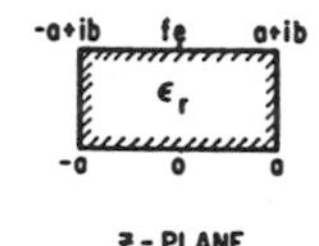

Fig. 4. Conformal mapping transformation of the upper half-plane of a CPW into the interior of a rectangle.

Fig. 3, serving both as a common ground and a protective cover. Because of the high dielectric constant of the substrate, most of the RF energy is stored in the dielectric and the loading effect of the grounded cover is negligible if it is more than two slot widths away from the surface.

Characteristic impedances Z_0 of CPWs fabricated on dielectric half-planes with relative dielectric constants ϵ_r have been calculated as a function of the ratio a_1/b_1, where $2a_1$ is the width of the center strip and $2b_1$ is the distance between two ground electrodes. A zeroth-order quasi-static approximation is employed. The dielectric half-plane Z_1 in Fig. 4 may be transformed to the interior of a rectangle in the Z-plane by conformal mapping [2]:

$$\frac{dZ}{dZ_1} = \frac{A}{(Z_1^2 - a_1^2)^{1/2}(Z_1^2 - b_1^2)^{1/2}} \qquad (1)$$

where A is a constant. The ratio a/b of the rectangle in the Z plane may be evaluated by multiplying both sides of (1) by dZ_1 and carrying out the integration

$$a + jb = \int_0^{b_1} \frac{A\,dZ_1}{(Z_1^2 - a_1^2)^{1/2}(Z_1^2 - b_1^2)^{1/2}}\,. \qquad (2)$$

As a result

$$\frac{a}{b} = \frac{K(k)}{K'(k)} \qquad (3)$$

where

$k = a_1/b_1$,
$K(k)$ = complete elliptical integral of the first kind [3],
$K'(k) = K(k')$,
$k' = (1-k^2)^{1/2}$.

If the relative dielectric constant of the material filling the rectangle in the Z plane of Fig. 4 is ϵ_r, a uniform electric field E is set up in the capacitor with the top and bottom plates charged up to opposite polarities and the capacitance per unit length of the line, including the empty space half-plane, is

$$C = (\epsilon_r + 1)\epsilon_0 \frac{2a}{b}\,. \qquad (4)$$

A zeroth-order quasi-static approximation has been employed to estimate the phase velocities and the characteristic impedances of CPWs. The approximation simply treats the CPW as a transmission line totally immersed in a dielectric with effective dielectric constant $(\epsilon_r+1)/2$. The resulting phase velocity is

$$v_{ph} = \left(\frac{2}{\epsilon_r + 1}\right)^{1/2} c \qquad (5)$$

where c is the velocity of light in free space and the characteristic impedance is

$$Z_0 = \frac{1}{Cv_{ph}}\,. \qquad (6)$$

The ratio v_{ph}/c is shown in Fig. 5 as a function of ϵ_r. In Fig. 6 the characteristic impedance Z_0 is shown as a function of a_1/b_1 with the relative dielectric constant ϵ_r as a parameter ranging from unity to 250. Experimental confirmation has been obtained at three points shown on the same figure with transmission lines fabricated on substrates of relative dielectric constant $\epsilon_r = 9.5$, 16, and 130, respectively. The calculated characteristic impedance of a parallel-strip coplanar line, the dual of the CPW, is shown in Fig. 7 as a function of a_1/b_1 with the relative dielectric constant ϵ_r of the substrate as a parameter [4]. The configuration of this line can be found on the same figure.

The thickness of the dielectric substrate becomes less critical with higher relative dielectric constants. In the CPW configuration the characteristic impedance increases by less than 10 percent when the thickness of the substrate is reduced from infinity to $b_1 - a_1$, the width of the slots, for infinitely large ϵ_r. In practice, the thickness t of the substrate should be one or two times the width of the slots. It is obvious that the finite thickness of the substrate will influence the dispersion characteristics of the transmission line but no estimate has been made on the extent of the effect. The experimentally measured dispersion characteristic of a coplanar waveguide fabricated on a single-crystal rutile substrate is shown on the frequency versus βL plot of Fig. 8. It is not known what portion of the dispersion is caused by the crystal anisotropy of the substrate and what portion is attributed to the inherent characteristics of the CPW mode of propagation on a dielectric half-space.

At microwave frequencies up to X band the attenuation of a CPW is due mainly to the copper loss of the conductors if the loss tangent of the dielectric is 0.001 or smaller. Measurements made on a 16.6-ohm CPW fabricated on a rutile substrate ($\epsilon_r \cong 130$, $k = \frac{1}{3}$, center conductor width is 0.025

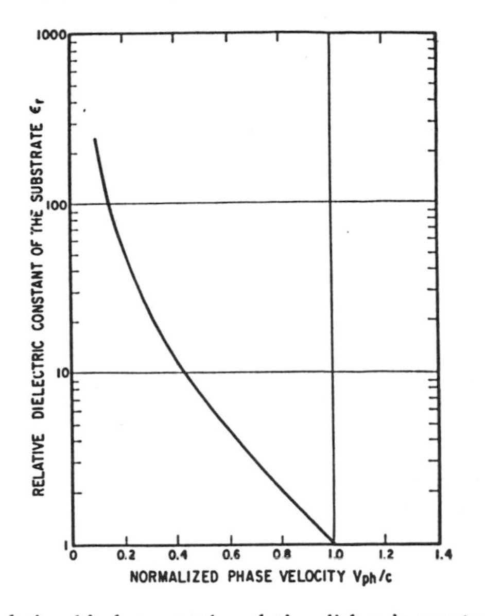

Fig. 5. Relationship between the relative dielectric constant ϵ_r of the substrate and the normalized phase velocity v_{ph}/c in a CPW.

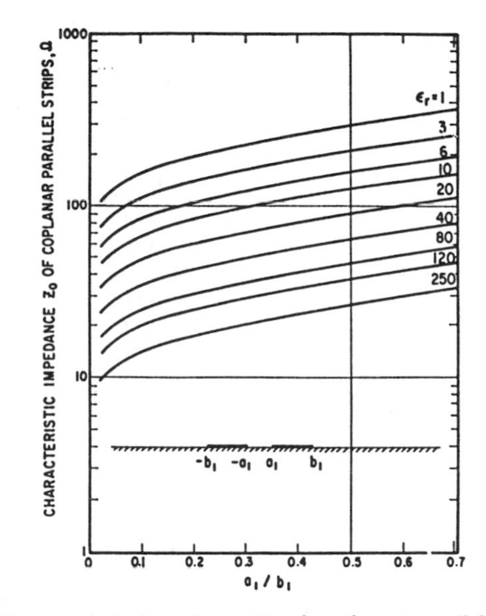

Fig. 7. Characteristic impedance Z_0 of coplanar parallel strips as a function of the ratio a_1/b_1 with the relative dielectric constant ϵ_r as a parameter.

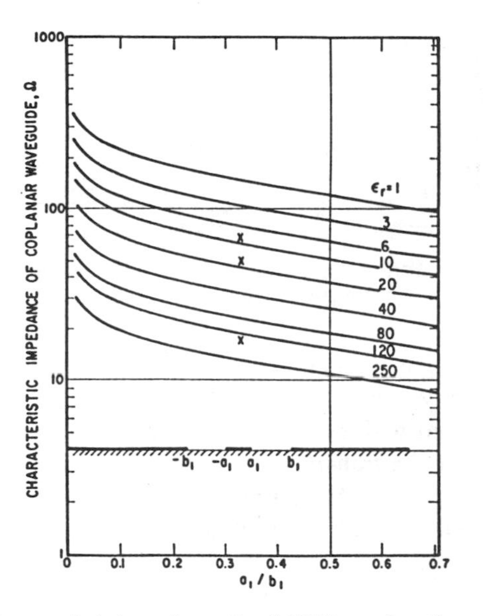

Fig. 6. Characteristic impedance Z_0 of CPW as a function of the ratio a_1/b_1 with the relative dielectric constant ϵ_r as a parameter

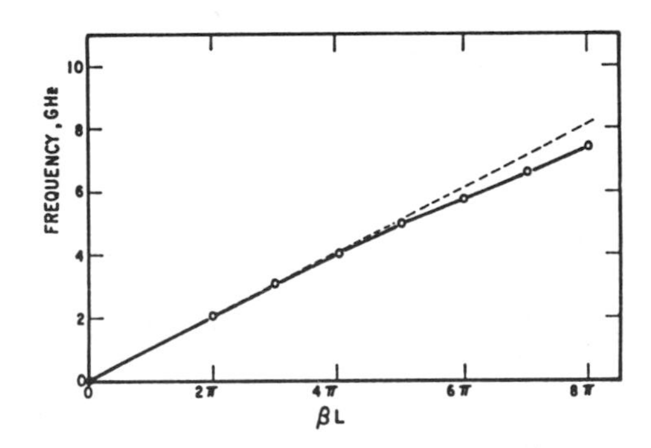

Fig. 8. Dispersion characteristics of a CPW on a TiO_2 substrate.

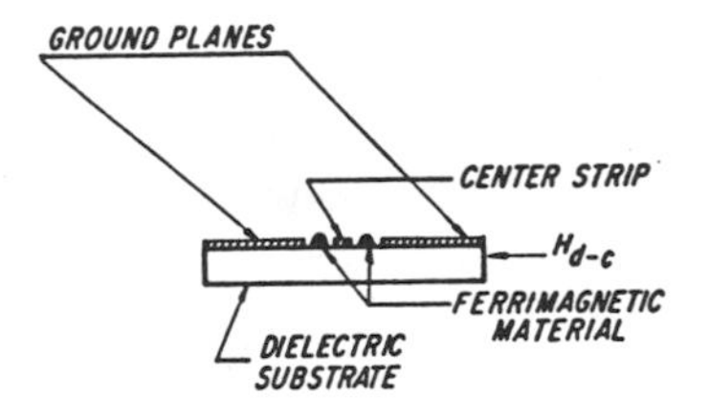

Fig. 9. Resonant isolator or differential phase shifter in CPW configurations.

inch, thickness of gold film is 2 microns, thickness of the dielectric substrate is 0.025 inch), yielded a Q of 173 at 4 GHz, corresponding to an attenuation of only 0.158 dB per wavelength.

Other design considerations include radiation problems when the total distance from one ground plane to another approaches $\lambda/2$. This will pose a limit on the dielectric constant of the substrate employed at a given frequency and the transverse dimension of the transmission line. CPWs can be easily adapted to coaxial systems using OSM connectors. Broad-band (2–12 GHz) matching has been achieved with a 50-ohm CPW deposited on a magnesium oxide substrate with $\epsilon_r = 16$.

Nonreciprocal gyromagnetic devices such as resonant isolators and differential phase shifters have been fabricated by attaching ferrimagnetic slabs at the air–dielectric interface between the conductors, as shown in Fig. 9. A differential phase shifter fabricated on an all-magnetic garnet substrate has also been built. A transverse dc magnetic field parallel to the surface of the substrate is required to provide appropriate bias conditions. As shown in Fig. 10, a CPW ferrimagnetic-resonance isolator built on the surface perpendicular to the c axis of a single rutile crystal provides 37-dB isolation at the center frequency of 6 GHz while the forward attenuation is below 2 dB. Overall length of the line is 0.8 inch including a quarter-wave transformer at each end. The center conductor width is 0.030 inch, $k = 0.33$, and the substrate is 0.025 inch thick. Strips of Trans-Tech G-1000 YIG

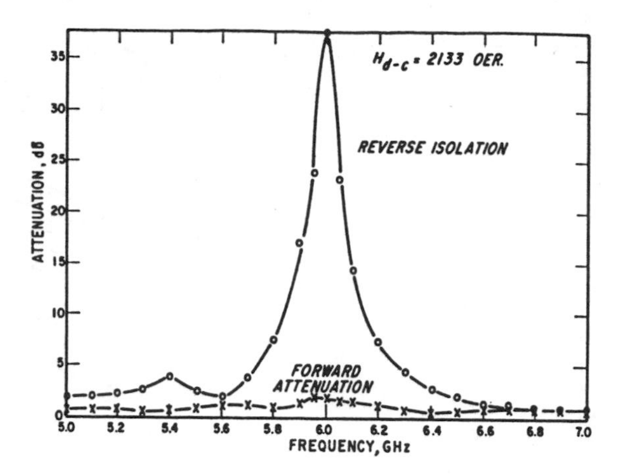

Fig. 10. Performance of a CPW ferrimagnetic-resonant isolator on a single-crystal rutile substrate.

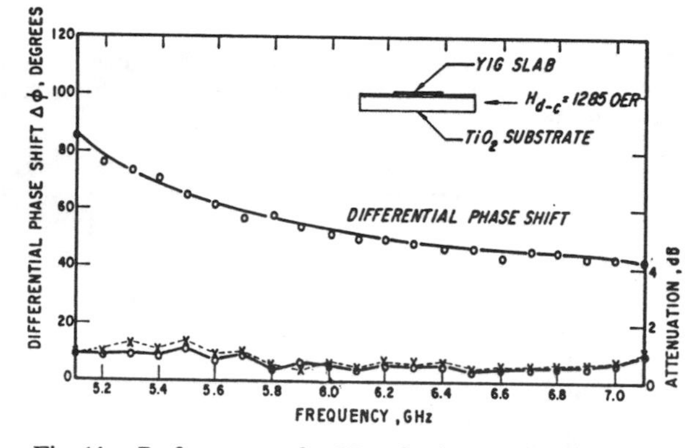

Fig. 11. Performance of a CPW ferrimagnetic differential phase shifter on a single-crystal rutile substrate.

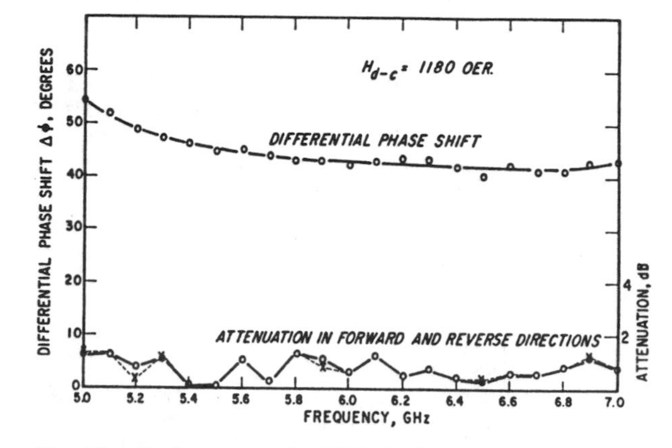

Fig. 12. Performance of a CPW ferrimagnetic differential phase shifter on a YIG substrate.

(0.010 inch×0.005 inch×0.60 inch) are attached by low-loss cement to the rutile surface with the center line of the YIG 0.010 inch from the ground planes. The performance of a phase shifter whose configurations are similar to the previously discussed isolator, are shown in Fig. 11. No attempt has been made to equalize the amount of differential phase shift across the band of frequencies. Average differential phase shift is over 45° while the insertion loss in either direction is less than 1 dB between 5.6 GHz and 7.1 GHz. Higher loss is observed at the lower frequencies which are neaı the ferrimagnetic resonance. Differential phase shift and loss data for a CPW fabricated on a YIG substrate is shown in Fig. 12. A tapered ceramic piece (0.800 inch long, 10° tapering on both ends, $\epsilon_r = 83$) is placed on top to provide the circularly polarized magnetic vector in the ferrite. The amount of differential phase shift varies little with frequency beyond 5.5 GHz. Insertion loss of 1.3 dB or less has been measured from 5.0 GHz to 7.0 GHz without special effort to match the device to the 50-ohm test system. These preliminary results demonstrate the gyromagnetic nonreciprocal device capabilities of coplanar waveguides.

In summary, the practicality of a novel microwave integrated circuit transmission line suitable for nonreciprocal gyromagnetic device applications has been demonstrated. Some preliminary calculations for characteristic impedance, phase velocity, and attenuation characteristics are presented and compared with experimental results. The coplanar configuration of the transmission system not only permits easy shunt connections of external elements, it also adapts well to the fabrication of monolithic integrated circuits.

Acknowledgment

The author wishes to thank H. Davis for his able assistance in preparation and during experimentation and R. Goodrich for the deposition of thin film circuits. Constant encouragement and valuable suggestions from B. Hershenov and L. S. Napoli are gratefully acknowledged.

References

[1] B. J. Duncan, L. Swern, K. Tomiyasu, and J. Hannwacker, "Design considerations for broad-band ferrite coaxial line isolators," *Proc. IRE*, vol. 45, pp. 483–490, April 1957.

[2] W. R. Smythe, *Static and Dynamic Electricity*. New York: McGraw-Hill, 1950.

[3] E. Jahnke and F. Emde, *Tables of Functions with Formulae and Curves*, 4th ed. New York: Dover, 1945.

[4] R. F. Frazita, "Transmission line properties of coplanar parallel strips on a dielectric sheet," M.S. thesis, Polytechnic Institute of Brooklyn, Brooklyn, N. Y., 1965.

Analysis of Coupled Slots and Coplanar Strips on Dielectric Substrate

JEFFREY B. KNORR, MEMBER, IEEE, AND KLAUS-DIETER KUCHLER, STUDENT MEMBER, IEEE

***Abstract*—A frequency-dependent hybrid-mode analysis of single and coupled slots and coplanar strips is presented. The dispersion characteristic and characteristic impedance of the structures are obtained by applying a Fourier transform technique and evaluating the resulting expressions numerically using the method of moments. Numerical results are presented and compared with results published by other investigators. The experimental performance of a slot-line coupler is compared with predicted performance based upon the results presented here for coupled slots. Excellent agreement has been obtained in all cases.**

I. INTRODUCTION

COPLANAR transmission lines have been studied by many investigators, mainly because they are easily adaptable to shunt-element connections without the need to penetrate the dielectric substrate as in the case of microstrip lines. Cohn [1] investigated the slot line and found an approximate analytic expression for the dispersion characteristic and the characteristic impedance by converting the slot line into a rectangular waveguide configuration. Recently this transmission line was analyzed by a new method proposed by Itoh and Mittra [2], but only to the extent that the dispersion characteristics were computed. To the authors' knowledge, there has been no other analysis published for the characteristic impedance of slot line besides Cohn's method [1], [3].

In connection with the increased interest in coplanar transmission lines, the need for an analysis of coupled slot lines or coplanar waveguide (CPW) structures is obvious. Wen [4] studied this transmission line with the assumption that the dielectric substrate is thick enough to be considered infinite for conformal mapping purposes. He also shows some theoretical results for two coplanar parallel strips, again on an infinitely thick dielectric substrate. For large values of ϵ_r, the relative permittivity of the substrate, this assumption may be practical, but it appears impractical for relatively small values of ϵ_r and for thin substrates. An alteration of this method used by Wen is given by Davis *et al.* [5] and takes the finite thickness of the dielectric substrate into account but also uses a quasi-static approximation, and thus lacks any frequency-dependent information on the behavior of phase velocity and characteristic impedance.

The purpose of this paper is to outline a new approach which was first suggested by Itoh and Mittra [2] and then extended by the authors to yield the characteristic impedance of the slot line as well as the dispersion characteristic and characteristic impedances of a pair of coupled slot lines in the odd and the even modes. During the development of the mathematical formulation, it was an easy extension to derive also the characteristics of a pair of coplanar parallel strips. The method is quite general and has also been used to analyze microstrip, although the results will not be presented here.

II. DISPERSION RELATIONSHIP ANALYSIS FOR A SINGLE SLOT LINE

A wave propagation problem on a slot transmission line, shown in Fig. 1, means, in general, the solution to the wave equation in an inhomogeneous medium with inhomogeneous boundary conditions. Moreover, the electric field in the slot is not known, and rather than finding the Green's function in a closed form, the investigator is forced to find an approximate solution. This led Cohn to his approach of using the infinite orthogonal set of waveguide modes, in other words, a complete set of functions, and a conversion from the slot-line configuration into a waveguide configuration by the use of electric and magnetic walls.

Itoh and Mittra [2] introduced a new technique for the analysis of the slot-line dispersion characteristic. To obtain a full understanding of the methods used in this paper, some reiterations from [2] are necessary.

It is known that all hybrid-field components can be obtained from a superposition of TE and TM modes which are related to the two scalar potential functions $\phi^e(x,y)$ and $\phi^h(x,y)$, where the superscripts e and h denote electric and magnetic, respectively. The axial components of TM and TE modes are then

Manuscript received October 1, 1974; revised February 19, 1975. This work was supported in part by the Office of Naval Research through the U. S. Naval Postgraduate School Foundation Research Program.

J. B. Knorr is with the Department of Electrical Engineering, U. S. Naval Postgraduate School, Monterey, Calif. 93940.

K.-D. Kuchler is with the Federal German Navy. He is currently assigned to the U. S. Naval Postgraduate School, Monterey, Calif. 93940.

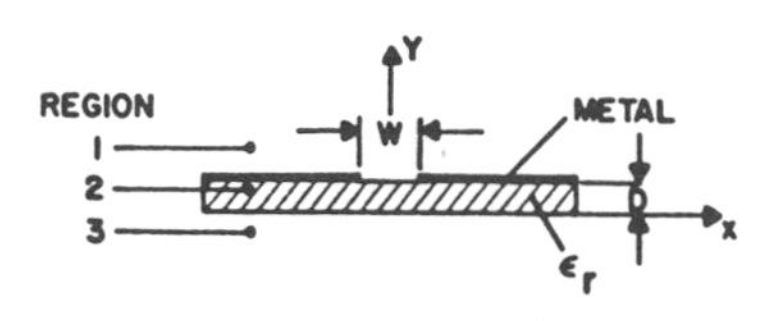

Fig. 1. Slot line.

Reprinted from *IEEE Trans. Microwave Theory Tech.*, vol. MTT-23, no. 7, pp. 541–548, July 1975.

$$E_z = k_c^2\phi^e(x,y)\exp(\pm j\beta z)$$

and

$$H_z = k_c^2\phi^h(x,y)\exp(\pm j\beta z)$$

respectively, where β is the propagation constant, assuming no losses, and

$$k_c^2 = k_i^2 - \beta^2$$

with

$$k_i = \omega(\epsilon_i\mu_i)^{1/2}, \qquad i = 1,2,3$$

defining the spatial region as shown in Fig. 1.

Both scalar potential functions satisfy the Helmholtz equation which is transformed into the Fourier domain thus converting a second-order partial differential equation into an ordinary differential equation. The solutions to these two ordinary differential equations can then be written as

$$\begin{aligned}\Phi_1^e(\alpha,y) &= A^e(\alpha)\exp[-\gamma_1(y-D)]\\ \Phi_2^e(\alpha,y) &= B^e(\alpha)\sinh\gamma_2 y + C^e(\alpha)\cosh\gamma_2 y\\ \Phi_3^e(\alpha,y) &= D^e(\alpha)\exp(\gamma_1 y)\end{aligned}\tag{1}$$

$$\begin{aligned}\Phi_1^h(\alpha,y) &= A^h(\alpha)\exp[-\gamma_1(y-D)]\\ \Phi_2^h(\alpha,y) &= B^h(\alpha)\sinh\gamma_2 y + C^h(\alpha)\cosh\gamma_2 y\\ \Phi_3^h(\alpha,y) &= D^h(\alpha)\exp(\gamma_1 y)\end{aligned}\tag{2}$$

where

$$\gamma_i^2 = \alpha^2 + \beta^2 - k_i^2\tag{3}$$

and the subscript defines the region. The preceding equations are found in [2, eqs. (2) and (3)]. It is important to observe that in region 2, $\gamma_2^2 < 0$ for small values of α, which means that the hyperbolic functions are replaced by trigonometric functions.

The eight unknown coefficients A^e through D^h are related to the horizontal electric- and magnetic-field components at the interfaces $y = 0$ and $y = D$ by the continuity conditions, and can be related also to the surface current density on the metal and the electric field in the slot at $y = D$.

If we denote the Fourier transforms of the x- and z-directed current-density and electric-field components by

$$\mathcal{E}_x(\alpha) = \mathcal{F}\{E_x(x)\} \qquad \mathcal{E}_z(\alpha) = \mathcal{F}\{E_z(x)\}$$

$$\mathcal{J}_x(\alpha) = \mathcal{F}\{j_x(x)\} \qquad \mathcal{J}_z(\alpha) = \mathcal{F}\{j_z(x)\}$$

we obtain a set of coupled equations of the form

$$\begin{bmatrix} M_1(\alpha,\beta) & M_2(\alpha,\beta)\\ M_3(\alpha,\beta) & M_4(\alpha,\beta)\end{bmatrix}\begin{bmatrix}\mathcal{J}_x(\alpha)\\ \mathcal{J}_z(\alpha)\end{bmatrix} = \begin{bmatrix}\mathcal{E}_x(\alpha)\\ \mathcal{E}_z(\alpha)\end{bmatrix}\tag{4}$$

where the elements of the M-matrix are the Fourier transforms of dyadic Green's function components.

If the M-matrix is inverted, we obtain a new matrix N and a second set of coupled equations

$$\begin{bmatrix} N_1(\alpha,\beta) & N_2(\alpha,\beta)\\ N_3(\alpha,\beta) & N_4(\alpha,\beta)\end{bmatrix}\begin{bmatrix}\mathcal{E}_x(\alpha)\\ \mathcal{E}_z(\alpha)\end{bmatrix} = \begin{bmatrix}\mathcal{J}_x(\alpha)\\ \mathcal{J}_z(\alpha)\end{bmatrix}.\tag{5}$$

This last formulation is equivalent to [2, eq. (4)].

Up to this point, the formulation of the problem is exact; no approximation has been made. If, however, the electric-field and current-density components are expanded in infinite series using a complete set of basis functions, and Galerkin's method [6] is applied, a homogeneous system of linear equations can be found [2, eqs. (7) and (8)]. An iteration scheme for β can be used to find a nontrivial solution for this set of equations.

The remaining question is what kind of basis functions to choose. The choice of this complete set of basis functions is arbitrary in a mathematical sense, since as long as this set is complete, any closed form of the field components can be represented by it. However, the rate of convergence of the series representation will depend on how well the first few terms approximate the closed form. In general, this requires some *a priori* knowledge of the true distributions.

In order to determine the sensitivity of the previously outlined method to the choice of basis functions, an investigation of various one-term approximations was made. The electric-field components were assumed to be of the form

$$e_x = \begin{cases} 1, & |x| < W/2\\ 0, & \text{elsewhere}\end{cases}\tag{6a}$$

$$e_z = \begin{cases} -1, & -W/2 < x < 0\\ 1, & 0 < x < W/2\\ 0, & \text{elsewhere.}\end{cases}\tag{6b}$$

Another choice which certainly approximates the fields more closely is

$$e_x = \begin{cases} [(W/2)^2 - x^2]^{-1/2}, & |x| \le W/2\\ 0, & \text{elsewhere}\end{cases}\tag{7a}$$

$$e_z = \begin{cases} x[(W/2)^2 - x^2]^{1/2}, & |x| \le W/2\\ 0, & \text{elsewhere}\end{cases}\tag{7b}$$

which was used in [2].

The use of x- and z-directed field components is called the second-order approximation. A reduction of computer time is possible by assuming $E_z = 0$, which we will refer to as the first-order approximation. The problem then reduces to the evaluation of a single integral instead of four during each iteration for β. In Fig. 2, the results from these four different approximations are shown, namely the first- and second-order approximations with either basis-function set (6) or (7), and are compared with results from [3]. Although no indication about the rate of convergence and hence the accuracy of the two series

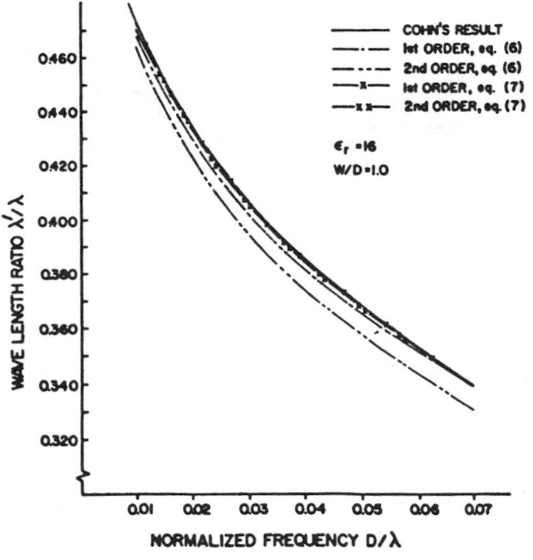

Fig. 2. Dispersion characteristic of a single slot.

for a general problem can be obtained, a comparison for this particular problem shows that the largest deviation between the different approximations is less than 4 percent. It is to be noted that the basis set given by (7) is superior to that given by (6) and that the first- and second-order solutions based on (7) agree very well with Cohn's results. The first- and second-order solutions based upon (6) give less accurate results with the second-order approximation being the poorer of the two due to the physically impossible discontinuities of (6b).

Using one specific set of parameters in the 1–3-GHz range, a comparison of the magnitudes of the x and z components of electric field was made. The x-directed electric-field component was found to have a magnitude greater than ten times that of the z-directed component. This provides further justification for use of the more efficient first-order approximation ($E_z = 0$).

III. CHARACTERISTIC IMPEDANCE OF A SINGLE SLOT LINE

The definition of the characteristic impedance for an ideal transmission line is uniquely defined by static quantities, but is somewhat arbitrary for the slot transmission line due to the non-TEM nature of this problem. One possible choice is to define it as

$$Z_0 = \frac{V_0^2}{2P_{\text{avg}}} \tag{8}$$

where V_0 is the slot voltage and P_{avg} is the time-averaged power flow on the slot line which is given by

$$P_{\text{avg}} = \tfrac{1}{2}\,\text{Re} \iint_s E \times H^* \cdot a_z \, dx\, dy. \tag{9}$$

The field components in this integral can be expressed in terms of the scalar potential functions by the use of Maxwell's equations.

Since the slot line is an open-boundary structure, the limits of integration in (9) are infinite, which makes the use of Paseval's theorem feasible. By this method (9) is transformed into the spectral domain where again a double integral is obtained with the variables of integration α and y instead of x and y. Equation (9) is then of the form

$$P_{\text{avg}} = \frac{1}{4\pi}\,\text{Re} \int_{-\infty}^{\infty}\int_{-\infty}^{\infty} f\left[\Phi^e(\alpha),\, \Phi^h(\alpha),\, \frac{\partial \Phi^e(\alpha)}{\partial y},\, \frac{\partial \Phi^h(\alpha)}{\partial y},\, \beta,\, \alpha,\, y\right] d\alpha\, dy. \tag{10}$$

The integral in (10) is a function of the Fourier transformed scalar potentials which were given in (1) and (2). Since the dispersion problem was already solved and the dependence of the coefficients $A^e(\alpha)$ through $D^h(\alpha)$ on the electric-field distributions is known, the integral of (10) can be evaluated.

Equations (1) and (2) show a simple functional dependence on the variable y, and thus integration with respect to y may be accomplished analytically in (10). One then obtains a single integral of the form

$$P_{\text{avg}} = \frac{1}{4\pi}\,\text{Re} \int_{-\infty}^{\infty} g(\alpha,\beta)\, d\alpha. \tag{11}$$

This integral has to be separately evaluated for the three spatial regions since the solution to the wave equation differs in each of these, as does the integrand of (11). Finally, to get all the necessary coefficients for (8), the voltage V_0 has to be computed. This involves simply the integration of the assumed electric-field distribution across the slot and can be done analytically. The evaluation of (11) must be done numerically on a digital computer. Although the limits of integration are infinite, the rapid decay of the integrand for large values of α and its well-behaved form make this computational task routine.

It should be noted that the amount of algebraic complexity in (11) is quite large and lengthy, and for this reason, the details are not presented here. For numerical purposes, the complexity is somewhat reduced by neglecting the z-directed electric-field component which means that the spatial phase shift between E_x and E_z requires no algebraic manipulation. It will be shown in the comparison between this and Cohn's method that use of the first-order solution results in good agreement. A computer program was developed which first computes the dispersion characteristic and then the characteristic impedance. Although the first- and second-order approximations for the dispersion calculation, as well as the two distributions (6) and (7) were investigated, the following results were computed using a second-order solution for the dispersion calculations [see (7)] and a first-order solution for characteristic-impedance calculation.

The comparison of these results with the values obtained by Mariani *et al.* [3] is shown in Fig. 3 and indicates a very close agreement for the two arbitrarily chosen sets of parameters. The average computation time on the IBM

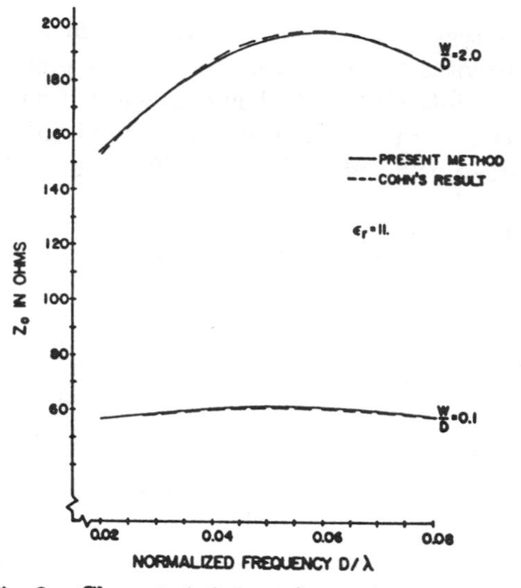

Fig. 3. Characteristic impedance of a single slot.

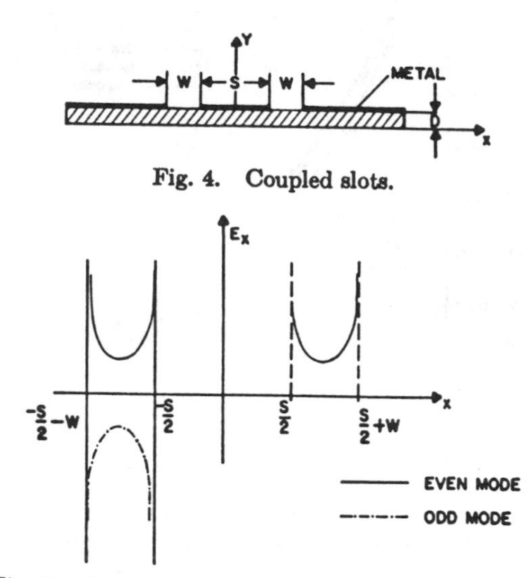

Fig. 4. Coupled slots.

Fig. 5. Assumed field distribution for coupled slots.

360 computer was 27 s for each value of D/λ. A first-order solution for the dispersion characteristic used in connection with distribution (6) required only 2.5 s and produced results which differed by less than 10 percent from those obtained using the second-order solution.

IV. CHARACTERISTICS OF A PAIR OF COUPLED SLOT LINES OF EQUAL WIDTH

Since the mathematical method developed so far produced results in agreement with those obtained by Cohn (which have been confirmed by several experiments) the method was applied to the analysis of two coupled slot lines, the geometry of which is shown in Fig. 4. This extension is straightforward for the following reason. The coefficients N_1 through N_4 are functions of α, β, ϵ_r, and D, and are independent of the slot configuration. The slot configuration enters into the calculation only through coefficients in the assumed field distributions or basis functions. Thus it is necessary only to modify (7) such that the mathematical description of the basis functions corresponds to the physical configuration and field distributions of the coupled slots.

Fig. 5 shows the two assumed distributions of electric field e_x for the even and odd modes, respectively. The change in the field component e_z is similar. In the Fourier domain, one simply applies the shifting theorem to the transforms of (7) to obtain the new transforms. Furthermore, (8) is changed to

$$Z_0 = \frac{V_0^2}{P_{\text{avg}}} \tag{12}$$

since the total time-averaged power surrounding the transmission lines is now due to two lines. Beyond this change, no major modifications were necessary to use the existing computer program. Figs. 6 and 7 show the results for several values of ϵ_r where Z_{0e} and Z_{0o} denote the even- and odd-mode characteristic impedances, respectively. A first-order approximation was used to calculate the dispersion and the characteristic impedances to reduce the

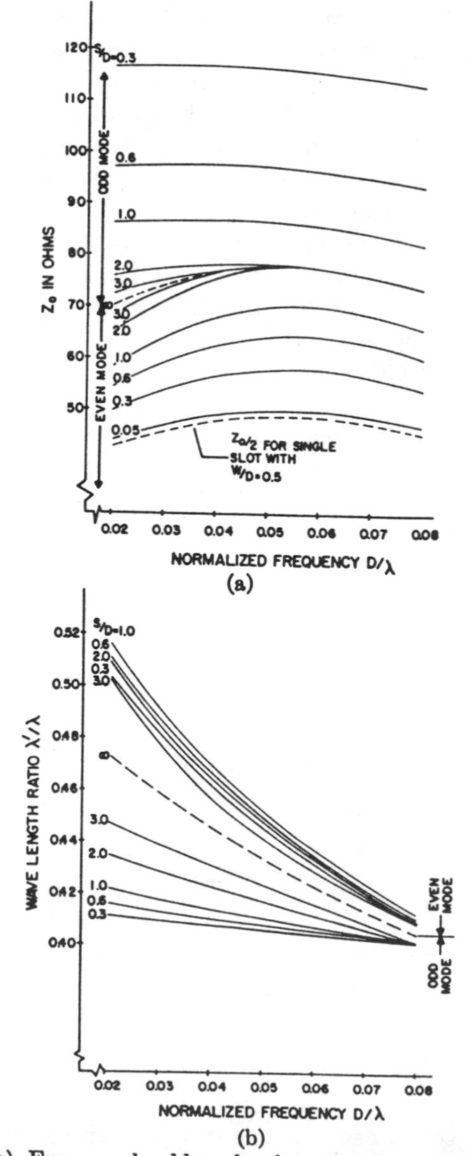

Fig. 6. (a) Even- and odd-mode characteristic impedances for coupled slots with $W/D = 0.25$, $\epsilon_r = 11$. (b) Even- and odd-mode dispersion characteristics for coupled slots with $W/D = 0.25$, $\epsilon_r = 11$.

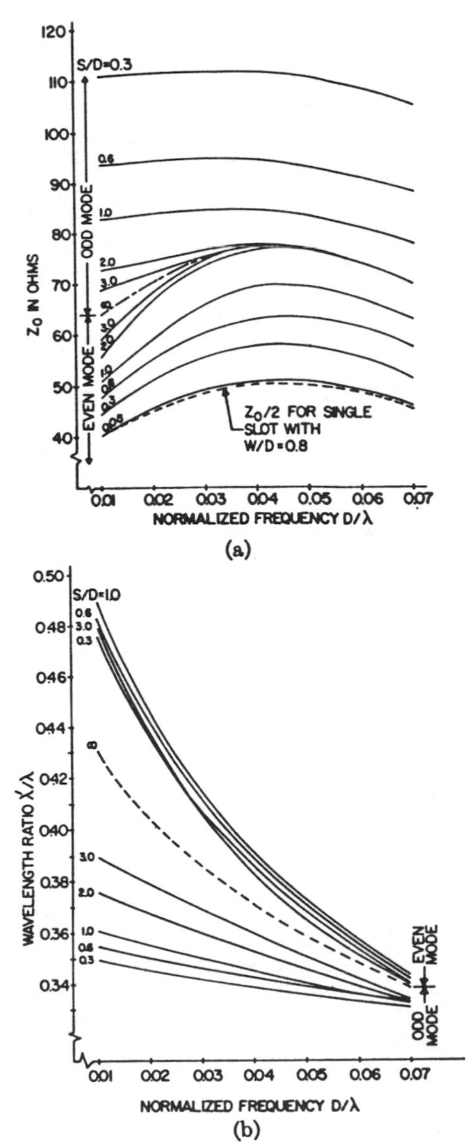

Fig. 7. (a) Even- and odd-mode characteristic impedances for coupled slots with $W/D = 0.4$, $\epsilon_r = 16$. (b) Even- and odd-mode dispersion characteristics for coupled slots with $W/D = 0.4$, $\epsilon_r = 16$.

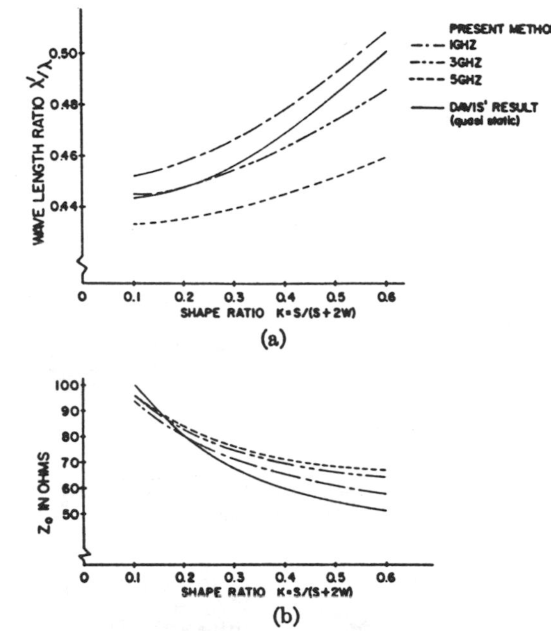

Fig. 8. (a) Dispersion characteristic and (b) characteristic impedance for CPW with $W/D = 1$, $\epsilon_r = 10$.

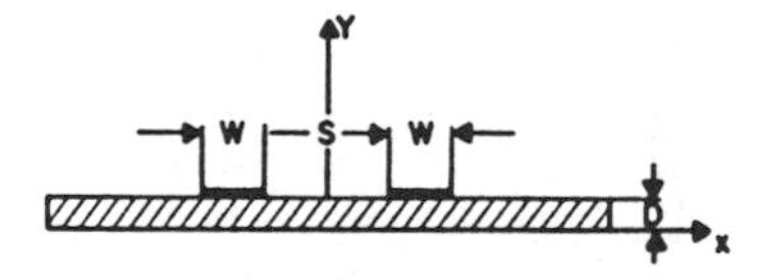

Fig. 9. Coplanar strip transmission line.

computer time which was, on the average, 80 s for both the odd and even modes. To the authors' knowledge, these results are basically new in the literature and can be only partially compared to other results. A quantitative comparison between these results for the odd-mode characteristic impedance and the quasi-static CPW impedance calculated by Davis *et al.* [5] can be made and is shown in Fig. 8 for $W/D = 1$ (or in the notation of [5], $t = s$) at frequencies of 1, 3, and 5 GHz. Reasonable agreement exists for the lower frequency range. Note that the characteristic impedance of the CPW is one half of Z_{0o} for equal slot widths. Another qualitative check on these results can be made by investigating the even-mode characteristic impedance in the limiting case as S, the separation between slots, becomes very small. One expects that Z_{0e} will be close to one half of Z_0, where Z_0 is the characteristic impedance of a single slot whose width is twice the width of the slot in the coupled structure. The dispersion characteristic for both structures, however, should be the same. This comparison is made in Figs. 6(a) and 7(a). It is interesting to observe the coupling and decoupling between the waves in the two slots as the frequency varies for large values of S/D. As the frequency increases, the waves become more closely bound to the slot which means there is less interaction between the two waves. In this case, Z_{0e} and Z_{0o} converge to Z_0, the characteristic impedance of a single slot with no coupling.

Another interesting phenomenon is the fact that for a fixed D/λ the ratio λ'/λ in the even mode first increases and then decreases as S/D increases from very small to larger values as shown in Figs. 6(b) and 7(b). An explanation may be given as follows. For small enough values of S/D, the metal strip between the slots has little effect on the propagating wave, and the wave propagates as if it were in a slot of width $2(W/D) + S/D$. Increasing the separation between the slots effectively increases slot width (the metal separation still has little effect), and the ratio λ'/λ increases. As S/D continues to increase, the two waves start to decouple and behave more as two waves on two slot lines which will finally be totally decoupled. Each wave then propagates on a slot line with width W/D, hence λ'/λ decreases.

V. CHARACTERISTICS OF COPLANAR STRIP LINE

A configuration of a pair of coplanar strip lines is shown in Fig. 9. The dispersion characteristic and the characteristic impedance can be found by again using Galerkin's method in the Fourier transform domain. Since an ap-

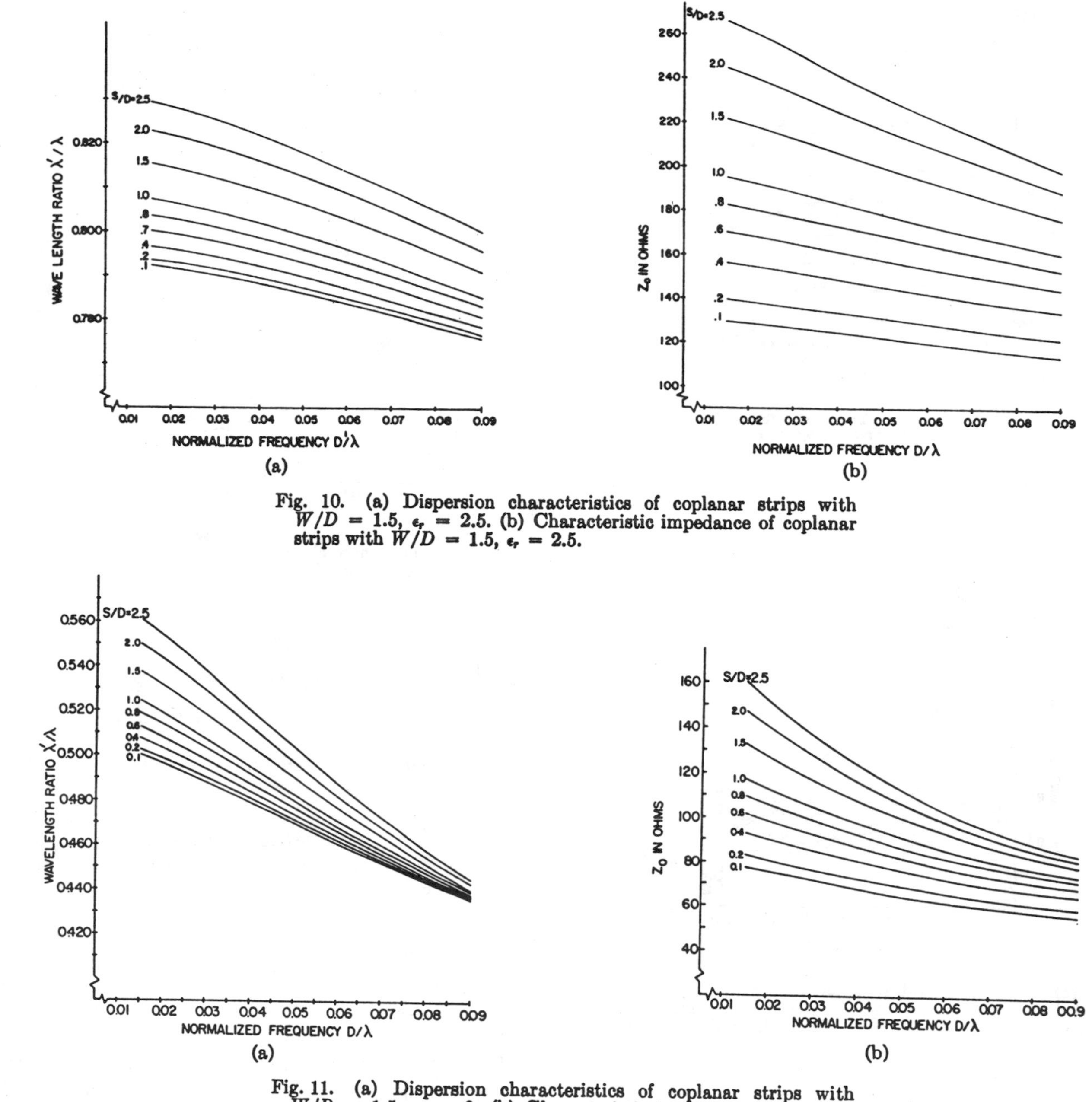

Fig. 10. (a) Dispersion characteristics of coplanar strips with $W/D = 1.5$, $\epsilon_r = 2.5$. (b) Characteristic impedance of coplanar strips with $W/D = 1.5$, $\epsilon_r = 2.5$.

Fig. 11. (a) Dispersion characteristics of coplanar strips with $W/D = 1.5$, $\epsilon_r = 9$. (b) Characteristic impedance of coplanar strips with $W/D = 1.5$, $\epsilon_r = 9$.

proximation of the current density across each strip is more feasible than an approximation of the electric field at $y = D$, the equation set (4) was used to determine the dispersion characteristic. A first-order solution was obtained assuming that the surface current in the x direction was negligible and that the z-directed surface current was of the form

$$j_s(x) = \begin{cases} \dfrac{\pm 1}{\{(W/2)^2 - [x \pm (S+W)/2]^2\}^{1/2}}, & S/2 < |x| < S/2 + W \\ 0, & \text{elsewhere} \end{cases} \tag{13}$$

over each strip. The characteristic impedance was calculated as

$$Z_0 = 2P_{avg}/I_0^2 \tag{14}$$

where I_0 is the total current on one strip. Any further necessary formulations were very similar to the previously outlined procedure for slot lines. Three representative graphs for the dispersion characteristics and characteristic impedances are shown in Figs. 10–12. Reasonable agreement for the impedances is found by comparing the present values with the results by Wen [4]. However, as one might expect, the present method yields somewhat larger values for the impedance due to the finite dielectric substrate.

VI. EXPERIMENTAL RESULTS

Although all previous comparisons showed reasonable agreement with existing results, some effort was devoted to obtaining experimental verification of this work. One

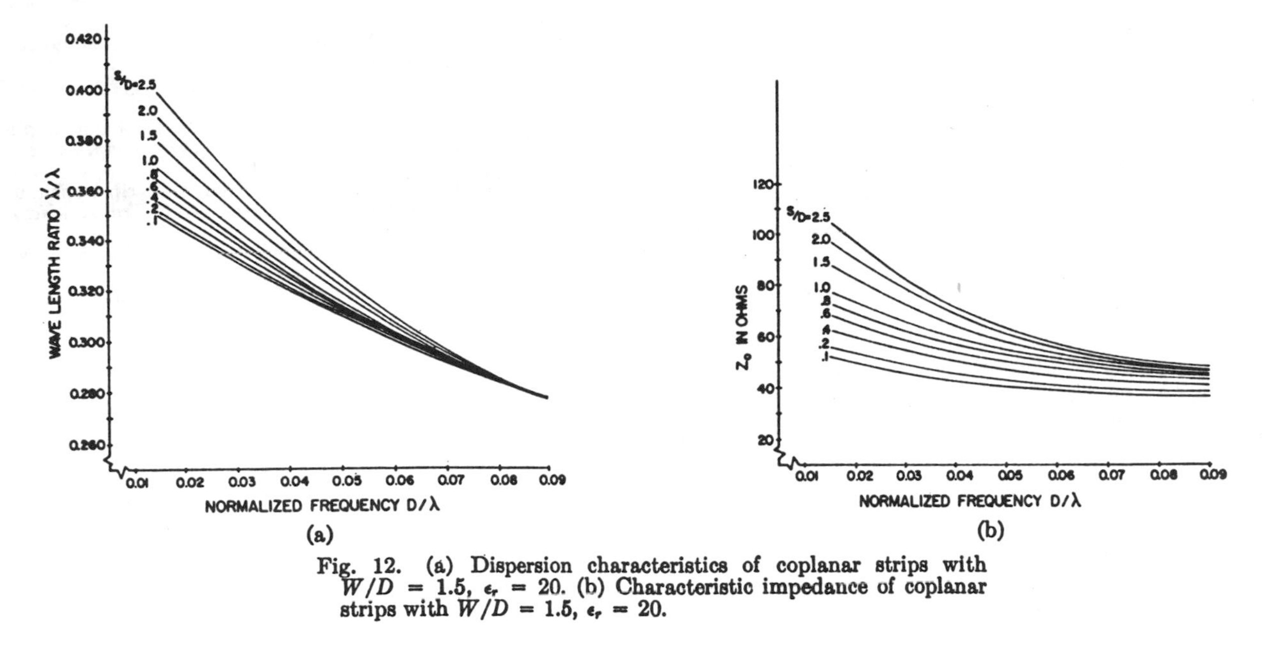

Fig. 12. (a) Dispersion characteristics of coplanar strips with $W/D = 1.5$, $\epsilon_r = 20$. (b) Characteristic impedance of coplanar strips with $W/D = 1.5$, $\epsilon_r = 20$.

experimental check is provided in the work of Luna [7] who measured the characteristics of coplanar strips and found an agreement of better than 5 percent between theory and experiment. The authors further confirmed the accuracy of the results for coupled slots by constructing a slot-line coupler.

Coupled slots ($W/D = 0.470$, $S/D = 1.08$) were etched onto one side of a 1-oz copper surface on a dielectric substrate with $D = 0.125$ in, and $\epsilon_r = 16$. Microstrip-to-slot transitions [8] were used at three ports, and the fourth was terminated with a chip resistor. A center frequency of 3 GHz ($l = 1.6$ cm) was chosen.

The work of Jones and Bolljahn [11] has been extended by Zysman and Johnson [9] who have derived the impedance matrix of dispersive coupled lines. Using the results presented earlier, the elements of the impedance matrix of the coupler were evaluated, and its performance as a function of frequency was calculated. The theoretical performance is compared with experiment in Fig. 13. Good correlation is evident. The deterioration of directivity noted experimentally at the band edges is in all probability due to the rising VSWR of the microstrip–slot transitions at these frequencies. It should be noted that the behavior of this coupler is different than that of the contradirectional (nondispersive) strip-line coupler. Here the difference in the phase velocities of the odd and even modes results in codirectional coupling as in waveguide (see also Mariani and Agrios [10]). Further investigations into the behavior of dispersive couplers have been undertaken and preliminary results have been reported elsewhere [12].

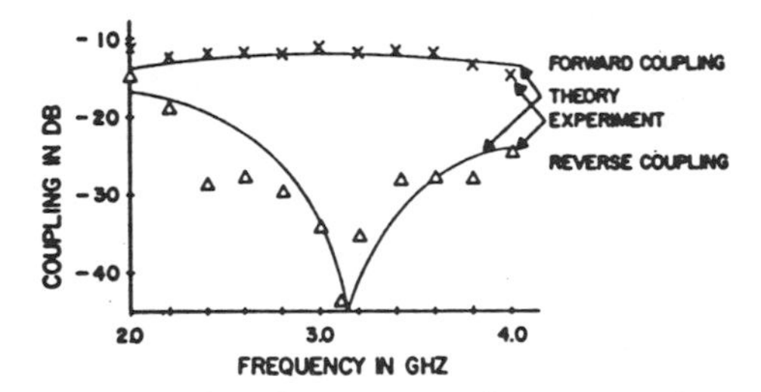

Fig. 13. Theoretical and experimental response for a slot-line coupler.

VII. CONCLUSIONS

An efficient numerical method has been presented for obtaining the dispersion characteristics and the characteristic impedances for a single slot line, two parallel slot lines of equal width, and two parallel coplanar strips. Solutions to the wave propagation problem were obtained in the Fourier transform domain. Numerical results obtained by this method have been presented and compared or related to other existing data and to experiment. In all cases, good agreement was obtained.

The transform technique is relatively straightforward in concept, but extensive algebraic manipulation of the resulting equations is required to achieve computational efficiency. The labor involved should not be underestimated. For this reason, the equations have not been presented here in detail. The authors expect to make this information available in a technical report in the near future. It is also anticipated that design curves will be made available in a technical report.

ACKNOWLEDGMENT

The authors wish to acknowledge the support provided by the Naval Postgraduate School Foundation Research Program and the W. R. Church Computer Center, also of the Naval Postgraduate School.

REFERENCES

[1] S. B. Cohn, "Slot line on a dielectric substrate," *IEEE Trans. Microwave Theory Tech.*, vol. MTT-17, pp. 768–778, Oct. 1969.
[2] T. Itoh and R. Mittra, "Dispersion characteristics of slot lines," *Electron. Lett.*, vol. 7, pp. 364–365, July 1971.
[3] E. A. Mariani, C. P. Heinzman, J. P. Agrios, and S. B. Cohn, "Slot line characteristics," *IEEE Trans. Microwave Theory Tech. (1969 Symp. Issue)*, vol. MTT-17, pp. 1091–1096, Dec. 1969.
[4] C. P. Wen, "Coplanar waveguide: A surface strip transmission line suitable for nonreciprocal gyromagnetic device applica-

tions," *IEEE Trans. Microwave Theory Tech.* (*1969 Symp. Issue*), vol. MTT-17, pp. 1087–1090, Dec. 1969.
[5] M. E. Davis, E. W. Williams, and A. C. Celestini, "Finite-boundary corrections to the coplanar waveguide analysis," *IEEE Trans. Microwave Theory Tech.* (Short Papers), vol. MTT-21, pp. 594–596, Sept. 1973.
[6] D. S. Jones, *The Theory of Electromagnetism.* New York: Pergamon, 1964.
[7] A. E. Luna, "Parallel coplanar strips on a dielectric substrate," M.E.E. thesis, Dep. Eng., Naval Postgraduate School, Monterey, Calif., 1973.
[8] J. B. Knorr, "Slot-line transitions," *IEEE Trans. Microwave Theory Tech.* (Short Papers), vol. MTT-22, pp. 548–554, May 1974.
[9] G. I. Zysman and A. K. Johnson, "Coupled transmission line networks in an inhomogeneous dielectric medium," *IEEE Trans. Microwave Theory Tech.*, vol. MTT-17, pp. 753–759, Oct. 1969.
[10] E. A. Mariani and J. P. Agrios, "Slot-line filters and couplers," *IEEE Trans. Microwave Theory Tech.* (*1970 Symp. Issue*), vol. MTT-18, pp. 1089–1095, Dec. 1970.
[11] E. M. T. Jones and J. T. Bolljahn, "Coupled-strip-transmission-line filters and directional couplers," *IRE Trans. Microwave Theory Tech.*, vol. MTT-4, pp. 75–81, Apr. 1956.
[12] K.-D. Kuchler and J. B. Knorr, "Coupler design using dispersive transmission lines," Proc. 8th Asilomar Conf. Circuits, Systems and Computers (Pacific Grove, Calif.), Dec. 3–5, 1974.

DISPERSION CHARACTERISTICS OF ASYMMETRIC COUPLED SLOT LINES ON DIELECTRIC SUBSTRATES

Indexing terms: Microwave devices and components, Transmission lines

A modified coplanar waveguiding structure, an asymmetric coupled slot line, on a dielectric substrate is considered. The dispersion characteristics of asymmetric coupled slot lines and asymmetric coplanar waveguides are found in terms of the numerically efficient spectral domain technique.

Introduction: A modified coplanar transmission line, an asymmetric coupled slot line (ACSL), shown in Fig. 1, is analysed in terms of the efficient exact spectral domain technique. This structure is believed to be potentially interesting for many practical applications in microwave integrated circuits. Moreover, the asymmetric coplanar waveguide (ACPW) case is simultaneously obtained due to the equivalence with the antiphase excitation of the ACSL structure, which for the symmetrical case has been extremely widely used for various passive and active microwave circuits. The developed formulation covers the symmetrical case also.

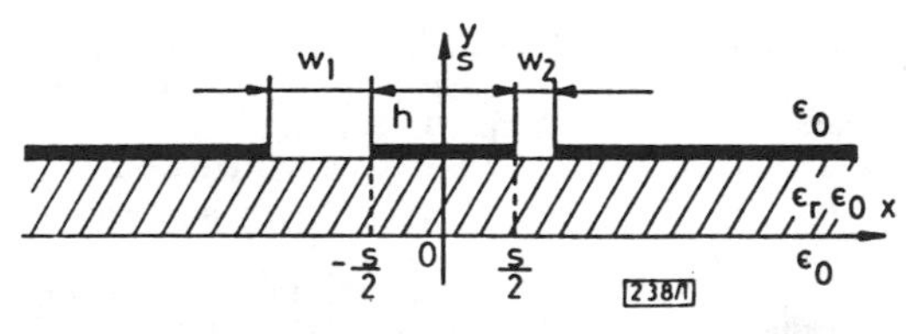

Fig. 1 *Cross-section of asymmetric coupled slot lines structure*

Coplanar transmission lines with plane symmetry have been studied in recent years by many investigators[1-3] due to their compatibility with microwave integrated circuits. They are much more easily adaptable to shunt-element connections than, for example, microstrip lines. It is the authors' belief that structural asymmetry of ACSL and ACPW structures should give additional possibilities for MIC designers as it should be useful especially for mounting solid-state devices of various dimensions in microwave integrated circuits. The additional advantages can be expected in filling the ACSL or ACPW lines with anisotropic materials where in symmetrical cases interesting field effects have been observed.[4] As previously mentioned, ACSL and ACPW can be considered as a modification of a well known coupled slot line structure with equal slot widths, or coplanar waveguide, respectively. These simple cases have been analysed and described by a number of authors.[1-8] The analysis covers the influence of various structural parameters on the line behaviour. In this letter, the general form of coupled slot lines is described, which has no symmetry in respect to the plane perpendicular to the surface substrate.

Analysis: Owing to the non-TEM propagation in the slots, the full-wave hybrid-mode analysis is necessary to determine the dispersion properties of ACSL, which are described in terms of π- and C-modes corresponding to anti- and in-phase excitation of the slots, respectively. The analysis is based on the spectral domain approach and Galerkin's procedure, first proposed by Itoh and Mittra.[9] To date, however, this method has been employed for calculating planar structures with symmetry plane except solutions[10] for asymmetric coupled microstrip lines. Our formulation, however, does not introduce any integral relation between electric field components in the slots, gives efficiency and versatility for the investigators and covers the simultaneously symmetrical case. Applying the second-order approximation, namely using the series representation for both tangential electric field components in the slots, and following Borburgh,[11] we choose trigonometric functions with Maxwell's terms as the basis function series. These

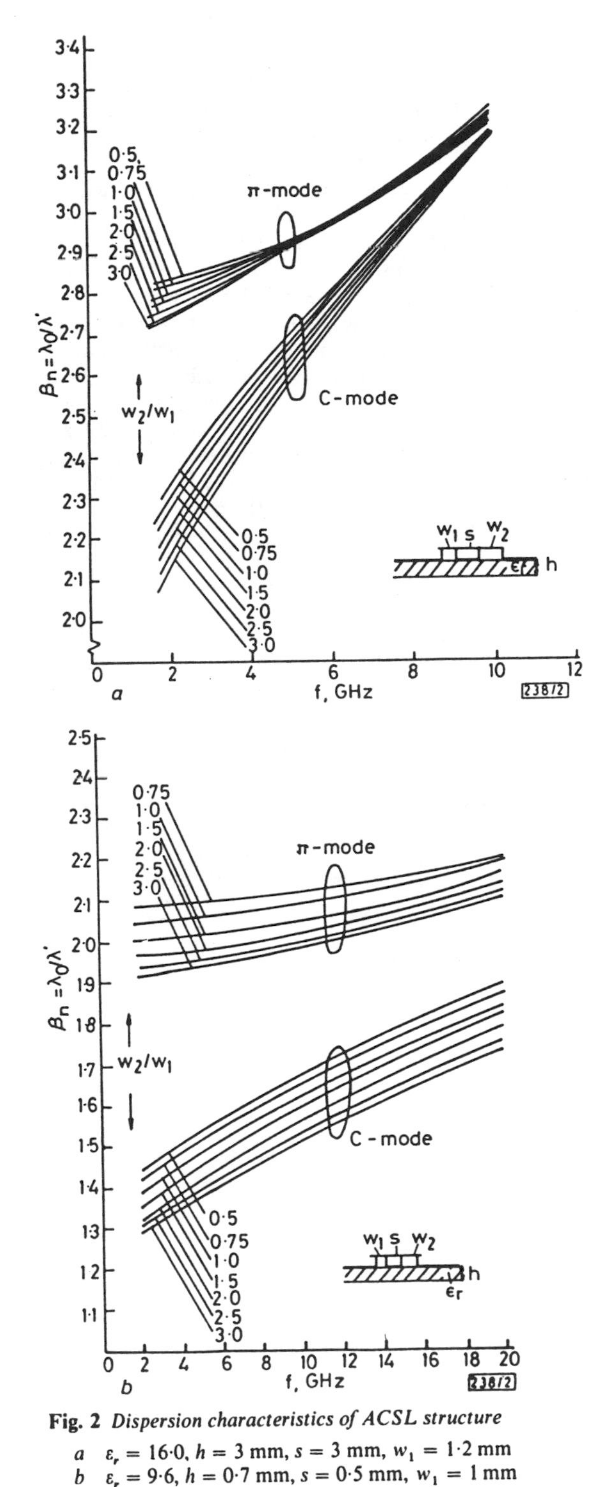

Fig. 2 *Dispersion characteristics of ACSL structure*

a $\varepsilon_r = 16{\cdot}0$, $h = 3$ mm, $s = 3$ mm, $w_1 = 1{\cdot}2$ mm
b $\varepsilon_r = 9{\cdot}6$, $h = 0{\cdot}7$ mm, $s = 0{\cdot}5$ mm, $w_1 = 1$ mm

Reprinted with permission from *Electronics Letters*, vol. 19, no. 3, pp. 91-92, February 1983.

functions can be done as follows:

$$E_x^{(i)}(y = h, x) = \begin{cases} \dfrac{2}{\pi} \sum_{n=0}^{\infty} b_n \dfrac{\cos\left(n\pi \dfrac{2x + (-1)^{i+1}(s + 2w_i)}{2w_i}\right)}{\sqrt{\left[1 - \left(\dfrac{2x + (-1)^{i+1}(s + w_i)}{w_i}\right)^2\right]}} & \text{for } \dfrac{s}{2} \le |x| \le \dfrac{s}{2} + w_i \\ 0 & \text{elsewhere} \end{cases}$$

$$E_z^{(i)}(y = h, x) = \begin{cases} j \sum_{n=1}^{\infty} a_n \sin\left(n\pi \dfrac{2x + (-1)^{i+1}(s + 2w_i)}{2w_i}\right) & \text{for } \dfrac{s}{2} \le |x| \le \dfrac{s}{2} + w_i \\ 0 & \text{elsewhere} \end{cases}$$

with the superscript i denoting the number of the slot. The total electric field at the $x = h$ interface is then obtained as a superposition of E_x^i and E_z^i components.

Numerical results: The developed analysis has been used for the calculation of the dispersion characteristics of the ACSL structure on dielectric substrate. At first, a symmetrical case was taken to check the validity of presented numerical solutions. The dispersion curves obtained from general analysis and from the symmetrical case were indistinguishable, confirming in this way the validity of presented numerical solutions. Two terms of the basis function series have been used in the computations for both considered cases. Next, the behaviour of the dispersion curves was analysed for two arbitrarily chosen sets of parameters against the asymmetry ratio w_2/w_1. It was found that for moderate values of the w_2/w_1 ratio, two terms in the series representation give the correct solution for the phase constant. In the case of very wide slots with respect to the substrate thickness, it is necessary to increase the number of basis function terms. It can be seen in Fig. 2 that phase velocities of C- and π-modes have straightforward dependence on the asymmetry ratio. In particular, the difference of the phase velocity values decreases with decreasing w_2/w_1; this can be profitable, for example, in the directional couplers design. Moreover, the dispersion of the π-mode is very close to the CPW behaviour, and is relatively less sensitive with respect to the asymmetry ratio. Thus, quasi-TEM approaches for determining the characteristic impedance of the ACPW can be applied for reasonable sets of parameters. In our opinion this case should be very interesting, and the above conclusions are important for the designers of various microwave integrated circuits. One can expect that proper choice of the surface permittivity and the line dimensions could extend the coplanar line usability range up to millimetre-wave frequencies.

M. KITLINSKI
B. JANICZAK

Technical University of Gdańsk
Telecommunication Institute
Majakowskiego 11/12
80-952 Gdańsk, Poland

14th December 1982

References

1 WEN, C. P.: 'Coplanar waveguide: a surface strip transmission line suitable for nonreciprocal gyromagnetic device applications', *IEEE Trans.*, 1969, **MTT-17,** pp. 1091–1096
2 KNORR, J. B., and KUCHLER, K.-D.: 'Analysis of coupled slots and coplanar strips on dielectric substrate', *ibid.*, 1975, **MTT-23,** pp. 541–548
3 DAVIES, J. B., and MIRSHEKAR-SYAHKAL, D.: 'Spectral domain solution of arbitrary coplanar transmission line with multilayer substrate', *ibid.*, 1977, **MTT-25,** pp. 143–146
4 KITLINSKI, M., and JANICZAK, B.: 'Dispersion characteristics of single and coupled open slot lines filled with ferrite medium'. Proc. 12th European Microwave Conference, Helsinki, 1982, pp. 759–764
5 PREGLA, R., and PINTZOS, S. G.: 'Determination of the propagation constants in coupled microslots by a variational method'. Proc. 5th Colloquium on microwave communication, Budapest, 1974, pp. MT491–500
6 HATSUDA, T.: 'Computation of coplanar-type strip-line characteristics by relaxation method and its application to microwave circuits', *IEEE Trans.*, 1975, **MTT-23,** pp. 795–802
7 FUJIKI, Y., SUZUKI, M., KITAZAWA, T., and HAYASHI, Y.: 'Higher order modes in coplanar-type transmission lines', *Electron. & Commun. Jpn.*, 1975, **58-B,** pp. 74–80
8 SAHA, P. K.: 'Dispersion in shielded planar transmission lines on two-layer composite substrate', *IEEE Trans.*, 1977, **MTT-25,** pp. 907–911
9 ITOH, T., and MITTRA, R.: 'A new technique for the analysis of the dispersion characteristics of microstrip lines', *ibid.*, 1971, **MTT-19,** pp. 47–56
10 JANSEN, R. H.: 'Fast accurate hybrid mode computation of non-symmetrical coupled microstrip characteristics'. Proc. 7th European Microwave Conference, Copenhagen, 1977, pp. 135–139
11 BORBURGH, J.: 'Theoretische Untersuchung der Dispersion und Feldverteilung von Wellentypen einer Microstreifenleitung mit gyrotropen Substrat'. Thesis, Erlangen, 1976

ANALYSIS OF CONDUCTOR-BACKED COPLANAR WAVEGUIDE

Indexing terms: Waveguides, Spectral analysis

The transmission properties of a coplanar waveguide printed on conductor-backed substrates are analysed using the spectral-domain technique. For a fixed substrate thickness, the characteristic impedance and the phase constant may be varied independently by simply adjusting the widths of the centre strip and the slots in the transmission line.

Introduction: Recently, remarkable progress has been achieved in GaAs monolithic microwave integrated circuits (MMIC). It is, therefore, necessary to acquire the knowledge of propagation properties in various structures appearing in GaAs MMICs.

The structure in Fig. 1 is a modification of the coplanar waveguide in that it has an additional ground plane. The latter is useful to increase mechanical strength of the circuit as the GaAs substrate is typically thin and fragile. To date, no analytical results have been reported on the propagation characteristics for such a structure. In this letter, the analysis based on the spectral-domain technique is described and the resulting propagation constants and characteristic impedances are reported.

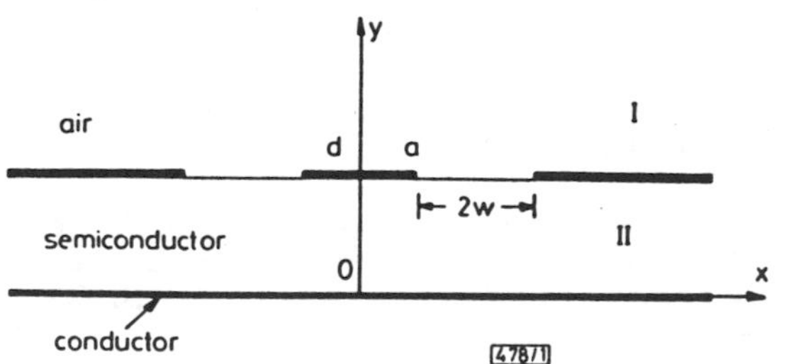

Fig. 1 *Cross-sectional view of conductor-backed coplanar waveguide*
a $\varepsilon_r = 13$ $d = 150\ \mu m$ $a/d = 1/3$

Method: The spectral-domain technique has been successfully applied to analyse a number of planar transmission lines.[1–3] A simple method for formulating the dyadic Green's functions in the spectral domain has been proposed by Itoh,[4] which is based on the transverse equivalent transmission lines for a spectral wave and on a simple co-ordinate transformation. By following this method, a set of coupled equations result:

$$\begin{bmatrix} \tilde{Y}_{xx}(\alpha, \beta) & \tilde{Y}_{xz}(\alpha, \beta) \\ \tilde{Y}_{zx}(\alpha, \beta) & \tilde{Y}_{zz}(\alpha, \beta) \end{bmatrix} \begin{bmatrix} \tilde{E}_x(\alpha) \\ \tilde{E}_z(\alpha) \end{bmatrix} = \begin{bmatrix} \tilde{J}_x(\alpha) \\ \tilde{J}_z(\alpha) \end{bmatrix} \tag{1}$$

where $\tilde{Y}_{xx}$, $\tilde{Y}_{xz}$, $\tilde{Y}_{zx}$ and $\tilde{Y}_{zz}$ are dyadic Green's functions similar to those derived in Reference 4, and $\tilde{E}_x$, $\tilde{E}_z$, $\tilde{J}_x$ and $\tilde{J}_z$ are the Fourier transforms of x- and z-components of the electric fields (E_x and E_z) and current densities (J_x and J_z), respectively. Alpha (α) is the Fourier transform variable and β is the phase constant that we seek. The solution is then found by means of Galerkin's method in the spectral domain, in which the slot field components $E_x(x)$ and $E_z(x)$ are expanded in terms of complete sets of known basis functions:

$$E_x(x) = \sum_{n=1}^{\infty} c_n E_{xn}$$

$$E_z(x) = \sum_{n=1}^{\infty} d_n E_{zn}$$

where c_n and d_n are unknown coefficients.

After taking Fourier transforms of these expressions, they are substituted into eqn. 1. Then the inner products of the resultant equations with each of the basis functions are performed and result in homogeneous linear simultaneous equations as the right-hand side becomes identically zero by the inner product process.[1,2] By equating the determinant to zero, the eigenvalue β is obtained. The relative magnitudes of c_n and d_n are then obtained, and provide information necessary to calculate the modal field and the characteristic impedance.

In the study, the following sets of basis functions are chosen for the fields in each slot:

$$E_{xn} = \frac{\cos\left(\frac{n\pi}{2w}x\right)}{\sqrt{(w^2 - x^2)}}, \quad n = 0, 2, 4, \ldots$$

$$\frac{\sin\left(\frac{n\pi}{2w}x\right)}{\sqrt{(w^2 - x^2)}}, \quad n = 1, 3, 5, \ldots$$

$$E_{zn} = \frac{\cos\left(\frac{n\pi}{2w}x\right)}{\sqrt{(w^2 - x^2)}}, \quad n = 1, 3, 5, \ldots$$

$$\frac{\sin\left(\frac{n\pi}{2w}x\right)}{\sqrt{(w^2 - x^2)}}, \quad n = 2, 4, 6, \ldots$$

These sets of basis functions satisfy the edge conditions at the edge and their Fourier sine and cosine transformations are simple zero-order Bessel functions of the first kind.

Numerical results: The conductor-backed coplanar waveguide has been studied for a wide range of geometric parameters. Fig. 2 shows typical results of the dispersion property, assuming GaAs as the substrate ($\varepsilon_r = 13{\cdot}0$). Aspect ratios a/d and w/d are varied as parameters. Since the present structure is a mixture of a microstrip line and a coplanar waveguide, properties of either become predominant depending on the structural parameters. As the slot width increases for a fixed substrate thickness, the characteristics approach that of microstrip line. This behaviour is shown in Fig. 2*a*. On the other hand, when the slot width is fixed and the thickness increases, the behaviour approaches the coplanar waveguide case, as depicted in Fig. 2*b*.

For a moderate aspect ratio (e.g. $a/d = 1/3$, $w/d = 1/3$), the transmission line becomes less dispersive than the corresponding microstrip line with the same aspect ratio ($a/d = 1/3$). This suggests that we may define the characteristic impedance of the transmission line as $Z_c = Z_0/\sqrt{\varepsilon_{eff}}$, where ε_{eff} is the effective dielectric constant defined as $(\beta/k)^2$ and Z_0 is the characteristic impedance when $\varepsilon_r = 1$. The impedance Z_c computed in this manner is plotted in Fig. 3. It is observed that the characteristic impedance decreases as the slot widths decrease, while the phase constant is relatively unaffected. Therefore, by adjusting both centre strip width and slot widths, independent control of the phase constant and the characteristic impedance may be obtained.

The number of basis functions required for accuracy is largely dependent on the aspect ratios. For small w/a and w/d, only E_{x0} is required since both the slot coupling effect and the effect of the ground-backing are small. For larger w/a, E_{x1} is needed to represent the stronger coupling effect between the slots. Finally, when the thickness becomes comparable to the slot width, more basis functions are required for accurate results. In this study, up to seven basis functions have been used

Reprinted with permission from *Electronics Letters*, vol. 18, no. 12, pp. 538–540, June 1982.

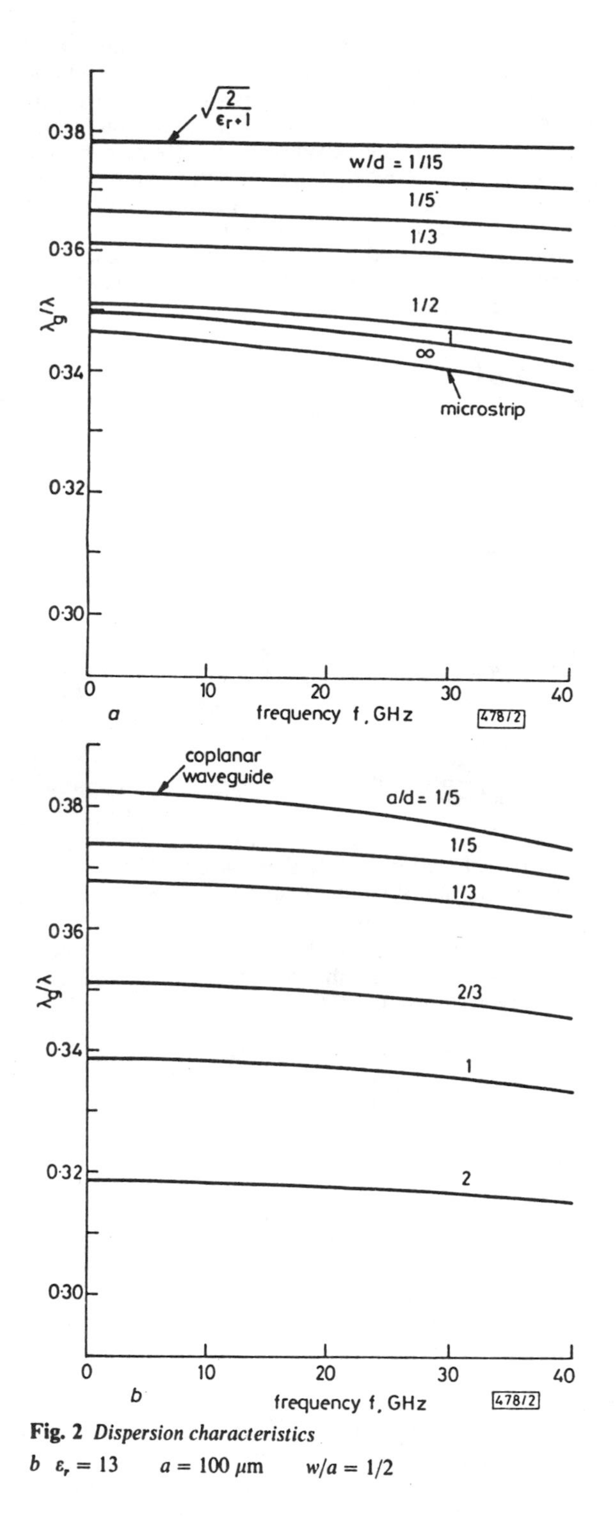

Fig. 2 *Dispersion characteristics*

b $\varepsilon_r = 13$ $a = 100\,\mu m$ $w/a = 1/2$

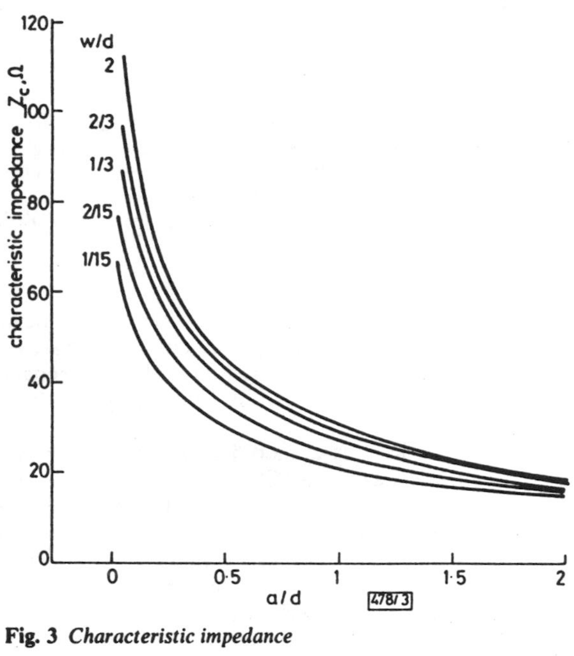

Fig. 3 *Characteristic impedance*

$\varepsilon_r = 13$ $d = 150\,\mu m$

for some extreme cases. The computation time for such a solution is about 5 s on a CDC Dual Cyber 170/750.

Acknowledgment: This work was supported by the Office of Naval Research N00014-79-0053, Joint Services Electronics Program F49620-79-C-0101 and US Army Research Office contract DAAG29-81-K-0053.

Y. C. SHIH
T. ITOH

Department of Electrical Engineering
The University of Texas at Austin
Austin, TX 78712, USA

30th April 1982

References

1 ITOH, T., and MITTRA, R.: 'Spectral-domain approach for calculating the dispersion characteristics of microstrip lines', *IEEE Trans.*, 1973, **MTT-21,** pp. 496–499
2 ITOH, T., and MITTRA, R.: 'Dispersion characteristics of slot lines', *Electron. Lett.*, 1971, **7,** pp. 364–365
3 KNORR, J., and KUCHLER, K.-D.: 'Analysis of coupled slots and coplanar strips on dielectric substrate', *IEEE Trans.*, 1975, **MTT-23,** pp. 541–548
4 ITOH, T.: 'Spectral domain immittance approach for dispersion characteristics of generalized printed transmission lines', *ibid.*, 1980, **MTT-28,** pp. 733–736

Losses in Coplanar Waveguides

A. GOPINATH

Abstract **—Conductor losses in coplanar waveguides have been calculated using a quasi-static Green's function approach. These calculations and others on dielectric and radiation losses are used to compute the quality factor of half-wavelength resonators, and comparison with measurements show good agreement.**

I. Introduction

The coplanar waveguide (CPW) is a planar guiding structure (Fig. 1) sometimes suggested as an alternative to microstrip lines [1], [2] in both hybrid and monolithic circuits. Losses in these guides have been investigated by several authors [3]–[6] over a limited range of parameters.

The present paper calculates the conductor losses in a coplanar waveguide, using a quasi-static approach, and provides a comprehensive set of results. The quality factors of half-wavelength resonators including radiation losses are calculated and the results are compared with experimental measurements on resonators on substrates with a relative dielectric constant ϵ_r of 13.0.

II. Calculation of Coplanar Waveguide Parameters

The various parameters of the coplanar waveguide are calculated for different strip width to ground plane spacing ratios $(2a/2b)$ (see Fig. 1), assuming quasi-static TEM propagation. The Green's function for a line charge distance p from the surface of dielectric slab of thickness $2h$ is derived using the method of images [7]. For the observation point P at (x, y), in the same air space as the line charge at $(0, h+p)$ of q C/m (see Fig. 2), the potential at P is given by

$$V(x,y) = -\frac{q}{4\pi\epsilon_0}\log\frac{(y-h-p)^2+x^2}{R^2} - \frac{Kq}{4\pi\epsilon_0} \cdot \log\frac{(y-h+p)^2+x^2}{R^2} + \frac{(1-K^2)q}{4\pi\epsilon_0} \cdot \sum_{n=1}^{\infty} K^{2n-1}\log\frac{(y+(4n-1)h+p)^2+x^2}{R^2}$$

where

$$K = \frac{\epsilon_0 - \epsilon_1}{\epsilon_0 + \epsilon_1}$$

$$R^2 = (x-x_p)^2 + (y-y_p)^2$$

$\epsilon_1 = \epsilon_r\epsilon_0$ the substrate dielectric constant

(x_p, y_p) = coordinates of the point with constant reference potential. (1)

The quantity R is the distance from the observation point to the reference potential point. In the present instance, this reference point was chosen to be the center of the conducting strip.

Manuscript received September 27, 1981; revised February 23, 1982. This work was sponsored by the Department of the Army. The U.S. Government assumes no responsibility for the information presented.

The author was with M.I.T. Lincoln Laboratory, Lexington, MA. He is now with the Department of Electronics, Chelsea College, University of London, Pulton Place, London SW6 5PR, England.

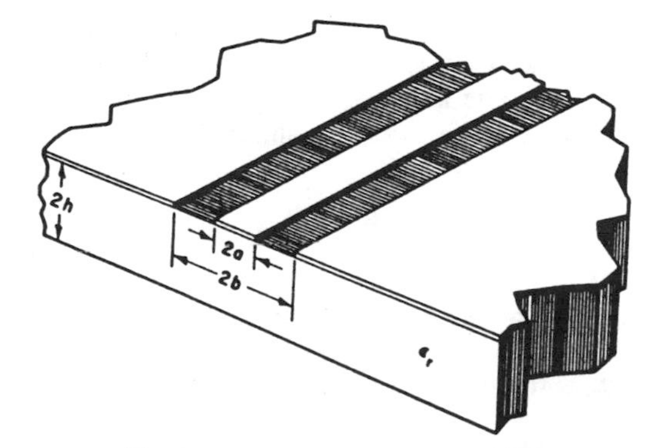

Fig. 1. Geometry of coplanar waveguide.

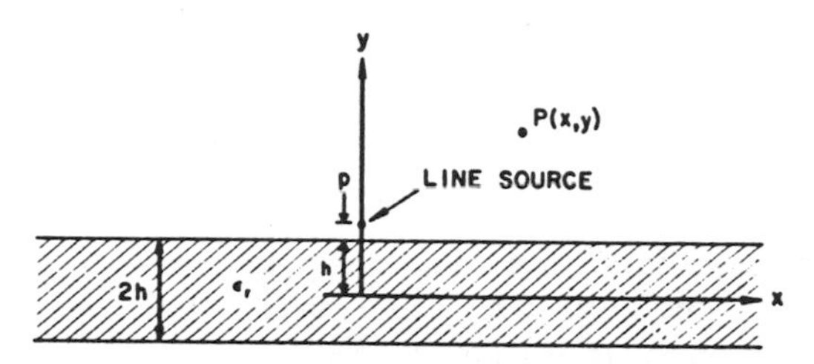

Fig. 2. Observation point P and line source over substrate, used in derivation of the quasi-static Green's function.

The ground planes on either side of the strip are assumed to be of finite width over the infinitely wide substrate of finite thickness $2h$. The strip and ground planes are subdivided into substrips and the charge distribution over each substrip is assumed to be uniform, but unknown in magnitude. The problem is solved by the usual matrix equation derived from the equation

$$V = \int \sigma_l G\, dl \tag{2}$$

where V is the potential on the strip and ground planes, and σ_l is the unknown charge distribution over the substrips. The strip and ground-plane potentials are assumed to be unknown and set at +1 and 0 V, respectively.

It is necessary, however, to impose the condition that the total charge on the strip is equal and opposite to the sum of the total charges on the ground planes. This condition is imposed by using the null matrix technique discussed in detail elsewhere [8]. In brief, this condition gives rise to a rectangular matrix equation, the null space of which represents a new basis-function set, which satisfies this equation and can be used as trial functions in (2).

The charge patterns are obtained without and with the substrate, and from these the capacitances C_o and C_d are obtained and hence the impedance, velocity factor, and effective dielectric constant $\epsilon_{reff}\epsilon_0$ are calculated.

III. Conductor and Dielectric Loss Constants

The charge pattern without the dielectric substrate corresponds to the longitudinal current distribution pattern and from this the conductor loss factor α_c is evaluated from

$$\alpha_c = \frac{R_s}{2Z_0I^2}\int_{-a}^{+a} J_s^2\, dx + 2\int_b^{b_{\max}} J_{gp}^2\, dx \text{ Np/m} \tag{3}$$

where J_s is the strip longitudinal current linear density, J_{gp} is the ground-plane current linear density, Z_0 is the characteristic im-

Reprinted from *IEEE Trans. Microwave Theory Tech.*, vol. MTT-30, no. 7, pp. 1101–1104, July 1982.

pedance of the CPW, I is the total strip or ground-plane current, and R_s is the metal surface resistivity in ohms.

Since the substrate loss tangent is generally small, the plane-wave approach may be used to evaluate the dielectric loss constant α_d. Thus

$$\alpha_d = \frac{q\epsilon_r \tan\delta}{\epsilon_{reff}\lambda_g}\ \text{Np/m} \tag{4}$$

where ϵ_r is the substrate relative dielectric constant, ϵ_{reff} is the relative effective dielectric constant of the guide, λ_g is the guide wavelength, and q is the guide filling factor, $(\epsilon_{reff}-1)/(\epsilon_r-1)$.

IV. Quality Factors of Half-Wavelength Resonators

The stored energy U in a $\lambda_g/2$ resonator with a voltage distribution $V \sin \beta_g z$ is given by

$$U = \frac{V^2}{8Z_0 f}. \tag{5}$$

The losses in the resonator arise from conductor dissipation, dielectric loss dissipation, and radiation. The conductor and dielectric losses are given by

$$W_c = \frac{1}{4}\frac{V^2}{Z_0}\lambda_g(\alpha_c + \alpha_d) \tag{6}$$

and the circuit quality factor Q_c is given by

$$Q_c = \frac{2\pi f U}{W_c} = \frac{\pi}{\lambda_g(\alpha_c+\alpha_d)}. \tag{7}$$

The radiation quality factor is given by

$$Q_r = \frac{2\pi f U}{W_r} \tag{8}$$

where W_r is the radiated power, estimated below. The total quality factor Q_t is given by

$$\frac{1}{Q_t} = \frac{1}{Q_r} + \frac{1}{Q_c}. \tag{9}$$

V. Radiation Loss From a Half-Wavelength Resonator

Radiation losses may only be estimated for finite lengths of line. The exact calculation of radiation loss from a half-wavelength CPW resonator has not been performed, and therefore an approximate method was used.

The two slots on either side of the $\lambda_g/2$ resonator between the strip and the ground planes are assumed to have magnetic currents that radiate into the air half-space and the half-space comprising substrate and the lower air space, if any. When the substrate fills this half-space, then the calculation is greatly simplified, and the radiation from the magnetic currents, whose distributions are known, may be readily estimated into each half-space. However, since the guide wavelength is longer than the dielectric free-space wavelength, the radiation in this half-space is very efficient, and in fact such calculation for the finite thickness substrate resonators overestimate the radiated power.

When the substrate is of finite thickness, the calculation of radiated power in this half-space is difficult. An approximation adopted here assumes that the half-space is filled with dielectric whose relative permittivity is ϵ_{reff}. The combination of this with

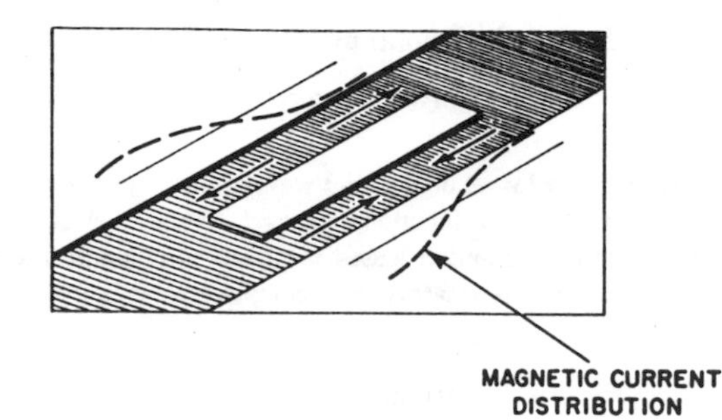

Fig. 3. Distribution of magnetic currents in the strip to ground-plane gaps for radiation calculations.

the other half-space filled with air provide an estimate of radiated power which may be acceptable.

The coplanar waveguide $\lambda_g/2$ resonators may be either open circuit or short circuit at their ends. Since the former case has smaller radiation losses and therefore higher radiation quality factors, it is calculated here; the latter case follows on similar lines. Fig. 3 shows the distribution of magnetic currents in the slot regions and for the open-circuit $\lambda_g/2$ resonator the currents have a $\pm \sin \beta_g z$ variation. The antiphase excitation and the change in direction of the current along the element reduces the radiated power.

For a vertical conducting plane with an open circuit CPW $\lambda_g/2$ resonator with two slots, each of width given by g (equal to $b-a$), and spaced d (equal to $b+a$) apart, the total E-field in any of the half-spaces is given by

$$E_T = \bar{a}_\phi \frac{k_{mm}^2 g E_m}{\pi r} \frac{\sin 2\theta \cos\left(\frac{k_m l}{2}\cos\theta\right)}{\left(\beta_g^2 - k_m^2\cos^2\theta\right)} \cdot \sin\left(\frac{k_m d}{2}\sin\theta\cos\phi\right)^{e^{-jkmr}} \tag{10}$$

where the voltage across the guide $V = gE_m$, and k_m is the propagation constant of the medium, either air space or effective dielectric constant space. The radiated power in each half-space is given by

$$W_{a,\,deff} = \frac{1}{2\eta_{a,\,deff}} \int_0^\pi \int_0^\pi (E_T)^2 r^2 \sin\phi \, d\theta \, d\theta \tag{11}$$

where $\eta_{a,\,deff}$ is the impedance of the appropriate half-space. Note that η_{deff} is the impedance of the half-space of dielectric constant ϵ_{reff}. The integrations are performed numerically for each half-space in turn and the total radiated power is equal to their sum.

VI. Computation of Losses and Quality Factors

A computer program has been written to calculate the impedance, wave velocity, and effective dielectric constant of coplanar waveguides for finite substrate thickness $2h$, and finite metallization thickness $2t$. The substrips are assumed to lie at the top and bottom of the strip and each ground plane, and the charge on each of these is obtained from the solution of the matrix equation. To increase the accuracy of the computation, the substrips are subdivided to be narrow in width close to the edges of the strip and inner edge of the ground planes and wider elsewhere. The conductivity of the upper and lower metalization layers corresponding to the respective substrips may also be specified as being different. The corresponding conductor loss constant and the dielectric loss constant are calculated based on these factors. At the specified frequencies, the circuit, radiation, and total

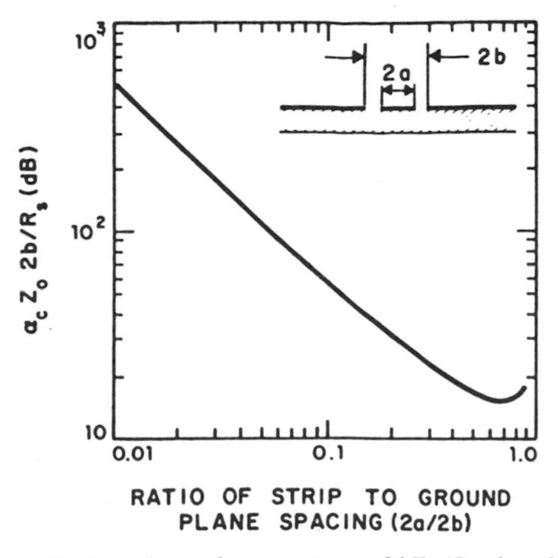

Fig. 4. Normalized conductor loss constant $\alpha_c 2bZ_0/R_s$ plotted against the strip width to ground-plane spacing ratio $2a/2b$.

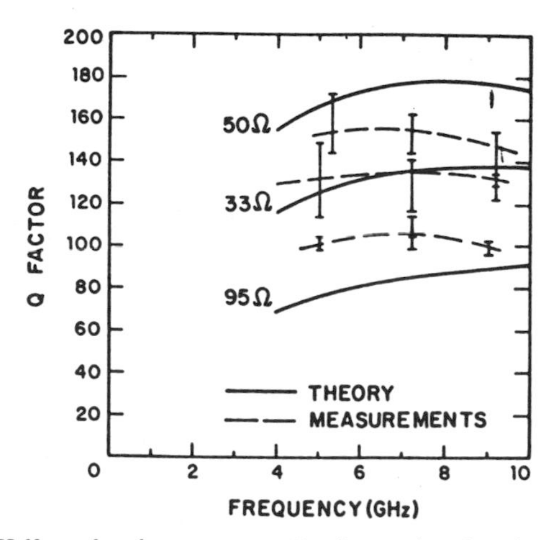

Fig. 5. Half-wavelength resonator quality factor plotted against frequency for resonators on substrates with $\epsilon_r = 13.0$, thickness 0.635 mm, and ground-plane spacing $2b = 1.2$ mm.

quality factors of $\lambda_g/2$ resonators are also calculated. In these calculations $b_{\max}$ was equal to $20b$.

A recent paper [9] has questioned the validity of using (3) in such loss calculations. The present calculations have been performed using a finite conductivity, finite (i.e., constant) basis functions, and uses the "thick strip" approach, all of which lead to a bounded solution. Increasing the substrip numbers from 24 to 72 metalized region has also resulted in the normalized conductor loss constant converging to such bounded solutions. It would therefore appear that comments from the above paper are not applicable to the present calculations.

VII. Computational Results and Comparison with Experiments

As there is adequate information in the literature [1]–[6] regarding the variation of Z_0 and ϵ_{reff} with a/b for coplanar waveguides, these results are not included here. The normalized conductor loss curve is given in Fig. 4, and from this, conductor losses for any ground-plane spacing may be obtained.

Quality factors of $\lambda_g/2$ resonators were calculated for $\epsilon_r = 13.0$ substrate, 0.635 mm thick, and ground plane spacing of 1.2 mm. Experimental measurements were performed on open-circuit resonators on Trans Tech D-13 substrates 0.635 mm thick for this ground-plane spacing of 1.2 mm. To extract the unloaded Q, the loss in the feed line was also taken into account in the usual manner. The comparison between theory and experiment is shown in Fig. 5, where the best agreement is for the 33-Ω resonators. Measurements of the Q-factors of resonators with the wider ground-plane spacing of 1.75 mm (not shown here) were lower by about 20 percent than predicted by the theory. The conductor-loss constants calculated here show reasonable agreement with other measurements [5] and predictions [6].

VIII. Conclusions

The losses in coplanar waveguides have been calculated and a normalized conductor loss curve is provided. The quality factors of half-wavelength resonators including radiation effects have been computed and show agreement with measured values.

References

[1] C. P. Wen, "Coplanar waveguides: A surface strip transmission line suitable for nonreciprocal gyromagnetic device application," *IEEE Trans. Microwave Theory Tech.*, vol. MTT-17, pp. 1087–1090, 1969.

[2] M. E. Davies, E. W. Williams, and A. C. Celestine, "Finite boundary corrections to coplanar waveguide analysis," *IEEE Trans. Microwave Theory Tech.*, vol. MTT-23, pp. 795–802, 1975.

[3] B. E. Spielman, "Computer-aided analysis of dissipation losses in isolated and coupled transmission lines for microwave and millimeter wave integrated circuit applications," Naval Research Laboratory, Washington, DC, NRL Rep 8009, 1979.

[4] J. B. Davies and D. Mirshekar-Syakhal, "Spectral domain solution of arbitrary coplanar transmission line with multilayer substrates," *IEEE Trans. Microwave Theory Tech.*, vol. MTT-25, pp. 143–146, 1977.

[5] J. A. Higgins, A. Gupta, G. Robinson, and D. R. Ch'en, "Microwave GaAs FET monolithic circuits," in *Int. Solid State Circuits Conf. Dig.*, 1979, pp. 20–21.

[6] K. C. Gupta, R. Garg, and I. J. Bahl, *Microstrip Lines and Slot Lines.* Dedham, MA: Artech House, 1979.

[7] P. Silvester, "TEM properties of microstrip transmission lines," in *Proc. Inst. Elec. Eng.*, vol. 115, 1968, pp. 42–49.

[8] A. Gopinath and P. Silvester, "Calculation of inductance of finite length strips and its variation with frequency," *IEEE Trans. Microwave Theory Tech.*, vol. MTT-21, pp. 380–386, 1973.

[9] R. Pregla, "Determination of conductor losses in planar waveguide structures," *IEEE Trans. Microwave Theory Tech.*, vol. MTT-28, pp. 433–434, 1980.

COPLANAR WAVEGUIDE VS. MICROSTRIP FOR MILLIMETER WAVE INTEGRATED CIRCUITS

R.W. Jackson

Department of Electrical and Computer Engineering
University of Massachusetts
Amherst, MA 01003

ABSTRACT

Using a full wave analysis, coplanar waveguide transmission line is compared to microstrip in terms of conductor loss, dispersion and radiation into parasitic modes. It is shown that, on standard .1 mm semiconductor at 60 GHz, the dimensions of coplanar waveguide can be chosen to give better results in terms of conductor loss and dispersion than microstrip. Curves are presented comparing the microstrip open end and the coplanar waveguide short circuit in terms of parasitic mode generation.

INTRODUCTION

For many years monolithic microwave integrated circuits have predominately used microstrip transmission lines. At microwave frequencies microstrip is well understood and flexible in that a large number of circuit elements can be made with it. However, for integrated circuits operating at millimeter wave frequencies it may not be the medium of choice. One disadvantage is that via holes are required to ground active devices. At millimeter wave frequencies these vias can introduce significant inductance and degrade circuit performance.

Coplanar waveguide has been suggested as an alternate to microstrip [1] but has not been widely used due to the mistaken assumption that it has inherently higher conduction loss than microstrip. Its principle advantage is that it is well suited for use with field effect transistors, especially at millimeter wave frequencies where R.F. grounding must be close to the device. Via holes are not necessary and fragile semiconductors need not be made excessively thin.

This paper compares coplanar waveguide to microstrip in terms of conductor and dielectric loss, dispersion characteristics and radiation of parasitic modes. A full wave technique is used to determine the currents in the vicinity of infinite lines on open substrate. These currents are then used to calculate loss [2] and impedance. Radiation at the aforementioned discontinuities is also determined using a full wave analysis [3] which is described briefly. A standard .1 millimeter thick GaAs substrate is assumed throughout as well as a 60 GHz operating frequency. The conclusions with regard to the comparison are not particularly sensitive to this choice of frequency.

INFINITE LINE ANALYSIS

Full wave analysis techniques are well known for infinite transmission lines [4], [5]. The techniques presented in this section outline the methods used for calculation of the loss, impedance and dispersion in coplanar waveguide (CPW). The same techniques were used to obtain the same quantities in microstrip.

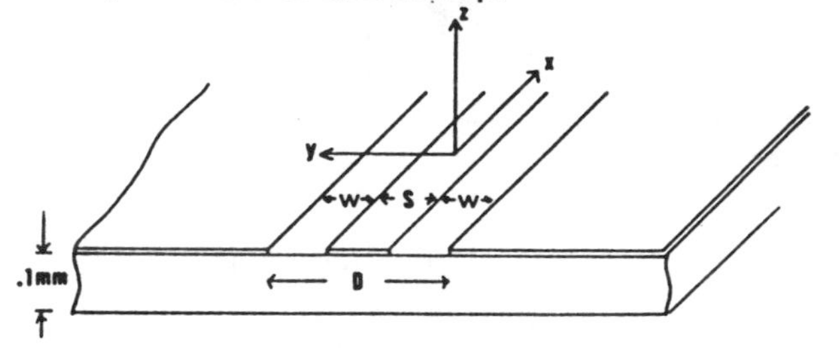

Figure 1. Structure of the coplanar waveguide transmission line.

For coplanar waveguide, currents on the z=0 plane (see Figure 1) are determined by the slot fields according to the relation

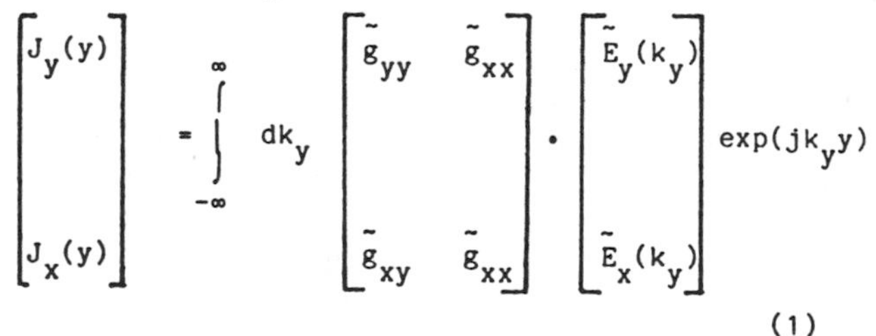

$$\begin{bmatrix} J_y(y) \\ J_x(y) \end{bmatrix} = \int_{-\infty}^{\infty} dk_y \begin{bmatrix} \tilde{g}_{yy} & \tilde{g}_{xx} \\ \tilde{g}_{xy} & \tilde{g}_{xx} \end{bmatrix} \cdot \begin{bmatrix} \tilde{E}_y(k_y) \\ \tilde{E}_x(k_y) \end{bmatrix} \exp(jk_y y) \quad (1)$$

where an $\exp(-j\beta x)$ dependence has been suppressed. The g_{ij} (see appendix) are the fourier transform of the Green's function and are dependent upon β and k_y. Following references [4] and [5] we expand $E_y(y)$ in terms of the functions shown below.

$$E_y(y) = \sum_{n=0} A_n \frac{\cos(n\pi[u/w+1/2])}{\sqrt{1-(2u/w)^2}},$$

$$E_x(y) = \sum_{n=0} B_n \frac{\sin(n\pi[u/w+1/2])}{\sqrt{1-(2u/w)^2}}, \quad (2)$$

$$u=|y|-(S+W)/2, \quad -w/2<u<w/2$$

The constants β, A_n and B_n are determined such that weighted moments of J_x and J_y in the slot are zero. The chosen weighting functions are the same as the

Reprinted from *IEEE MTT-S Digest*, pp. 699-702, 1986.

expansion functions. As usual β is varied until nontrivial values of A_n and B_n can exist. The A_n and B_n can then be used with (1) and (2) to determine J_x and J_y everywhere on the conductors.

Characteristic impedance is defined using a voltage-current definition,

$$Z_c = V \cdot \left[2 \int_0^s dy \; J_x(y) \right]^{-1}$$

where V is the voltage across the gap. This definition is not the only one possible, but it is felt that it reflects the character of the local fields more accurately than the power-voltage definition and the local fields are most significant to circuit modelers. Microstrip impedances are also defined using a voltage-current model where the current is the x-directed current on the strip and the voltage is the line integral of the E_z field under the strip average over the strip width.

Conductor loss is determined from the formula

$$\alpha = C \frac{Z_c}{V^2} \left[\int_0^{S/2-\Delta} dy + \int_{S/2+w+\Delta}^{\infty} dy \right] \cdot \left[|J_y^u|^2 + |J_x^u|^2 + |J_y^\ell|^2 + |J_x^\ell|^2 \right] \qquad (4)$$

where Δ=t/290, t is the conductor thickness, and C is a constant which includes surface resistivity. The choice of Δ is made according to Lewin's work [2] in order to avoid the nonintegrable singularity at the edge of the zero thickness conductor assumed in the full wave analysis. The u and ℓ superscript on the current refer to the current on the upper and lower sides of the conductor. These currents were calculated using equation 1 with different Green's functions (see appendix). The y integration in (4) and the k_y integration in (1) were performed numerically. When computing currents near the conductor edges the integration in (1) is very slow to converge and one must subtract the asymptotic value of the integrand and evaluate it analytically.

Dielectric loss is calculated using standard formulas and is only a small part of the total loss.

INFINITE LINE COMPARISON

Figure 2 shows conductor and dielectric loss plotted against impedance for microstrip and coplanar waveguide on a substrate with a permittivity of 12.8. A conductor thickness of 3 microns is chosen. Copper conductor was assumed, but any other conductor would result in the same conclusions as far as comparisons are concerned. The coplanar waveguide impedance is varied by keeping a constant cross-section (D) and changing the center conductor width. For microstrip the only free parameter is the strip width. Strip widths were constrained to be between 10 and approximately 300 microns.

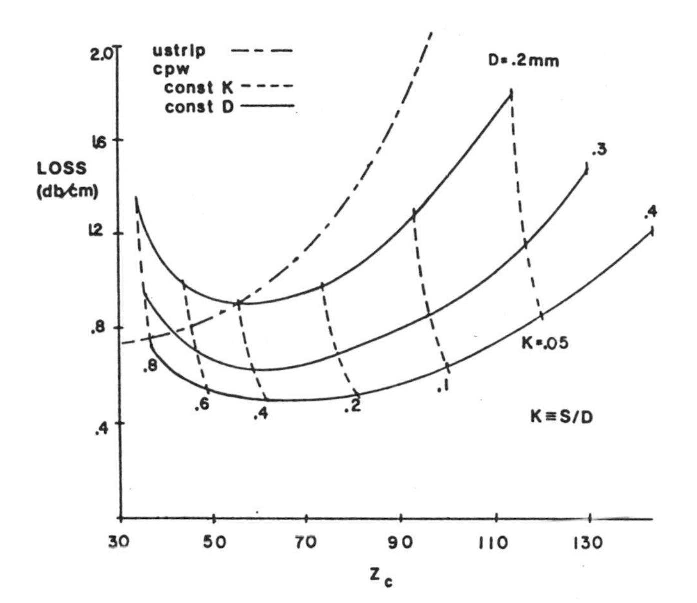

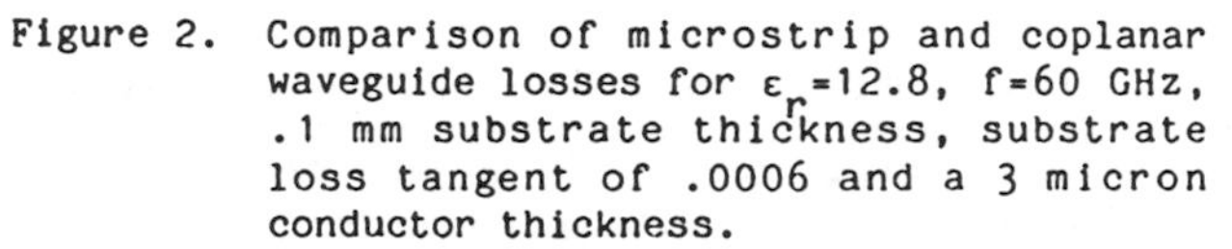

Figure 2. Comparison of microstrip and coplanar waveguide losses for ε_r=12.8, f=60 GHz, .1 mm substrate thickness, substrate loss tangent of .0006 and a 3 micron conductor thickness.

The figure shows that coplanar waveguide can have significantly less loss than microstrip over a broad range of impedances but especially at higher impedances. The impedance for minimum coplanar waveguide loss appears to be about 60 ohms for any of the chosen cross-sections. The microstrip width at minimum loss is about 300 microns whereas the smallest coplanar waveguide cross-section which will give the same loss is about 250 microns. So sizes are comparable.

The loss calculation shows the size that coplanar waveguide must be in order to compete with microstrip. For these sizes figure 3 shows a comparison of dispersion for microstrip and coplanar waveguide. The fractional change in effective dielectric constant per fractional change in frequency is plotted against impedance. For frequencies near 60GHz the figure shows that coplanar waveguides with cross-sections between 200 and 300 microns have as much or less dispersion than microstrip.

ANALYSIS OF PARASITIC RADIATION LOSS

One way of comparing microstrip and coplanar waveguide in terms of parasitic radiation is to compare radiation (surface wave and space wave) from a coplanar waveguide short circuit and from a microstrip open circuit using the dimensions given in section 1.

The analysis of losses from a microstrip open end and coplanar waveguide short circuit has been reported previously [3]. In that work the authors used a moment method technique to calculate the slot fields (strip currents) at the end of a coplanar waveguide (microstrip). These fields (currents) were assumed to be transverse (longitudinal). The analysis which is presented

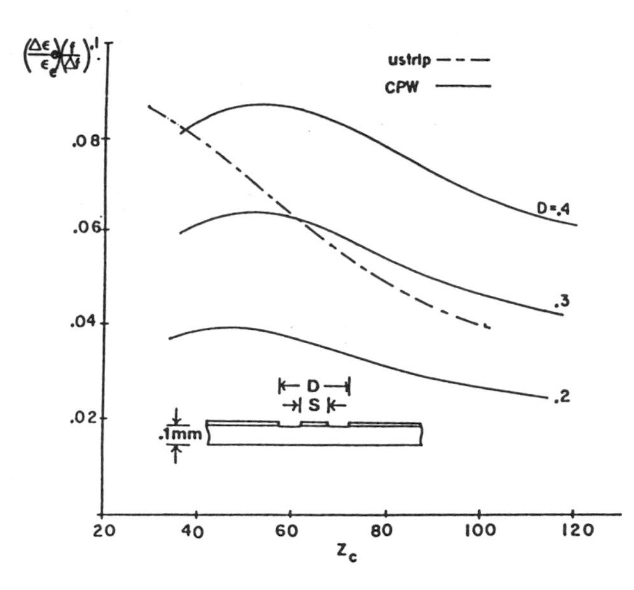

Figure 3. Comparison of microstrip and coplanar waveguide dispersion.

here also includes the longitudinal fields (transverse currents). In the previous analysis of the coplanar waveguide short, the slot fields were assumed to be symmetric around the slot centers. For tightly coupled slots, this can be inaccurate as Jansen [4] points out. The possible asymmetry is allowed in the analysis which is now outlined. Only the differences between this calculation and the calculations in reference 3 will be discussed.

The coordinate system used is shown in Figure 1 except that the slots only exist x<0. Slot fields are related to z=0 surface currents by an equation similar to (1) except that $\vec{J}$ is a function of x and y, $g_{ij}(\beta,k_y)$ is replaced by $G_{ij}(k_x,k_y)$, $\vec{E}$ is a function of k_x and k_y and the inverse fourier transform is with respect to both k_x and k_y. The slot fields are expanded in terms of known functions multiplied by unknown constants,

$$E_\alpha(x,y)=g^\alpha(y)[(1+R)f_c(x)+j(1-R)f_s(x)+\sum_{n+1}^{N} A_n^\alpha f(x-x_n)],\quad \alpha=x \text{ or } y. \qquad (5)$$

The functions f_c and f_s are cosine and sine functions which are several periods in length, f(x) is a piecewise sinusoidal function. R, the reflection coefficient, and A_n^α are unknown constants. For the coplanar waveguide the transverse dependence of the fields $g^y(y)$ and $g^x(y)$ is made up of three pulse and three triangle functions, respectively (see figure 4). The amplitudes of all six functions are related to each other and and are computed prior to the discontinuity calculation. This relationship as well as the propagation constant used in f_c was obtained by using pulse and triangle functions as expansion modes in the infinite line solution which was described in the first section of this paper. Due to the similarity of the expansion functions to one another, the amplitudes and propagation constant can be calculated very quickly. The propagation constants so calculated are within 1% of those calculated using modes in equation 2 for all impedances of interest. By using this type of function for $g^\alpha(y)$, the asymmetry in the slot fields can be included in the analysis of the short circuit discontinuity.

Weighting functions are chosen to be the piecewise sinusoidal parts of equation 5, $f(x-x_n)g^x(y)$ for x-directed fields and $f(x-x_n)g^y(y)$ for y-directed fields. All piecewise modes were located in the range $0>x\geq-\lambda_g/4$. Sufficient expansion modes were used such that the end resistance and length extension converge to within an estimated 5% of their final value. Although length extension was not a goal of this calculation, this method produced results which were within 10% of those calculated by Jansen [4].

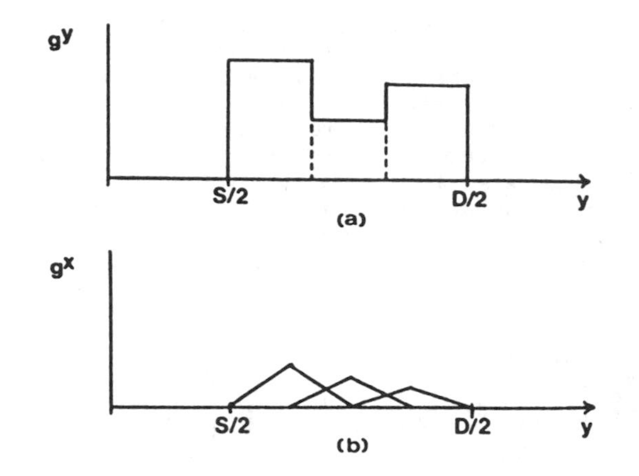

Figure 4. Transverse variation of slot fields, for y>0, (a) E_y and (b) E_x.

Loss from an open ended microstrip was calculated in a similar manner except that the transverse variation of the x and y directed strip currents were, respectively,

$$\frac{1}{\sqrt{1-(2y/w)^2}} \quad \text{and} \quad \frac{\sin(\pi[y/w+1/2])}{\sqrt{1-(2y/w)^2}} \quad .$$

COMPARISON OF DISCONTINUITY LOSS

Figure 5 compares power lost due to space wave and surface wave radiation from a microstrip open end and a coplanar waveguide short circuit. These values are obtained from end impedance calculations by the relationship

$$\frac{P_{rad}}{4P_{incidence}} = \begin{cases} GZ_c, & \text{open end microstrip} \\ RY_c, & \text{short circuit CPW} \end{cases}$$

where G (R) is a the real part of the end admittance (impedance). Coplanar waveguide size (D) is chosen in accordance with the conductor loss results in the second section. The coplanar waveguide evidently radiates much less energy.

CONCLUSIONS

It has been shown that, at millimeter wave frequencies, coplanar waveguide can be equal to or better than microstrip when loss and dispersion on GaAs substrate are used as a basis for comparison. Minimum loss for a given coplanar waveguide cross-section occurs at about a 60 ohm impedance whereas the minimum loss for microstrip occurs at about 25

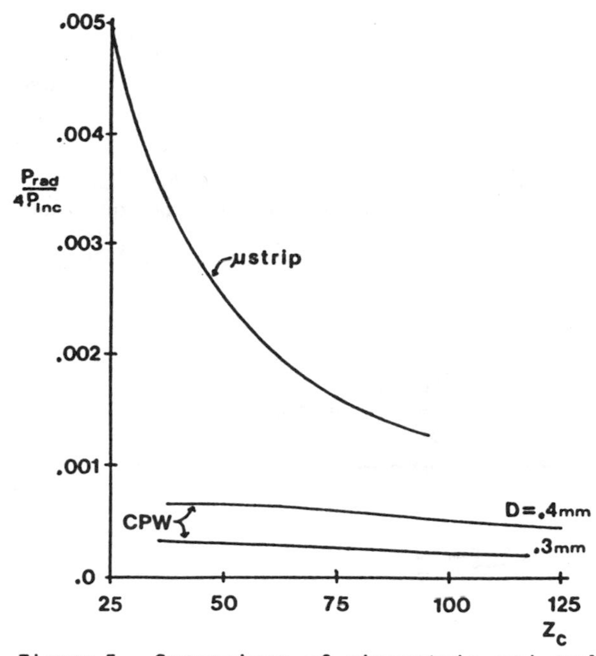

Figure 5. Comparison of microstrip and coplanar waveguide discontinuity loss.

ohms. The physical sizes at these minimum loss impedances are similar. This is important when long runs of transmission line are being contemplated. For higher impedances coplanar waveguide can give much smaller loss but will take up more space than the same impedance microstrip line.

Using the coplanar waveguide sizes required to made conductor loss comparable to that of microstrip, we have calculated discontinuity radiation loss for the two types of lines. A full wave analysis which includes space wave and surface wave radiation shows that coplanar waveguide discontinuities radiate much less energy than microstrip discontinuities.

The disadvantages of coplanar waveguide compared to microstrip are; size, the possibility that an even mode can be excited at non symmetric discontinuities and possibly heat transfer for active devices.

The advantages of coplanar waveguide include; easier construction using thicker substrates and no via holes, good grounding for integrated active devices, less radiation at discontinuities and, in some cases, lower conductor loss.

ACKNOWLEDGEMENT

This work was supported by the General Electric Company and by the Airforce Office of Scientific Research under contract #F49620-82-C-0035.

APPENDIX

The following functions are the Fourier transform of the Green's function which relates an infinitesimal slot electric field (magnetic current) at x=y=z=0 to the electric current on the lower side of a conductor on the z=0 plane.

$$\tilde{G}'_{xy}(k_x,k_y,\varepsilon_r)=\frac{-1}{\omega\mu}\frac{k_x^2k_1(1-\varepsilon_r)}{TM\cdot TE}+\frac{k_xk_y(k_1\cos k_1d+j\varepsilon_r\sin k_1d)}{k_1TM},$$

$$\tilde{G}'_{xy}(k_x,k_y,\varepsilon_r)=\frac{-1}{\omega\mu}\frac{k_xk_yk_1(1-\varepsilon_r}{TM\cdot TE}+\frac{k_xk_y(k_1\cos k_1d+j\varepsilon_rk_2\sin k_1d)}{k_1TM},$$

$$TE = k_1\cos k_1d+jk_2\sin k_1d, \quad TM = \varepsilon_rk_2\cos k_1d+jk_1\sin k_1d$$

$$k_2 = k_o^2 - \beta^2, \; k_1 = \varepsilon_rk_o^2 - \beta^2, \quad \beta^2 = k_x^2 + k_y^2$$

Except for loss calculations, only the total current from both the lower and upper sides is of interest,

$$\tilde{G}_{yy}(k_x,k_y)=\tilde{G}'_{yy}(k_x,k_y,\varepsilon_r)+\tilde{G}'_{yy}(k_x,k_y1),$$

$$\tilde{G}_{xy}(k_x,k_y)=\tilde{G}'_{xy}(k_x,k_y,\varepsilon_r)+\tilde{G}'_{xy}(k_x,k_y,1),$$

$$\tilde{G}_{yx}=\tilde{G}_{xy}, \quad \tilde{G}_{xx}(k_x,k_y)=\tilde{G}_{yy}(k_x\to k_y,k_y\to k_x).$$

For an infinite line the x variation is determined by the propagation constant, β, and therefore the Green's functions used in equation 1 are,

$$\tilde{g}_{yy}(\beta,k_y) = \tilde{G}_{yy}(\beta,k_y), \quad \tilde{g}_{yx}(\beta,k_y) = \tilde{G}_{xy}(\beta,k_y),$$

$$\tilde{g}_{xy}(\beta,k_y) = \tilde{G}_{xy}(\beta,k_y), \quad \tilde{g}_{xx}(\beta,k_y) = \tilde{G}_{yy}(k_y,\beta).$$

REFERENCES

[1] R. A. Pucel, "Design Considerations for Monolithic Microwave Circuits", IEEE Trans. MTT-29, pp. 513-534, April 1981.

[2] L. Lewin, "A Method of Avoiding the Edge Current Divergence in Perturbation Loss Calculations", IEEE Microwave Theory and Tech., MTT-32, No. 7, pp. 717-719, July 1984.

[3] R. W. Jackson and D. M. Pozar, "Surface Wave Losses at Discontinuities in Millimeter Wave Integrated Transmission Lines", IEEE MicrowaveSymposium Digest, MTT-S, pp. 563-565, June 1985.

[4] R. Jansen, "Unified User-oriented Computation of Shielded, Covered and Open Planar Microwave and Millimeter-wave Transmission-line Characteristics", Microwave, Optics and Acoustics, Vol. 3, No. 1, pp. 14-22, January 1979.

[5] T. Itoh, "Spectral-domain Inmittance Approach for Dispersion Characteristics of Generalized Printed Transmission Lines", IEEE Trans. Microwave Theory Tech., Vol. MTT-28, p. 733-736, July 1980.

[6] R. Jansen, "Hybrid Mode Analysis of End Effects of Planar Microwave and Millimeter Transmission Lines", IEEE Proc., Vol. 128, PT.H, No. 2, pp. 77-86, 1981.

ANALYTICAL FORMULAS FOR COPLANAR LINES IN HYBRID AND MONOLITHIC MICs

Indexing terms: Integrated circuits, Microwave circuits and systems

Some analytical formulas for the parameters of coplanar lines are discussed and validated; a chart is given for the design of coplanar waveguides on GaAs. The formulas discussed here, together with those presented previously by us (1983) represent a suitable set for the design of coplanar lines for hybrid and monolithic MICs (microwave integrated circuits).

Owing to the increasing popularity of coplanar waveguides (CPW, Fig. 1) and coplanar striplines (CPS, Fig. 3) for the design of hybrid and monolithic microwave integrated circuits, it is important to have a set of reliable analytical formulas for their quasi-TEM electrical parameters (characteristic impedance Z_ϕ and effective dielectric constant ε_{eff}). While exact formulas are available for lines having infinitely thick substrates, the analytical expressions published so far for lines having finite substrates are approximate in principle and should be validated by a more rigorous (numerical) analysis. The purpose of the present letter is to assess the validity of some formulas published in the literature, also suggesting possible improvements where results of questionable correctness are found. Finally, an example of design chart is presented for integrated coplanar waveguides on GaAs substrate.

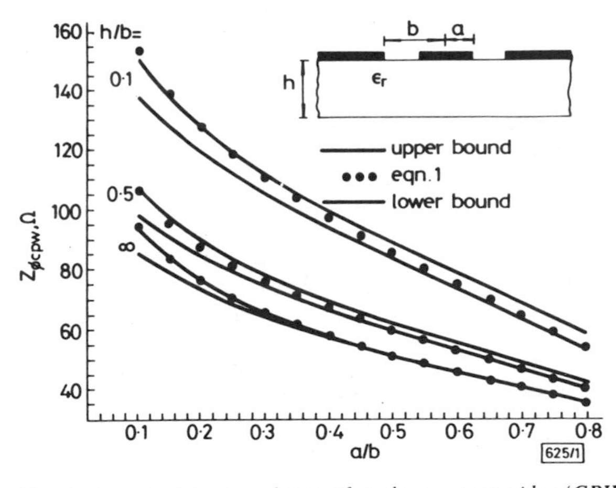

Fig. 1 *Characteristic impedance of coplanar waveguides (CPW) as a function of the shape ratio $k = a/b$, taking the normalised substrate thickness h/b as a parameter*

$\varepsilon_r = 10$

Analytical expressions for coplanar waveguides are found in References 1 and 2. Gupta's formulas are obtained by fitting the data published in Reference 3, and only hold for moderately thick substrates ($h > b - a$), while Fouad-Hanna's formulas are based on an approximated theory whose general validity is questionable. However, they can deal with lines having arbitrarily thin substrates and yield the correct limit for $h \to 0$. A comparison has been carried out between the results obtained through a spectral-domain variational analysis (Figs. 1 and 2) and Fouad-Hanna's expressions:

$$Z_{\phi(CPW)} = \frac{30\pi}{\sqrt{\varepsilon_{eff}}} \frac{K(k')}{K(k)} \tag{1}$$

$$\varepsilon_{eff} = 1 + \frac{\varepsilon_r - 1}{2} \frac{K(k')K(k_1)}{K(k)K(k'_1)} \tag{2}$$

where

$$k = a/b \tag{3a}$$

$$k_1 = \sinh(\pi a/2h)/\sinh(\pi b/2h) \tag{3b}$$

K is the complete elliptic integral of the first kind, and $k' = \sqrt{(1 - k^2)}$. As a result, it has been found that eqns. 1 and 2 are satisfactorily accurate in a wide range of substrate thicknesses and approach the variational upper limit for small

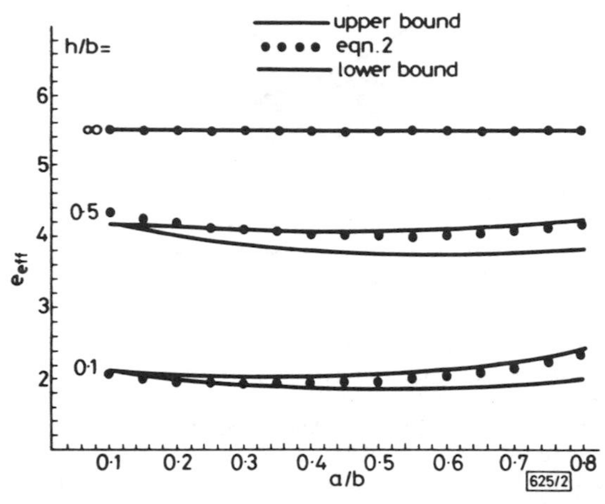

Fig. 2 *Same as Fig. 1: effective dielectric permittivity*

shape ratios ($k = a/b$) and the lower limit for large shape ratios, as could be expected from the variational trial functions used in the analysis; besides, Fouad-Hanna's formulas yield practically the same results as Gupta's for $h > b - a$, though Gupta's expressions tend to slightly underestimate the characteristic impedance, with respect to the variational lower limit, for rather thin substrates. Good agreement with published experimental values is found to exist for noncritical cases (i.e. for h large), as it happens for Gupta's formulas,[2] whereas for very thin substrates the available experimental data are of uncertain interpretation. It ought to be noticed that the formulas for CPW with lower ground plane, also published in Reference 1, are of dubious validity. This has been pointed out in a previous letter,[4] where a different set of formulas for this structure is proposed.

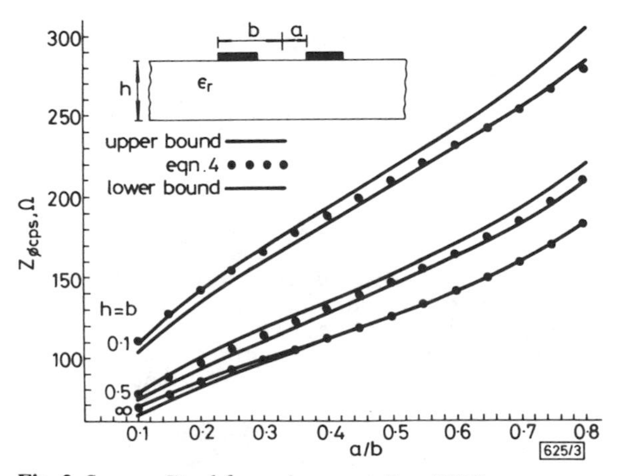

Fig. 3 *Same as Fig. 1 for coplanar striplines (CPS)*

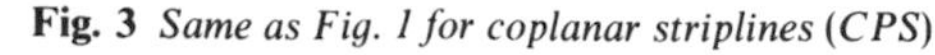

Reprinted with permission from *Electronics Letters*, vol. 20, no. 4, pp. 179–181, February 1984.

An application of Fouad-Hanna's method to the CPS has been tried in Reference 5, while Gupta's formulas for the same structure are reported in Reference 2. Unfortunately, the analysis published in Reference 5 seems to lead to uncorrect results; in particular, the conclusion according to which the impedance of the line increases when the substrate thickness decreases is unacceptable, and leads to absurd consequences in the limit $h \to 0$. However, reliable formulas for the CPS can be obtained through the assumption that the phase velocities of complementary lines (as CPW and CPS) are equal. Although this is strictly valid only for lines in homogeneous or semi-infinite media, the aforementioned assumption approximately holds for TEM lines on finite substrates as well, and is also used in Gupta's formulas. The proposed set for the parameters of CPS with finite substrate is therefore

$$Z_{\phi(CPS)} = \frac{120\pi}{\sqrt{\varepsilon_{eff}}} \frac{K(k)}{K(k')} \tag{4}$$

while ε_{eff} is always given by eqn. 2. A comparison with the variational upper and lower bounds for the impedance is shown in Fig. 3; for the effective permittivity, the results coincide with those reported in Fig. 2.

As a possible application for the analytical formulas discussed before, a design chart (Fig. 4) is presented for the design of CPW on GaAs substrate ($\varepsilon_r = 13$). The thickness of the substrate is typical of MIC applications ($h = 200\ \mu m$); the range of impedances considered covers the interval 40–60 Ω. One clearly sees that, in order to realise lines having constant impedance (e.g. 50 Ω) with different ground-plane spacings ($2b$) also the ratio $a/b = k$ has to be changed, whereas for semi-infinite substrates keeping k constant is enough to this purpose.

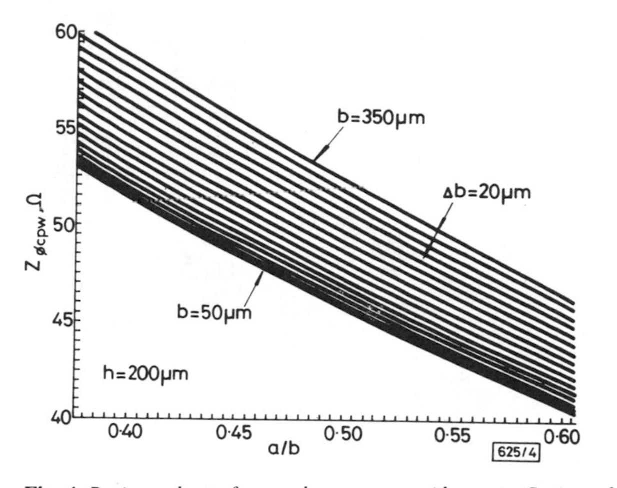

Fig. 4 *Design chart for coplanar waveguides on GaAs substrate* ($\varepsilon_r = 13$)

Characteristic impedance is given as a function of shape ratio $k = a/b$, taking half of the ground-plane spacing (b) as a parameter. The substrate thickness is $h = 200\ \mu m$

G. GHIONE
C. NALDI

17th January 1984

Department of Electronics
Politecnico di Torino
Corso Duca degli Abruzzi 24, Torino, Italy

References

1 VEYRES, C., and FOUAD-HANNA, V.: 'Extension of the application of conformal mapping techniques to coplanar lines with finite dimensions', *Int. J. Electron.*, 1980, **48**, pp. 47–56

2 GUPTA, K. G., GARG, R., and BAHL, I. J.: 'Microstrip lines and slotlines' (Artech House, 1979), Par. 7.3.1

3 DAVIS, M. E., WILLIAMS, E. W., and CELESTINI, C.: 'Finite-boundary corrections to the coplanar waveguide analysis', *IEEE Trans.*, 1973, **MTT-21**, pp. 594–596

4 GHIONE, G., and NALDI, C.: 'Parameters of coplanar waveguides with lower ground plane', *Electron. Lett.*, 1983, **19**, pp. 734–735

5 FOUAD-HANNA, V.: 'Finite boundary corrections to coplanar stripline analysis', *ibid.*, 1980, **16**, pp. 604–606

Part IX
Finlines and Slot Lines

FINLINES and slot lines treated in this part have modal field configurations substantially different from those in quasi-TEM lines such as microstrip lines and coplanar waveguides. The slot line was invented by S. B. Cohn as a planar waveguide having a region of elliptic polarization of the field. Initial realizations typically used a substrate with a high permittivity such as alumina. Due to elliptic polarizations, the slot line has been considered useful for applications involving ferrite materials. Compared to other planar transmission lines, the slot lines have been used less frequently. On the other hand, the finline invented by P. J. Meier has proven to be a very popular configuration for both analysis and practice at millimeter-wave frequency regions. This structure is similar to a slot line inserted along the E-plane of a TE_{10} waveguide. It can also be viewed as a printed form of ridge waveguide. Hence, the finline has a single mode bandwidth larger than that of the corresponding rectangular waveguide with interior dimensions identical to those of the housing of the finline. Three types of finlines are conceived: unilateral, bilateral, and antipodal. The unilateral configuration has fins on one side of the substrate while the bilateral uses both sides of the substrate for metallization, often symmetrically. The antipodal finline has one fin on one side and another on the other side of the substrate. The fins of the antipodal configuration are connected to the opposite broad walls of the housing. In some cases, the substrate is completely removed. Such a metal-only fin has been originally proposed by Konishi *et al.** Finline configurations are often called E-plane structures because the principal transverse electric field between the edges of the fins is along the E-plane of the waveguide housing. In many applications, the finlines are combined with suspended strip structures. One of the problems of finlines is that the characteristic impedance is quite high. A low impedance can be realized with an antipodal configuration in which two fins on both sides of the substrate can have an overlapping region to increase a distributed capacitance. The antipodal finline can be used as a transition from a strip-type transmission line to a finline configuration. One of the antipodal fins can be tapered down to a strip and another to a ground plane by gradually increasing the width.

The papers in this part cover a number of subjects related to the slot line and finline. It should be noted that there is no low-frequency approximation such as the quasi-TEM analysis applicable to these structures. In addition to several papers dealing with the rigorous full-wave analysis of the characteristic impedance and the propagation constants, papers analyzing the attenuation, the effect of metallization thickness, etc., are included as well. Note that the definition of the characteristic impedance is somewhat arbitrary. Often, the voltage-power definition is employed where the voltage is the electric field integrated from the edge of one fin to another (Knorr and Shayda; Schmidt and Itoh; Mirshekar-Syahkal and Davies). An example of the empirical design formula of finlines is also included (Sharma and Hoefer). Important subjects of finline discontinuities have been treated by an increasing number of researchers recently. Several examples of these contributions appear in the last three papers in this part.

* Y. Konishi, K. Uenakada, N. Yazawa, N. Hoshino, and T. Takahashi, "Simplified 12-GHz Low-Noise Converter with Mounted Planar Circuit in Waveguide," *IEEE Trans. Microwave Theory and Techniques,* vol. MTT-22, pp. 451–454, April 1974.

Slot Line on a Dielectric Substrate

SEYMOUR B. COHN, FELLOW, IEEE

***Abstract*—Slot line consists of a narrow gap in a conductive coating on one side of a dielectric substrate, the other side of the substrate being bare. If the substrate's permittivity is sufficiently high, such as $\epsilon_r = 10$ to 30, the slot-mode wavelength will be much smaller than free-space wavelength, and the fields will be closely confined near the slot. Possible applications of slot line to filters, couplers, ferrite devices, and circuits containing semiconductor elements are discussed. Slot line can be used either alone or with microstrip line on the opposite side of the substrate. A "second-order" analysis yields formulas for slot-line wavelength, phase velocity, group velocity, characteristic impedance, and effect of adjacent electric and magnetic walls.**

Manuscript received February 17, 1969; revised May 15, 1969. This work was performed for Stanford Research Institute as a part of their study program for U. S. Army Electronics Command, Contract DAAB07-68-c-0088. This paper covers material presented at the 1968 G-MTT Symposium, Detroit, Mich., May 20–22, 1968.

The author is a consultant to Stanford Research Institute, Menlo Park, Calif.

I. Introduction

SLOT LINE consists of a slot or gap in a conductive coating on a dielectric substrate, as shown in Fig. 1. Both resonant and propagating slots in metal sheets have been used as radiating antenna elements (see [1], [2], and [3], and their bibliographies). For slot line to be practical as a transmission line, radiation must be minimized. This is accomplished through the use of a high permittivity substrate, which causes the slot-mode wavelength λ' to be small compared to free-space wavelength λ, and thereby results in the fields being closely confined to the slot with negligible radiation loss. For example, if $\epsilon_r = 20$, then $\lambda'/\lambda \approx \frac{1}{3}$, and analysis shows the slot-mode fields to be sharply attenuated at a distance $r/\lambda = \frac{1}{8}$, or $r = 0.5$ inch at 3 GHz.

Fig. 2(a) shows the slot-mode fields in a cross-sectional view. A voltage difference exists between the slot edges. The

Reprinted from *IEEE Trans. Microwave Theory Tech.*, vol. MTT-17, no. 10, pp. 768–778, October 1969.

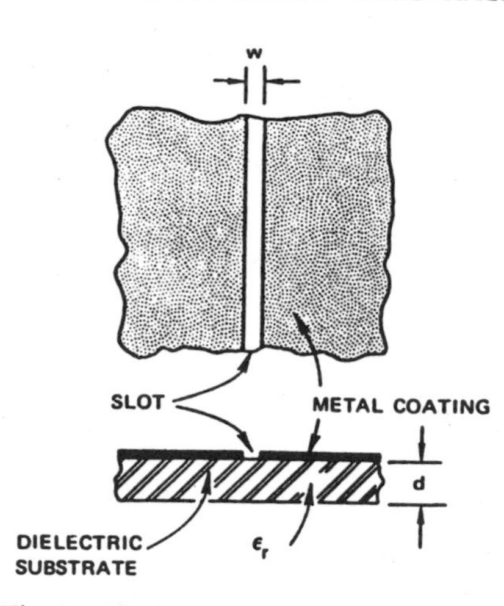

Fig. 1. Slot line on a dielectric substrate.

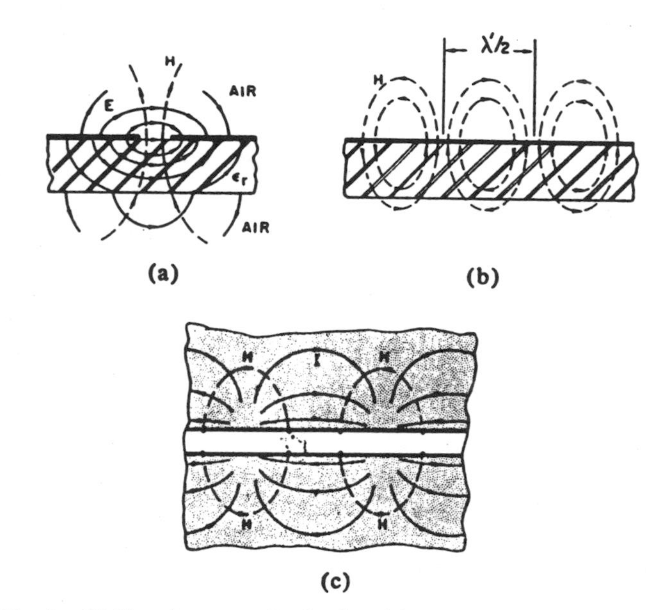

Fig. 2. Field and current distribution. (a) Field distribution in cross section. (b) *H* field in longitudinal section. (c) Current distribution on metal surface.

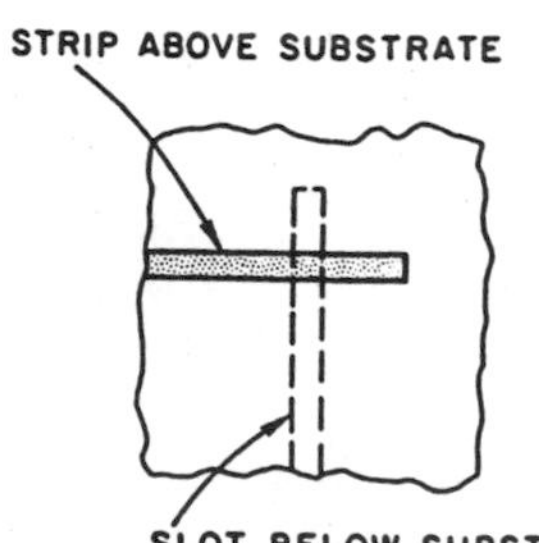

Fig. 3. Simple transition between slot line and microstrip.

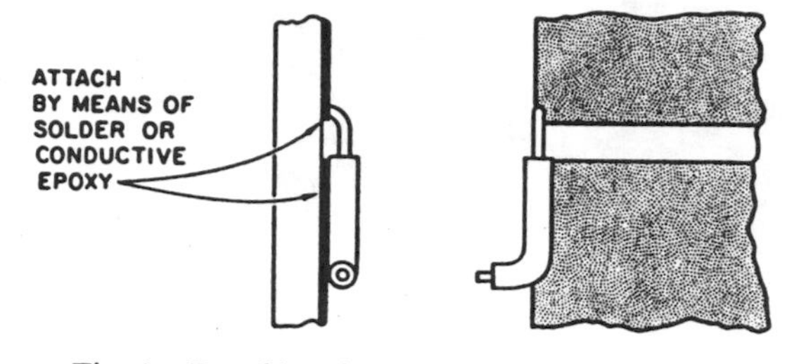

Fig. 4. Broad-band transition between slot line and miniature semirigid coaxial line.

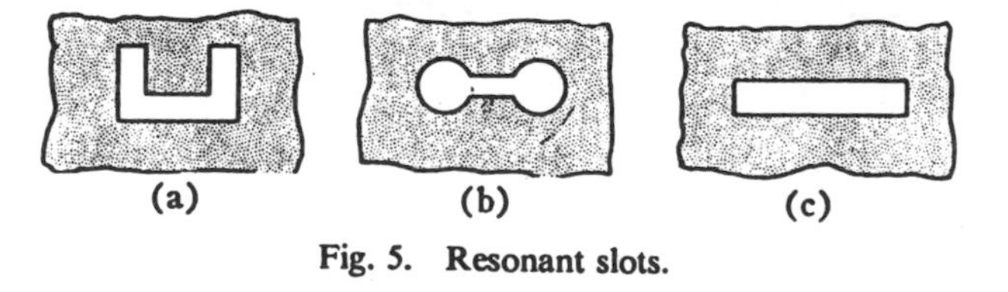

Fig. 5. Resonant slots.

electric field extends across the slot; the magnetic field is perpendicular to the slot. Because the voltage occurs across the slot, the configuration is especially convenient for connecting shunt elements such as diodes, resistors, and capacitors. The longitudinal view in Fig. 2(b) shows that in the air regions the magnetic field lines curve and return to the slot at half-wavelength intervals. Consequently, a propagating wave has elliptically polarized regions that can be usefully applied in creating certain ferrite components. The current paths on the conducting surface are shown in Fig. 2(c). The surface-current density is greatest at the edges of the slot and decreases rapidly with distance from the slot. A propagating wave has regions of elliptically polarized current and magnetic field in this view, also.

An interesting possibility for microwave integrated circuits is the use of slot lines on one side of a substrate and microstrip lines on the other. When close to each other, coupling between the two types of lines will exist, and when sufficiently far apart they will be independent. Coupling between a slot and a strip can be used intentionally in certain components. For example, parallel lengths of slot and strip can be made to act as a directional coupler. If a slot and strip cross each other at right angles, as in Fig. 3, coupling will be especially tight, and a transition covering approximately 30-percent bandwidth can be achieved when the strip and slot widths are optimally related, and when the strip and slot are extended approximately one-quarter wavelength beyond the point of crossing. With matching techniques, a bandwidth of an octave or so should be feasible.

Fig. 4 shows one way that a wide-band transition between miniature-cross-section coaxial line and slot line can be made. Additional structural details may be needed to obtain optimum matching and to prevent radiation loss.

A half-wavelength slot, as shown in Fig. 5(a), can be used as a resonator. If desired, the resonant slot may be made more compact by capacitively loading its center, as in Fig. 5(b), or by bending it, as in Fig. 5(c). Applications to bandstop and bandpass filters are shown in Fig. 6. In the bandstop example, the terminating lines are microstrip, and in the bandpass example they are slot lines. Many other filter configurations are feasible, using slots alone or slots with strips on the opposite side of the substrate.

The basic electrical parameters of slot line are the characteristic impedance Z_0 and the phase velocity v. Because of the non-TEM nature of the slot-line mode these parameters are not constant, but vary with frequency at a rather slow rate per octave. This behavior contrasts with quasi-TEM

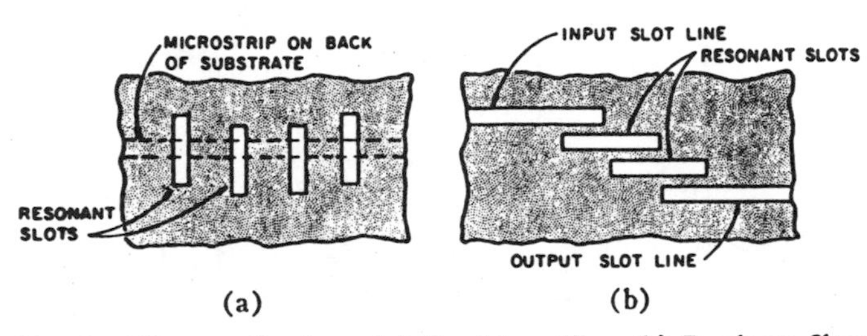

Fig. 6. Filter applications. (a) Bandstop filter. (b) Bandpass filter.

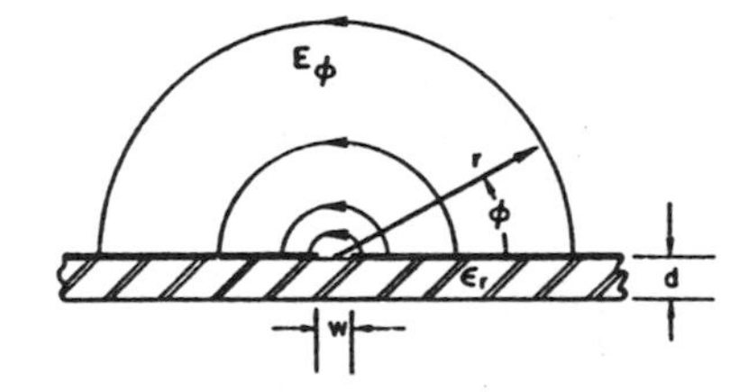

Fig. 7. Cylindrical coordinates with axis on center line of slot.

microstrip line, whose Z_0 and v are first-order independent of frequency. On the other hand, slot line differs from waveguide in that it has no cutoff frequency. Propagation along the slot occurs at all frequencies down to $f=0$, where, if the metal-coated substrate is assumed infinite in length and width, v/c approaches unity and Z_0 approaches zero. Other parameters treated in this paper are the ratio of phase and group velocities v/v_g and the effect of adjacent walls. Attenuation has not yet been treated, but data thus far indicates it to be about the same as for microstrip on the same substrate [20].

II. Approximations for Slot Line

Several references on slot antennas were found in which the presence of a substantial amount of dielectric material in or near the slot was taken into account. Strumwasser, Short, Stegen, and Miller [4] have studied experimentally the effects of filling a slot in a thick metal plate with dielectric material. They give data on resonant-length reduction and radiation resistance coupled into an air-dielectric TEM line. Bailey [5] has measured resonant length and radiation conductance of a slot in a waveguide wall covered by a protective layer of dielectric material. Galejs [6] has analyzed theoretically a slot in a zero-thickness, perfectly conducting sheet separating free space from a lossy dielectric medium of infinite extent. For example, a slot radiator in a wire mesh on the surface of the ground would be simulated by this model.

Galejs utilizes an integral-equation method to obtain complex expressions for radiation efficiency and other parameters of the slot antenna. His *zero-order* solution for the propagation constant along the slot can be easily modified into the following simple formula for relative wavelength:

$$\frac{\lambda'}{\lambda} = \sqrt{\frac{2}{\epsilon_r + 1}}\,. \tag{1}$$

Since wavelength is inversely proportional to the square root of permittivity, an effective permittivity of a uniform medium replacing the two different dielectric half spaces may be defined as

$$\epsilon_r' = \frac{\epsilon_r + 1}{2}\,. \tag{2}$$

The *second-order* solution for slot line derived in this paper shows that (1) is a fair approximation for slot line, yielding values within about 10 percent in typical slot-line cases. The second-order solution shows quantitatively how λ'/λ varies with the parameters d, w, ϵ_r, and λ.

The field components on the air side of the slot can be computed quite easily as a function of λ, λ', and distance r from the slot (Fig. 7). If we assume $w/\lambda \ll 1$, then the electric voltage across the slot may be replaced by an equivalent line source of magnetic current. At a distance r at least several times larger than w the longitudinal component of magnetic field is given by [7]

$$H_z = AH_0^{(1)}(k_c r) \tag{3}$$

where $H_n^{(1)}(x)$ is the Hankel function of first kind, order n, and argument x. The coefficient k_c is

$$k_c = \sqrt{\gamma_z^2 + k^2} = j\frac{2\pi}{\lambda}\sqrt{\left(\frac{\lambda}{\lambda'}\right)^2 - 1} \tag{4}$$

since $\gamma_z = j2\pi/\lambda_z = j2\pi/\lambda'$ and $k = 2\pi/\lambda$.

By (1), a zero-order value of k_c is

$$k_c = j\frac{2\pi}{\lambda}\sqrt{\frac{\epsilon_r - 1}{2}}\,. \tag{5}$$

The other field components are H_r and E_ϕ. They are related to H_z by [7]

$$H_r = -\frac{\gamma_z}{k_c^2}\frac{\partial H_z}{\partial r} = \frac{A}{\sqrt{1 - \left(\frac{\lambda'}{\lambda}\right)^2}} H_1^{(1)}(k_c r) \tag{6}$$

$$E_\phi = \frac{j\omega\mu}{k_c^2}\frac{\partial H_z}{\partial r} = \frac{-\eta(\lambda'/\lambda)A}{\sqrt{1 - \left(\frac{\lambda'}{\lambda}\right)^2}} H_1^{(1)}(k_c r) \tag{7}$$

where the identity $d[H_0^{(1)}(x)]/dx = -H_1^{(1)}(x)$ was used.

The Hankel function of imaginary argument, $H_n^{(1)}(j|x|)$, approaches zero proportional to $e^{-|x|}/\sqrt{|x|}$ for $|x|$ large. Equation (4) shows that the argument $k_c r$ is imaginary for $\lambda'/\lambda < 1$. Hence a relative wavelength ratio less than unity is a sufficient condition to ensure decay of the slot-mode field with radial distance. As λ'/λ is decreased, the decay becomes sharper and the fields become more tightly bound to the slot.

A radius r_{cp} of circular polarization of the magnetic field requires $|H_z/H_r| = 1$. By means of (3) and (6), r_{cp} must satisfy

$$\left|\frac{H_1^{(1)}(k_c r_{cp})}{H_0^{(1)}(k_c r_{cp})}\right| = \sqrt{1 - \left(\frac{\lambda'}{\lambda}\right)^2}\,. \tag{8}$$

However, tables show that $|H_1^{(1)}(j|x|)| > |H_0^{(1)}(j|x|)|$ for all $|x|$ [8]. Since the right-hand side of (8) is less than one, a solution for r_{cp} does not exist. Nevertheless, elliptical polarization occurs for all r, and low axial ratios occur for r sufficiently large.

Also of interest is the ratio of voltage along a semicircular path at constant radius divided by the voltage directly across the slot. This ratio is

$$\frac{V(r)}{V} = \frac{k_c r H_1^{(1)}(k_c r)}{\lim\limits_{|x|\to 0} [|x| H_1^{(1)}(j|x|)]} = \frac{\pi}{2} k_c r |H_1^{(1)}(k_c r)|. \tag{9}$$

As an example of field decay, let $\epsilon_r = 16$, $f = 3$ GHz, and $\lambda = 4$ inches. The zero-order value of λ'/λ is 0.343, and of $k_c r$ is 4.30 r, where r is in inches. At $r=0$, 0.7, 1.0, and 1.3 inches, $V(r)/V = 1$, 0.120, 0.0382, and 0.0118, respectively. If a plane metal wall is positioned perpendicular to the radius vector at distance $r/2$ from the slot, an image of the slot will appear at distance $r/2$ behind the wall. The effect will be that of two equally excited parallel slots spaced by r. Thus if the metal wall is at distance 0.5 inch in the above example, $r=1$ inch and $V(r)/V$ of one slot is 0.0382, or -28.36 dB, at the other slot. Coupling between slots is even weaker than this, since only part of the total voltage $V(r)$ of one slot affects the other slot. Therefore, a wall or other perturbing object can be as close as $r=0.5$ inch with little effect on λ' or Z_0, and the fields and stored energy of the slot mode are mainly confined within this same radial distance. These conclusions have been verified both by experiment and by computations using the second-order solution.

III. Basis of Second-Order Solution

An analytical approach for slot line offering high accuracy is described in this section. The solution obtained by this approach will be referred to as *second-order*. A first-order solution offering intermediate complexity and accuracy between zero-order (1) and second-order equations would also be useful, but has not yet been completed.

The key feature of the approach used in the second-order solution is the introduction of boundary walls permitting the slot-line configuration to be treated as a rectangular-waveguide problem rather than as a problem in cylindrical coordinates. Thus the infinite orthogonal sets of relatively simple rectangular-waveguide modes apply rather than sets of cylindrical modes embodying all orders of Hankel functions.

Parameters evaluated in this paper are: relative-wavelength ratio λ'/λ, characteristic impedance Z_0, ratio of phase velocity to group velocity v/v_g, and the effect of nearby electric and magnetic walls.

Future plans are to adapt the second-order solution to yield the even- and odd-mode characteristic impedances and velocities of parallel slots, the effect of metal-coating thickness greater than zero, attenuation per unit length, unloaded

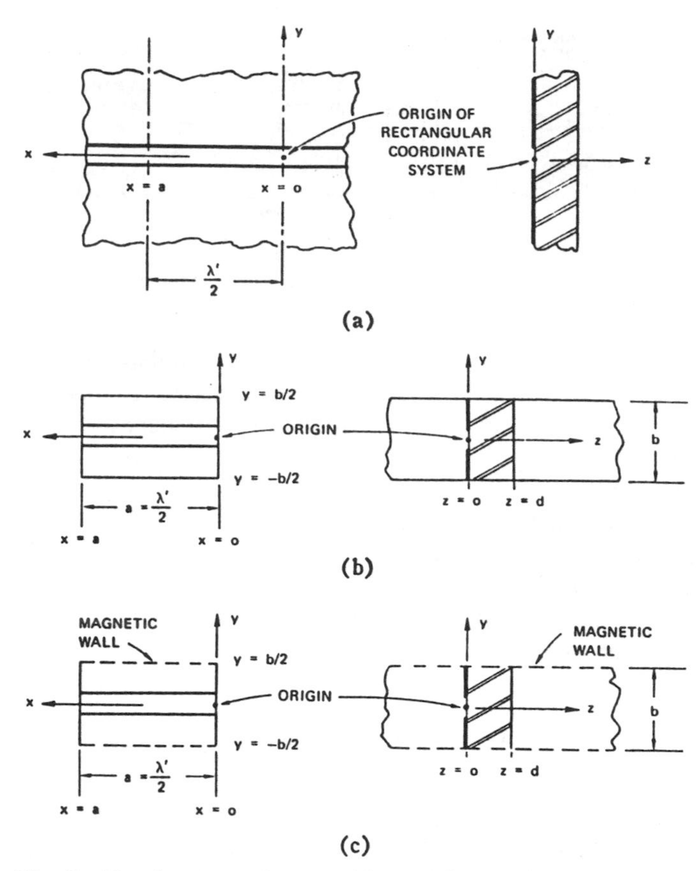

Fig. 8. Development of waveguide models for slot-line solution. (a) Insertion of transverse conducting planes at $x=0$ and a. (b) Insertion of conducting planes at $y=\pm b/2$. (c) Insertion of magnetic walls at $y=\pm b/2$.

Q of a resonant slot, and presence of more than one dielectric substrate on one or both sides of the slotted conductive sheet.

Conversion of the slot-line configuration into a rectangular-waveguide problem is illustrated in Fig. 8. First assume that slot waves of equal amplitude are traveling in the $+x$ and $-x$ directions. Then transverse planes spaced by $\lambda'/2$ exist where the transverse E field and normal H field cancel to zero. Let two such planes occur at $x=0$ and $x=\lambda'/2=a$ in Fig. 8(a). Conducting (or electric) walls of infinite extent may be inserted in these planes without disturbing the field components between the planes, and the semi-infinite regions at $x<0$ and $x>a$ may be eliminated. The section of slot line between the transverse planes supports a resonant slot-wave mode with no loss of energy, if the dielectric substrate and the conducting walls are assumed dissipationless.

Next, conducting walls are inserted in planes parallel to the slot and perpendicular to the substrate at $y=\pm b/2$. The region separated out of the original infinite space has the rectangular-waveguide boundary shown in Fig. 8(b). Since the fields are tightly bound to the vicinity of the slot, the walls at $y=\pm b/2$ will have negligible effect for b sufficiently large (typically one inch at 3 GHz), yet they serve the important function of enabling the use of rectangular-waveguide mode sets, thereby greatly simplifying the analysis.

Magnetic walls may be placed at $y \doteq \pm b/2$ instead of electric walls. The result is the boundary shown in Fig. 8(c), where two magnetic walls spaced by b and two electric walls spaced by a are used.

Images of the slot in the electric or magnetic walls at $y=\pm b/2$ result in an infinite array of parallel slots in the $z=0$ plane having center-to-center spacing of b. Therefore, the effect of adjacent slots may be computed for both electric- and magnetic-wall imaging, allowing the even- and odd-mode characteristic impedances and wavelengths to be evaluated for a slot in an infinite array. For one pair of slots rather than an array, these even- and odd-mode quantities are given approximately by the $b\to\infty$ values modified by one half the change computed for the infinite array.

Thus, the introduction of walls in Fig. 8 has created the configuration of a capacitive iris in a rectangular waveguide, with air and dielectric regions as indicated. Consider the metal-walled case in Fig. 8(b). All waveguide modes must have the $\lambda'/2$ variation of the slot wave in the x direction. Also, because of the symmetry of the structure, all modes must have an E-field maximum at the center of the slot. Therefore, the full set of modes satisfying the boundary conditions are TE_{10}, TE_{12}, TE_{14}, $\cdots$, and TM_{12}, TM_{14}, $\cdots$; that is, $TE_{1,2n}$ for n an integer ≥ 0 and $TM_{1,2n}$ for $n\geq 1$.

For the slot wave, $\lambda'<\lambda$ and hence $a<\lambda/2$. Therefore, the TE_{10} and all higher modes are cut off, or nonpropagating, in the air regions. In the dielectric region the TE_{10} mode is propagating, and the first few higher modes may propagate or all higher modes may be cut off, depending on the size of b. Since all modes are cut off in the two air regions, the energy of the resonant slot-wave mode is trapped near the slot. The amplitude of each mode in each region must be such that when the full set of modes are superimposed, the boundary conditions in the iris plane at $z=0$ will be met, and all field components on either side of the dielectric-to-air interface at $z=d$ will be matched. An alternative but equivalent condition is that transverse resonance occur; that is, the sum of the susceptances at the iris plane must equal zero. This sum includes the susceptances of the TE_{10} mode looking in the $-z$ and $+z$ directions, and the capacitive-iris susceptance representing higher modes on the $-z$ and $+z$ sides of the iris.

A formula for characteristic impedance is derived in Appendix II by a method utilizing the total-susceptance formula. Because of the non-TEM nature of the slot wave, definition of characteristic impedance is somewhat arbitrary. The reasonable and useful definition chosen here is $Z_0=V^2/2P$, where $V=-\int E_y dy$ is peak voltage amplitude across the slot and P is average power flow of the wave. As in waveguide, this definition does not necessarily yield the best match in a transition to coaxial or microstrip line. In fact, experimental results indicate that a 50 ohm coaxial or microstrip line requires about 75 ohm slot impedance computed by the voltage power definition [19], [20]. Reactive discontinuity effects may also play a part in this discrepancy.

IV. Parameter Formulas—Second-Order Solution

Define as an independent variable $p=\lambda/2a$. At the transverse-resonance frequency, $a=\lambda'/2$ and hence $p=\lambda/\lambda'$ for $B_t=0$ where B_t is the total susceptance at the plane of the slot. Once $p=\lambda/\lambda'$ has been determined as the solution of $B_t=0$ for a given set of parameters ϵ_r, w, d, b, and a, the wavelength and frequency for this solution are simply $\lambda=2a(\lambda/\lambda')$ and $f=c/\lambda$.

The formula for B_t is as follows for the case of electric walls at $y=\pm b/2$:

$$\eta B_t = \frac{a}{2b}\left[-v + u\tan\left(\frac{\pi d u}{ap} - \tan^{-1}\frac{v}{u}\right)\right] + \frac{1}{p}\left\{\left(\frac{\epsilon_r+1}{2} - p^2\right)\ln\frac{2}{\pi\delta} + \frac{1}{2}\cdot\sum_{n=1,2,3,\cdots}\left[v^2\left(1-\frac{1}{F_n}\right) + M_n\right]\frac{\sin^2(\pi n\delta)}{n(\pi n\delta)^2}\right\}, \tag{10}$$

and as follows for magnetic walls at $y=\pm b/2$:

$$\eta B_t = \frac{1}{p}\left\{\left(\frac{\epsilon_r+1}{2} - p^2\right)\ln\frac{8}{\pi\delta} + \frac{1}{2}\cdot\sum_{n=\frac{1}{2},\frac{3}{2},\frac{5}{2},\cdots}\left[v^2\left(1-\frac{1}{F_n}\right) + M_n\right]\frac{\sin^2(\pi n\delta)}{n(\pi n\delta)^2}\right\} \tag{11}$$

where $\eta=\sqrt{\mu_0/\epsilon_0}=376.7$ ohms, $\delta=w/b$, and

$$u = \sqrt{\epsilon_r - p^2}, \qquad v = \sqrt{p^2-1} \tag{12}$$

$$F_n = \sqrt{1+\left(\frac{b}{2an}\cdot\frac{v}{p}\right)^2}, \qquad F_{n1} = \sqrt{1-\left(\frac{b}{2an}\cdot\frac{u}{p}\right)^2}. \tag{13}$$

For F_{n1} real, M_n is

$$M_n = \frac{\epsilon_r\tanh r_n - p^2F_{n1}^2\coth q_n}{\left[1+\left(\frac{b}{2an}\right)^2\right]F_{n1}} - u^2 \tag{14}$$

where

$$r_n = \frac{2\pi n d F_{n1}}{b} + \tanh^{-1}\left(\frac{F_{n1}}{\epsilon_r F_n}\right) \tag{15}$$

$$q_n = \frac{2\pi n d F_{n1}}{b} + \coth^{-1}\left(\frac{F_n}{F_{n1}}\right). \tag{16}$$

For F_{n1} imaginary, M_n is

$$M_n = \frac{\epsilon_r\tan r_n' - p^2|F_{n1}|^2\cot q_n'}{\left[1+\left(\frac{b}{2an}\right)^2\right]|F_{n1}|} - u^2 \tag{17}$$

where

$$r_n' = \frac{2\pi n d|F_{n1}|}{b} + \tan^{-1}\left(\frac{|F_{n1}|}{\epsilon_r F_n}\right) \tag{18}$$

$$q_n' = \frac{2\pi n d|F_{n1}|}{b} + \cot^{-1}\left(\frac{F_n}{|F_{n1}|}\right). \tag{19}$$

Equations (10) and (11) for B_t are valid for $\delta = w/b \leq 0.15$, $w < \lambda/(4\sqrt{\epsilon_r})$, and $w \leq d$. These ranges are more than adequate for usual slot-line dimensions.

The procedure in using the above equations is to substitute a set of values ϵ_r, w, d, b, and $a = \lambda'/2$, and then to solve (10) or (11) for the value of p at which $\eta B_t = 0$. This p is equal to λ/λ'. Wavelength and frequency are then given by $\lambda = 2a(\lambda/\lambda')$ and $f = c/\lambda$. Solution of the above equations is most conveniently accomplished with an electronic computer, arriving at $B_t = 0$ by an iterative process.

The group velocity is $v_g = d\omega/d\beta$, where $\omega = 2\pi f$ and β is the slot-wave phase constant in radians per unit length. From this definition of v_g, one may obtain the following relations for the ratio of phase velocity to group velocity:

$$\frac{v}{v_g} = 1 + \frac{f}{\lambda'/\lambda} \cdot \frac{-\Delta(\lambda'/\lambda)}{\Delta f} = 1 + \frac{f}{\lambda/\lambda'} \cdot \frac{\Delta(\lambda/\lambda')}{\Delta f} \tag{20}$$

where $\Delta(\lambda'/\lambda)$, $\Delta(\lambda/\lambda')$, and Δf are computed from two separate solutions of $\eta B_t = 0$ for fixed values of ϵ_r, w, d, and b, and for two slightly different values of $a = \lambda'/2$ incremented plus and minus from the desired a. The frequency f may be assumed to lie midway in the Δf interval.

The slot-wave characteristic impedance is obtained as follows from the ηB_t formulas:

$$Z_0 = 376.7 \frac{v}{v_g} \frac{\pi}{p} \cdot \frac{\Delta p}{-\Delta \eta B_t} \text{ ohms.} \tag{21}$$

In this equation $\Delta \eta B_t$ is computed from (10) or (11) with ϵ_r, w, d, b, and a held constant, and with p incremented slightly plus and minus from the value $p = \lambda/\lambda'$ at $\eta B_t = 0$; v/v_g is obtained from (20) for the same parameters. Equations (20) and (21) may be used with (10) to yield values for electric walls at $y = \pm b/2$, or with (11) for magnetic walls.

V. Computed Data

The second-order solution equations in Section IV were programmed on an electronic computer and the parameters λ'/λ, v/v_g, and Z_0 were computed for various slot-line cases.[1]

Fig. 9 shows the effect of wall spacing b on the parameters for both electric and magnetic walls. In this computation, $\epsilon_r = 20$, $d = 0.137$ inch, $w = 0.025$ inch, and λ' is constant at 1.360 inches. The corresponding curves for electric and magnetic walls merge together for $b > 1.5$ inches and are only slightly separated for $b = 1$ inch. As b decreases below 1 inch, the λ'/λ curves diverge more and more rapidly. Since λ' is held constant, λ varies inversely and f proportionally with λ'/λ. The ratio v/v_g decreases with b for magnetic walls and increases for electric walls. The behavior of the characteristic impedance is more complex, the electric-wall curve rising and then falling sharply as b decreases, and the magnetic-wall curve falling and then rising sharply. This reversal of direction is caused by the v/v_g factor in (21). The quantity $Z_0/(v/v_g)$ is virtually constant for the two types of walls for

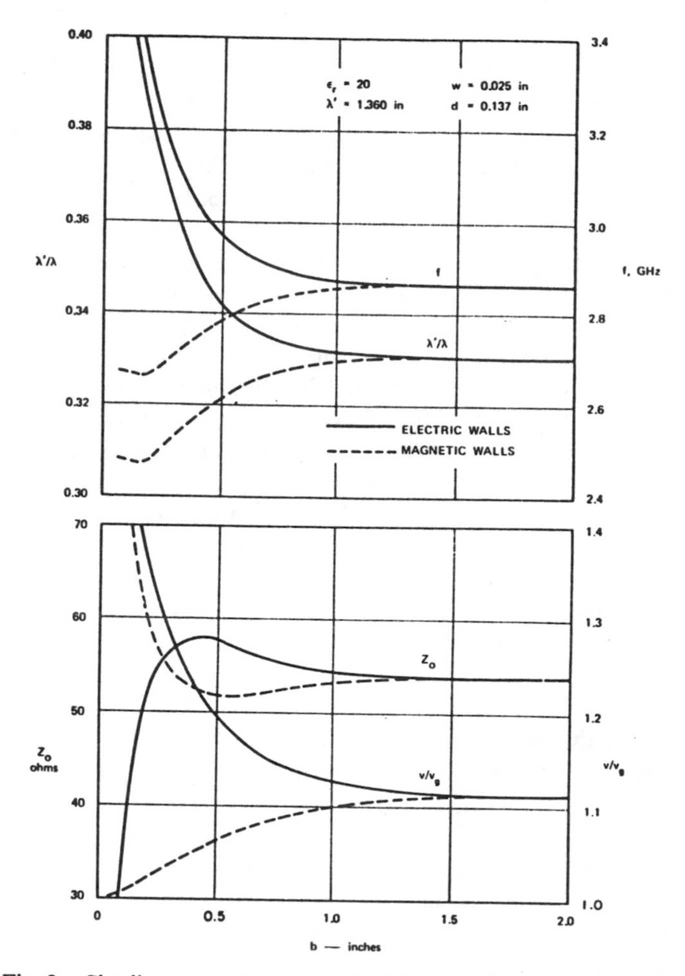

Fig. 9. Slot-line parameters versus b with ϵ_r, w, d and $a = \lambda'/2$ fixed.

b as low as 0.6 inch, while for b smaller, this quantity diverges without changing direction.

The variation of Z_0 and λ'/λ with frequency is shown in Fig. 10 for $\epsilon_r = 20$, $d = 0.137$ inch, $w = 0.025$ inch, and for $\lambda' = 2a$ stepped from 0.5 to 150 inches in the computation. Curves are plotted for $b = 3$ inches and $b \to \infty$ (that is, b sufficiently large at each λ' value such that Z_0 and λ'/λ are essentially independent of b). The ratio λ'/λ decreases monotonically in the plotted range. For increasing f, λ'/λ falls toward $1/\sqrt{\epsilon_r} = 0.224$, while for f approaching zero λ'/λ rises to 0.390 for magnetic walls at $b = 3$ inches and to 1.0 for the two other curves. The zero-order value of λ'/λ calculated from (1) is indicated for comparison.

The characteristic impedance plotted in Fig. 10 has a broad maximum, varying by only ± 7 percent between 1 and 6 GHz. As f approaches zero, Z_0 approaches 40.3 ohms for magnetic walls at $b = 3$ inches, and approaches 0 ohms for the other two cases.

A few experimental λ'/λ points are shown in Fig. 10. The substrate used in the test piece had a specified ϵ_r of 20. Length, width, and thickness were approximately 5 by 1.75 by 0.137 inches. For the circled points one side was covered with Scotch Brand No. 51 aluminum tape, about 0.0007 inch thick, including 0.0001 inch of adhesive. A slot 0.025 inch wide and 3 inches long was left bare. A probe terminating a

[1] Normalized graphs of λ'/λ and Z_0 versus d/λ for $w/d = 0.02$, 0.1, 0.2, 0.4, and 0.6, and for $\epsilon_r = 13$, 16, and 20 have been prepared and are included in another paper [19].

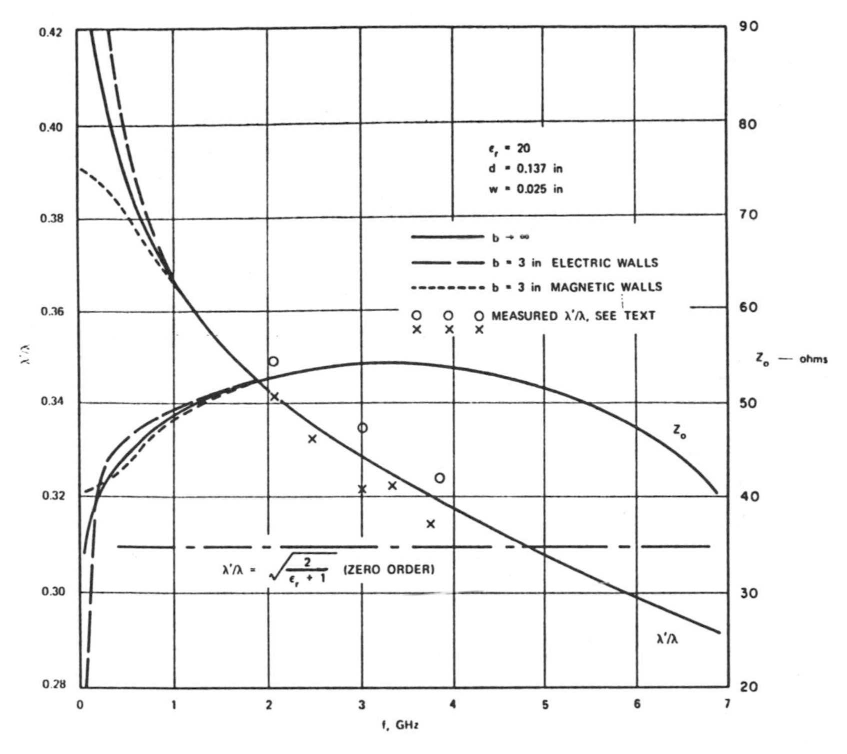

Fig. 10. Graph of λ'/λ and Z_0 versus frequency.

reflectometer coupler was coupled to the slot near one end. Resonance was indicated by a dip in the reflected wave. The slot length was varied by sliding a thin steel scale over the slot from the end opposite the probe. At each of the test frequencies the overlapping length of steel scale was varied until resonance was found. Several lengths for resonance were obtained differing in $\lambda'/2$ steps, thus yielding λ' and λ'/λ. The circled points lie parallel to the theoretical λ'/λ curve, but about 2.3 percent above it. For the crossed points, the same substrate sample was tested with copper plating directly applied. The thickness was about 0.0008 inch and slot width was 0.0235 inch. These points lie about 1.1 percent below the theoretical curve. The difference between the two sets of points may be attributed to the thin layer of adhesive on the aluminum tape. Additional tests with other permittivities and other w, d, and λ yield similar agreement with λ'/λ computations [19].

The effect of varying slot width w on λ'/λ and Z_0 is shown in Fig. 11 for two different cases of ϵ_r and d. In each case λ' is held constant. The ratio λ'/λ increases with w, but only slightly. The characteristic impedance Z_0 increases substantially, although far less than in proportion to w.

Measurements of Z_0 have not yet been made.[2] However, the computed Z_0 values for the magnetic-wall case can be checked when $b/\lambda' \ll 1$, since in the static limit this slot-line cross section becomes equivalent to a TEM-mode transmission line containing an equivalent dielectric medium $\epsilon_r' = (\lambda/\lambda')^2$. The Z_0 formula for this TEM line may be obtained by a straightforward modification of a case treated in [18]:

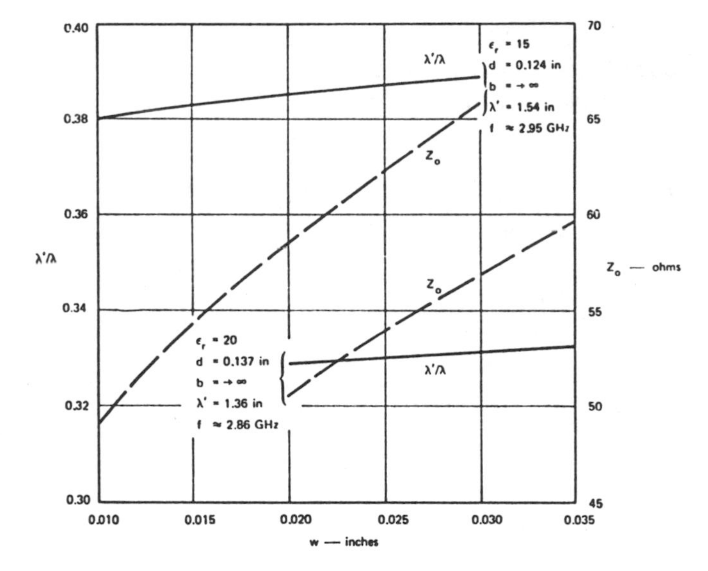

Fig. 11. Graph of λ'/λ and Z_0 versus slot width.

$$Z_0 = \frac{591.7}{\sqrt{\epsilon_r'}\ \ln\left(\frac{8b}{\pi w}\right)} = \frac{591.7(\lambda'/\lambda)}{\ln\left(\frac{8b}{\pi w}\right)} \text{ ohms} \qquad \begin{array}{l} b/w > 3 \\ b/\lambda' \to 0. \end{array} \tag{22}$$

[2] Good transitions have been achieved at C and X bands between slot line and 50 ohm coaxial and microstrip lines with $\epsilon_r = 16$, $d = 0.055$ inch, and $w = 0.021$ inch [20]. Another good transition at S band had $\epsilon_r = 16$, $w = 0.031$ inch, and $d = 0.062$ inch [19]. See the discussion at the end of Section III on the arbitrary nature of slot-line Z_0 definition.

TABLE I

COMPARISON OF SECOND-ORDER AND STATIC SOLUTIONS FOR Z_0

Second-Order Solution			Eq. (22)
b (inches)	λ'/λ	Z_0 (Ω)	Z_0 (Ω)
0.10	0.30752	78.38	78.40
0.14	0.30707	68.35	68.37
0.20	0.30763	60.45	60.39
0.30	0.31230	54.60	54.04
0.40	0.31640	52.42	50.50
0.60	0.32384	51.77	46.59

In Table I, (22) is tested against data computed for the curves of Fig. 9 ($\epsilon_r=20$, $d=0.137$ inch, $w=0.025$ inch, $\lambda'=1.360$ inches, $f\approx 2.670$ GHz). Agreement is excellent for $b\leq 0.20$ inch or $b/\lambda'\leq 0.15$. A second test is afforded by a computed point used in plotting the graph in Fig. 10. For $\epsilon_r=20$, $d=0.137$ inch, $w=0.025$ inch, $b=3.00$ inches, and $\lambda'=150.0$ inches, the second-order solution gives $\lambda'/\lambda=0.38987$, $f=30.67$ MHz, and $Z_0=40.33$ ohms. Equation (22) yields 40.32 ohms.

APPENDIX I

DERIVATION OF B_t FORMULAS

Fig. 12 shows a longitudinal yz-plane view through the waveguide model for the slot-line analysis. The waveguide boundaries in this case are conducting, or electric, walls. The TE_{10}-mode susceptance at the iris plane ($z=0$) is B_d looking to the right into the dielectric slab and B_a looking to the left into the air region. The total susceptance in the iris plane is $B_t=B_d+B_a$.

A formula for B_d will be derived first and then modified to yield B_a. The approach is based on a previously published analysis of a waveguide filter consisting of an alternating series of steps between two cross-section heights [9], [10], and on earlier analyses applied to other discontinuities [11], [12]. Fig. 13 shows the basic structure treated in [9] and [10], with notation modified for slot line. A waveguide of height b is driven by a waveguide of height w, with an abrupt step at $z=0$. A reactive plane is placed transverse to the waveguide at $z=l$. This terminating plane can be either an electric or magnetic wall.

Since all walls and the dielectric material are assumed lossless, the admittance looking to the right at $z=0$ is a pure susceptance, B_d. Because of symmetry, only $TE_{1,2n}$ (with $n\geq 0$) and $TM_{1,2n}$ (with $n\geq 1$) modes are present. In a transverse plane the E_y and H_x components of each mode are proportional to $\sin \pi x/a \cos 2\pi ny/b$. Thus the total E_y and H_x fields at the $z=0$ plane and $x=a/2$ are functions of y as follows:

$$E_y = R_0 + \sum_{n>0} R_n \cos\frac{2\pi ny}{b} \tag{23}$$

$$H_x = -y_{i0}R_0 - \sum_{n>0} y_{in}R_n \cos\frac{2\pi ny}{b} \tag{24}$$

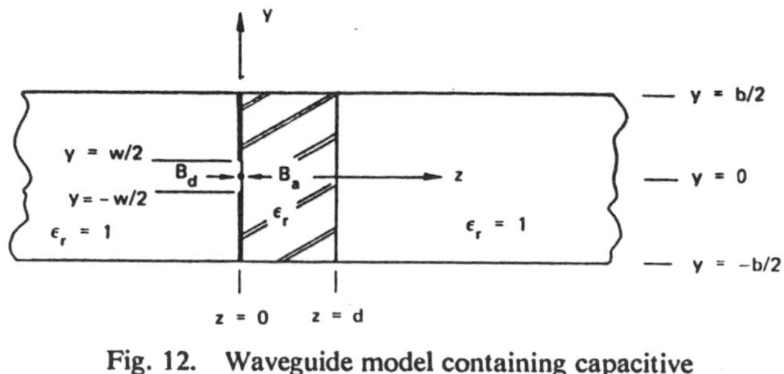

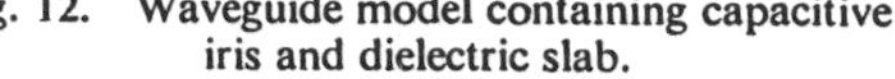

Fig. 12. Waveguide model containing capacitive iris and dielectric slab.

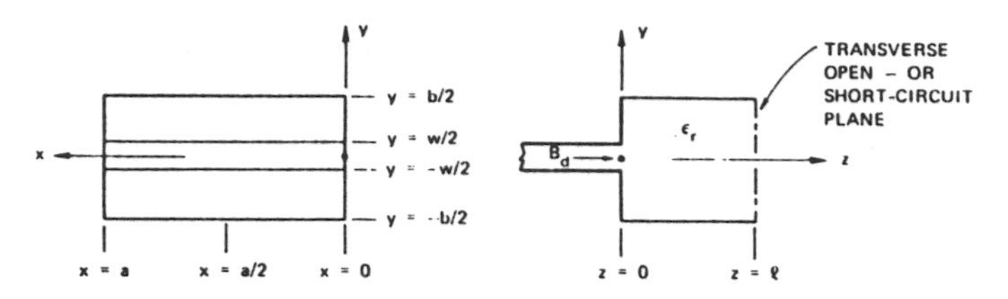

Fig. 13. Waveguide of height b driven by waveguide of height w.

where $n=1, 2, 3, \cdots$, R_0 and R_n are constants, and input wave admittances y_{i0} and y_{in} are defined by

$$y_{i0} = -\left(\frac{H_x}{E_y}\right)_{TE_{10}} \tag{25}$$

$$y_{in} = -\left(\frac{(H_x)_{TE_{1,2n}} + (H_x)_{TM_{1,2n}}}{(E_y)_{TE_{1,2n}} + (E_y)_{TM_{1,2n}}}\right). \tag{26}$$

The analysis will be simplified by assuming that only the TE_{10} mode is present in the waveguide of height w at $z=0$. Therefore, E_y and H_x are constants as follows:

$$E_y = \begin{cases} C_0 & \text{for } |y| \leq w/2 \\ 0 & \text{for } w/2 < |y| \leq b/2 \end{cases} \tag{27}$$

$$H_x = -C_0 y_i' \quad \text{for } |y| \leq w/2 \tag{28}$$

where y_i' is input wave admittance in the waveguide of height w at the plane $z=0$. The error in the analysis due to neglecting higher modes in the region $z<0$ is very small for $\delta=w/b$ small, as would be usual for slot line.

Equation (23) has the form of a Fourier series. Let this series equal E_y as defined by (27). Then R_0 and R_n are determined as follows:

$$R_0 = C_0\delta, \qquad R_n = 2C_0\delta\frac{\sin \pi n\delta}{\pi n\delta}\cdot \tag{29}$$

Next, set the right-hand sides of (24) and (28) equal, and integrate with respect to y over the opening of height w to obtain

$$C_0 y_i' = y_{i0}R_0 + \sum_{n>0} y_{in}R_n\frac{\sin \pi n\delta}{\pi n\delta} \tag{30}$$

and, by (29),

$$(b/w)y_i' = y_{i0} + 2\sum_{n>0} y_{in}\left(\frac{\sin \pi n\delta}{\pi n\delta}\right)^2. \tag{31}$$

At this point replace the wave admittances by guide admittances defined on the TE_{10}-mode voltage power basis in the complete waveguide cross section [13]; that is, replace y_{in} by $Y_{in}=(a/2b)y_{in}$ and y_i' by $Y_i=(a/2w)y_i'$. Then

$$Y_i = jB_d = Y_{i0} + 2\sum_{n>0} Y_{in}\left(\frac{\sin \pi n\delta}{\pi n\delta}\right)^2. \tag{32}$$

With reference to Fig. 12, Y_{i0} is the admittance seen by a TE_{10} wave directed into a dielectric-filled waveguide region of length d terminated by an infinite air-filled region:

$$Y_{i0} = jY_{o1}\tan\left(\beta_1 d + \tan^{-1}\frac{Y_0}{jY_{o1}}\right). \tag{33}$$

In the notation used in (33), symbols with the subscript 1 apply to the dielectric-filled region between $z=0$ and $z=d$; without this subscript they apply to the air regions. Y_{o1} and Y_o are TE_{10}-mode characteristic admittances and $\beta_1=\gamma_1/j=2\pi/\lambda_{g1}$ is the TE_{10}-mode phase constant.

$$Y_0 = \frac{a\gamma}{j2b\eta k}, \quad Y_{o1} = -j\frac{a\gamma_1}{2b\eta_1 k_1} = \frac{a}{2b\eta}\cdot\frac{\lambda}{\lambda_{g1}} \tag{34}$$

where $\eta=\sqrt{\mu_0/\epsilon_0}=376.7$ ohms and $\eta_1=\eta/\sqrt{\epsilon_r}$; γ and γ_1 are the TE_{10}-mode z-directed propagation constants; $k=2\pi/\lambda$ and $k_1=2\pi\sqrt{\epsilon_r}/\lambda$ are the plane-wave constants; and λ_{g1} is the TE_{10}-mode guide wavelength in the ϵ_r region.

For $n>0$, both $TE_{1,2n}$ and $TM_{1,2n}$ modes are present. For each n, the corresponding TE and TM amplitudes must be chosen so that E_x exactly cancels at $z=0$. In this way the total E_x field will be zero at $z=0$, as is required by the boundary conditions in that transverse plane. From (26) we obtain for the nth mode

$$Y_{in} = \frac{Y_{iTMn} + Y_{iTEn}D_n}{1 + D_n} \tag{35}$$

where

$$D_n = \frac{(E_y)_{TE1,2n}}{(E_y)_{TM1,2n}} \text{ when } (E_x)_{TE1,2n} + (E_x)_{TM1,2n} = 0. \tag{36}$$

From the field-component formulas for the TE and TM modes [7], [9] we obtain

$$D_n = \left(\frac{b}{2an}\right)^2. \tag{37}$$

The input admittances Y_{iTMn} and Y_{iTEn} for each n are[3]

$$Y_{iTMn} = Y_{o1TMn}\tanh\left(\gamma_{n1}d + \tanh^{-1}\frac{Y_{oTMn}}{Y_{o1TMn}}\right) \tag{38}$$

$$Y_{iTEn} = Y_{o1TEn}\coth\left(\gamma_{n1}d + \coth^{-1}\frac{Y_{oTEn}}{Y_{o1TEn}}\right) \tag{39}$$

where the characteristic admittances and propagation constants are

$$Y_{o1TMn} = \frac{jak_1}{2b\eta_1\gamma_{n1}}, \quad Y_{oTMn} = \frac{jak}{2b\eta\gamma_n} \tag{40}$$

$$Y_{o1TEn} = \frac{a\gamma_{n1}}{j2b\eta_1 k_1}, \quad Y_{oTEn} = \frac{a\gamma_n}{j2b\eta k} \tag{41}$$

$$\gamma_{n1} = \sqrt{\left(\frac{2\pi n}{b}\right)^2 + \left(\frac{\pi}{a}\right)^2 - \left(\frac{2\pi}{\lambda}\right)^2\epsilon_r} = \frac{2\pi n}{b}\sqrt{1-\left(\frac{b}{n\lambda_{g1}}\right)^2} \tag{42}$$

$$\gamma_n = \frac{2\pi n}{b}\sqrt{1+\left(\frac{b\gamma}{2\pi n}\right)^2}. \tag{43}$$

Then (35) and (37) through (41) yield

$$\eta Y_{in} = j\frac{ak}{2b\gamma_{n1}}\cdot\frac{\epsilon_r\tanh r_n - \left(\frac{b\gamma_{n1}}{2ank}\right)^2\coth q_n}{1+\left(\frac{b}{2an}\right)^2} \tag{44}$$

$$r_n = \gamma_{n1}d + \tanh^{-1}\left(\frac{\gamma_{n1}}{\epsilon_r\gamma_n}\right) \tag{45}$$

$$q_n = \gamma_{n1}d + \coth^{-1}\left(\frac{\gamma_n}{\gamma_{n1}}\right). \tag{46}$$

Now let $p=\lambda/2a$ and

$$u = \frac{\lambda}{\lambda_{g1}} = \frac{\gamma_1}{jk} = \sqrt{\epsilon_r - p^2} \tag{47}$$

$$v = \frac{\gamma}{k} = \sqrt{p^2-1} \tag{48}$$

$$F_{n1} = \frac{b\gamma_{n1}}{2\pi n} = \sqrt{1-\left(\frac{b}{n\lambda_{g1}}\right)^2} = \sqrt{1-\left(\frac{bu}{2anp}\right)^2} \tag{49}$$

$$F_n = \frac{b\gamma_n}{2\pi n} = \sqrt{1+\left(\frac{bv}{2anp}\right)^2}. \tag{50}$$

Equations (32) through (34) and (44) through (50) yield

$$\eta B_d = \frac{au}{2b}\tan\left(\frac{\pi ud}{ap} - \tan^{-1}\frac{v}{u}\right) + \frac{1}{2p} \cdot \sum_{n=1,2,\cdots}\left[\frac{\epsilon_r\tanh r_n - p^2F_{n1}^2\coth q_n}{\left[1+\left(\frac{b}{2an}\right)^2\right]F_{n1}}\right]\cdot\frac{\sin^2\pi n\delta}{n(\pi n\delta)^2}. \tag{51}$$

The first term is the TE_{10}-mode susceptance; the second term is the discontinuity susceptance representing the effect of higher modes. The rate of convergence of the series is very slow but will be improved by the following procedure. If we let $d\to\infty$ and $f\to 0$, then $\lambda'=2a\to\infty$ and each term S_n of the series becomes

$$S_n' = \frac{u^2\sin^2\pi n\delta}{n(\pi n\delta)^2}. \tag{52}$$

[3] In (38) and (39), the pair of functions tanh and $\tanh^{-1}$ is interchangeable with coth and $\coth^{-1}$. The selection here is appropriate since $Y_{oTMn}/Y_{o1TMn}<1$ and $Y_{oTEn}/Y_{o1TEn}\geq 1$.

The original series may be replaced by

$$\sum_{n=1,2,\cdots} S_n = \sum_{n=1,2,\cdots} (S_n - S_n') + \sum_{n=1,2,\cdots} S_n'. \quad (53)$$

The first series on the right of the equal sign converges rapidly compared to the series in (51), while the second may be summed in closed form by the following identity [11], [15]:

$$\lim_{\delta\to 0} \sum_{n=1,2,3,\cdots} \frac{\sin^2 \pi n\delta}{n(\pi n\delta)^2} = \ln \frac{1}{2\pi\delta} + \frac{3}{2} = \ln\frac{1}{\delta} - 0.3379. \quad (54)$$

The accuracy of this is excellent for $\delta \leq 0.15$. At $\delta = 0.15$ the error is only 0.4 percent. Therefore, the susceptance of both sides of a capacitive iris is approximately

$$\eta B_C \approx \frac{u^2}{p}\left(\ln\frac{1}{\delta} - 0.3379\right). \quad (55)$$

However, the exact formula in the limit $\lambda_{g1} \to \infty$ is [14]

$$B_C = \frac{4bY_{o1}}{\lambda_{g1}} \ln \csc \frac{\pi\delta}{2} = \frac{u^2}{p} \ln \csc \frac{\pi\delta}{2}, \quad \delta \leq 1$$
$$\approx \frac{u^2}{p} \ln \frac{2}{\pi\delta} = \frac{u^2}{p}\left(\ln\frac{1}{\delta} - 0.452\right), \quad \delta \ll 1. \quad (56)$$

The small difference between the constants in (55) and (56) is the result of assuming in the analysis that E_y and H_x are constant across the slot instead of being the correct functions of y. At this point (53) may be written

$$\sum_{n=1,2,\cdots} S_n = \sum_{n=1,2,\cdots} (S_n - S_n') + u^2 \ln \frac{2}{\pi\delta} \quad (57)$$

where (56) is used instead of (55) for better accuracy. Equation (51) for the susceptance looking into the dielectric section is now

$$\eta B_d = \frac{au}{2b} \tan\left(\frac{\pi ud}{ap} - \tan^{-1}\frac{v}{u}\right) + \frac{u^2}{2p}\ln\frac{2}{\pi\delta} + \frac{1}{2p} \cdot \sum_{n=1,2,\cdots}\left[\frac{\epsilon_r \tanh r_n - p^2F_{n1}^2 \coth q_n}{\left[1+\left(\frac{b}{2an}\right)^2\right]F_{n1}} - u^2\right]\cdot\frac{\sin^2 \pi n\delta}{n(\pi n\delta)^2}. \quad (58)$$

The susceptance B_a looking toward the left from the iris in Fig. 12 may be obtained from (58) by letting $\epsilon_r = 1$ or $d = 0$. With either substitution, the following equation results:

$$\eta B_a = -\frac{av}{2b} - \frac{v^2}{2p}\ln\frac{2}{\pi\delta} + \frac{1}{2p}\sum_{n=1,2,\cdots} v^2\left(1 - \frac{1}{F_n}\right)\frac{\sin^2 \pi n\delta}{n(\pi n\delta)^2}. \quad (59)$$

When (58) and (59) are added, the total susceptance B_t given by (10) is obtained for electric walls at $y = \pm b/2$.

The susceptance B_t for the case of magnetic walls at $y = \pm b/2$ may be obtained easily from the electric-wall solution. Let the cross section be as shown in Fig. 8(c). Modes excited in the waveguide are TE_{11}, TE_{13}, TE_{15}, $\cdots$ and TM_{11}, TM_{13}, TM_{15}, $\cdots$; that is, $TE_{1,2n}$ and $TM_{1,2n}$ where $n = \frac{1}{2}, \frac{3}{2}, \frac{5}{2}$, etc. The TE_{10} mode cannot exist within this boundary. Careful examination of each step of the above analysis for electric walls shows that (51) applies when the first term representing the TE_{10} contribution is dropped, and when the summation is performed for $n = \frac{1}{2}, \frac{3}{2}, \frac{5}{2}, \cdots$ rather than $n = 1, 2, 3, \cdots$. Evaluation of (54) with $n = \frac{1}{2}, \frac{3}{2}, \frac{5}{2}, \cdots$, yields [15]

$$\lim_{\delta\to 0} \sum_{n=\frac{1}{2},\frac{3}{2},\cdots} \frac{\sin^2 \pi n\delta}{n(\pi n\delta)^2} = \ln\frac{4}{\delta} - 0.3379. \quad (60)$$

Therefore, in (58) for ηB_d and (59) for ηB_a drop the first term, replace $\ln 2/\pi\delta$ by $\ln 8/\pi\delta$, and change the summation index from $n = 1, 2, 3, \cdots$ to $n = \frac{1}{2}, \frac{3}{2}, \frac{5}{2}, \cdots$. The resulting expression giving ηB_t for magnetic walls at $y = \pm b/2$ is (11).

Appendix II

Derivation of Formulas for Z_0 and v/v_g

Define the slot-line characteristic impedance Z_0 by

$$Z_0 = \frac{V_+^2}{2P_+} \quad (61)$$

where P_+ is the average power flow of a slot wave traveling in the $+x$ direction (Fig. 8) and V_+ is the peak amplitude of the voltage across the slot. Now assume a resonant length $\lambda'/2$ of slot line having waves of equal power P^+ and P^- traveling in the $+x$ and $-x$ directions. The total stored energy in this length is $W_t = (P^+ + P^-)(\lambda'/2v_g) = P^+\lambda'/v_g = (2\pi P^+/\omega')(v/v_g)$, where v_g is group velocity, or velocity of energy transport along the slot line [16]. Let V_0 be the maximum voltage at the center of the resonant length of slot. Since the waves in the $+x$ and $-x$ directions have equal voltages $V_+ = V_-$, the maximum voltage is $V_0 = 2V_+$, and $Z_0 = (V_0/2)^2/2P^+ = \pi V_0^2 v/4\omega W_t v_g$.

The following general relation holds at a port of a cavity at resonance [16]: $W_t = (V^2/4)(dB/d\omega)$. In the case of the $\lambda'/2$ resonant slot, we shall assume the port to be at the iris plane $z = 0$ in Fig. 8. We shall set B equal to the total waveguide susceptance B_t at that plane and V equal to the slot voltage V_0 at $x = \lambda'/4$. These choices of B and V are consistent with the waveguide impedance and slot impedance definitions, both of which are on a voltage power basis. Therefore,

$$Z_0 = \frac{\pi}{\omega(dB_t/d\omega)} \cdot \frac{v}{v_g} \quad (62)$$

where $dB_t/d\omega$ is evaluated at the resonant frequency; that is, at $B_t = 0$. Let $\omega = 2\pi c/\lambda$ and $p = \lambda/2a$. Then $\omega = \pi c/ap$ and $d\omega = -(\pi c/ap^2)dp = -\omega(dp/p)$. Substitution of these relations in (62) yields (21).

The ratio v/v_g will now be evaluated. Phase and group velocity are given by $v = \omega/\beta_x = f\lambda'$ and $v_g = d\omega/d\beta_x$ [16], [17] where β_x is the slot-wave phase constant. Since $\beta_x = 2\pi/\lambda'$, we obtain $v_g = -\lambda'^2/(d\lambda'/df)$. Differentiate λ'/λ as follows:

$$\frac{d(\lambda'/\lambda)}{df} = \frac{1}{\lambda}\frac{d\lambda'}{df} - \frac{\lambda'}{\lambda^2}\frac{d\lambda}{df}.$$

Solve this for $d\lambda'/df$ and substitute $d\lambda/df = d(c/f)/df = -\lambda/f$. The resulting relations yield the first part of (20). The second part is obtained in a similar manner.

Acknowledgment

The author wishes to thank E. G. Cristal of Stanford Research Institute for preparing the computer program of the second-order solution, and J. P. Agrios, C. Heinzman and E. A. Mariani of U. S. Army Electronics Command for their experimental studies of slot line. The above people, and also L. A. Robinson and L. Young of Stanford Research Institute, participated in numerous discussions with the author.

References

[1] H. Jasik, Ed., *Antenna Engineering Handbook*. New York: McGraw-Hill, 1961, chs. 8, 9.

[2] A. F. Harvey, *Microwave Engineering*. New York: Academic Press, 1963, pp. 633–638.

[3] W. H. Watson, *Waveguide Transmission and Antenna Systems*. New York: Oxford University Press, 1947.

[4] E. Strumwasser, J. A. Short, R. J. Stegen, and J. R. Miller, "Slot study in rectangular TEM transmission line," Hughes Aircraft Company, Tech. Memo 265, Air Force Contract AF 19(122)-454, January 1952.

[5] M. C. Bailey, "Design of dielectric-covered resonant slots in a rectangular waveguide," *IEEE Trans. Antennas and Propagation*, vol. AP-15, pp. 594–598, September 1967.

[6] J. Galejs, "Excitation of slots in a conducting screen above a lossy dielectric half space," *IRE Trans. Antennas and Propagation*, vol. AP-10, pp. 436–443, July 1962.

[7] S. Ramo and J. R. Whinnery, *Fields and Waves in Modern Radio* 2nd ed. New York: Wiley, 1953, pp. 357–358.

[8] E. Jahnke and F. Emde, *Tables of Functions with Formulae and Curves*. New York: Dover, 1943, pp. 236–243.

[9] S. B. Cohn, "A theoretical and experimental study of a waveguide filter structure," Cruft Lab., Harvard University, Cambridge, Mass., Tech. Rept. 39, Contract N5 ORI-76, Task Order 1, April 25, 1948.

[10] ——, "Analysis of a wide-band waveguide filter," *Proc. IRE*, vol. 37, pp. 651–656, June 1949.

[11] W. C. Hahn, "A new method for the calculation of cavity resonators," *J. Appl. Phys.*, vol. 12, p. 62, 1941.

[12] J. R. Whinnery and H. W. Jamieson, "Equivalent circuits for discontinuities in transmission lines," *Proc. IRE*, vol. 32, pp. 98–115, February 1944.

[13] S. A. Schelkunoff, *Electromagnetic Waves*. Princeton, N. J.: Van Nostrand, 1943.

[14] N. Marcuvitz, *Waveguide Handbook*, M.I.T. Rad. Lab. Ser., vol. 10. New York: McGraw-Hill, 1951, pp. 218–219.

[15] R. E. Collin, *Field Theory of Guided Waves*. New York: McGraw-Hill, 1960. Use $\Sigma_{1,2,3,\cdots} e^{inx}/n^3$ on p. 579 and $\Sigma_{1,3,5,\cdots} e^{inx}/n^3$ on p. 580.

[16] C. G. Montgomery, R. H. Dicke, and E. M. Purcell, *Principles of Microwave Circuits*, M.I.T. Rad. Lab. Ser., vol. 8. New York: McGraw-Hill, 1948. See p. 230, eq. (40) for relation between W_t, V, and $dB_t/d\omega$; see p. 53, eqs. (103) and (104) for v and v_g.

[17] L. Brillouin, *Wave Propagation in Periodic Structures*. New York: Dover, 1953, pp. 72–76.

[18] F. Oberhettinger and W. Magnus, *Anwendung der Elliptischen Funktionen in Physik und Technik*. Berlin: Springer, 1949, pp. 63, 114–116.

[19] E. Mariani, C. Heinzman, J. Agrios, and S. B. Cohn, "Measurement of slot-line characteristics," presented at the 1969 IEEE G-MTT Internatl. Symp., Dallas, Tex., May 5–7, 1969, to be published in *IEEE Trans. Microwave Theory and Techniques*.

[20] G. H. Robinson and J. L. Allen, "Applications of slot line to miniature ferrite devices," presented at the 1969 IEEE G-MTT Internatl. Symp., Dallas, Tex., May 5–7, 1969.

Integrated Fin-Line Millimeter Components

PAUL J. MEIER, SENIOR MEMBER, IEEE

Abstract—**This paper reviews the characteristics of integrated fin-line, a low-loss transmission line which is compatible with batch-processing techniques and superior to microstrip in several respects at millimeter wavelengths. Relative to microstrip, fin-line can provide less stringent tolerances, greater freedom from radiation and higher mode propagation, better compatibility with hybrid devices, and simpler interfaces with waveguide instrumentation. Examples of solid-state and passive components are presented which illustrate the potential of integrated fin-line at millimeter wavelengths. The examples include a p-i-n attenuator which has demonstrated the capability of constructing low-loss semiconductor mounts in fin-line. A four-pole bandpass filter, which performs in close agreement with theory, is also discussed.**

INTRODUCTION

INCREASED activity in the spectrum above 30 GHz has recently stirred interest in the development of millimeter integrated circuits. Much of the enthusiasm associated with integrated circuits can be traced to the clear advantages that such circuits provide below 3 GHz, namely, reduced size, weight, and cost combined with improved electrical performance, production uniformity, and reliability. However, those who have worked with integrated circuits at centimeter wavelengths (3–30 GHz) have encountered some fundamental problems which have limited the utility of such circuits. These problems include the critical tolerances and questionable production uniformity that can occur when miniaturization is carried too far. Although enhanced performance is possible in centimeter integrated circuits through the reduction of parasitics and the elimination of superfluous interfaces, poorer overall performance is also possible. The fundamental microstrip problems, which generally increase in severity as the operating frequency is raised, include radiation loss, spurious coupling, dispersion, and higher mode propagation. Attempts to integrate a large number of components in a single housing have generally demonstrated the need for "mode barriers" and "box-resonance absorbers." Radiation and related problems can be controlled by choosing progressively thinner substrates as the operating frequency is raised, but this only serves to degrade the Q factor, compound tolerance problems, and restrict the range over which the characteristic impedance can be varied.

Although standard microstrip techniques can be applied to millimeter components [1], [2], the problems listed

Manuscript received May 8, 1974; revised August 30, 1974.
The author is with AIL, a Division of Cutler-Hammer, Melville, N. Y. 11746.

Reprinted from *IEEE Trans. Microwave Theory Tech.*, vol. MTT-22, no. 12, pp. 1209–1216, December 1974.

previously can be expected to become more severe. As the operating frequency is raised and the microstrip dimensions are decreased, a limit will be reached where the strip width is no longer compatible with chip and beam-lead devices. In addition, millimeter integrated circuits must be tailored to requirements that are generally different from those which apply at lower frequencies. For example, the ability to construct a simple transition to waveguide becomes important at millimeter wavelengths where coaxial instrumentation is not practical. Moreover, the miniaturization that proved to be an asset at centimeter wavelengths can become a liability in millimeter applications. Designers of millimeter equipment have historically selected quasi-optical approaches to *increase* the physical size of components and thereby ease tolerance problems and improve performance [3]. Thus the ideal transmission line for millimeter integrated circuits is one which avoids miniaturization and yet offers the potential for low-cost production through batch-processing techniques. Integrated fin-line [4]–[6] is such a transmission line.

After introducing the symbols to be utilized throughout this paper, the body of the text will review the characteristics of integrated fin-line and demonstrate how this transmission line can be profitably applied to solid-state and passive millimeter components. Conclusions will then be presented regarding the potential application of integrated fin-line to future millimeter systems.

NOMENCLATURE

a	Major inner dimension of fin-line housing.
b	Minor inner dimension of fin-line housing.
c	Thickness of dielectric substrate.
d	Gap between fins.
s	Equivalent ridge thickness.
B	Susceptance magnitude.
k_e	Equivalent dielectric constant.
Q	Unloaded quality factor.
Y_0	Characteristic admittance of fin-line.
Z_0	Characteristic impedance of fin-line $= 1/Y_0$.
R_0	Resistance of load matched to Z_0.
Z_∞	Characteristic impedance at infinite frequency.
λ_g	Guide wavelength.
λ_0	Free-space wavelength.
λ_c	Cutoff wavelength of equivalent ridged waveguide.
C_j,R_j	Capacitance and resistance of p-i-n junction.
L_s,R_s	Series parasitic inductance and resistance of p-i-n diode.
C_p	Capacitance of diode package.
L_{s1}	Series inductance external to diode package.
W	Width of inductive strip.

CHARACTERISTICS OF INTEGRATED FIN-LINE

In an integrated fin-line structure [4], metal fins are printed on a dielectric substrate which bridges the broad walls of a rectangular waveguide. This adaptation of ridge-loaded waveguide permits circuit elements to be photoetched by low-cost batch techniques. The dimensions of practical fin-line circuits remain compatible with chip and beam-lead devices throughout the millimeter spectrum, thereby offering great potential for the construction of active and passive hybrid integrated circuits. Examples of such circuits and a simple low-reflection transition to rectangular waveguide will be presented.

In addition to serving as the bonding areas for hybrid devices, the printed fins increase the separation between the first two modes of propagation thereby providing a wider useful bandwidth than conventional waveguide. Owing to the similarity between integrated fin-line and conventional ridged waveguide, considerable design information [7], [8] is available. For thin substrates of moderate permittivity, the dielectric will have a minor effect [9] and single-mode bandwidth and attenuation of integrated fin-line may be estimated from existing data [8]–[10]. Such estimates lead to the conclusion that integrated fin-line can provide bandwidths in excess of an octave with less attenuation than microstrip. Larger single-mode bandwidths are feasible at the expense of attenuation.

The low-loss properties of integrated fin-line have been experimentally demonstrated through cavity tests at centimeter wavelengths [5] and transmission tests at millimeter wavelengths [11]. Preliminary tests were conducted at X band where wide-band sweepers were more readily available. The substrate thickness, however, was chosen to be sufficiently large (0.062 in) to permit direct scaling to higher frequencies.

Two different fin-line configurations, as shown in Fig. 1, were evaluated. In the configuration of Fig. 1(a), the fins are printed on opposite sides of a single dielectric substrate. Since the fins are directly grounded to the metal waveguide walls, this configuration is applicable only to passive devices. In the configuration of Fig. 1(b), however, the fins are printed on the mating surfaces of two dielectric substrates. Since the fins are insulated from the waveguide at dc, bias may be introduced for active components. RF continuity between the fins and the waveguide wall is obtained by choosing the thickness of the broad walls to be a quarter wavelength in the dielectric medium and by selecting $c \ll a$. The resultant choke prevents radiation or TEM propagation across a bandwidth in excess of an octave.

Teflon-fiber glass (3M type FL-GT) substrates with normalized gaps ranging from $d/b = 0.1$ to 1.0 were evaluated in the cavity whose aspect ratio (b/a) was 0.45. The center frequency and 3-dB bandwidth were recorded for each sample and the results were analyzed to determine the manner in which the guide wavelength and unloaded Q varied.

Fig. 2 shows how the measured Q (including the effect of losses within the irises) varies with respect to the gap between the fins. For moderate loading ($d/b \geq 0.6$), the measured Q of integrated fin-line exceeds 400 for insulated fins, and is better than 700 for the grounded-fin configura-

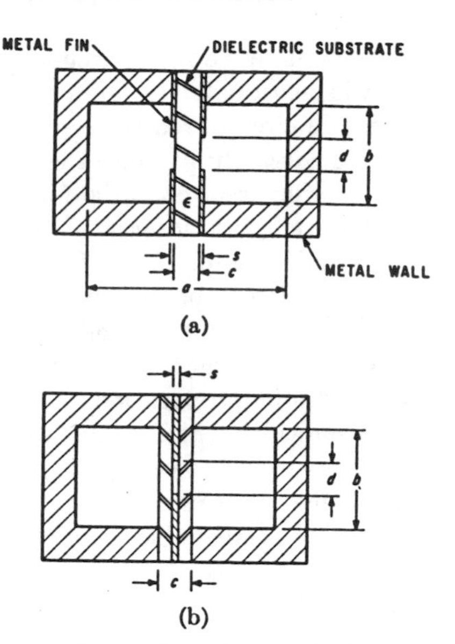

Fig. 1. Fin-line cavity configurations. (a) Grounded fins. (b) Insulated fins.

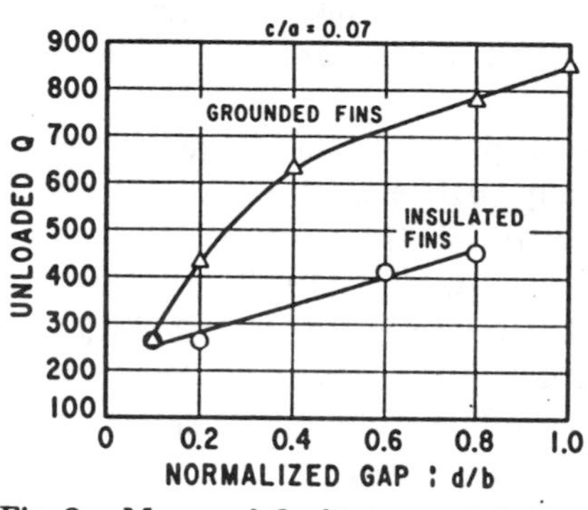

Fig. 2. Measured Q of integrated fin-line.

tion. These values compare favorably with those reported for practical microstrip lines at centimeter wavelengths [12].

To supplement the earlier measurements of unloaded Q performed in the 3-cm band, the insertion loss of integrated fin-line has been measured for a comparable fin-line structure in the 9-mm band [11]. A copper-clad laminate (Duroid 5880) with a thickness of 0.020 in was selected to maintain the ratio of c/a at 0.07. By accurately measuring the insertion loss of two fin-line transmission fixtures of different lengths, the loss was found to be 0.1 dB/wavelength for grounded fins with $d/b = 0.1$. This corresponds to an unloaded Q of 273, which is in agreement with the X-band cavity measurement of 260. Since the latter figure is somewhat pessimistic owing to loss in the coupling irises, it may be concluded that the loss in integrated fin-line holds fairly constant throughout the frequency range of 10–40 GHz. Although the loss can be expected to increase gradually as the frequency is raised above 40 GHz, the general utility of fin-line should be preserved throughout the millimeter spectrum.

Another important property of integrated fin-line is the guide-wavelength variation which can be calculated from the simple equation

$$\lambda_g = \frac{\lambda_0}{[k_e - (\lambda_0/\lambda_c)^2]^{1/2}} \tag{1}$$

where k_e is the equivalent dielectric constant, λ_0 is the free-space wavelength, and λ_c is the cutoff wavelength for an air-filled ridge-loaded waveguide of the same dimensions [8]. The value of k_e may be experimentally determined for a given fin-line configuration from a single cavity test [5]. It has been shown that the value of k_e varies little with respect to d/b when the fins are grounded directly to the waveguide. When d/b approaches unity, the value of k_e determined from the cavity agrees with the published value [10] for both the grounded and insulated cases. As the gap between the insulated fins becomes smaller, however, the value of k_e approaches the dielectric constant of the substrate material. Representative values of k_e appear in Table I.

To confirm the validity of (1) and demonstrate the capability of integrated fin-line at millimeter wavelengths, a transmission fixture was developed for operation in the frequency range of 26.5–40.0 GHz. Fig. 3 shows the cross section of this structure which is suitable for mounting semiconductor devices. In this structure, the upper fin is insulated from the housing at dc by a dielectric gasket, but is grounded at RF by choosing the thickness of the broad walls to be a quarter wavelength in the dielectric medium. The lower fin is grounded directly by a metal gasket to provide a dc return in solid-state applications. Nylon screws hold the halves of the housing together and align the substrate and its associated gaskets. Nylon is preferable to metal as the former has little effect on the RF choke formed within the broad walls.

The guide wavelength in the millimeter fin-line fixture has been measured across the band of 26.5–40.0 GHz by a sliding short-circuit technique. The fin-line was connected to a conventional slotted line through a low-reflection transition and the fin-line was internally terminated in a sliding short circuit of special design. The short consisted of a forked structure which straddled the dielectric substrate and provided a high standing wave ratio (SWR) by means of noncontacting chokes. The measurement was performed by recording the distance through which the short was moved in order to repeat nulls observed at a fixed location in the slotted line. Fig. 4 is a plot of the measured guide wavelength as a function of frequency for a typical structure ($d/b = 0.13$) suitable for mounting semiconductor devices. The value of k_e was determined by substituting into (1) the measured λ_g at 31 GHz ($\lambda_0 = 0.3807$ in) and the published value of λ_c [8] for an air-filled ridged waveguide of the same dimensions ($s/a \approx 0$, $a = 0.280$ in). Based upon this single-frequency measurement, the value of k_e was determined to be 1.31. Also shown in Fig. 4 is the curve calculated by applying (1) throughout the band for $k_e = 1.31$.The measured values of λ_g agree with the calculation ± 2 percent across the 40-percent instrumentation band.

To predict the behavior of semiconductor devices in fin-line, it is helpful to know the absolute value of the

TABLE I

EQUIVALENT DIELECTRIC CONSTANT

Dielectric Constant of Substrate Material	Fin-Line Configuration	s/a	d/b	Equivalent Dielectric Constant k_e
2.5	Figure 1A	0.07	0.1	1.50
2.5	Figure 1A	0.07	1.0	1.25
2.5	Figure 1B	0.07	0.1	2.10
2.5	Figure 1B	0.07	1.0	1.25
2.2	Figure 3	0.036	0.08	1.33
2.2	Figure 3	0.036	0.13	1.31

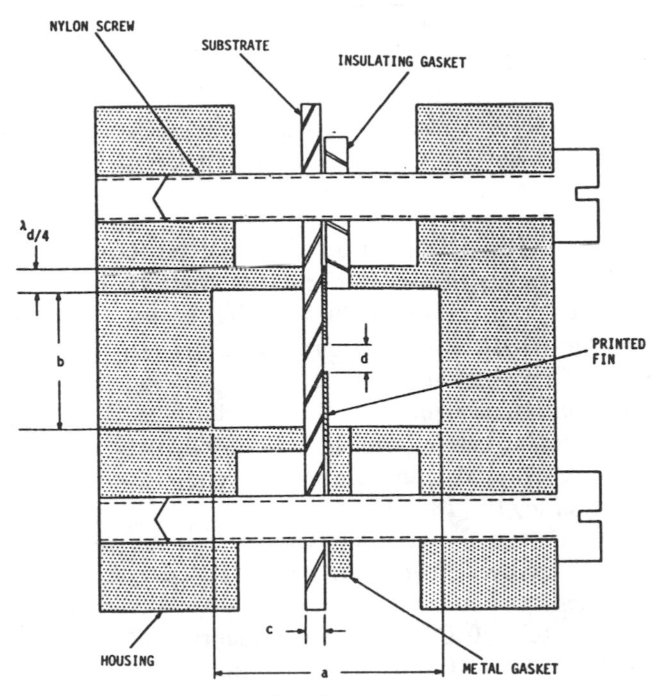

Fig. 3. Integrated fin-line semiconductor mount.

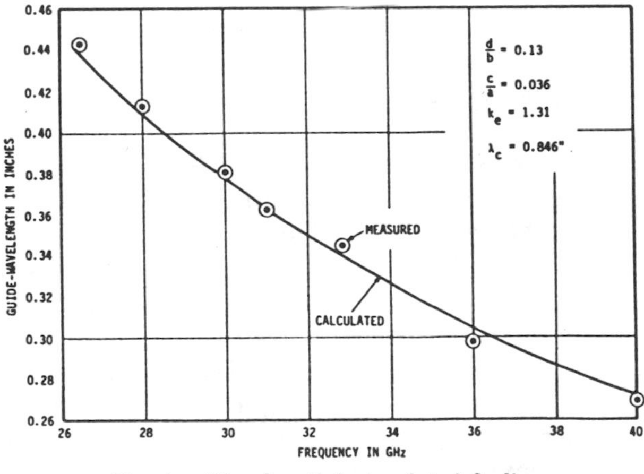

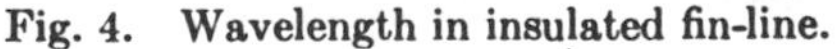
Fig. 4. Wavelength in insulated fin-line.

characteristic impedance ($Z_0 = 1/Y_0$). Values of the characteristic admittance of ridge-loaded waveguide, based upon a power-voltage definition, have been published for a wide range of parameters [8]. It has been found that these data are directly applicable to fin-line structures containing thin substrates ($c/a < 0.1$) of moderate permittivity. The characteristic impedance of fin-line may be approximated by

$$Z_0 \approx \frac{Z_\infty}{[k_e - (\lambda_0/\lambda_c)^2]^{1/2}} \tag{2}$$

where Z_∞ is the characteristic impedance of a ridge-loaded waveguide of the same dimensions at infinite frequency (that is, the reciprocal of the published admittance values). Appropriate factors may be applied to account for other aspect ratios (b/a), configurations (single versus double ridge), and impedance definitions [13].

Based upon a knowledge of the basic characteristics discussed previously, one may design a wide range of solid-state and passive components in integrated fin-line. The following sections will present examples of such components which serve to illustrate the advantages of fin-line construction techniques.

APPLICATION TO SOLID-STATE DEVICES

A millimeter demonstration of the compatibility of integrated fin-line with semiconductors has been performed with the aid of the previously discussed transmission fixture (Fig. 3). Additional features of this fixture appear in Fig. 5, wherein the fully assembled fixture is shown adjacent to a duplicate set of components. The housing mates directly with two standard WR28 waveguides and has identical inner dimensions (0.140 × 0.280 in). The substrate is cut from a 0.010-in sheet of laminate (Duroid 5880) and includes six mounting holes and two stepped edges. The latter protrude into the abutting WR28 waveguides and serve as quarter-wave transformers. After establishing a low-reflection transition between the WR28 waveguide and a slotted waveguide loaded by a dielectric slab ($d/b = 1.0$), the substrate metallization is tapered until the desired gap between the fins is obtained. The illustrated transitions at the ends of the substrate are each three wavelengths long near the middle of the instrumentation band (33 GHz). The measured VSWR of each transition is 1.2 or better across the 26.5–40-GHz band. It is believed that a more compact transition is feasible with quarter-wave transformers substituted for the fin-line taper. The substrate metallization also includes an RF blocking network connected to the upper fin.

To demonstrate the compatibility of integrated fin-line with semiconductor devices, two beam-lead p-i-n diodes (Alpha D5840B) were mounted across the fins near the center of the previously discussed substrate ($d/b = 0.13$). The diodes were spaced a quarter wavelength apart at a frequency near the upper end of the instrumentation band. The measured insertion loss of the p-i-n fin-line attenuator is plotted in Fig. 6 as a function of bias with frequency as a parameter. At the lower end of the band, where parasitics play a minor role, the reversed-bias insertion loss of the attenuator is only 0.3 dB, thereby demonstrating the capability of constructing low-loss semiconductor mounts in fin-line. As the bias is varied in the forward direction,

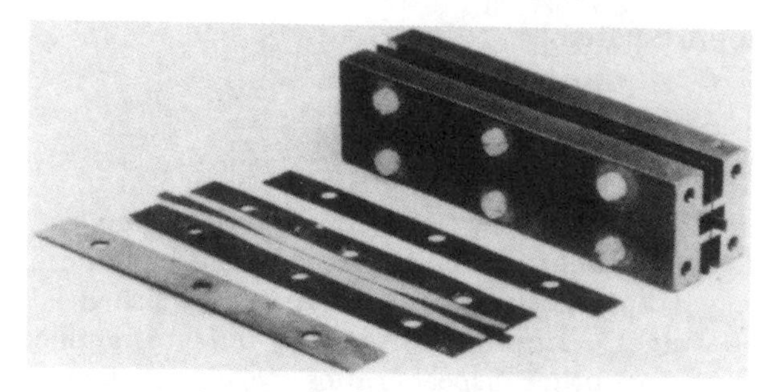

Fig. 5. Fin-line test fixture.

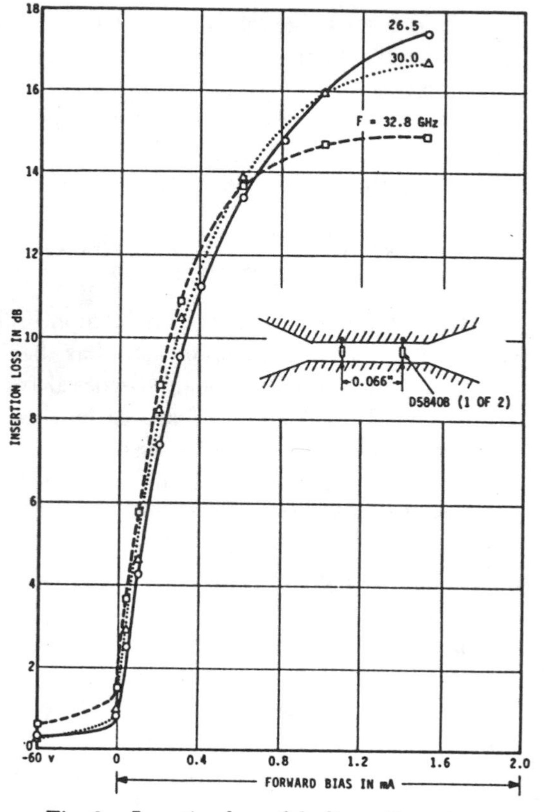

Fig. 6. Insertion loss of fin-line attenuator.

the attenuation varies smoothly over a 14-dB range throughout a 20-percent band.

Although the preliminary results of Fig. 6 are gratifying, it may be observed that the minimum loss and maximum attenuation deteriorate as the frequency is raised. To obtain a better understanding of the frequency limitations of this design, a program was launched to obtain an equivalent circuit for the attenuator, including the parasitic elements associated with the diode and its mounting structure. The SWR, as a function of bias and frequency, was measured for a single diode mounted across a matched fin-line. A computer-aided technique was then applied to find the parasitic elements which best reconciled the measured data with the general equivalent circuit shown in Fig. 7. Based upon the manufacturer's published data, a value of 1 Ω was assigned to R_s and R_j was represented as

$$R_j = 14.1 I^{-0.83} \quad (3)$$

where I is the diode current in milliamperes ($0.02 \leq I \leq 2$).

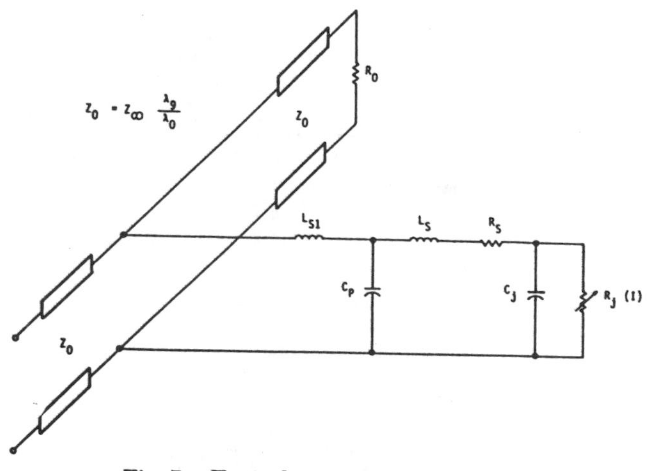

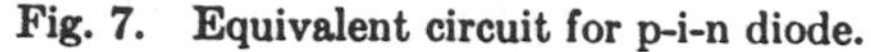

Fig. 7. Equivalent circuit for p-i-n diode.

All other parasitics were assigned plausible initial values and then optimized by an iterative process. Based upon a power-voltage definition of Z_0, the fit between the measured and calculated values of SWR proved to be only fair after exhaustive iterations. Moreover, the optimum value of C_j was found to be well below the specified value of 0.03 pF. Consequently, other definitions of Z_0 were examined; the best overall fit was obtained by adopting the power-current definition and thereby reducing Z_∞ by a factor of $\pi^2/16$ [13].

Table II lists the values of the parasitic elements that provided the best fit (rms error equals 4 percent) with 21 SWR measurements at three equally spaced frequencies from 26.5 to 40.0 GHz and seven logarithmically spaced bias currents from 0 to 1.5 mA. This simple technique provided an unambiguous determination of the parasitics without complex impedance measurements (which prove difficult to reference at the plane of the diode). The validity of the equivalent circuit was confirmed by accurately predicting the total insertion loss of the diode which is generally greater than that due to the observed SWR alone.

The key parasitic element, which limits the high-frequency performance in the existing test fixture, is the series inductance L_s. The inductive reactance at 40 GHz is 35 Ω, which is significant relative to the forward-biased junction resistance and the characteristic impedance of the line ($Z_0 = 73.5$ Ω). This accounts for the reduced attenuation under forward bias. For reverse bias, the series inductance tends to resonate with the junction capacitance, thereby introducing a reflection which increases the minimum insertion loss.

The high-frequency response of the p-i-n attenuator may be improved by several methods. The series inductance can be lowered by decreasing the gap between the fins; however, this further reduces Z_0 and compounds the problem under forward bias. A better approach is to raise Z_0 by constricting the width of the fin-line housing and thereby increasing the ratios λ_g/λ_0 and b/a. This minimizes the effect of the series inductance under forward bias and increases the maximum attenuation. Raising Z_0 will also

TABLE II

P-I-N EQUIVALENT CIRCUIT PARAMETERS

L_s	= 0.14 nH
L_{s1}	≈ 0
C_p	= 0.002 pF
C_j	= 0.037 pF
Z_∞	= 80.6 ohms [a]

[a] Power current definition.

increase the reflection of a single diode under reverse bias, but this can be offset over a moderate bandwidth by spacing two diodes such that their reflections cancel. The upper frequency limitation then becomes the series resonance between L_s and C_j which is 70 GHz for the parasitics listed in Table II. Substituting a diode with a smaller junction capacitance (such as the D5840E) can move this series resonance well above 100 GHz. Further increases in the operating frequency and bandwidth are feasible with a more complex fin-line mount which tunes out the inductance L_s.

A p-i-n attenuator has been selected for measurements and analysis to demonstrate the potential of a broad class of semiconductor fin-line components. Other solid-state millimeter components which are amenable to fin-line construction include mixers, oscillators, amplifiers, limiters, switches, and phase shifters. A fin-line Gunn oscillator, which provides a power output of 45 mW in the 9-mm band, has been described elsewhere [14]. In addition to the existing examples of one- and two-port components, fin-line is well suited to the fabrication of E-plane Y junctions. Consequently, three-port networks, such as parametric amplifiers, phase modulators, and switching matrices, are feasible. For the reasons presented previously, integrated fin-line is preferred to microstrip for the construction of solid-state components and systems throughout the millimeter spectrum.

APPLICATION TO PASSIVE COMPONENTS

Although fin-line configurations with insulated fins and small gaps are best suited to solid-state applications, lower loss can be obtained with grounded fins and larger gaps. It has been shown, in connection with the discussion of Fig. 2, that an unloaded Q of 700–850 is possible for grounded fins with moderated loading ($d/b > 0.6$). Although another planar configuration [15] will provide higher values of Q, many millimeter applications exist where fin-line should prove to be adequate.

To demonstrate the capability of fin-line components fabricated according to the cross section of Fig. 1(a), various filter components were printed on 0.020-in Duroid. Test samples include the one- and four-pole inductively coupled filters shown at the center of Fig. 8, surrounded by four substrates, each printed with a single inductive element. By measuring the insertion loss of these elements across the 26.5–40-GHz band, families of design curves were generated to present the shunt susceptance as a function of strip width, with free-space wavelength as a parameter.

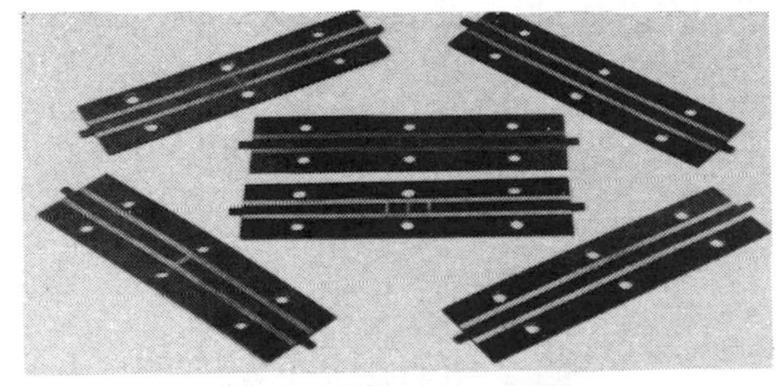

Fig. 8. Filter components.

Fig. 9 shows how the shunt susceptance (normalized to the characteristic admittance of the fin-line) varies with respect to the strip width (normalized with respect to the wide inner dimension of the housing) for the special case of $d/b = 1.0$. Larger values of B/Y_0 can be obtained for a given w/a by reducing d/b but only at the expense of the unloaded resonator Q. Additional one-pole resonator measurements have been performed to determine the unloaded resonator Q and the reference plane at which the inductive strip can be modeled as a pure shunt susceptance. Based upon the characterization of fin-line filter elements and published design curves [16], a four-pole equal-element filter has been constructed and tested. Fig. 10 compares the calculated response with measurements. The calculation was performed by a computer-aided technique which solves for the overall $ABCD$ matrix of the entire filter network including:

1) shunt susceptance of each filter element and its variation with respect to frequency;

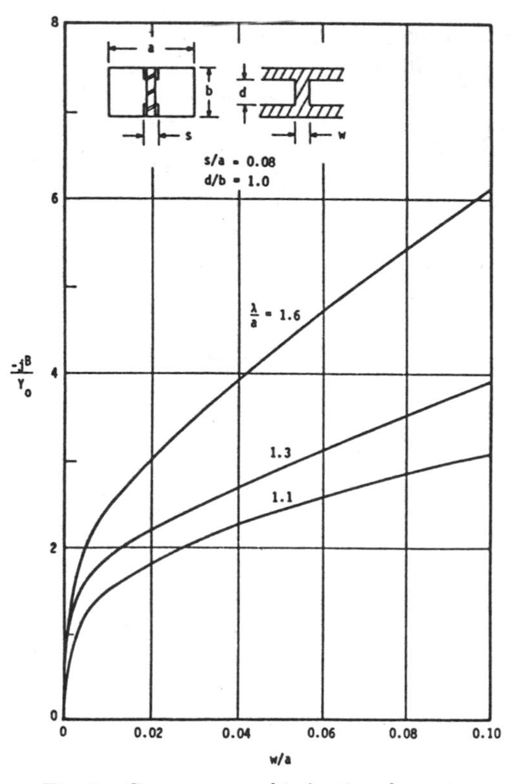

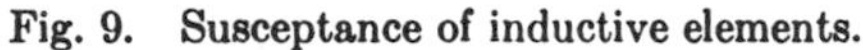
Fig. 9. Susceptance of inductive elements.

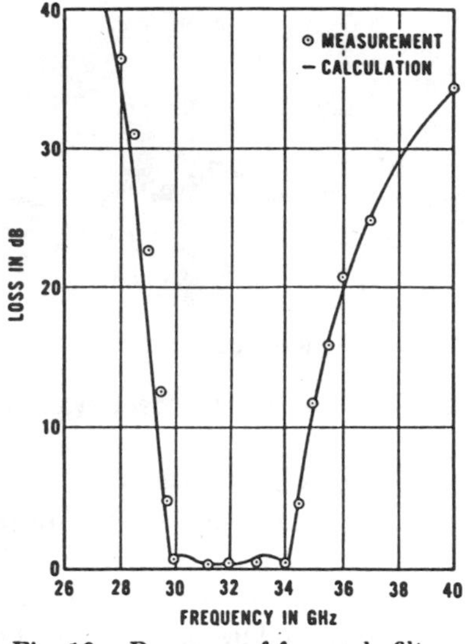

Fig. 10. Response of four-pole filter.

2) offset between the electrical and mechanical centerlines of the inductive elements;

3) variation of guide wavelength with respect to frequency, which is known accurately for $d/b = 1.0$ [5], [10].

4) unloaded resonator Q of 350 for center sections (based upon one-pole measurement).

The good agreement that has been obtained between measurements and theory serves to illustrate the straightforward application of fin-line techniques to the design of millimeter filters.

To explore the possible advantages of fin-line filters with $d \neq b$, the susceptance of a more generalized fin-line inductor has been measured. Table III lists the susceptance measured at three frequencies across the instrumentation band for two inductor configurations. For the first configuration, the fins do not load the waveguide and a fairly wide strip ($w/a = 0.2$) is required to establish a high susceptance suitable for narrow-band filters. In the second configuration, the fins provide heavy loading to the waveguide and a higher susceptance can be obtained with a moderately narrow strip ($w/a = 0.1$). Heavy fin loading not only provides a larger susceptance but one whose magnitude is less dependent upon frequency. Consequently, inductors in heavily loaded fin-line should prove to be of great value and deserve further study.

CONCLUSIONS

The application of standard microstrip techniques at millimeter wavelengths can result in problems which include excessive radiation loss, spurious coupling, dispersion, and higher mode propagation. These problems can be controlled by choosing progressively thinner substrates as the operating frequency is raised but this only serves to degrade the Q, increase tolerance problems, restrict the range over which Z_0 can be varied, and complicate the design of transitions to waveguide instrumentation. Integrated fin-line is superior to microstrip at millimeter wavelengths as the former provides eased production tolerances, compatibility with chip and beam-lead devices throughout the millimeter spectrum, and single-mode octave bandwidths, combined with the ability to construct simple transitions to waveguide. Fin-line avoids excessive miniaturization and offers low loss combined with the potential for low-cost production through batch-processing techniques.

The basic characteristics of integrated fin-line have been reviewed previously. It has been demonstrated that practical fin-line structures can provide an unloaded Q of 260–850 and that the guide wavelength can be accurately predicted by a simple equation.

Examples of solid-state and passive components have been presented which serve to illustrate the potential of integrated fin-line at millimeter wavelengths. These examples have included a p-i-n attenuator which has demonstrated the capability of constructing low-loss semiconductor mounts in fin-line. An analysis has shown that existing beam-lead devices in a simple fin-line mount can be useful well beyond 100 GHz.

Based upon the analysis and examples presented, it is clear that fin-line construction techniques can be applied to a broad class of millimeter components including mixers, oscillators, amplifiers, and phase shifters. Fin-line is also suited to the fabrication of E-plane Y junctions and is therefore compatible with three-port networks such as parametric amplifiers, phase modulators, and switching matrices.

Since integrated fin-line can provide high performance combined with low-cost batch processing, it is ideally suited to a wide range of millimeter applications including communication systems, high-resolution imaging, collision-avoidance radars, and environmental monitoring equipment.

TABLE III

SUSCEPTANCE OF FIN-LINE INDUCTORS

Frequency (GHz)	λ_o/a	Normalized Susceptance (B/Y_o) No Fin Loading $d/b = 1.0$ $w/a = 0.2$	Normalized Susceptance (B/Y_o) Heavy Fin Loading $d/b = 0.1$ $w/a = 0.1$
26.5	1.59	13.2	18.1
33.0	1.28	6.8	14.7
40.0	1.05	4.9	12.0

ACKNOWLEDGMENT

This program was sponsored by AIL under the direction of K. S. Packard, M. T. Lebenbaum, and J. J. Whelehan. D. Fleri and J. J. Taub formulated the initial program plan and made technical contributions throughout the program. Technical assistance was provided by R. Chew, R. Gibbs, L. Hernandez, A. Kunze, and W. Reinheimer.

REFERENCES

[1] D. Dobramysl, "Integrated mixer for 18 and 26 GHz," in *IEEE 1971 G-MTT Symp. Digest* (Washington, D. C.), May 16–19, 1971, pp. 18–19.

[2] T. H. Oxley, K. J. Ming, G. H. Swallow, B. J. Climer, and M. J. Sisson, "Hybrid microwave integrated circuits for millimeter wavelengths," in *IEEE 1972 G-MTT Symp. Digest* (Arlington Heights, Ill.), May 22–24, 1972, pp. 224–226.

[3] J. J. Taub, "The status of quasi-optical waveguide components for millimeter and submillimeter wavelengths," *Microwave J.*, pp. 57–62, Nov. 1970.

[4] P. J. Meier, "Two new integrated-circuit media with special advantages at millimeter wavelengths," in *IEEE 1972 G-MTT Symp. Digest* (Arlington Heights, Ill.), May 22–24, 1972, pp. 221–223.

[5] ——, "Equivalent relative permittivity and unloaded Q factor of integrated fin-line," *Electron. Lett.*, vol. 9, pp. 162–163, Apr. 1973.

[6] ——, "Microwave transmission line," U. S. Patent 3 825 863, July 1974.

[7] S. B. Cohn, "Properties of ridge wave guide," *Proc. IRE*, vol. 35, pp. 783–788, Aug. 1947.

[8] S. Hopfer, "The design of ridged waveguides," *IRE Trans. Microwave Theory Tech.*, vol. MTT-3, pp. 20–29, Oct. 1955.

[9] F. E. Gardiol, "Higher-order modes in dielectrically loaded rectangular waveguides," *IEEE Trans. Microwave Theory Tech.*, vol. MTT-16, pp. 919–924, Nov. 1968.

[10] P. H. Vartanian, W. P. Ayres, and A. L. Helgesson, "Propagation in dielectric slab loaded rectangular waveguide," *IRE Trans. Microwave Theory Tech.*, vol. MTT-6, pp. 215–222, Apr. 1958.

[11] D. Fleri, "AIL project R035 status report," Internal Memo, Oct. 1973.

[12] R. A. Pucel, D. J. Massé, and C. P. Hartwig, "Losses in mircostrip," *IEEE Trans. Microwave Theory Tech.*, vol. MTT-16, pp. 342–350, June 1968.

[13] G. C. Southworth, *Principles and Applications of Waveguide Transmission*. Princeton, N. J.: Van Nostrand, 1950, pp. 104–105.

[14] D. A. Fleri *et al.*, "A low cost X-band MIC paramp," presented at the IEEE Int. Microwave Symp., Atlanta, Ga., June 1974.

[15] Y. Konishi, K. Uenakada, N. Yazawa, N. Hoshino, and T. Takahashi, "Simplified 12-GHz low-noise converter with mounted planar circuit in waveguide," *IEEE Trans. Microwave Theory Tech.* (Short Paper), vol. MTT-22, pp. 451–454, Apr. 1974.

[16] J. J. Taub, "Design of minimum loss band-pass filters," *Microwave J.*, Nov. 1963.

Dispersion of Planar Waveguides for Millimeter-Wave Application

by Holger Hofmann *

Report from AEG-Telefunken, Fachgebiet Hochfrequenztechnik, Ulm (Donau)

The dispersion of the phase coefficient and the wave impedance of some planar and quasi-planar structures such as slot-line, coplanar-strip-line, suspended substrate-line, and different fine-line geometries are calculated applying Galerkin's method. The formulation is adapted to the various structures under the aspect of minimizing computational time in the numerical evaluation.

Dispersion von planaren Wellenleitern für Millimeterwellen-Anwendungen

Die Dispersion des Phasenkoeffizienten und des Wellenwiderstandes einiger planarer und quasi-planarer Strukturen, wie Schlitzleitung, Coplanarleitung, „Suspended Substrate"-Leitung und verschiedene „Fin"-Leitungen, wird mit der Methode von Galerkin berechnet. Die Formulierung wird den verschiedenen Strukturen angepaßt unter dem Gesichtspunkt, minimale Rechenzeit bei der numerischen Auswertung zu erhalten.

1. Introduction

Different planar waveguide structures have been proposed for millimeter-wave applications. The slot-line and the coplanar-strip-line have been treated sufficiently (e.g. [1], [2]). Fin-lines have been applied only recently by Meyer (e.g. [3], [4]) but no theoretical approach is available to date. For the suspended substrate line only quasi-static results have been published (e.g. [5], [6]). However, thinking of the surrounding rectangular waveguide of the fin-line as an electric shield, it is possible to calculate fin-line, suspended substrate-line, slot-line and coplanar-strip-line, in one treatment, if the shield dimensions are allowed to become very large. Six

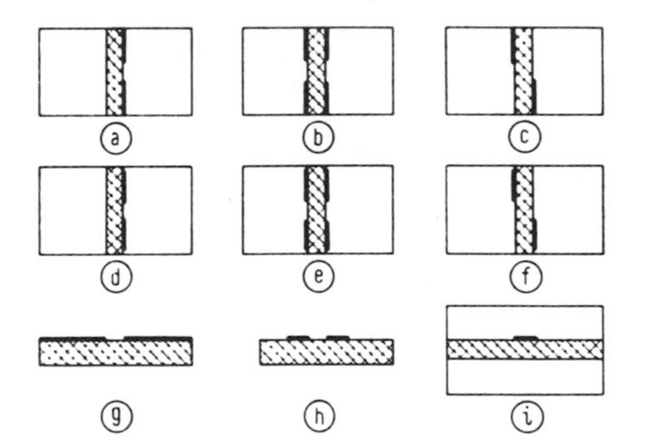

Fig. 1. Fin-line geometries and related structures to be treated in one uniform formulation: (a) unilateral fin, (b) bilateral fin, (c) antipodal fin; (d), (e), (f) insulated counterparts of (a), (b), (c); (g) slot, (h) coplanar strips, (i) suspended substrate.

different fin-line geometries will be discussed. Fig. 1, which shall be titled unilateral-, bilateral-, and antipodal-fin-line, and their insulated counterparts. Fin-lines and suspended substrate-lines are considered quasi-planar since components are modelled by strip structures, keeping the hollow waveguide unchanged.

* Dr.-Ing. H. Hofmann, c/o AEG-Telefunken, Postfach 1730, D-7900 Ulm (Donau).

2. Fourier Representation of the Fields

Using the reduced field quantities

$$\begin{aligned} \boldsymbol{E} &= \sqrt{\varepsilon_0}\,\boldsymbol{E}', & \boldsymbol{D} &= 1/\sqrt{\varepsilon_0}\,\boldsymbol{D}', \\ \boldsymbol{H} &= -\,\mathrm{j}\sqrt{\mu_0}\,\boldsymbol{H}', & \boldsymbol{B} &= 1/\mathrm{j}\sqrt{\mu_0}\,\boldsymbol{B}', \end{aligned} \tag{1}$$

the primes indicating the classical notification, the fields shall be calculated via the E_x- and H_x-components only:

$$\begin{gathered} \left(\frac{\partial^2}{\partial x^2} + \frac{\partial^2}{\partial y^2} + (k_0^2\,\varepsilon_r - \beta^2)\right)\begin{pmatrix} E_x \\ H_x \end{pmatrix} = 0\,, \\ \begin{pmatrix} \boldsymbol{E}_t \\ \boldsymbol{H}_t \end{pmatrix} = \frac{1}{k_0^2\,\varepsilon_r - k_x^2}\begin{pmatrix} \partial/\partial x & -k_0\,\boldsymbol{e}_x\times \\ -k_0\,\varepsilon_r\,\boldsymbol{e}_x\times & \partial/\partial x \end{pmatrix}\nabla_t\begin{pmatrix} E_x \\ H_x \end{pmatrix}, \\ k_x^2 + k_y^2 + \beta^2 = k_0^2\,\varepsilon_r\,, \qquad k_0 = \omega\sqrt{\varepsilon_0\,\mu_0}\,, \\ \nabla_t = \partial/\partial y\,\boldsymbol{e}_y - \mathrm{j}\,\beta\,\boldsymbol{e}_z\,, \end{gathered} \tag{2}$$

understanding wave propagation of the form $\exp\{\mathrm{j}(\omega t - \beta z)\}$.

The relevant cross-sections are divided into subregions as Fig. 2 shows. The electric and magnetic walls account for the symmetries of the fundamental mode. In Fig. 2c the field space of the antipodal-fin line can be further reduced to $x \geqq 0$ by postulating

$$\begin{aligned} E_x(-x, 2b - y) &= E_x(x, y)\,, \\ H_x(-x, 2b - y) &= H_x(x, y)\,. \end{aligned} \tag{3}$$

In any case the metallic fins are considered to be infinitely thin and the dielectric and metal loss-free.

A complete solution to eq. (2) in the case of the bilateral-fin line, holding for the two subregions is

$$\begin{aligned} E_x^{(1)} &= \sum_{n=1}^{\infty} A_n^{(1)} \sin\bar{\alpha}_n(x - \alpha)\sin\alpha_n y\,, \\ E_x^{(2)} &= \sum_{n=1}^{\infty} A_n^{(2)} \sin\tilde{\alpha}_n(x - a)\sin\alpha_n y\,, \end{aligned} \tag{4}$$

Reprinted with permission from *AEU,* Band 31, Heft 1, pp. 40–44, 1977.

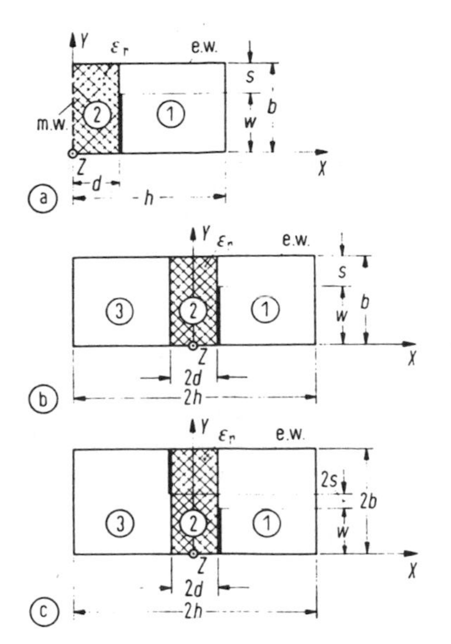

Fig. 2. Division of the field space into subregions for the different geometries; e.w.: electric wall, m.w.: magnetic wall; (a) bilateral fin, (b) unilateral fin, (c) antipodal fin.

with the transversal phase coefficients

$$\begin{aligned}\alpha_n &= n\pi/b\,,\\ \bar{\alpha}_n &= \sqrt{k_0^2-\beta^2-\alpha_n^2}\,,\\ \tilde{\alpha}_n &= \sqrt{k_0^2\varepsilon_r-\beta^2-\alpha_n^2}\,.\end{aligned} \tag{5}$$

The components H_x are obtained by replacing $\sin\to\cos$, $\cos\to\sin$ and naming the Fourier coefficients B_n, starting the series at $n=0$. The solution for the unilateral-fin line is achieved by taking in subregion ① the same $E_x^{(1)}$ as above, in subregion ③ by replacing $(x-a)\to(x+a)$ of $E_x^{(1)}$ and by writing in subregion ②

$$E_x^{(2)} = \sum_{n=1}^{\infty} A_n^{(2)} \sin\tilde{\alpha}_n x \sin\alpha_n y + \bar{A}_n^{(2)} \cos\tilde{\alpha}_n x \sin\alpha_n y\,. \tag{6}$$

Finally the series for the antipodal-fin case will be explained. Taking $E_x^{(1)}$ as in eq. (4) with α_n replaced by

$$\alpha_n = n\pi/2b\,, \tag{7}$$

and choosing

$$A_n^{(3)} = -(-1)^n A_n^{(1)}\,,\quad B_n^{(3)} = -(-1)^n B_n^{(1)} \tag{8}$$

renders the complete solutions in subregions ① and ③, accounting for the symmetric condition of eq. (3). In the same way

$$E_x^{(2)} = \sum_{n=1}^{\infty} A_n^{(2)}(\delta_e \cos\tilde{\alpha}_n x \sin\alpha_n y + \delta_o \sin\tilde{\alpha}_n x \sin\alpha_n y)\,, \tag{9}$$

$$\delta_e = \begin{cases}0 & \text{for } n \text{ even}\\ 1 & \text{for } n \text{ odd}\,,\end{cases} \qquad \delta_o = \begin{cases}0 & \text{for } n \text{ odd}\\ 1 & \text{for } n \text{ even}\end{cases}$$

supplies the solution in subregion ②.

Common to all Fourier-series representations of above is, that any of the boundary conditions on the electric and magnetic walls are satisfied.

3. Continuity Conditions

Applying all continuity conditions

$$\begin{aligned}\boldsymbol{E}_{\tan}^{(i)} - \boldsymbol{E}_{\tan}^{(j)} &= 0\,,\\ (\boldsymbol{H}_{\tan}^{(i)} - \boldsymbol{H}_{\tan}^{(j)}) \times \boldsymbol{e}_x &= \boldsymbol{I}\,,\end{aligned} \tag{10}$$

where $\boldsymbol{I}$ is a surface current density in the interface, allows all Fourier coefficients to be expressed in terms of $\boldsymbol{I}$, that is

$$A_n^{(i)},\, B_n^{(i)} = f\{L_{1n}(I_z),\, L_{2n}(I_y)\}\,, \tag{11}$$

where the linear operator L applied to any function in its domain is given by

$$\begin{aligned}L_{1n}(\xi(\tau)) &= \int_0^{\hat{y}} \xi(\tau)\sin\alpha_n\tau\,\mathrm{d}\tau\,,\\ L_{2n}(\xi(\tau)) &= \int_0^{\hat{y}} \xi(\tau)\cos\alpha_n\tau\,\mathrm{d}\tau\,,\end{aligned} \tag{12}$$

with $\hat{y} = \begin{cases} b & \text{if } \alpha_n \text{ of eq. (5) is taken}\\ 2b & \text{if } \alpha_n \text{ of eq. (7) is taken.}\end{cases}$

Now the remaining two boundary conditions

$$\boldsymbol{E}_{\tan}^{(1)} = 0\,,\quad x = d\,,\quad 0 \leqq y \leqq w\,, \tag{13}$$

with eq. (11) substituted for the $A_n^{(i)}$, $B_n^{(i)}$ deliver two operator equations in I_y, I_z

$$\begin{aligned}\sum_n \Gamma_{11} L_{2n}(I_y)\cos\alpha_n y + \sum_n \Gamma_{12} L_{1n}(I_z)\cos\alpha_n y &= 0\,,\\ \sum_n \Gamma_{21} L_{2n}(I_y)\sin\alpha_n y + \sum_n \Gamma_{22} L_{1n}(I_z)\sin\alpha_n y &= 0\,,\end{aligned} \tag{14}$$

where the Γ_{ij} are well defined functions rendered by the continuity conditions.

Eqs. (14) are solved applying Galerkin's method: the current densities are represented by appropriate complete Fourier series

$$\begin{aligned}I_z &= \begin{cases}\sum_m C_m i_m(y), & \text{for } 0 \leqq y \leqq w\\ 0 & \text{for } w \leqq y \leqq \hat{y},\end{cases}\\ I_y &= \begin{cases}\sum_m D_m j_m(y) & \text{for } 0 \leqq y \leqq w\\ 0 & \text{for } w \leqq y \leqq \hat{y}.\end{cases}\end{aligned} \tag{15}$$

Then eqs. (14) are tested with $i_m(y)$ and $j_m(y)$ respectively, supplying a doubly-infinite system of homogeneous equations in C_m and D_m. Vanishing of the system's determinant is condition for the determination of the phase coefficient β.

For the numerical evaluation the field series are truncated after N_f and the current series after N_i terms. The matrix size depends on N_i only, while each matrix coefficient is a sum over N_f terms. On account of the edge-condition [7] the indices have to be chosen as

$$N_i/N_f = w/\hat{y}\,. \tag{16}$$

On approximating the slot-line by choosing $b \gg b-w$, N_i will be equal to N_f, thus delivering very poor convergence as shall be shown later. In this case it is advantageous to formulate the problem in the slot-fields only instead of in the current densities [2].

4. Slot-field Solution

Instead of taking boundary condition (13) write

$$\begin{aligned} E_y^{(1)} &= f(y)\,, \\ E_z^{(1)} &= g(y)\,, \end{aligned} \qquad x = d,\ 0 \leq y \leq \hat{y}, \tag{17}$$

where f, g represent the fields in the slot, and are zero elsewhere. Eq. (17) allows the Fourier coefficients of region ① to be expressed as

$$A_n^{(1)},\ B_n^{(1)} = f\{L_{2n}(f),\ L_{1n}(g)\}\,. \tag{18}$$

On eliminating $A_n^{(1)}$, $B_n^{(1)}$ eqs. (11) and (18) can be rearranged to

$$\begin{aligned} A_{11}\,L_{2n}(f) + A_{12}\,L_{1n}(g) &= L_{2n}(I_y)\,, \\ A_{21}\,L_{2n}(f) + A_{22}\,L_{1n}(g) &= L_{1n}(I_z)\,, \end{aligned} \tag{19}$$

where A_{ij} are again well defined functions. Now sum eqs. (19) as follows:

$$\begin{aligned} \sum_n 1/\sigma_n\,L_{2n}(I_y)\cos\alpha_n y &= 1/2\,\hat{y}\,I_y(y)\,, \\ \sum_n L_{1n}(I_z)\sin\alpha_n y &= 1/2\,\hat{y}\,I_z(y) \end{aligned} \tag{20}$$

with

$$\sigma_n = \begin{cases} 2 & \text{for } n = 0 \\ 1 & \text{for } n \neq 0 \end{cases}$$

or in other words

$$\begin{aligned} &\sum_n 1/\sigma_n\,A_{11}\,L_{2n}(f)\cos\alpha_n y + \\ &\quad + \sum_n A_{12}\,L_{1n}(g)\cos\alpha_n y = 1/2\,\hat{y}\,I_y(y)\,. \\ &\sum_n A_{21}\,L_{2n}(f)\sin\alpha_n y + \\ &\quad + \sum_n A_{22}\,L_{1n}(g)\sin\alpha_n y = 1/2\,\hat{y}\,I_z(y)\,. \end{aligned} \tag{21}$$

Galerkin's method is applied to eqs. (21) in writing

$$\begin{aligned} f(y) &= \begin{cases} \sum_m C_m f_m(y) & \text{for } w \leq y \leq \hat{y} \\ 0 & \text{for } 0 \leq y \leq w\,, \end{cases} \\ g(y) &= \begin{cases} \sum_m D_m g_m(y) & \text{for } w \leq y \leq \hat{y} \\ 0 & \text{for } 0 \leq y \leq w \end{cases} \end{aligned} \tag{22}$$

and testing with f_m and g_m respectively. The right hand sides will vanish when testing:

$$\int_0^{\hat{y}} I_y(y)\,f_n(y)\,\mathrm{d}y = \int_0^{\hat{y}} I_z(y)\,g_n(y)\,\mathrm{d}y = 0\,, \tag{23}$$

since I_y, I_z and f_n, g_n are defined in complementary field spaces. Thus again a doubly-infinite system of homogeneous equations is sustained.

If now N_i is the truncation index of the interface-fields the ratio

$$N_i/N_f = (\hat{y} - w)/\hat{y} \tag{24}$$

is small resulting in a much faster convergence of the solution, as shall be shown beyond.

For completeness the used Fourier components are stated following. For the earthed fins

$$\begin{aligned} i_m = \sin\eta_m y\,, \quad j_m &= \cos\eta_m y\,, \\ \eta_m &= (m - 1/2)\,\pi/w \end{aligned} \tag{25}$$

is used, while in the insulated fin case

$$\begin{aligned} i_m = \cos\eta_m(y - y_0)\,, \quad j_m &= \sin\eta_m(y - y_0)\,, \\ \eta_m &= m\,\pi/w \end{aligned} \tag{26}$$

is applied, where y_0 provides for a finite spacing of the fin from the waveguide, thus the fin extending from y_0 to $y_0 + w$. The slot fields in turn are represented by

$$\begin{aligned} f_m = \cos\eta_m(y - w), \quad g_m &= \sin\eta_m(y - w), \\ \eta_m &= m\,\pi/(\hat{y} - w)\,. \end{aligned} \tag{27}$$

It should be mentioned that any of the examined problems presents us with a very similar final system of equations (as does, as a matter of fact, the microstrip problem too), so that one FORTRAN program with only little modifications can be used for all cases. The coefficient matrix of the system is all real and symmetric.

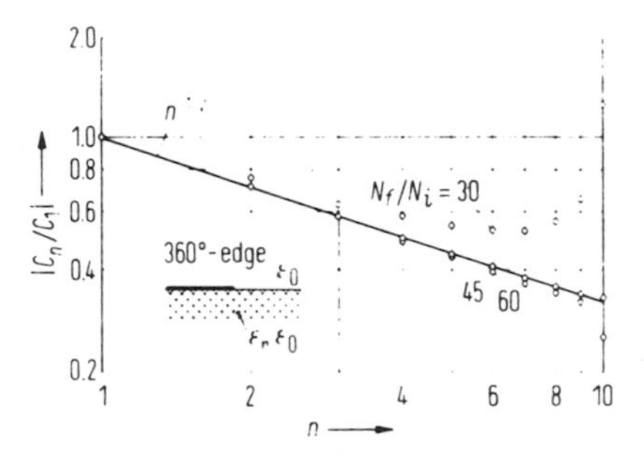

Fig. 3. Asymptotic behaviour of the Fourier coefficients of the transverse electric field in the interface; $N_i = 10$, cover-to-slot ratio $b/s = 30$, $d/\lambda = 0.01$.

5. Relative Convergence and Edge Condition

Though wellknown, there should be some remarks made upon the problem of relative convergence, which will of course emerge in the above calculations. It has been proved [7] that the edge condition supplies the correct ratio N_i/N_f of truncation indices as stated in eqs. (16) and (24). This means, only this ratio will render the correct behaviour of the fields at the edge as N_i, N_f tend to infinity. But the edge condition has also been shown to be approximated much better for a different ratio N_i/N_f if N_i, N_f are finite [8]. Convergence can thus be speeded up extremely.

In Fig. 3 the approximation of the edge condition is shown for the slot-field solution. The Fourier coefficients C_n have to be proportional to $n^{-1/2}$ for large n if the edge condition is to be fulfilled. Best approximation is achieved with

$$N_f/N_i = 1.5\,\hat{y}/(\hat{y} - w)\,. \tag{28}$$

This ratio also gives fastest convergence as is proved in Fig. 4. With $N_i = 1$ for the slot-field solution the relative error of $\varepsilon_{\text{eff}} = (\beta/k_0)^2$ is about $1^0/_{00}$. Fig. 4 also reveals the fact that for large shielding the current formulation is not appropriate, since it gives only very poor convergence.

The distribution of the slot-fields and the current densities is plotted in Fig. 5. Good approximation of the edge behaviour can be recognized in this picture.

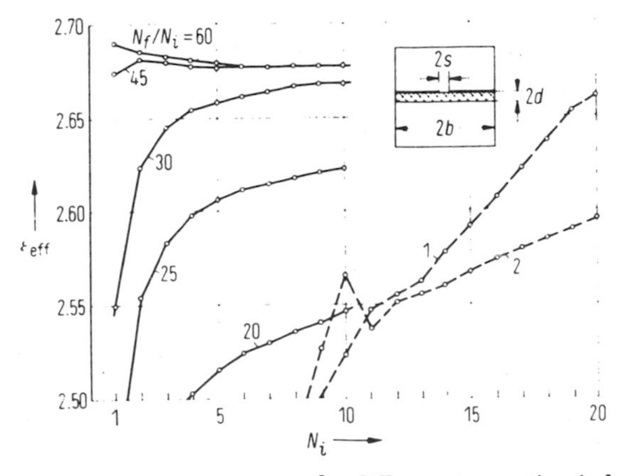

Fig. 4. Relative convergence for different truncation-index ratios;
—— calculation via the slot-fields,
---- calculation via the current densities; cover-to-slot ratio $b/s = 30$, $d/\lambda = 0.01$, $s/d = 1$.

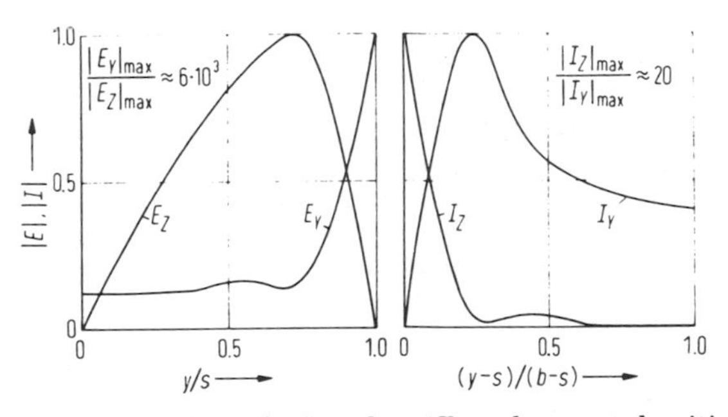

Fig. 5. Slot-fields in the interface (E) and current densities on the conductor (I) versus transverse coordinate; $2s$: slot width, $2b$: cover width.

6. Slot-line and Coplanar-strip-line Solutions

To check the validity of the calculation comparisons with published data of slot-line [1] and coplanar-strip-line [2] are depicted in Fig. 6. Very good agreement is evident. It should be stated that a cover to slot ratio or cover to strip ratio respectively of about 30 is sufficient for not too small values of d/λ. In contrast to the slot-line the configuration of Fig. 1a has a low frequency cut-off so that the shield has to be made very large for d/λ approaching zero if the open structure is to be approximated.

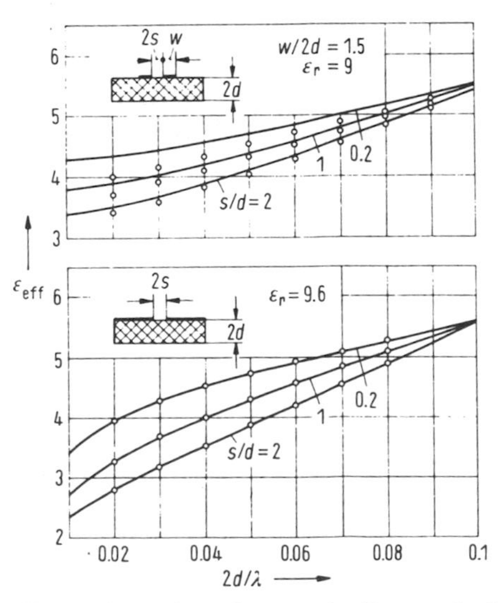

Fig. 6. Dispersion of coplanar-strip line and slot-line: —— this theory, o theory of [2] and [1], respectively.

In the case of the coplanar strips the deviation of this theory to the one in [2] grows with decreasing s/d ratio. It is believed that the herein presented results are more accurate since the quasi-static current approximation of [2] is inferior to the complete Fourier series of eq. (26).

7. Fin-line Solutions

In Fig. 7 dispersion curves of the bilateral and antipodal fin-lines are presented. Dispersion of the unilateral fin-line is within the plotting accuracy identical to that of the bilateral fin-line. (The latter has only halve the attenuation though.) A Kapton-

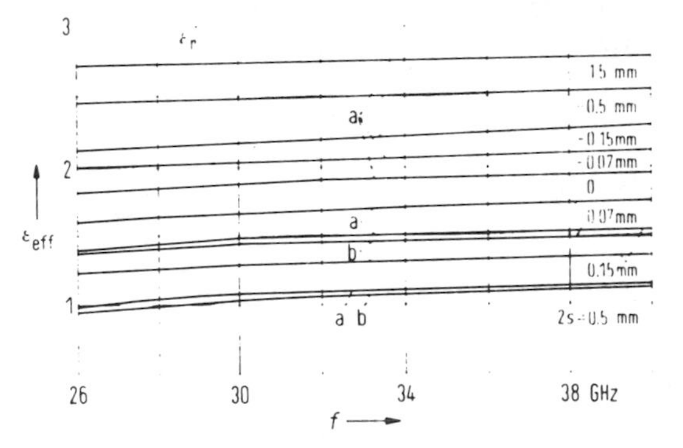

Fig. 7. Dispersion of the bilateral (curves b) and the antipodal (curves a) fin-line. Substrate: Kapton ($\varepsilon_r = 3$, thickness 0.125 mm) within WR 28 waveguide (compare Fig. 8). Negative fin spacing $2s$ implies overlapping of the antipodal fins.

substrate within WR 28 waveguide is taken as an example. Dispersion of the fin-lines is almost linear in the range of Ka-band frequencies, the antipodal fin-line rendering higher ε_{eff} than the other types.

Though very arbitrary, a wave impedance has been defined as

$$Z_L = U/I\,, \tag{29}$$

where I is the integral over the longitudinal current density and U is the voltage sustained by integrating the E-field along the interface from fin to fin.

Values of this wave impedance compare favourably to those of [1], ranging up to 40% higher. Cohn stated that his results are about 30% high compared to measurement. In Fig. 8 the wave impedance is plotted versus the fin-spacing. Here the advantage of the newly presented antipodal fin-line emerges. While the low limit of the wave impedances of the other two fine-line geometries is given by a minimum etchable slot width, the antipodal fine-line can easily be realized in the very wide range of about 10 to 300 Ω wave impedance by letting the fins overlap. In the vicinity of zero spacing a transition

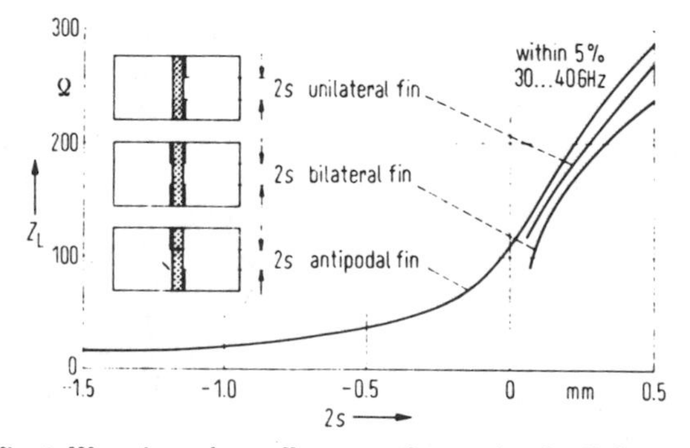

Fig. 8. Wave impedance Z_L versus fin-spacing $2s$. Substrate: Kapton ($\varepsilon_r = 3$, thickness 0.125 mm) within WR 28 waveguide. Negative fin spacing $2s$ implies overlapping of the antipodal fins.

of fine-line behaviour to parallel-plate-line behaviour takes place. For large negative spacing (i.e. overlapping) the wave impedance of this latter line is approached (e.g. $2s = -1.5$ mm renders 18 Ω as compared to 16 Ω with the fin-line). Dispersion of

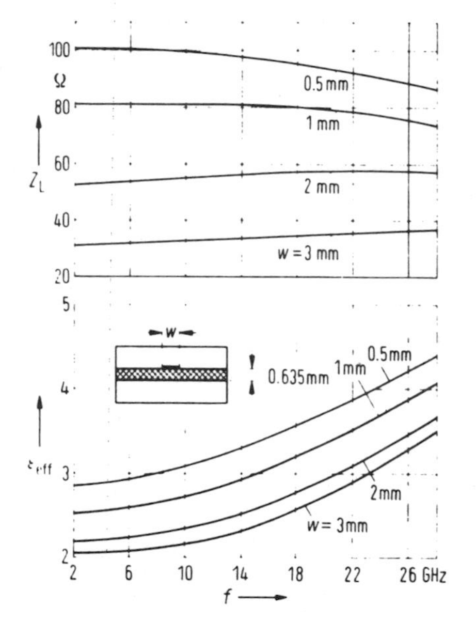

Fig. 9. Wave impedance Z_L and effective dielectric constant ε_{eff} of the suspended substrate line; substrate: $\varepsilon_r = 9.6$, shield: WR 28 waveguide.

the wave impedance is so little that the results are true within 5% in the range of 30 to 40 GHz.

If in the case of the insulated fins a large spacing from the waveguide walls is allowed, the suspended-substrate line can be calculated. Fig. 9 is a plot giving the dispersion of ε_{eff} and Z_L for this line. Alumina substrate inside a WR 28 waveguide was employed as an example.

8. Conclusion

A theory has been presented to calculate the dispersion of the phase coefficient and the wave impedance for a variety of planar structures suitable for mm-wave applications. Included are slot-line, coplanar-strip-line, suspended-substrate line, and unilateral, bilateral, and antipodal fin-lines. Particularly the antipodal fin-line promises interesting applications since its wave impedance can easily be varied in a wide range.

(Received July 14th, 1976.)

References

[1] Mariani, E. A., Heinzmann, C. P., Agrios, J. P., and Cohn, S. B., Slot line characteristics. Transact. Inst. Elect. Electron. Engrs. MTT-17 [1969], 1091–1096.
[2] Knorr, J. B. and Kuchler, K.-D., Analysis of coupled slots and coplanar strips on dielectric substrate. Transact. Inst. Elect. Electron. Engrs. MTT-23 [1975], 541–548.
[3] Meier, P. J., Equivalent relative permittivity and unloaded Q-factor of integrated finline. Electron. Letters 9 [1973], 162–163.
[4] Meier, P. J., New developments with integrated fin-line and related printed millimeter circuits. Inst. Elect. Electron. Engrs. Microwave Symp. Digest [1975], 143–145.
[5] Brenner, H. E., Use a computer to design suspended-substrate ICs. Microwaves 7 [1968], No. 9, 38–45.
[6] Gish, D. L. and Graham, O., Characteristic impedance and phase velocity of a dielectric-supported air strip transmission line with side walls. Transact. Inst. Elect. Electron. Engrs. MTT-18 [1970], 131–148.
[7] Mittra, R. and Lee, S. W., Analytical techniques in the theory of guided waves. MacMillan, New York 1971, pp. 4–14, 38–41.
[8] Jansen, R., Zur numerischen Berechnung geschirmter Streifenleiterstrukturen. AEÜ 29 [1975], 241–247.

Millimeter-Wave Fin-Line Characteristics

JEFFREY B. KNORR, MEMBER, IEEE, AND PAUL M. SHAYDA, MEMBER, IEEE

Abstract—**This paper presents an analysis of the fin line. The spectral-domain technique is used to determine both wavelength and characteristic impedance. Numerical results are compared with known data for ridged waveguide slab-loaded waveguide, and slotline. Design curves are presented for practical millimeter-wave fin line structures.**

I. Introduction

IN THE PAST several years, fin line has been increasingly used as a medium for constructing millimeter wave circuits [1]–[4]. Fig. 1 shows a cross-sectional view of a fin line. The structure may be viewed as a slotline with a shield, a ridged waveguide with dielectric or a slab loaded waveguide with fins. A major attribute of the structure is its compatability with rectangular waveguide and the ease with which the fin pattern may be realized using printed circuit techniques.

In practice, when fin lines are constructed, the dielectric material is often allowed to pass through the broad wall of the shield [1]. An additional dielectric spacer may be used to provide complete dc isolation of one fin from the shield to allow biasing of solid-state devices mounted between the fins. In this case, the wall thickness is made equal to one quarter of a wavelength thereby causing an RF short circuit to appear between the fin and the inner wall of the shield. Thus, a practical fin line will have electrical characteristics similar to the idealized structure shown in Fig. 1.

Several recent papers [5]–[7] have described methods for the determination of fin line wavelength and an approximate method for calculating characteristic impedance is suggested in [1]. This paper presents an analysis of fin line using the spectral domain technique. The analysis covers both wavelength and characteristic impedance.

The paper first discusses the application of the spectral domain technique to fin line. A new matrix approach to the implementation of this method is described. Numerical results are then presented and compared with known data for ridged waveguide, slab loaded waveguide, and slotline. These comparisons establish the accuracy of the numerical results and illustrate the applicability of the method for the full range of structure parameters. It also provides perspective on the substructures which result in the various limits. Finally, several families of design curves are presented for a practical choice of fin-line parameters.

Manuscript received November 30, 1979; revised February 7, 1980. This work was supported in part by the Naval Ocean Systems Center, San Diego, CA, and in part by the Foundation Research Program, Naval Postgraduate School, Monterey, CA.

J. Knorr is with the Department of Electrical Engineering, Naval Postgraduate School, Monterey, CA 93940.

P. Shayda is with the 3D Radar Division, Surveillance System Subgroup, Naval Sea Systems Command, Washington, DC.

II. Fin-Line Analysis

The analysis of various structures using the spectral-domain technique has been discussed previously in several papers by one of the authors as well as other researchers [7]–[12]. Thus the application of the spectral-domain technique to the analysis of a fin line will be presented here in an abbreviated fashion.

The fin line supports a hybrid field. All transverse field components may be found from

$$E_z = k_c^2 \phi^e(x,y) e^{\Gamma z} \tag{1a}$$

$$H_z = k_c^2 \phi^h(x,y) e^{\Gamma z} \tag{1b}$$

where the scalar potential functions ϕ^e, ϕ^h satisfy the Helmholtz equation, and we assume lossless propagation so that $\gamma = \pm j\beta$. Further

$$k_{ci}^2 = k_i^2 - \beta^2 \tag{2}$$

with

$$k_i^2 = \omega^2 \mu_i \epsilon_i, \qquad i = 1,2,3 \tag{3}$$

for each of the three regions defined in Fig. 1. We will assume here that $\epsilon_1 = \epsilon_3 = \epsilon_0$ and $\epsilon_2 = \epsilon_0 \epsilon_r$.

The second-order partial differential equations for the unknown potentials ϕ^e, ϕ^h may be Fourier transformed with respect to x to obtain ordinary differential equations. When this is done and boundary conditions at the shield walls are applied, we obtain the solutions

$$\Phi_1^e(\alpha_n, y) = A^e(\alpha_n) \sinh \gamma_1 (D + h_1 - y) \tag{4a}$$

$$\Phi_2^e(\alpha_n, y) = B^e(\alpha_n) \sinh \gamma_2 y + C^e(\alpha_n) \cosh \gamma_2 y \tag{4b}$$

$$\Phi_3^e(\alpha_n, y) = D^e(\alpha_n) \sinh \gamma_3 (h_2 + y) \tag{4c}$$

$$\Phi_1^h(\alpha_n, y) = A^h(\alpha_n) \cosh \gamma_1 (D + h_1 - y) \tag{4d}$$

$$\Phi_2^h(\alpha_n, y) = B^h(\alpha_n) \sinh \gamma_2 y + C^h(\alpha_n) \cosh \gamma_2 y \tag{4e}$$

$$\Phi_3^h(\alpha_n, y) = D^h(\alpha_n) \cosh \gamma_3 (h_2 + y) \tag{4f}$$

where

$$\gamma_i^2 = \alpha_n^2 - k_{ci}^2 \tag{5}$$

and

$$\alpha_n = \begin{cases} n2\pi/b, & \phi^h \text{ even} \quad (6a) \\ (2n-1)\dfrac{\pi}{b}, & \phi^h \text{ odd} \quad (6b) \end{cases}.$$

Reprinted from *IEEE Trans. Microwave Theory Tech.*, vol. MTT-28, no. 7, pp. 737–743, July 1980.

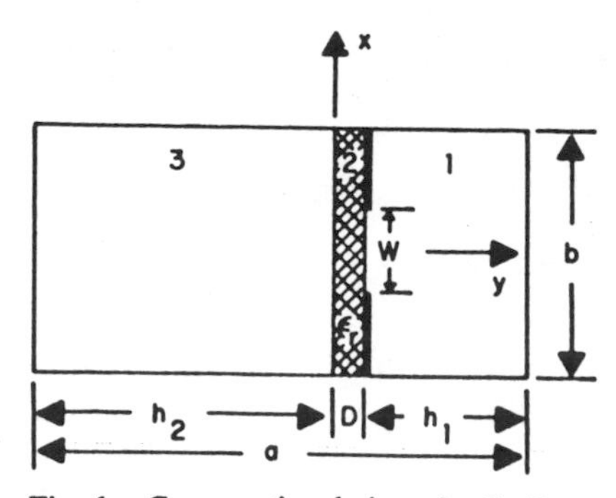

Fig. 1. Cross-sectional view of a fin line.

Through the application of boundary conditions at $y=0$ and $y=D$, the eight coefficients A^e through D^h may be related to each other, to the fin surface current, $j(x)$, and to the slot field $e(x)$. The resulting set of linear equations (really a set of transformed boundary equations) may be written in matrix form as follows:

$$[\mathfrak{M}_{\mathcal{E}}]\begin{bmatrix} A^e \\ B^e \\ \vdots \\ D^h \end{bmatrix} = \begin{bmatrix} 0 \\ 0 \\ \vdots \\ \mathcal{E}_x \\ \mathcal{E}_z \end{bmatrix} \tag{7a}$$

$$[\mathfrak{M}_{\mathcal{J}}]\begin{bmatrix} A^e \\ B^e \\ \vdots \\ D^h \end{bmatrix} = \begin{bmatrix} 0 \\ 0 \\ \vdots \\ \mathcal{J}_x \\ \mathcal{J}_z \end{bmatrix} \tag{7b}$$

where $\mathcal{E}_i(\alpha_n)$ is the transform of the slot field and $\mathcal{J}_i(\alpha_n)$ is the transform of the current. The matrices $[\mathfrak{M}_{\mathcal{E}}]$ and $[\mathfrak{M}_{\mathcal{J}}]$ differ in only the last two rows. Each is a square 8×8 matrix. Using (7a) and (7b), we may write

$$[\mathfrak{M}_{\mathcal{J}}][\mathfrak{M}_{\mathcal{E}}^{-1}]\begin{bmatrix} 0 \\ 0 \\ \vdots \\ \mathcal{E}_x \\ \mathcal{E}_z \end{bmatrix} = \begin{bmatrix} 0 \\ 0 \\ \vdots \\ \mathcal{J}_x \\ \mathcal{J}_z \end{bmatrix}. \tag{8}$$

From (8), using the four elements in the lower right-hand corner of the matrix $[\mathfrak{M}_{\mathcal{J}}][\mathfrak{M}_{\mathcal{E}}^{-1}]$, we obtain

$$[\mathcal{G}]\begin{bmatrix} \mathcal{E}_x \\ \mathcal{E}_z \end{bmatrix} = \begin{bmatrix} \mathcal{J}_x \\ \mathcal{J}_z \end{bmatrix} \tag{9}$$

where the 2×2 matrix $[\mathcal{G}]$ is the transform of the dyadic Green's function for this structure.

Previously, (9) has been obtained by extensive algebraic manipulation of the boundary equations. In the approach described here, the matrices (7) are defined directly from the boundary equations and (9) is arrived at by numerical computation. There is some sacrifice in numerical efficiency, but the formulation of the problem is more straightforward.

Equation (9) is exact, no approximations having been made so far. A solution to (9) is obtained using the Method of Moments as discussed elsewhere [7]–[12]. For this problem, we have chosen to approximate the field between the fins as

$$e_x(x) = \begin{cases} 1, & |x| < W/2 \\ 0, & \text{elsewhere} \end{cases} \tag{10a}$$

$$e_z = 0. \tag{10b}$$

This choice has been shown to give accurate results for slotlines with $W/D < 2$ [11], and for dielectric-loaded waveguide, $W/b = 1$, it is exact. The results to be presented shortly indicate that (10) is a good choice for the fin line as well.

The dispersion problem is now reduced to the form

$$\sum_n g_{11}(\alpha_n,\beta)|\mathcal{E}_x(\alpha_n)|^2 = 0. \tag{11}$$

A numerical search for the value of β which satisfies (11) yields the propagation constant for the dominant fin-line mode.

The characteristic impedance of the fin line is not uniquely defined since the field is non-TEM. A useful definition, however, is

$$Z_0 = \frac{V_0^2}{2P_{\text{avg}}} \tag{12}$$

where V_0 is the voltage between the fins and P_{avg} is the time-averaged power flow. V_0, the voltage across the interface, is determined by (10a) and

$$P_{\text{avg}} = \tfrac{1}{2}\text{Re}\int_S\int (E_x H_y^* - E_y H_x^*)\,dx\,dy. \tag{13}$$

Parseval's theorem is applied to (13) to obtain

$$P_{\text{avg}} = \tfrac{1}{2}\text{Re}\frac{1}{b}\sum_{n=-\infty}^{+\infty}\int_{-h_2}^{D+h_1}[\mathcal{E}_x(\alpha_n,y)\mathcal{H}_y^*(\alpha_n,y) - \mathcal{E}_y(\alpha_n,y)\mathcal{H}_x^*(\alpha_n,y)]\,dy. \tag{14}$$

The integration with respect to y can be carried out analytically in (14). This leaves an equation of the form

$$P_{\text{avg}} = \frac{1}{2b}\sum_{n=-\infty}^{+\infty} f(\alpha_n,\beta) \tag{15}$$

which is evaluated numerically in each of the three regions.

III. Numerical Results

A computer program was developed to solve the equations presented in the previous section. To check the accuracy of the numerical results, they were compared with data available in the literature as described below.

A. Ridged Waveguide

If $\epsilon_r = 1$, the fin line becomes a ridged waveguide with zero-thickness ridges. The wavelength and impedance in this case are given by

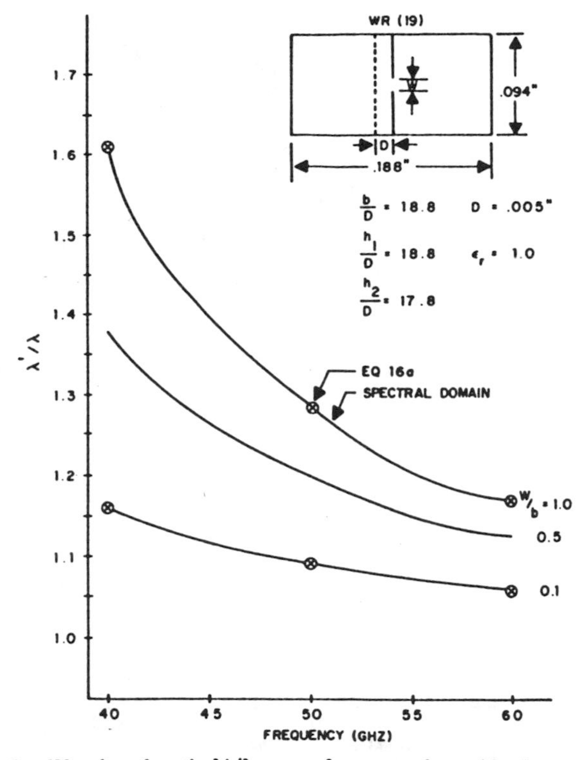

Fig. 2. Wavelength ratio λ'/λ versus frequency for a ridged waveguide. Ridges are centered and have zero thickness.

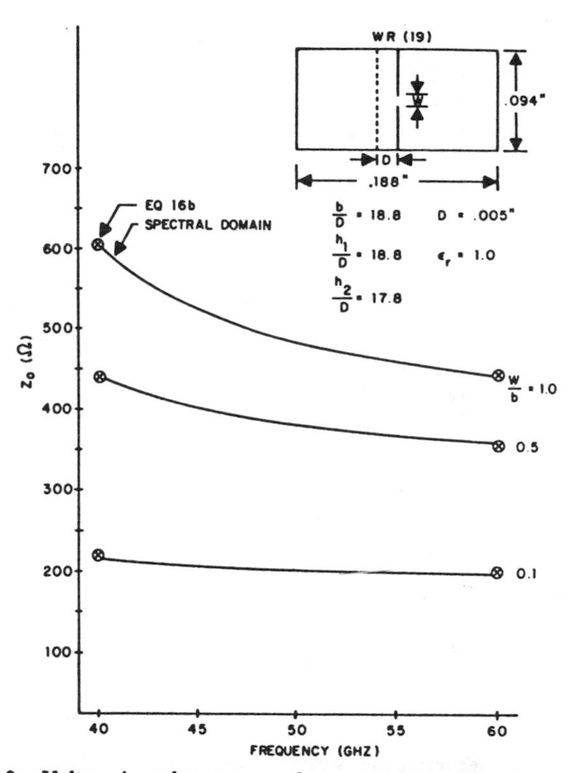

Fig. 3. Voltage impedance versus frequency for a ridged waveguide. Ridges are centered and have zero thickness.

$$\lambda'/\lambda = \frac{1}{\left[1-\left(\frac{\lambda}{\lambda_c}\right)^2\right]^{1/2}} \tag{16a}$$

$$Z_0 = \frac{Z_{0\infty}}{\left[1-\left(\frac{\lambda}{\lambda_c}\right)^2\right]^{1/2}}. \tag{16b}$$

The ridged waveguide has been treated by Hopfer [13] and Lagerlöf [14]. The cutoff wavelength λ_c was determined from [13] for several values of W/b. Equation (16a) was then used to calculate λ'/λ and these values were compared with results obtained using the spectral-domain method as shown in Fig. 2. It can be seen that the two results agree to within 0.5 percent. A similar comparison of impedance [14] as determined by both methods appears in Fig. 3. Again, the agreement is very good, with the difference being only 2 percent.

B. Slab-Loaded Waveguide

If $W/b=1$ and $\epsilon_r>1$, the fin line becomes a slab-loaded waveguide. The wavelength may easily be determined analytically by the transverse resonance procedure [15]. Fig. 4 shows the variation of λ'/λ with dielectric thickness D for several values of ϵ_r. The edge of the slab is placed at the midpoint of the broad wall of a WR(19) guide operating at 40 GHz. The agreement between the results obtained by the two methods is excellent.

The slab-loaded waveguide has also been studied by Vartanian *et al.* [16]. They consider a guide with the slab

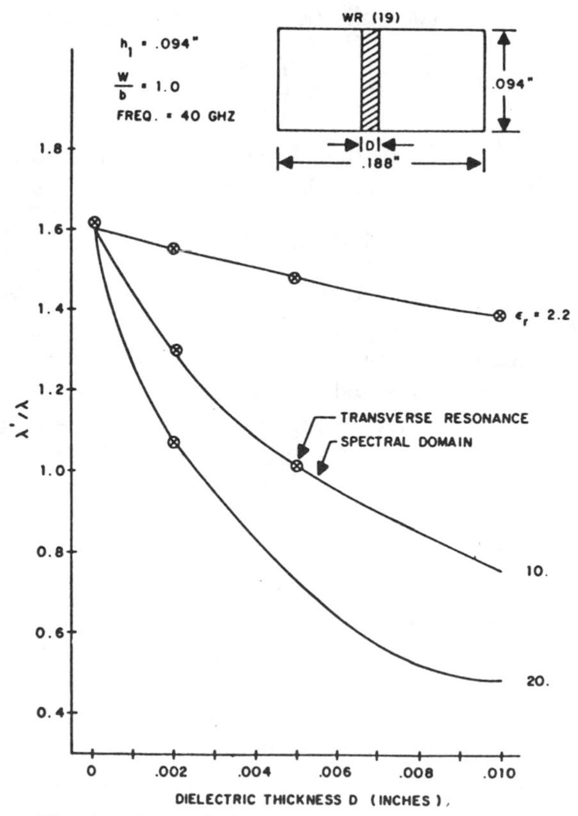

Fig. 4. Wavelength ratio λ'/λ versus dielectic thickness D for a slab-loaded WR(19) guide. Edge of slab is centered.

centered and give an analytical expression for the (voltage) impedance Z_{pv} (using notation of [16]) at the center of the slab. The impedance at the edge of the slab (as it is

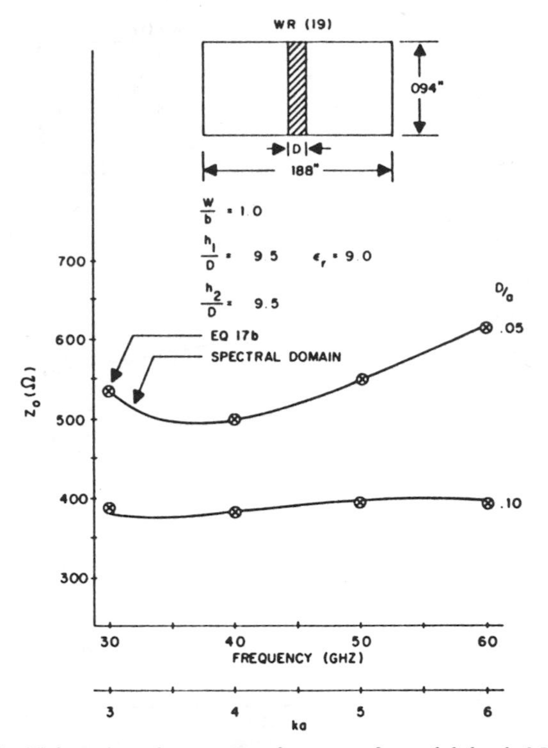

Fig. 5. Voltage impedance versus frequency for a slab-loaded WR(19) waveguide. Slab is centered.

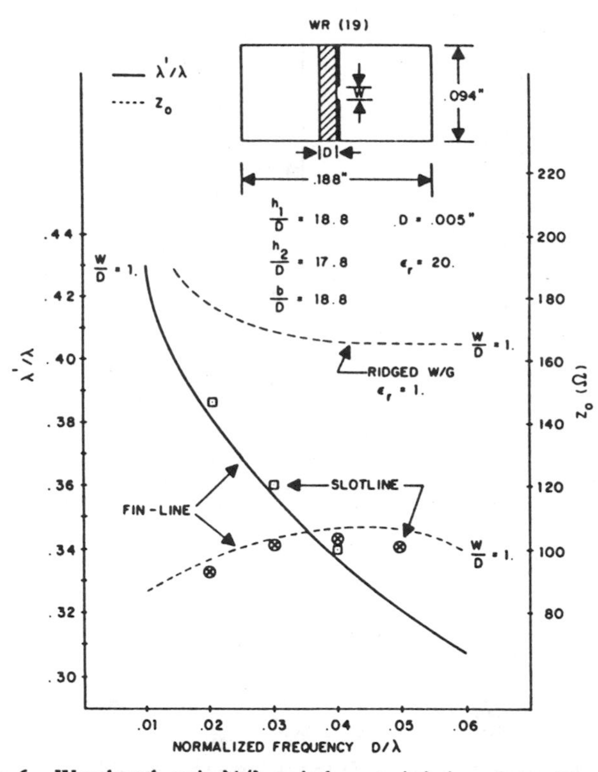

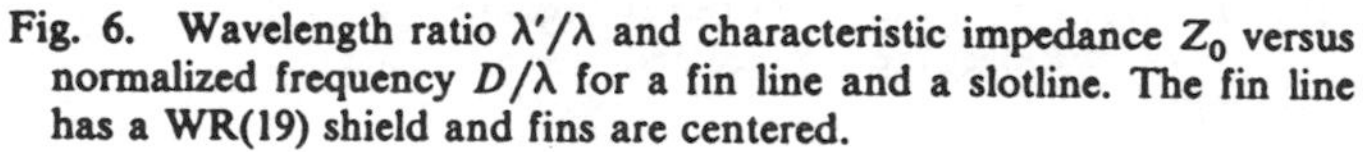

Fig. 6. Wavelength ratio λ'/λ and characteristic impedance Z_0 versus normalized frequency D/λ for a fin line and a slotline. The fin line has a WR(19) shield and fins are centered.

defined here) may easily be obtained from Z_{pv} as

$$Z_0 = Z_{pv}\left(\frac{E_x^{\text{edge}}}{E_x^{\text{ctr}}}\right)^2 \tag{17a}$$

where E_x^{edge} is the field at the edge of the slab and E_x^{ctr} is the field at the center of the slab. Thus

$$Z_0 = Z_{pv}\cos^2\frac{qc}{2s} \tag{17b}$$

where the various quantities on the right-hand side of (17b) are defined in [16]. The characteristic impedance of a slab-loaded guide was computed using both the spectral-domain method and Vartanian's equations. The results are compared in Fig. 5 where the discrete points are calculated from (17a). It can be seen that the results agree very closely.

C. Slotline

If $W/D < 2$ and D/λ and ϵ_r are sufficiently large, the field is tightly bound to the slot. For this condition, the presence of the shield will have little effect if the walls are sufficiently far removed from the slot. In this case, the fin line will behave like a slotline. This behavior is illustrated in Fig. 6 where wavelength and characteristic impedance of a fin line with $W/D = 1$, $\epsilon_r = 20$ have been plotted. The discrete points are the values of the same parameters for a slotline with equal W/D as obtained from [17]. It can be seen that the results agree closely for the two structures. The agreement is within 1 percent for wavelength and 5 percent for impedance. The impedance of the ridged-waveguide substructure has also been plotted for reference.

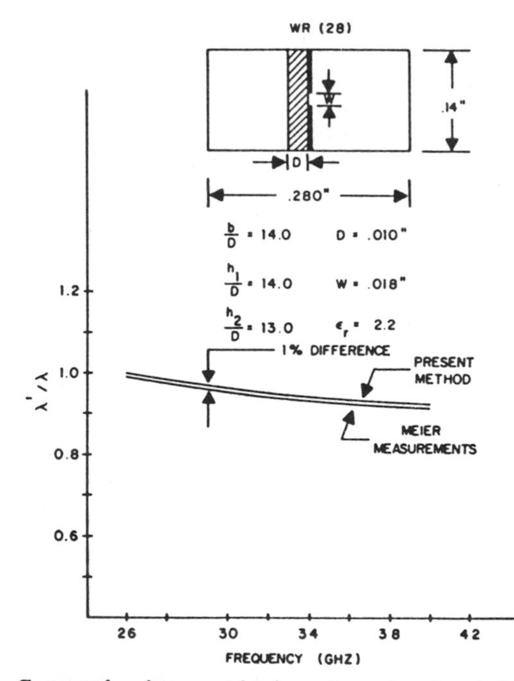

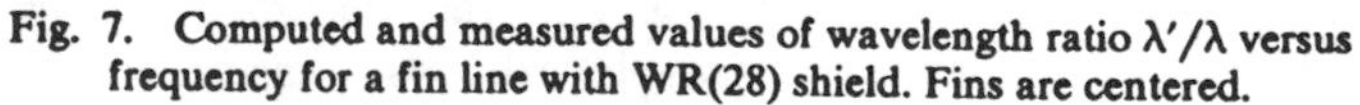

Fig. 7. Computed and measured values of wavelength ratio λ'/λ versus frequency for a fin line with WR(28) shield. Fins are centered.

D. Measurement

Fin-line wavelength measurements have been reported by Meier [1]. A computer run was made for a structure with the physical parameters reported in [1], and the results were compared with the measured values. This comparison appears in Fig. 7, where the values are seen to be in agreement to within about 1 percent.

E. Approximations

It has been suggested [1] that for low-dielectric-constant substrates the fin-line wavelength and impedance are approximated well by the relations

$$\lambda'/\lambda \doteq \frac{1}{\sqrt{k_e - \left(\frac{\lambda}{\lambda_c}\right)^2}} \tag{18a}$$

$$Z_0 \doteq \frac{Z_{0\infty}}{\sqrt{k_e - \left(\frac{\lambda}{\lambda_c}\right)^2}} \tag{18b}$$

where k_e is the effective dielectric constant and λ_c and $Z_{0\infty}$ are the cutoff wavelength and high-frequency limit impedance of the ridged-waveguide substructure. k_e is assumed invariant with frequency through the waveguide band and may be determined experimentally.

To check the accuracy of this approximation, k_e was determined for a fin line in WR(28) guide at $f=33$ GHz. This was accomplished using the numerical value of λ'/λ from Fig. 8 and the value of λ_c from [13] in (18a). The value of k_e thus determined was then assumed constant and (18) were used to calculate λ'/λ and Z_0 at $f=40$ GHz. These values were, in turn, compared with those in Fig. 8 for $W/b=0.1$, 0.5, and 1.0. Wavelength ratios agreed to within 0.2 percent while the difference in impedance values was 7–9 percent. The approximate expressions are thus useful for fin lines using low-dielectric-constant substrates. It is clear from Figs. 5 and 6, however, that there is no possibility that (18b) can be used to approximate the impedance if ϵ_r is high.

IV. Fin-Line Design Curves

In general, the publication of design curves for a structure like the fin line is problematic since there are several independently variable parameters which describe the structure. Practical considerations alleviate this difficulty to some extent, however. First, fin lines are generally enclosed with a shield that is compatible with the dimensions of the standard rectangular waveguides for the millimeter wavebands. Above 22 GHz (WR(34)), all these guides have aspect ratios $b/a=0.5$. Further, the fins are most often centered in the guide and printed using $D=$ 0.005-in substrate with $\epsilon_r=2.2$. It thus seems that it would be useful to provide design curves for structures with these parameters. This has been done for the 26.5–40-GHz, 40–60-GHz, and 60–90-GHz waveguide bands. The results appear in Figs. 8–10.

An inspection of Figs. 8–10 reveals that even with (centered) fin separations of a few mils, it is difficult to achieve low values of characteristic impedance. The lower limit for the parameters chosen here is in the range of 125–150 Ω. In some applications, it is desirable to have a lower impedance which raises the question as to how this might be achieved. One method is to increase the dielectric constant. This is clear from a comparison of Figs. 9 and 6. For a $W=0.005$-in slot, use $W/D=1$ in Fig. 6 and $W/b=0.053$ in Fig. 9 to find that the impedance is decreased from about 165 to 100 Ω when the dielectric constant is increased from $\epsilon_r=2.2$ to $\epsilon_r=20$. This approach has a disadvantage since the wavelength is decreased by a factor of 3 making already small circuit dimensions even smaller.

Another possibility for achieving a lower Z_0 might be to relocate the fins toward the sidewall of the guide. Fig. 11 illustrates the characteristics of a fin line with the fins located midway between the sidewall and the center of a WR(19) guide. Comparing Fig. 11 with Fig. 9, it can be seen that the relocation of the fins causes a small increase in the wavelength and pronounced changes in the impedance. The change in wavelength, for example, is 4 percent at 40 GHz and 2 percent at 60 GHz for $W/b=1$. The change is less for smaller values of W/b. With regard to characteristic impedance, it can be seen that the values are relatively unchanged for small values of W/b. For this situation, the line behaves like a slotline and there is little effect produced by the presence of the shield. For $W/b=1$, however, the impedance decreases significantly, from about 500 Ω for the centered configuration to about 275 Ω for the off-center configuration. In this case, the guide is also loaded, and since the low dielectric constant causes little change in the empty guide fields, the result agrees with expectations based on the voltage-impedance definition for an empty guide. The interesting feature of the results here is that the impedance for $W/b=1$ falls below that for $W/b=0.5$. Thus for the off-center configurations, the impedance first increases with W/b, then reaches a maximum, and decreases until $W/b=1$. Overall, it seems clear that relocation of the fins toward the side of the guide does not result in lower structure impedance when W/b is small. It should also be noted that the ridged waveguide data necessary for the evaluation of the approximate expressions given by (18) are not readily available in the literature.

Probably the best way to realize a lower structure impedance is to use a single-fin structure. If a single fin were separated from the bottom wall of the shield by a distance $W/2$, then for small values of W/b the impedance would be approximately one half that for a normal fin line with fin spacing W. For example, Fig. 8 shows that the impedance of a WR(28) fin line with $W/b=0.05$ ($W=0.007$ in) is approximately $Z_0=160$ Ω. Thus a single fin in the same shield would result in a structure impedance of approximately $Z_0=80$ Ω if located a distance $W=0.0035$ in from the bottom wall of the shield. This configuration has the additional advantage that the bottom wall of the shield may be used as a heat sink.

V. Conclusions

A transform method of obtaining the wavelength and characteristic impedance for the dominant mode of the fin line has been presented. It has been shown that a matrix formulation of the problem permits the elements of the

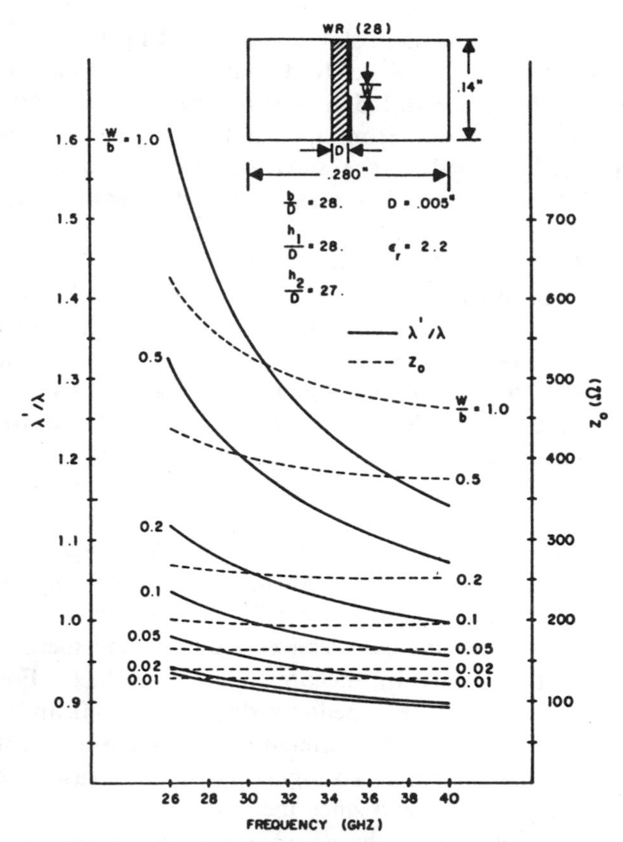

Fig. 8. Wavelength ratio λ'/λ and characteristic impedance Z_0 versus frequency for a fin line with WR(28) shield. Fins are centered.

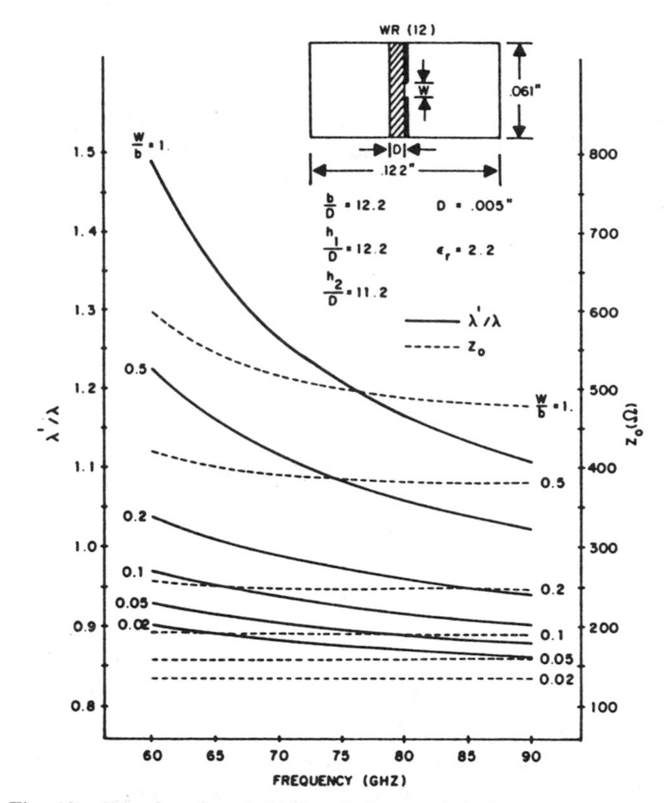

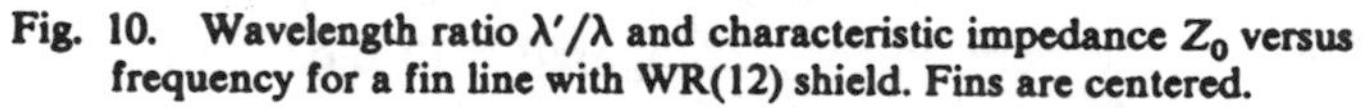
Fig. 10. Wavelength ratio λ'/λ and characteristic impedance Z_0 versus frequency for a fin line with WR(12) shield. Fins are centered.

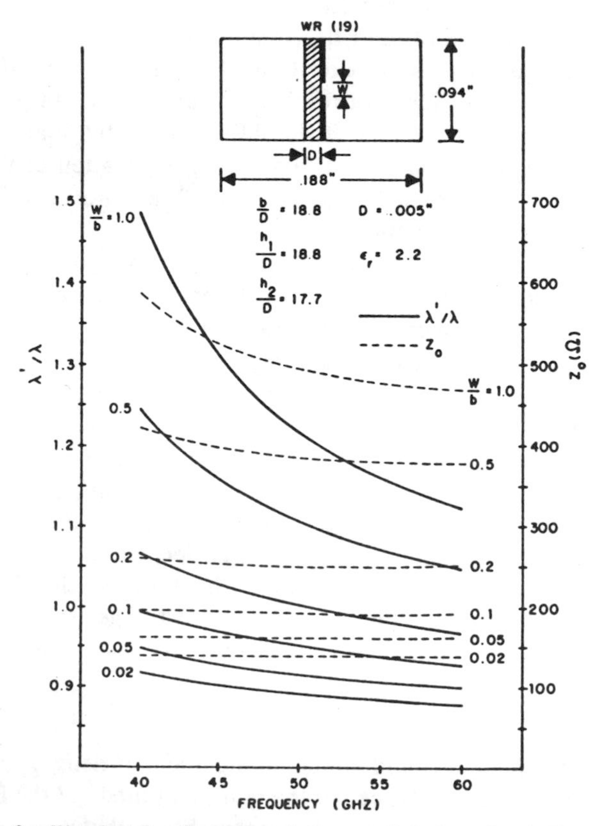

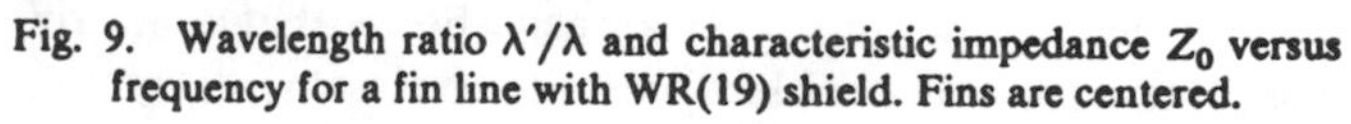
Fig. 9. Wavelength ratio λ'/λ and characteristic impedance Z_0 versus frequency for a fin line with WR(19) shield. Fins are centered.

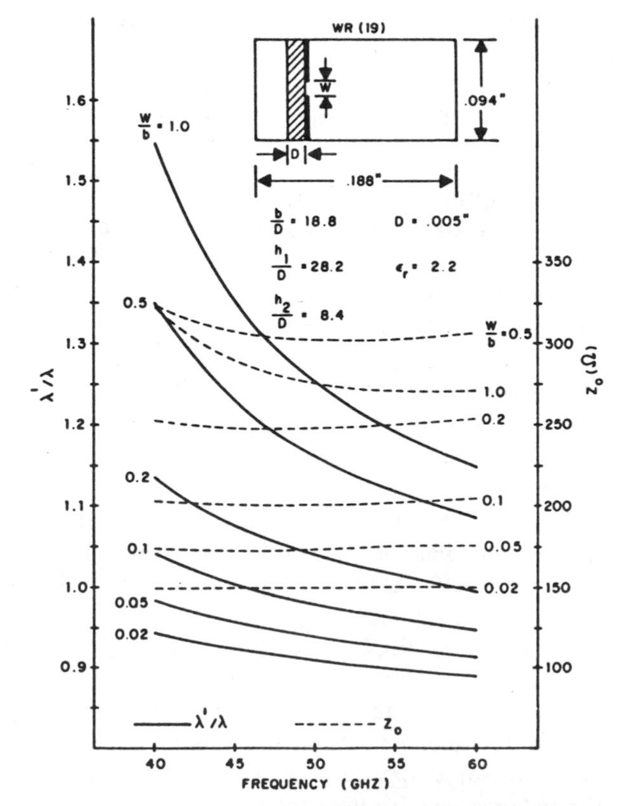

Fig. 11. Wavelength ratio λ'/λ and characteristic impedance Z_0 versus frequency for a fin line with WR(19) shield. Fins are located halfway between side wall and guide center.

dyadic Green's function to be calculated, and circumvents the extensive algebraic manipulation associated with formulations described previously. Numerical results obtained using this method have been presented and compared to other existing data. Good agreement was obtained in all cases thus establishing the accuracy and applicability of the method for the full range of structure parameters. Design curves have been included here for millimeter-wave fin lines of practical interest. Both center and off-center fin locations have been discussed, and the off-center location was shown to result in no significant change in impedance for small values of W/b. Lower impedance may be realized, however, by using a single-fin configuration.

It is clear from the results presented here that the fin line may exhibit the characteristics of a ridged waveguide, slotline, or dielectric slab-loaded waveguide, depending upon the values of the various fin-line parameters. All of these structures are fin-line substructures.

References

[1] P. J. Meier, "Integrated fin-line millimeter components," *IEEE Trans. Microwave Theory Tech.* vol. MTT-22, pp. 1209–1216, Dec. 1974.

[2] ——, "Millimeter integrated circuits suspended in the *E*-plane of rectangular waveguide," *IEEE Trans. Microwave Theory Tech.*, vol. MTT-26, pp. 726–733, Oct. 1978.

[3] ——, "Printed circuit balanced mixer for the 4 and 5 mm bands," in *IEEE MTT-S Symp. Dig.*, pp. 84–86, May 1979.

[4] W. Kpodzo *et al.*, "A quadriphase modulator in fin-line technique," in *IEEE MTT-S Symp. Dig.*, pp. 119–121, May 1979.

[5] A. M. K. Saad and K. Schunemann, "A simple method for analyzing fin-line structures," *IEEE Trans. Microwave Theory Tech.*, vol. MTT-26, pp. 1002–1007, Dec. 1978.

[6] W. Hoefer, "Fin-line parameters calculated with the TLM method," in *IEEE MTT-S Symp. Dig.*, pp. 341–343, May 1979.

[7] T. Itoh, "Spectral domain analysis of dominant and higher order modes in fin-lines," in *IEEE MTT-S Symp. Dig.*, pp. 344–345, May 1979.

[8] T. Itoh and R. Mittra, "Dispersion characteristics of slot lines," *Electron. Lett.*, vol. 7, pp. 364–365, July 1971.

[9] ——, "Spectral-domain approach for calculating the dispersion characteristics of microstrip lines," *IEEE Trans. Microwave Theory Tech.*, (Short Papers), vol. MTT-21, pp. 496–499, July 1973.

[10] ——, "A technique for computing dispersion characteristics of shielded microstrip lines," *IEEE Trans. Microwave Theory Tech.* (Short Papers), vol. MTT-22, pp. 896–898, Oct. 1974.

[11] J. B. Knorr and K -D. Kuchler, "Analysis of coupled slots and coplanar strips on dielectric substrate," *IEEE Trans. Microwave Theory Tech.*, vol. MTT-23, pp. 541–548, July 1975.

[12] J. B. Knorr and A. Tufekcioglu, "Spectral-domain calculation of microstrip characteristic impedance," *IEEE Trans. Microwave Theory Tech.*, vol. MTT-23, pp. 725–728, Sept. 1975.

[13] S. Hopfer, "The design of ridged waveguides," *IRE Trans. Microwave Theory Tech.*, vol. MTT-3, pp. 20–29, Oct. 1955.

[14] R. O. Lagerlöf, "Ridged waveguide for planar microwave circuits," *IEEE Trans. Microwave Theory Tech.*, vol. MTT-21, pp. 499–501, July 1973.

[15] R. Collin, *Field Theory of Guided Waves*. New York: McGraw-Hill, 1960.

[16] P. H. Vartanian *et al.*, "Propagation in dielectric slab loaded rectangular waveguide," *IRE Trans. Microwave Theory Tech.*, vol. MTT-6, pp. 215–222, Apr. 1958.

[17] E. A. Mariani *et al.*, "Slot line characteristics," *IEEE Trans. Microwave Theory Tech.* (1969 Symp. Issue), vol. MTT-17, pp. 1091–1096, Dec. 1969.

Spectral Domain Analysis of Dominant and Higher Order Modes in Fin-Lines

LORENZ-PETER SCHMIDT AND TATSUO ITOH, SENIOR MEMBER, IEEE

Abstract—The spectral domain analysis is applied for deriving dispersion characteristics of dominant and higher order modes in fin-line structures. In addition to the propagation constant, the characteristic impedance is calculated based on the power–voltage definition. Numerical results are compared for different choices of basis functions and allow to estimate the accuracy of the solution.

I. Introduction

THE FIN-LINE structure is a special printed transmission line proposed for millimeter wave integrated circuits in 1973 by Meier [1]. Since then, a number of millimeter-wave components have been developed in the fin-line form (e.g., [2]). The single-mode range of frequency is relatively wide, as the fin-line somewhat resembles the ridged waveguide. Propagation characteristics of fin-line structures have been investigated by a number of workers such as Hofmann [3], Hoefer [4], [5], and Saad and Begemann [10]. In [3], which is based on Galerkin's method in the space domain, sinusoidal functions are used as expansion functions. Hence a comparatively large number of expansion functions is required to obtain accurate results and, in addition, relative convergence problems occur and have to be handled carefully. On the other hand, some engineering approximations are involved in the work in [4].

In the present paper, fin-line structures are analyzed using the spectral domain technique, which has been developed for the analysis of various printed transmission lines for microwave integrated circuits [6], [7]. In this method, the information on the propagation constant at a given frequency is extracted from algebraic equations that relate Fourier transforms of the currents on the fins to those of the electric field in the dielectric–air interface. These equations are *discrete* Fourier transforms[1] of coupled integral equations one would obtain if the formulation is done in the space domain. Obviously, algebraic equations are much easier to handle in numerical processing. In addition to standard features of the spectral domain technique, the present work contains the following provisions.

1) The accuracy of the method is checked by comparing results obtained from three different choices of basis functions. A convergence check is also performed by increasing the number of basis functions for one of these sets.

2) In addition, dispersion curves for higher order modes are presented. For practical applications, the knowledge of higher order modes is important, because often a single mode operation is required.

3) Another important quantity for design purposes is the characteristic impedance of the dominant mode. By applying a definition suitable to fin-line structures, useful results for the characteristic impedance could be obtained and are presented in this paper.

II. Formulation of the Eigenvalue Problem

Since the details of the spectral domain method itself have been reported in [6] and [7], only the key steps will be given here. The method of using alternative sets of basis functions for accuracy checks has recently been

Manuscript received March 17, 1980; revised April 15, 1980. This work was supported in part by the U.S. Army under Grant DAAG29-78-G-0145.

L.-P. Schmidt is with AEG-Telefunken N1 E32, Ulm, West Germany, on leave from the Department of Electrical Engineering, The University of Texas at Austin, Austin, TX 78712.

T. Itoh is with the Department of Electrical Engineering, The University of Texas at Austin, Austin, TX 78712.

[1] Henceforth referred to as Fourier transform.

Reprinted from *IEEE Trans. Microwave Theory Tech.*, vol. MTT-28, no. 9, pp. 981–985, September 1980.

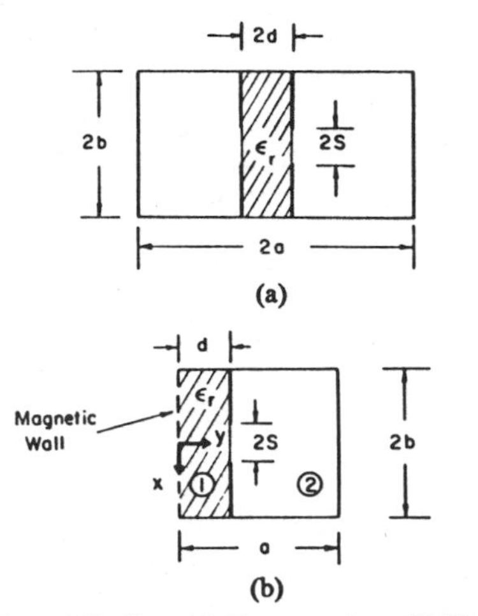

Fig. 1. Bilateral fin-line. (a) Cross section. (b) Equivalent structure.

employed for a higher order mode analysis of microstrip lines [8]. The significance of the modifications will be pointed out in this paper.

Several versions of fin-lines have been proposed, including bilateral, unilateral, and antipodal fin arrangements. Although the present method is applicable to other types of fin-line structures, we will formulate the problem for the bilateral fin-line, the cross section of which is shown in Fig. 1(a). Because of the symmetry, we only need to consider the half-structure given in Fig. 1(b).

Since the modal field in the fin-line is of hybrid type, the fields in regions 1 (dielectric) and 2 (air) can be derived from two scalar potentials $\phi_i(x,y)$ and $\psi_i(x,y)$, $i=1,2$, for instance

$$E_{zi}=j\frac{k_i^2-\beta^2}{\beta}\phi_i(x,y) \tag{1}$$

$$H_{zi}=j\frac{k_i^2-\beta^2}{\beta}\psi_i(x,y) \tag{2}$$

where $i=1,2$ signifies the region, k_i is the wavenumber in region i, and β is the propagation constant of the mode in the z direction. The time and z dependence of the field $\exp(j\omega t-j\beta z)$ is omitted throughout the paper. All other field components are derivable from Maxwell's equations.

In the spectral domain approach, the potentials ϕ_i and ψ_i as well as all the field quantities are Fourier transformed via

$$\tilde{\phi}_i(n,y)=\int_{-b}^{b}\phi_i(x,y)\exp(j\hat{k}_n x)\,dx \tag{3}$$

where $\hat{k}_n=n\pi/b$ for all odd (in E_z) modes including the dominant one and $\hat{k}_n=(n-1/2)\pi/b$ for the even modes. $\tilde{\phi}_i(n,y)$ and $\tilde{\psi}_i(n,y)$ now satisfy Helmholtz equations, e.g.,

$$\left(\frac{d^2}{dy^2}-\gamma_i^2\right)\tilde{\phi}_i(n,y)=0 \tag{4}$$

where

$$\gamma_1=\sqrt{\hat{k}_n^2+\beta^2-\epsilon_r k_0^2} \qquad \gamma_2=\sqrt{\hat{k}_n^2+\beta^2-k_0^2}$$

and k_0 is the free space wavenumber. Since the tangential electric fields E_{z2} and E_{x2} must be zero at $y=a, |x|<b$, and the tangential magnetic fields H_{z1} and H_{x1} be zero at $y=0, |x|<b$, appropriate solutions to the above equations are

$$\tilde{\phi}_1(n,y)=A_n^e\cosh\gamma_1 y \qquad \tilde{\phi}_2(n,y)=B_n^e\sinh\gamma_2(a-y)$$

$$\tilde{\psi}_1(n,y)=A_n^h\sinh\gamma_1 y \qquad \tilde{\psi}_2(n,y)=B_n^h\cosh\gamma_2(a-y).$$

By applying the interface conditions at $y=d$, the unknown coefficients A_n^e, A_n^h, B_n^e, and B_n^h can be eliminated, and we obtain two coupled algebraic equations

$$Y_{xx}\tilde{E}_x+Y_{xz}\tilde{E}_z=j\omega\mu_0\tilde{J}_x \tag{5}$$

$$Y_{zx}\tilde{E}_x+Y_{zz}\tilde{E}_z=j\omega\mu_0\tilde{J}_z \tag{6}$$

where

$$Y_{xx}=(\epsilon_r k_0^2-\beta^2)\frac{\tanh\gamma_1 d}{\gamma_1}+(k_0^2-\beta^2)\frac{\coth\gamma_2 h}{\gamma_2} \tag{7}$$

$$Y_{xz}=Y_{zx}=\beta\hat{k}_n\left(\frac{\tanh\gamma_1 d}{\gamma_1}+\frac{\coth\gamma_2 h}{\gamma_2}\right) \tag{8}$$

$$Y_{zz}=(\epsilon_r k_0^2-\hat{k}_n^2)\frac{\tanh\gamma_1 d}{\gamma_1}+(k_0^2-\hat{k}_n^2)\frac{\coth\gamma_2 h}{\gamma_2} \tag{9}$$

$$h=a-d$$

all are known. $\tilde{E}_x$, $\tilde{E}_z$, and $\tilde{J}_x$, $\tilde{J}_z$ are Fourier transforms of the unknown tangential electric field in the gap $(y=d, |x|<s)$ and the unknown current components on the fins $(y=d, s<|x|<b)$. Up to this stage the method of analysis is exact. In the following we present a solution based on Galerkin's method.

To this end, the unknown aperture fields $\tilde{E}_x$ and $\tilde{E}_z$ are expanded in terms of known basis functions $\tilde{\xi}_i$, $\tilde{\eta}_i$

$$\tilde{E}_x(\hat{k}_n)=\sum_{i=1}^{M}c_i\tilde{\xi}_i(\hat{k}_n) \tag{10}$$

$$\tilde{E}_z(\hat{k}_n)=\sum_{j=1}^{N}d_j\tilde{\eta}_j(\hat{k}_n) \tag{11}$$

where $\tilde{\xi}_i$ and $\tilde{\eta}_j$ are Fourier transforms of $\xi_i(x)$ and $\eta_j(x)$, which are chosen to be zero except for $|x|<s$.

Now (10) and (11) are substituted into (5) and (6) and the inner products of the resulting equations with $\tilde{\xi}_i$ and $\tilde{\eta}_j$, respectively, are obtained. This results in a homogeneous matrix equation for the unknown expansion coefficients c_i and d_j

$$\sum_{i=1}^{M}K_{pi}^{xx}c_i+\sum_{j=1}^{N}K_{pj}^{xz}d_j=0, \qquad p=1,\cdots,M \tag{12}$$

$$\sum_{i=1}^{M}K_{qi}^{zx}c_i+\sum_{j=1}^{N}K_{qj}^{zz}d_j=0, \qquad q=1,\cdots,N \tag{13}$$

where

$$K_{pi}^{xx}=\sum_{n=0}^{\infty}\tilde{\xi}_p(\hat{k}_n)Y_{xx}(\beta,\hat{k}_n)\tilde{\xi}_i(\hat{k}_n) \tag{14a}$$

$$K_{pj}^{xz}=\sum_{n=0}^{\infty}\tilde{\xi}_p(\hat{k}_n)Y_{xz}(\beta,\hat{k}_n)\tilde{\eta}_j(\hat{k}_n) \tag{14b}$$

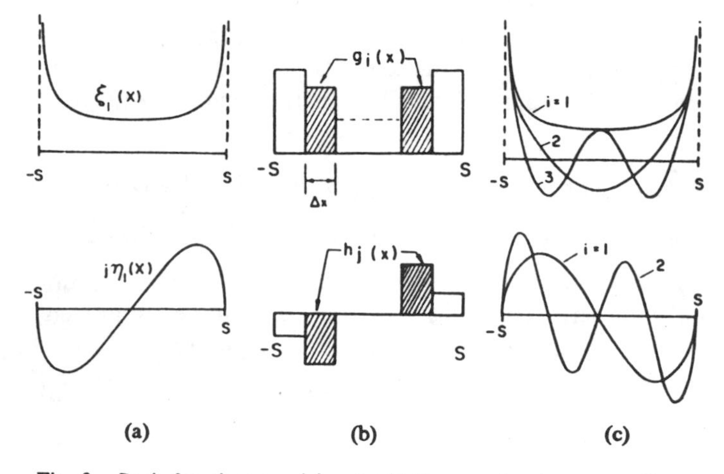

Fig. 2. Basis functions used for slot-field expansion (corresponding to (15)–(20)).

$$K_{qi}^{zx}=\sum_{n=0}^{\infty}\tilde{\eta}_q(\hat{k}_n)Y_{zx}(\beta,\hat{k}_n)\tilde{\xi}_i(\hat{k}_n) \quad (14c)$$

$$K_{qj}^{zz}=\sum_{n=0}^{\infty}\tilde{\eta}_q(\hat{k}_n)Y_{zz}(\beta,\hat{k}_n)\tilde{\eta}_j(\hat{k}_n). \quad (14d)$$

Equating the determinant of the coefficient matrix associated with (12) and (13) to zero, we finally obtain the eigenvalue equation, and its solutions are the desired propagation constants of the dominant and higher order modes.

III. Choices of Basis Functions

One of the features of the spectral domain method applied in the described manner is that quite accurate solutions result even if an extremely small size matrix such as $M=N=1$ is used. This takes place because certain qualitative natures such as the edge condition of the aperture electric field can be incorporated in the choice of basis functions.

In the present case three different choices of basis functions have been used, all of them being readily Fourier transformed analytically.

1) Qualified one-term expansions satisfying the edge condition at $|x|=s$ (see Fig. 2(a)):

$$\xi_1(x)=\frac{1}{\sqrt{s^2-x^2}} \quad (15)$$

$$\eta_1(x)=x\cdot\sqrt{s^2-x^2}\,. \quad (16)$$

2) Trains of rectangular pulses with unknown amplitudes (Fig. 2(b)):

$$\xi_i(x)=\begin{cases}1, & (i-1)\cdot\Delta x<|x|<i\cdot\Delta x\\ 0, & \text{elsewhere}\end{cases} \quad (17)$$

$$\eta_j(x)=\begin{cases}1, & (j-1)\cdot\Delta x'<x<j\cdot\Delta x'\\ -1, & -j\cdot\Delta x'<x<(j-1)\cdot\Delta x'\\ 0, & \text{elsewhere}\end{cases} \quad (18)$$

and $\Delta x=s/M$, $\Delta x'=s/N$.

3) Sinusoidal functions modified by an "edge condition" term (Fig. 2(c)):

$$\xi_i(x)=\frac{\cos\{(i-1)\pi(x/s+1)\}}{\sqrt{1-(x/s)^2}} \quad (19)$$

$$\eta_j(x)=\frac{\sin\{j\pi(x/s+1)\}}{\sqrt{1-(x/s)^2}}. \quad (20)$$

The second set of basis functions is numerically less advantageous than the others, because it requires an inherently larger matrix order, and the edge condition cannot be directly incorporated. However, it is very flexible and general and the expansion coefficients c_i, d_j are adjusted automatically to represent the aperture field distributions.

IV. The Characteristic Impedance

In addition to the propagation constants of the propagating modes, the characteristic impedance of the dominant mode is an important quantity for the design of microwave and millimeter-wave integrated circuits. Three definitions are possible for the impedance Z_c: they are V_x/I, $2P/I^2$, and $V_x^2/(2P)$ where P is the transmitted power, V_x the slot voltage, and I the current on the fins and waveguide walls. For reasons pointed out earlier [9], an impedance definition via slot voltage and transported power

$$Z_c=\frac{V_x^2}{(2P)} \quad (21)$$

was applied to the bilateral fin-line considered here. The calculation of the slot voltage

$$V_x=\int_{-s}^{+s}E_x(x,d)\,dx \quad (22)$$

can directly be accomplished, since the x dependence of the integral is known from the slot-field series expansion. The expansion coefficients c_i and d_j result from inverting the eigenvalue matrix equation, after the propagation constant has been computed. The transported power

$$P=\mathrm{Re}\int_{-b}^{+b}\int_0^a(E_xH_y^*-E_yH_x^*)\,dy\,dx \quad (23)$$

can be transformed to the spectral domain using Parseval's theorem

$$\tilde{P}=\mathrm{Re}\int_0^a(\tilde{E}_x\tilde{H}_y^*-\tilde{E}_y\tilde{H}_x^*)\,dy\,. \quad (24)$$

Since the spectral field components have already been derived in the course of formulating the eigenvalue problem, this integral can directly be solved and, thus, the calculation of the characteristic impedance completed.

It should be noted that the impedance defined in (21)–(24) is for one half of the bilateral fin-line in Fig. 1(b). The impedance thus defined may be called the one associated with one of the slots. The *total* impedance is one half of the values computed by the definition above, because the total power is twice that of (23); whereas, V_x given by (22)

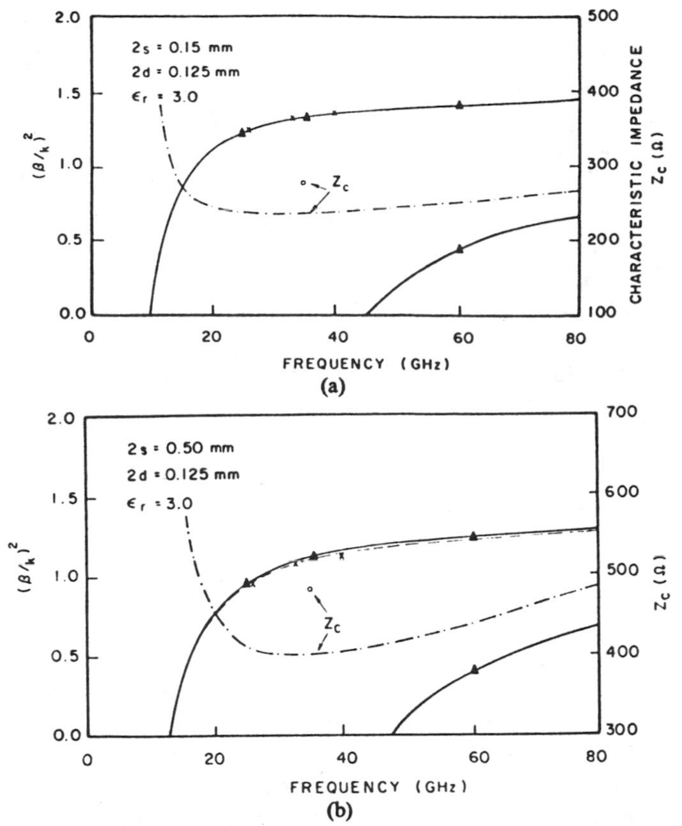

Fig. 3. Dispersion characteristics of dominant and first higher order mode and characteristic impedance of dominant mode. (200 spectral terms). (a) $2S=0.15$ mm. (b) $2S=0.50$ mm.——one-term expansion according to (15), (16), respectively, (19), (20), and $M=N=1$. ----- two-term expansion according to (19), (20) and $M=N=2$. (Indistinguishable from ——in (a)). ▲▲ expansion with rectangular pulses according to (17), (18) and $M=N=7$. × × Hofmann's results [3]. ∘ Impedance by Hofmann [3]. –·– Characteristic impedance for $M=N=1$ and 2.

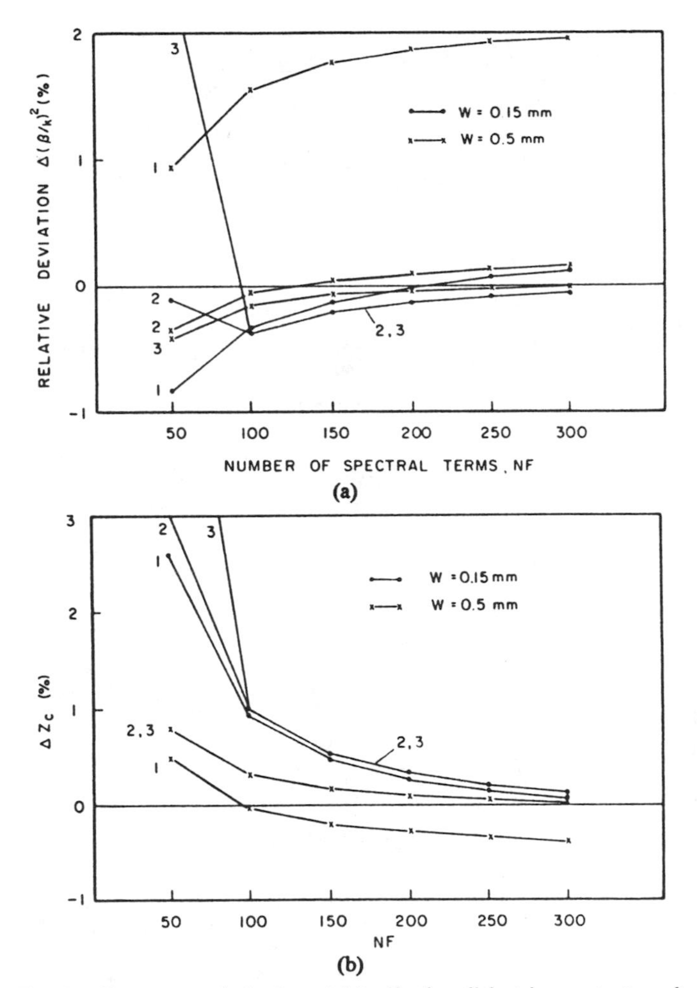

Fig. 4. Convergence behavior of (a) effective dielectric constant, and (b) characteristic impedance for $M=N=1,2,3$. (Expansion according to (19), (20); reference value: $NF=500$, $M=N=4$; $f=30$ GHz; $\epsilon_r=3$).

is identical to both slots. It is readily shown that total impedances based on other definitions are also one half of those for one slot.

V. Numerical Results

Dispersion characteristics for different choices of basis functions have been computed, including the effective dielectric constant $\epsilon_{eff}=\beta^2/k_0^2$ of the dominant and the first higher order mode and the characteristic impedance Z_c of the dominant mode (Fig. 3). Note that this definition of ϵ_{eff} is different from the one used in [1]. The latter provides information as to what kind of effects the dielectric substrate has in reference to the air-filled fin-line. Our present definition is simply square of the normalized propagation constant and hence, ϵ_{eff} here compares the guided wave in the structure with the plane wave in free space. Since the fin-line is not a quasi-TEM waveguide, the present ϵ_{eff} may be less than unity. This definition has also been used by Hofmann [3] and Jansen [9].

The dimensions of the shielding walls were chosen to coincide with the WR-28 waveguide usually utilized for 26.5 to 40-GHz operation. It is clearly seen that the dominant fin-line mode is not quasi-TEM but resembles somewhat the ridged waveguide dominant mode as pointed out earlier [1], [4].

Comparing the results for different sets of basis functions, a good conformity of all solutions can be testified for a small slotwidth. For broader slots (Fig. 3(b)), however, a one-term expansion results in relative errors of up to 2 percent. This fact is confirmed in Fig. 4, where the convergence behavior of ϵ_{eff} and Z_c is presented for one-, two-, and three-term expansions and a varying number of spectral terms. The solution for ϵ_{eff}, using a one-term expansion, obviously deviates about 2 percent from the reference value for a broad slot, whereas, the deviation of Z_c is not so significant. Further investigations have shown that for all normally used slotwidths, a two-term expansion according to (19), (20) and a number of 250 spectral terms are sufficient for very accurate solutions.

Fig. 5 shows modal dispersion characteristics for two more substrates and several slotwidths. In addition, in Fig. 5(b) the existence of modes resulting from an electric wall symmetry at $y=0$ is pointed out. Though modes with this kind of symmetry may not be excited by the dominant H_{10} mode of the empty waveguide, they have to be taken into account at all discontinuities that are not symmetric with respect to the $y=0$ plane.

Figs. 3 and 5 confirm Hofmann's statement in [3] that the characteristic impedance is fairly constant in the frequency range 30–40 GHz. Additionally these figures

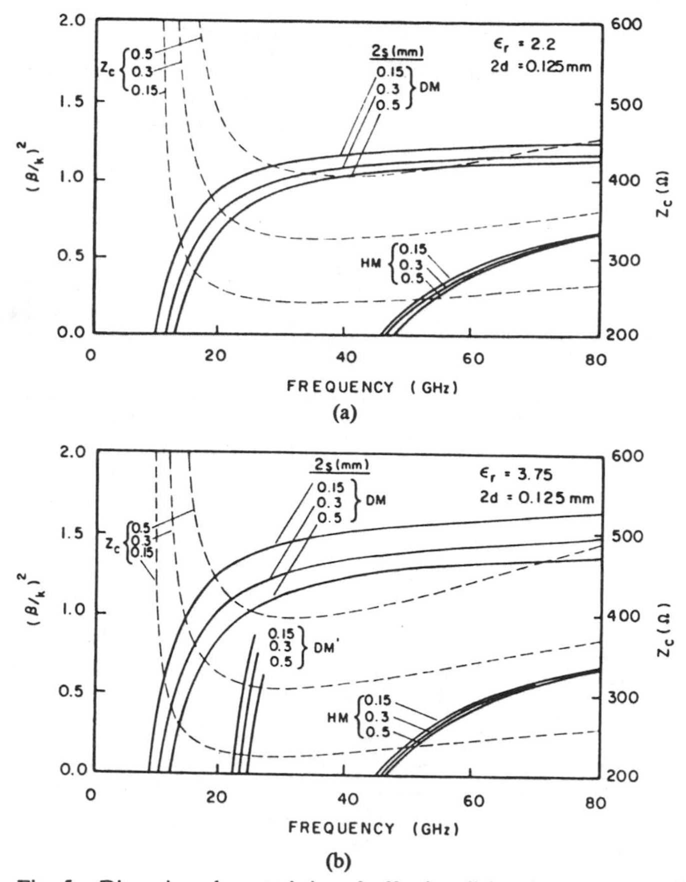

Fig. 5. Dispersion characteristics of effective dielectric constant and characteristic impedance for other substrates and different slotwidths $2s$. (*DM*-dominant mode, *HM*-higher order mode, *DM'*-dominant mode for the structure in Fig. 1(b) with magnetic wall replaced by an electric wall).

show a broad minimum of Z_c in this frequency range. Since we found that this phenomenon, which was not observed in Hofmann's work [3], is not due to numerical instability or convergence, we conjecture as follows. The main tendency of increasing impedance at higher frequencies can be interpreted with the help of the fact that the fields concentrate more and more in the vicinity of the slot as the frequency gets higher. This results in higher values of slot voltage V_x if the power P is set constant. Hence, $Z_c = V_x^2/(2P)$ becomes larger. Near the cutoff, however, another mechanism takes place. Since the power transmitted P in the z direction is zero at cutoff, Z_c becomes infinite. Since this cannot happen discontinuously, Z_c takes higher values again as the cutoff is approached, resulting in a broad minimum at intermediate frequencies. Since Hofmann [3] takes only the current on the fins into account in his definition $Z_c = V_x/I$, a different mechanism takes place for his impedance. At higher frequencies, the fin current increases as the fields concentrate near the slot, and the portion of current flowing on the waveguide walls decreases. Since the latter is neglected, Hofmann's impedance decreases at higher frequencies.

Hofmann's impedance values plotted in the figure are twice those found in his original paper. This is because he used the total impedance described earlier. Even after this factor of two is introduced, Hofmann's results deviate considerably from the present data. The deviations may be due to the fact that he used an impedance definition via slot voltage and fin current which seems to be questionable for broader slots, because the portion of the current on the waveguide wall which is neglected in his analysis becomes more important.

VI. Conclusions

Derived from a thorough dynamic spectral domain analysis, a wide variety of useful information about the bilateral fin-line has been given in this paper, including important aspects for practical application, as well as theoretical considerations about suitable choices of basis functions and the convergence behaviors of the solutions. Obviously, this method is applicable to other fin-line configurations as well.

References

[1] P. J. Meier, "Equivalent relative permittivity and unloaded *Q*-factor of integrated fin-line," *Electron. Lett.*, vol. 9, no. 7, pp. 162–163, Apr. 1973.

[2] H. Hofmann, H. Meinel, and B. Adelseck, "Integration of millimeter wave components," (in German), *NTZ* (*Commun. J.*) vol. 31, pp. 752–757, 1978.

[3] H. Hofmann, "Dispersion of planar waveguides for millimeter-wave application," *Arch. Elek. Ubertragung.*, vol. 31, no. 1, pp. 40–44, Jan. 1977.

[4] W. J. R. Hoefer, "Fin line design made easy," in *1978 IEEE MTT Symp.* (Ottawa, Canada), June 1978.

[5] W. J. R. Hoefer and A. Ros, "Fin-line parameters calculated with the TLM-method," in *1979 IEEE MTT Symp.*, (Orlando, FL), Apr. 1979.

[6] T. Itoh and R. Mittra, "Dispersion characteristics of slot lines," *Electron. Lett.*, vol. 7, no. 13, pp. 364–365, July 1971.

[7] T. Itoh, "Analysis of microstrip resonators," *IEEE Trans. Microwave Theory Tech.*, vol. MTT-22, pp. 946–952, Nov. 1974.

[8] C. Chang, MS thesis, Univ. Kentucky, Lexington, KY, 1978.

[9] R. Jansen, "Unified user-oriented computation of shielded, covered and open planar microwave and millimeter-wave transmission-line characteristics," *Microwaves Opt. Acoust.* vol. 3, no. 1, pp. 14–22, Jan. 1979.

[10] A. M. K. Saad and G. Begemann, "Electrical performance of fin-lines of various configurations," *Microwaves Opt. Acoust.*, vol. 1, no. 2, pp. 81–88, Jan. 1977.

An Accurate, Unified Solution to Various Fin-Line Structures, of Phase Constant, Characteristic Impedance, and Attenuation

D. MIRSHEKAR-SYAHKAL AND J. BRIAN DAVIES, MEMBER, IEEE

***Abstract*—The analysis of several fin-line configurations (unilateral fin-line, bilateral fin-line, antipodal fin-line, and coupled fin-lines) has been completed accurately. In this unified method, propagation constant is achieved via the generalized spectral domain technique where the basis functions for the bounded and unbounded fields are chosen to be trigonometric functions and Legendre polynomials, respectively. The conduction loss and dielectric loss solution for the first time are found through a perturbation method. The conductor loss so derived is believed to be sufficiently accurate for practical purposes. Characteristic impedances of these transmission lines using tentative definitions have been presented. The CPU time on an IBM 360/65 for calculation of the mentioned parameters does not exceed five seconds if the fourth-order solution in the spectral analysis gives the required accuracy. The programs are also capable of detection of higher order modes.**

Manuscript received July 29, 1981; revised March 3, 1982.

The authors are with the Department of Electronic and Electrical Engineering, University College London, London, England, WC1E 7JE.

Reprinted from *IEEE Trans. Microwave Theory Tech.,* vol. MTT-30, no. 11, pp. 1854–1861, November 1982.

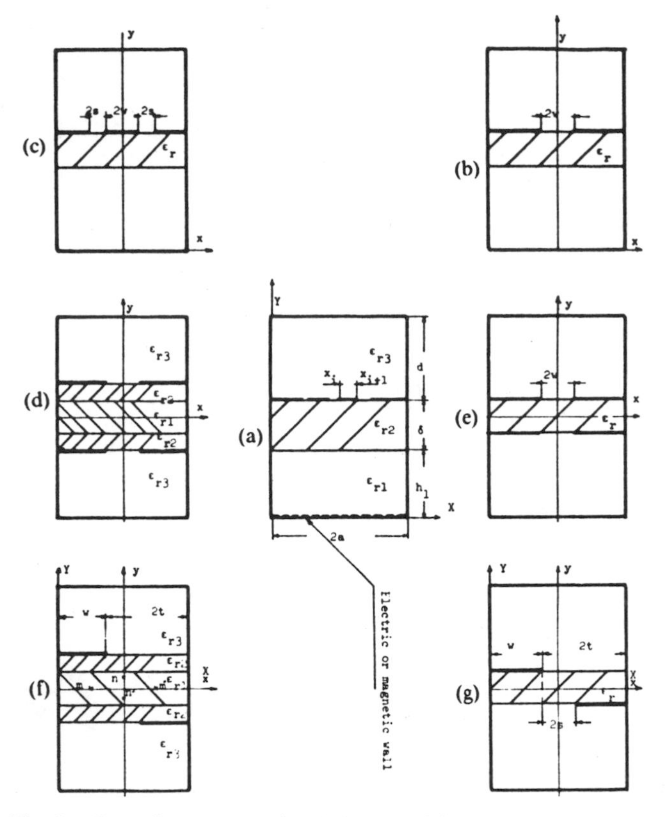

Fig. 1. General structure and various type of finlines. Numerical examples are performed for structures (a), (b), (e), and (g), where the outside cover is WG-22 waveguide and the dielectric is placed in the middle of its broad wall. In all the mentioned cases, dielectric thickness = 0.127 mm, $\epsilon_r = 2.22$, $\tan\delta_i = 2\times10^{-4}$, and $\rho = 3\times10^{-8}$ $\Omega\cdot$m.

I. Introduction

WORK AT millimeter bands has presented a need for a new low-loss transmission line capable of being interfaced with other integrated components. Fin-lines, viz. metallic strips etched on plastic or ceramic substrate embodied in the trunk of the standard rectangular waveguides have been the first guiding structures fulfilling these engineering requirements. Many descriptions of these lines can be found in the literature [1], [2]. Alternative transmission lines at millimeter band are image lines [3]. Although they present lower losses than fin-line, difficulty in making them easily adapted to conventional microwave components has left them second in application to fin-lines.

In spite of scattered results in the literature on the phase constant and characteristic impedance of some fin-lines, no account of losses has yet been given. Therefore, the problem of optimum design of the fin-line, involving the conductor and dielectric losses has remained unsolved. One must bear in mind that, as the gap in fin-lines becomes smaller and smaller, the conductor loss may even exceed the microstrip loss which can be achieved at the same band as fin-line.

The technique implemented to get the phase constants of all fin-lines, including unilateral, bilateral, antipodal, and coupled fin-line, Fig. 1, is the generalized spectral domain approach developed by the authors [4], [5]. The basis functions as before [5] are chosen to be Legendre polynomials for the unbounded field components near the sharp edges of fins and/or strips. The reason for the choice of Legendre polynomials is not just because they enjoy a closed-form Fourier transformation, but they are also advantageous in computing conductor losses. This point will be clarified in Section II-B. A perturbation solution for conductor loss and dielectric loss has been given in [6] for a general microwave integrated coplanar transmission line. This is used directly in fin-lines for the same purposes.

In this analysis of fin-lines, results are compared with accurate and reliable sources of information, if available. The presented techniques are very efficient in computing, and it is believed they could be programmed for fin-lines onto a small computer, with about 30K-byte store.

II. Theoretical Background

As the method is a development of that described in [4], [6] for multilayer dielectric multistrips, a very short resumé is given below to highlight the remarks made later for individual fin-line configurations.

Consider Fig. 1(a), where the right-hand side wall (i.e., $Y=0$), can be assumed as a magnetic wall or electric wall, and conductors, either strips or fins, are placed on one side of the dielectric as shown. This structure is a substructure of the general planar configuration shown in [4], [6]. Therefore, in view of lossless conductors and homogeneous dielectrics the following relations hold true [4], [6]:

$$\begin{bmatrix} A_2 \\ B_2 \\ C_2 \\ D_2 \end{bmatrix} = [T_{2,1}] \begin{bmatrix} A_1 \\ B_1 \\ C_1 \\ D_1 \end{bmatrix} \tag{1}$$

where

$$[T_{2,1}] = [\gamma_2 h_1]^{-1} [\gamma_1 h_1] \tag{2}$$

and A_2, B_2, C_2, D_2, A_1, B_1, C_1, and D_1 are the coefficients of potential functions in the finite Fourier domain given as follows:

$$\tilde{\psi}_i^e = A_i \sinh\gamma_{i,n} Y + B_i \cosh\gamma_{i,n} Y \tag{3}$$

$$\tilde{\psi}_i^h = C_i \sinh\gamma_{i,n} Y + D_i \cosh\gamma_{i,n} Y, \qquad i=1,2 \tag{4}$$

where

$$\gamma_{i,n}^2 = \alpha_n^2 + \beta^2 - k_i^2 \tag{5}$$

α_n, $j\gamma_{i,n}$, β are the X, Y, Z components of wavenumber k_i in each region of dielectrics, and α_n denotes the Fourier spectrum constant given by

$$\alpha_n = \frac{n\pi}{2a}, \quad n = 0,1,2,\cdots \tag{6}$$

for unsymmetrical structures. For a structure with magnetic wall or electric wall at $X=a$, α_n becomes

$$\alpha_n = (n+1/2)\pi/a \quad \text{or} \quad \alpha_n = n\pi/a \tag{7}$$

respectively.
From [4] and [6], the matrix $[G]$ relating E_X, E_Z, J_X, and J_Z at $y = h_1 + \delta$ and in the Fourier domain is specified by

$$\begin{bmatrix} G_{1,1} & G_{1,2} \\ G_{2,1} & G_{2,2} \end{bmatrix} \begin{bmatrix} \tilde{J}_X \\ \tilde{J}_Z \end{bmatrix} = \begin{bmatrix} \tilde{E}_Z \\ \tilde{E}_X \end{bmatrix} \tag{8}$$

where $\tilde{E}$ and $\tilde{J}$ are E and J in the Fourier domain. As numerically proved, different arrangements of (8) submit different computing efficiencies [6]. Efficiency is measured by the number of basis functions required in each arrangement to reach the same precision in a solution. An instance of this can be seen in the microstrip problem where (8) is much more efficient than (9), while in slot relation (9) is superior

$$[G]^{-1}\begin{bmatrix}\tilde{E}_Z\\ \tilde{E}_X\end{bmatrix}=\begin{bmatrix}\tilde{J}_X\\ \tilde{J}_Z\end{bmatrix}. \tag{9}$$

It is believed that to achieve the optimum CPU time, the arrangement should be chosen in which currents and/or electric fields are to be approximated over the shortest intervals. In the proceeding work, it is thus much easier to work with (9) since usually E_X and E_Z have to be expanded over a small interval, within the gap for unilateral (Fig. 1(b) and (c)) and bilateral (Fig. 1(d) and (e)) fin-line. For some gap dimensions, the performance of (9) may not be the most efficient, but since continuity of numerical solutions is of concern, (9) has been used for all gap dimensions. In antipodal fin-line (Fig. 1(f) and (g)), (8) is advantageous for fin dimensions less than 'a', but as the fin size becomes larger than 'a', (9) yields a solution in less CPU time.

Assuming G_{ij} in (9) are known, the next step is to select expansion functions for E_X and E_Z. The expansion functions chosen throughout are Legendre polynomials for the unbounded field E_X, and sinusoids for the bounded field E_Z. The explicit expressions for the basis functions are given for each fin-line structure later in the paper. The choice of Legendre polynomials for approximation of E_X is because a) they have closed-form Fourier transform, and b) although required to approximate the unbounded singularity of E_X [7], the individual Legendre polynomials are bounded. The latter fact is actually advantageous in the computation of conductor loss where, due to the unbounded fields for an infinitely thin perfect conductor, the use of unbounded basis functions leads to unbounded conductor loss. This point is expanded later in Section II-B.

Transforming the basis functions into the Fourier domain and substituting in (9), then applying Galerkin's method and Parseval's identity, one obtains a set of homogeneous linear equations whose nontrivial solution yields phase constant of the structure in question [8].

A. Dielectric Loss

Calculation of dielectric loss is based on the assumption of low-loss dielectric materials. Therefore, a perturbation formulation developed in [6] for a general shape planar structure can be directly implemented for all fin-line configurations and the final expression is given by

$$\alpha_d=\frac{\omega\sum_i\epsilon_i\tan\delta_i\int_{S_i}|\vec{E}_0|^2\,ds}{2\,\mathrm{Re}\int_S\vec{E}_0\times\vec{H}_0^*\,d\vec{s}}. \tag{10}$$

In (10), $\vec{E}_0$ and $\vec{H}_0$ are electric and magnetic fields before introduction of dielectric loss, $\tan\delta_i$. S and S_i represent the whole cross section and dielectric sections, respectively. Since all fields are available in the transformed domain, both integrals in (10) are reduced to truncated series, greatly simplifying the computations of α_d [6].

B. Conductor Loss

The conventional formula for conductor loss of a transmission line with high conductivity conductor is given by [9]

$$\alpha_c=\frac{R_s\int_c|\vec{H}_t|^2\,dl}{2\,\mathrm{Re}\int_S\vec{E}_0\times\vec{H}_0^*d\vec{s}} \tag{11}$$

where R_s is surface resistance and $\vec{H}_t$ is the tangential magnetic field around the conductor periphery for lossless case. The only problem encountered in calculation of α_c is the integral

$$\int_c|\vec{H}_t|^2\,dl. \tag{12}$$

This integral is unbounded at the edge of an infinitely thin lossless strip [10], and (11) has to be used cautiously with conductor edges.

In fin-line, microstrip, and similar planar guides, field components E_X and J_Z will have $r^{-1/2}$ type singularities near any thin perfectly conducting edge. Substitution into (11) gives an unbounded value—an infinity that is a mathematical artefact. This is because, consequent on the assumption of a finite conductivity, it immediately follows that fields are everywhere bounded—even in geometric limit of the zero-thickness strip with sharp edges. Bounded fields will yield a bounded value for attenuation, (assuming only a nonzero net power flow).

Of course, the assumption of a finite conductivity together with a zero-thickness conductor must be interpreted carefully at microwave frequencies. It must be assumed that the conductor thickness is much greater than the skin depth; this is usually true in practice at microwave frequencies, and needs to be true for the perturbation approach of (11) to be valid. For validity of the field analysis between the conductor regions, 'zero thickness' must be taken as meaning that the conductor thickness is much less than any structural dimensions.

To describe the procedure used for the computation of conductor loss, suppose that either E_X or J_Z is expanded as a truncated series of Legendre polynomials, (in Section III-A, (18) gives the choice for unilateral fin-line; J_Z would be similarly expanded for microstrip)

$$E_x=\sum_{m=1}^{P}a_mP_{2(m-1)}(x/w). \tag{13}$$

Limiting the number of terms to P means that spatial discrimination in the X direction is limited to w/P; the essential singularity of

$$\int_0^w\frac{dx}{(w-x)}$$

is replaced by the finite

$$\int_0^{w(1-1/P)} \frac{dx}{(w-x)}.$$

One can then ask the question: How does the computed result for attenuation depend on P the number of terms? Tests with microstrip have shown that increasing the highest polynomial order from 4 to 12 gives an attenuation higher by about 10 percent. The structure chosen was the microstrip in [5, fig. 6(b)], with $w/d=2$ and using 150 Fourier terms, where the resulting attenuation increases from 0.59 to 0.64 dB/m. Although not conclusive, it does indicate the reasonable convergence, and so the usefulness, of the conductor-loss calculation. Formula (11) does, of course, depend on satisfying the inequalities

skin depth $\ll$ strip thickness $\ll$ other structural dimensions.

C. Characteristic Impedance

Due to hybrid wave propagation in fin-lines, a unique definition for characteristic impedance does not exist. However, the most common and useful definition is based on the voltage/power relationship, where the voltage is measured from one fin edge to another. In the configuration analyzed, for single fin-line, the characteristic impedance Z_0 is given by

$$Z_0 = V^2/P \tag{14}$$

while for coupled lines

$$Z_0 = 2V^2/P \tag{15}$$

where

$$V = \int_{\text{over gap}} E_x dx \tag{16}$$

$$P = \mathrm{Re} \int_S \vec{E}_0 \times \vec{H}_0^* \, d\vec{s} \tag{17}$$

where S is the whole cross section of the guide.

III. Applications

The technique described above is now applied to various examples of fin-line, outlined as follows.

A. Unilateral Fin-Line

Fig. 1(b) shows a generic cross section of unilateral fin-line where h_1 may be different from d (see Fig. 1(a)). This case is similar to shielded slot analyzed by the authors in [5]. Therefore, all the relations given for G_{ij} can be directly implemented in a computer program to analyze the unilateral fin-line. Notice that for this structure, the G_{ij} elements of (8) correspond to consideration of an electric wall at $y=0$; in relation (1) $B_1 = C_1 = 0$. As discussed above, basis functions are given by

$$\left.\begin{aligned} E_x &= \sum_{m=1}^{P} a_m P_{2(m-1)}(x/w) \\ E_z &= \sum_{m=1}^{Q} b_m \sin m\pi(x/w) \end{aligned}\right\}, \quad |x|<w \quad \text{at} \quad y=h_1+\delta \tag{18}$$

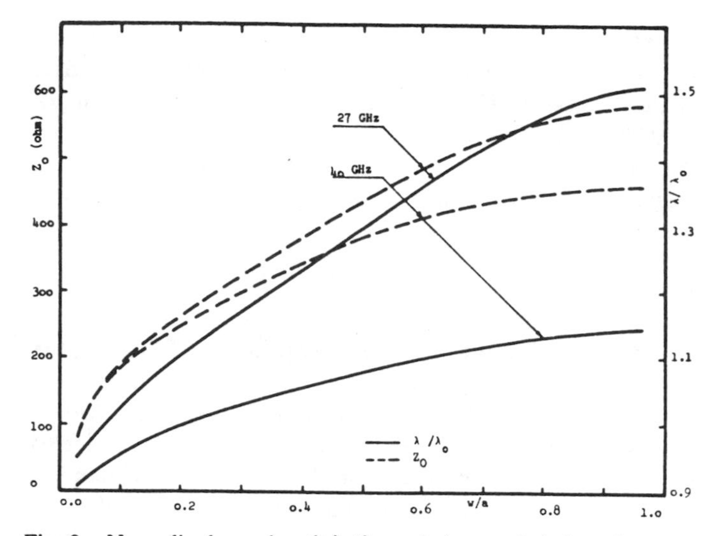

Fig. 2. Normalized wavelength λ/λ_0 and characteristic impedance Z_0 of unilateral finline (Fig. 1(b)).

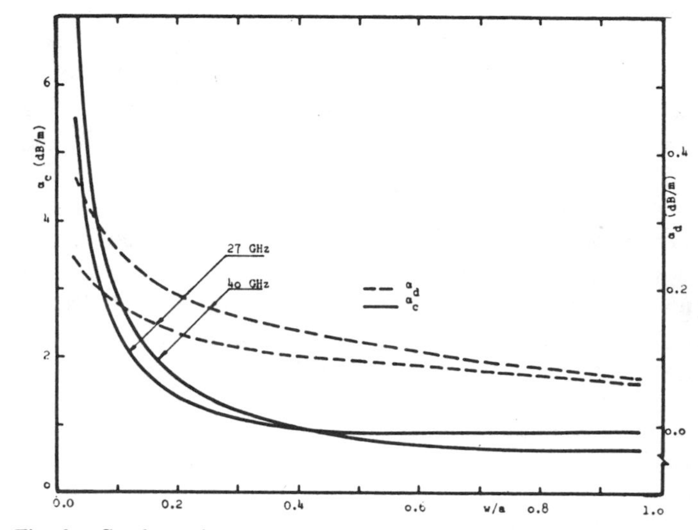

Fig. 3. Conductor loss α_c and dielectric loss α_d of unilateral finline (Fig. 1(b)).

where $P_{2(m-1)}$ represents Legendre polynomials of degree $2(m-1)$.

The computer results for propagation constant, characteristic impedance, dielectric loss, and conductor loss for a fin-line whose dielectric substrate is symmetrically placed within WG-22 waveguide are given in Fig. 2 and Fig. 3 over *Ka*-band. To check the speed of convergence and the accuracy of the applied method, it was tested against [11] where a very close agreement after considering $P=Q=4$ was achieved.

For unilateral fin-line with very low dielectric constant, $\epsilon \simeq 1.0003$ and with large gap $w/a \simeq 0.98$, the normalized wavelength, characteristic impedance, and conductor loss are compared with those of empty guide supporting the TE_{10} mode (Table I). The close agreement of values of those two guides indeed supports the accuracy of the presented numerical technique.

From Fig. 3, it is seen that, due to heavy concentration of fields near the small gap, the conductor loss and dielectric loss are not very low. Nevertheless, for a 400-μm gap, commonly etched, the total loss is around 2.5 dB/m over the whole *Ka*-band.

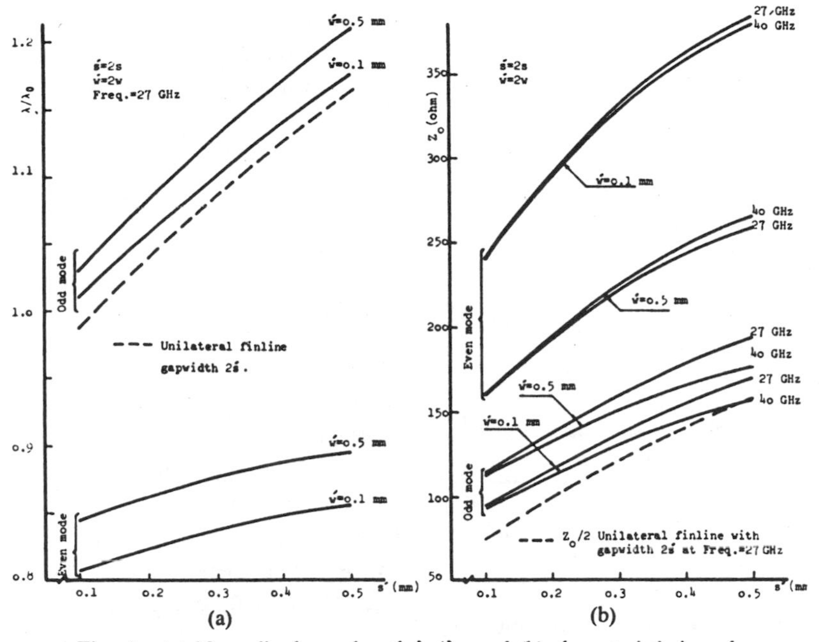

Fig. 4. (a) Normalized wavelength λ/λ_0 and (b) characteristic impedance Z_0 of even and odd mode of coupled finline (Fig. 1(c)).

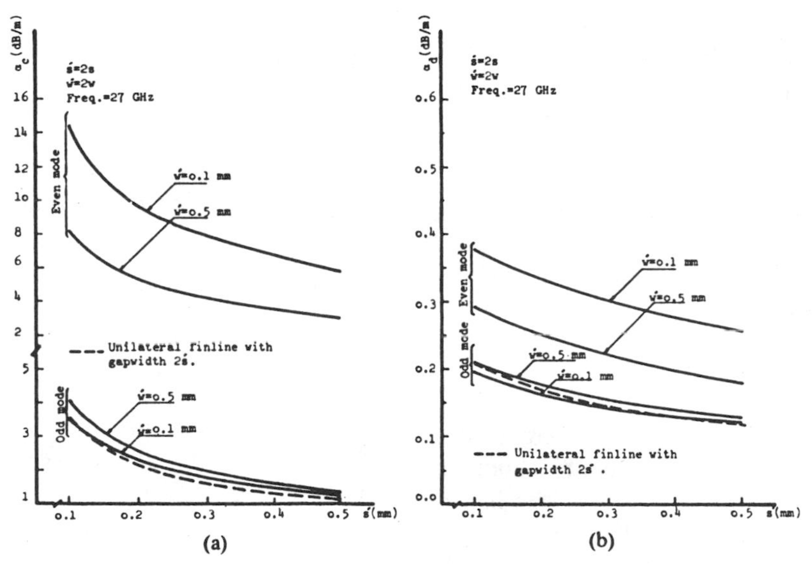

Fig. 5. (a) Conductor loss α_c and (b) dielectric loss α_d of even and odd mode of coupled finline (Fig. 1(c)).

TABLE I
A COMPARISON BETWEEN EMPTY WG-22 WAVEGUIDE PARAMETERS (*) AND THOSE OF UNILATERAL FIN-LINE WITH $w/a = 0.98$, $\epsilon_r = 1.0003$ AND $\rho = 3\times10^{-8}\Omega\cdot$m

Freq. GHz	λ_g/λ_0	$Z_0(\Omega)$	α_c dB/m
	1.598*	599.76*	0.937*
27	1.603	604.24	0.946
	1.176*	440.54*	0.675*
40	1.177	443.78	0.671

B. Coupled Coplanar Fin-Lines

This structure, Fig. 1(c), has recently been used in making a quadriphase fin-line modulator [12] where the parameters have been obtained empirically. However, with the introduced general technique, determination of the required parameters of the coupled coplanar fin-lines is not difficult. The only alteration to be made in the unilateral fin-line analysis is the change of basis functions given by

$$\left.\begin{aligned} E_x &= \sum_{m=1}^{P} a_m P_{m-1}[(x-w-s)/s] \\ E_z &= \sum_{m=1}^{Q} b_m \sin[m\pi(x-w)/2s] \end{aligned}\right\},$$

$$w < x < w+2s \quad \text{at} \quad y = h_1+\delta. \qquad (19)$$

For the 'odd mode' (defined here as having E_z an odd function of x and E_x an even function) the basis functions are correspondingly defined for negative x. The 'even

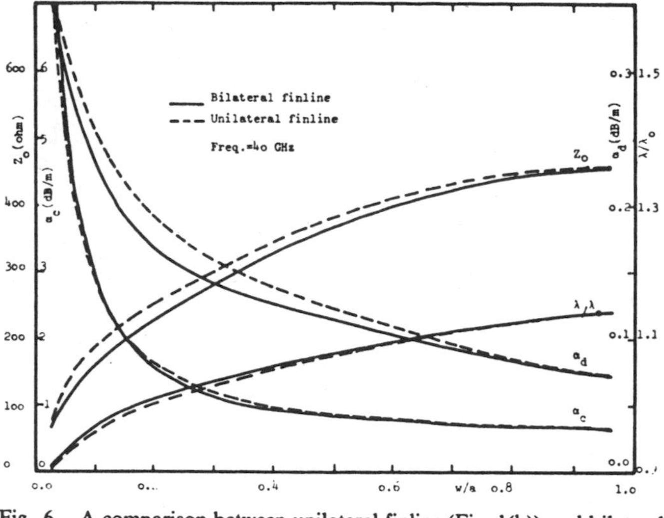

Fig. 6. A comparison between unilateral finline (Fig. 1(b)) and bilateral finline (Fig. 1(e)).

mode' has E_z an even function of x and E_x an odd function.

Figs. 4 and 5 illustrate the results of λ/λ_0, Z_0, α_c, and α_d versus slot width for two different separations of slots at a frequency of 27 GHz. The coupled coplanar fin-line as before is considered in WG-22 waveguide. A comparison of the parameters of the odd mode is carried out with those of unilateral fin-line whose slot width is twice the value of each slot. At the limit, as the separation between the two slots becomes zero, the coupled coplanar fin-line approaches the mentioned unilateral fin-line. From the given curves, the approach of parameters of the coupled line to the mentioned single line is not always monotonic but one can see their general consistency.

C. Unilateral Fin-Line and Bilateral Fin-Line as Coupled Line

Conventional and more general forms of bilateral fin-line are shown in Fig. 1(e) and (d). By symmetry about the x-axis, the analysis can be carried out just for one half of the structure. Considering a magnetic wall at $y=0$, the $[G]$ matrix in (8) is obtained by substituting $A_1 = D_1 = 0$ in (1). This change in (1) corresponds to exchange of sinh by cosh and vice versa for all field relations in the first dielectric region, for the case of electric wall symmetry plane at $y=0$, equivalent to Fig. 1(b) [4]. Therefore, the computer program developed for unilateral fin-line can be directly applied to the bilateral fin-line with some minor modification. The computer results for a conventional bilateral fin-line embodied in WG-22 waveguide have been illustrated in Fig. 6. A comparison of this line with unilateral fin-line of the same category, Figs. 2 and 3, reveals only a small change in the values of the associated parameters from one line to another. However, this observation may not be true for a substrate with greater thickness. Another feature of bilateral fin-line is its flexibility in being used as a pair of coupled lines, i.e., two coupled unilateral fin-lines. For the analysis of this aspect of unilateral fin-line, programs developed for the unilateral fin-line and bilateral fin-line can be called into service provided that the characteristic impedances for the even and odd modes are now defined through relation (15).

D. Antipodal Fin-Line

Cross sections of antipodal fin-lines have been shown in Fig. 1(g) and (f), where—in the latter—the structure is more general than Fig. 1(g). Note that in these two figures, there are distinct x, y axes and X, Y axes; this simplifies the equation presentation.

It is assumed that both structures have 180° rotational symmetry. This symmetry permits the solution for the fields to be expressed in a particular arrangement of even and odd symmetry, although the antipodal fin-line structure does not enjoy the simpler reflection symmetries about a plane, like those in unilateral and bilateral fin-lines. To find the solution, consider the potential functions in the space domain for the first dielectric region which, (from (3), (4), and (6)), are given as follows:

$$\psi^e(X,Y) = \sum_{n=0}^{\infty} (A_1 \sinh \gamma_{1,n} Y + B_1 \cosh \gamma_{1,n} Y) \sin \alpha_n X \tag{20}$$

$$\psi^h(X,Y) = \sum_{n=0}^{\infty} (C_1 \sinh \gamma_{1,n} Y + D_1 \cosh \gamma_{1,n} Y) \cos \alpha_n X. \tag{21}$$

For two points m and m' equally distant from 0 and on the X-axis, the following relations hold (Fig. 1(f)):

$$\psi^{(e,h)}(X,0) = \psi^{(e,h)}(2a - X,0) \tag{22}$$

provided that in (20) and (21) B_1 and D_1 are given by

$$B_1 = 0 \qquad \text{for } n = 0,2,4,6,\cdots$$
$$D_1 = 0 \qquad \text{for } n = 1,3,5,7,\cdots. \tag{23}$$

A similar argument for equally distant points n and n' on the y-axis leads to the following conclusion in (20) and (21):

$$A_1 = 0 \qquad \text{for } n = 1,3,5,7,\cdots$$
$$C_1 = 0 \qquad \text{for } n = 0,2,4,6,\cdots. \tag{24}$$

Therefore, as n takes $0,1,2,\cdots$, the right-hand side of (1) changes alternately between the following matrices:

$$\begin{vmatrix} 0 \\ B_1 \\ C_1 \\ 0 \end{vmatrix} \quad \text{and} \quad \begin{vmatrix} A_1 \\ 0 \\ 0 \\ D_1 \end{vmatrix}. \tag{25}$$

Hence, having found $[G]$ matrices corresponding to the above situations, their combination gives the required $[G]$ matrix of the antipodal fin-line. But the $[G]$ matrices corresponding to (25) are already available as the $[G]$ matrix of unilateral fin-line and $[G]$ matrix of bilateral fin-line with magnetic wall considered at the y-axis. Therefore, the solution of the antipodal fin-line by combining the two programs of the previous structures are now accessible provided that the basis functions are chosen as below

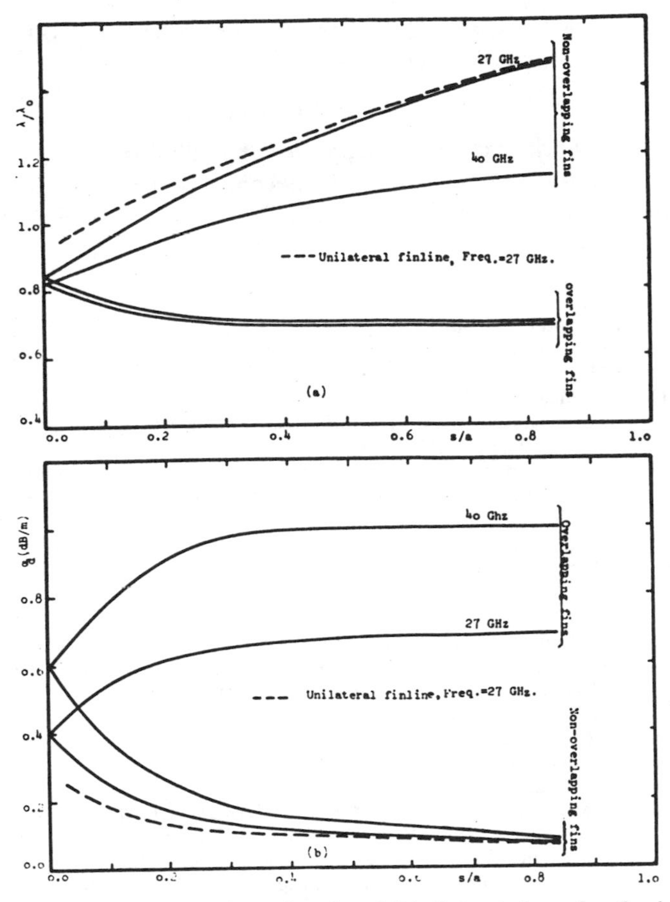

Fig. 7. (a) Normalized wavelength and (b) dielectric loss of antipodal finline (Fig. 1(g)).

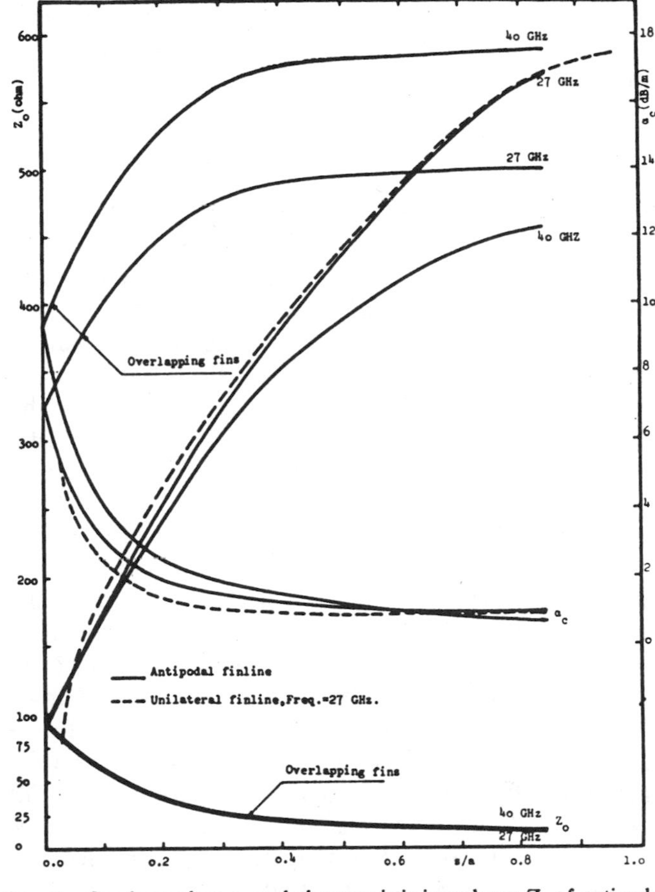

Fig. 8. Conductor loss α_c and characteristic impedance Z_0 of antipodal finline (Fig. 1(g)).

TABLE II
$2s = 3$ mm, $\delta = 0.127$ mm, $\epsilon_r = 2.22$, $\rho = 3\times10^{-8}\Omega$m, and $\tan\delta_i = 2\times10^{-4}$

Freq. GHz	TEM Approx. Z_0	α_c	α_d	Spectral Domain Z_0	α_c	α_d
27	10.77	15.0	0.73	10.05	14.0	0.69
40	10.77	18.5	1.08	10.51	17.5	1.00

(Fig. 1(f)):

$$\left.\begin{aligned} E_X &= \sum_{m=1}^{P} a_m P_{m-1}[(X-w-t)/t] \\ E_Z &= \sum_{m=1}^{Q} b_m \sin\frac{m\pi}{2t}(X-w) \end{aligned}\right\},$$

$$w < X < 2a \quad \text{at} \quad Y = h_1 + \delta. \quad (26)$$

By means of this developed method, α_c, α_d, λ/λ_0, and Z_0 of an antipodal fin-line with $h_1 = 0$ (Fig. 1(g)) have been computed and shown in Figs. 7 and 8. From these results and their comparison with unilateral fin-line results, Figs. 2 and 3, it is concluded that without overlap of the two fins, parameters of the antipodal, unilateral, and bilateral fin-lines are not significantly different. But as the fins start overlapping each other, the antipodal parameters vary substantially and are like those of parallel plate waveguide in which a TEM wave propagates. To substantiate the latter argument, α_c, α_d, Z_0 of the antipodal fin-line at large overlapping have been contrasted to those of parallel plate waveguide with width equal to $2s$ (Table II). In this comparison, parameters of finite width parallel plate waveguide have been derived, assuming that the fringing fields are negligible and it supports the TEM wave with the following relations [9]:

$$Z_0 = 1/v_p C \quad (27)$$

$$\alpha_c = 4.23 R_s / Z_0 s \quad (28)$$

$$\alpha_d = 4.23\omega \tan\delta_i / v_p \quad (29)$$

where

$$C = 2\epsilon_0\epsilon_r s/\delta \quad (30)$$

$$v_p = 1/\sqrt{\mu_0\epsilon_0\epsilon_r}. \quad (31)$$

Indeed, a close agreement is seen from Table II between the accurate solutions of the given antipodal fin-line and its approximate TEM model as s/a is large. Thus relations (27)–(31) may be used for approximate solution of antipodal fin-line with largely overlapped fins. As far as characteristic impedance is concerned, in the antipodal fin-lines the definition (14) has been used. For nonoverlapped fins, the voltage is measured from one fin edge to another, and for the overlapped fins just $E_y(x, y)$ at $x=0$ is integrated over the dielectric thickness as the substitution for the voltage.

IV. Conclusions

A unified method for the calculation of phase constant, characteristic impedance, dielectric loss, and conductor loss of fin-lines has been introduced. The method is in fact an application of the general analysis given by the authors for multilayer multiconductor planar transmission line [4]. Since the technique includes use of the spectral domain in determination of electric and magnetic fields, the computations of all parameters for any fin-line presented in this paper does not take more than five seconds on an IBM 360/65 if an accuracy of 0.1 percent for phase constant is acceptable. This is typical for $P=Q=4$ in (18), (19) or (26). For the first time, dielectric loss and conductor loss, (taking account of dispersion), have been computed, in fact using a perturbation technique. For maximum advantage, particularly for achieving the conductor loss, the basis functions are chosen to be Legendre polynomials for fields that are singular near the 180° conductor edges and trigonometric functions for fields that are bounded at the same edge. This is in contrast to the use of singular basis functions [13] where loss has not been considered. The parameters of unilateral, bilateral, antipodal, and coupled fin-lines built in WG-22 waveguide were computed and compared against each other. Very close agreement was seen as the method was tested against [11] for the phase constant of unilateral fin-line. For characteristic impedance and conductor loss, the results of unilateral fin-line with very small fins were contrasted with those of empty waveguide and as expected a close connection between them was seen. An approximate solution to the antipodal fin-line with large overlapped fins has also been presented, showing agreement with the general techniques. Specially, this comparative study supports the approximations assumed for the calculation of conductor loss. The higher order mode in each discussed structure can be obtained easily without any change in programming. The developed programs for the analyzed fin-lines could be adapted to other fin structure by further modifications.

Acknowledgment

The work has been supported by Philips Research Laboratory, Redhill, England, in connection with a program sponsored by D. C. V. D., Procurement Executive, Ministry of Defence.

References

[1] P. J. Meier, "Integrated fin-line millimeter components," *IEEE Trans. Microwave Theory Tech.*, vol. MTT-22, pp. 1209–1216, Dec. 1974.

[2] P. J. Meier, "Millimeter integrated circuits suspended in the *E*-plane of rectangular waveguide," *IEEE Trans. Microwave Theory Tech.*, vol. MTT-26, pp. 726–733, Oct. 1978.

[3] R. M. Knox, "Dielectric waveguide microwave integrated circuits—An overview," *IEEE Trans. Microwave Theory Tech.*, vol. MTT-24, pp. 806–814, Nov. 1976.

[4] J. B. Davies and D. Mirshekar-Syahkal, "Spectral domain solution of arbitrary coplanar transmission line with multilayer substrate," *IEEE Trans. Microwave Theory Tech.*, vol. MTT-25, pp. 143–146, Feb. 1977.

[5] D. Mirshekar-Syahkal and J. B. Davies, "Accurate solution of microstrip and coplanar structures for dispersion and for dielectric and conductor losses," *IEEE Trans. Microwave Theory Tech.*, vol. MTT-27, pp. 694–699, July 1979.

[6] D. Mirshekar-Syahkal, "Analysis of uniform and tapered transmission lines for microwave integrated circuits," Ph.D. thesis, University of London, 1979.

[7] R. Mittra and S. W. Lee, *Analytical Techniques in the Theory of Guided Waves.* New York: Macmillan, 1971.

[8] T. Itoh and R. Mittra, "A technique for computing dispersion characteristics of shielded microstrip lines," *IEEE Trans. Microwave Theory Tech.*, vol. MTT-22, pp. 889–891, Oct. 1974.

[9] R. F. Harrington, *Time Harmonic Electromagnetic Fields.* New York: McGraw-Hill, 1968.

[10] R. Pregla, "Determination of conduction losses in planar waveguide structures," *IEEE Trans. Microwave Theory Tech.*, vol. MTT-28, pp. 433–434, Apr. 1980.

[11] E. Yamashita and K. Atsuki, "Analysis of microstrip-like transmission lines by nonuniform discretization of integral equations," *IEEE Trans. Microwave Theory Tech.*, vol. MTT-24, pp. 195–200, Apr. 1976.

[12] E. Kpodzo, K. Shunemann, and G. Begmann, "A quadriphase fin-line modulator," *IEEE Microwave Theory Tech.*, vol. MTT-28, pp. 747–752, July 1980.

[13] L. P. Schmidt, Tatsuo Itoh, and Holgar Hofman, "Characteristics of unilateral fin-line structures with arbitrary located slots," *IEEE Trans. Microwave Theory Tech.*, vol. MTT-29, pp. 352–355, Apr. 1981.

Accurate Hybrid-Mode Analysis of Various Finline Configurations Including Multilayered Dielectrics, Finite Metallization Thickness, and Substrate Holding Grooves

RÜDIGER VAHLDIECK

Abstract — **An accurate analysis of various finline configurations is introduced. The method of field expansion into suitable eigenmodes used considers the effects of finite metallization thickness as well as waveguide wall grooves to fix the substrate. Especially for millimeter-wave range applications, the propagation constant of the fundamental mode is found to be lower than by neglecting the finite thickness of metallization. For increasing groove depth in cases of asymmetrical and "isolated finline," higher order mode excitation reduces the monomode bandwidth significantly. In contrast to hitherto known calculations, this parameter only causes negligible influence on a fundamental mode if the groove depth is lower than half of the waveguide height.**

I. Introduction

MILLIMETER WAVE application of finline structures is of increasing importance for *E*-plane integrated circuit designs [1]–[5]. Real structure parameters, like finite metallization thickness, and waveguide grooves to fix the inset, considerably influence circuit behavior, especially in the higher frequency range. As for the metallization thickness, this influence has already been demonstrated by the example of low-insertion-loss finline and metal-insert filters [5], [6].

Hitherto known design theories [8], [11]–[18], however, widely neglect the influence of these parameters, and are considered, therefore, to yield appropriate finline circuit designs only for the lower frequency range. The unilateral earthed finline investigations by Beyer [19] reveal that both parameters have relatively high influence on fundamental mode behavior. In [19], the Ritz–Galerkin method is used and the continuity of the odd TE-mode field at the interfaces is applied as an example.

A comparison to the propagation constant of an idealized finline structure, given by Hofmann [9], seems to be in good agreement with [19] only when comparing finite strip thickness (70 μm) and zero groove depth or finite groove depth (0.326 mm) and zero strip thickness. It appears that this problem has not yet been solved completely as evi-

Manuscript received November 7, 1983; revised June 4, 1984.

The author was with the Microwave Department, University of Bremen, Federal Republic of Germany. He is now with the Department of Electrical Engineering, University of Ottawa, Ottawa, Ont., Canada K1N 6N5.

Reprinted from *IEEE Trans. Microwave Theory Tech.*, vol. MTT-32, no. 11, pp. 1454–1460, November 1984.

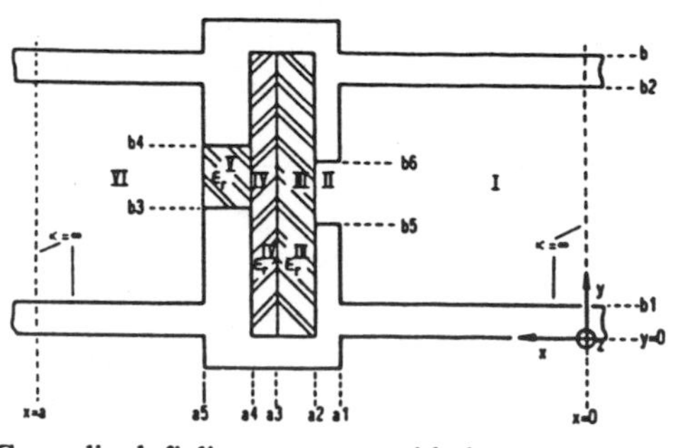

Fig. 1. Generalized finline structure with homogeneous and lossless dielectric in each subregion ($\mu_r = 1$).

denced by the absence of thorough investigation of higher order mode behavior with respect to the technologically conditioned circuit dimensions.

The purpose of this paper is to describe an accurate hybrid mode theory of a generalized finline configuration (Fig. 1) which includes finite strip thicknesses, substrate supporting grooves, asymmetric structures, and more than one dielectric region.

The higher order mode analysis presented utilized a generalized transverse-resonance relation which has already been successfully applied at microstrip structures [7] to reduce the size of the characteristic matrix equation considerably compared with the mode-matching technique in [10], [19]. In contrast to [19], the hybrid mode description used in this paper automatically involves the coupling of TE- and TM-waves. It will be shown that, only for finlines with homogeneous cross section ($\epsilon_r = 1$, ridged waveguide) or at cut-off frequencies both types of waves are decoupled. For the generalized case, therefore, inclusion of the TE- and TM-wave coupling effects on the field to be matched at the interfaces is necessary.

The transverse-resonance hybrid mode theory used in this paper yields a very efficient computer program to evaluate the normalized propagation constant. Numerical results are presented, especially for the E and T band and comparison with available results [4], [8]–[12], [14] establishes the accuracy of the numerical solutions in some special cases. In addition to the hybrid mode dispersion characteristics given, the effects due to finite metallization thickness and waveguide grooves are investigated in detail in order to provide design information for the practical choice of finline structural parameters.

II. Theory

The generalized finline structure given in Fig. 1 can be regarded as a transversal inhomogeneous parallel-plate line subdivided into partial homogeneous crossection ($\nu \epsilon$ I, II, III, IV, V, and VI). The electromagnetic field

$$\vec{E}^{(\nu)} = \nabla \times \nabla \times \vec{\Pi}_e^{(\nu)} - j\omega\mu \nabla \times \vec{\Pi}_m^{(\nu)} \tag{1}$$

$$\vec{H}^{(\nu)} = \nabla \times \nabla \times \vec{\Pi}_m^{(\nu)} + j\omega\epsilon^{(\nu)} \nabla \times \vec{\Pi}_e^{(\nu)} \tag{2}$$

in each subregion is expanded in an infinite series of eigenmodes and derived from the axial z components of the two independent vector potentials $\vec{\Pi}_m$ and $\vec{\Pi}_e$, which are assumed to be a sum of suitable eigenfunctions

$$\Pi_{mz}^{(\nu)} = \sum_{n=0}^{\infty} fc_{(n)}^{(\nu)}(y) \cdot Im_{(n)}^{(\nu)}(x) \cdot e^{-jkz \cdot z} \tag{3}$$

$$\Pi_{ez}^{(\nu)} = \sum_{n=1}^{\infty} fs_{(n)}^{(\nu)}(y) \cdot Ue_{(n)}^{(\nu)}(x) \cdot e^{-jkz \cdot z} \tag{4}$$

satisfying the boundary conditions and the scalar wave equation.

Considering the y-dependent boundary condition, the abbreviation $fs_{(n)}^{(\nu)}(y)$ and $fc_{(n)}^{(\nu)}(y)$ are given by

$$fc_{(n)}^{(\nu)}(y) = \frac{\cos k\tilde{y}_{(n)}^{(\nu)}}{\sqrt{1+\delta_{on}}}$$

$$fs_{(n)}^{(\nu)}(y) = \sin k\tilde{y}_{(n)}^{(\nu)}$$

$$k\tilde{y}_{(n)}^{(\nu)} = \frac{n \cdot \pi}{f^{(\nu)}} \cdot q^{(\nu)}$$

$$q^{(\nu)\prime} = [y - b_1, y - b_5, y, y, y - b_3, y - b_1]$$

$$f^{(\nu)\prime} = [b_2 - b_1, b_6 - b_5, b, b, b_4 - b_3, b_2 - b_1]$$

with the Kronecker delta δ_{on}. $Im_{(n)}^{(\nu)}(x)$ and $Ue_{(n)}^{(\nu)}(x)$ denote the partial wave amplitudes explained in the Appendix. The transverse resonance principle [7] is applied to reduce the size of the characteristic matrix equation for determining the normalized propagation constant kz/ko. For that reason, partial wave amplitudes

$$Im_{(n)}^{(\nu)}(x), Ue_{(n)}^{(\nu)}(x)$$

and

$$Ie_{(n)}^{(\nu)}(x) = \frac{1}{jkx_{(n)}^{(\nu)}} \frac{dUe_{(n)}^{(\nu)}(x)}{dx}$$

$$Um_{(n)}^{(\nu)}(x) = \frac{dIm_{(n)}^{(\nu)}(x)}{dx}$$

summarized as vectors $U^{(\nu)}$ and $I^{(\nu)}$ at the left-side boundary ($x = xo$) of each subregion (ν) are determined by the amplitudes of its right-side boundary ($x = xu$). The transfer matrix $R^{(\nu)}$ gives the relation between $I^{(\nu)}$ and $U^{(\nu)}$ at the two coordinates in the following manner

$$\underset{(x=xo)}{\begin{bmatrix} U^{(\nu)} \\ I^{(\nu)} \end{bmatrix}} = \underbrace{\begin{bmatrix} Rc^{(\nu)} & Rs^{(\nu)} \\ Rs^{(\nu)\prime} & Rc^{(\nu)} \end{bmatrix}}_{R^{(\nu)}} \cdot \underset{(x=xu)}{\begin{bmatrix} U^{(\nu)} \\ I^{(\nu)} \end{bmatrix}} \tag{5}$$

with the diagonal matrices $Rc^{(\nu)}$, $Rs^{(\nu)}$, and $Rs^{(\nu)\prime}$ given in the Appendix.

To satisfy the continuity condition at each common interface, modified continuity equations lead to the following expressions

$$Ey:\quad \frac{\partial \Pi_{mz}^{(\nu+1)}}{\partial x} = \frac{\partial \Pi_{mz}^{(\nu)}}{\partial x} - K_{\mu}^{(\nu)}(\omega, kz)\frac{\partial \Pi_{ez}^{(\nu)}}{\partial y} \tag{6}$$

$$Ez:\quad \Pi_{ez}^{(\nu+1)} = K^{(\nu)}(\omega, kz)\Pi_{ez}^{(\nu)} \tag{7}$$

$$Hz:\quad \Pi_{mz}^{(\nu+1)} = K^{(\nu)}(\omega, kz)\Pi_{mz}^{(\nu)} \tag{8}$$

$$Hy:\quad \frac{\partial \Pi_{ez}^{(\nu+1)}}{\partial x} = \frac{\epsilon_r^{(\nu)}}{\epsilon_r^{(\nu+1)}}\cdot\frac{\partial \Pi_{ez}^{(\nu)}}{\partial x} + K_{\epsilon}^{(\nu)}(\omega, kz)\frac{\partial \Pi_{mz}^{(\nu)}}{\partial y} \tag{9}$$

with

$$K^{(\nu)}(\omega, kz) = \frac{\epsilon_r^{(\nu)} - (kz/ko)^2}{\epsilon_r^{(\nu+1)} - (kz/ko)^2}$$

$$K_{\mu}^{(\nu)}(\omega, kz) = \frac{kz}{\omega\mu}\left(1 - K^{(\nu)}(\omega, kz)\right)$$

for $K_{\epsilon}^{(\nu)}(\omega, kz)$ replace $\epsilon_o \cdot \epsilon_r^{(\nu+1)}$ by μ.

The coupling between TE and TM waves is automatically involved in (6)–(9). Only in cases of finline configurations with homogeneous cross-sectional propagation media (e.g., ridged waveguide) or at cut-off frequency ($kz/ko = 0$), both types of waves are decoupled since $K\mu^{(\nu)}(\omega, kz)$ and $K\epsilon^{(\nu)}(\omega, kz)$ vanish.

The left-side $(\nu+1)$ x-dependent partial wave amplitudes in (6)–(9) are separated by multiplying with appropriate orthogonal functions which lead directly to the transition matrix $V^{(\nu)}$ of each interface section and hence combine the adjacent amplitude vectors $I^{(\nu)}$ and $I^{(\nu+1)}$ as well as $U^{(\nu)}$ and $U^{(\nu+1)}$

$$\underset{(x=a_\nu)}{\begin{bmatrix} Um^{(\nu+1)} \\ Ue^{(\nu+1)} \\ Im^{(\nu+1)} \\ Ie^{(\nu+1)} \end{bmatrix}} = \underbrace{\begin{bmatrix} \tilde{y}^{(\nu)}ec^{(\nu)} & -\tilde{y}^{(\nu)}K_{\mu}^{(\nu)}(\omega, kz)\tilde{e}c^{(\nu)} & 0 & 0 \\ 0 & \tilde{y}^{(\nu)}K^{(\nu)}(\omega, kz)\tilde{e}s^{(\nu)} & 0 & 0 \\ 0 & 0 & \tilde{y}^{(\nu)}K^{(\nu)}(\omega, kz)\tilde{e}c^{(\nu)} & 0 \\ 0 & 0 & -K_{\epsilon}^{(\nu)}(\omega, kz)\tilde{e}s^{(\nu)} & \tilde{y}^{(\nu)}\frac{\epsilon_r^{(\nu)}}{\epsilon_r(\nu+1)}es^{(\nu)} \end{bmatrix}}_{V^{(\nu)}} \cdot \underset{(x=a_\nu)}{\begin{bmatrix} Um^{(\nu)} \\ Ue^{(\nu)} \\ Im^{(\nu)} \\ Ie^{(\nu)} \end{bmatrix}}. \tag{10}$$

The abbreviations are explained in the Appendix.

Successively applying the transfer matrix $R^{(\nu)}$ to the corresponding transition matrix $V^{(\nu)}$ finally leads to a relation between amplitude vectors at the left ($x = a$) and the right boundary ($x = 0$)

$$\underset{(x=a)}{\begin{bmatrix} U^{(\mathrm{VI})} \\ I^{(\mathrm{VI})} \end{bmatrix}} = \underbrace{R^{(\mathrm{VI})}\cdot \prod_{\nu=\mathrm{V}}^{\nu=\mathrm{I}} V^{(\nu)}\cdot R^{(\nu)}}_{G} \underset{(x=0)}{\begin{bmatrix} U^{(\mathrm{I})} \\ I^{(\mathrm{I})} \end{bmatrix}}. \tag{11}$$

For the numerical solution, the infinite set of equations for U_L and I_L is truncated by the end index $L = 2N-1$; so, the matrix size of G is of order $4N-2$ and keeps constant, even for an increasing number of discontinuities. If N is the number of summation terms in (3) and (4) for all investigated finline structures, a maximum of $N = 17$ has turned out to be sufficient.

The transverse resonance relation requires the boundary conditions at the metallic surface $x = 0, x = a$ (Fig. 1), to be considered now. With respect to the relation $U \sim 0$, consequently the characteristic matrix equation is reduced to the upper right quarter of the matrix product in (11)

$$\underset{(x=a)}{\begin{bmatrix} 0 \\ 0 \end{bmatrix}} = \begin{bmatrix} G_{12}^{hh} & G_{12}^{he} \\ G_{12}^{eh} & G_{12}^{ee} \end{bmatrix} \cdot \underset{(x=0)}{\begin{bmatrix} Im^{(\mathrm{I})} \\ Ie^{(\mathrm{I})} \end{bmatrix}}. \tag{12}$$

G_{12} is of size $2N-1$ and the zeros of the determinant

$$\det(G_{12}) = 0 \tag{13}$$

which is a transcendent function of

$$kx_{(n)}^{(\nu)} = ko\sqrt{\epsilon_r^{(\nu)} - \left(\frac{n\cdot\pi}{f^{(\nu)}\cdot ko}\right)^2 - \left(\frac{kz}{ko}\right)^2} \tag{14}$$

$$ko = \omega\sqrt{\mu_o \epsilon_o}$$

provide the desired dispersion characteristic.

III. Results

Dispersion characteristics of the dominant and first higher order modes are given in Fig. 2 for the bilateral finline. The structure shielded by a *Ka*-band waveguide has the same dimensions as that used by Hofmann [8] and Schmidt [12]. Considering finite thickness of metallization ($t1 = t2 = 1$ μm) and a dielectric between the fins ($\epsilon_r^{\mathrm{V}} = 3.0 = \epsilon_r^{\mathrm{II}}$, see Fig. 1), results are in good agreement with these authors. Without a slot dielectric ($\epsilon_r^{\mathrm{V}} = \epsilon_r^{\mathrm{II}} = 1.0$), however, evaluated data are 3 percent less, on the average. This discrepancy can be explained by the fact that, due to an infinitely thin metallization [8], [12], the field between the fins is mainly concentrated within the dielectric and thus causes a higher value of propagation constant.

In order to demonstrate the fundamental mode behavior in finlines with considerable metallization thickness at 75 GHz (Fig. 3(a)), unilateral and bilateral finlines with thin substrates are considered. In the case of a bilateral finline with relatively thick substrate in a *T*-band waveguide mount (Fig. 3(b)), the influence of the finite strip thickness on the fundamental mode is smaller and deviations from values with negligible strip thickness given in [4] are not as high as expected in view of Fig. 3(a).

Fig. 4 shows the influence of the groove depth e on the propagation constant for the unilateral finline centered in the waveguide (E band). Comparing the results with Beyer [19] and using his definition of ϵ_{eff}, which gives a relation between the cut-off wavelength of an equivalent air-filled

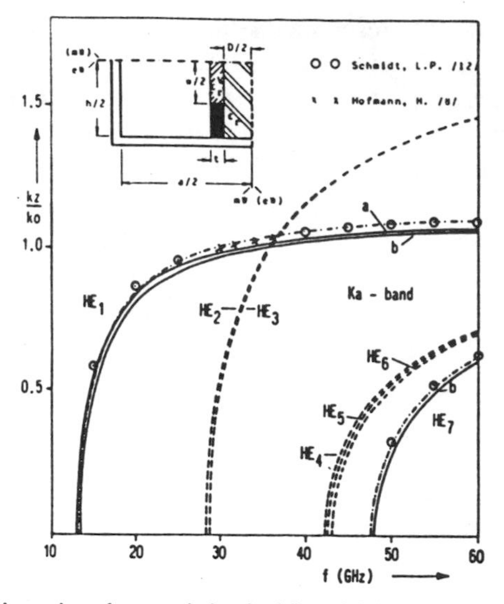

Fig. 2. Dispersion characteristic of a bilaterial finline (eW: electric wall, mw: magnetic wall; $a = 7.112$ mm, $h = a/2$; $\epsilon_r = 3.0$; $D = 125$ μm; $w = 0.5$ mm; (- - - - - -) $t = 1.0$ μm, $\epsilon_r^V = 3.0$; (———) $\epsilon_r^V = 1.0$, a: $t = 5$ μm, b: $t = 1$ μm; (- - - -) nonexcited modes, $t = 1.0$ μm, groove depth $e = 0$).

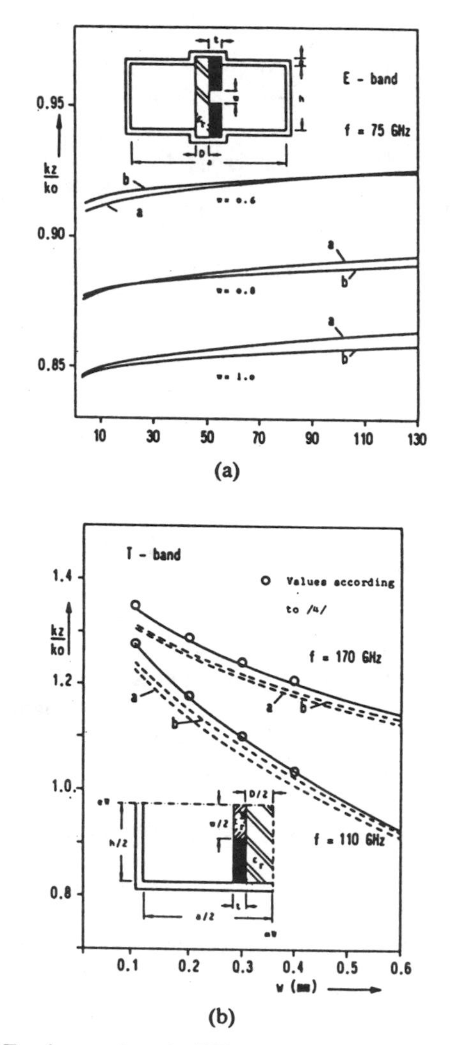

Fig. 3. (a) Fundamental mode (HE_1) versus the metallization thickness $t(\mu m)$ in a unilateral finline b with several slot widths w(mm). a a bilateral finline for comparison ($a = 3.1$ mm, $h = a/2$; $D = 50.$ μm, $\epsilon_r = 3.0$, $e = 0.326$ mm). (b) Fundamental mode (HE_1) versus the slot width in a bilateral finline ($a = 1.65$ mm, $h = a/2$; $D = 110$ μm, $\epsilon_r = 3.75$; $e = 0$; (———) $t = 0.5$ μm, $\epsilon_r^V = 3.75$; (- - - -) $\epsilon_r^V = 1.0$, a: $t = 0.5$ μm, b: $t = 5$ μm).

ridged waveguide and the finline wavelength at 75 GHz, the disagreement is significant (Fig. 4(a)). Since the field is concentrated mainly in the slots, one expects that the depth of the grooves has only a minor influence on the dominant mode. This fact has been confirmed with the present method for all investigated finlines with various slot widths and groove depths lower than half of the waveguide height. However, the second-order mode propagation which limits the practically most interesting monomode range is highly influenced. This behavior is illustrated in Fig. 4(b) containing the bandwidth which is a ratio of the nearly constant first cut-off wavelength λ_{C1} and the next higher mode cut-off wavelength λ_{C2} versus the groove depths. Although the first higher order mode for the bilateral finline reveals the highest dependence on groove depth, this mode is not excited by an H_{10}-wave of the empty waveguide. In contrast to other finline structures, the next excitable higher order mode (HE_7, Fig. 2) indicates a negligible influence on groove depth. So, this configuration provides the highest monomode range. If the grooves are deeper than half of the waveguide height, for a unilateral finline with the same dimension given in Fig. 3(b), Fig. 4(c) shows an obvious interaction between the fundamental and second higher order mode and results in a significant deviation of dispersion characteristics (Fig. 5) from those given in [4]. This interaction effect is also evident for a generalized finline structure with multilayered dielectrics and different metallization thicknesses (Fig. 6). This structure obviously reveals the necessity to consider all important higher order modes.

Similar observations are possible in other asymmetrical configurations, but will be of minor interest for small substrate thicknesses with low dielectric constant and small groove depths related to the waveguide height. This is demonstrated (Fig. 7) for an antipodal finline with nonoverlapping fins. The bandwidth behavior, as well as the dispersion characteristic, resembles somewhat that of a comparable unilateral structure, but all higher order modes are excited.

Modal dispersion curves of a so-called "isolated finline" are presented in Fig. 8. In practical applications for active components, one or both fins are isolated by a gasket which allows a dc voltage to be developed across the fins. RF continuity between the fins and the waveguide wall is achieved by using a choke section of a quarter wavelength groove in the waveguide broadwall at center frequency. Neglecting the groove as well as the finite metallization thickness shows a good agreement with the cut-off wavelength obtainable from [14]. In practical realization, however, this configuration provides TEM behavior which is also neglected in [14], [16] and leads to a somewhat different theoretical procedure.

Furthermore, it should be noticed that, in contrast to the unilateral and antipodal finline, where the next excitable higher order modes result originally from the H_{2o}-, respec-

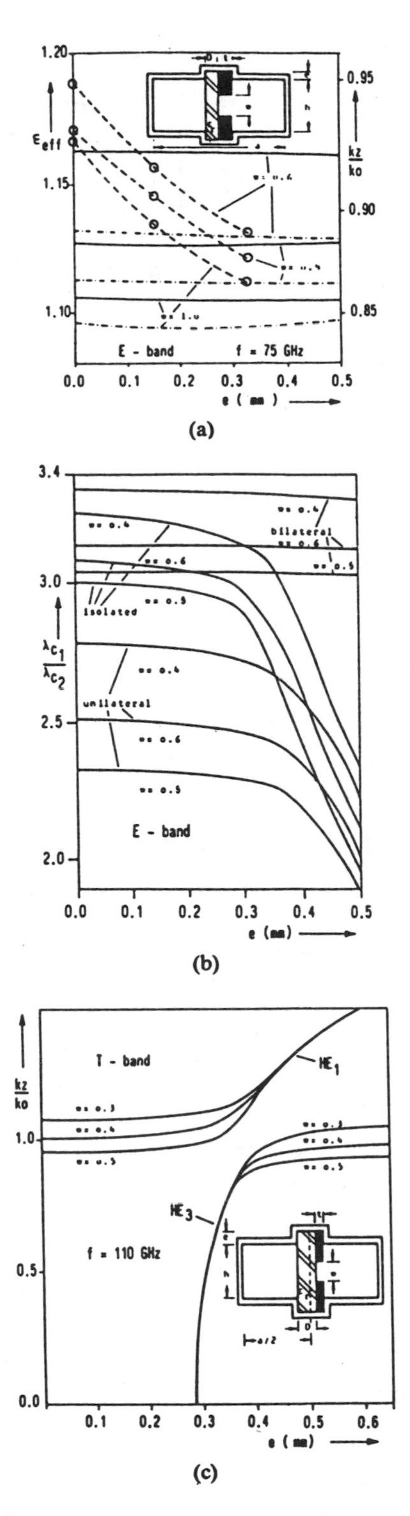

Fig. 4. (a) Fundamental mode (HE_1) versus the groove depths e in a unilateral finline with several slot widths w(mm) ($a = 3.1$ mm, $h = a/2$; $D = 50.0\ \mu$m, $\epsilon_r = 3.0$, $t = 70.0\ \mu$m)

∘∘∘ ┐
- - - - - - - - - ┘ values according to [19] $\Big\} \epsilon_{eff} = \lambda_0^2\left(\frac{1}{\lambda_g^2} + \frac{1}{\lambda_c^2}\right)$

λ_0—free space wavelength, λ_g—finline wavelength (75 GHz), λ_c—cutoff wavelength of an equivalent ridge waveguide. (——) normalized phase constant kz/ko). (b) Monomode bandwidth for a bilateral, unilateral and isolated finline with several slot widths. For the isolated finline $D/2 = 25.0\ \mu$m (see the structure in Fig. 8) (λc_1: cutoff-wavelength of the dominant mode, λc_2: cutoff-wavelength of the next higher order mode excited by an incident H_{10}-wave). (c) Normalized phase constant versus the groove depth in a unilateral structure with several slot widths ($a = 1.65$ mm, $h = a/2$; $D = 110.0\ \mu$m, $\epsilon_r = 3.75$, $t = 5.0\ \mu$m).

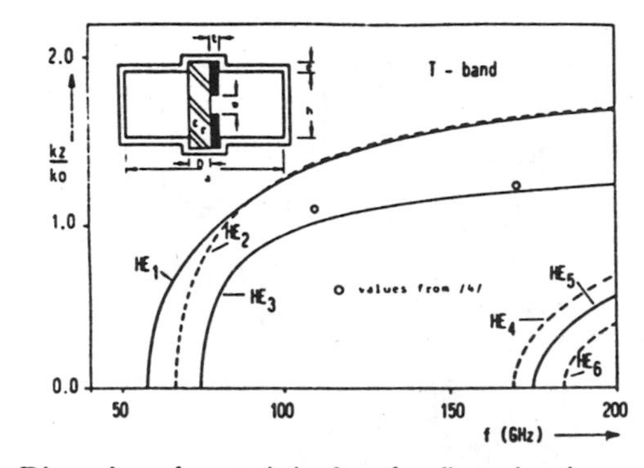

Fig. 5. Dispersion characteristic for the first six eigenmodes in a unilateral finline $w = 0.3$ mm, $e = 0.5$ mm, all other dimensions according to Fig. 4(c). (----) these modes are not excited by an incident H_{10}-wave.

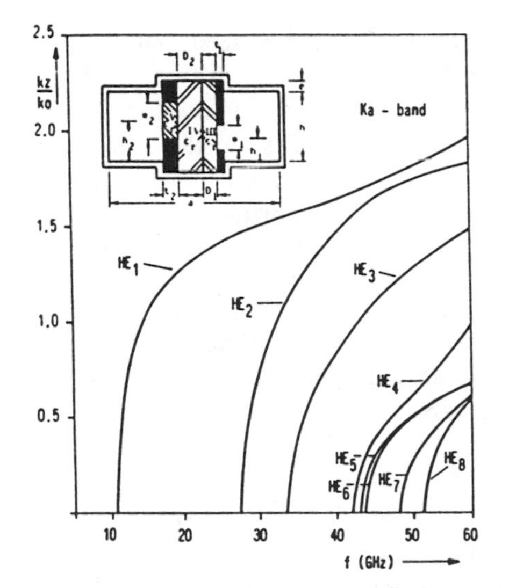

Fig. 6. Dispersion characteristic of an asymmetrical bilateral finline in the Ka - band range ($a = 7.112$ mm, $h = a/2$, $h1 = 1.0$ mm, $h2 = 2.0$ mm; $D1 = 125.0\ \mu$m, $\epsilon_r^{III} = 3.0$, $D2 = 254.0\ \mu$m, $\epsilon_r^{IV} = 9.6$, $\epsilon_r^{V} = 3.75$; $t1 = 10.0\ \mu$m, $t2 = 50.0\ \mu$m; $w1 = 1.5$ mm, $w2 = 2.5$ mm; $e = 0.5$ mm).

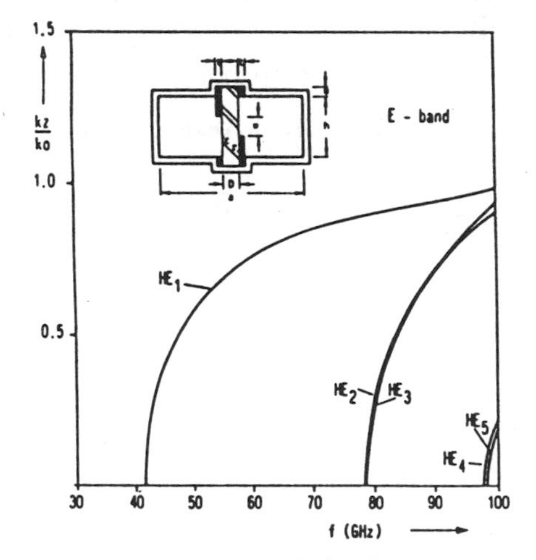

Fig. 7. Dispersion characteristic of the first five eigenmodes in an antipodal finline ($a = 3.1$ mm, $h = a/2$; $D = 50.0\ \mu$m, $\epsilon_r = 3.0$; $t = 10.0\ \mu$m; $w = 0.8$ mm; $e = 0.5$mm).

tively, H_{o1}-mode of the empty waveguide, the HE_4-mode in Fig. 8 results originally from the H_{3o}-mode and shows a high dependency on groove depth (Fig. 4(b)) which is not observed for the same mode in the case of a symmetric bilateral finline.

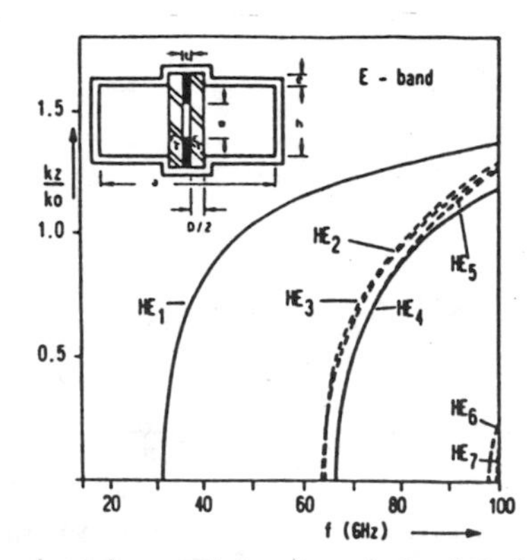

Fig. 8. Dispersion characteristic of an isolated finline ($a = 3.1$ mm, $h = a/2$; $D = 220.0$ μm, $\epsilon_r = 3.75$; $w = 0.6$ mm; $t = 5.0$ μm; $e = 0.5$ mm).

Finally, it must be emphasized that, for all these finline modes, the field distribution differs widely from those of an empty waveguide. The computation time for most symmetrical structures is 1–2 s, on the average, on a Siemens 7880. This corresponds to a 0.5- percent accuracy in the calculation of one phase constant for one frequency sample point.

IV. Conclusion

A method has been described which allows an accurate analysis of various finline structures, including all important parameters of real dimensions. Comparison between often used finline configurations indicates that the bilateral finline behaves best and provides the largest monomode bandwidth due to its virtual insensitivity to the groove depth. In asymmetrical structures, higher order modes cut-off frequencies are considerably reduced by the depth of the groove, which is much more critical than the strip thickness. Hence, it follows that, at high frequencies, the grooves supporting the substrate cannot be neglected and their influence is more significant than the effect of finite metallization thickness.

Appendix I

Abbreviations for the partial wave amplitudes in (3) and (4)

$$Im_{(n)}^{(\nu)}(x) = A_{(n)}^{(\nu)} e^{jkx_{(n)}^{(\nu)}\cdot x_a} + B_{(n)}^{(\nu)} e^{-jkx_{(n)}^{(\nu)}\cdot x_a}$$

$$Ue_{(n)}^{(\nu)}(x) = \frac{1}{jkx_{(n)}^{(\nu)}}\left(C_{(n)}^{(\nu)} e^{jkx_{(n)}^{(\nu)}\cdot x_a} - D_{(n)}^{(\nu)} e^{-jkx_{(n)}^{(\nu)}\cdot x_a}\right)$$

$$x_a = xo - xu$$

and for the diagonal matrices in (5)

$$Rc_{(n)}^{(\nu)} = \begin{bmatrix} \cosh\left(jkx_{(n)}^{(\nu)}\cdot x_a\right) & 0 \\ 0 & \cosh\left(jkx_{(n)}^{(\nu)}\cdot x_a\right) \end{bmatrix}$$

$$Rs_{(n)}^{(\nu)} = \begin{bmatrix} jkx_{(n)}^{(\nu)} \sinh\left(jkx_{(n)}^{(\nu)}\cdot x_a\right) & 0 \\ 0 & \dfrac{1}{jkx_{(n)}^{(\nu)}} \sinh\left(jkx_{(n)}^{(\nu)}\cdot x_a\right) \end{bmatrix}$$

$$Rs_{(n)}^{(\nu)\prime} = \begin{bmatrix} \dfrac{1}{jkx_{(n)}^{(\nu)}} \sinh\left(jkx_{(n)}^{(\nu)}\cdot x_a\right) & 0 \\ 0 & jkx_{(n)}^{(\nu)} \sinh\left(jkx_{(n)}^{(\nu)}\cdot x_a\right) \end{bmatrix}.$$

Appendix II

Abbreviations in the transition matrix equation (10)

$$\tilde{y}^{(\nu)\prime} = \left[\frac{b_2 - b_1}{2}, \frac{2}{b}, 1, \frac{b}{2}, \frac{2}{b_1 - b_2}\right]$$

$$\bar{y}^{(\nu)\prime} = \left[\frac{2}{b_6 - b_5}, \frac{b_6 - b_5}{2}, 1, \frac{2}{b_4 - b_3}, \frac{b_4 - b_3}{2}\right]$$

$$\tilde{ec}^{(\nu)} = \begin{cases} Fc^{(\mathrm{II})}\cdot p \cdot \dfrac{1}{b_6 - b_5} \\ E \cdot p \cdot \dfrac{1}{b} \\ \left(Fc^{(\mathrm{IV})}\right)^{-1}\cdot p \cdot \dfrac{1}{2} \\ Fc^{(\mathrm{V})}\cdot p \cdot \dfrac{1}{b_2 - b_1} \end{cases}$$

$$\tilde{es}^{(\nu)} = \begin{cases} \left(Fs^{(\mathrm{II})T}\right)^{-1}\cdot p \cdot \dfrac{1}{2} \\ E \cdot p \cdot \dfrac{1}{b} \\ Fs^{(\mathrm{IV})^T}\cdot p \cdot \dfrac{1}{b} \\ Fs^{(\mathrm{V})^T}\cdot p \cdot \dfrac{1}{2} \end{cases}$$

where $\tilde{es}^{(\nu)}$ receives an additional zero column, p is the diagonal matrix with the elements $i\cdot\pi$ $(i = 1, 2, 3 \cdots, N)$, and $Fc^{(\nu)}$, $Fs^{(\nu)}$ are the coupling integrals

$$Fc_{(n,i)}^{(\mathrm{I})} = \int_{y=b_5}^{y=b_6} fc_{(n)}^{(\mathrm{I})}(y)\cdot fc_{(i)}^{(\mathrm{II})}(y)\, dy$$

$$Fc_{(n,i)}^{(\mathrm{II})} = \int_{y=b_5}^{y=b_6} fc_{(n)}^{(\mathrm{III})}(y)\cdot fc_{(i)}^{(\mathrm{II})}(y)\, dy$$

$$Fc_{(n,i)}^{(\mathrm{III})} = \int_{y=b_3}^{y=b_4} fc_{(n)}^{(\mathrm{IV})}(y)\cdot fc_{(i)}^{(\mathrm{V})}(y)\, dy$$

$$Fc_{(n,i)}^{(\mathrm{IV})} = \int_{y=b_3}^{y=b_4} fc_{(n)}^{(\mathrm{VI})}(y)\cdot fc_{(i)}^{(\mathrm{V})}(y)\, dy.$$

For the coupling integrals $Fs^{(\nu)}$, replace $fs^{(\nu)}(y)$ instead of $fc^{(\nu)}(y)$

$$ec^{(\nu)} = \begin{cases} \left(Fc^{(\mathrm{I})}\right)^{-1} \\ Fc^{(\mathrm{II})} \\ E \\ \left(Fc^{(\mathrm{IV})}\right)^{-1} \\ Fc^{(\mathrm{V})} \end{cases} \qquad es^{(\nu)} = \begin{cases} Fs^{(\mathrm{I})^T} \\ \left(Fs^{(\mathrm{II})^T}\right)^{-1} \\ E \\ Fs^{(\mathrm{IV})^T} \\ \left(Fs^{(\mathrm{V})^T}\right)^{-1} \end{cases}$$

$$\bar{ec}^{(\nu)} = \begin{cases} Fc^{(\mathrm{I})T} \\ \left(Fc^{(\mathrm{II})^T}\right) - 1 \\ E \\ Fc^{(\mathrm{IV})^T} \\ \left(Fc^{(\mathrm{V})^T}\right)^{-1} \end{cases} \qquad \bar{es}^{(\nu)} = \begin{cases} \left(Fs^{(\mathrm{I})}\right)^{-1} \\ Fs^{(\mathrm{II})} \\ E \\ \left(Fs^{(\mathrm{IV})}\right)^{-1} \\ Fs^{(\mathrm{V})} \end{cases}.$$

where E denotes the unit matrix.

Acknowledgment

The author wishes to thank Prof. Dr.-Ing. F. Arndt who supported this work with helpful discussions.

References

[1] P. T. Meier, "Integrated fin-line millimeter components," *IEEE Trans. Microwave Theory Tech.*, vol. MTT-22, pp. 1209–1216, Dec. 1974.

[2] P. T. Meier "Millimeter integrated circuits suspended in the *E*-plane of rectangular waveguides," *IEEE Trans. Microwave Theory Tech.*, vol. MTT-26, pp. 726–733, Oct. 1978.

[3] R. N. Bates, S. J. Nightingale, and P. M. Ballard, "Millimeter-wave *E*-plane components and subsystems," *Radio Electron. Eng.*, vol. 57, pp. 506–512, Nov./Dec., 1982.

[4] B. Adelseck, H. Callsen, H. Hofmann, H. Meinel, and B. Rembold, "Neue millimeterwellenkomponenten in quasiplanarer leitungstechnik," *Frequenz*, vol. 35, pp. 118–123, 1981.

[5] F. Arndt, J. Bornemann, D. Grauerholz, and R. Vahldieck, "Theory and design of low-insertion loss fin-line filter," *IEEE Trans. Microwave Theory Tech.*, vol. MTT-30, pp. 155–163, Feb. 1982.

[6] R. Vahldieck, J. Bornemann, F. Arndt, and D. Grauerholz, "Optimized waveguide *E*-plane metal insert filters for millimeter-wave applications," *IEEE Trans. Microwave Theory Tech.*, vol. MTT-31, pp. 65–69, Jan. 1983.

[7] F. Arndt and G. U. Paul, "The reflection definition of the characteristic impedance of microstrips," *IEEE Trans. Microwave Theory Tech.*, vol. MTT-27, pp. 724–731, Aug. 1979.

[8] H. Hofmann, "Dispersion of planar waveguides for millimeter-wave application," *Arch. Elek. Übertragung*, vol. 31, pp. 40–44, 1977.

[9] H. Hofmann, H. Meinel, and B. Adelseck, "Möglichkeiten der Integration von Millimeter-Wellen-Komponenten," *Nachrichtentechn. Zeitschrift (NTZ)*, vol. 31, pp. 752–757, 1978.

[10] J. Siegel, "Phasenkonstante und Wellenwiderstand einer Schlitzleitung mit rechteckigem Schirm und endlicher Metallisierungsdicke," *Frequenz*, vol. 31, pp. 216–220, July 1977.

[11] J. Siegel "Grundwelle und höhere wellentypen bei fin-leitungen," *Frequenz*, vol. 34, pp. 196–200, July 1980.

[12] L.-P. Schmidt and T. Itoh, "Spectral domain analysis of dominant and higher order modes in fin-lines," *IEEE Trans. Microwave Theory Tech.*, vol. MTT-28, pp.981–985, Sept. 1980.

[13] L.-P. Schmidt, T. Itoh, and H. Hofmann, "Characteristics of unilateral fin-line structures with arbitrarily located slots," *IEEE Trans. Microwave Theory Tech.*, vol. MTT-29, pp. 352–355, Apr. 1981.

[14] Y.-C. Shih and W. J. R. Hoefer, "Dominant and second-order mode cutoff frequencies in fin-lines calculated with a two-dimensional TLM program," *IEEE Trans. Microwave Theory Tech.*, vol. MTT-24, pp. 1443–1448, Dec. 1980.

[15] A. K. Sharma and W. J. R. Hoefer, "Empirical expression for fin-line design," *IEEE Trans. Microwave Theory Tech.*, vol. MTT-31, pp. 350–356, Apr. 1983.

[16] A. M. K. Saad and G. Begemann, "Electrical performance of fin-lines of various configuration," *Microwaves, Opt. Acoust.*, vol. 1, no. 2, pp. 81–88, Jan. 1977.

[17] A. M. K. Saad and K. Schünemann, "A simple method for analyzing fin-line structures," *IEEE Trans. Microwave Theory Tech.*, vol. MTT-26, pp. 1007–1011, Dec. 1978.

[18] A. M. K. Saad and K. Schünemann, "Efficient eigenmode analysis for planar transmission lines," *IEEE Trans. Microwave Theory Tech.*, vol. MTT-30, pp. 2125–2132, Dec. 1982.

[19] A. Bayer, "Analysis of the characteristics of an earthed fin-line," *IEEE Trans. Microwave Theory Tech.*, vol. MTT-29, pp. 676–680, July 1981.

Empirical Expressions for Fin-Line Design

ARVIND K. SHARMA, MEMBER, IEEE, AND WOLFGANG J. R. HOEFER, SENIOR MEMBER, IEEE

Abstract —This paper presents empirical expressions in closed form for the design of unilateral and bilateral fin-lines. The guided wavelength and the characteristic impedance calculated with these expressions agree, typically, within ± 2 percent with values obtained using numerical techniques in the normalized frequency range $0.35 \leqslant b/\lambda \leqslant 0.7$, which is suitable for most practical applications.

I. Introduction

FIN-LINES FIND frequent applications in millimeter-wave integrated-circuit design. This is attributed to their favorable properties, such as low dispersion, broad single-mode bandwidth, moderate attenuation, and compatibility with semiconductor devices. Among various possible configurations, unilateral and bilateral fin-lines are of particular interest (see Fig. 1).

To this date, the propagation characteristics of fin-lines have been obtained with various methods. An early paper by Meier [1] described the propagating mode as a variation of the dominant mode in ridged waveguide. His procedure requires a test measurement to determine the equivalent dielectric constant of the fin-line structure. This is both expensive and time consuming. The analysis procedures by Saad and Begemann [2] and Hoefer [3] are based on ridged waveguide theory, and provide only an approximate solution. On the other hand, an accurate description of propagation in fin-lines, such as presented by Hofmann [4], and

Manuscript received November 23, 1981; revised December 8, 1982.

A. K. Sharma is with the Microwave Technology Center, RCA Laboratories, David Sarnoff Research Center, Princeton, NJ 08540.

W. J. R. Hoefer is with the Department of Electrical Engineering, University of Ottawa, Ottawa, Ontario, K1N6N5, Canada.

Reprinted from *IEEE Trans. Microwave Theory Tech.*, vol. MTT-31, no. 4, pp. 350–356, April 1983.

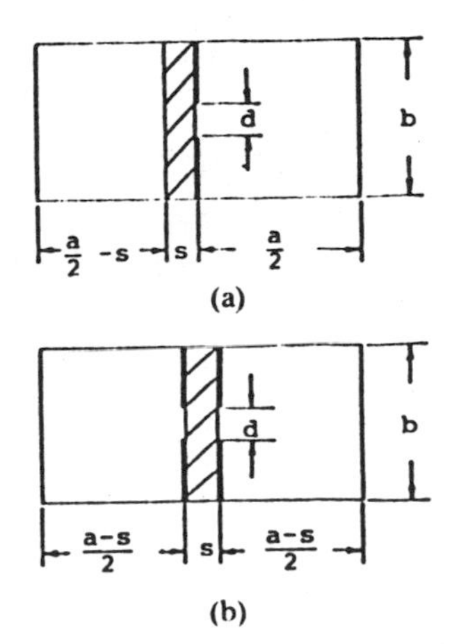

Fig. 1. Fin-line configurations. (a) Unilateral fin-line. (b) Bilateral fin-line.

recently by Knorr and Shayda [5], Schmidt and Itoh [6], and Beyer and Wolff [7], demands considerable analytical efforts and invariably leads to complicated computer programming. It is, therefore, desirable to have a design method which combines the flexibility of analytical expressions with the accuracy of numerical techniques. With this in mind, we have developed the following empirical formulas.

We believe that Meier's expressions describe the dispersion in fin-lines well enough for most practical applications. But, in order to circumvent the inconvenience of the required test measurement, we have developed empirical expressions for the equivalent dielectric constant k_e, the cutoff wavelength λ_{ca} in the equivalent ridged waveguide, as well as for the cutoff wavelength λ_{cf} in fin-lines. We present the basic approach in Section II, and the detailed empirical expressions in subsequent sections.

II. The Derivation of the Design Expressions

Meier's expressions for guided wavelength λ_g and characteristic impedance Z_0 in fin-line are [1]

$$\lambda_g = \lambda\left[k_e - (\lambda/\lambda_{ca})^2\right]^{-1/2} \tag{1}$$

and

$$Z_0 = Z_{0\infty}\left[k_e - (\lambda/\lambda_{ca})^2\right]^{-1/2} \tag{2}$$

where k_e is the equivalent dielectric constant, and λ is the free-space wavelength. λ_{ca} and $Z_{0\infty}$ are the cutoff wavelength and the characteristic impedance at infinite frequency of a ridged waveguide of identical dimensions. In Meier's expressions (1) and (2), the term k_e is regarded as a constant and is determined by a single test measurement. Strictly speaking, it characterizes a fictitious ridged waveguide uniformly filled with a dielectric of relative permittivity k_e. This first-order approximation is satisfactory only if the relative dielectric constant of the fin-line substrate is close to unity, and if the substrate occupies only a very small fraction of the guide cross section. If, however, ϵ_r is larger than 2.5, k_e must be considered frequency dependent, and we assume it to have the following general form [8]:

$$k_e = k_c \cdot F(d/b, s/a, \lambda, \epsilon_r). \tag{3}$$

k_c is the equivalent dielectric constant at cutoff given by

$$k_c = (\lambda_{cf}/\lambda_{ca})^2 \tag{4}$$

where λ_{cf} and λ_{ca} are the cutoff wavelength in the fin-line and in the equivalent ridged waveguide, respectively. The correction factor F is determined such that (1) and (2) yield the results obtained with the rigorous numerical techniques [4]–[9].

In the millimeter-wave range, standard waveguides have an aspect ratio $b/a = 1/2$. Furthermore, the substrates most frequently used in this range have a relative dielectric constant $\epsilon_r = 2.22$ (RT – Duroid) or $\epsilon_r = 3$ (Kapton). Expressions in this paper have therefore been derived for these parameters in the normalized frequency range of $0.35 \leqslant b/\lambda \leqslant 0.7$ which is suitable for most practical applications.

III. Numerical Evaluation of the Normalized Cutoff Frequencies

The accurate numerical evaluation of the normalized cutoff frequencies in fin-lines is accomplished with the hybrid mode formulation of the spectral domain technique [4], [5], [9]. In this technique, the Fourier transform of the dyadic Green's functions are related to the transform of the current densities on the conductors and the electric fields in the region complementary to the conductors, via the equation

$$\begin{bmatrix} \tilde{H}_{11}(\alpha_n, \beta, k_0) & \tilde{H}_{12}(\alpha_n, \beta, k_0) \\ \tilde{H}_{21}(\alpha_n, \beta, k_0) & \tilde{H}_{22}(\alpha_n, \beta, k_0) \end{bmatrix} \begin{bmatrix} \tilde{E}_x(\alpha_n) \\ \tilde{E}_z(\alpha_n) \end{bmatrix} = \begin{bmatrix} \tilde{J}_x(\alpha_n) \\ \tilde{J}_z(\alpha_n) \end{bmatrix} \tag{5}$$

where α_n is the Fourier-transform variable, β is the propagation constant, and k_0 is the free-space wavenumber. $\tilde{E}_x$, $\tilde{E}_z$, $\tilde{J}_x$, and $\tilde{J}_z$ are the electric fields in the aperture and the current densities on the conductors, respectively.

With the application of Galerkin's procedure and Parseval's theorem, we obtain a set of algebraic equations in terms of unknown constants of the basis functions. At cutoff, a nontrivial solution for the wavenumber k_0 is obtained by setting the determinant of the coefficient matrix equal to zero and finding the root of the equation. The numerical values for the normalized cutoff frequencies evaluated for the dielectric constants $\epsilon_r = 2.22$ and 3 are displayed in Tables I and II for unilateral and bilateral fin-lines, respectively. These values serve as a reference for other methods of fin-line analysis and are also utilized to derive the empirical expressions.

IV. Empirical Expressions for the Normalized Cutoff Frequencies

Meier's expressions require the knowledge of the cutoff wavelength λ_{ca} in an equivalent ridged waveguide, obtained by setting $\epsilon_r = 1$. However, in order to keep the analytical expressions for λ_{ca} as simple as possible, we

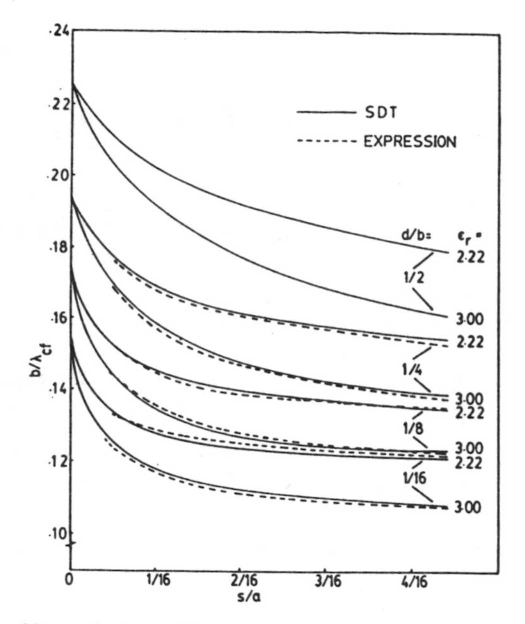

Fig. 2. Normalized cutoff frequencies in unilateral fin-lines. $b/a = 0.5$, $\epsilon_r = 2.22$ and 3.

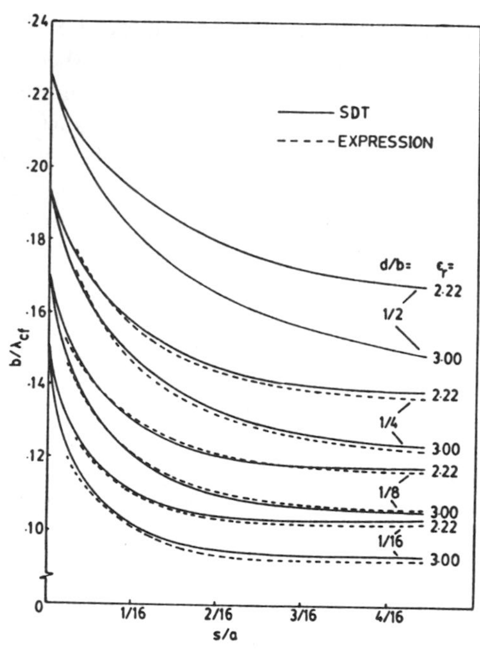

Fig. 3. Normalized cutoff frequencies in bilateral fin-lines. $b/a = 0.5$, $\epsilon_r = 2.22$ and 3.

assume that the equivalent ridged waveguide is obtained by letting the substrate thickness tend toward zero, which leads to the same expression for λ_{ca} in the unilateral and bilateral case. The normalized cutoff frequency (b/λ_{ca}) is then given by

$$b/\lambda_{ca} = 0.245(d/b)^{0.173} \tag{6}$$

which is valid for $1/16 \leqslant d/b \leqslant 1/4$ and is accurate to ± 1 percent.

For unilateral and bilateral fin-lines, the general expression for the normalized cutoff frequency (d/λ_{cf}) can be written in terms of the d/b and s/a

$$b/\lambda_{cf} = A(d/b)^{p}(s/a)^{q}. \tag{7}$$

TABLE I

NORMALIZED CUTOFF FREQUENCY b/λ_c OF THE DOMINANT MODE IN UNILATERAL FIN-LINES

Normalized Thickness s/a	Normalized Gap Width d/b	Cutoff frequency b/λ_c of the Dominant Mode: $\epsilon_r = 2.22$ SDT	$\epsilon_r = 2.22$ Expression	$\epsilon_r = 3.0$ SDT	$\epsilon_r = 3.0$ Expression
1/4	1/2	0.18070	-	0.16269	-
	1/4	0.15457	0.15340	0.13908	0.13840
	1/8	0.13505	0.13561	0.12146	0.12233
	1/16	0.12096	0.11988	0.10874	0.10814
1/8	1/2	0.19210	-	0.17706	-
	1/4	0.16140	0.16040	0.14756	0.14673
	1/8	0.13942	0.14085	0.12684	0.12800
	1/16	0.12396	0.12369	0.11244	0.11167
1/16	1/2	0.20248	-	0.19114	-
	1/4	0.16925	0.16755	0.15799	0.15640
	1/8	0.14499	0.14609	0.13410	0.13502
	1/16	0.12796	0.12738	0.11755	0.11657
1/32	1/2	0.21049	-	0.20275	-
	1/4	0.17698	0.17603	0.16881	0.16766
	1/8	0.15139	0.15283	0.14285	0.14364
	1/16	0.13286	0.13268	0.12409	0.12306

TABLE II

NORMALIZED CUTOFF FREQUENCY b/λ_c OF THE DOMINANT MODE IN BILATERAL FIN-LINES

Normalized Thickness s/a	Normalized Gap Width d/b	Cutoff frequency b/λ_c of the Dominant Mode: $\epsilon_r = 2.22$ SDT	$\epsilon_r = 2.22$ Expression	$\epsilon_r = 3.0$ SDT	$\epsilon_r = 3.0$ Expression
1/4	1/2	0.16833	-	0.15079	-
	1/4	0.13814	0.13695	0.12401	0.12292
	1/8	0.11779	0.11876	0.10576	0.10689
	1/16	0.10387	0.10298	0.09325	0.09296
1/8	1/2	0.17973	-	0.16531	-
	1/4	0.14489	0.14365	0.13254	0.13149
	1/8	0.12058	0.12154	0.10976	0.11087
	1/16	0.10409	0.10283	0.09443	0.09348
1/16	1/2	0.19279	-	0.18168	-
	1/4	0.15732	0.15577	0.14706	0.14567
	1/8	0.12955	0.13073	0.11999	0.12118
	1/16	0.11014	0.10972	0.10123	0.10082
1/32	1/2	0.20399	-	0.19623	-
	1/4	0.16925	0.16987	0.16159	0.16270
	1/8	0.14110	0.14183	0.13332	0.13409
	1/16	0.11941	0.11842	0.11160	0.11051

In the following, the unknown constants appearing in (7) are given for the range of structural parameters $1/16 \leqslant d/b \leqslant 1/4$ and $1/32 \leqslant s/a \leqslant 1/4$.

For Unilateral Fin-Lines ($\epsilon_r = 2.22$)

$$A = 0.1748$$

$$p = \begin{cases} 0.16(s/a)^{-0.07}, & 1/32 \leqslant s/a \leqslant 1/20 \\ 0.16(s/a)^{-0.07} \\ \quad -0.001\ln[(s/a)-(1/32)], & 1/20 \leqslant s/a \leqslant 1/4 \end{cases}$$

$$q = -0.0836. \tag{8}$$

For Unilateral Fin-Lines ($\epsilon_r = 3$)

$$A = 0.1495$$

$$p = \begin{cases} 0.1732(s/a)^{-0.073}, & 1/32 \leqslant s/a < 1/10 \\ 0.1453(s/a)^{-0.1463}, & 1/10 \leqslant s/a \leqslant 1/4 \end{cases}$$

$$q = -0.1223. \tag{9}$$

For Bilateral Fin-Lines ($\epsilon_r = 2.22$)

$$A = 0.15$$

$$p = \begin{cases} 0.225(s/a)^{-0.042}, & 1/32 \leqslant s/a \leqslant 1/10 \\ 0.149(s/a)^{-0.23}, & 1/10 \leqslant s/a \leqslant 1/4 \end{cases}$$

$$q = -0.14. \tag{10}$$

For Bilateral Fin-Lines ($\epsilon_r = 3$)

$$A = 0.1255$$

$$p = \begin{cases} 0.21772(s/a)^{-0.07155}, & 1/32 \leqslant s/a \leqslant 1/15 \\ 0.2907 - 0.3568(s/a), & 1/15 \leqslant s/a \leqslant 1/4 \end{cases}$$

$$q = -0.1865. \tag{11}$$

Tables I and II compare the above expressions and numerical results obtained with the spectral domain technique. Results agree within ± 1 percent, which inspires confidence in the above expressions. Figs. 2 and 3 display these results graphically.

V. Equivalent Dielectric Constant

Given the cutoff frequencies in fin-lines and ridged waveguides of identical dimensions, the equivalent dielectric constant k_c at cutoff is calculated with (3). k_e is then obtained by multiplying k_c with a correction factor F. The expressions for F are as follows.

For Unilateral Fin-Lines ($\epsilon_r = 2.22$)

$$F = \begin{cases} [1.0 + 0.43(s/a)](d/b)^{p_1}, & 1/32 \leqslant s/a \leqslant 1/8 \\ [1.02 + 0.264(s/a)](d/b)^{p_1}, & 1/8 \leqslant s/a \leqslant 1/4 \end{cases} \tag{12}$$

where

$$p_1 = 0.096(s/a) - 0.007.$$

For Unilateral Fin-Lines ($\epsilon_r = 3$)

$$F = F' + 0.25308(b/\lambda) - 0.135$$

$$F' = \begin{cases} 1.368(s/a)^{0.086}(d/b)^{p_1}, & 1/32 \leqslant s/a \leqslant 1/8 \\ [1.122 + 0.176(s/a)]^{p_2}, & 1/8 < s/a \leqslant 1/4 \end{cases} \tag{13}$$

where

$$p_1 = 0.375(s/a) - 0.0233$$

$$p_2 = 0.032 - 3.0[(s/a) - (3/16)]^2.$$

For Bilateral Fin-Lines ($\epsilon_r = 2.22$)

$$F = \begin{cases} 0.78(s/a)^{-0.098}(d/b)^{0.109}, & 1/32 \leqslant s/a < 1/8 \\ [1.04 - 0.2(s/a)](d/b)^{p_1}, & 1/8 \leqslant s/a \leqslant 1/4 \end{cases} \tag{14}$$

where

$$p_1 = 0.152 - 0.256(s/a).$$

For Bilateral Fin-Lines ($\epsilon_r = 3$)

$$F = F' + 0.08436(b/\lambda) - 0.045$$

$$F' = \begin{cases} 0.975(s/a)^{-0.026}(d/b)^{p_1}, & 1/32 \leqslant s/a < 1/8 \\ [1.0769 - 0.2424(s/a)](d/b)^{p_2}, & 1/8 \leqslant s/a \leqslant 1/4 \end{cases} \tag{15}$$

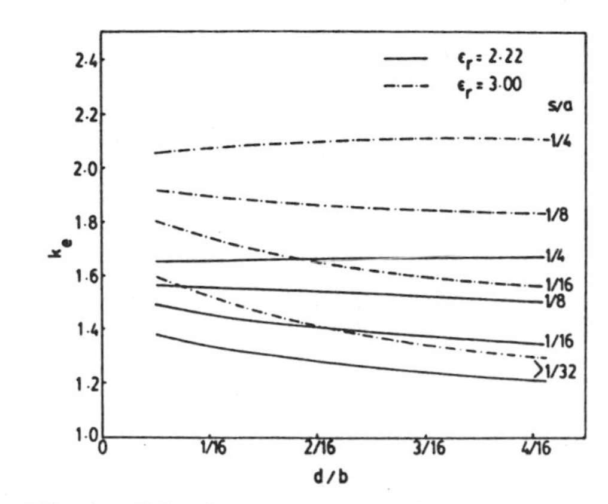

Fig. 4. Effective dielectric constant k_e in unilateral fin-lines. $b/a = 0.5$, $\epsilon_r = 2.22$ and 3, $b/\lambda = 0.3556$.

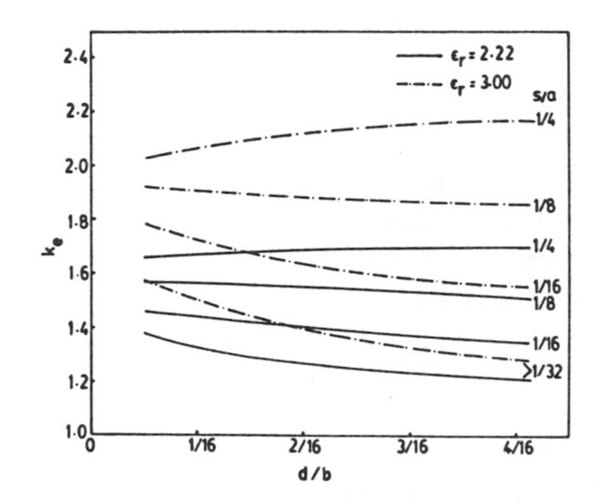

Fig. 5. Effective dielectric constant k_e in bilateral fin-lines. $b/a = 0.5$, $\epsilon_r = 2.22$ and 3, $b/\lambda = 0.3556$.

where

$$p_1 = 0.089 + 0.288(s/a)$$

$$p_2 = 0.16 - 0.28(s/a).$$

Figs. 4 and 5 show the typical values of k_e computed with the above expressions at $b/\lambda = 0.3556$.

VI. Characteristic Impedance

The characteristic impedance of the fin-line has been presented by Meier [1] in terms of the asymptotic value $Z_{0\infty}$ (2), that is, the impedance at infinite frequency of an equivalent ridged waveguide structure. This impedance can be defined in many different ways. The choice of the definition depends on the application. For instance, in Meier's expression (2), $Z_{0\infty}$ is defined on a power-voltage basis. However, Meinel and Rembold [10] have found that in the design of fin-line switches it is appropriate to define characteristic impedance in terms of a voltage and current, that is

$$Z_0 = \frac{V_0}{I_l} \tag{16}$$

where V_0 is the line integral over the electric field between the fins taken along the shortest path on the substrate surface, and I_l is the total longitudinal surface current in the structure. This definition was proposed by Hofmann [4].

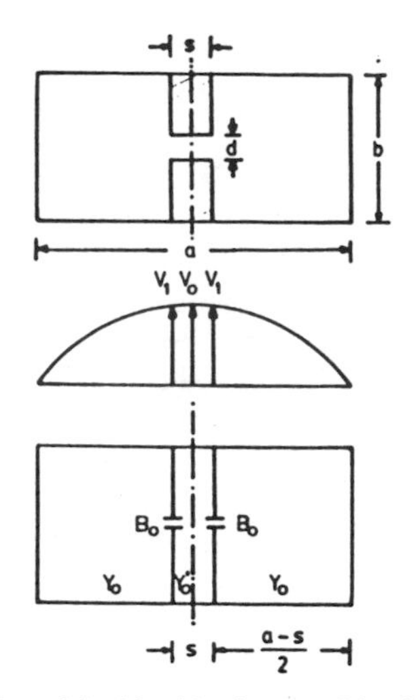

Fig. 6. Cross section of double-ridged waveguide with transverse equivalent network showing the voltage distribution.

In this section, we shall derive an analytical expression for the characteristic impedance of an equivalent ridged waveguide structure at infinite frequency. To that end, we shall calculate the total longitudinal current with a procedure similar to that of Cohn [11], taking into account the current on the edges of the ridge.

The longitudinal current is equal to the sum of the respective currents in the three regions of the double-ridged waveguide structure shown in Fig. 6. The transverse equivalent network and the voltage distribution in the TE_{10} mode are also shown there.

A. The Longitudinal Current Between the Ridges

(Region 1): Following the notation of Fig. 6, the voltage decreases cosinusoidally outwards from the center and can be expressed as

$$V(l) = V_0 \cos 2\pi l/\lambda_t \tag{17}$$

where V_0 is the magnitude of the voltage at the center, l is the variable distance from the center, and λ_t is the wavelength in the transverse direction, which is equivalent to the cutoff wavelength λ_{ca} of the air-filled ridged waveguide given by

$$\lambda_t = \lambda\left[1-(\lambda/\lambda_g)^2\right]^{-1/2} = \lambda_{ca}. \tag{18}$$

The voltage at the step is

$$V_1 = V_0 \cos \pi s/\lambda_{ca} \tag{19}$$

which is obtained setting $l = s/2$ in (17). Thus, the longitudinal linear current density at the top wall is

$$J(l) = \frac{V_0}{d\eta} \cos 2\pi l/\lambda_{ca} \tag{20}$$

where

$$\eta = \sqrt{\frac{\mu_0}{\epsilon_0}}\,\frac{\lambda_g}{\lambda} \tag{21}$$

is the characteristic wave impedance of the TE_{10} mode in the structure. The longitudinal current is then derived as

$$I_{l1} = \frac{2}{d}\int_0^{\frac{s}{2}} \frac{V_0}{\eta} \cos 2\pi l/\lambda_{ca}\, dl$$

$$= \frac{V_0 \lambda_{ca}}{\pi \eta d} \sin \pi s/\lambda_{ca}. \tag{22}$$

B. The Longitudinal Current in the Discontinuity Plane

(Region 2): Assuming that the discontinuity region can be represented by a shunt capacitance C_s per unit length subject to the voltage

$$V_1 = V_0 \cos \pi s/\lambda_{ca} \tag{23}$$

we can imagine it as a parallel plate capacitor of plate distance h and width l in the transverse direction

$$C_s = \epsilon_0 l/h. \tag{24}$$

The electric field strength in the capacitor is then

$$E_c = V_1/h = \frac{V_0}{h} \cos \pi s/\lambda_{ca}. \tag{25}$$

The current in the top plate is

$$I_t = \frac{V_0 C_s}{\eta \epsilon_0} \cos \pi s/\lambda_{ca}. \tag{26}$$

The total discontinuity current, taking into account both halves of the cross section, is then

$$I_{l2} = \frac{2V_0}{\eta\omega\epsilon_0}\,\frac{\omega C_s}{Y_{0t}} \cdot Y_{0t} \cos \pi s/\lambda_{ca} \tag{27}$$

with

$$Y_{0t} = \frac{\epsilon_0}{\mu_0}\,\frac{1}{b}. \tag{28}$$

After some further modifications, this current becomes, for a finite real λ_g in the longitudinal direction

$$I_{l2} = \frac{V_0 \lambda_{ca}}{\pi \eta b}(B_0/Y_0) \cos \pi s/\lambda_{ca}. \tag{29}$$

The expression for B_0/Y_0 is presented later in (34).

C. The Longitudinal Current in the Lateral Parts

(Region 3): In the lateral parts of the cross section, the voltage variation in the transverse direction is

$$V(l) = V_0 \frac{\cos \pi s/\lambda_{ca}}{\sin \pi(a-s)/\lambda_{ca}} \sin 2\pi l/\lambda_{ca} \tag{30}$$

where l is now the variable distance inward from the side walls. The longitudinal current density in the top wall becomes

$$J(l) = \frac{V_0 \cos \pi s/\lambda_{ca}}{b\eta \sin \pi(a-s)/\lambda_{ca}} \sin 2\pi l/\lambda_{ca} \tag{31}$$

and the expression for longitudinal current is given by

$$I_{l3} = 2\int_0^{(a-s)/2} \frac{V_0 \cos \pi s/\lambda_{ca}}{b\eta \sin \pi(a-s)/\lambda_{ca}} \sin 2\pi l/\lambda_{ca}\, dl$$

$$= \frac{V_0 \lambda_{ca}}{\pi \eta b} \cos(\pi s/\lambda_{ca}) \tan\left[\pi(a-s)/2\lambda_{ca}\right]. \tag{32}$$

With the three components of the total longitudinal current derived above, the characteristic impedance at in-

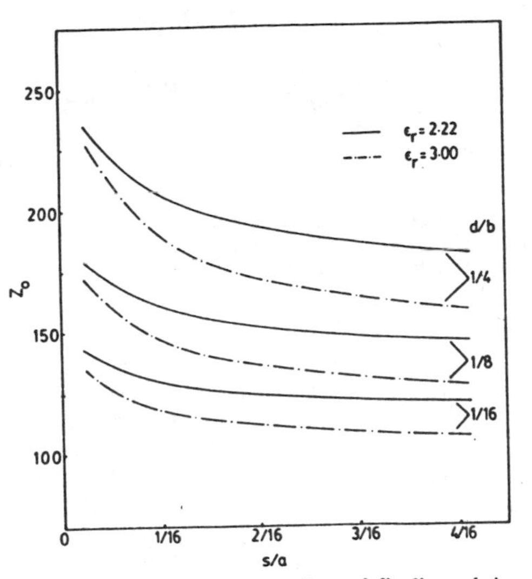

Fig. 7. Characteristic impedance of unilateral fin-lines. $b/a = 0.5$, $\epsilon_r = 2.22$ and 3, $b/\lambda = 0.3556$.

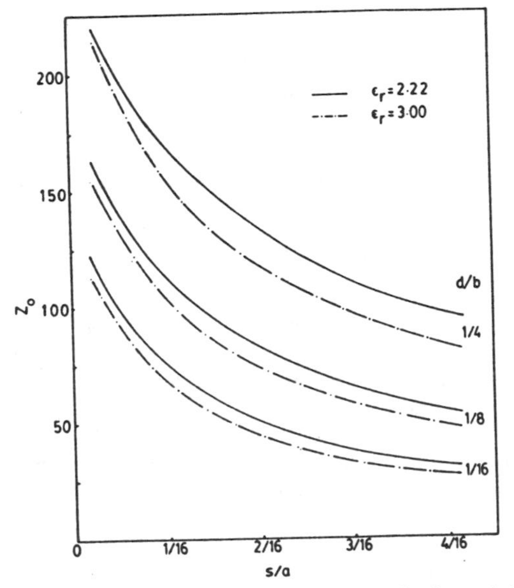

Fig. 8. Characteristic impedance of bilateral fin-lines. $b/a = 0.5$, $\epsilon_r = 2.22$ and 3, $b/\lambda = 0.3556$.

finite frequency is given by

$$Z_{0\infty VI} = \frac{120\pi^2(b/\lambda_{ca})}{\dfrac{b}{d}\sin\dfrac{\pi s}{\lambda_{ca}} + \left[\dfrac{B_0}{Y_0} + \tan\dfrac{\pi(a-s)}{2\lambda_{ca}}\right]\cos\dfrac{\pi s}{\lambda_{ca}}} \tag{33}$$

with

$$\frac{B_0}{Y_0} = \frac{2b}{\lambda_{ca}}\left\{\ln\csc\left(\frac{\pi d}{2b}\right) + \frac{Q\cos^4\left(\frac{\pi d}{2b}\right)}{1+Q\sin^4\left(\frac{\pi d}{2b}\right)} + \frac{1}{16}\left(\frac{b}{\lambda_{ca}}\right)^2\left[1-3\sin^2\left(\frac{\pi d}{2b}\right)\right]^2\cos^4\left(\frac{\pi d}{2b}\right)\right\} \tag{34}$$

and

$$Q = \left[1-(b/\lambda_{ca})^2\right]^{-1/2} - 1. \tag{35}$$

The characteristic impedance Z_0 is computed using (33) and (2). The value of s in (33) is set equal to zero in the case of unilateral fin-line, and it is set equal to the substrate thickness in the case of bilateral fin-line. Z_0 is shown in Figs. 7 and 8 for unilateral and bilateral fin lines as a function of s/a for various values of d/b at $b/\lambda = 0.3556$. These values agree within ± 2 percent with Hofmann's results [12].

VII. Results and Conclusions

In the foregoing sections, we have presented expressions for the evaluation of the cutoff wavelength, guided wavelength, and characteristic impedance of unilateral and bilateral fin-lines. These expressions are directly applicable to the design of fin-line circuits. The expressions for the cutoff wavelength agree within ± 1 percent and those of guided wavelength agree within ± 2 percent with the results obtained with the spectral domain technique. These expressions may look slightly complicated at a first glance, but when applied to a practical problem, they reduce to a very simple expression of the form $y = Ax^B$. This is so because the designer initially fixes the thickness of the substrate and chooses a given waveguide size, thus fixing the values of b/a and s/a. The only remaining variable parameter is then the normalized gap width d/b. The expression for the characteristic impedance agrees also within ± 2 percent with the values given by Hofmann [12]. This definition of characteristic impedance is appropriate in impedance matching problems and in the characterization of discontinuities in fin-line structures [13].

References

[1] P. J. Meier, "Integrated fin-line millimeter components," *IEEE Trans. Microwave Theory Tech.*, vol. MTT-22, pp. 1209–1216, Dec. 1974.

[2] A. M. K. Saad and G. Begemann, "Electrical performance of fin-lines of various configurations," *Microwaves, Opt. Acoust.*, vol. 1, pp. 81–88, Jan. 1977.

[3] W. J. R. Hoefer, "Fin-line design made easy," in 1978 IEEE MTT-S Int. Microwave Symp., (Ottawa, Canada), p. 471.

[4] H. Hofmann, "Dispersion of planar waveguides for millimeter-wave applications," *Arch. Elek. Ubertragung.*, vol. 31, pp. 40–44, Jan. 1977.

[5] J. B. Knorr and P. M. Shayda, "Millimeter-wave fin-line characteristics," *IEEE Trans. Microwave Theory Tech.*, vol. MTT-28, pp. 737–743, July 1980.

[6] L.-P. Schmidt and T. Itoh, "Spectral domain analysis of dominant and higher order modes in fin-lines," *IEEE Trans. Microwave Theory Tech.*, vol. MTT-28, pp. 981–985, Sept. 1980.

[7] A. Beyer and I. Wolff, "A solution of the earthed fin-line with finite metallization thickness," in *1980 IEEE MTT-S Int. Microwave Symp. Dig.*, (Washington, DC), pp. 258–260.

[8] A. K. Sharma and W. J. R. Hoefer, "Empirical analytical expressions for the fin-line design," in *1981 IEEE MTT-S Int. Microwave Symp. Dig.*, (Los Angeles, CA), pp. 102–104.

[9] A. K. Sharma, G. I. Costache, and W. J. R. Hoefer, "Cutoff in fin-lines evaluated with the spectral domain technique and with the finite element method," in *1981 IEEE AP-S Int. Antenna Propagation Symp. Dig.*, (Los Angeles, CA), pp. 308–311.

[10] H. Meinel and B. Rembold, "New millimeter-wave fin-line attenuators and switches," in *1979 IEEE MTT-S Int. Microwave Symp. Dig.*, (Orlando, FL), pp. 249–252.

[11] S. B. Cohn, "Properties of ridge waveguide," *Proc. IRE*, vol. 35, pp. 783–788, Aug. 1947.

[12] H. Hofmann, private communication.

[13] E. Pic and W. J. R. Hoefer, "Experimental characterization of fin-line discontinuities using resonant techniques," in *1981 IEEE MTT-S Int. Microwave Symp. Dig.*, (Los Angeles, CA), pp. 108–110.

Impedance transformation in fin lines

Hadia El Hennawy, Ph.D., Mem. I.E.E.E., and Prof. Klaus Schünemann, Dr.-Ing., Mem. I.E.E.E., M.V.D.E.

Indexing terms: *Semiconductor devices and materials, Microwave circuits and networks, Microwave components, Numerical analysis*

Abstract: Various discontinuities in the slot width of fin lines are accurately analysed. These include a step in the slot width, an inductive notch, a small capacitive strip and a longitudinal stripe for mounting a semiconductor diode. Equivalent circuits are derived and related to the geometrical parameters. The results can directly be applied to a design of impedance transforming and filter networks.

1 Introduction

Various passive and active components have been realised in integrated fin line throughout a broad frequency spectrum ranging from 10 to 140 GHz. These include oscillators [1, 2], mixers [3–5], digital modulators [6], switches and attenuators [7], detectors [8], couplers [6, 9], filters [10, 11] and tapers [12]. Nearly all of these components have been developed by trial and error, except for the filters and tapers reported in References 11 and 12. Theoretical investigations published so far deal almost exclusively with the calculation of the propagation constants and wave impedances of the fundamental and higher-order modes: this topic has been solved by approximations (leading to closed-form solutions [13]) and by rigorous calculations [14, 15]. There is, however, a lack of fundamental theoretical research concerning an analytical description of almost all passive circuit elements used for either mounting semiconductor devices or impedance transformation, or for both. It is hoped that the present work will be a first step in filling this gap.

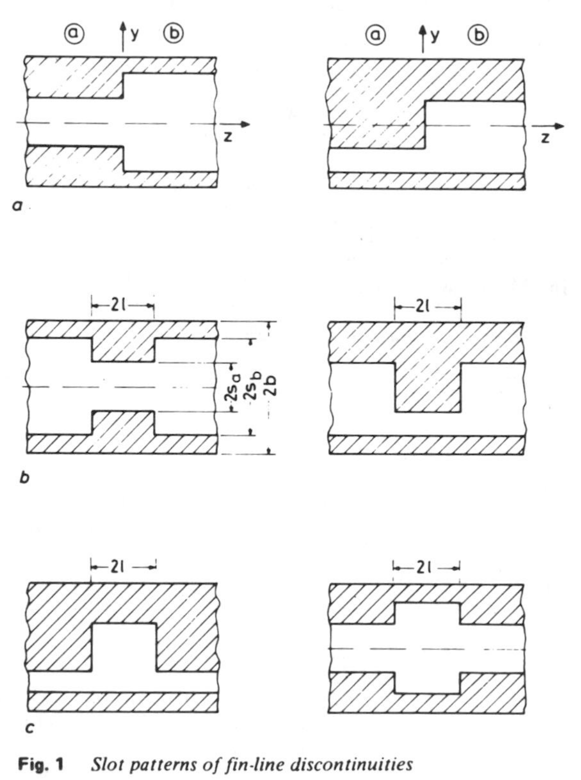

Fig. 1 *Slot patterns of fin-line discontinuities*

a step, *b* strip, *c* notch

Paper 2234H, received 8th December 1981

The authors are with the Institut fuer Hochfrequenztechnik, Technische Universitaet, Postfach 3329, D-3300 Braunschweig, West Germany

The structures to be investigated are shown in Fig. 1. The basic building block is a step discontinuity in the slot width (Fig. 1*a*) which may or may not be symmetric with respect to a plane $y =$ constant. This discontinuity acts as an impedance transformer shunted by a susceptance. Two such steps can be combined to form either a strip (Fig. 1*b*) or a notch (Fig. 1*c*). Potential applications depend on the electrical length of the strip or of the notch. A short strip can be used for realising a capacitance, whereas a short notch can be modelled by an inductance [6]. A long, unsymmetric strip (which will be called a longitudinal stripe) is often used for mounting a semiconductor device [1]. This mount resembles a post mount in rectangular waveguides. A narrow slit in one fin, which can be utilised as a DC separation in a complex structure, can be regarded as a special case of the notch: it forms a very short, unsymmetric notch with a slot width equal to the height of the waveguide housing. Finally, strips and notches can be cascaded for the design of a lowpass filter.

2 Modal analysis

The step discontinuity in the slot width of Fig. 1*a* will be treated by a modal analysis [16]. In Reference 16, two uniform cylindrical waveguides *a* and *b* with different cross-sections and distributions of enclosed electrical properties are considered which are joined end to end forming a junction. The boundary conditions are satisfied by a suitable infinite series of modes appropriate to each side of the junction. We will assume, now, that the modes in both waveguides and the scattering coefficients of successive discontinuities are known. The properties of the junction can then be calculated, if one finds out how power is distributed between the various scattered modes.

The solution of this basic problem will first be described with the help of a modal analysis. The co-ordinate system is chosen so that $z = 0$ is the plane of the junction, and x and y are transverse co-ordinates. The transverse fields of each mode are written as

$$e_i(x,y,z) = a_i\,\boldsymbol{e}_i(x,y)\,e^{\pm\gamma_i z}$$
$$h_i(x,y,z) = a_i\,\boldsymbol{h}_i(x,y)\,e^{\pm\gamma_i z} \qquad (1)$$

$\boldsymbol{e}_i$ and $\boldsymbol{h}_i$ are transverse vector functions of the electric and magnetic fields, respectively. Expanding the transverse fields in terms of these modes just to the left of the junction (waveguide *a*), yields

$$\boldsymbol{E}_a = (1+\rho)\,a_1\,\boldsymbol{e}_{a1} + \sum_{i=2}^{\infty} a_i \boldsymbol{e}_{ai}$$

$$\boldsymbol{H}_a = (1-\rho)\,a_1\,\boldsymbol{h}_{a1} - \sum_{i=2}^{\infty} a_i \boldsymbol{h}_{ai} \qquad (2)$$

Similarly, we can write, just to the right of the junction

Reprinted with permission from *IEE Proceedings,* vol. 129, Pt. H, no. 6, pp. 342–350, December 1982.

(waveguide b),

$$E_b = \sum_{j=1}^{\infty} b_j(e_{bj} + \sum_{k=1}^{\infty} s_{jk}\, e_{bk})$$

$$H_b = \sum_{j=1}^{\infty} b_j(h_{bj} - \sum_{k=1}^{\infty} s_{jk} h_{bk}) \qquad (3)$$

The subscripts a and b denote the individual waveguides, ρ means reflection coefficient of the incident mode, which is characterised by $i = 1$ and amplitude a_1. s_{jk} are the scattering coefficients of the next discontinuity, which is located at $z > 0$. The magnetic and electric fields have been expressed independently, because a unique wave impedance cannot be defined for a fin line.

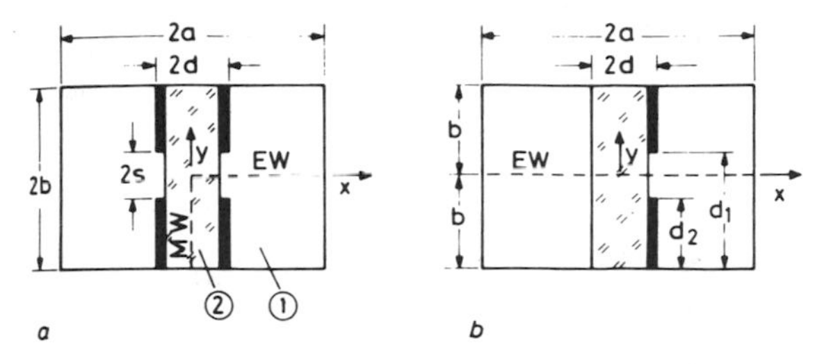

Fig. 2 *Cross-section of bilateral and unilateral fin lines*
EW = electric wall, MW = magnetic wall
a Bilateral fin line, *b* unilateral fin line

The following calculations will be illustrated for a bilateral fin line with a symmetrically located slot as an example. Its cross-section is shown in Fig. 2a. It is obvious that only modes which are symmetrical with respect to $x = 0$ will be excited. One may hence assume a magnetic wall (MW) at $x = 0$ and an electric wall (EW) at $y = 0$. The transverse vector functions e_i and h_i can then be composed of the transverse field components in subregions 1 and 2 of Fig. 2a at $z = 0$, according to:

$$E_{ti}^{(1)} = \sum_{n=1}^{\infty} A_n^{(1)} \cos \boldsymbol{\alpha}_n (x - a) \sin \alpha_n y \cdot u_x$$

$$+ \sum_{n=0}^{\infty} S_n^{(1)} \sin \boldsymbol{\alpha}_n (x - a) \cos \alpha_n y \cdot u_y$$

$$E_{ti}^{(2)} = \sum_{n=1}^{\infty} A_n^{(2)} \sin \tilde{\boldsymbol{\alpha}}_n x \sin \alpha_n y \cdot u_x$$

$$+ \sum_{n=0}^{\infty} S_n^{(2)} \cos \tilde{\boldsymbol{\alpha}}_n x \cos \alpha_n y \cdot u_y \qquad (4)$$

and

$$H_{ti}^{(1)} = \sum_{n=0}^{\infty} B_n^{(1)} \sin \boldsymbol{\alpha}_n (x - a) \cos \alpha_n y \cdot u_x$$

$$+ \sum_{n=1}^{\infty} M_n^{(1)} \cos \boldsymbol{\alpha}_n (x - a) \sin \alpha_n y \cdot u_y$$

$$H_{ti}^{(2)} = \sum_{n=0}^{\infty} B_n^{(2)} \cos \tilde{\boldsymbol{\alpha}}_n x \cos \alpha_n y \cdot u_x$$

$$+ \sum_{n=1}^{\infty} M_n^{(2)} \sin \tilde{\alpha}_n x \sin \alpha_n y \cdot u_y \qquad (5)$$

u_x and u_y are unit vectors in the x- and y-directions, respectively. Owing to the dielectric substrate, the fields are neither TE nor TM but hybrid. Further details concerning the fin line eigenmodes are given in Appendix 7.2.

The boundary conditions for the junction of Fig. 1a (continuity of the transverse fields through all apertures and zero tangential electric field at conducting obstacles), which can be formulated with eqns. 2 and 3, are now satisfied in the following way: The crossproduct of the electric-field component of eqn. 2 with h_{am} is taken and integrated over the cross-section of waveguide a. For the unknown aperture field on the left-hand side, E_b is inserted from eqn. 3. The crossproduct of the magnetic-field component of eqn. 2 with e_{bn} is also taken and integrated over the cross-section of waveguide b. The unknown aperture field H_b is inserted from eqn. 3. Both operations yield:

$$(1 + \rho) a_1 \int_{(a)} e_{a1} \times h_{am} \cdot u_z\, dxdy$$

$$+ \sum_{i=2}^{\infty} a_i \int_{(a)} e_{ai} \times h_{am} \cdot u_z\, dxdy = \sum_{j=1}^{\infty} \left[b_j \left(\left(\int_{(b)} e_{bj} \times h_{am} \right.\right.\right.$$

$$\left.\left.\left. \cdot\, u_z\, dxdy + \sum_{k=1}^{\infty} s_{jk} \int_{(b)} e_{bk} \times h_{am} \cdot u_z\, dxdy \right)\right)\right] \qquad (6a)$$

$$(1 - \rho) a_1 \int_{(b)} e_{bn} \times h_{a1} \cdot u_z\, dxdy - \sum_{i=2}^{\infty} a_i \int_{(b)} e_{bn} \times h_{ai}$$

$$\cdot\, u_z\, dxdy = \sum_{j=1}^{\infty} \left[b_j \left(\left(\int_{(b)} e_{bn} \times h_{bj} \cdot u_z\, dxdy \right.\right.\right.$$

$$\left. - \sum_{k=1}^{\infty} s_{jk} \int_{(b)} e_{bn} \times h_{bk} \cdot u_z\, dxdy \right) \qquad (6b)$$

u_z is the unit vector in the z-direction. Both equations assure the continuity of the transverse fields. Because E_a exists over the aperture only and vanishes elsewhere, the integrals on the right-hand side of eqn. 6a must only be taken over waveguide b. The solutions of the integrals in eqns. 6 are given in Appendix 7.1.

The next task is to calculate the scattering coefficients s_{jk}. In a waveguide system with more than one discontinuity, interactions occur between dominant (fundamental) and higher-order modes. In our case of an abrupt transition between waveguides a and b, it is necessary to know the scattering properties of the termination of waveguide b, before eqns. 6 can be solved. It is sufficient to consider simple terminations such as a matched load, a short-circuit, or an open circuit, which cause an independent reflection of each mode incident upon it regardless of the amplitude and phase of any other mode. In the special case of a single discontinuity, waveguide b can be thought to be terminated with a matched load; hence $s_{jk} = 0$. In the other case shown in Fig. 1 (two abrupt transitions in cascade), we can take advantage of the symmetry of the structure with respect to the plane $z = l$. The equivalent circuit of the symmetric strip or notch is usually found from a symmetrical and an antisymmetrical excitation of both ports, which produces either a short-circuit or an open circuit in the plane of symmetry. Hence $s_{jk} = 0$ for $j \neq k$ and

$$s_{jj} = (1 - y_{bj})/(1 + y_{bj}) = \pm \exp(-2j\beta_{bj} l) \qquad (7)$$

y_{bj} is the normalised input admittance of the jth mode in waveguide b measured at $z = 0$, β_{bj} is the propagation constant of this mode. The solution to the general case (more than two discontinuities or two unsymmetrical transitions) can be composed as sketched in Reference 16.

We are now in a position to solve the system of eqns. 6. It is convenient to normalise the wave amplitudes on a_1. Setting

$$a'_i = a_i/a_1,\ b'_j = b_j/a_1,\ I'_{ambj} = I_{ambj}/(F_{am}F_{bj}) \tag{8}$$

with the Is and Fs taken from Appendix 7.1 we obtain, for the case that the exciting wave a_1 impinges on waveguide b from waveguide a,

$$\rho I'_{a1bn} + \sum_{i=2}^{L} a'_i I'_{aibn} - \sum_{j=1}^{L} b'_j (1+s_{jj}) I'_{bjbn} = -I'_{a1bn}$$

$$\rho I'_{ama1} + \sum_{i=2}^{L} a'_i I'_{amai} + \sum_{j=1}^{L} b'_j (1-s_{jj}) I'_{ambj} = -I'_{ama1} \tag{9}$$

The slot width of waveguide b is assumed to be larger than that of waveguide a: $s_b > s_a$. This junction (which corresponds to a boundary enlargement problem in the terminology of Reference 16), in this case, will be called a step-up junction. In the complementary case called a step-down junction (excited from waveguide b) $s_b > s_a$ (boundary reduction problem), we obtain

$$\rho I'_{b1bm} + \sum_{i=2}^{L} a'_i I'_{bibm} - \sum_{j=1}^{L} b'_j (1+s_{jj}) I'_{ajbm} = -I'_{b1bm}$$

$$\rho I'_{anb1} + \sum_{i=2}^{L} a'_i I'_{anbi} + \sum_{j=1}^{L} b'_j (1-s_{jj}) I'_{anaj} = I'_{anb1} \tag{10}$$

The infinite series are truncated at $i = j = L$, thus taking into account L modes in each waveguide. Then eqns. 9 or 10 generate $2L$ linear equations for $1 + (L-1) + L$ unknowns.

The discontinuities are characterised either by a transformer (Fig. 3*a*) or by a π-network (Fig. 3*b*), for a notch, and by a T-network (Fig. 3*c*), for a strip. The former is calculated from the reflection coefficient ρ, the latter from B_{sc} (X_{sc}), the short-circuit admittance (impedance), and B_{oc} (X_{oc}), the open-circuit input admittance (impedance). Strictly speaking, the equivalent circuit for the step discontinuity (Fig. 3*a*) must be completed by two series reactances. The numerical calculations have, however, shown that both are extremely small and can be neglected for any parameter combination.

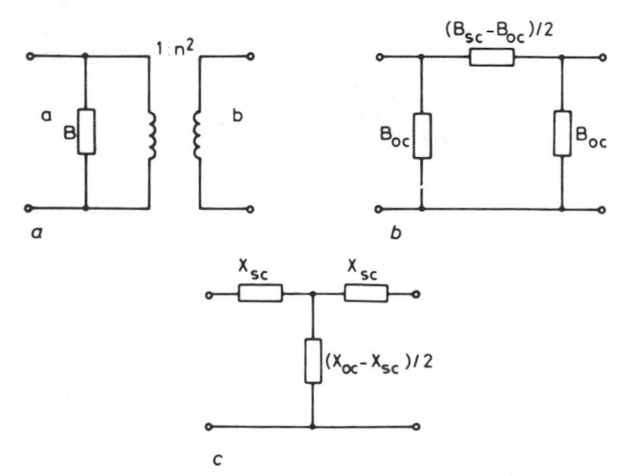

Fig. 3 *Equivalent circuits of a step, a notch and a strip discontinuity*
a **Step discontinuity,** ***b*** **notch discontinuity,** ***c*** **strip discontinuity**

3 Results

The impedance transforming potential of a single step, a notch and a strip will be illustrated in the following (some preliminary results have already been given in Reference 22): The characteristics of a single step in both unilateral and bilateral fin lines are shown in Figs. 4 and 5. The shunt susceptance is of course capacitive; it increases against the slot-width ratio s_b/s_a. Furthermore, it is larger for a bilateral than for a unilateral fin line, while the turns ratio behaves in an opposite way. The normalised series reactances (which have been neglected in the simplified equivalent circuit of Fig. 3*a*) are shown against frequency in Fig. 5*b*. They are indeed very small, except near the cutoff frequency and far above cutoff. The normalised shunt susceptance also shows a pole at cutoff. This can be attributed to the rapid increase of the wave impedance. All three parasitic elements increase almost linearly with frequency well above cutoff.

The characteristics of various notches have been displayed in Figs. 6–8. The series element of the π-equivalent circuit

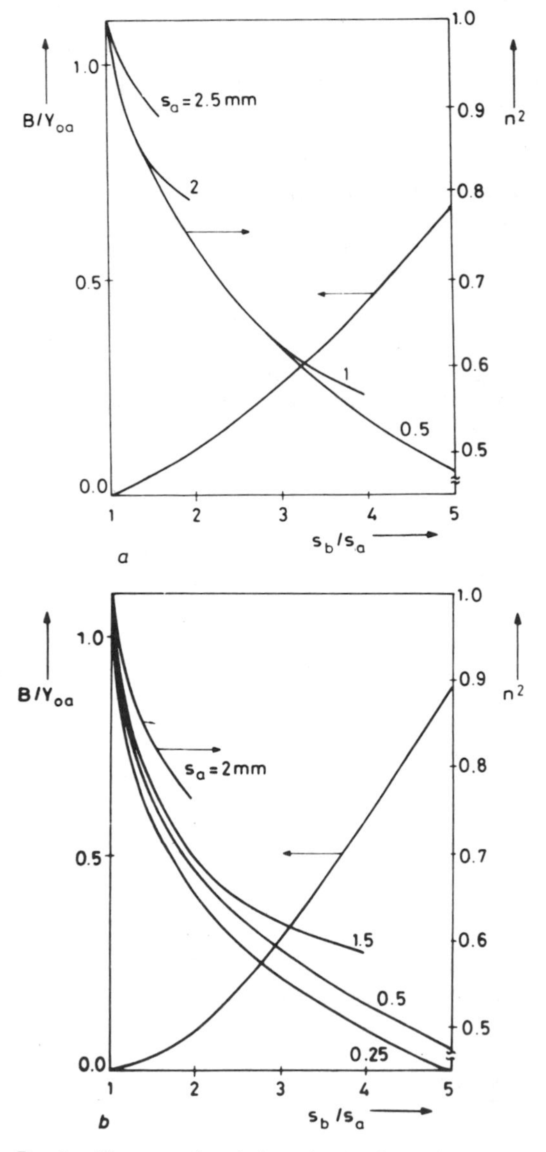

Fig. 4 *Elements of equivalent circuit of step discontinuity (Fig. 3a) against slot width ratio s_b/s_a*

a **unilateral fin line,** ***b*** **bilateral fin line**
Common parameters: symmetrically located slots of width $2s$, f = 15 GHz, ϵ_r = 2.22 (RT-Duroid), $2a$ = 15.8 mm, $2b$ = 7.9 mm, $2d$ = 0.254 mm, Y_{oa} = wave admittance of waveguide *a*

turns out to be always inductive, as long as the notch length is less than half a wavelength, while the shunt element is capacitive. This behaviour can be explained if higher-order mode coupling between the two step discontinuities of the notch is neglected. Then the equivalent circuit can be thought to be composed of the equivalent circuit of the two steps with a combined transmission line of length l. Hence the series element $B_{sc} - B_{oc}$ is proportional to $n^2/\sin(2\beta l)$, with β as the propagation constant of the fundamental mode. This relation easily explains the shape of the curves in Figs. 6–8.

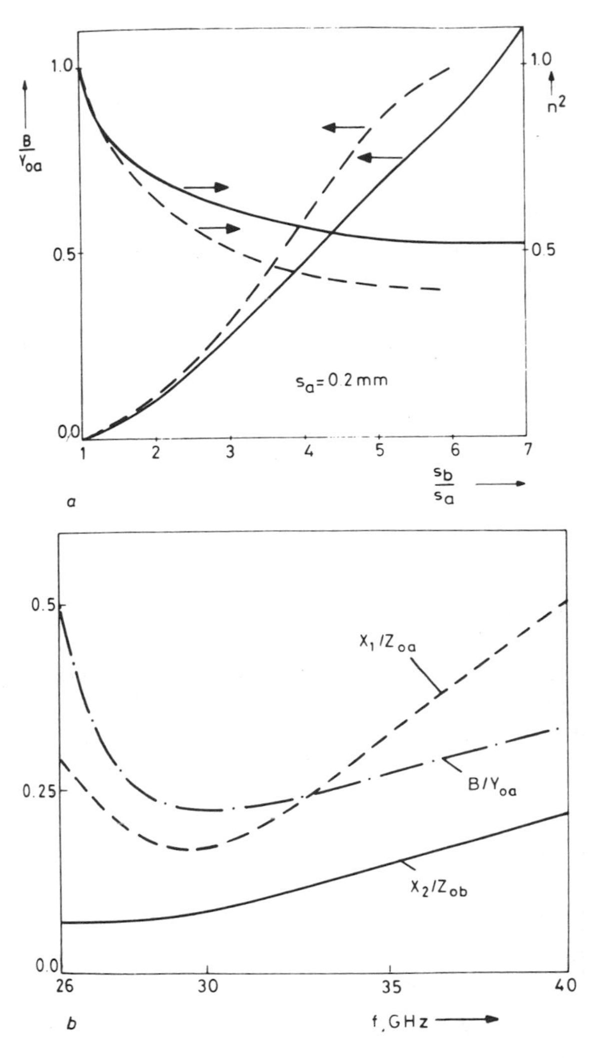

Fig. 5 *Elements of equivalent circuit of step discontinuity (Fig. 3a)*

a Unilateral (solid lines) and bilateral fin line (dashed lines) at 30 GHz
b Shunt susceptance B and series reactances X_1 (port *a*) and X_2 (port *b*), for $s_b/s_a = 1.778$ and $2s_a = 1$ mm
Common parameters: symmetrically located slots of width $2s$, $\epsilon_r = 2.22$, $2a = 7.112$ mm, $2b = 3.556$ mm, $2d = 0.254$ mm, $Y_0 (X_0)$ wave admittance (impedance)

The inductive susceptance increases with increasing ratio of the slot widths, because the turns ratio increases similarly. The susceptance is very large for both $l \to 0$ and $l \to \lambda/4$, where λ is the wavelength of the fundamental mode. In between these two extremes, we observe a flat response against l. The shunt element B_{oc} is proportional to $n^2\{B + Y_{ob} \tan(\beta l)\}$. Hence the susceptance increases against l up to a maximum (pole) at $l = \lambda/4$. Increasing the ratio of the slot widths means increasing B (see Fig. 4), and thus increasing B_{oc}. The overlapping of the curves in Fig. 7 can be understood, if we take into account the fact that the wavelength λ decreases when the slot width s_b, within the notch, decreases. Then, $l = \lambda/4$ must be achieved for a decreasing ratio l/λ_0 also.

Higher-order mode coupling between the step discontinuities can be neglected if $\gamma_i l > 2$, where γ_i is the real propagation constant of the ith mode. The numerical calculations yielded as a rule of thumb that this inequality is always fulfilled for $l/\lambda_0 > 0.05$. The error introduced by the neglect of higher-order mode coupling is then less than 10%.

The characteristics of an unsymmetrically located notch whose slot width equals the waveguide height are depicted in Fig. 8. A very short notch of this type ($l \to 0$) is of practical interest, because it can be used for DC separation in a circuit containing semiconductor devices; then the shunt susceptance should be small and the series susceptance large. The curves in Fig. 8 show that extremely short notches must be realised in order not to influence the high frequency performance.

The characteristics of various strips have been displayed in Figs. 9–11. The series element is always inductive while the shunt element is capacitive for $l/\lambda_0 < 1/4$. As for the notch, impedances within a large range can be realised. The shape of the curves can again be explained with the help of a simplified equivalent circuit neglecting higher-order

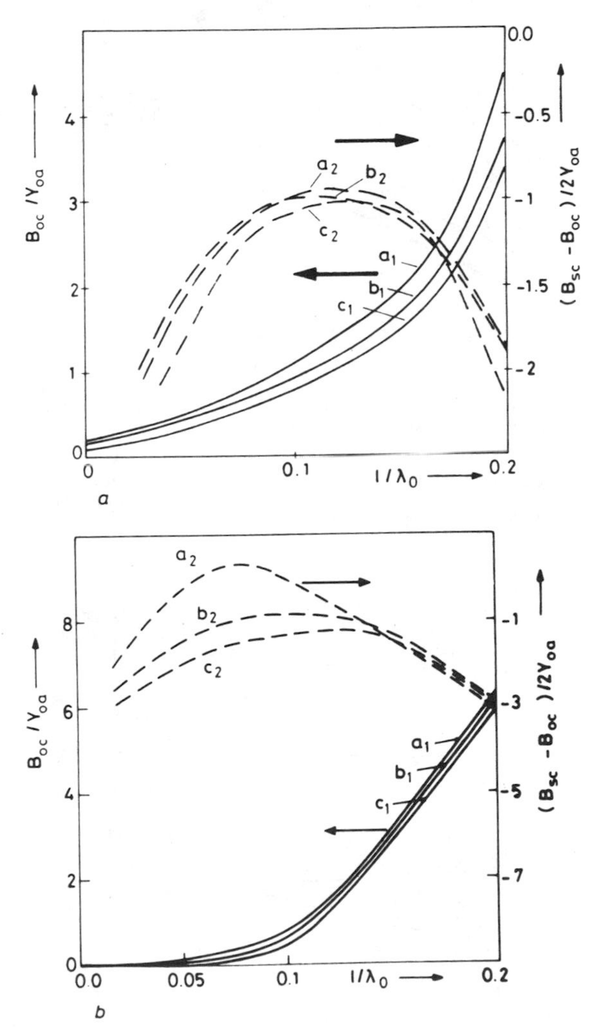

Fig. 6 *Elements of equivalent circuit of symmetrically located notch (Fig. 3b) against normalised notch length*

a Unilateral fin line, *b* bilateral fin line
Common parameters as in Fig. 4. Furthermore, $2s_a = 1.5$ mm and $2s_b = 3$ mm (curves a_1, a_2), 2.5 mm (curves b_1, b_2) and 2 mm (curves c_1, c_2),
λ_0 = free-space wavelength

mode coupling between the two step discontinuities of the strip. Then the series element X_{sc} is proportional to

$$X_{sc} = 1/[(Y_{oa}/n^2)\cot(\beta l) - B]$$

whereas the shunt element $(X_{oc} - X_{sc})/2$ is proportional to

$$(X_{oc} - X_{sc})/2 = -Y_{oa}/[(Y_{oa}^2/n^2 - n^2B^2)\sin(2\beta l) + 2BY_{oa}\cos(2\beta l)]$$

The series element represents the length of the strip. It increases against l, while its variation with the slot-width ratio is small. The shunt element, on the other hand, varies strongly with the slot-width ratio, whereas its response against the strip length is flat.

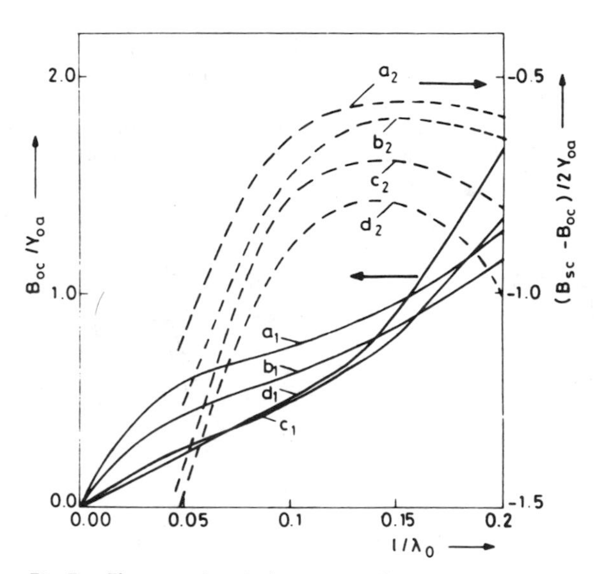

Fig. 7 *Elements of equivalent circuit of symmetrically located notch (Fig. 3b) against normalised notch length for a bilateral fin line at 30 GHz*

Common parameters as in Fig. 5. Furthermore, $2s_a = 0.5$ mm and $2s_b = 1.786$ mm (curves a_1, a_2), 1.5 mm (curves b_1, b_2), 1 mm (curves c_1, c_2) and 0.665 mm (curves d_1, d_2),
λ_0 = free-space wavelength

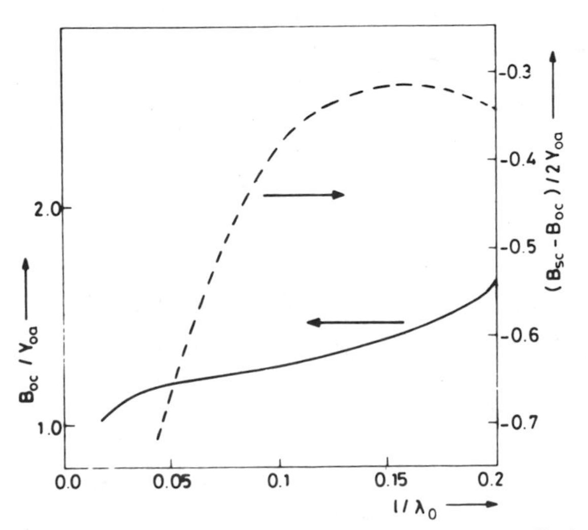

Fig. 8 *Elements of equivalent circuit of unsymmetrically located notch (Fig. 3b) against normalised notch length for a unilateral fin line at 30 GHz*

$s_a = 0.5$ mm, $s_b = 3.56$ mm, WR-28 waveguide, λ_0 = free-space wavelength

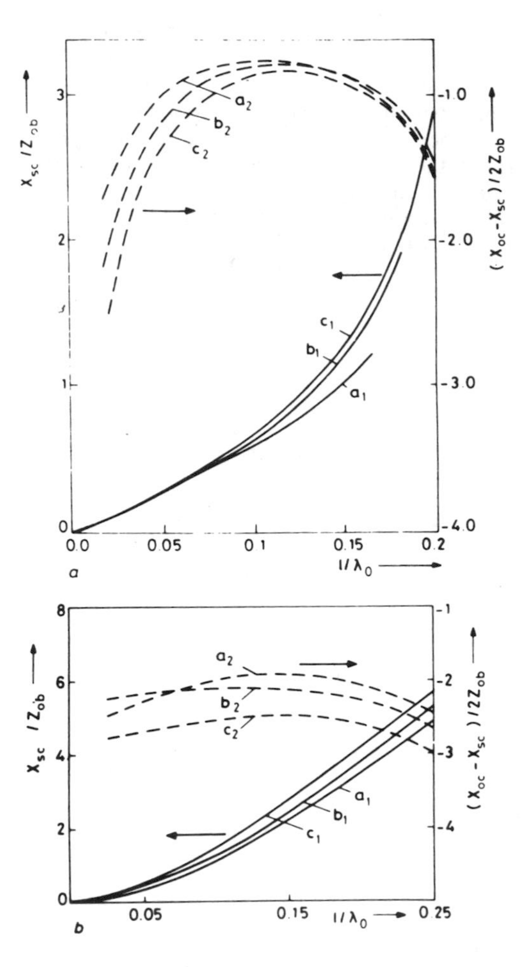

Fig. 9 *Elements of equivalent circuit of symmetrically located strip (Fig. 3c) against normalised strip length*

a Unilateral fin line, *b* bilateral fin line
Common parameters as in Fig. 4. The curves have been labelled as in Fig. 6.

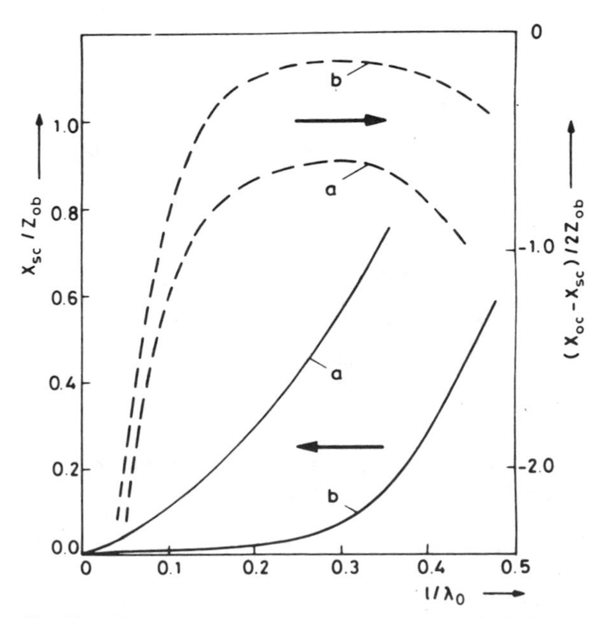

Fig. 10 *Elements of equivalent circuit of symmetrically located strip (Fig. 3c) against normalised strip length for a bilateral fin line at 30 GHz*

Common parameters as in Fig. 5. Furthermore, $s_b = 1.778$ mm with $s_a = 1$ mm (curves *a*) and $s_a = 0.5$ mm (curves *b*)

Relatively short strips are used for narrowing a fin-line slot to solder a beam-lead diode directly on the metallic fins. The length of these diodes amounts to some 0.1 mm in the lower millimetre-wave region. The characteristics of such a strip are shown in Fig. 11. The series element cannot be neglected except for extremely short strips. In practice, such a diode mount will also perform an impedance transformation, so that it must be taken into account in the circuit design.

Finally, Fig. 12 shows that the computed results agree well with measurements, which have been performed in K_u-band in order to use a network analyser.

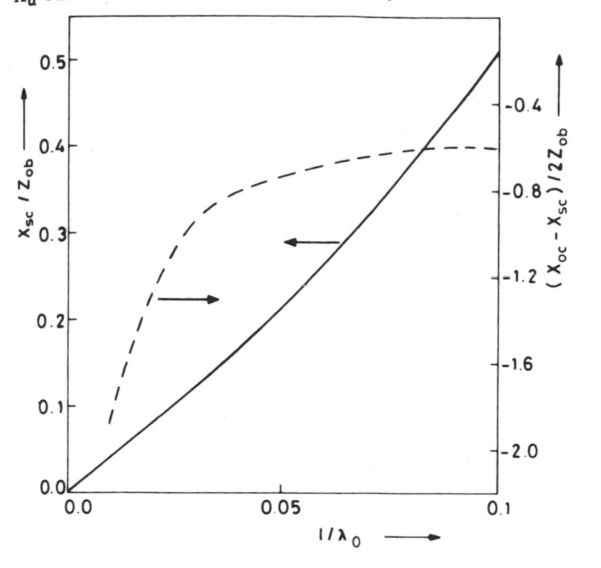

Fig. 11 *Elements of equivalent circuit of unsymmetrically located strip (Fig. 3c) against normalised strip length for a unilateral fin line at 30 GHz*

s_a = 0.2 mm, s_b = 0.6 mm, WR-28 waveguide, λ_o = free-space wavelength

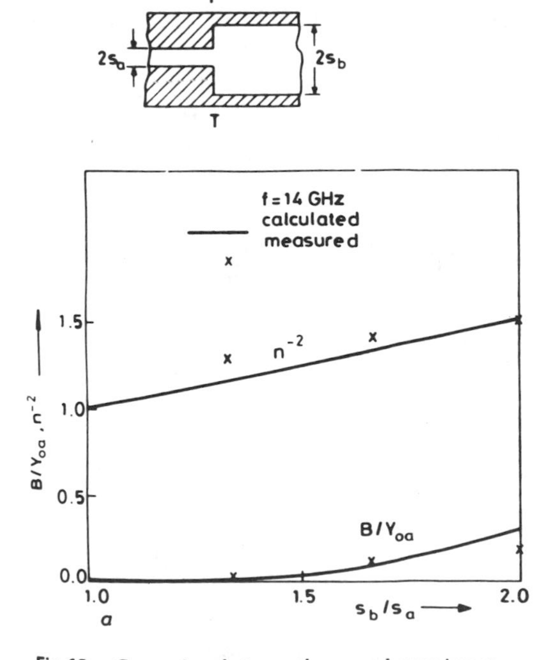

Fig. 12 *Comparison between theory and experiment*

a **Step discontinuity, $2s_a$ = 3.0 mm**
b **Notch with B_{in} = Im (Y_{in}), $2s_a$ = 3.0 mm, $2s_b$ = 6.0 mm**
c **Strip with B_{in} = Im (Y_{in}), $2s_a$ = 3.0 mm, $2s_b$ = 6.0 mm**
Common parameters as in Fig. 4
(WR-62 waveguide), unilateral fin line

4 Conclusions

A modal analysis has been applied to various discontinuities in the slot width of unilateral and bilateral fin lines. Equivalent circuits have been calculated for a single step discontinuity and for composite structures like notches, slits and strips. These structures offer a large flexibility to the circuit engineer for impedance transformation and semiconductor mounts. A computer-aided circuit design should now be possible, based on the numerous illustrative examples given in this paper.

5 Acknowledgment

The authors thank the Deutche Forschungsgemeinschaft for financial support.

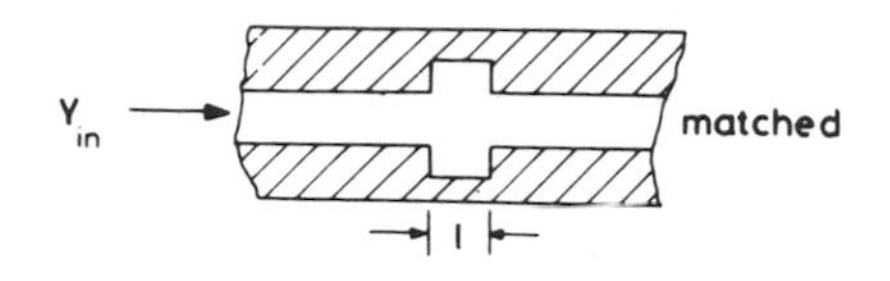

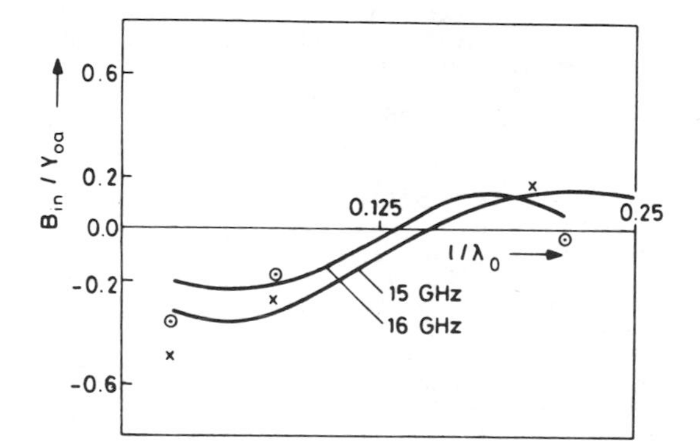

b

x 15 GHz
⊙ 16 GHz

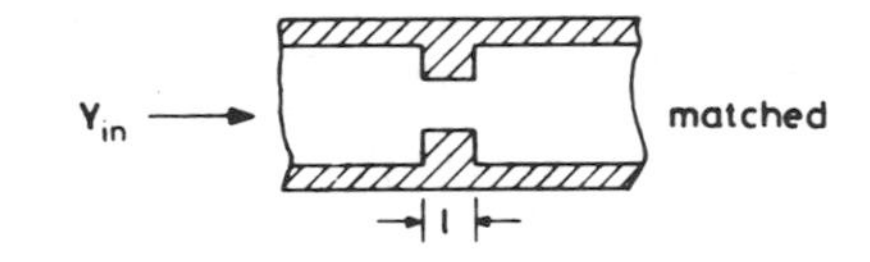

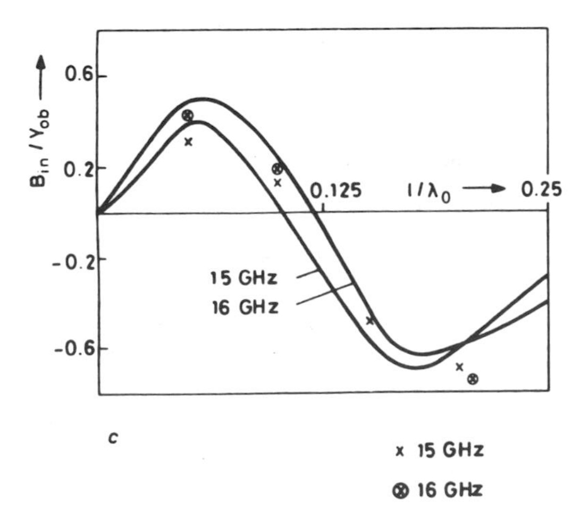

c

x 15 GHz
⊗ 16 GHz

6 References

1 KNÖCHEL, R.: 'Design and performance of microwave oscillators in integrated fin-line technique', *IEE J. Microwaves, Opt. & Acoust.*, 1979, **3**, (3), pp. 115–120

2 COHEN, L.D.: 'Advances in printed millimeter-wave oscillator circuits'. IEEE MTT Symposium digest, Washington DC, USA, 1980, pp. 264–266

3 BEGEMANN, G.: 'An X-band balanced fin-line mixer', *IEEE Trans.*, 1978, **MTT-26**, pp. 1007–1011

4 TROOST, A.: 'Fortschritte auf dem Gebiet der Millimeterwellen im Bereich 35 bis 90 GHz', *Nach. Elektron.*, 1976, **12**, pp. 295–297

5 KNÖCHEL, R., and SCHLEGEL, A.: 'Octave-band double-balanced integrated fin-line mixers at mm-wavelengths'. Proceedings of the 10th European Microwave Conference, Warsaw, 1980, pp. 722–726

6 KPODZO, E., SCHÜNEMANN, K., and BEGEMANN, G.: A quadriphase fin-line modulator'. *IEEE Trans.*, 1980, **MTT-28**, pp. 747–752

7 MEINEL, H., and REMBOLD, B.: 'New mm-wave fin-line attenuators and switches'. IEEE MTT Symposium digest, Orlando, USA, 1979, pp. 249–252

8 MEINEL, H., and SCHMIDT, L.-P.: 'High sensitivity mm-wave detectors using fin-line technology'. Fifth international conference on Infrared and Millimeter Waves, Würzburg, Deutschland, 1980

9 COHEN, L.D., and MEIER, P.J.: 'Advances in E-plane printed millimeter-wave circuits', IEEE MTT Symposium digest, Ottawa, Canada, 1978, pp. 27–29

10 MEIER, P.J.: 'Integrated fin-line millimeter components', *IEEE Trans.*, 1974, **MTT-22**, pp. 1209–1216

11 SAAD, A.M.K., and SCHÜNEMANN, K.: 'Design and performance of fin-line bandpass filters'. Proceedings of the 9th European Microwave Conference, Brighton, England, 1979, pp. 397–401

12 BEYER, A., and WOLFF, I.: 'Calculation of the transmission properties of inhomogeneous fin-lines'. Proceedings of the 10th European Microwave Conference, Warsaw, 1980, pp. 322–326

13 SAAD, A.M.K., and BEGEMANN, G.: 'Electrical performance of finlines of various configurations', *IEE J. Microwaves, Opt. & Acoust.*, 1977, **1**, (2), pp. 81–88

14 SCHMIDT, L.P., ITOH, T., and HOFMANN, H.: 'Characteristics of unilateral finline structures with arbitrarily located slots'. IEEE MTT Symposium digest, Washington DC, USA, 1980, pp. 255–257

15 BEYER, A., and WOLFF, I.: 'A solution of the earthed fin-line with finite metallization thickness'. IEEE MTT Symposium digest, Washington DC, USA, 1980, pp. 258–260

16 WEXLER, A.: 'Solution of waveguide discontinuities by modal analysis', *IEEE Trans.*, 1967, **MTT-15**, pp. 508–517

17 HOFMANN, H.: 'Dispersion of planar waveguides for millimetrewave applications', *Arch. Elektron. & Uebertragungstech.*, 1977, **31**, pp. 40–44

18 MITTRA, R., ITOH, T., and LI, T.: 'Analytical and numerical studies of the relative convergence phenomenon arising in the solution of an integral equation by the moment method', *IEEE Trans.*, 1972, **MTT-20**, pp. 96-104

19 EL-SHERBINY, A., and SAID, M.: 'Spectral domain calculation of the fin-line dispersion characteristics', 3rd international congress for Statistics and Computer Science, Cairo, Egypt, 1978

20 CHANG, C., and ITOH, T.: 'Spectral domain analysis of dominant and higher-order modes in fin-lines', IEEE MTT Symposium digest, Orlando, USA, 1979, pp. 344–346

21 SCHMIDT, L.P., and ITOH, T.: 'Spectral domain analysis of dominant and higher-order modes in fin-lines', *IEEE Trans.*, 1980, **MTT-28**, pp. 981–985

22 EL HENNAWY, H., and SCHÜNEMANN, K.: 'Analysis of fin-line discontinuities'. Proceedings of the 9th European Microwave Conference, Brighton, 1979, pp. 448–452

7 Appendix

7.1 Some integrals

For solving the various integrals appearing in eqns. 6, the transverse vector functions have to be inserted from eqns. 4 and 5. This yields

$$I_{aibj} = \int_{(a,b)} \boldsymbol{e}_{ai} \times \boldsymbol{h}_{bj} \cdot \boldsymbol{u}_z \, dxdy$$

$$= \frac{b}{2} \sum_{n=0}^{\infty} \Bigg\{ [\delta_n A_{na}^{(1)} M_{nb}^{(1)} - \eta_1 \epsilon_n B_{nb}^{(1)} S_{na}^{(1)}] \frac{\sin(\alpha_{nai} - \alpha_{nbj})(a-d)}{2(\alpha_{nai} - \alpha_{nbj})}$$

$$+ [\delta_n A_{na}^{(1)} M_{nb}^{(1)} + \eta_1 \epsilon_n B_{nb}^{(1)} S_{na}^{(1)}] \frac{\sin(\alpha_{nai} + \alpha_{nbj})(a-d)}{2(\alpha_{nai} + \alpha_{nbj})} + [\eta_2 \delta_n A_{na}^{(2)} M_{nb}^{(2)} - \epsilon_n B_{nb}^{(2)} S_{na}^{(2)}] \frac{\sin(\tilde{\alpha}_{nai} - \tilde{\alpha}_{nbj})d}{2(\tilde{\alpha}_{nai} - \tilde{\alpha}_{nbj})}$$

$$- [\eta_2 \delta_n A_{na}^{(2)} M_{nb}^{(2)} + \epsilon_n B_{nb}^{(2)} S_{na}^{(2)}] \frac{\sin(\tilde{\alpha}_{nai} + \tilde{\alpha}_{nbj})d}{2(\tilde{\alpha}_{nai} + \tilde{\alpha}_{nbj})} \Bigg\} \tag{11}$$

with

$$\eta_1 = \begin{cases} 1 \text{ for } \alpha_{nbj} \text{ real} \\ -1 \text{ for } \alpha_{nbj} \text{ imaginary} \end{cases} \qquad \delta_n = \begin{cases} 0 \text{ for } n = 0 \\ 1 \text{ for } n \neq 0 \end{cases}$$

$$\eta_2 = \begin{cases} 1 \text{ for } \tilde{\alpha}_{nbj} \text{ real} \\ -1 \text{ for } \tilde{\alpha}_{nbj} \text{ imaginary} \end{cases} \qquad \epsilon_n = \begin{cases} 1 \text{ for } n \neq 0 \\ 2 \text{ for } n = 0 \end{cases} \tag{12}$$

$$F_{ai}^2 = \int_{(a,b)} \boldsymbol{e}_{ai} \times \boldsymbol{e}_{ai}^* \cdot \boldsymbol{u}_z \, dxdy$$

$$= \frac{b}{2} \sum_{n=0}^{\infty} \Bigg\{ [\delta_n |A_n^{(1)}|^2 + \eta_1 \epsilon_n |S_n^{(1)}|^2] \frac{a-d}{2}$$

$$+ [\delta_n |A_n^{(1)}|^2 - \eta_1 \epsilon_n |S_n^{(1)}|^2] \frac{\sin 2\alpha_n (a-d)}{4\alpha_n}$$

$$+ [\delta_n \eta_2 |A_n^{(2)}|^2 + \epsilon_n |S_n^{(2)}|^2] \frac{d}{2}$$

$$- [\eta_2 \delta_n |A_n^{(2)}|^2 - \epsilon_n |S_n^{(2)}|^2] \frac{\sin 2\tilde{\alpha}_n d}{4\tilde{\alpha}_n} \Bigg\} \tag{13}$$

$$I_{aiai} = \int_{(a,b)} \boldsymbol{e}_{ai} \times \boldsymbol{h}_{ai} \cdot \boldsymbol{u}_z \, dxdy$$

$$= \frac{b}{2} \sum_{n=0}^{\infty} \Bigg\{ [\delta_n A_{na}^{(1)} M_{na}^{(1)} - \eta_1 \epsilon_n S_{na}^{(1)} B_{na}^{(1)}] \frac{a-d}{2}$$

$$+ [\delta_n A_{na}^{(1)} M_{na}^{(1)} + \eta_1 \epsilon_n S_{na}^{(1)} B_{na}^{(1)}] \frac{\sin 2\alpha_n (a-d)}{4\alpha_n}$$

$$+ [\eta_2 \delta_n A_{na}^{(2)} M_{na}^{(2)} - \epsilon_n B_{na}^{(2)} S_{na}^{(2)}] \frac{d}{2}$$

$$- [\eta_2 \delta_n A_{na}^{(2)} M_{na}^{(2)} + \epsilon_n B_{na}^{(2)} S_{na}^{(2)}] \frac{\sin 2\tilde{\alpha}_n d}{4\tilde{\alpha}_n} \Bigg\} \tag{14}$$

$$I_{aiaj} = \int_{(a,b)} \boldsymbol{e}_{ai} \times \boldsymbol{h}_{aj} \cdot u_z \, dxdy$$

$$= \frac{b}{2} \sum_{n=0}^{\infty} \left\{ [\delta_n A^{(1)}_{nai} M^{(1)}_{naj} - \epsilon_n B^{(1)}_{naj} S^{(1)}_{nai}] \frac{\sin(\boldsymbol{\alpha}_{ni} - \boldsymbol{\alpha}_{nj})(a-d)}{2(\boldsymbol{\alpha}_{ni} - \boldsymbol{\alpha}_{nj})} \right.$$

$$+ [\delta_n A^{(1)}_{nai} M^{(1)}_{naj} + \epsilon_n B^{(1)}_{naj} S^{(1)}_{nai}] \frac{\sin(\boldsymbol{\alpha}_{ni} + \boldsymbol{\alpha}_{nj})(a-d)}{2(\boldsymbol{\alpha}_{ni} + \boldsymbol{\alpha}_{nj})} + [\delta_n A^{(2)}_{nai} M^{(2)}_{naj} - \epsilon_n B^{(2)}_{naj} S^{(1)}_{nai}] \frac{\sin(\tilde{\boldsymbol{\alpha}}_{ni} - \tilde{\boldsymbol{\alpha}}_{nj}) d}{2(\tilde{\boldsymbol{\alpha}}_{ni} - \tilde{\boldsymbol{\alpha}}_{nj})}$$

$$- [\delta_n A^{(2)}_{nai} M^{(2)}_{naj} + \epsilon_n B^{(2)}_{naj} S^{(2)}_{nai}] \frac{\sin(\tilde{\boldsymbol{\alpha}}_{ni} + \tilde{\boldsymbol{\alpha}}_{nj}) d}{2(\tilde{\boldsymbol{\alpha}}_{ni} + \tilde{\boldsymbol{\alpha}}_{nj})} \qquad (15)$$

Some of the indices in eqns. 11 to 15 have been omitted, as they were not necessary for a unique characterisation. The integral F^2_{ai} of eqn. 13 is later used for normalisation. The infinite series in the above formulas should be truncated at $N = Mb/s_b$ for $s_b > s_a$ and at $N = Mb/s_a$ for $s_b < s_a$, in order to make accurate use of the edge condition [18]. M is the number of slot field components used in eqn. 25. (We have chosen $M = 4$.)

7.2 Fin-line eigenmodes

The eigenmodes have been calculated by modifying the analysis given in Reference 17 in such a way that the edge conditions [18] can be taken into account without considerably increasing the computation time. This has been achieved by using the slot field distribution suggested in Reference 19 for the first time. (More recently, the method has also been described in Reference 20.) The procedure will be briefly described in the following:

The fin-line fields are derived from the E_x- and H_x-components which must satisfy the Helmholtz equation

$$\left[\frac{\partial^2}{\partial x^2} + \frac{\partial^2}{\partial y^2} + (k_0^2 \epsilon_r - \beta^2)\right] \begin{pmatrix} E_x \\ H_x \end{pmatrix} = 0 \qquad (16)$$

The tangential field components are calculated from

$$\begin{pmatrix} E_t \\ H_t \end{pmatrix} = \frac{1}{k_0^2 \epsilon_r - k_x^2} \begin{pmatrix} \partial/\partial x & -k_0 u_x x \\ -k_0 \epsilon_r u_x x & \partial/\partial x \end{pmatrix} \nabla_t \begin{pmatrix} E_x \\ H_x \end{pmatrix} \qquad (17)$$

with $\nabla_t = \partial/\partial y(u_y) - j\beta u_z$, β = the propagation constant, and $k_0^2 = \omega^2 \mu_0 \epsilon_0$.

E_x and H_x are given in eqns. 4 and 5 for the odd modes of a bilateral fin line with symmetrically located slot. The various constants read

$$\boldsymbol{\alpha}_n = \sqrt{k_0^2 - \beta^2 - \alpha_n^2} \qquad \tilde{\boldsymbol{\alpha}}_n = \sqrt{\epsilon_r k_0^2 - \beta^2 - \alpha_n^2}$$

$$\alpha_n = n\pi/b \qquad (18)$$

Applying eqn. 17 to the x-components of the electric- and magnetic-field strength yields relations between the Fourier coefficients of the x- and y-components:

$$S^{(1)}_n = (-A^{(1)}_n \alpha_n \boldsymbol{\alpha}_n - j k_0 \beta B^{(1)}_n)/(\beta^2 + \alpha_n^2)$$
$$S^{(2)}_n = (A^{(2)}_n \alpha_n \tilde{\boldsymbol{\alpha}}_n - j k_0 \beta B^{(2)}_n)/(\beta^2 + \alpha_n^2)$$
$$M^{(1)}_n = (-B^{(1)}_n \alpha_n \boldsymbol{\alpha}_n - j k_0 \beta A^{(1)}_n)/(\beta^2 + \alpha_n^2)$$
$$M^{(2)}_n = (B^{(2)}_n \alpha_n \tilde{\boldsymbol{\alpha}}_n - j k_0 \epsilon_r \beta A^{(2)}_n)/(\beta^2 + \alpha_n^2) \qquad (19)$$

It follows from the continuity of the tangential field components that:

$$A^{(2)}_n = \frac{-\boldsymbol{\alpha}_n}{\tilde{\boldsymbol{\alpha}}_n} \frac{\sin \boldsymbol{\alpha}_n(d-a)}{\cos \tilde{\boldsymbol{\alpha}}_n d} A^{(1)}_n \qquad B^{(2)}_n = \frac{\sin \boldsymbol{\alpha}_n(d-a)}{\cos \tilde{\boldsymbol{\alpha}}_n d} B^{(1)}_n \qquad (20)$$

The remaining two coefficients $A^{(1)}_n$ and $B^{(1)}_n$ can be expressed in terms of the slot field components $E_y(y)$ and $E_z(y)$. This is done, in the spectral domain, by Fourier transforming all field components and applying the interface conditions. One obtains the field coefficients:

$$A^{(1)}_n = -2(\alpha_n L_{2n} + j\beta L_{1n})/[\boldsymbol{\alpha}_n b \sin \boldsymbol{\alpha}_n (d-a)]$$
$$B^{(1)}_n = 2(\alpha_n L_{1n} + j\beta L_{2n})/[k_0 b \sin \boldsymbol{\alpha}_n (d-a)] \qquad (21)$$

and the coupled algebraic equations

$$\sum_{n=0}^{\infty} \Lambda_{11} L_{2n} \cos\alpha_n y + \sum_{n=1}^{\infty} \Lambda_{12} L_{1n} \cos\alpha_n y = \frac{b}{2} I_y(y)$$

$$\sum_{n=1}^{\infty} \Lambda_{21} L_{2n} \sin\alpha_n y + \sum_{n=1}^{\infty} \Lambda_{22} L_{1n} \sin\alpha_n y = \frac{b}{2} I_z(y) \qquad (22)$$

L_{1n} and L_{2n} are Fourier transforms of the slot field components

$$L_{1n} = \int_0^s E_z \sin\alpha_n y \, dy \qquad L_{2n} = \int_0^s E_y \cos\alpha_n y \, dy \qquad (23)$$

and I_y and I_z are transforms of the current on the fins. These two unknowns are of no importance for the solution of eqns. 22, because they can easily be eliminated.

The various constants in eqns. 22 read:

$$\Lambda_{11} = -\frac{[\alpha_n^2 k_0 H_{1n}/(\boldsymbol{\alpha}_n \tilde{\boldsymbol{\alpha}}_n) + \beta^2/k_0 H_{2n}]}{(\alpha_n^2 + \beta^2) \sin \boldsymbol{\alpha}_n (d-a)}$$

$$\Lambda_{12} = -\Lambda_{21} = \frac{-j\alpha_n \beta k_0 H_{1n}/(\boldsymbol{\alpha}_n \tilde{\boldsymbol{\alpha}}_n) + j\alpha_n \beta H_{2n}/k_0}{(\alpha_n^2 + \beta^2) \sin \boldsymbol{\alpha}_n (d-a)}$$

$$\Lambda_{22} = -\frac{[\alpha_n^2 H_{2n}/k_0 + \beta^2 k_0 H_{1n}/(\boldsymbol{\alpha}_n \tilde{\boldsymbol{\alpha}}_n)]}{(\alpha_n^2 + \beta^2) \sin \boldsymbol{\alpha}_n (d-a)} \qquad (24)$$

The abbreviations H_{1n} and H_{2n} will be given below.

The coupled equations (eqns. 22) for the transformed-slot-field components are solved by applying Ritz-Galerkin's method. To this end, the fields are expanded in terms of known basis functions which have been chosen according to Reference 19:

$$E_y(y) = \sum_{m=0}^{M} c_m y^{2m} (d^2 - y^2)^{-1/2} \quad \text{for } |y| \leqslant s$$

$$E_z(y) = \sum_{m=1}^{M} d_m y^{(2m-1)} (d^2 - y^2)^{+1/2} \quad \text{for } |y| \leqslant s \qquad (25)$$

Substituting these values into eqns. 23 yields

$$L_{1n}^{(m)} = \pi/2\, s^{(2m+1)} (-1)^m \left[\left(\frac{\partial}{\partial \alpha_n s}\right)^{2m-1} J_0(\alpha_n s) + \left(\frac{\partial}{\partial \alpha_n s}\right)^{2m+1} J_0(\alpha_n s) \right]$$

$$L_{2n}^{(m)} = \pi/2\, s^{2m} (-1)^m \left(\frac{\partial}{\partial \alpha_n s}\right)^{2m} J_0(\alpha_n s) \qquad (26)$$

with J_0 Bessel function of order zero.

The Fourier transforms of the assumed-slot-field components are substituted into eqns. 22. Eliminating the fin currents by taking the inner products of these equations with L_{1n} and L_{2n}, one obtains a linear system of homogeneous equations for the unknown expansion coefficients c_m and d_m:

$$\sum_{k=0}^{M} c_k \sum_{n=0}^{N} \Lambda_{11} L_{2n}^{(m)} L_{2n}^{(k)} + \sum_{k=1}^{M} d_k \sum_{n=1}^{N} \Lambda_{12} L_{2n}^{(m)} L_{1n}^{(k)} = 0$$

$$\sum_{k=0}^{M} c_k \sum_{n=1}^{N} \Lambda_{21} L_{1n}^{(m)} L_{2n}^{(k)} + \sum_{k=1}^{M} d_k \sum_{n=1}^{N} \Lambda_{22} L_{1n}^{(m)} L_{1n}^{(k)} = 0$$

$$m = 0, 1, 2 \ldots M \qquad (27)$$

M is the total number of basis-function components and N is the number of fringing-field components. They are both related to each other via the edge conditions. $M = 4$ has been chosen here, which gives accurate results up to the 30th mode. N has been specified at the end of Appendix 7.1.

The dispersion relation for the propagation constant of a particular mode is obtained from eqns. 27 by equating the determinant of the coefficient matrix to zero.

The basis functions have to be modified for the case when the fin-line slot is not symmetrically located in the cross-section (i.e. it is not around $y = 0$). Then, we set:

$$E_y(y) = \sum_{m=0}^{M} c_m y^m (s^2 - y'^2)^{-1/2}$$

$$E_z(y) = \sum_{m=0}^{M} d_m y^m (s^2 - y'^2)^{+1/2} \qquad (28)$$

Here $y' = y - (d_1 + d_2)/2$ and d_1 and d_2 are defined in Fig. 2. The Fourier transforms are given by:

$$L_{1n}^{(m)} = \frac{\pi}{2} s^2 \left(\frac{\partial}{\partial \alpha_n}\right)^m \left[\sin\left(\alpha_n \frac{d_1 + d_2}{2} + \frac{m\pi}{2}\right) [J_0(\alpha_n s) + J_2(\alpha_n s)] \right]$$

$$L_{2n}^{(m)} = \pi \left(\frac{\partial}{\partial \alpha_n}\right)^m \left[\cos\left(\alpha_n \frac{d_1 + d_2}{2} + \frac{m\pi}{2}\right) J_0(\alpha_n s) \right] \qquad (29)$$

Up to now, we have only treated a bilateral fin line with both a symmetrically and an unsymmetrically located slot. Both cases differ only in the choice of basis functions for the slot fields. These functions are given by eqns. 25 for a symmetrically located slot and by eqns. 28 for an unsymmetrically located slot. The results must still be completed by defining the symbols H_{1n} and H_{2n} in eqns. 24. They read, for a bilateral fin line,

$$H_{1n} = [\tilde{\alpha}_n \cos \alpha_n (d - a) \cos \tilde{\alpha}_n d - \epsilon_r \alpha_n \times \sin \tilde{\alpha}_n d \sin \alpha_n (d - a)] / \cos \tilde{\alpha}_n d$$

$$H_{2n} = [\alpha_n \cos \tilde{\alpha}_n d \cos \alpha_n (d - a) + \tilde{\alpha}_n \sin \tilde{\alpha}_n d \sin \alpha_n (d - a)] / \cos \tilde{\alpha}_n d \qquad (30)$$

for the odd modes, and

$$H_{1n} = [\tilde{\alpha}_n \cos \alpha_n (d - a) \sin \tilde{\alpha}_n d - \epsilon_r \alpha_n \times \cos \tilde{\alpha}_n d \sin \alpha_n (d - a)] / \sin \tilde{\alpha}_n d$$

$$H_{2n} = [\alpha_n \cos \alpha_n (d - a) \sin \tilde{\alpha}_n d - \tilde{\alpha}_n \cos \tilde{\alpha}_n d \sin \alpha_n (d - a)] / \sin \tilde{\alpha}_n d \qquad (31)$$

for the even modes.

For a unilateral fin line, one must also distinguish between a symmetrically and an unsymmetrically located slot. All results given above, for the bilateral fin line, remain valid except for H_{1n} and H_{2n}. They are given for a unilateral fin line, as

$$H_{1n} = \frac{-2\epsilon_r \alpha_n \tilde{\alpha}_n + [\tilde{\alpha}_n^2 \cot \alpha_n (d - a) - \epsilon_r^2 \alpha_n^2 \tan \alpha_n (d - a)] \tan 2\tilde{\alpha}_n d}{\epsilon_r \alpha_n \tan \alpha_n (a - d) + \tilde{\alpha}_n \tan 2\tilde{\alpha}_n d}$$

$$H_{2n} = \frac{-2\alpha_n \tilde{\alpha}_n - [\alpha_n^2 \cot \alpha_n (a - d) - \tilde{\alpha}_n^2 \tan \alpha_n (a - d)] \tan 2\tilde{\alpha}_n d}{\alpha_n \tan 2\tilde{\alpha}_n d - \tilde{\alpha}_n \tan \alpha_n (d - a)} \qquad (32)$$

For an antipodal fin line [17], the basis functions are given by eqns. 25 and the symbols H_{1n} and H_{2n} are given by

$$H_{1n} = \left[\tilde{\alpha}_n \cot \alpha_n (d - a) + \epsilon_r \alpha_n \frac{\delta_0 \sin \tilde{\alpha}_n d + \delta_e \cos \tilde{\alpha}_n d}{\delta_0 \cos \tilde{\alpha}_n d - \delta_e \sin \tilde{\alpha}_n d} \right] \sin \alpha_n (d - a)$$

$$H_{2n} = \left[\alpha_n \cot \alpha_n (d - a) + \tilde{\alpha}_n \frac{\delta_0 \sin \tilde{\alpha}_n d - \delta_e \cos \tilde{\alpha}_n d}{\delta_0 \cos \tilde{\alpha}_n d - \delta_e \sin \tilde{\alpha}_n d} \right] \sin \alpha_n (d - a) \qquad (33)$$

with $\delta_0 = 1$, $\delta_e = 0$, for $n = 0, 2, 4 \ldots$, and $\delta_0 = 0$, $\delta_e = 1$, for $n = 1, 3, 5 \ldots$.

A detailed description of the eigenmode analysis for a bilateral fin line has recently been published in Reference 21.

Transverse Resonance Analysis of Finline Discontinuities

ROBERTO SORRENTINO, MEMBER, IEEE, AND TATSUO ITOH, FELLOW, IEEE

Abstract —A method of analysis is proposed for characterizing finline discontinuities. Two conducting or magnetic planes are inserted at some distances away from the discontinuity so as to obtain a closed resonant structure. A transverse resonance technique is then used to compute the resonant frequencies and, from these, the equivalent circuit parameters of the discontinuity. In the particular case when the discontinuity is removed, the method can be used to characterize uniform finlines.

I. Introduction

FINLINES are now recognized as a suitable technology of millimeter-wave integrated circuits. While much theoretical work has been done concerning the analysis and characterization of uniform finline structures [1], [2], a relatively small number of analyses of finline discontinuities have been developed [3], [4].

This paper presents a new method of analysis and characterization of both uniform finlines and finline discontinuities. The method consists of computing the resonant frequencies of a resonator obtained by inserting two conducting or magnetic planes apart from the discontinuity; using a transverse resonance technique, the electromagnetic (EM) fields are expanded in terms of longitudinal section magnetic (LSM) and electric (LSE) modes of the rectangular waveguide. With respect to other approaches based on the field expansion in terms of finline modes [3], [4], the present one has the advantage of a substantial reduction of computer time. In this paper, this new method is applied to the simple step discontinuity as well as to the cascade of step discontinuities.

II. Characterization of the Discontinuity

The characterization of a finline discontinuity is obtained with the resonant frequencies of resonators which are obtained by introducing two shorting planes at some distances away from the discontinuity. The resultant structure is shown in Fig. 1 along with the dimensions and the coordinate system.

As long as the frequency is such that only dominant modes can propagate in the two finline sections and the higher order modes excited at the discontinuity have negligible amplitudes at the shorting planes, the discontinuity can be modeled as an equivalent two-port network, as shown in Fig. 2.

Manuscript received April, 26 1984; revised July 31, 1984. This work was supported in part by the U.S. Army Research Office under Contract DAAG 29-81-K-0053, and in part by the Joint Services Electronics Program.

R. Sorrentino is with the Department of Electronics, University of Rome La Sapienza, Via Eudossiana 18, 00184 Rome, Italy.

T. Itoh is with the Department of Electrical Engineering, University of Texas at Austin, Austin, TX 78712.

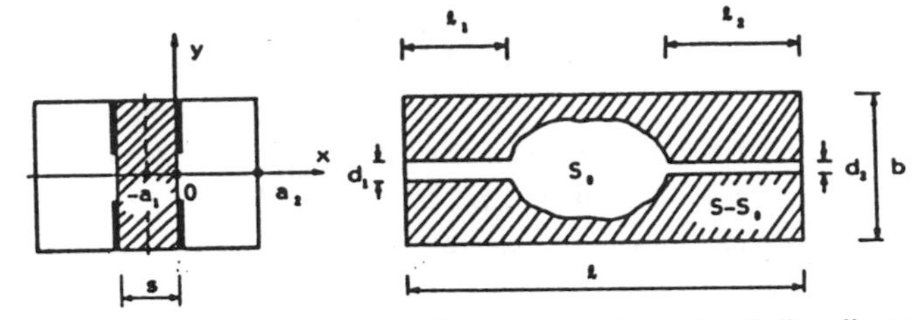

Fig. 1. Transverse and longitudinal cross sections of a finline discontinuity in a shorted cavity.

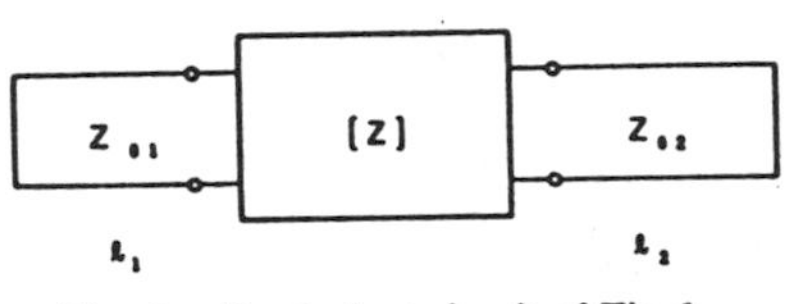

Fig. 2. Equivalent circuit of Fig. 1.

The resonance condition in terms of the impedance parameters of the discontinuity is

$$(Z_{11}+Z_1)(Z_{22}+Z_2)-Z_{12}^2=0 \tag{1}$$

where

$$Z_i = jZ_{oi}\tan(\beta_i l_i), \qquad i=1,2.$$

Z_{oi} is the characteristic impedance of the ith finline, and β_i is the corresponding phase constant. Alternatively, (1) can also be formulated in terms of the scattering parameters of the discontinuity.

If the same resonant frequency ω_r rad/s is obtained for three different pairs of l_1, l_2, (1) allows the evaluation of the three impedance parameters of the discontinuity at ω_r.

In the absence of the discontinuity, $\beta_1=\beta_2=\beta$; (1) then reduces to

$$\beta(\omega_r)=n\pi/l$$

with $l=l_1+l_2$. Thus, the length l corresponding to the resonant frequency ω_r yields the phase constant of a uniform finline at ω_r.

With simple modifications, the above procedure can be applied to other types of finline discontinuity problems, such as end effects in open- or short-circuited finline sections. In such cases, the equivalent circuit will consist of a line section terminated at one end with an unknown reactance. Its value can be computed by way of the resonant frequencies of a resonator obtained by short-circuit-

Reprinted from *IEEE Trans. Microwave Theory Tech.*, vol. MTT-32, no. 12, pp. 1633–1638, December 1984.

ing the waveguide at some distance away from the line termination.

III. Computation of the Resonant Frequencies

The method for computing the resonant frequencies of the structure will be illustrated in the case of bilateral finline, shown in Fig. 1. The metallic fins are assumed to be infinitely thin, although the method can easily be modified to account the finite thickness of metallization.

Because of symmetry, a longitudinal magnetic plane can be inserted at the symmetric plane $x = -a_1$, so that only the region $x \geqslant -a_1$ has to be analyzed. The extension to nonsymmetrical structures, such as unilateral finlines, is straightforward and will not be considered here.

The EM field in the dielectric region (region 1: $-a_1 \leqslant x \leqslant 0$) and in the air region (region 2: $0 \leqslant x \leqslant a_2$) can be expanded in terms of TE and TM modes of a rectangular waveguide with inner dimensions l and b. We obtain the following expressions for the transverse E- and H-field components in the two regions:

Dielectric Region: $-a_1 \leqslant x \leqslant 0$

$$\begin{aligned} \boldsymbol{E}_{t1} &= \sum_{mn} A'_{mn} \cos k'_{mn}(x+a_1)\hat{x} \times \nabla_t \psi_{mn} \\ &\quad + \frac{1}{j\omega\epsilon_0\epsilon_r} \sum_{mn} B'_{mn} k'_{mn} \cos k'_{mn}(x+a_1) \nabla_t \varphi_{mn} \\ \boldsymbol{H}_{t1} &= \frac{-1}{j\omega\mu_0} \sum_{mn} A'_{mn} k'_{mn} \sin k'_{mn}(x+a_1) \nabla_t \psi_{mn} \\ &\quad + \sum_{mn} B'_{mn} \sin k'_{mn}(x+a_1) \nabla_t \varphi_{mn} \times \hat{x}. \end{aligned} \tag{2}$$

Air Region: $0 \leqslant x \leqslant a_2$

$$\begin{aligned} \boldsymbol{E}_{t2} &= \sum_{mn} A_{mn} \sin k_{mn}(x-a_2)\hat{x} \times \nabla_t \psi_{mn} \\ &\quad - \frac{1}{j\omega\epsilon_0} \sum_{mn} B_{mn} k_{mn} \sin k_{mn}(x-a_2) \nabla_t \varphi_{mn} \\ \boldsymbol{H}_{t2} &= \frac{1}{j\omega\mu_0} \sum_{mn} A_{mn} k_{mn} \cos k_{mn}(x-a_2) \nabla_t \psi_{mn} \\ &\quad + \sum_{mn} B_{mn} \cos k_{mn}(x-a_2) \nabla_t \varphi_{mn} \times \hat{x} \end{aligned} \tag{3}$$

where

$$\begin{aligned} \psi_{mn} &= P_{mn} \cos\frac{m\pi z}{l} \cos\frac{n\pi y}{b} \\ \varphi_{mn} &= P_{mn} \sin\frac{m\pi z}{l} \sin\frac{n\pi y}{b} \\ P_{mn} &= \sqrt{\frac{\delta_m \delta_n}{lb}} \frac{1}{\gamma_{mn}} \qquad \delta_i = \begin{cases} 1, & i=0 \\ 2, & i \neq 0 \end{cases} \\ \gamma_{mn}^2 &= \left(\frac{m\pi}{l}\right)^2 + \left(\frac{n\pi}{b}\right)^2 \\ k_{mn}^2 &= k_0^2 - \gamma_{mn}^2 \qquad k'^2_{mn} = k_0^2 \epsilon_r - \gamma_{mn}^2 \\ k_0^2 &= \omega^2 \mu_0 \epsilon_0 \end{aligned} \tag{4}$$

where ψ_{mn} and φ_{mn} are the TE and TM scalar potentials, and m and n are integers with starting values of 0 or 1, depending on whether the TE or TM mode is being considered. P_{mn} are determined from the normalization conditions for ψ_{mn}, φ_{mn}

$$\int_S |\nabla_t \psi_{mn}|^2 \, dS = 1$$

$$\int_S |\nabla_t \varphi_{mn}|^2 \, dS = 1.$$

Equations (2)–(4) already satisfy the boundary conditions at $x = -a_1$ and $x = a_2$. The boundary conditions at $x = 0$ are

$$\boldsymbol{E}_{t1} = \boldsymbol{E}_{t2} = \begin{cases} \boldsymbol{E}_{t0}, & \text{on } S_0 \\ 0, & \text{on } S - S_0 \end{cases} \tag{5}$$

$$\boldsymbol{H}_{t1} = \boldsymbol{H}_{t2} = \boldsymbol{H}_{t0}, \qquad \text{on } S_0 \tag{6}$$

where $\boldsymbol{E}_{to}$ and $\boldsymbol{H}_{to}$ are unknown functions of z, y. These functions are expanded in terms of a set of orthonormal vector functions $\boldsymbol{e}_\nu$, or $\boldsymbol{h}_\mu$ defined over aperture region S_o (see Appendix)

$$\boldsymbol{E}_{to} = \sum_\nu V_\nu \boldsymbol{e}_\nu \tag{7}$$

$$\boldsymbol{H}_{to} = \sum_\mu I_\mu \boldsymbol{h}_\mu. \tag{8}$$

Inserting (2), (3), (7), and (8) into (5) and (6), and making use of the orthogonal properties of $\psi_{mn}, \varphi_{mn}, \boldsymbol{e}_\nu$, and $\boldsymbol{h}_\mu$, we obtain a homogeneous system of equations in terms of unknown coefficients V_ν

$$\begin{aligned} \sum_\nu V_\nu \Bigg[& \xi_{mn\nu} \zeta_{mn\mu} \left(k'_{mn} \tan k'_{mn} a_1 - k_{mn} \operatorname{cotan} k_{mn} a_2 \right) \\ & + \chi_{mn\nu} \theta_{mn\mu} k_0^2 \left(\epsilon_r \frac{\tan k'_{mn} a_1}{k'_{mn}} - \frac{\operatorname{cotan} k_{mn} a_2}{k_{mn}} \right) \Bigg] = 0, \\ & \mu = 1, 2 \cdots \end{aligned} \tag{9}$$

where

$$\begin{aligned} \xi_{mn\nu} &= \int_{S_0} \hat{x} \times \nabla_t \psi_{mn} \cdot \boldsymbol{e}_\nu \, dS \qquad \chi_{mn\nu} = \int_{S_0} \nabla_t \varphi_{mn} \cdot \boldsymbol{e}_\nu \, dS \\ \zeta_{mn\mu} &= \int_{S_0} \nabla_t \psi_{mn} \cdot \boldsymbol{h}_\mu \, dS \qquad \theta_{mn\mu} = \int_{S_0} \nabla_t \varphi_{mn} \times \hat{x} \cdot \boldsymbol{h}_\mu \, dS. \end{aligned} \tag{10}$$

The condition for nontrival solutions determines the characteristic equation of the given structure. This equation may be regarded as a real function of ω, l_1, and l_2 equated to zero

$$f(\omega, l_1, l_2) = 0. \tag{11}$$

For a given value of $\omega = \omega_r$, (11) can be solved to evaluate three different pairs of l_1 and l_2 yielding the same resonant frequency ω_r. These values of l_1 and l_2 can be used for computing the discontinuity parameters discussed in the previous section.

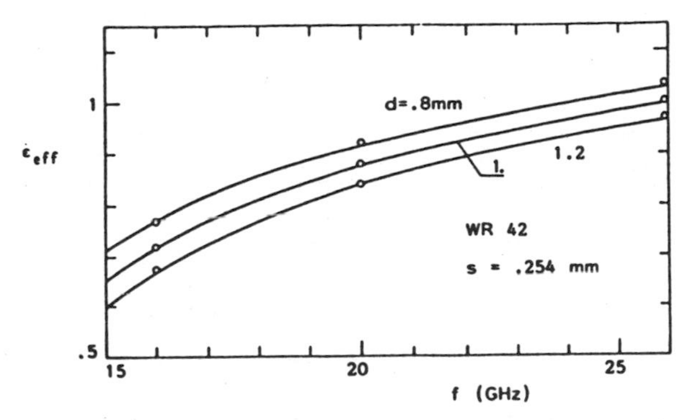

Fig. 3. Effective permittivity of a uniform finline. ∘ Spectral-domain method.

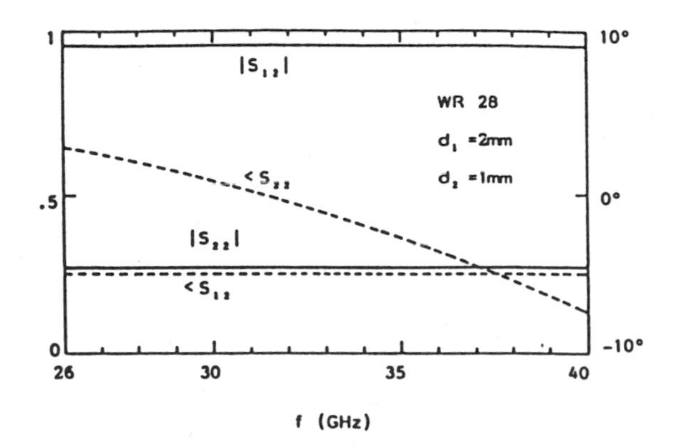

Fig. 5. Scattering parameters of the step discontinuity of Fig. 4.

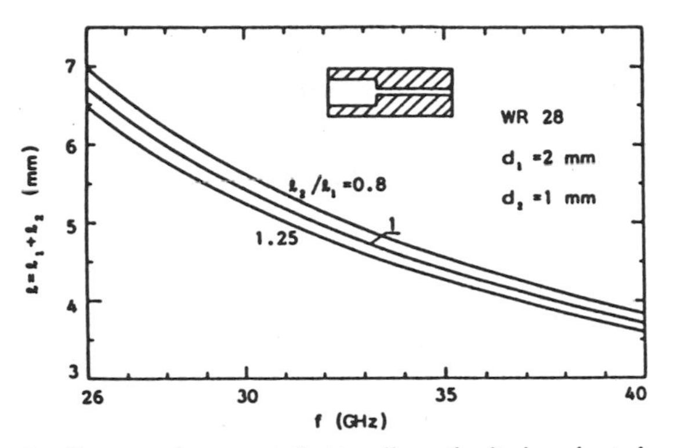

Fig. 4. Resonant frequency of a step discontinuity in a shorted cavity.

IV. Computed Results

According to the above-described technique, the EM fields in the air region, in the dielectric region, and in the aperture region of the resonator are expressed in terms of the series expansions (2), (3), (7), and (8). In the numerical computations, only a finite number of terms of each series can be retained. In order to obtain a proper convergent behavior of the solution, the number of terms in adjacent regions was chosen in such a way that the highest spatial frequencies of the EM field were about the same in the two regions [5], [6]. The method was first tested for computing the propagation characteristics of a uniform finline. In the absence of the discontinuity, the vector basis functions $\boldsymbol{e}_\nu$ and $\boldsymbol{h}_\mu$ on the aperature region (see Appendix) simply reduce to the transverse components of the normal modes of a rectangular waveguide with inner dimensions l and b.

The computed frequency behavior of the effective permittivity

$$\epsilon_{\text{eff}} = (\beta/\beta_0)^2$$

for different gap widths is shown in Fig. 3. Increasing the number of basis functions from 1 to 10, only small differences (less than 1 percent) have been obtained. The time required for computing one resonant frequency using four basis functions was typically 0.3 s on a Univac 1100 computer. The comparison with the results obtained using the spectral-domain approach is quite satisfactory.

In the presence of a step discontinuity, the vector basis functions $\boldsymbol{e}_\nu$ and $\boldsymbol{h}_\mu$ required to represent the EM field over the aperture, have a more complicated spatial distribution, and were evaluated as shown in Appendix. This required some additional computer time.

Fig. 4 shows the resonant frequency of the finline resonator containing a step discontinuity as a function of the total length $l = l_1 + l_2$, with the ratio l_2/l_1 as a parameter. Utilizing these data, the scattering parameters of the discontinuity have been computed using the procedure outlined in Section II, and are shown in Fig. 5. The computed scattering parameters of a unilateral finline discontinuity are compared in Fig. 6 with those computed by Schmidt [5] using the mode-matching procedure.

Although the procedure described above applies to a more complicated discontinuity structure, a certain simplification can be introduced if the discontinuity is longitudinally symmetric, such as the cascaded step discontinuities shown in Fig. 7. For instance, because of the symmetry, the analysis of the structure in Fig. 7(a) is reduced to the two equivalent structures containing a single step terminated by either a magnetic wall or an electric wall, as shown in Fig. 8. The equivalent circuits of the original and the two reduced structures are also shown there.

With obvious modifications of expressions (4) for ψ_{mn} and φ_{mn}, and of the basis functions $\boldsymbol{e}_\nu$ and $\boldsymbol{h}_\mu$ (see Appendix), the field analysis procedure described in Section III can be applied to the case of magnetic walls to obtain Z_{11}, Z_{22}, and Z_{21} by way of the resonant frequencies.

Fig. 9 shows the computed results at 26 GHz for the capacitive strips. The normalized reactance parameters of the equivalent T-network are shown as a function of the fin gap d_2 and the distance h. As expected, the capacitance associated with the shunt branch X_{12} increases with both h and the ratio d_1/d_2. On the contrary, the series branches have an inductive reactance whose value is much less sensitive to variations with respect to d_1/d_2. It can be shown that increasing h or d_1/d_2 results in an increase in the magnitudes of the reflection coefficient s_{11}. The phase of s_{11} varies almost linearly with h.

The dual case of inductive notches is shown in Fig. 10, where the normalized admittance parameters of the equivalent π-network are shown as functions of h and d_2/d_1. In this case, the inductance associated with the series branch

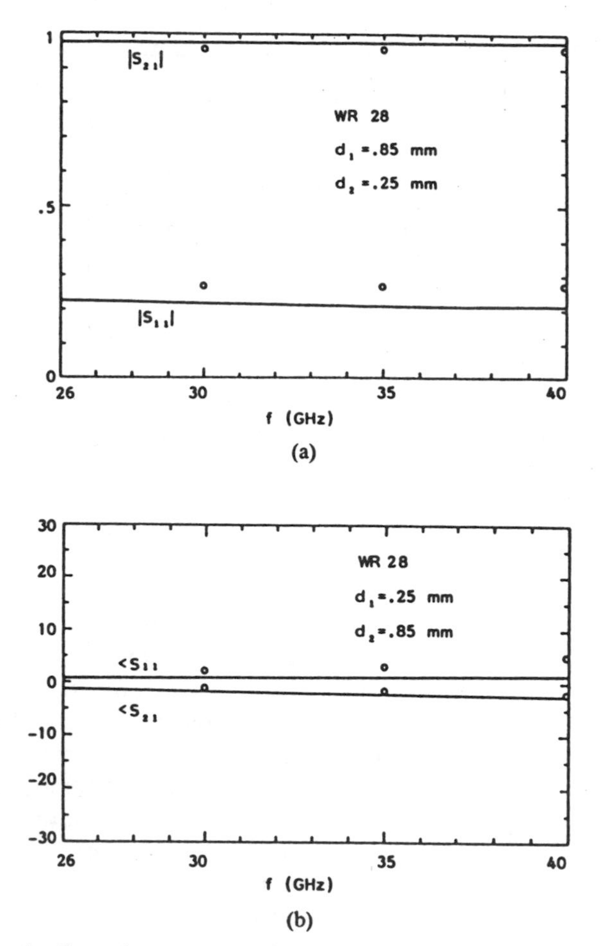

Fig. 6. Scattering parameters of a unilateral finline step discontinuity. ∘ Schmidt [7].

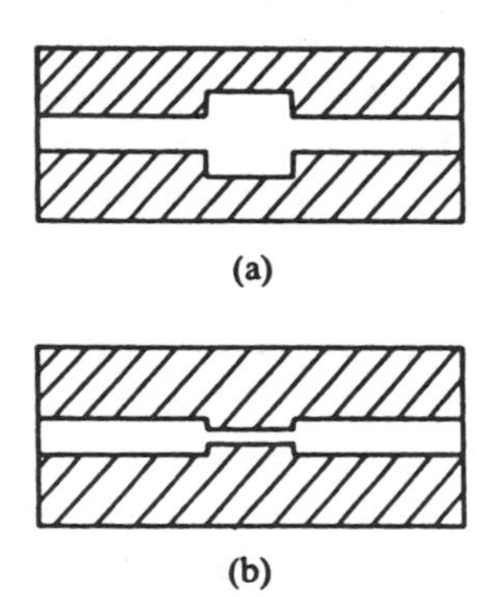

Fig. 7. Cascaded step discontinuities: (a) inductive notch and (b) capacitive strip.

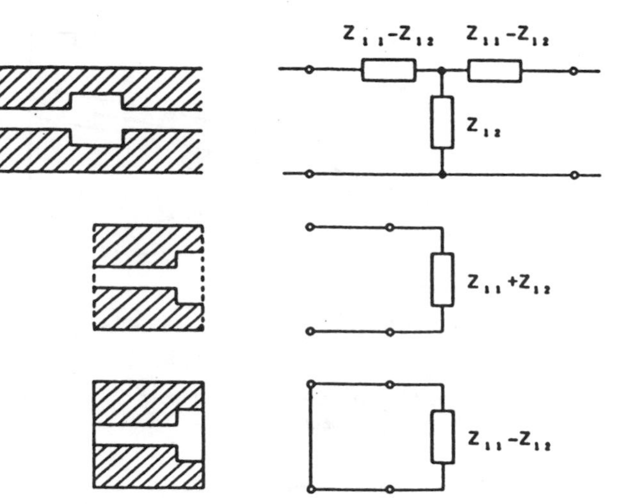

Fig. 8. Evaluation of the Z-parameters of a longitudinally symmetric discontinuity.

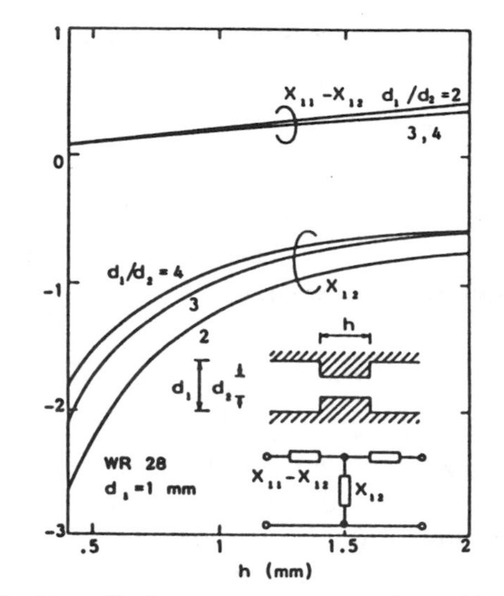

Fig. 9. Normalized reactance parameters of capacitive strips.

increases with h and d_2/d_1, while the capacitance of the shunt branches increases only slightly as a function of these parameters.

V. Conclusions

A new method of analysis has been proposed for the characterization of uniform finlines and finline discontinuities. The method is based on the computation of the resonant frequencies of a resonator obtained by short- (or open-) circuiting a finline section containing the discontinuity. The analysis procedure consists of a field expansion in terms of LSM and LSE modes of the rectangular waveguide. These expressions are matched with the field distribution in the plane of the fins. With respect to other approaches based on the field expansion in terms of finline modes, this procedure reduces computer time. The results are in good agreement with the numerical values obtained with other techniques.

Appendix

The two sets of orthonormalized vector functions $\boldsymbol{e}_\nu, \boldsymbol{h}_\mu$ used in (7) and (8) for expanding the EM field at $x = 0$ in the aperture region are derived in this Appendix in the case of a step discontinuity between two finline sections of different slot widths. Because of symmetry considerations, a longitudinal electric plane can be placed at $y = 0$ (see Fig. 1), so reducing the longitudinal section to that of Fig. 11.

The aperture region $S_0 \equiv (S_1 \cup S_2)$ may be viewed as the cross section of a waveguide having a stepped cross section. We can therefore expand the EM-field components $\boldsymbol{E}_{t0}, \boldsymbol{H}_{t0}$ lying in the yz plane in terms of the TE and TM scalar

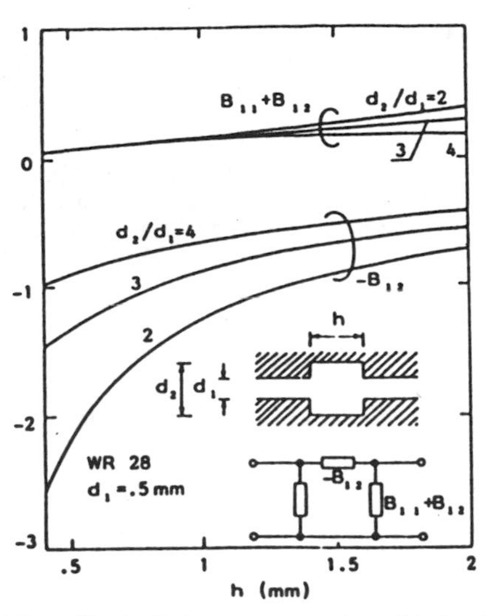

Fig. 10. Normalized admittance parameters of inductive notches.

potentials

$$E_{t0} = \sum_n V_n \hat{x} \times \nabla_t \psi_n + \sum_\nu V_\nu \nabla_t \varphi_\nu \quad \text{(A1)}$$

$$H_{t0} = \sum_n I_n \nabla_t \psi_n + \sum_\nu I_\nu \nabla_t \varphi_\nu \times \hat{x}. \quad \text{(A2)}$$

ψ_n and φ_ν represent the transverse potentials for TE and TM modes, respectively, satisfying the eigenvalue equations

$$\nabla_t^2 \psi_n + k_{cn}^2 \psi_n = 0 \quad \text{(A3)}$$

$$\nabla_t^2 \varphi_\nu + k_{c\nu}^2 \varphi_\nu = 0 \quad \text{(A4)}$$

in S_0 together with proper boundary conditions.

For the sake of brevity, only the solution of (A3) will be illustrated. Moreover, in order to simplify the notation, the index n will be dropped.

In order to solve (A3), the function ψ can be expressed as follows:

$$\psi = \begin{cases} \psi_1 = \sum_r A_r \psi_r^{(1)}, & \text{in } S_1 \\ \psi_2 = \sum_s B_s \psi_s^{(2)}, & \text{in } S_2 \end{cases} \quad \text{(A5)}$$

where

$$\psi_r^{(1)} = \cos k_{1r}(z + l_1) \cos \frac{r\pi y}{d_1/2} \quad \text{(A6)}$$

$$\psi_s^{(2)} = \cos k_{2s}(z - l_2) \cos \frac{s\pi y}{d_2/2} \quad \text{(A7)}$$

$$k_{ir}^2 = k_c^2 - \left(\frac{r\pi}{d_i/2}\right)^2, \qquad i = 1, 2. \quad \text{(A8)}$$

Expressions (A5)–(A8) are such that (A3) is satisfied together with the boundary conditions at $z = -l_1, l_2$ and $y = 0, d_1/2, d_2/2$. The boundary conditions at $z = 0$

$$\psi_1 = \psi_2, \qquad 0 \leqslant y \leqslant d_2/2 \quad \text{(A9)}$$

$$\frac{\partial \psi_1}{\partial z} = \begin{cases} \dfrac{\partial \psi_2}{\partial z}, & 0 \leqslant y \leqslant d_2/2 \\ 0, & d_2/2 \leqslant y \leqslant d_1/2 \end{cases} \quad \text{(A10)}$$

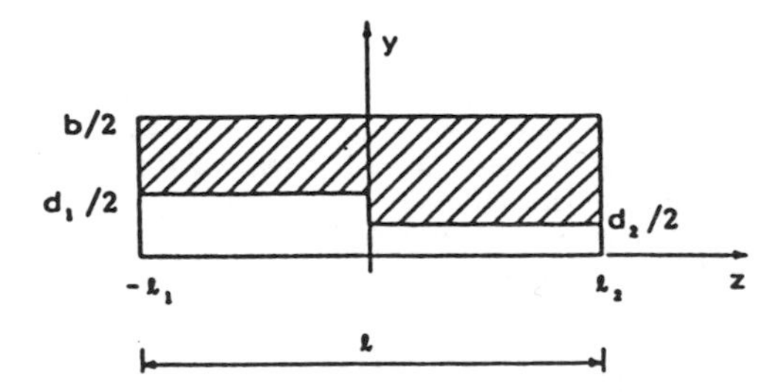

Fig. 11. Reduced geometry of the step discontinuity.

through the orthogonal properties of cosine functions, lead to a homogeneous system of equations in the expansion coefficients A_r, B_s

$$\sum_r A_r f_{rs} \cos k_{1r} l_1 - \frac{d_2}{2\delta_s} B_s \cos k_{2s} l_z = 0, \qquad s = 0, 1, 2 \cdots \quad \text{(A11)}$$

$$\frac{d_1}{2\delta_r} A_r k_{1r} \sin k_{1r} l_1 + \sum_s B_s f_{rs} k_{2s} \sin k_{2s} l_2 = 0, \qquad r = 0, 1, 2 \cdots \quad \text{(A12)}$$

where

$$\delta_r = \begin{cases} 1, & r = 0 \\ 2, & r \neq 0 \end{cases}$$

$$f_{rs} = \int_0^{d_2/2} \cos \frac{r\pi y}{d_1/2} \cos \frac{s\pi y}{d_2/2}\, dy.$$

The condition for nontrivial solutions of (A11)–(A12) constitutes the characteristic equation from which the eigenvalues k_c^2 can be computed. For each k_c^2, the expansion coefficients A_r, B_s are determined using (A11)–(A12) and imposing the normalization condition

$$\int_{S_0} |\nabla_t \psi|^2\, dS = 1.$$

Finally, it can be easily demonstrated that the ψ_n's so obtained satisfy the orthogonality condition

$$\int_{S_0} \nabla_t \psi_n \cdot \nabla_t \psi_m\, dS = 0, \qquad n \neq m$$

even if, for numerical reasons, the series in (A5) will be truncated to a finite number of terms.

A similar procedure can be applied to the evaluation of the φ_ν's. The right-hand side of (A1) and (A2) finally provide the required expansions in terms of orthonormal vector functions.

If the resonator is terminated at $z = -l_1, l_2$ by magnetic walls, (A6) and (A7) are modified corresponding, in order to satisfy the open-circuit boundary conditions. Moreover, the eigenfunction φ_0, corresponding to the eigenvalue $k_c^2 = 0$, must also be included in expansions (A1) and (A2). This eigenfunction corresponds to the TEM mode of the stepped waveguide with mixed conducting and magnetic boundaries.

References

[1] H. Hofmann, "Dispersion of planar waveguides for millimeter-wave application," *Arch. Elek. Übertragung.*, vol. 31, pp. 40–44, Jan. 1977.

[2] L.-P. Schmidt and T. Itoh, "Spectral domain analysis of dominant and higher order modes in fin-lines," *IEEE Trans. Microwave Theory Tech.* vol. MTT-28, pp. 981–985, Sept. 1980.

[3] A. Beyer, "Calculation of discontinuities in grounded finlines taking into account the metallization thickness and the influence of the mount-slits," in *Proc. of the 12th European Microwave Conf.* (Helsinki, Finland), 1982, pp. 681–686.

[4] H. El Hennaway and K. Schunemann, "Analysis of fin-line discontinuities," in *Proc. of the 9th European Microwave Conf.* (Brighton, England), 1979, pp. 448–452.

[5] S. W. Lee, W. R. Jones, and J. J. Campbell, "Convergence of numerical solutions of iris-type discontinuity problems," *IEEE Trans. Microwave Theory Tech.*, vol. MTT-19, pp. 528–536, June 1971.

[6] Y. C. Shih and K. G. Gray, "Convergence of numerical solutions of step-type waveguide discontinuity problems by modal analysis," in *IEEE MTT-S Int. Symp. Dig.* (Boston, MA), 1983, pp. 233–235.

[7] L.-P. Schmidt, private communication.

Theoretical and Experimental Investigation of Finline Discontinuities

MARYLINE HELARD, JACQUES CITERNE, ODILE PICON, AND VICTOR FOUAD HANNA

Abstract —**The dominant and the first-five higher order modes in a unilateral finline are precisely described from a thorough spectral-domain approach. Then, using the modal analysis, coupling coefficients between eigenmodes at a discontinuity that have to be introduced into the scattering matrix formulation are directly computed in the spectral-domain, and, consequently, the equivalent circuit parameters of the discontinuity are determined. Finally, finline discontinuities often used for impedance transformation are investigated and a good agreement between theoretical and experimental results is reported.**

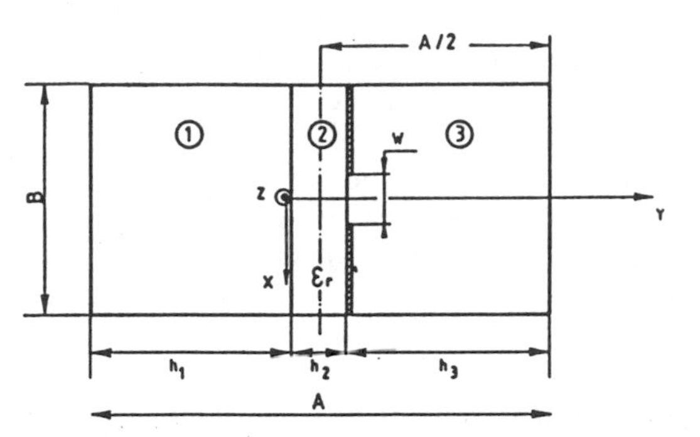

Fig. 1. Cross-section parameters of a unilateral finline. In the *Ku*-band (12–18 GHz), $A = 15.8$ mm, $B = 7.9$ mm, $h_2 = 0.635$ mm, and $\epsilon_r = 9.6$ (Alumina). In the *Ka*-band (26–40 GHz), $A = 7.112$ mm, $B = 3.556$ mm, $h_2 = 0.254$ mm, and $\epsilon_r = 2.22$ (Duroid 5880).

I. Introduction

THE FIELD THEORETICAL solution of finline discontinuities presents a complex problem. This explains the small number of concise and rigorous attempts of analyses reported in the recent literature [1].

In fact, the major complexity for efficient and accurate analysis of single or complex discontinuities in finline structures lies in the necessary treatment of hybrid eigenmodes to which there is no closed-form solution.

Firstly, the modal analysis [2] has been applied [3] to single and double steps in the finline slot width, while the characterization of uniform finline structures was obtained by the moment mode-matching technique [4].

This pioneering work provided an insight into CAD potentialities of integrated millimeter-wave finline circuits [5] that such a field theoretical solution could provide. However, a hasty model in terms of equivalent circuits prevents an unbiased evaluation of computed intermediate field parameters.

A. Beyer and I. Wolff [6] used a nearly similar approach of the finline treatment and solved the discontinuity problem by a flexible and elegant combination of the moment mode-matching technique and the modal analysis.

In solving the problem of the rectangular waveguide finline tapers, use was made, for the first time, of the powerful method of coupled modes [7] accompanied by the spectral-domain approach. It was shown that such a combination offers a very efficient computational scheme of the return and insertion losses of the "back-to-back" tapering transition. Unfortunately, experiments were unable to validate the presumed exact fields computations.

Recently, the transverse resonance techniques [8] were used to compute the resonant frequencies of a resonant structure containing the finline discontinuity and in consequence to determine the equivalent-circuit parameters of the discontinuity.

This paper presents a precise field theoretical solution of finline discontinuities using a combination of the spectral-domain approach and the modal analysis.

The aim of the paper is threefold: 1) to give an accurate evaluation of six eigenmodes of the unilateral finline via a thorough spectral-domain approach; 2) to perform calculations of the generalized scattering matrix of a single-step slot width discontinuity by combining the direct modal analysis and the spectral-domain approach [9] and, consequently, to determine the equivalent-circuit parameters of this discontinuity; 3) to compare the theory and experiments in the *Ku*- and *Ka*-bands on complex discontinuities [10], as this is the only method that enables the direct comparison of measured and calculated scattering parameters of either the simple or the complex discontinuity.

II. Spectral-Domain Approach of a Unilateral Finline

The axial field components $E_{z,i}(x,y)$ and $H_{z,i}(x,y)$ in the ith region ($i = 1,2,3$) are expanded in Fourier series (Fig. 1) within their domain

$$-B/2 \leqslant x \leqslant B/2.$$

Manuscript received February 12, 1985; revised May 31, 1985.

M. Helard is with the Centre Commun d'Etudes des Télécommunications et Télédivision, 35510 Cesson Sevigne, France.

J. Citerne is with the Laboratories Structures Rayonnantes, Department Genie Electrique, INSA, 35031 Rennes Cedex, France.

O. Picon and V. Fouad Hanna are with the Division Espace et Transmission Radioélectrique, Centre National d'Etudes des Télécommunications, 92131 Issy-Les-Moulineaux, France.

Reprinted from *IEEE Trans. Microwave Theory Tech.*, vol. MTT-33, no. 10, pp. 994–1003, October 1985.

For odd waves, the following expansions of $E_{z,i}(x,y)$ and $H_{z,i}(x,y)$ are valid:

$$E_{zi}(x,y)=\sum_{m=1}^{\infty}\tilde{E}_{zi}(m,y)\sin\alpha_m x$$

$$H_{zi}(x,y)=\sum_{m=0}^{\infty}\tilde{H}_{zi}(m,y)\cos\alpha_m x \tag{1}$$

where quantities with the sign (~) designate the line amplitude (i.e., the mth term of the Fourier series) associated with the space harmonic $\alpha_m = 2m\pi/B$.

The partial differential equations for the axial field components $E_{z,i}(x,y)$ and $H_{z,i}(x,y)$ are also Fourier expanded with respect to x; ordinary differential equations are derived for the mth line amplitudes $\tilde{E}_{z,i}(m,y)$ and $\tilde{H}_{z,i}(m,y)$, respectively.

For a unilateral finline (Fig. 1), these mth line amplitudes are given in Appendix I.

Through the application of boundary conditions at $y=0$ and $y=h_2$, which are also Fourier expanded, the spectral coefficients are related to each other, to the mth line amplitudes of fin surface current components denoted $\tilde{J}_x(m,h_2)$ and $J_z(m,h_2)$, and to the mth line amplitude of the slot aperture field components denoted $\tilde{E}_x(m,h_2)$ and $\tilde{E}_z(m,h_2)$. Extensive algebraic manipulations of these boundary conditions then yield functional equations relating the mth line amplitude of the fin surface current components to the mth line amplitude of the slot aperture field components. For a unilateral finline, the standard computational scheme uses the following admittance matrix representations of these functional equations, namely:

$$\begin{bmatrix} G_{11} & G_{12} \\ G_{21} & G_{22} \end{bmatrix}\begin{bmatrix} \tilde{E}_x(m,h_2) \\ \tilde{E}_z(m,h_2) \end{bmatrix}=\begin{bmatrix} \tilde{J}_x(m,h_2) \\ \tilde{J}_z(m,h_2) \end{bmatrix}. \tag{2}$$

Closed-form expressions of the $\bar{\bar{G}}$ matrix elements are listed in Appendix II.

Now it is the moment to start the numerical part through the application of the Galerkin's form of the general method of moments to the functional equations (2). The slot aperture field components $E_x(x,h_2)$ and $E_z(x,h_2)$ are expanded in terms of two complete sets of R and S basis functions denoted $\mathscr{E}_{x,r}(x,h_2)(r=1,\cdots,R)$ and $\mathscr{E}_{z,s}(x,h_2)(s=1,\cdots,S)$

$$E_x(x,h_2)=\sum_{r=1}^{R}a_r\mathscr{E}_{x,r}(x,h_2)$$

$$E_z(x,h_2)=\sum_{s=1}^{S}b_s\mathscr{E}_{z,s}(x,h_2). \tag{3}$$

Obviously, the basis functions in (3) have nonzero values in the interval $-W/2\leqslant x\leqslant W/2$ of the cell $-B/2\leqslant x\leqslant B/2$.

The expansions (3) are displayed in Fourier series to bring the mth line amplitude of the aperture field components and the mth line amplitude of basis functions into relationship denoted by

$$\tilde{E}_x(m,h_2)=\sum_{r=1}^{R}a_r\tilde{\mathscr{E}}_{x,r}(m,h_2)$$

$$\tilde{E}_z(m,h_2)=\sum_{s=1}^{S}b_s\tilde{\mathscr{E}}_{z,s}(m,h_2). \tag{4}$$

The coefficients a_r and b_s involved in (3) and (4) are the first true constants of the waveguide problem.

By using an inner product consistent with the Parseval theorem, the Galerkin's procedure is directly applied to the matrix form (2) in the Fourier domain. A set of $R+S$ homogeneous and linear equations, for which the $R+S$ unknowns are precisely the constants a_r and b_s, is obtained.

Nontrivial solutions of this set of equations occur for zero values of its matrix determinant. The real roots (i.e., $\beta^2>0$) determine the propagating eigenmodes, whereas the imaginary roots (i.e., $\beta^2<0$) determine evanescent ones.

For any given root β, the associated mth line amplitudes of the aperture field components are expressed in terms of the $R+S$ coefficients a_r and b_s involved in (4). By a substitution process of $\tilde{E}_x(m,h_2)$ and $\tilde{E}_z(m,h_2)$ in boundary conditions, the coefficients $A(m)$ through $H(m)$ are determined, as well as the mth line amplitude of the axial field components. The summation of Fourier series gives finally the axial field components of eigenmodes anywhere in the waveguide cross section.

Eigenmode normalization is not a necessary task as far as the waveguide treatment is concerned. However, this operation can be found relevant in the further discontinuity problem. So, applying again the Parseval's theorem, the orthogonality of eigenmodes in lossless waveguides can be expressed directly in the spectral domain as

$$\frac{B}{2}\sum_{m=0}^{\infty}\int_{-h_1}^{h_2+h_3}\Big[\tilde{E}_{x,p}(m,y)\tilde{H}^*_{y,p'}(m,y) - \tilde{E}_{y,p}(m,y)\tilde{H}^*_{x,p'}(m,y)\Big]\,dy = s\frac{\beta_p}{|\beta_p|}P\delta_{pp'} \tag{5}$$

where δ_{pp} represents the Kroneker delta and the coefficient s can take the values $+1$ or -1 depending on the propagating or evanescent nature of the eigenmode labelled p. Line amplitudes of transverse field components involved in (5) are derived from longitudinal components given by (1). The eigenmode normalization is carried out if the $R+S$ unknowns a_r and b_r in (3) and (4) associated with a given root β are reevaluated to ensure a power flow value P equal to 1 W in (5).

III. Modal Analysis of Single- and Multiple-Step Slot Width Discontinuities

The problem to be treated is how an incident accessible power entering each side of the junction is apportioned between the various scattered eigenmodes. The solution lies in the derivation of the generalized scattering matrix $[S]$ of the junction.

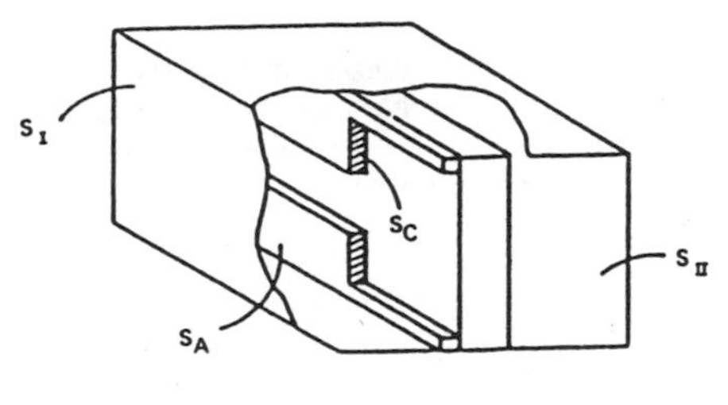

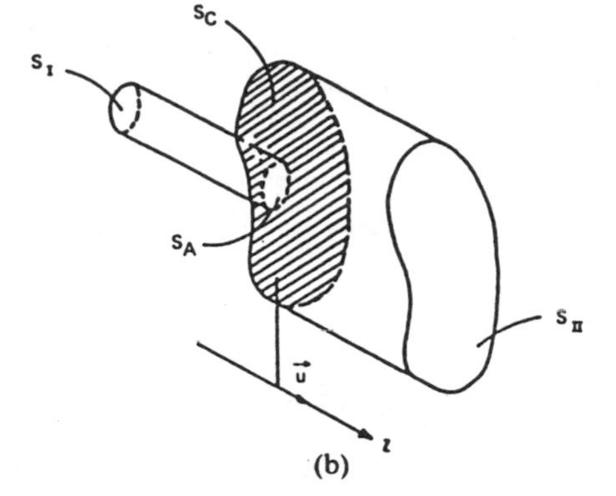

Fig. 2. (a) A single-step slot width junction between two unilateral finlines. (b) Model for modal analysis.

A single slot width discontinuity as that shown in Fig. 2(a) can be modeled by a junction between two cylindrical closed waveguides as shown in Fig. 2(b). We have to notice that the conducting fins in Fig. 2(a) have negligible thickness. They have been drawn so for a better comprehension of the model of Fig. 2(b).

The transverse electric and magnetic field to the left of the junction plane $z = 0$ (waveguide I side) can be expressed as

$$\vec{E}_T^{\mathrm{I}} = \sum_{p=1}^{P} \left(A_p^{\mathrm{I}} + B_p^{\mathrm{I}}\right) \vec{e}_{p,T}^{\,\mathrm{I}}$$

$$\vec{H}_T^{\mathrm{I}} = \sum_{p=1}^{P} \left(A_p^{\mathrm{I}} - B_p^{\mathrm{I}}\right) \vec{h}_{p,T}^{\,\mathrm{I}}. \tag{6}$$

In a similar manner, to the right of the junction plane $z = 0$, the expansion of the transverse electric and magnetic fields is written as

$$\vec{E}_T^{\mathrm{II}} = \sum_{q=1}^{Q} \left(A_q^{\mathrm{II}} + B_q^{\mathrm{II}}\right) \vec{e}_{q,T}^{\,\mathrm{II}}$$

$$\vec{H}_T^{\mathrm{II}} = \sum_{q=1}^{Q} \left(B_q^{\mathrm{II}} - A_q^{\mathrm{II}}\right) \vec{h}_{q,T}^{\,\mathrm{II}}. \tag{7}$$

The above expansions involve normalized transverse electric and magnetic fields $\vec{e}_{p,T}^{\,\mathrm{I}}, \vec{h}_{p,T}^{\,\mathrm{I}}$ (resp.: $\vec{e}_{q,T}^{\,\mathrm{II}}, \vec{h}_{q,T}^{\,\mathrm{II}}$) associated with the eigenmode p (resp.: q) of the waveguide I (resp.: II). Eigenmodes p and q are forward traveling waves in the two waveguides I and II, while the modal amplitudes $A_p^{\mathrm{I}}, B_p^{\mathrm{I}}$ (resp. $A_q^{\mathrm{II}}, B_q^{\mathrm{II}}$) are referred to incident and reflected waves in the waveguide I (resp. II) at the junction plane.

If $S_{\mathrm{I}} < S_{\mathrm{II}}$ (this is the case outlined in Fig. 2), the boundary conditions at the junction plane $z = 0$ can be expressed as

$$\vec{E}_T^{\mathrm{I}} = \vec{E}_T^{\mathrm{II}} \tag{8a}$$

$$\vec{H}_T^{\mathrm{I}} = \vec{H}_T^{\mathrm{II}} \tag{8b}$$

on the aperture surface S_A and as

$$\vec{E}_T^{\mathrm{II}} = 0 \tag{8c}$$

$$\vec{H}_T^{\mathrm{II}} + \vec{J}_T^{\mathrm{I}} \Lambda \vec{u} = 0 \tag{8d}$$

on the transverse conducting wall S_C.

The junction surfaces S_A and S_C are related to waveguide cross sections S_{I} and S_{II} by $S_A = S_{\mathrm{I}}$ and $S_A + S_C = S_{\mathrm{II}}$.

The surface current $\vec{J}_T^{\mathrm{I}}$ on the conducting wall S_C is labeled with a subscript I to indicate a relationship with waveguide I.

The next step is to transform pairs of boundary functional equations (8a) and (8b) into an equivalent set of linear equations involving the $2(P+Q)$ modal amplitudes A_p^{I}, B_p^{I}, A_q^{II}, and B_q^{II} to be determined.

There is a unique procedure to derive such an equivalent linear set of $2(P+Q)$ equations that is closely related to the basic assumption $S_{\mathrm{I}} < S_{\mathrm{II}}$. This point has never been clarified enough in the literature, especially according to the uniqueness of the solution; therefore, it is summarized briefly in Appendix III.

Thus, the generalized scattering matrix of the simple discontinuity can be constructed from this system of equivalent linear equations which can be written in a matrix form as

$$\begin{aligned} {}^{T}[P]([A^{\mathrm{I}}]+[B^{\mathrm{I}}]) &= [L]([A^{\mathrm{II}}]+[B^{\mathrm{II}}]) \\ [K]([A^{\mathrm{I}}]-[B^{\mathrm{I}}]) &= [N]([B^{\mathrm{II}}]-[A^{\mathrm{II}}]) \end{aligned} \tag{9}$$

with

$$[A^{\mathrm{I}}] = \begin{bmatrix} A_1^{\mathrm{I}} \\ A_2^{\mathrm{I}} \\ \vdots \\ A_p^{\mathrm{I}} \\ \vdots \\ A_P^{\mathrm{I}} \end{bmatrix} \quad [B^{\mathrm{I}}] = \begin{bmatrix} B_1^{\mathrm{I}} \\ B_2^{\mathrm{I}} \\ \vdots \\ B_p^{\mathrm{I}} \\ \vdots \\ B_P^{\mathrm{I}} \end{bmatrix}$$

$$[A^{\mathrm{II}}] = \begin{bmatrix} A_1^{\mathrm{II}} \\ A_2^{\mathrm{II}} \\ \vdots \\ A_q^{\mathrm{II}} \\ \vdots \\ A_Q^{\mathrm{II}} \end{bmatrix} \quad [B^{\mathrm{II}}] = \begin{bmatrix} B_1^{\mathrm{II}} \\ B_2^{\mathrm{II}} \\ \vdots \\ B_q^{\mathrm{II}} \\ \vdots \\ B_Q^{\mathrm{II}} \end{bmatrix} \tag{10}$$

where

$$[P] \text{ matrix } P\times Q;\ P_{p,q}=\int_{S_I}\vec{e}^{\,I}_{p,T}\Lambda\vec{h}^{\,II*}_{q,T}\cdot\vec{u}\,dS$$

$$[L] \text{ diagonal matrix } Q\times Q;\ L_{q,q}=S_q\frac{\beta_q}{|\beta_q|}$$

$$[K] \text{ diagonal matrix } P\times P;\ K_{p,p}=S_p\frac{\beta_p}{|\beta_p|}$$

$$[N] \text{ matrix } Q\times P;\ N_{q,p}=\int_{S_I}\vec{e}^{\,I*}_{p,T}\Lambda\vec{h}^{\,II}_{q,T}\cdot\vec{u}\,dS. \tag{11}$$

From (9), the generalized scattering matrix can be written in the form

$$[B]=[S]\cdot[A] \tag{12}$$

where

$$[B]=\begin{bmatrix}[B^{I}]\\ [B^{II}]\end{bmatrix}\quad [A]=\begin{bmatrix}[A^{I}]\\ [A^{II}]\end{bmatrix}. \tag{13}$$

The generalized scattering matrix of the junction assembles two reflection blocks denoted $[S_{11}]$ and $[S_{22}]$ and two transmission blocks denoted $[S_{21}]$ and $[S_{12}]$ arranged as

$$[S]_{S_I<S_{II}}=\begin{bmatrix}[S_{11}]_{S_I<S_{II}} & [S_{12}]_{S_I<S_{II}}\\ [S_{21}]_{S_I<S_{II}} & [S_{22}]_{S_I<S_{II}}\end{bmatrix}. \tag{14}$$

Each block has a size $(P+Q)\times(P+Q)$ and can be determined separately from matrices $[P]$, $[L]$, $[K]$, and $[N]$ after extensive algebraic manipulation of (9). Results are listed below:

$$\begin{aligned}
[S_{11}]_{S_I<S_{II}} &= \left([K]+[N]\cdot[L]^{-1}\cdot{}^{T}[P]\right)^{-1}\\
&\quad\cdot\left([K]-[N]\cdot[L]^{-1}\cdot{}^{T}[P]\right)\\
[S_{12}]_{S_I<S_{II}} &= 2\left([K]+[N]\cdot[L]^{-1}\cdot{}^{T}[P]\right)^{-1}\cdot[N]\\
[S_{21}]_{S_I<S_{II}} &= 2\left([L]+{}^{T}[P]\cdot[K]^{-1}\cdot[N]\right)^{-1}\cdot{}^{T}[P]\\
[S_{22}]_{S_I<S_{II}} &= -\left([L]+{}^{T}[P]\cdot[K]^{-1}\cdot[N]\right)^{-1}\\
&\quad\cdot\left([L]-{}^{T}[P]\cdot[K]^{-1}\cdot[N]\right).
\end{aligned} \tag{15}$$

The above $[S]$ matrix is labeled with subscripts $S_I<S_{II}$ to recall that it is addressed only to the case $S_I<S_{II}$. As mentioned in Appendix III, the $[S]$ matrix addressed to the alternative case $S_{II}<S_I$ is derived in a quite different way. However, the results can be related to the previously studied case: $S_I<S_{II}$. This relation can be written as

$$[S]_{S_I>S_{II}}=\begin{bmatrix}[S_{22}]_{S_I<S_{II}} & [S_{21}]_{S_I<S_{II}}\\ [S_{12}]_{S_I<S_{II}} & [S_{11}]_{S_I<S_{II}}\end{bmatrix}. \tag{16}$$

IV. Combination of the Spectral-Domain Approach and the Modal Analysis

The spectral-domain approach evaluates both the phase constants and the associated line amplitudes of eigenmode field components.

TABLE I
BASIS FUNCTIONS AND ASSOCIATED LINE AMPLITUDES OF THE APERTURE FIELD CORRESPONDING TO ODD MODES

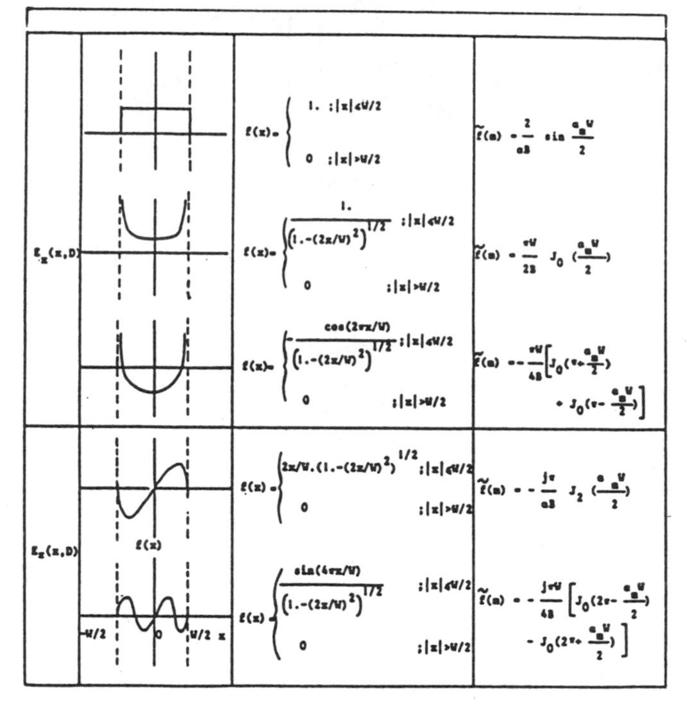

As a result of the eigenmode normalization, the calculation of elements L_{qq} and K_{pp} of matrices $[L]$ and $[K]$ is reduced to the determination of the propagating or evanescent nature of eigenmodes labeled p and q in waveguides I and II, respectively.

The computation of elements $P_{p,q}$ and $N_{q,p}$ of matrices $[P]$ and $[Q]$ can be quickly accomplished as it is done directly in the spectral domain. For example, the $P_{p,q}$ coefficient can be expressed from (11) as

$$P_{p,q}=B\sum_{m=0}^{\infty}\int_{-h_1}^{h_2+h_3}\left[\tilde{e}^{I}_{p,x}(m,y)\tilde{h}^{II*}_{q,y}(m,y)-\tilde{e}^{I}_{p,y}(m,y)\tilde{h}^{II*}_{q,x}(m,y)\right]dy. \tag{17}$$

V. Computed Results

A. Unilateral Finline Analysis

The sets of basis functions selected to describe the aperture field are represented in Table I. As shown in [11], the dominant mode can be described precisely by means of a single $E_x(x,D)$ basis function: the unit rectangular pulse denoted $f_1(x)$ in Table I. However, to describe both the dominant and the higher order modes, the aperture field expanded with the basis functions denoted $f_1(x)$, $f_2(x)$, $f_3(x)$, and $f_4(x)$ in Table I appears as a more judicious choice. Checks of this aperture field have been made in two limit cases for the unilateral finline: the standard rectangular waveguide ($\epsilon_r\to1$; $W/B\to1$) and the rectangular waveguide loaded symmetrically by a dielectric slab ($W/B\to1$). They allowed conclusions to be made about the completeness of at least the first six eigenmodes.

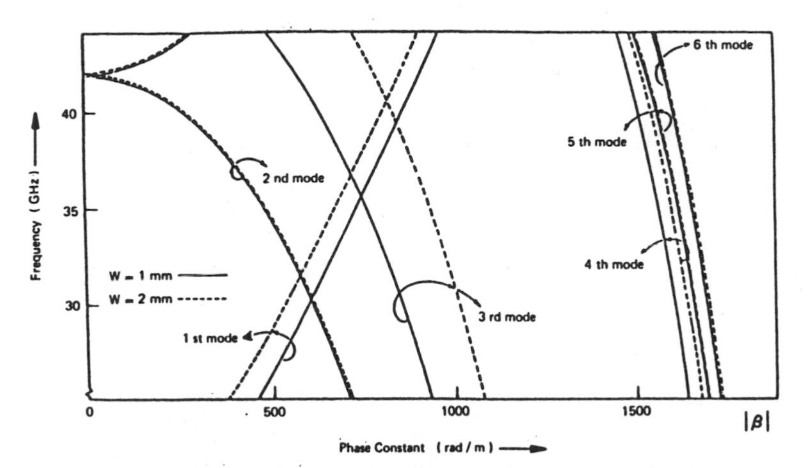

Fig. 3. Dispersion characteristics of eigenmodes in a unilateral finline in the *Ku*-band ($\epsilon_r = 2.22$, $h_2 = 0.254$ mm).

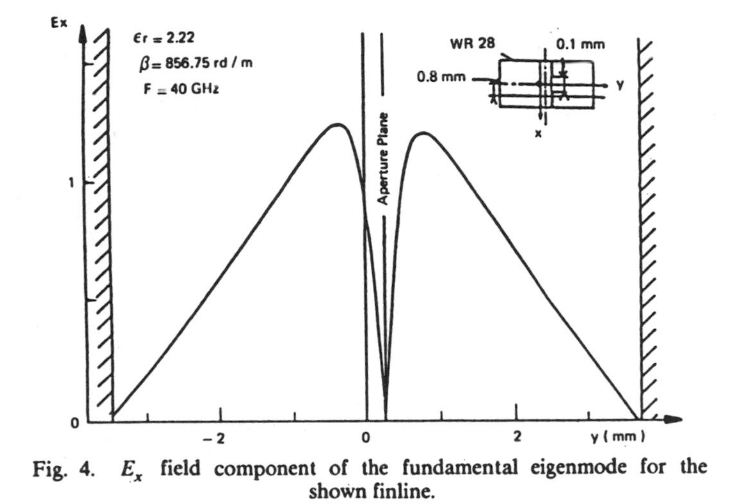

Fig. 4. E_x field component of the fundamental eigenmode for the shown finline.

Dispersion characteristics of the first six eigenmodes in a unilateral finline in the *Ka*-band are plotted in Fig. 3 for two values of the slot width. They show that the frequency band for single-mode operation in a unilateral finline is quite identical to those of the standard WR28 rectangular waveguide. Another check of the aperture field is to compute the eigenmode distribution to be sure of the boundary conditions as shown in Fig. 4 for the E_x component of the fundamental mode as a function of y at $x = 0.8$ mm.

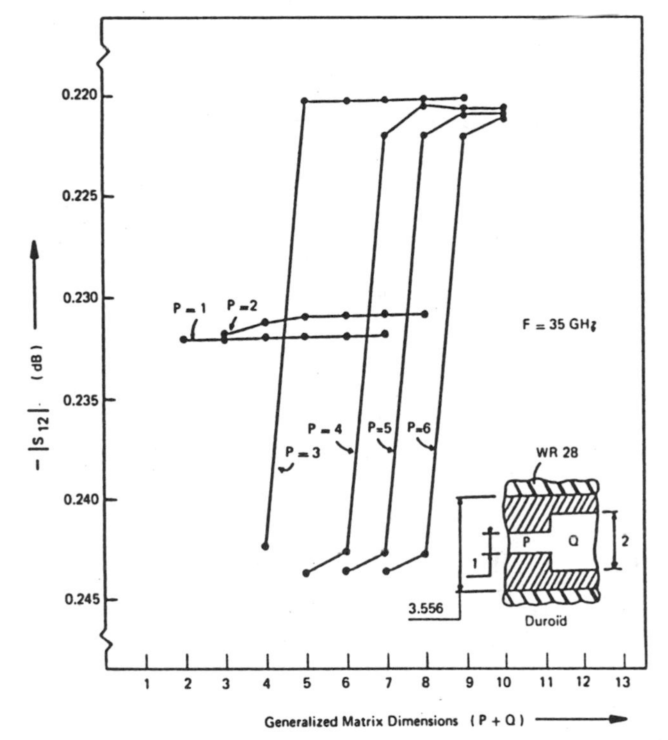

Fig. 5. Convergence test of the amplitude of the transmission coefficient for a step slot width discontinuity between two unilateral finlines (dimensions are in millimeters).

B. Scattering Parameters of Single-Step Slot Discontinuity in a Unilateral Finline

The most critical factor in the modal analysis is the convergence of reflection and transmission coefficients as a function of the number of modes taken into consideration when writing the boundary condition equations (8) at the junction plane. A relative convergence towards wrong values may be obtained if the number of modes is not sufficiently high.

Convergence tests on the moduli and the phases of reflection and transmission coefficients are performed making use of the description possibilities that are offered by the spectral-domain approach for the finline. The results of these tests applied on an abrupt junction between two finlines, as well as on an abrupt junction between a finline and a rectangular waveguide, are respectively given in Figs. 5 and 6. Here, the total number of modes did not exceed 14. The highest level of the curves drawn in Figs. 5 and 6 is supposed to be the real convergence level. Fig. 7 shows the computed scattering parameters of a unilateral finline discontinuity as a function of frequency compared with these computed by Schmidt [8] using the mode-matching procedure and with these computed by Sorrentino and Itoh [8] using the transverse resonance technique. Our results are in excellent agreement with Schmidt's results and in good agreement with Sorrentino's.

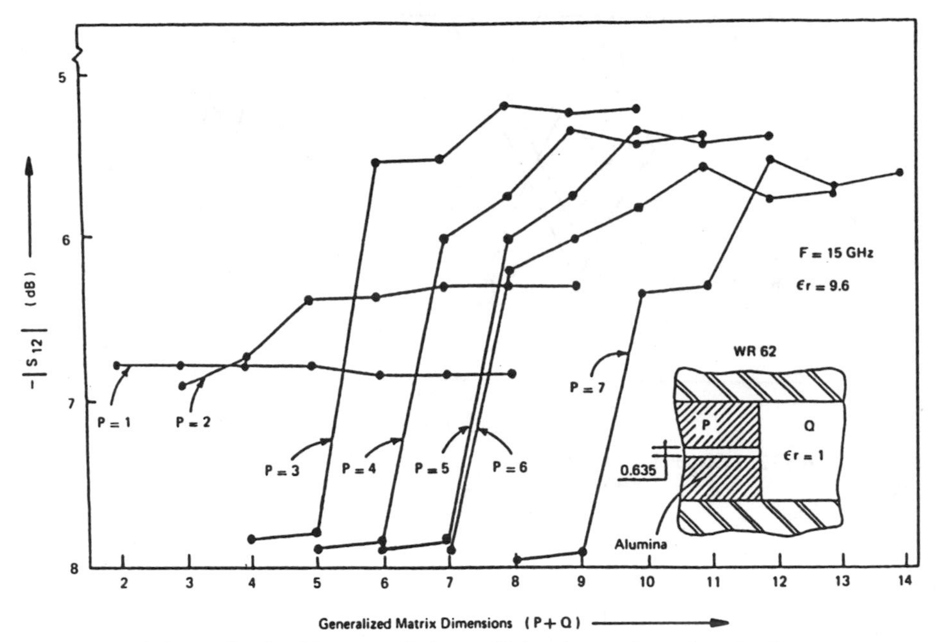

Fig. 6. Convergence test of the amplitude of the transmission coefficient for an abrupt junction between an empty rectangular waveguide and a unilateral finline (dimensions are in millimeters).

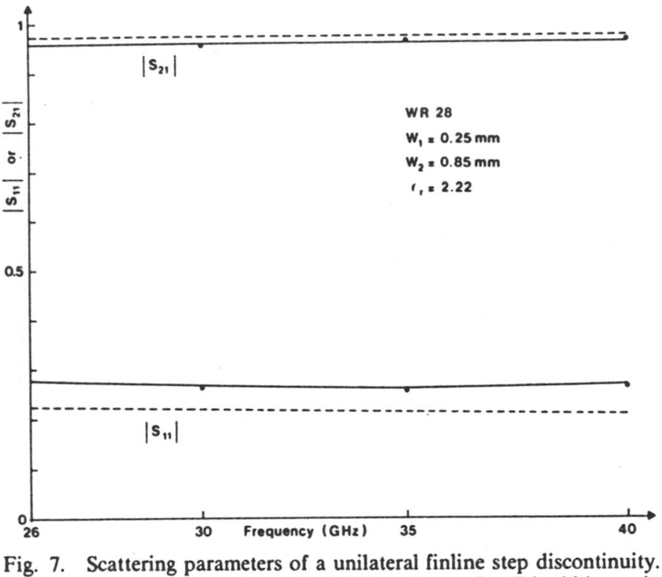

Fig. 7. Scattering parameters of a unilateral finline step discontinuity. —— Our theory, ----- Sorrentino's [8] results, and ∘ Schmidt's results.

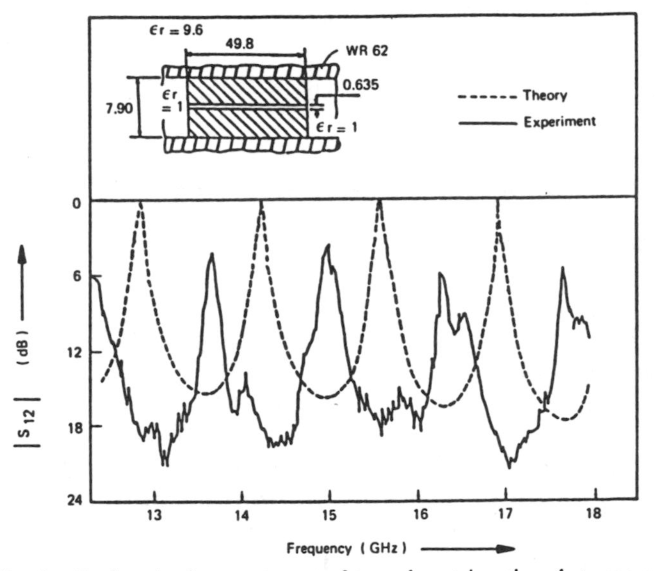

Fig. 8. Back-to-back arrangement of two abrupt junctions between an empty rectangular waveguide and a unilateral finline (dimensions are in millimeters.

C. Comparison Between Theory and Experiments

In order to evaluate objectively the effectiveness of the direct modal analysis for computer-aided design of finline circuits containing different discontinuities, the following three circuits have been fabricated, their scattering parameters have been calculated, and finally measured. The calculation of each circuit includes the effect of the discontinuity created by the narrow face of the substrate. The first circuit is that of two simple rectangular waveguide-unilateral finline junctions as shown in Fig. 8. The agreement can be judged satisfactory for the module of the transmission coefficient. The frequency shift of about 500 MHz can be explained by the mechanical constraints in the fabrication process (e.g., positioning grooves). The second circuit represents a pair of complex transitions each composed of three single simple transitions as shown in Fig. 9. The agreement between theory and experiment is still good in spite of the systematic frequency shift like that observed in the results of the first circuit. The third circuit is that given in Fig. 10, which represents a gradual transition operating in the *Ku*-band. The calculations are performed after dividing the transition into nine single simple discontinuities (Fig. 11). The agreement between theory and experiment is considered definitively satisfactory.

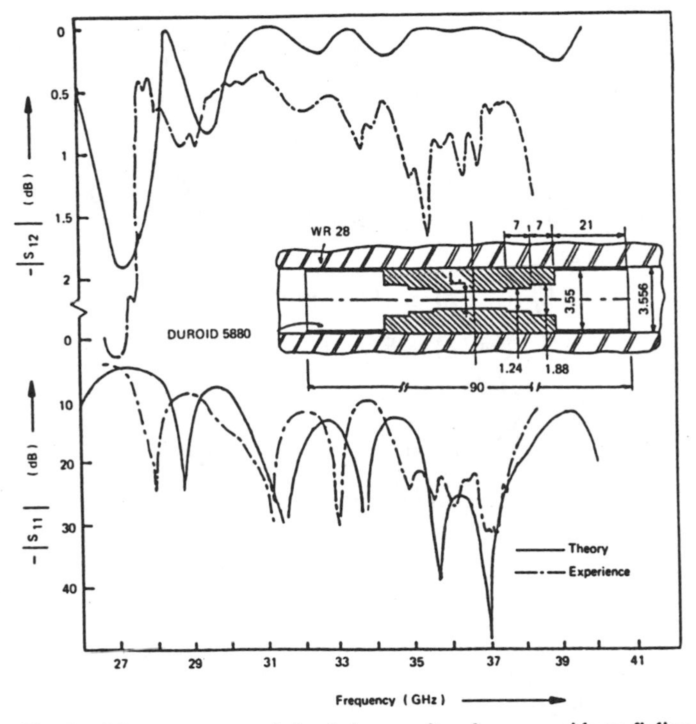

Fig. 9. Measurements and simulation results of a waveguide to finline complex transition in the *Ka*-band (dimensions are in millimeters).

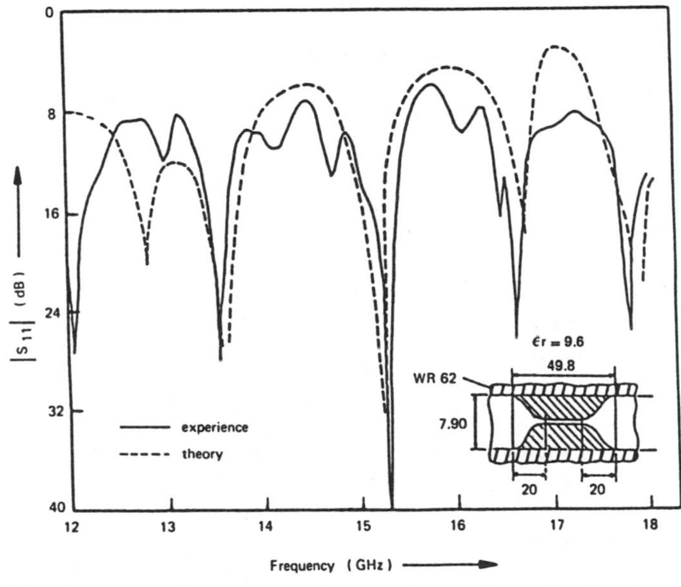

Fig. 10. Comparison between theory and experiment for the shown tapered transition (dimensions are in millimeters).

VI. Equivalent-Circuit Parameters of Simple Finline Discontinuities

The knowledge of both the reflection coefficient S_{11} and the transmission coefficient S_{21} in a given frequency band allows one to construct the equivalent circuit of the junction.

Each line having an access to the junction is considered as a lossless transmission line for which the effective dielectric constant and the characteristic impedance are those corresponding to the fundamental mode. The spectral-domain approach of a finline allows one to calculate these parameters directly in the spectral domain with excellent

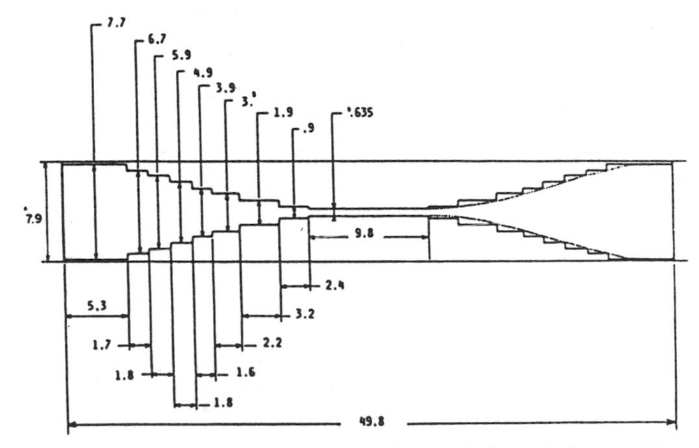

Fig. 11. Theoretical cutting used in the simulation of the transition reported in Fig. 10 (dimensions are in millimeters).

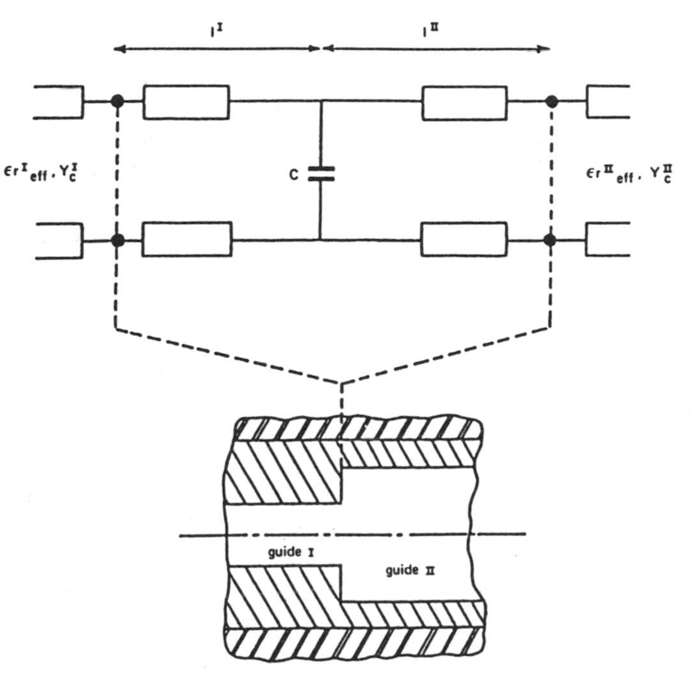

Fig. 12. Equivalent-circuit parameters of an elementary discontinuity.

accuracy. It is worth mentioning that the characteristic impedance is calculated according to a definition that relates the power flow of the fundamental mode and the potential that is induced between the slot edges.

As far as the single-step slot discontinuity is concerned, an equivalent circuit like that shown in Fig. 12 can be selected. The parameters C, l^{I}, and l^{II} of this equivalent circuit can be calculated by comparison of its scattering matrix to that of the single simple discontinuity restricted to four coefficients which represent the reflection and the transmission on only the fundamental eigenmodes of the two finlines having an access on the junction.

Examples of the results are reported in Figs. 13 and 14.

VII. Conclusion

The spectral-domain approach combined with the direct modal analysis appears as a very promising technique for calculation of scattering matrix elements of finline discontinuities.

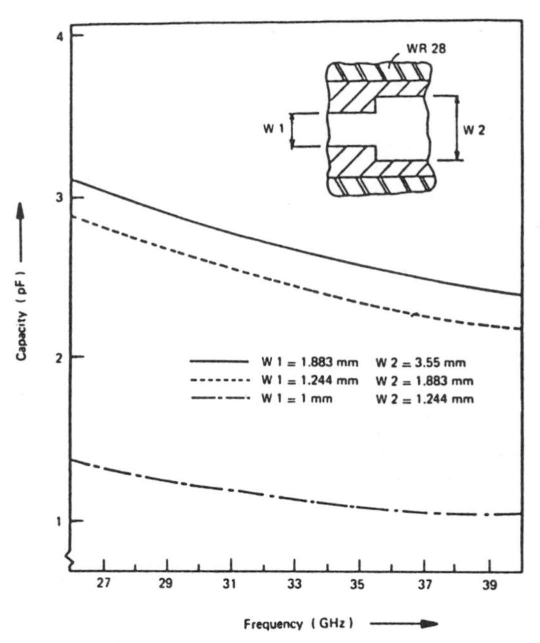

Fig. 13. Values of the lumped capacitance of the equivalent circuit of an elementary discontinuity as a function of frequency ($\epsilon_r = 2.22$, $h_2 = 0.254$ mm).

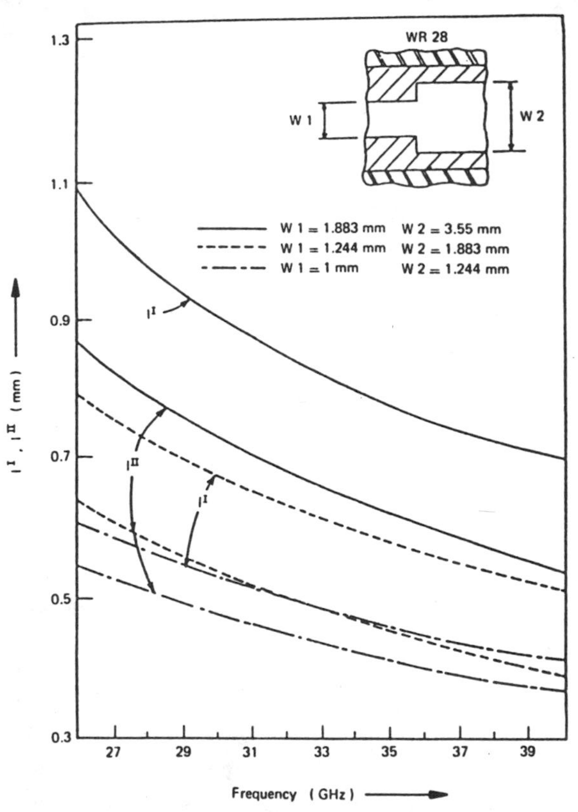

Fig. 14. Reference plane localization as a function of the frequency for the given elementary discontinuity ($\epsilon_r = 2.22$, $h_2 = 0.254$ mm).

Convergence is indeed necessary both in the waveguide eigenmode evaluation and in the waveguide discontinuity problem. In the case of a unilateral configuration, results show that a relatively simple aperture field allowed an unambiguous identification of six eigenmodes. Moreover, they show that, within such an identification, a satisfactory convergence on scattering parameters of a step slot width discontinuity can be achieved.

The agreement between theory and measurements performed on three different complex finline discontinuities is quite satisfactory.

Appendix I

The mth line amplitudes $\tilde{E}_{z,i}(m, y)$ and $\tilde{H}_{z,i}(m, y)$ can be written in the case of unilateral finline in the following form:

$$\tilde{E}_{z,1}(m, y) = A(m)\sinh\gamma_1(y + h_1)$$
$$\tilde{H}_{z,1}(m, y) = B(m)\cosh\gamma_1(y + h_1) \qquad \text{(A1a)}$$

in region (2)

$$\tilde{E}_{z,2}(m, y) = C(m)\sinh\gamma_2 y + D(m)\cosh\gamma_2 y$$
$$\tilde{H}_{z,2}(m, y) = E(m)\sinh\gamma_2 y + F(m)\cosh\gamma_2 y \qquad \text{(A1b)}$$

and in region (3)

$$\tilde{E}_{z,3}(m, y) = G(m)\sinh\gamma_3(h_2 + h_3 - y)$$
$$\tilde{H}_{z,3}(m, y) = H(m)\cosh\gamma_3(h_2 + h_3 - y) \qquad \text{(A1c)}$$

where coefficients γ_i ($i = 1$, 2, and 3) are defined as

$$\gamma_1^2 = \gamma_3^2 = \alpha_m^2 - k_1^2$$
$$\gamma_2^2 = \alpha_m^2 - k_2^2 \qquad \text{(A2a)}$$

with

$$k_1^2 = \omega^2\mu_0\epsilon_0 - \beta^2$$
$$k_2^2 = \omega^2\mu_0\epsilon_0\epsilon_r - \beta^2 \qquad \text{(A2b)}$$

represent the transverse wavenumbers of the mth line amplitude of the guided wave inside each region.

In (A2b), β denotes the phase constant of the guided wave to be determined at any given angular frequency ω.

Appendix II

The elements of the admittance matrix representation [G] given in (2) can be written as

$$G_{11} = -j\left[k_2^2\frac{\tanh(\gamma_2 h_2)}{\mu_0\omega\gamma_2} - \epsilon_0 F\left(k_2^2\frac{\gamma_1}{\tanh(\gamma_1 h_1)} + k_1^2\frac{\epsilon_r\gamma_2}{\tanh(\gamma_2 h_2)}\right) + k_1^2\frac{\coth(\gamma_1 h_3)}{\mu_0\omega\gamma_1}\right] \qquad \text{(A3)}$$

$$G_{12} = G_{21} = j\alpha\beta\left[\frac{\tanh(\gamma_2 h_2)}{\mu_0\omega\gamma_2} - \epsilon_0 F\left(\frac{\gamma_1}{\tanh(\gamma_1 h_1)} + \frac{\epsilon_r\gamma_2}{\tanh(\gamma_2 h_2)}\right) + \frac{\coth(\gamma_1 h_3)}{\mu_0\omega\gamma_1}\right] \qquad \text{(A4)}$$

$$G_{22} = j\left[B_2\left(\frac{\tanh(\gamma_2 h_2)}{\mu_0\omega\gamma_2} - F\frac{\epsilon_0\gamma_1}{\tanh(\gamma_1 h_1)}\right) + B_1\left(\frac{\coth(\gamma_1 h_3)}{\mu_0\omega\gamma_1} - F\frac{\epsilon_0\epsilon_r\gamma_2}{\tanh(\gamma_2 h_2)}\right)\right] \qquad \text{(A5)}$$

where

$$k_2^2 = \omega^2 \mu_0 \epsilon_0 \epsilon_r - \beta^2$$

$$k_1^2 = \omega^2 \mu_0 \epsilon_0 - \beta^2$$

$$B_2 = \alpha^2 - \mu_0 \epsilon_0 \epsilon_r \omega^2$$

$$B_1 = \alpha^2 - \mu_0 \epsilon_0 \omega^2$$

$$F = \frac{k_1^2 k_2^2}{\cosh^2(\gamma_2 h_2)} \cdot \frac{\omega}{\delta}$$

$$\delta = (\alpha\beta)^2 (k_1^2 - k_2^2)^2 - \epsilon_0 \mu_0 \omega^2 \left[k_1^4 \epsilon_r \gamma_2^2 + k_2^4 \gamma_1^2 + (k_1 k_2)^2 \gamma_1 \gamma_2 \left(\frac{\tanh(\gamma_2 h_2)}{\tanh(\gamma_1 h_1)} + \epsilon_r \frac{\tanh(\gamma_1 h_1)}{\tanh(\gamma_2 h_2)} \right) \right].$$

Appendix III

Equations (8a) and (8c) are scalarly multiplied by $\vec{h}^{\mathrm{II}*}_{q,T} \Lambda \vec{u}$ and $e^{\mathrm{I}*}_{p,T} \Lambda \vec{u}$, respectively, and then integrated over S_A. Since $S_{\mathrm{I}} < S_{\mathrm{II}}$, $S_A = S_{\mathrm{I}}$, thus the orthogonality properties of eigenmodes in waveguide I can be used to obtain the following set of $P + Q$ linear equations:

$$\sum_{p=1}^{P} \left(A_p^{\mathrm{I}} + B_p^{\mathrm{I}} \right) \int_{S_{\mathrm{I}}} \vec{e}^{\,\mathrm{I}}_{p,T} \Lambda \vec{h}^{\mathrm{II}*}_{q',T} \vec{u}\, dS =$$

$$\sum_{q=1}^{Q} \left(A_q^{\mathrm{II}} + B_q^{\mathrm{II}} \right) \int_{S_{\mathrm{I}}} \vec{e}^{\,\mathrm{II}}_{q,T} \Lambda \vec{h}^{\mathrm{II}*}_{q',T} \vec{u}\, dS, \qquad q' = 1, \cdots, Q \quad \text{(A6)}$$

$$S_{p'} \frac{\beta^{\mathrm{I}}_{p'}}{|\beta_{p'}|} \left(A^{\mathrm{I}}_{p'} - B^{\mathrm{I}}_{p'} \right) = \sum_{q=1}^{Q} \left(B_q^{\mathrm{II}} - A_q^{\mathrm{II}} \right) \int_{S_{\mathrm{I}}} \vec{e}^{\,\mathrm{I}*}_{p',T} \Lambda \vec{h}^{\mathrm{II}}_{q,T} \vec{u}\, dS,$$

$$p' = 1, \cdots, P \quad \text{(A7)}$$

which are found quite equivalent to boundary functional equations (8).

Similarly, the scalar multiplication of (8b) by $h^{\mathrm{II}*}_{q,T} \Lambda \vec{u}$ and (8d) by $e^{\mathrm{I}*}_{p,T} \Lambda \vec{u}$ and the surface integration over S_C provides the set of $P + Q$ equations

$$\sum_{q=1}^{Q} \left(A_q^{\mathrm{II}} + B_q^{\mathrm{II}} \right) \int_{S_C} \vec{e}^{\,\mathrm{II}}_{q,T} \Lambda h^{\mathrm{II}*}_{q,T} \cdot \vec{u}\, dS = 0,$$

$$q' = 1, \cdots, Q \quad \text{(A8)}$$

$$\sum_{q=1}^{Q} \left(B_q^{\mathrm{II}} - A_q^{\mathrm{II}} \right) \int_{S_C} \vec{e}^{\,\mathrm{I}*}_{p',T} \Lambda \vec{h}^{\mathrm{II}}_{q,T} \cdot \vec{u}\, dS = 0,$$

$$p' = 1, \cdots, P. \quad \text{(A9)}$$

During derivation, use has been made of the following equation:

$$\int_{S_C} \left(\vec{e}_{p',T} \Lambda \vec{u} \right) \cdot \left(\vec{J}^{\mathrm{I}}_T \Lambda \vec{u} \right) dS = \int_{S_C} \vec{J}^{\mathrm{I}}_T \cdot e^{\mathrm{I}*}_{p'T}\, dS = 0$$

due to the above-mentioned connection of the surface current $\vec{J}^{\mathrm{I}}_T$ with waveguide I.

Clearly, the set (A8) can be imbedded in set (A6) after enlarging the surface integration at its right-hand side from S_{I} to $S_{\mathrm{II}} = S_A + S_C$. Such a widening is done without altering the left-hand side of set (A6), and, moreover, use can then be made of the orthogonality in waveguide II.

Now, as far as the set (A7) is concerned, another look at the set (A9) shows that its right-hand side already includes it.

Considering now the case $S^{\mathrm{I}} > S^{\mathrm{II}}$, the functional boundary conditions of (8) still hold except those over S_C ((8c) and (8d)), which must be replaced by

$$E^{\mathrm{I}}_T = 0 \qquad \text{(A10)}$$

$$\vec{H}^{\mathrm{I}}_T = \vec{J}^{\mathrm{II}}_T \Lambda \vec{u} \qquad \text{(A11)}$$

where the superscript II of the surface current notation $\vec{J}^{\mathrm{II}}_T$ indicates now a relationship with the waveguide II. There, another unique procedure must be employed to derive the linear set of $P + Q$ equations equivalent to the boundary functional equations (8), (A9), and (A10).

This alternative procedure starts from the scalar multiplication of (8a) and (A6) by $h^{\mathrm{I}*}_{p',T} \Lambda \vec{u}$ and of (8b) and (A7) by $e^{\mathrm{II}*}_{q'T} \Lambda \vec{u}$, followed by surface integration where use must be made of the basic assumptions $S_A = S_{\mathrm{II}}$ and $S_A + S_C = S_{\mathrm{I}}$.

References

[1] K. Solbach, "The status of printed millimeter-wave *E*-plane circuits," *IEEE Trans. Microwave Theory Tech.*, vol. MTT-31, pp. 107–121, Feb. 1983.

[2] A. Wexler, "Solution of waveguide discontinuities by modal analysis," *IEEE Trans. Microwave Theory Tech.*, vol. MTT-15, pp. 508–517, Sept. 1967.

[3] H. El Hennaway and K. Schunemann, "Analysis of finline discontinuities," in *Proc. 9th Eur. Microwave Conf.* (Brighton, England), 1979, pp. 448–452.

[4] H. Hofmann, "Dispersion of planar waveguides for millimeter-wave application," *Arch. Elek. Übertragung.*, vol. 31, pp. 40–44, Jan. 1977.

[5] H. El Hennaway and K. Schunemann, "Computer-aided design of finline detectors, modulators and switches," *Arch. Elek. Übertragung.*, vol. 36, pp. 49–56, Feb. 1982.

[6] A. Beyer and I. Wolff, "Calculation of transmission properties of inhomogeneous finlines," in *Proc. 10th Eur. Microwave Conf.* (Warsaw, Poland), 1980, pp. 322–326.

[7] D. Mirshekar-Syahkal and J. B. Davies, "Accurate analysis of tapered planar transmission lines for microwave integrated circuits," *IEEE Trans. Microwave Theory Tech.*, vol. MTT-29, pp. 123–128, Feb. 1981.

[8] R. Sorrentino and I. Itoh, "Transverse resonance analysis of finline discontinuities," *IEEE Trans. Microwave Theory Tech.*, vol. MTT-32, pp. 1633–1638, Dec. 1984.

[9] M. Helard, J. Citern, O. Picon, and V. Fouad Hanna, "Exact calculation of scattering parameters of a step slot width discontinuity in a unilateral finline," *Electron. Lett.*, vol. 14, pp. 537–539, July 1983.

[10] M. Helard, J. Citern, O. Picon, and V. Fouad Hanna, "Solution of finline discontinuities through the identification of its first four higher order modes," in *IEEE MTT-S Int. Microwave Symp. Dig.*, May 1983, pp. 387–389.

[11] J. B. Knorr and P. M. Shayda, "Millimeter wave finline characteristics," *IEEE Trans. Microwave Theory Tech.*, vol. MTT-28, pp. 737–742, July 1980.

Part X
Slow Wave Structures

THE subject in this part is somewhat different from the earlier parts of the book. The topic of the slow wave planar transmission lines is treated. These structures can be in the form of a microstrip line or a coplanar waveguide. An essential difference from the conventional planar transmission lines lies in the substrate. The substrate typically contains a lossy semiconductor layer. For instance, the simplest slow wave microstrip takes the form of an MIS (metal-insulator-semiconductor) configuration. The substrate is made of a lossy or doped semiconductor (such as silicon) on which an insulating layer (such as SiO_2) is deposited. In such a structure, three types of guided waves are possible depending on the material and structural parameters (Hasegawa, Furukawa, and Yanai). If the conductivity of the lossy layer is relatively low, the lossy layer can be treated as an imperfect dielectric, and hence the guided wave may be called the dielectric quasi-TEM mode. On the other hand, if the conductivity is high, the semiconductor layer behaves like a lossy metal and the guided wave may be called the skin effect mode. The slow wave mode exists between these two extremes. The guide wavelength of this mode is much smaller than that attainable by the wavelength reduction due to the relative permittivity of the substrate and is often only a few percent of the free space wavelength. The origin of the slow wave effect can be explained by the spatial separation of the electric and magnetic energy. With appropriate material and structural parameters for which the slow wave can exist, the electric field of the guided wave is essentially confined in a thin insulating layer immediately below the strip conductor. On the other hand, the magnetic field is relatively unaffected and penetrates deep into the lossy semiconductor region. Heuristically stated, one sees a tremendous increase of the line capacitance while the line inductance is rather unchanged from the lossless microstrip configuration. The result is an increased phase velocity while, as one of the undesirable features, the characteristic impedance is extremely low. Another major problem of the structure is its high attenuation characteristic. This is inherent because the structure contains a lossy layer.

The slow wave phenomena have been predicted by Guckel *et al.** The first extensive study was done by Hasegawa, *et al.*, as reported in the paper included in this part. Instead of an MIS configuration, a Shottky contacted microstrip can be used as a slow wave structure. In this case the depleted region replaces the insulating layer in the MIS configuration. Furthermore, in place of a microstrip configuration, a coplanar waveguide can be used for realization of slow wave structures either in a microstrip form or a coplanar waveguide form. Several papers in this part present rigorous field-theoretical analysis of the slow wave propagations. Analytical models have also been presented by a number of researchers including Hasegawa.

* H. Guckel, P. A. Brennan, and I. Palocz, "A Parallel-Plate Waveguide Approach to Microminiaturized, Planar Transmission Lines for Integrated Circuits," *IEEE Trans. Microwave Theory and Techniques*, vol. MTT-15, no. 8, pp. 468–476, August 1967.

Properties of Microstrip Line on $Si-SiO_2$ System

HIDEKI HASEGAWA, MEMBER, IEEE, MIEKO FURUKAWA,
AND HISAYOSHI YANAI, SENIOR MEMBER, IEEE

Abstract—A parallel-plate waveguide model for the microstrip line formed on the $Si-SiO_2$ system is analyzed theoretically and the results are compared with the experiment. The experiment has been performed over wide ranges of frequency, substrate resistivity, and strip width. Existence of three types of fundamental modes is concluded and the condition for the appearance of each mode is clarified. In particular, the slow-wave mode is found to propagate within the resistivity-frequency range suited to the monolithic circuit technology, and its propagation mechanism is discussed. Approximate analysis of the fringing effect is also made for the slow-wave mode.

Manuscript received November 19, 1970; revised March 29, 1971.
H. Hasegawa was with the Department of Electronic Engineering, University of Tokyo, Tokyo, Japan. He is now with the Department of Electrical Engineering, Hokkaido University, Sapporo, Japan.
M. Furukawa and H. Yanai are with the Department of Electronic Engineering, University of Tokyo, Tokyo, Japan.

I. Introduction

RECENTLY, planar metallizations on surface-passivated semiconductor substrates have become a familiar form of signal transmission with the rapid development of the monolithic integrated circuit technology now extending partly into the microwave frequency region. However, little is known at present concerning the transmission properties of such a composite structure. Interests of previous workers in the field of planar transmission lines [1]–[4] have been limited mostly to the properties of the quasi-TEM mode through lossless inhomogeneous media. As a matter of course, the above structure can be regarded lossless, when the conductivity of the semiconductor

Reprinted from *IEEE Microwave Theory Tech.*, vol. MTT-19, no. 11, pp. 869–881, November 1971.

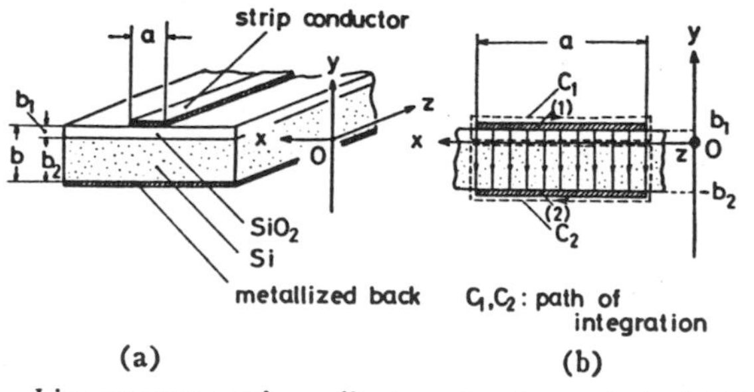

Fig. 1. Line structure and coordinate system for analysis. (a) Microstrip line on Si–SiO_2 system. (b) Its parallel-plate waveguide model.

substrate is decreased. However, this condition is not necessarily realized in the integrated circuit design, because the choice of the conductivity value of the semiconductor substrate in the monolithic circuit technology is closely related to ease of device fabrication; and in most cases, high conductivity values are preferred. Thus, in general, the effect of the finite conductivity of the substrate cannot at all be expected to be a small perturbation to lossless cases.

Of many configurations possible, the microstrip line on the Si–SiO_2 double-layer system illustrated in Fig. 1(a) appears the simplest and perhaps the most useful model for studying the high-frequency signal transmission along the planar metallizations in monolithic ICs. The purpose of the paper is to study the fundamental transmission properties of the microstrip line of this type, varying the conductivity of the semiconductor substrate, frequency, and linewidth over wide ranges.

Exact theoretical analysis of the wave propagation along this type of microstrip line requires the solution of an extremely complicated eigenvalue problem, owing to the presence of two boundary planes and the finite and not necessarily small conductivity of the substrate. For this reason, the line is replaced in Section II by the parallel-plate waveguide filled with the Si–SiO_2 layer [Fig. 1(b)], and the effects of the substrate conductivity and frequency are studied theoretically in this "wide-strip" limit. Existence of the three fundamental modes, i.e., "dielectric quasi-TEM mode," "skin-effect mode," and "slow-wave mode" is shown, and the condition for the appearance of each mode is clarified on the resistivity (of the substrate)–frequency plane. Physical mechanism of propagation is also discussed.

In Section III, the detailed experimental results upon the microstrip lines with finite linewidths are presented, and they are compared with the theory in the wide-strip limit. The experiment has been performed over the frequency range of 30–4000 MHz, using the samples which have the substrate resistivity of 10^{-3}–10^2 $\Omega \cdot$cm and the strip width of 70–1600 μm. The measured resistivity–frequency characteristics of the characteristic impedance and propagation constant agree fairly well with the theory. It is concluded that the above three fundamental modes also exist in the microstrip configuration and their behaviors can be explained qualitatively and, to some extent, quantitatively by the theory in the wide-strip limit.

In Section IV, the "fringing" effect caused by the finite linewidth is approximately analyzed in the case of the slow-wave mode, and the result is compared with the experiment.

Let us survey in the following the physical origin of the above three fundamental modes. Strictly speaking, they are the three types of limiting forms of propagation that are possessed by the lowest order mode of the microstrip line formed on the surface-passivated semiconductor substrate. These limiting forms come out of the different behaviors of the substrate, when frequency and its resistivity are varied. Physically, existence of the two limits would be evident as shown in the following. 1) When the product of the frequency and the resistivity of the Si-substrate is large enough to produce a small dielectric loss angle, the substrate acts like a dielectric as reported by Hyltin [5], and the line can be regarded as a microstrip line loaded with a double-layer dielectric consisting of silicon and silicon-dioxide. The fundamental mode would closely resemble to the TEM mode, so long as the wavelength is much larger than the thickness of the double layer. Electrostatic analysis of this quasi-TEM mode has been carried out by Yamashita [6]. With the further increase in frequency, the conventional TEM approximation becomes invalid, and the mode would undergo the similar dispersive effect as reported in the case of the standard microstrip lines [7], [8], although it is not discussed in the present paper. 2) When the product of the frequency and substrate conductivity is large enough to yield a small depth of penetration into silicon, the substrate would behave like a lossy conductor wall, and the line may be treated as microstrip line on the imperfect "ground plane" made of silicon. Essentially the same discussion as above would be applied as for the fundamental mode, if the depth of penetration is much smaller than the thickness of the silicon-dioxide layer. However, this condition is not usually satisfied, and the line behaves highly dispersive due to the skin effect.

In addition to the above "dielectric" and "metallic" limits, however, there exists another type of propagation. 3) When the frequency is not so high and the resistivity is moderate, the substrate acts like neither of the above, and a slow-surface wave propagates along the line. In contrast to the above limits, this mode might be recognized as corresponding to the "semiconductor" limit of the substrate. Existence of such a slow-wave mode was first predicted by Guckel *et al.* [9], and verified by the authors [10]. This mode is of particular interest, because it appears within the resistivity and frequency ranges suited for the monolithic circuit technology which has recently grown up to treat extremely high-speed pulses, such as subnanosecond ones. In addition, this mode suffers considerably smaller attenuation than what is expected from the conductivity value. Possible applications are suggested in Section V.

II. Analysis of Parallel-Plate Waveguide Model

The microstrip line shown in Fig. 1(a) is replaced by the idealized parallel-plate waveguide without fringing effect, and the latter is analyzed theoretically in this section. The model for analysis is illustrated in Fig. 1(b). The same model was already analyzed numerically by Guckel *et al.* [9], but the wave propagation is studied in this paper from a somewhat different point of view, introducing the effective permittivity and permeability to characterize the fundamental modes, and considering their relaxation behaviors. The electromagnetic field distributions are also studied.

A. Effective Permittivity and Permeability of Double Layer

The fundamental propagation mode of the waveguide shown in Fig. 1(b) is a TM wave. The longitudinal and transverse propagation constants of the wave should satisfy the following relation required by the wave equation:

$$\gamma_i^2 + \gamma^2 = -\omega^2 \epsilon_i^* \mu_i^*, \qquad i = 1, 2 \tag{1a}$$

where γ_1 and γ_2 denote the transverse propagation constants (y direction) in SiO_2 and Si layers, respectively, and γ, the longitudinal one (z direction). $\epsilon_i^* = \epsilon_i' - j\epsilon_i''$ and $\mu_i^* = \mu_i' - j\mu_i''$ denote the complex permittivity and permeability of each loading layer, respectively. Another relation is derived by the application of the transverse resonance method [11].

$$\sum_{i=1}^{2} Z_i \tanh \gamma_i b_i = 0 \tag{1b}$$

where $Z_i = \gamma_i/(j\omega\epsilon_i^*)$ and b_i is the thickness of each layer. Equation (1b) together with (1a) determines the allowed values of γ_1, γ_2, and γ.

The field components in each layer are calculated as follows in terms of γ_1, γ_2, and γ, using the Maxwell field equations:

$$\left[\begin{aligned} E_{yi} &= \mp \frac{\gamma}{\gamma_i} \cdot E_{z0} \cdot \frac{\cosh[\gamma_i(y \mp b_i)]}{\sinh \gamma_i b_i} \cdot e^{-\gamma z} \\ H_{xi} &= \mp \frac{j\omega\epsilon_i^*}{\gamma_i} \cdot E_{z0} \cdot \frac{\cosh[\gamma_i(y \mp b_i)]}{\sinh \gamma_i b_i} \cdot e^{-\gamma z} \\ E_{zi} &= \mp E_{z0} \cdot \frac{\sinh[\gamma_i(y \mp b_i)]}{\sinh \gamma_i b_i} \cdot e^{-\gamma z} \end{aligned}\right. \tag{2}$$

where $i = 1, 2$ (for $i = 1$, $-$sign and for $i = 2$, $+$sign) and $E_{z0} = E_z|z = 0, y = 0$; $i = 1$ represents the field distribution within the SiO^2 layer, and $i = 2$, that within the Si-layer, respectively.

The effective complex permittivity, ϵ^*_{eff}, and the effective complex permeability, μ^*_{eff}, of the double layer should be defined by

$$\left[\begin{aligned} \epsilon^*_{\text{eff}} &\equiv \epsilon'_{\text{eff}} - j\epsilon''_{\text{eff}} \equiv \left[\frac{1}{b} \sum_{i=1}^{2} \frac{1}{\epsilon_i^*} \frac{1}{\gamma_i} \tanh \gamma_i b_i\right]^{-1} \\ \mu^*_{\text{eff}} &\equiv \mu'_{\text{eff}} - j\mu''_{\text{eff}} \equiv \frac{1}{b} \sum_{i=1}^{2} \mu_i^* \frac{1}{\gamma_i} \tanh \gamma_i b_i \end{aligned}\right. \tag{3}$$

where $b = b_1 + b_2$, the total thickness of the double layer. From (1a), (1b), and (3), one can express γ as

$$\gamma = j\omega\sqrt{\epsilon^*_{\text{eff}}\mu^*_{\text{eff}}}. \tag{4}$$

In the next place, the line voltage, V, and line current, I, should be defined as follows, making reference to Fig. 1(b):

$$V \equiv -\int_{(2)}^{(1)} E_t dl \qquad I \equiv \oint_{C1} H_t dl \equiv -\oint_{C_2} H_t dl \tag{5}$$

where the subscript t represents the transverse component.

Then, the usual differential equation for the transmission line is derived after a little algebra. Using (2), (3), and (5), the characteristic impedance, Z_0, of the line can be expressed as

$$Z_0 = \frac{b}{a}\sqrt{\frac{\mu^*_{\text{eff}}}{\epsilon^*_{\text{eff}}}}. \tag{6}$$

Thus, it is seen from (4) and (6) that ϵ^*_{eff} and μ^*_{eff} are convenient quantities[1] to describe the transmission properties of the TM wave. Furthermore, ϵ^*_{eff} reduces to the classical Maxwell–Wagner permittivity for the interfacial polarization [12] in the limit of $\gamma_i b_i \to 0$ ($i = 1, 2$).

B. Three Fundamental Modes and Resistivity–Frequency Domain Chart

As stated in Section I, the fundamental TM wave exhibits three kinds of principal limiting forms, depending upon the frequency, f, and the resistivity, ρ_2, of the Si layer. They should be defined as the three fundamental modes. Each mode can be characterized in terms of the real parts of ϵ^*_{eff} and μ^*_{eff} as follows.

1) *Dielectric Quasi-TEM Mode:*

$$\epsilon'_{\text{eff}} = \epsilon_0 \epsilon_{s\infty} \qquad \mu'_{\text{eff}} = \mu_0$$

[1] ϵ^*_{eff} and μ^*_{eff} can be generalized to the case of N different loading layers ($N \geqq 3$) by letting $i = 1, 2, \cdots, N$ in (3). But this generalization is valid only for the case of $|\gamma_i b_i| \ll 1$ ($i = 1, 2, \cdots, N$), because for $N \geqq 3$ the exact transverse resonance equation becomes complicated due to the multiple reflection of plane waves in the transverse direction and is not given simply by

$$\sum_{i=1}^{N} Z_i \tanh \gamma_i b_i = 0.$$

where

$\epsilon_{S\infty}$ $= [(1/b)\ (b_1/\epsilon_{SiO_2}+b_2/\epsilon_{Si})]^{-1}$, optical value of Maxwell–Wagner permittivity;

ϵ_0, μ_0 = permittivity and permeability of vacuum;

$\epsilon_{SiO_2}, \epsilon_{Si}$ = relative permittivities of SiO_2 and Si.

2) *Skin-Effect Mode*:

$$\epsilon'_{eff} = \epsilon_0\epsilon_{s0} \qquad \mu'_{eff} = \mu_0 \frac{1}{b}\left(b_1 + \frac{\delta}{2}\right)$$

where

ϵ_{s0} $= \epsilon_{SiO_2}\cdot(b/b_1)$, static value of Maxwell–Wagner permittivity;

δ $= \sqrt{2\rho_2/\omega\mu_0}$, depth of penetration in Si-layer.

3) *Slow-Wave Mode*:

$$\epsilon'_{eff} = \epsilon_0\epsilon_{s0}, \qquad \mu'_{eff} = \mu_0.$$

Here, the permeabilities of the SiO_2 and Si layers are assumed to be equal to μ_0, and the dielectric loss in the SiO_2 layer is ignored. $2\pi f\sqrt{\epsilon_0\epsilon_{s\infty}\mu_0}\,b \ll 1$ is also assumed.

The first mode appears when the product of f and ρ_2 is large and the Si layer acts like a dielectric. In the practical situation of $\epsilon_{Si}\ b_1 \ll \epsilon_{SiO_2} b_2$, almost all the energy is transmitted through the Si layer with the velocity nearly equal to $1/\sqrt{\epsilon_{Si}}$ times the light velocity in vacuum. The second mode appears when the product of f and the substrate conductivity, $\sigma_2 = \rho_2^{-1}$, is large. In this mode, the substrate behaves like a lossy conductor wall. Note that the effective wall recession is equal to $\delta/2$, as pointed out by Wheeler [13]. In the practical situation of extremely thin SiO_2 layer, $b_1 \ll \delta/2$ is probable in most cases. The dispersion behavior is then governed by the skin effect in the silicon substrate, and becomes different from that of a TEM wave through the SiO_2 layer (see Table II). For this reason, the mode is called the skin-effect mode. The third mode appears when f is not so large and ρ_2 is moderate. In the practical situation of $b_1 \ll b_2$, a strong interfacial polarization occurs, and ϵ_{s0} becomes very large owing to the so-called "Maxwell–Wagner mechanism" [12]. For example, with $b_1 = 5000$ Å and $b_2 = 200\ \mu m$, ϵ_{s0} increases to as much as 1600. Correspondingly, the propagation velocity slows down and therefore this mode is called the slow-wave mode.

To clarify the domain of the resistivity–frequency plane in which each mode propagates, the various characteristic frequencies should be defined as summarized in Table I. In Table I, f_s is the relaxation frequency of the interfacial polarization in the case of $\epsilon_{Si}b_1 \ll \epsilon_{SiO_2}b_2$. f_δ is the frequency at which the depth of penetration, δ, in the Si layer becomes equal to its thickness, b_2.

Using the approximate solutions of (1a) and (1b), whose derivation is outlined in the Appendix, the frequency ranges of the three modes are derived as listed in Table II.[2] Since these three modes gradually change their forms and make transitions to each other in a continuous manner with the variation of f and ρ_2, the frequency ranges in the Table II are defined as follows. Inside the frequency range of each mode, the guide wavelength, λ_g, should be given by $\sqrt{\epsilon_0\mu_0/\epsilon'_{eff}\,\mu_{eff}}\ \lambda_0$ within a small error of about 5 percent. Here, λ_0 is the free-space wavelength, and for ϵ'_{eff} and μ'_{eff}, the values specified in the previously mentioned 1)–3) should be inserted for each mode. From these frequency ranges, the resistivity–frequency domain chart shown in Fig. 2 is obtained. In the figure, the domain boundaries for the case of $b_1 = 10^4$ Å and $b_1/b_2 = 5\times10^{-3}$ are drawn by thick lines for convenience of comparison with the experiment. The region other than the three mode domains is defined as the transition region.

TABLE I

Definitions of Various Characteristic Frequencies

f_e	$\equiv \frac{1}{2\pi}\frac{\sigma_2}{\epsilon_0\epsilon_{Si}}$	: dielectric relaxation frequency of Si-layer
f_s	$\equiv \frac{1}{2\pi}\frac{\sigma_2}{\epsilon_0\epsilon_{SiO_2}}\frac{b_1}{b_2}$	: relaxation frequency of interfacial polarization
f_δ	$\equiv \frac{1}{2\pi}\frac{2}{\mu_0\sigma_2 b_2^2}$	: characteristic frequency for skin-effect in Si-layer
f_0	$\equiv [f_s^{-1} + \frac{2}{3}f_\delta^{-1}]^{-1}$	: characteristic frequency of slow-wave mode
f_s	$\equiv f_s^{\,2}/f_\delta$	
f_m	$\equiv \sqrt{f_s \cdot f_e}$	

The characteristic impedance, Z_0, and the propagation constant, $\gamma \equiv \alpha + j\beta$ (α is the attenuation constant; β, the phase constant) are calculated using the approximate solutions of the eigenvalue equation together with (4) and (6), and are also shown in Table II. The equivalent circuit of each mode determined from Z_0 and γ is given in Fig. 3.[3]

Judging from the resistivity–frequency domain chart, the slow-wave mode appears of most importance for the resistivity and frequency ranges practically employed in IC's. It is seen from Tables I and II that the upper

[2] In the frequency range of the skin-effect mode, the second condition, $f \lesssim 10^{-2} f_{s\delta}$, assures $\epsilon^*_{eff} = \epsilon_0\epsilon_{s0}$, that is, the electric field in the "skin layer" can be ignored as compared with that in the SiO_2 layer.

[3] The equivalent circuits, (1) and (3), in Fig. 3 are simpler than the corresponding ones in [9], because the elements for the higher order corrections can be ignored within the resistivity–frequency domains of the modes. It can be also shown that when these higher order elements are to be included, the equivalent circuits in [9] become incomplete, because other elements already present in the equivalent circuits must also undergo the corrections of the same order.

TABLE II

Properties of Fundamental Modes

$(\epsilon_{Si} b_1 \ll \epsilon_{SiO_2} b_2, \quad b_1 \ll \delta/2)$

	slow-wave mode	dielectric quasi-TEM mode	skin-effect mode
frequency range	$f \lesssim 0.3 f_0$	$f \gtrsim 1.5 f_e$	$f \gtrsim 4 \cdot f_\delta$ $f \lesssim 10^{-2} f_{s\delta}$
Re (Z_0)	Z_{os}	$Z_{o\infty}$	$K_1 \cdot Z_{os} \cdot (f_\delta / f)^{\frac{1}{4}}$
Im (Z_0)	$\frac{1}{2} Z_{os} [(f/f_s) - \frac{2}{3}(f/f_\delta)]$	$\frac{1}{2} \cdot Z_{o\infty} \cdot (f_e / f)$	$-K_2 \cdot Z_{os} (f_\delta / f)^{\frac{1}{4}}$
$\beta / \beta_0 (= \lambda_0/\lambda_g)$	$\sqrt{\epsilon_{so}}$	$\sqrt{\epsilon_{s\infty}}$	$\sqrt{\epsilon_{so}} \cdot K_1 \cdot (f_\delta / f)^{\frac{1}{4}}$
α	$\alpha_0 (f/f_0)^2$ $= \alpha_s (f/f_s)^2 + \alpha_\delta (f/f_\delta)^2$	α_∞	$3 \cdot K_2 \cdot \alpha_\delta (f / f_\delta)^{\frac{3}{4}}$

$Z_{os} \equiv 120\pi \frac{1}{\sqrt{\epsilon_{so}}} \frac{b}{a}$ ohms, $\quad Z_{o\infty} \equiv 120\pi \frac{1}{\sqrt{\epsilon_{s\infty}}} \frac{b}{a}$ ohms

$\alpha_0 \equiv \pi f_0 \sqrt{\epsilon_0 \epsilon_{so} \mu_0}, \quad \alpha_s \equiv \pi f_s \sqrt{\epsilon_0 \epsilon_{so} \mu_0}, \quad \alpha_\delta \equiv \frac{2}{3} \cdot \pi f_\delta \sqrt{\epsilon_0 \epsilon_{so} \mu_0}$

$\alpha_\infty \equiv \pi f_e \sqrt{\epsilon_0 \epsilon_{s\infty} \mu_0}, \quad \beta_0 \equiv 2\pi f \sqrt{\epsilon_0 \mu_0}$

$K_1 \equiv \frac{\cos(\pi/8)}{\sqrt[4]{2}}, \quad K_2 \equiv \frac{\sin(\pi/8)}{\sqrt[4]{2}}$

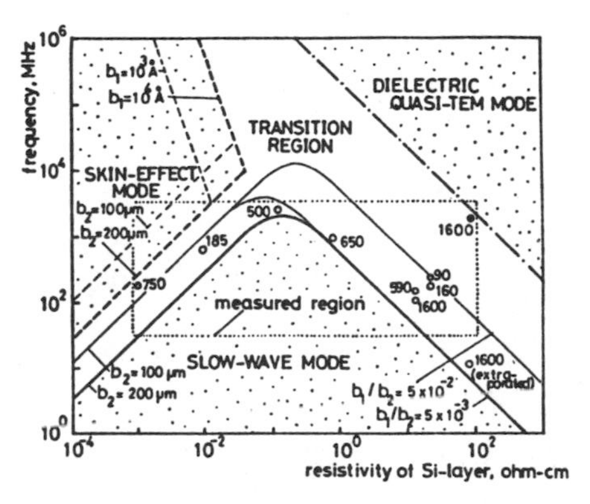

Fig. 2. Resistivity–frequency domain chart. $\epsilon_{SiO_2}=4$, $\epsilon_{Si}=12$. Thick lines represent $b_1=10^4$Å and $b_1/b_2=5\times10^{-3}$, and correspond to geometrical conditions of experimental samples. ○ and ● are measured slow-wave mode and dielectric quasi-TEM mode boundary points, respectively. Numbers attached to the points are the strip widths in μm.

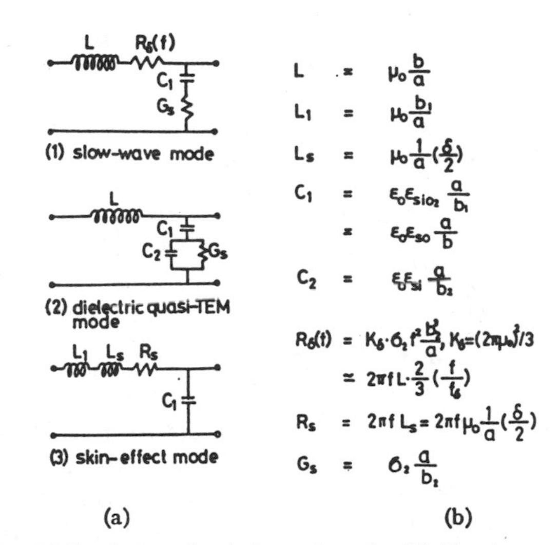

Fig. 3. (a) Equivalent circuit for each mode. (b) Circuit parameters.

frequency limit of the slow-wave mode becomes highest under the condition of

$$f_s^{-1} = \tfrac{2}{3} f_\delta^{-1}. \tag{7}$$

This condition is satisfied when ρ_2 is equal to the characteristic value, ρ_c, determined by

$$\rho_c = \sqrt{\mu_0 b_1 b_2 / (3 \epsilon_0 \epsilon_{SiO_2})}. \tag{8}$$

Im (Z_0) changes its sign according to $f_s^{-1} \gtrless (2/3) f_\delta^{-1}$, i.e., $\rho_2 \gtrless \rho_c$. Furthermore, the attenuation constant is in direct proportion to the square of frequency and the constant of proportionality becomes minimum under the condition of $\rho_2 = \rho_c$. Thus, ρ_c is an important characteristic quantity for the slow-wave mode. It should be noted here that the attenuation constant of the slow-wave mode is much smaller than that of the homogeneous parallel-plate waveguide filled with the silicon layer only. The propagation constant of the latter is simply given by $\gamma = \gamma_m \equiv \sqrt{j\omega\mu_0\sigma_2}$ for $f \ll f_e$.

Next, the transition region is considered. The frequency characteristics of Z_0 and γ in this region can be described by the relaxation behaviors of ϵ^*_{eff} and μ^*_{eff}. Consider first the transition from the slow-wave mode to the dielectric quasi-TEM mode under the condition of $f \ll f_\delta$. In this transition, μ^*_{eff} is held constant and ϵ^*_{eff} exhibits dielectric relaxation behavior. An approx-

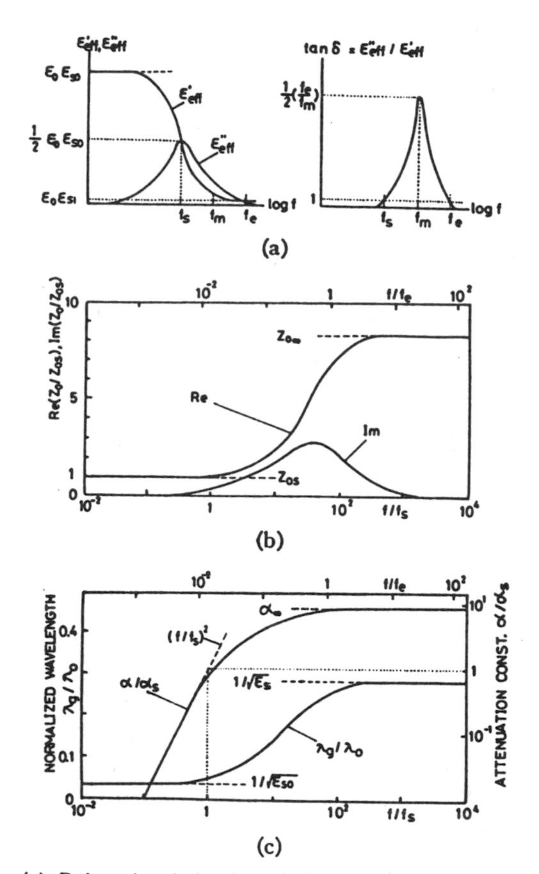

Fig. 4. (a) Relaxation behavior of ϵ^*_{eff} in the case of $f \ll f_\delta$. (b), (c) Corresponding frequency characteristics of Z_0 and γ. (For notations, see Tables I and II.)

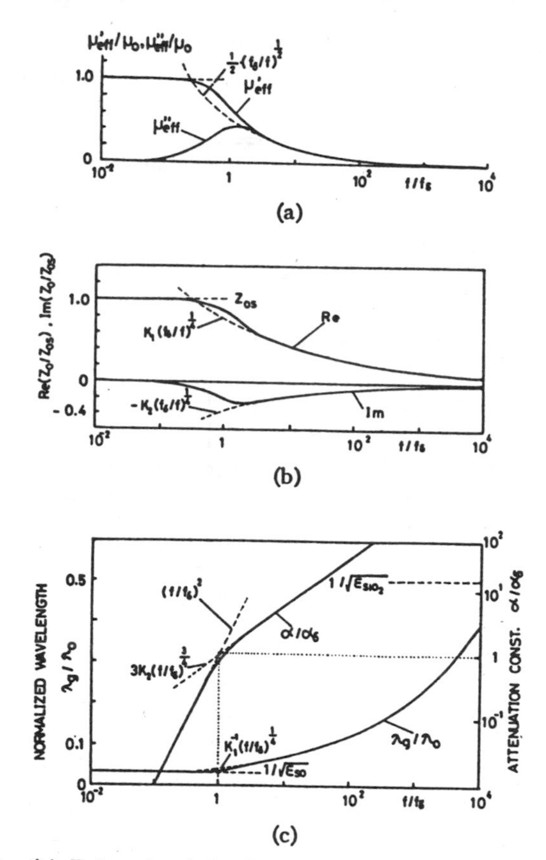

Fig. 5. (a) Relaxation behavior of μ^*_{eff} in the case of $f \ll f_s$. (b), (c) Corresponding frequency characteristics of Z_0 and γ. (For notations, see Tables I and II.)

imate expression for the relaxation can be derived from (3) as

$$\left[\begin{aligned} \epsilon'_{eff}/\epsilon_0 &= \epsilon_{s\infty} + \frac{\epsilon_{s0} - \epsilon_{s\infty}}{1 + (\omega\tau_{se})^2} \\ \epsilon''_{eff}/\epsilon_0 &= (\epsilon_{s0} - \epsilon_{s\infty}) \frac{\omega\tau_{se}}{1 + (\omega\tau_{se})^2} \end{aligned}\right. \tag{9}$$

where $\tau_{se} = (1/2\pi)(f_s^{-1} + f_e^{-1}.)$ Thus, ϵ^*_{eff} is formally subject to the Debye-type relaxation with the relaxation time constant, τ_{se}, which is very close to $(1/2\pi)f_s^{-1}$ in the case of $\epsilon_{Si}b_1 \ll \epsilon_{SiO_2}b_2$. The relaxation behavior of ϵ^*_{eff} and the corresponding frequency characteristics of Z_0 and γ are illustrated in Fig. 4. In the figure, the phase constant is expressed by the normalized wavelength, λ_g/λ_0. As the frequency is increased, the slow-wave mode changes its form into a diffusion-like TEM mode. γ approaches $\gamma_m \equiv \sqrt{j\omega\mu_0\sigma_2}$ in the vicinity of $f_m = \sqrt{f_s \cdot f_e}$, where the attenuation per wavelength becomes maximum. With further increase in frequency, the displacement current in the Si layer becomes dominant and the mode changes smoothly into the dielectric quasi-TEM mode.

For the transition from the slow-wave mode to the skin-effect mode, ϵ^*_{eff} is held constant and μ^*_{eff} shows a relaxation behavior owing to the skin effect. Since $\gamma_2 = \gamma_m$ is a good approximation for $f \ll f_e$, as shown in the Appendix, the following approximate expression for the relaxation is derived:

$$\mu^*_{eff}/\mu_0 = 1 - \frac{b_2}{b} F(f/f_\delta) \tag{10}$$

where

$$F(f/f_\delta) = 1 - \frac{1}{\gamma_m b_2} \tanh \gamma_m b_2$$

and

$$\gamma_m b_2 = (1 + j)\sqrt{f/f_\delta}.$$

The relaxation behavior and the corresponding frequency characteristics of Z_0 and γ are illustrated in Fig. 5.

C. Electromagnetic Field Distribution

The previous explanation of propagation modes in terms of ϵ^*_{eff} and μ^*_{eff} is somewhat phenomenological, especially in the case of the slow-wave mode. It is easily recognized that the permittivity looks very large due to the strong interfacial polarization, but what kind of wave "propagates" is still not clear. For this reason, the electromagnetic field distributions are now briefly discussed.

Since the field distribution of the quasi-TEM mode in a dielectric medium appears well known, only those of

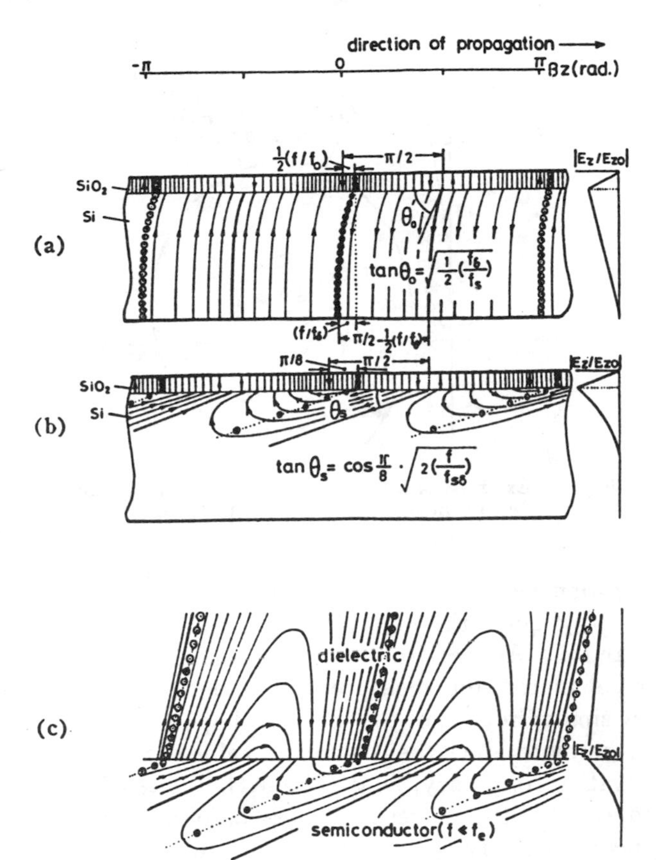

Fig. 6. Electromagnetic field distributions of (a) slow-wave mode, (b) skin-effect mode, and (c) Zenneck surface wave. ——— electric field lines; ⊙, ⊗ magnetic field lines corresponding to the phase of maximum magnetic field.

the slow-wave mode and the skin-effect mode are schematically illustrated in Fig. 6. For convenience, attenuation in the propagation direction is ignored. This figure is drawn by inserting the approximate solutions of the eigenvalue equation into (2). The field distribution of the Zenneck surface wave [14] that appears relevant to the previously mentioned two modes is also shown [Fig. 6(c)]. Comparison of field distributions reveals that the first two modes belong also to the category of the surface wave.

In the slow-wave mode, almost all the active power is transmitted through the SiO_2 layer, but a large amount of reactive power is exchanged between two layers across the Si–SiO_2 interface with the movement of charge at the interface. The average amount of magnetic energy stored in the waveguide is equal to that of electric energy stored in the SiO_2 layer, but the former is stored mainly within the Si layer because of $b_2 \gg b_1$. The sum of these energies is transmitted in z direction through the thin SiO_2 layer. The propagation velocity slows down owing to the energy transfer across the interface.

The part of the active power that flows into the Si layer is totally dissipated by the conduction current in the Si layer. As shown in Fig. 6(a), the conduction current flows making an angle, θ_0, at the interface. The magnitude of the angle is independent of frequency and depends on the substrate resistivity. Under the condition of (7), the power dissipated by the longitudinal component of the conduction current becomes equal to that dissipated by the transverse one, and θ_0 becomes 30°. Furthermore, the average phase difference between H_{x2} and E_{y1} vanishes under this condition, and the total reactive power flow in z direction resultantly vanishes.

Transition from the slow-wave mode to the skin-effect mode can be easily traced by Fig. 6(a) and (b).

TABLE III

Dimensions and Resistivities of Samples

Thickness of SiO_2-layer	:	$b_1 = 10^4$ Å
Thickness of Si-substrate	:	$b_2 = 200 - 250$ μm
Width of strip conductor	:	$a = 70, 160, 600, 1600$ μm
Length of the line	:	$\ell = 5, 10, 20, 30$ mm
Resistivity of n-type Si-substrate	:	$\rho_2 = 10^{-3}, 10^{-2}, 10^{-1}, 0.9$ 12.5, 20, 85 ohm·cm

III. Experimental Study of Microstrip Lines on Si–SiO_2 Systems

A. Measurement Method and Samples

The dimensions and resistivities of the samples employed in the experiment are listed in Table III. The strip conductors were fabricated by vacuum evaporation or beam-lead technology. The Si substrates were of n type (phosphor doped) for all the samples.

The characteristic impedance and the propagation constant were determined from 30 to 4000 MHz by measuring the short-circuit and open-circuit input impedances, Z_{sh} and Z_{op}, of the samples. The input impedances were measured by the coaxial impedance meter (Z–g diagraph, Rhode und Schwarz) from 30 to 600 MHz. For higher frequencies, a microstrip standing-wave detector was specially developed to avoid measurement errors introduced by the coaxial-to-microstrip transition; these errors are especially serious for the samples with lower characteristic impedances. The experimental setup is shown in Fig. 7(a). Mountings of the samples are illustrated in Fig. 7(b). Detailed preliminary experiments using lossless lines confirmed that the measurement accuracies were sufficient within the frequency range in concern.

B. Behavior of Input Impedances

The input impedances of this type of line behave quite differently versus frequency when the substrate resistivity is varied. Some of the experimental data are presented and briefly discussed below:

Fig. 8(a) shows the measured loci of Z_{sh} and Z_{op} for a sample with a fairly high resistivity of about 85 Ω · cm. Corresponding theoretical loci of Z_{sh}, Z_{op}, and Z_0 are shown in Fig. 8(b), which are calculated assuming the

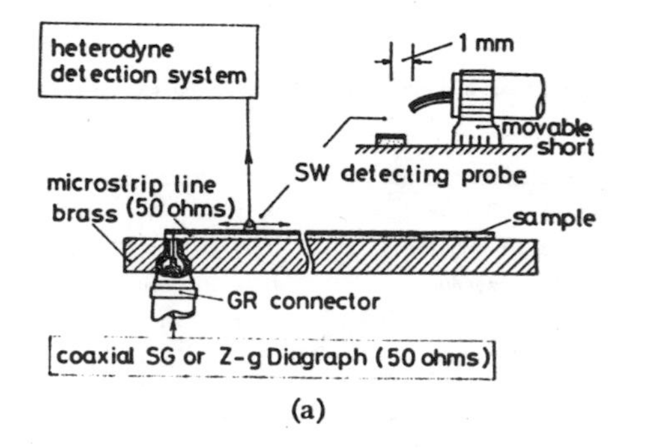

(a)

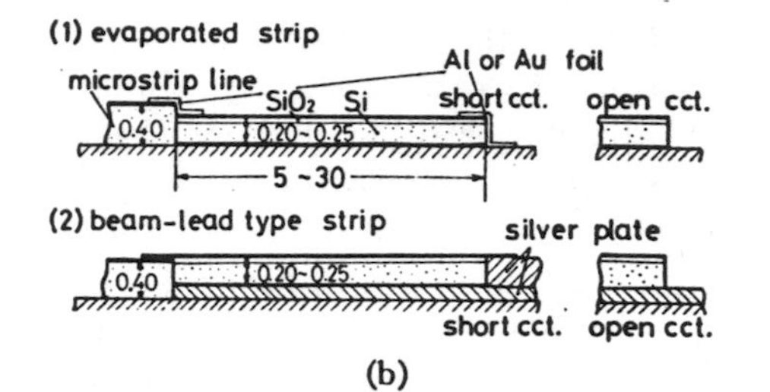

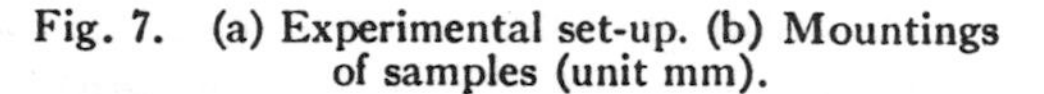

(b)

Fig. 7. (a) Experimental set-up. (b) Mountings of samples (unit mm).

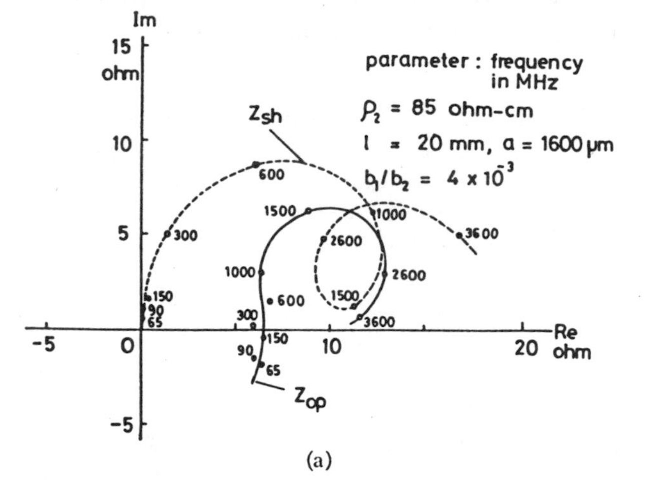

(a)

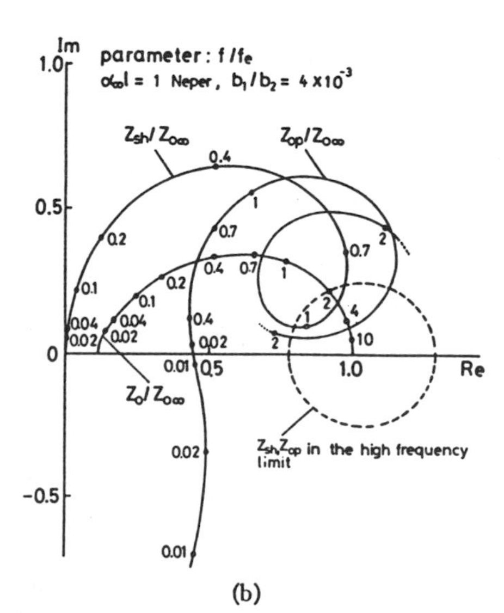

(b)

Fig. 8. (a) Example of frequency loci of Z_{op} and Z_{sh} for the sample with a fairly high resistivity. (b) Theoretical loci of Z_{op}, Z_{sh}, Z_0 assuming the relaxation behavior of (9).

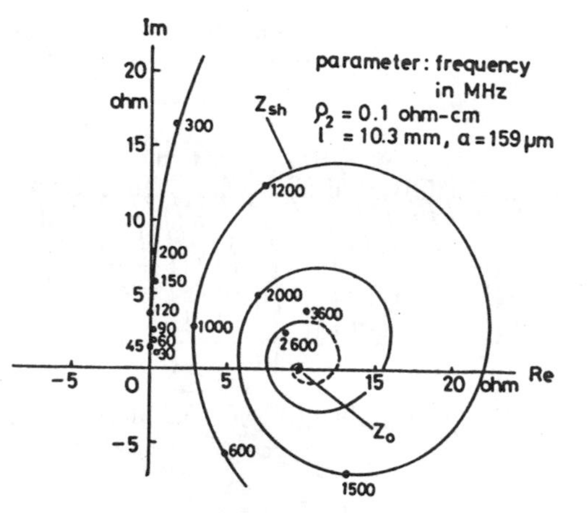

Fig. 9. Example of frequency locus of Z_{sh} for the sample with a moderate resistivity of 0.1 Ω·cm.

relaxation behavior given in (9). The dotted circle in Fig. 8(b)[4] gives the high-frequency limit of the loci of Z_{sh} and Z_{op}. As seen from Fig. 8 (a) and (b), the forms of the calculated and measured input impedance loci are very similar in both of the short- and open-circuit cases.[5] Furthermore, since the theoretical value of the dielectric relaxation frequency, f_e, is 1800 MHz for $\rho_2 = 85\ \Omega \cdot$ cm and $\epsilon_{Si} = 12$, the correspondences of parameter values appear reasonbly good between both figures.

Fig. 9 is an example of the short-circuit impedance locus[6] of the sample with an intermediate resistivity of $1.0 \times 10^{-1}\ \Omega \cdot$cm. This value of resistivity is roughly equal to ρ_c given by (8) (ρ_c is $1.5 \times 10^{-1}\ \Omega \cdot$cm in this case). Theoretically, for $\rho_2 = \rho_c$, a locus that rotates spirally around and converges to a point on the real axis corresponding to Z_0 is expected. The solid curve in the figure is drawn by taking this theoretical behavior into account. It should be noted that the electrical length of the line is much larger than the physical length in the figure; the former obtained from the resonant and anti-resonant frequencies is about 170 mm, while the latter is only 10.3 mm. This implies a small propagation velocity of about 1/16.5 of light velocity in vacuum, and proves the propagation of the slow-wave mode.

C. Characteristic Impedance and Propagation Constant

Fig. 10 (a)–(d), shows the measured frequency dependences[7] of the real and imaginary part of the characteristic impedance, and the phase and attenuation constants, respectively, for various resistivities of the

[4] The radius of the circle is given by $(1/2)(\coth \alpha_\infty l - \tanh \alpha_\infty l)$ and the center, by $(1/2)(\tanh \alpha_\infty l + \coth \alpha_\infty l) + j0$.

[5] Since $\alpha_\infty = 6.4 \times 10^{-2}$ Np/mm for $\rho_2 = 85\ \Omega \cdot$cm from Table II, the line length that satisfies $\gamma_\infty l = 1$ Np is 16 mm, and thus the condition for the line length is nearly the same in both figures.

[6] Since Z_0 was real almost over the whole measured frequency range, the open-circuit locus shows nearly the same behavior with Fig. 9, if represented in the form of input admittance.

[7] Small discontinuity reactances at the sample connection point were evaluated experimentally and corrected. Because of low-impedance measurement, only the parasitic series-inductance, which was typically of the order of 0.1 nH, was effective. It was found that it had an appreciable effect only on Im (Z_0) and effects on the other quantities was below 2 percent.

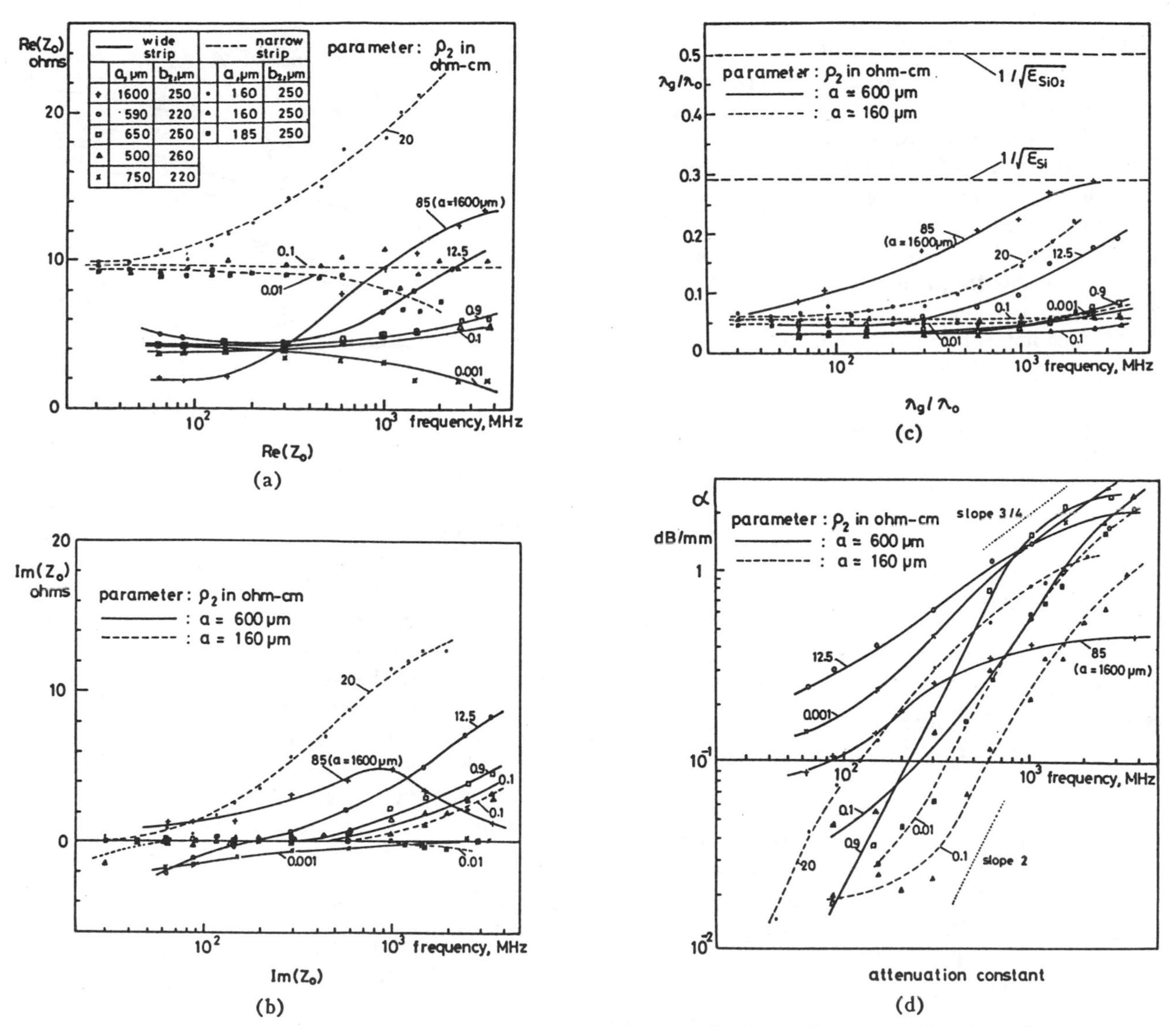

Fig. 10. Measured frequency characteristics of (a) real part of Z_0, (b) imaginary part of Z_0, (c) λ_g/λ_0, and (d) attenuation constant.

Si substrate. The linewidth is about 600 μm or 160 μm except the sample of 85 $\Omega\cdot$cm. Precise line dimensions are tabulated in Fig. 10(a). The phase constants are expressed in terms of the normalized wavelength, λ_g/λ_0, which is equal to the propagation velocity divided by the light velocity in vacuum.

As seen from the figures, the propagation velocity is constant and very slow in the lower frequency region, indicating the propagation of the slow-wave mode. The measured velocity is 1/15–1/30 of the light velocity in vacuum, and is essentially smaller than that of TEM propagation in silicon or silicon-dioxide. The upper frequency limit of the slow-wave mode propagation is strongly dependent upon the substrate resistivity, and becomes maximum for $\rho_2 = 10^{-1}\ \Omega\cdot$cm, where it extends into low-gigahertz region. The slow-wave mode boundary points are determined from the measured data by taking the points of a 5-percent rise of λ_g/λ_0 from the low-frequency values, according to the definition of the mode boundary explained in Section II. The points thus obtained are plotted in the resistivity–frequency domain chart shown in Fig. 2. In spite of the large differences in the strip width involved in the experiment, a fairly good agreement is observed between theoretical and experimental mode boundaries. Thus, the resistivity–frequency domain chart based on a somewhat idealized model, serves as a rough measure for the propagation modes along the actual lines. As for the attenuation constant, the measured values of some of the samples increase approximately in proportion to the square of frequency in the middle frequency range; in this range, the attenuation constant is minimum for $\rho_2 = 10^{-1}\ \Omega\cdot$cm, which is close to the characteristic resistivity, ρ_c. This behavior is in agreement with the theory in the wide-strip limit.

Outside the slow-wave mode domain, the measured characteristic impedance and propagation constant show quite different behavior, depending upon whether $\rho_2 > \rho_c$ or $\rho_2 < \rho_c$; for $\rho_2 > \rho_c$, they behave like the theoretical curves shown in Fig. 4, and for $\rho_2 < \rho_c$, like those shown in Fig. 5. It could be concluded from this result that the remaining two modes obtained in the wide-strip limit exist also in the case of finite strip width, and the above mentioned behaviors of the measured data are caused by the mode transitions.

Thus, most part of the measured resistivity–frequency characteristics can be explained qualitatively, and to some extent, quantitatively by the theory in the wide-strip limit. But, several disagreements are also present in the low-frequency region: Re (Z_0) becomes large, Im (Z_0) becomes negative, and the attenuation constant does not exhibit the quadratic dependence on frequency in some of the samples. These disagreements can be explained by the finite resistance of the line conductors ignored in the theoretical analysis, and this explanation has been quantitatively confirmed by the measurement of the sheet resistances of the strip conductors employed in the experiment. The measured dc resistance per unit length of the evaporated strips was $10^{-2}-2\times10^{-1}$ Ω·mm and that of the beam-lead-type ones was $10^{-3}-4\times10^{-1}$ Ω·mm.

IV. Strip Width Dependences of Slow-Wave Mode Properties

As seen in Fig. 10, the propagation constant and characteristic impedance are strongly dependent on the strip width, a, of the line. For example, the measured values of λ_0/λ_g of the slow-wave mode are plotted versus a/b in Fig. 11. Naturally, the idealized model in Section II cannot provide any quantitative information concerning the strip-width dependences of the properties of each mode. On the other hand, exact analysis of the wave propagation along the microstrip configuration is extremely difficult and beyond the scope of the present paper. As an alternative, the effects of strip width on the transmission properties are approximately analyzed in this section, using the quasistatic approach. Only the slow-wave mode is considered, because it covers the practically most important resistivity–frequency domain. Quasistatic analysis of the dielectric quasi-TEM mode has been carried out in [6].

A. Equivalent Circuit

From the physical propagation mechanism and the agreement between theory and experiment obtained in the previous section, it appears reasonable to assume the equivalent circuit shown in Fig. 3(a) also for the slow-wave mode along the microstrip configuration. The effect of the finite strip width is to be included as the fringing effects for the circuit parameters. The effect due to the finite resistance of the line conductors should be taken into account as well.

In order to check the above assumption experimentally, the series impedance, $Z(f)$, and shunt admittance, $Y(f)$, per unit length of the line, have been calculated as functions of frequency from the measured characteristic impedance and propagation constant through the relations, $Z(f)=Z_0\gamma$ and $Y(f)=Z_0^{-1}\gamma$. Then, their behavior versus frequency has been studied in detail. The result was precisely what was expected; the equivalent circuit in Fig. 3 (a) has been found correct with sufficient accuracies within the slow-wave mode domain in the resistivity–frequency domain chart and the deviation from it occurs slowly outside that domain. It has been also found that the deviation occurs so slowly that the equivalent circuit can be practically employed outside the domain under the rough limitations of $f \lesssim f_m$ and $f \lesssim f_\delta$. This is not in contradiction with the definition of the mode boundary, since the propagation velocity becomes different from the slow-wave mode value outside the domain, if calculated using that equivalent circuit.

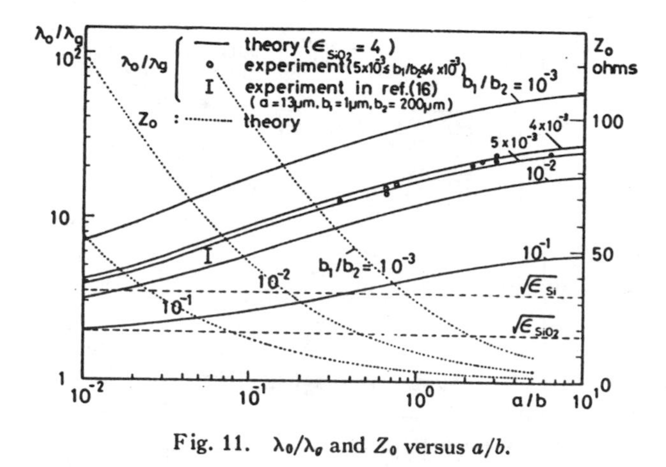

Fig. 11. λ_0/λ_g and Z_0 versus a/b.

B. Circuit Parameters

For the quantitative treatment, the effective height, $b^\dagger$, of the strip should be defined as

$$b^\dagger = az_0(b/a) \tag{11}$$

where $z_0(b/a)$ is the characteristic impedance of a microstrip line normalized to that of free space (120π Ω), and b is the height of a strip of width a, measured from the ground plane. Since $z_0(b/a)$ is always less than b/a, $b^\dagger$ is always smaller than b.

In addition, the effective permittivity, $\epsilon^\dagger$, of a dielectric loading sheet with thickness, b, and relative permittivity, ϵ_s, should be defined as

$$\epsilon^\dagger(b/a, \epsilon_s) = (z_0/z_0')^2 \tag{12}$$

where z_0 is the normalized characteristic impedance of the microstrip line without the sheet and z_0', that of the same line loaded with the sheet. $\epsilon^\dagger$ is always smaller than ϵ_s except for $\epsilon_s=1$.

Consider L and C_1 of the equivalent circuit shown in Fig. 3(a) in the first place. In the quasistatic limit, it can be assumed that electric and magnetic energies are stored independently from each other. Under this assumption, L and C_1 are given by

$$L = \mu_0 \frac{b^\dagger}{a} \qquad C_1 = \epsilon_0\epsilon_{s0}{}^\dagger \frac{a}{b^\dagger} \tag{13}$$

where

$$\epsilon_{s0}^{\dagger} = \epsilon^{\dagger}_{SiO_2} \frac{b^{\dagger}}{b_1^{\dagger}}, \qquad \text{with } \epsilon^{\dagger}_{SiO_2} = \epsilon^{\dagger}(b_1/a, \epsilon_{SiO_2}).$$

These expressions are formally the same as those given in Fig. 3(b). The characteristic impedance (real part) and λ_0/λ_g are obtained as

$$Z_0 = 120\pi \frac{1}{\sqrt{\epsilon_{s0}^{\dagger}}} \frac{b^{\dagger}}{a} \Omega \qquad \lambda_0/\lambda_g = \sqrt{\epsilon_{s0}^{\dagger}}. \tag{14}$$

In (13) and (14), $\epsilon_{s0}^{\dagger}$ can be interpreted as the static value of the effective permittivity of the loading double-layer in the microstrip configuration. Indeed, the capacitance is enhanced by a factor of $\epsilon_{s0}^{\dagger}$ from its vacuum value, $C_0 = \epsilon_0 a/b^{\dagger}$ by the insertion of the double layer.

As for the explicit forms of $z_0(b/a)$ and $\epsilon^{\dagger}(b/a, \epsilon_s)$, various works concerning the standard microstrip lines could be referred to. Among them, the analytical expressions for the fringing effect after Schneider [15] should be adopted here. In the present notation, they are given by

$$\begin{cases} z_0(r) = (1/2\pi) \log (8r + 4r^{-1}), & \text{for } r \equiv b/a \geqq 1 \\ \quad\;\; = [r^{-1} + 2.42 - 0.44r + (1 - r)^6]^{-1}, & \text{for } r \equiv b/a < 1 \end{cases} \tag{15}$$

$$\epsilon^{\dagger}(r, \epsilon_s) = \frac{1 + \epsilon_s}{2} - \frac{1 - \epsilon_s}{2} (1 + 10r)^{-1/2}, \quad \text{with } r \equiv b/a. \tag{16}$$

Fig. 12(a) compares the theoretical values of L and C_1 calculated from (13), (15), and (16) with those experimentally determined. Fig. 11 shows the calculated and measured values of λ_0/λ_g. In Fig. 11, the propagation delay data after Ho and Mullick [16] are also included. In both figures, agreement between theory and experiment is good. Theoretical curves for the characteristic impedance are also given in Fig. 11.

Next, the lossy elements in the equivalent circuit are considered. $R_\delta(f)$ in the circuit represents the loss caused by the longitudinal component of the conduction current in the Si substrate. Evaluation of its value in the microstrip case cannot be made by the present quasistatic approach, and appears to require a more sophisticated method. But it seems reasonable to assume that the loss tangent of the series arm of the equivalent circuit increases in proportion to frequency also in the microstrip configuration, i.e., $R_\delta(f)$ itself is proportional to the square of frequency. This quadratic dependence on frequency has been confirmed experimentally as shown in Fig. 12(b). In the figure, the measured values of $R_\delta(f)$ are plotted versus frequency. These values were determined by subtracting the measured dc resistance per unit length of the experimental strip from the measured total series resistance per unit length given by $\text{Re}(Z_0\gamma)$. The basic assumption is that the variation of the series resistance of the strip conductor itself caused

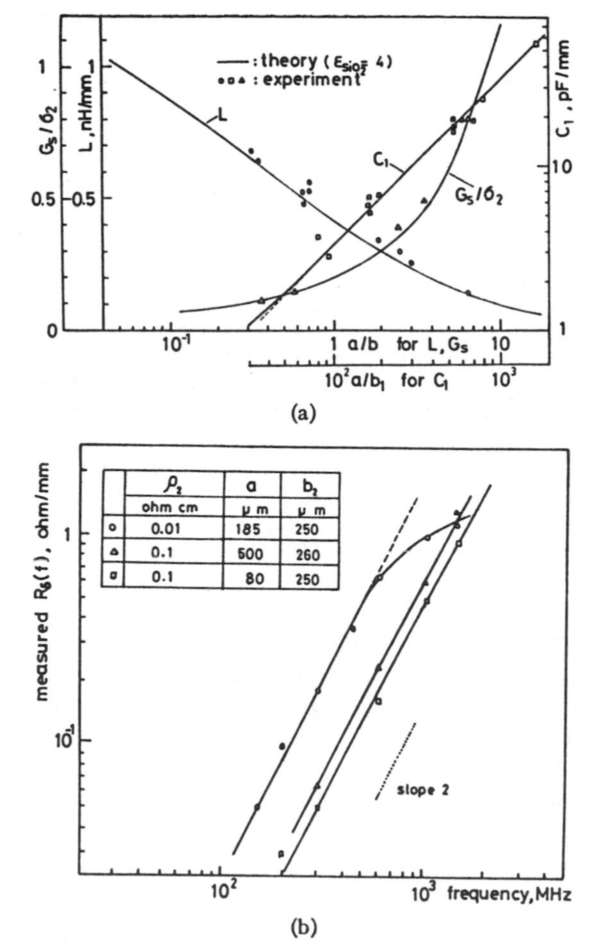

Fig. 12. (a) Theoretical and experimental values of L, C_1, and G_s. (b) Measured frequency dependency of $R_\delta(f)$.

by the skin effect is slow and small as compared with that of $R_\delta(f)$ over the frequency range in concern. To obtain a formal unification, $R_\delta(f)$ could be expressed as follows, introducing the effective longitudinal conductivity, $\sigma_\parallel$:

$$R_\delta(f) = K_\delta \sigma_\parallel f^2 b_2^{\dagger 3}/a, \qquad \text{with } K_\delta \equiv (2\pi\mu_0)^2/3. \tag{17}$$

Equation (17) should be understood as the definition of $\sigma_\parallel$, which can presumably be written as a product of σ_2 and an unknown geometrical factor.

G_s in the equivalent circuit represents the loss caused by the transverse component of the conduction current in the Si substrate. In the microstrip case, the effective complex permittivity of the loading double layer would not conform in general to the Debye-type relaxation with a single relaxation time constant. A shunt arm consisting of C_1 and G_s is merely a single-time-constant approximation for the low-frequency behavior. But, this approximation appears particularly good in the usual situation of the extremely thin SiO$_2$ layer, in which the condition, $b_1 \ll a$ is satisfied. Under this condition, G_s can be evaluated approximately by

$$G_s = \sigma_\perp \frac{a}{b_2^{\dagger}} \tag{18}$$

with

$$\sigma_{\perp} = \sigma_2\left[\frac{1}{2} + \frac{1}{2}\left(1 + 10\frac{b_2}{a}\right)^{-1/2}\right].$$

Here, the effective transverse conductivity, $\sigma_{\perp}$, is derived from (16), taking a proper limit. Fig. 12(a) compares the values of G_s calculated from (18) with those determined from the experimental data.

Equations (13), (17), and (18) have the same forms as the expressions given in Fig. 3(b). It follows from this that there exists also in the case of microstrip configuration, a certain resistivity value, $\rho_c^{\dagger}$, that maximizes the upper frequency limit of the slow-wave propagation and minimizes attenuation. $\rho_c^{\dagger}$ is determined by the condition

$$\frac{R_\delta(f)}{2\pi fL} = \frac{2\pi fC_1}{G_s}. \qquad (19)$$

This condition is reminiscent of the distortionless condition of the $LRCG$ distributed line, given by $R/(2\pi fL) = G/(2\pi fC)$. Though $\rho_c^{\dagger}$ would be an extremely complicated function of the geometrical parameters of the line, the experimental results indicate that its value does not largely differ from ρ_c determined by the simple formula of (8).

In conclusion, the equivalent circuit in Fig. 3(a) together with the circuit parameters given in (13), (17), and (18) represents sufficiently well the main features of the slow-wave mode along the microstrip line on the Si–SiO_2 system. The good agreement between the theory based on the quasistatic approach and the experiment proves the validity of the expectation that the slow-wave mode itself is the lowest order quasistatic mode whose main properties are determined in the first approximation by the quasistatic considerations. It should be noted, however, that this equivalent circuit is merely a low-frequency approximation and would become inadequate when the detailed dispersive behaviors of the mode at higher frequencies are concerned. Further investigations are required on this point as well as on the evaluation of $R_\delta(f)$ or $\sigma_{\parallel}$ in (17).

V. Conclusions

The fundamental transmission properties of the microstrip line formed on a surface-passivated silicon substrate are studied in the paper. An approximate analysis based on the parallel-plate waveguide model is presented, and is compared with the detailed experimental data. Existence of three fundamental modes is concluded, and the resistivity–frequency domain chart is shown to be useful. In particular, the slow-wave mode is found to propagate within the practically important resistivity–frequency range. The effect of the finite strip width is approximately analyzed for the case of the slow-wave mode, using the quasistatic approach.

One possible application of the slow-wave mode is a new type of slow-wave circuit for the solid-state traveling-wave amplifier, in which this mode is transmitted along planar periodic structure [17]. Furthermore, a substrate with an extremely large permittivity and fairly small loss could be obtained by constructing a multilayer structure in which semiconducting and dielectric layers are formed alternately with their thickness ratio being kept large. This is because the skin-effect loss could be reduced a great deal in such a structure. This type of substrate would possibly serve to reduce the size of the microwave integrated circuits in low gigahertz region, or enable one to construct distributed-type circuits in substantially lower frequency region.

Although the properties of the microstrip line of this type have been clarified qualitatively and, to some extent, quantitatively in the paper, further investigations are required concerning the detailed quantitative behaviors of the line, involving the strip-width dependences of the mode boundaries, detailed dispersion behaviors of the modes and mode transitions, evaluation of $R_\delta(f)$ of the slow-wave equivalent circuit, etc. In order to clarify these points, it appears necessary to treat the microstrip configuration in a more rigorous manner.

Appendix

Derivation of Approximate Solutions of (1a) and (1b) ($\epsilon_{Si}b_1 \ll \epsilon_{SiO_2}b_2$)

When $f \gg f_e$ or $f \ll f_\delta$, $|\gamma_1 b_1| \ll 1$ and $|\gamma_2 b_2| \ll 1$ can be assumed.

The approximate solution for this case is given by

$$\left[\begin{aligned} \gamma_1^2 &= -\omega^2\epsilon_1^* b_2 \frac{\epsilon_1^*\mu_1^* - \epsilon_2^*\mu_2^*}{\epsilon_1^* b_2 + \epsilon_2^* b_1} \\ \gamma_2^2 &= -\omega^2\epsilon_2^* b_1 \frac{\epsilon_2^*\mu_2^* - \epsilon_1^*\mu_1^*}{\epsilon_1^* b_2 + \epsilon_2^* b_1} \\ \gamma^2 &= -\omega^2\epsilon_1^*\epsilon_2^* \frac{\mu_1^* b_1 + \mu_2^* b_2}{\epsilon_1^* b_2 + \epsilon_2^* b_1}. \end{aligned}\right. \qquad (20)$$

From (20), one obtains for the dielectric quasi-TEM mode ($f \gg f_e$)

$$\gamma \simeq j2\pi f\sqrt{\epsilon_0\epsilon_{Si}\mu_0} + \pi f_e\sqrt{\epsilon_0\epsilon_{Si}\mu_0} \qquad (21)$$

and for the slow-wave mode ($f \ll f_e$ and $f \ll f_\delta$),

$$\left[\begin{aligned} &\gamma_1 = 2\pi f\sqrt{\epsilon_0\epsilon_{s0}\mu_0}\,[1 - j\tfrac{1}{2}(f/f_s)] \\ &\gamma_2 \simeq \gamma_m[1 - j\tfrac{1}{2}(f/f_s)] \\ &\gamma \simeq j2\pi f\sqrt{\epsilon_0\epsilon_{s0}\mu_0}\,[1 - j\tfrac{1}{2}(f/f_s)] \\ &(\gamma_m \equiv \sqrt{j\omega\mu_0\sigma_2}). \end{aligned}\right. \qquad (22)$$

When the skin effect becomes appreciable in the Si layer, $\gamma_2 b_2 \simeq \gamma_m b_2$ gets large, and (22) is no longer valid.

Setting $\gamma_2 \equiv \gamma_m(1-\Delta_m)$ and assuming $|\gamma_1 b_1| \ll 1$, one can evaluate Δ_m by returning to (1a) and (1b). The result is

$$\Delta_m = -j\tfrac{1}{2}(f/f_s) \cdot \frac{(b_1/b_2) + (\gamma_m b_2)^{-1} \tanh \gamma_m b_2}{1 + j(f/f_s)[1 + (\gamma_m b_2)^{-1} \tanh \gamma_m b_2 - \tanh^2 \gamma_m b_2]}. \tag{23}$$

Since $|(\gamma_m b_2)^{-1} \tanh \gamma_m b_2| \leqq 1$, $\gamma_2 = \gamma_m$ is a good approximation for $f \ll f_s$.

For the slow-wave mode, one notes that $(\gamma_m b_2)^{-1} \tanh \gamma_m b_2 \simeq 1 - \frac{1}{3}(\gamma_m b_2)^2 = 1 - j\frac{2}{3}(f/f_\delta)$. After substituting γ_2 into (1a) and retaining the first-order terms in (f/f_s) and (f/f_δ), (22) is modified as

$$\begin{bmatrix} \gamma_1 \simeq 2\pi f\sqrt{\epsilon_0 \epsilon_{s0} \mu_0}\,[1 - j\tfrac{1}{2}(f/f_0)] \\ \gamma_2 \simeq \gamma_m[1 - j\tfrac{1}{2}(f/f_s)] \\ \gamma \simeq j2\pi f\sqrt{\epsilon_0 \epsilon_{s0} \mu_0}\,[1 - j\tfrac{1}{2}(f/f_0)]. \end{bmatrix} \tag{24}$$

For the skin effect mode, one sets $\tanh \gamma_m b_2 = 1 + j0$ in (23) and obtains

$$\begin{bmatrix} \gamma_1 \simeq \dfrac{1}{\delta}\sqrt[4]{2(f/f_{s\delta})} \qquad \gamma_2 \simeq \gamma_m\left[1 - \dfrac{1+j}{4}\sqrt{f/f_{s\delta}}\right] \\ \gamma \simeq j2\pi f\sqrt{\epsilon_0 \epsilon_{SiO_2} \mu_0}\sqrt{\left(1 + \dfrac{\delta}{2b_1}\right) - j\left(\dfrac{\delta}{2b_1}\right)}. \end{bmatrix}$$

Acknowledgment

The authors wish to thank the members of the technical staff, IC division, Nippon Electric Co. Ltd., for their useful discussions and for provision of the experimental samples.

References

[1] H. A. Wheeler, "Transmission-line properties of parallel strips separated by dielectric sheet," *IEEE Trans. Microwave Theory Tech.*, vol. MTT-13, Mar. 1965, pp. 172–185.

[2] M. V. Schneider, "Computation of impedance and attenuation of TEM-lines by finite difference methods," *IEEE Trans. Microwave Theory Tech.*, vol. MTT-13, Nov. 1965, pp. 793–800.

[3] E. Yamashita and R. Mittra, "Variational method for the analysis of microstrip lines," *IEEE Trans. Microwave Theory Tech.*, vol. MTT-16, Apr. 1968, pp. 251–256.

[4] P. Silvester, "TEM wave properties of microstrip transmission lines," *Proc. Inst. Elec. Eng.*, vol. 115, Jan. 1968, pp. 43–48.

[5] T. M. Hyltin, "Microstrip transmission on semiconductor dielectrics," *IEEE Trans. Microwave Theory Tech.*, vol. MTT-13, Nov. 1965, pp. 777–781.

[6] E. Yamashita, "Variational method for the analysis of microstrip-like transmission lines," *IEEE Trans. Microwave Theory Tech.*, vol. MTT-16, Aug. 1968, pp. 529–535.

[7] C. P. Hartwig, D. Massé, and A. P. Pucel, "Frequency dependent behavior of microstrip," *1968 Int. Microwave Symp. Dig.*, Detroit, Mich., pp. 110–116.

[8] G. I. Zysman and D. Varon, "Wave propagation in microstrip transmission lines," *1969 Int. Microwave Symp. Dig.*, Dallas, Tex., pp. 3–9.

[9] H. Guckel, P. A. Brennan, and I. Palócz, "A parallel-plate waveguide approach to microminiaturized, planar transmission lines for integrated circuits," *IEEE Trans. Microwave Theory Tech.*, vol. MTT-15, Aug. 1967, pp. 468–476.

[10] H. Hasegawa, M. Furukawa, and H. Yanai, "Slow wave propagation along a microstrip line on Si-SiO₂ system," *Proc. IEEE* (Lett.), vol. 59, Feb. 1971, pp. 297–299.

[11] R. E. Collin, *Field Theory of Guided Waves.* New York: McGraw-Hill, 1960, pp. 227–228.

[12] A. R. von Hippel, *Dielectrics and Waves.* New York: Wiley, 1954, pp. 228–234.

[13] H. A. Wheeler, "Formulas for the skin effect," *Proc. IRE*, vol. 30, Sept. 1942, pp. 412–424.

[14] H. M. Barlow and A. L. Cullen, "Surface waves," *Proc. Inst. Elec. Eng.*, vol. 100, pt. III, Nov. 1953, pp. 329–347.

[15] M. V. Schneider, "Microstrip lines for microwave integrated circuits," *Bell Syst. Tech. J.*, vol. 48, May 1969, pp. 1421–1444.

[16] I. T. Ho and S. K. Mullick, "Analysis of transmission lines on integrated-circuit chips," *IEEE J. Solid-State Circuits*, vol. SC-2, Dec. 1967, pp. 201–208.

[17] H. Yanai, H. Hasegawa, H. Nomura, and M. Furukawa, "Slow-wave propagation along the transmission line on Si-SiO₂ system," presented at the 8th Int. Conf. on Microwave and Optical Generation and Amplification (MOGA 70), Amsterdam, The Netherlands, Sept. 7–11, 1970.

Slow-Wave Propagation Along Variable Schottky-Contact Microstrip Line

DIETER JÄGER

Abstract—Schottky-contact microstrip lines (SCML) are a special type of transmission line on the semiconducting substrate: the metallic-strip conductor is specially selected to form a rectifying metal–semiconductor transition while the ground plane exhibits an ohmic metallization. Thus the cross section of SCML is similar to that of a Schottky-barrier diode. The resulting voltage-dependent capacitance per unit length causes the nonlinear behavior of such lines.

In this paper a detailed analysis of the slow-wave propagation on SCML is presented, including the effect of metallic losses. Formulas for the propagation constant and characteristic impedance are derived and an equivalent circuit is presented. Conditions for slow-mode behavior are given, particularly taking into account the influence of imperfect conductors and defining the range of many interesting applications. Experimental results performed on Si-SCML are compared with theory.

I. Introduction

DUE to particular applications in microwave integrated circuits, microstrip lines on semiconductor substrates have been thoroughly investigated both in theoretical and experimental works during the last few years. The Schottky-contact microstrip line (SCML) is a special form of microstrip line on a semiconducting substrate: the cross section [Fig. 1(a)] is similar to a Schottky-barrier diode; i.e., the stripline forms a rectifying metal–semiconductor transition to the chip with a large-area ohmic-contact back metallization. At the Schottky-barrier contact a depletion layer arises, the depth of which depends strongly on the applied voltage. Thus the most interesting features of SCML's are caused by this voltage-dependent depletion-layer capacitance per unit length. Two modes of operation may be distinguished: the large-signal behavior, which is characterized by nonlinear wave propagation, and the small-signal properties, which are determined by bias-dependent transmission-line parameters.

The wave propagation on SCML's has been investigated recently, leading to several fundamental results: large-signal operation leads to distributed harmonic frequency generation and the possibility of parametric amplification [1], [2]. Under small-signal conditions a slow-wave propagation occurs, and the propagation constant and characteristic impedance may be changed by an external dc bias [3], [4]. In particular, it has been verified that bias-dependent phase delay gives rise to possible applications of SCML in variable IC microwave components, such as resonators, delay lines, phase shifters, or tunable filters [5]–[7].

To a certain extent the SCML resembles the microstrip line, which serves as the electrical-interconnection pattern in IC technology on MOS or MIS systems where an oxide layer insulates the semiconductor wafer from the metallic conductors. The high-frequency behavior has been investigated by several workers, since the propagation delay imposes a limitation upon signal velocity [8]–[11]. Introducing the voltage-dependent capacitance of the MIS system, a variable (nonlinear) MIS microstrip line results [12], [13]. The fundamental theoretical work on wave propagation along such transmission lines has been done by Guckel *et al.* [8], assuming perfect conductors and a large ratio $r = w/l$ of strip width w to substrate thickness l.

Until now an accurate calculation of the influence of imperfect metallic conductors has been neglected in theoretical analysis. The experimental results, however, show large deviations from theory, especially in the lower slow-mode region [4], [11] which exhibits the most interesting features for possible applications. The efficiency of harmonic-frequency conversion and parametric amplification in nonlinear SCML depends strongly on the metallic losses [1], [2], and the phase delay of the variable SCML is influenced by the additional attenuation. In this way, the influence of the metallic losses has become a central problem in the discussion of possible practical applications.

In this paper, a more detailed analysis of small-signal slow-wave propagation along variable SCML is presented, including the effect of imperfect conductors. The treatment

Manuscript received December 2, 1975; revised February 25, 1976. This work was supported by the Deutsche Forschungsgemeinschaft under Grant Ha 505/6/10.

The author is with the Institute for Applied Physics, University of Münster, 4400 Münster, Germany.

Reprinted from *IEEE Trans. Microwave Theory Tech.*, vol. MTT-24, no. 9, pp. 566–573, September 1976.

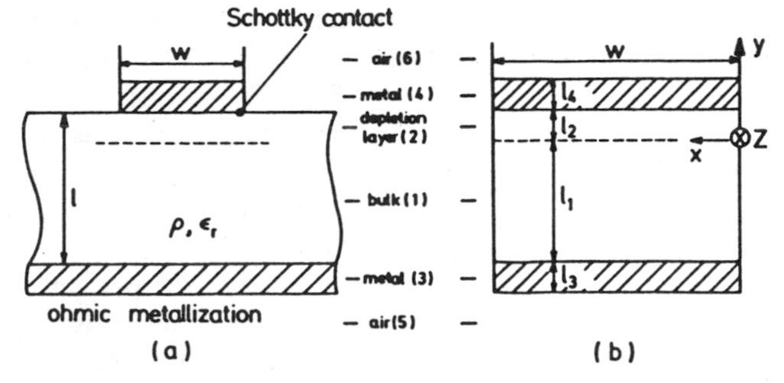

Fig. 1. Schottky-contact microstrip line. (a) Cross section. (b) Parallel-plate waveguide model.

is based upon the work of Guckel *et al.* [8] and confined to the slow-mode region. Formulas are derived in the limit of a parallel-plate waveguide approach and an equivalent circuit is introduced. The slow-mode region is restricted due to metallic losses, and the characteristic frequencies are given. Finally, the theoretical results are compared with experimental values. The influence of fringing fields in the case of real SCML are not considered, since it may be calculated as in [4].

II. Parallel-Plate Waveguide Model

Fig. 1(a) shows the cross section of the SCML. The ohmic back metallization forms the ground plane. The Schottky contact is realized by the strip conductor of width w in intimate contact with the semiconductor of thickness l, doping concentration n_D (here n-type), resistivity ρ, and relative permittivity ε_r. A depletion layer is generated a depth l_2, which depends on the applied reverse bias V according to the well-known equation [14]

$$l_2 = \sqrt{\frac{2\varepsilon_0\varepsilon_r(V + V_D)}{en_D}} \tag{1}$$

where e is the elementary charge, ε_0 the free-space permittivity, and V_D the barrier potential of the Schottky barrier. In the following, l_2 has to be considered bias-dependent and the RF amplitudes to be small enough to allow the use of (1).

Fig. 1(b) shows the parallel-plate waveguide model of the SCML assuming $w \gg l$, so the fringing fields may be neglected. The following six layers have to be considered. The first layer is the semiconducting bulk with thickness l_1 and resistivity ρ. The second layer is represented by the depletion layer, with thickness l_2 and a negligible conductivity at microwave frequencies, since the concentration of free charge carriers is considerably reduced. The two metallic layers are formed by the conductors with thicknesses l_3 and l_4 and conductivities σ_3 and σ_4. The fifth and sixth layers are given by the surrounding air. For such a multilayered transmission line, the fundamental mode of electromagnetic-wave propagation is a TM mode which is a surface wave. Propagation in the z direction is characterized by the complex propagation constant $\gamma = \alpha + j\beta$, according to $\exp(-\gamma z)$ and the time factor $\exp(j\omega t)$. By using the Maxwellian equations the field components of the ith layer are given by

$$H_{xi}(y) = a_{+i} \exp(\gamma_i y) + a_{-i} \exp(-\gamma_i y)$$

$$E_{yi}(y) = -\frac{\gamma}{j\omega\varepsilon_0\varepsilon_i} H_{xi}(y),$$

$$E_{zi}(y) = -\frac{\gamma_i}{j\omega\varepsilon_0\varepsilon_i}(a_{+i}\exp(\gamma_i y) - a_{-i}\exp(-\gamma_i y)),$$

$$i = 1,\cdots,6 \tag{2}$$

where the $a_{\pm i}$ are complex amplitudes, γ_i denotes the transverse propagation constants in the y direction, and $\varepsilon_i = \varepsilon_{ri} - j\sigma_i/(\omega\varepsilon_0)$, the complex permittivity of the ith layer. The separation conditions are formulated as follows:

$$\gamma_i^2 + \gamma^2 = -k_0^2\varepsilon_i, \qquad i = 1,\cdots,6 \tag{3}$$

where $k_0^2 = \omega/c$, with $c = 1/\sqrt{\mu_0\varepsilon_0}$ the free-space speed of light. At the boundaries $y = 0,\ l_2,\ l_2 + l_4,\ -l_1,\ -(l_1 + l_3)$ the electromagnetic continuity conditions must be satisfied, which, together with (3), gives

$$\frac{\gamma_1}{j\omega\varepsilon_0\varepsilon_1}\tanh\gamma_1 l_1 + \frac{\gamma_2}{j\omega\varepsilon_0\varepsilon_2}\tanh\gamma_2 l_2 + M(\gamma_1,\cdots,\gamma_6) = 0. \tag{4}$$

In this equation the quantity M signifies the "metal function"

$$M = \sum_{i=3}^{6} Z_i t_i + \sum_{\substack{(i,j,k)=1 \\ n_i<n_j<n_k}}^{6} \frac{Z_i Z_k}{Z_j} t_i t_j t_k + \sum_{\substack{(i,j,k,l,m)=1 \\ n_i<n_j<n_k<n_l<n_m}}^{6} \frac{Z_i Z_k Z_m}{Z_j Z_l} t_i t_j t_k t_l t_m \tag{5}$$

with $t_i = \tanh\gamma_i l_i$ for $i = 1,\cdots,4$, $t_5 = t_6 = 1$, and $n_i = i(-1)^i$. Z_i denotes the characteristic field impedance

$$Z_i = \frac{\gamma_i}{j\omega\varepsilon_0\varepsilon_i}, \qquad i = 1,\cdots,6. \tag{6}$$

The summation in (5) has to be done as follows [10]. In the second sum such triplets (i,j,k) are selected from the numbers $1,\cdots,6$, for which $n_i < n_j < n_k$ is fulfilled. The twenty terms are added up. The third sum is handled in a similar manner, yielding six terms.

The eigenvalue equation (4) is the same as in [8], except for the term M which vanishes in the case of perfect conductors. Thus, by use of the separation conditions, (4) may be rewritten as

$$\gamma^2 = \frac{-k_0^2\left[\frac{1}{\gamma_1}\tanh\gamma_1 l_1 + \frac{1}{\gamma_2}\tanh\gamma_2 l_2\right] + j\omega\varepsilon_0 M}{\frac{1}{\gamma_1\varepsilon_1}\tanh\gamma_1 l_1 + \frac{1}{\gamma_2\varepsilon_2}\tanh\gamma_2 l_2}. \tag{7}$$

In regard to the numerical solution of (4), the transverse propagation constants γ_i $(i = 2,\cdots,5)$ are expressed in terms of γ_1 with the help of (3). The zeros are then found by Newton's iteration procedure. The results are illustrated in Fig. 2(a) and (b) for attenuation constant α and normalized wavelength λ_0/λ, respectively, where λ_0 denotes the

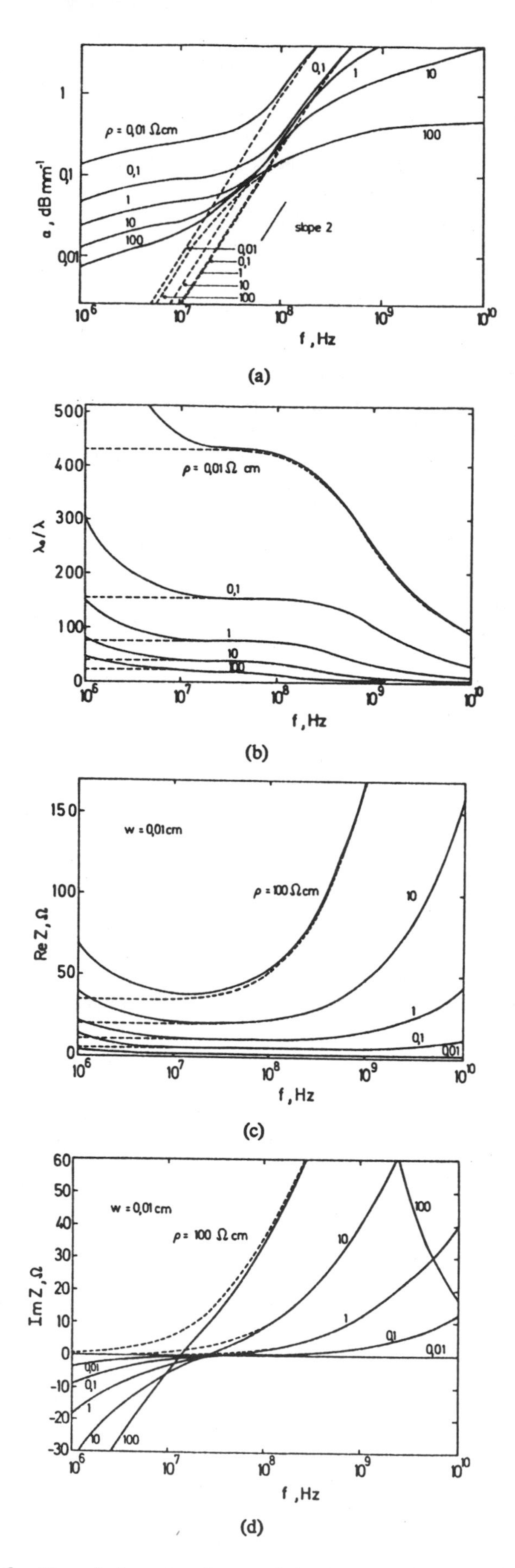

Fig. 2. Numerical results of transmission-line parameters. (a) Attenuation constant. (b) Slowing factor. (c) Real part. (d) Imaginary part of characteristic impedance. n–Si substrate, $l = 0.022$ cm, $l_3 = l_4 = 5\ \mu$m, $V + V_D = 0.8$ V, $\epsilon_r = 11.7$. Parameter is the resistivity of Si. ——: imperfect conductors with $\sigma_m = 3.5 \cdot 10^5\ \Omega^{-1}\ \text{cm}^{-1}$. - - - -: perfect conductors.

free-space wavelength. The semiconductor resistivity is a parameter. The dielectric constant and mobility are those of the Si substrate, needed to calculate the required values of the doping concentration in (1). A fixed value $V + V_D = 0.8$ V is chosen as the Schottky-contact potential corresponding to a typical barrier potential $V_D = 0.5$ V of n-type Si-Schottky barriers and a reverse bias $V = 0.3$ V. Thickness and conductivity of the metallic layers are common values for microminiaturized Al conductors. The dotted lines show the case where $M = 0$ for perfect conductors. Assuming lossless conductors, the slow-mode region is defined by a frequency-independent phase velocity v, which is very small as compared with c. In Fig. 2(b), $\lambda_0/\lambda > 400$ for the $\rho = 0.01\text{-}\Omega \cdot \text{cm}$ substrate. Within this frequency range the attenuation would be proportional to the square of the frequency; cf. Fig. 2(a). By this definition the slow-mode region is confined by an upper frequency limit only (for $M = 0$).

As can be seen from Fig. 2, the transmission-line parameters are strongly influenced by the metallic losses in the lower frequency range. The attenuation constant is considerably enhanced and the phase velocity exhibits dispersion. It may be concluded therefore that the slow-mode region as previously defined must be restricted by a lower frequency limit.

Equation (7) may be interpreted on the basis of transmission-line theory [5]. Then the numerator represents the series impedance W' and the denominator the shunt admittance Y' per unit length of a T-equivalent circuit. Thus the characteristic impedance Z of the SCML can be evaluated from

$$Z = (W'/Y')^{1/2} \tag{8}$$

in a way similar to that previously shown.

The numerical results of Re Z and Im Z are plotted in Fig. 2(c) and (d), respectively. Again the transmission-line parameters are strongly influenced by imperfect conductors within the lower frequency range: Re Z becomes frequency dependent and Im Z negative. In the case of lossless conductors, the slow-mode region is characterized by frequency independent Re Z.

III. Influence of Imperfect Conductors

In the following some approximate expressions for the transmission-line parameters under slow-mode conditions are derived. In particular, the term M in (7) has to be calculated. Since it is difficult to make approximations from (5) with the help of the previous treatment, an analogous procedure, based upon the transverse resonant method [15] which has already been employed in [8], [10], is used.

In the six-layer configuration of Fig. 1(b), the characteristic impedance $Z_0 = 377\ \Omega$ of the air layers is first transformed through the metallic layers by common transmission-line theory. This yields the y-directed field impedances Z_t at the boundaries $y = -l_1$ and $y = l_2$. In the following, for simplicity, conductors identical with $l_3 = l_4 = l_m$ and $\sigma_3 = \sigma_4 = \sigma_m$ are assumed. If $l_m \gtrsim 0.1$

μm, Z_t is approximately

$$Z_t(y = -l_1, l_2) = Z_m \coth \gamma_m l_m \tag{9}$$

with

$$\gamma_m = \frac{1+j}{\delta_m}$$

and

$$Z_m = \frac{1+j}{\delta_m \sigma_m} \tag{10}$$

where $\delta_m = [2/(\mu_0 \sigma_m \omega)]^{1/2}$ is the skin depth of the metallic layers. Equation (9) is a reasonably good approximation for typical values of σ_m. In this case, the influence of the air layers has disappeared and a four-layered configuration results. Now Z_t is transformed through the depletion layer and the bulk material to the plane $y = 0$. Then the resonance condition finally yields

$$\sum_{i=1}^{2} Z_i \tanh \gamma_i l_i + Z_m \coth \gamma_m l_m (2 - \tanh^2 \gamma_1 l_1) = 0 \tag{11}$$

if

1) $|Z_2| \gg |Z_m \coth \gamma_m l_m \tanh \gamma_2 l_2|$;
2) $l_m \gtrsim 0.1$ μm;
3) $\omega < \omega_s = \dfrac{2}{\mu_0 \sigma l_1^2}$;
4) $\sigma l_1 \lesssim 0.1 \sigma_m \min(l_m, \delta_m)$. (12)

The first and second conditions are satisfied in the case of the typical SCML for microelectronic applications. The third and fourth conditions ensure that the longitudinal component of the current within the semiconductor is less than the current through the strip conductor. ω_s is the frequency at which l_1 equals the skin depth of the semiconductor.

Equation (11) yields the desired expression for M of (4)

$$M = W_m (2 - \tanh^2 \gamma_1 l_1) \tag{13}$$

with

$$W_m = Z_m \coth \gamma_m l_m = \begin{cases} \dfrac{1}{\sigma_m l_m} + j \dfrac{l_m \mu_0 \omega}{3}, & \text{for } \delta_m \gtrsim 2 l_m \\ \dfrac{1+j}{\sigma_m \delta_m}, & \text{for } 2\delta_m \lesssim l_m. \end{cases}$$

The slow-mode region may be defined for frequencies [4]

$$\omega \lesssim 0.3\omega_c = 0.3\sqrt{3}\,(\omega_p^{-1} + \tfrac{2}{3}\omega_s^{-1})^{-1} \tag{14}$$

with

$$\omega_p = \omega_d \frac{l_2}{l_1}$$

and

$$\omega_d = \frac{\sigma}{\varepsilon_r \varepsilon_0}. \tag{15}$$

ω_d designates the dielectric relaxation frequency and ω_p the characteristic frequency limit for interfacial polarization

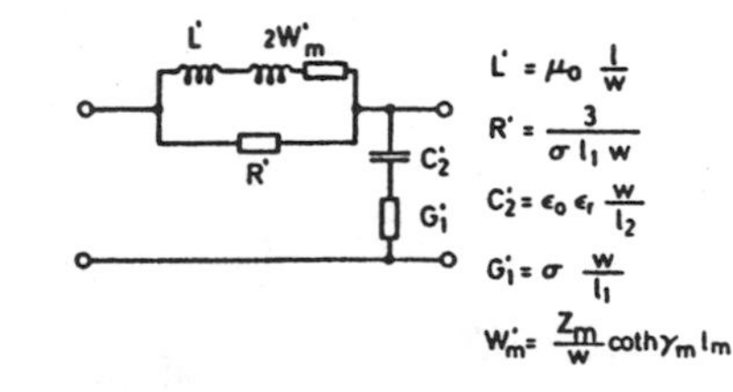

Fig. 3. Equivalent circuit and circuit parameters.

of the two-layered semiconductor substrate. Within the slow-mode region of (14) the hyperbolic functions in (7) may be expanded, assuming $l_1 \gg l_2$, [5], [8]

$$\gamma^2 = \frac{j\omega\mu_0 l(1 - W_m \sigma l_1) + \frac{1}{3}\omega^2 \mu_0^2 \sigma l_1^3 (1 - 2W_m \sigma l_1) + 2W_m}{l_1 \sigma^{-1} + (j\omega\varepsilon_0\varepsilon_r)^{-1} l_2}. \tag{16}$$

Finally, from the viewpoint of transmission-line theory, this equation yields the equivalent circuit in Fig. 3 and the following approximations, if $|W_m \sigma l_1| \ll 1$, is used:

$$\gamma = jk_0 \sqrt{\varepsilon_r \frac{l}{l_2}} \left[1 - j\left(\sqrt{3}\frac{\omega}{\omega_c} + \frac{2W_m}{\omega\mu_0 l}\right)\right]^{1/2} \tag{17}$$

$$Z = \frac{Z_0}{w} \sqrt{\frac{l l_2}{\varepsilon_r}} \left[1 + j\left(\frac{\omega}{\omega_p} - \frac{2}{3}\frac{\omega}{\omega_s} - \frac{2W_m}{\omega\mu_0 l}\right)\right]^{1/2}. \tag{18}$$

L' of Fig. 3 represents the common inductance per unit length of the parallel-plate waveguide and $2W_m'$ the influence of the two metallic conductors. The resistance R' per unit length accounts for longitudinal losses in the semiconductor due to the electric-field component in the direction of propagation. The effective height $l_1/3$ is caused by a current density not being constant in the transversal plane: from (2), (3), and (17) it can be verified that E_{z1} varies linearly with y [5], [11], a sufficiently good approximation under slow-mode conditions. C_2' is identical with the depletion-layer capacitance and G_1' with the semiconductor bulk conductance per unit length. A more general equivalent circuit is obtained if L' is replaced by $L'(1 - W_m \sigma l_1)$, (16), provided that the longitudinal current through the semiconductor is not neglected in comparison with that through the adjacent metallic ground plane. Moreover, the parallel circuit of L' and R' in Fig. 3 is in accordance with the physical mechanism of current transport along such layered lines. Thus it is evident that the attenuation due to series losses in the semiconductor is proportional to the square of frequency, a fact that gave rise to much confusion in the literature [8], [11].

The desired slow-wave properties of the SCML may now be deduced from (17) and (18), if W_m may be neglected

$$\alpha \sim \omega^2, \qquad v = c\left[\frac{2\varepsilon_0 (V + V_D)}{\varepsilon_r l^2 e n_D}\right]^{1/4} \tag{19}$$

$$\text{Re}\, Z = \frac{Z_0}{w} \sqrt{\frac{1}{\varepsilon_r}} \left[\frac{2\varepsilon_r \varepsilon_0 (V + V_D)}{e n_D}\right]^{1/4}. \tag{20}$$

The attenuation becomes a minimum for

$$\sigma = \sigma_{\min} = \left(\frac{9\varepsilon_0 \varepsilon_r}{2\mu_0^2 l_1^2 (V + V_D)\mu_n}\right)^{1/3}. \tag{21}$$

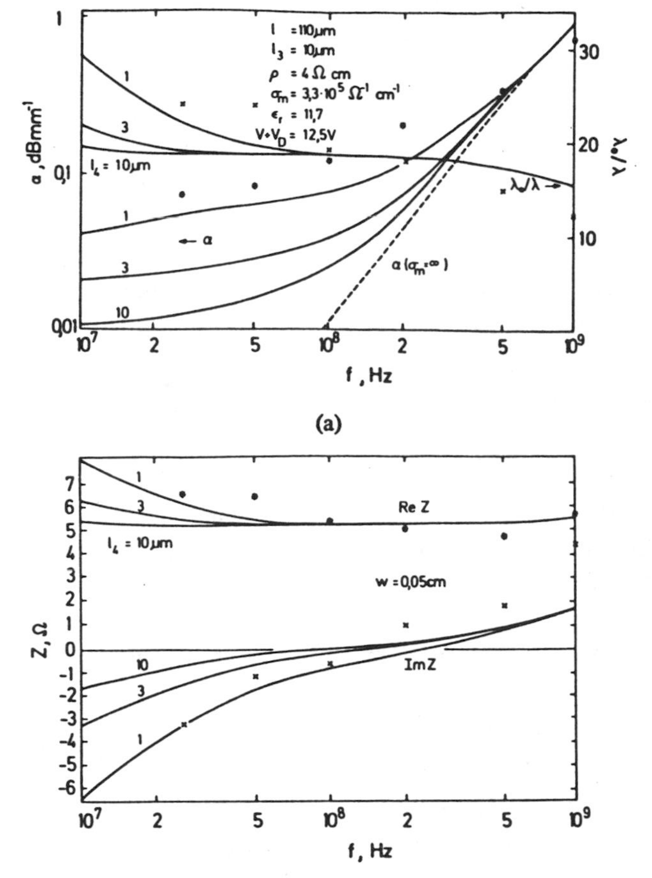

Fig. 4. Transmission-line parameters versus frequency; parameter is the strip-conductor thickness. (a) Attenuation constant and slowing factor. (b) Characteristic impedance, experimental points of SCML-4 [4].

Altogether, ω_c gets a maximum value and Im $Z \approx 0$; cf. Fig. 2. These are the most attractive features of the SCML: the phase velocity v is very small compared with c, and v and Re Z are voltage dependent and exhibit no dispersion.

Fig. 4 shows detailed plots of transmission-line parameters influenced by the finite metallic conductivity calculated as in Section II. As can be seen, the slow-mode properties, as defined by (19) and (20), can only be obtained for frequencies $\omega_{m1} \lesssim \omega \lesssim 0.3\omega_c$. This lower frequency limit ω_{m1} may be evaluated from (17) and (18)

$$\omega_{m1} \approx \begin{cases} \left[\dfrac{9}{\sigma_m \mu_0 l l_m} \omega_c\right]^{1/2}, & \text{for } \delta_m \gtrsim 2l_m \\ \left[\dfrac{40}{\sigma_m \mu_0 l^2} \omega_c^2\right]^{1/3}, & \text{for } l_m \gtrsim 2\delta_m. \end{cases} \quad (22)$$

For the practical example $f_c = 1$ GHz, $l = 0.2$ mm, and $\sigma_m = 3.3 \cdot 10^5\ \Omega^{-1}\ \text{cm}^{-1}$, one obtains $f_{m1} \approx 150$ MHz if $l_m \gtrsim 14\ \mu$m so that $l_m > 2\delta_m$. From (17) and (18) a second lower frequency limit may be derived so that for $\omega_{m2} \lesssim \omega \lesssim 0.3\omega_c$ only v and Re Z exhibit slow-mode properties

$$\omega_{m2} = \begin{cases} \omega_{m1}^2/\omega_c, & \text{for } \delta_m \gtrsim 2l_m \\ \omega_{m1}^3/\omega_c^2, & \text{for } l_m \gtrsim 2\delta_m. \end{cases} \quad (23)$$

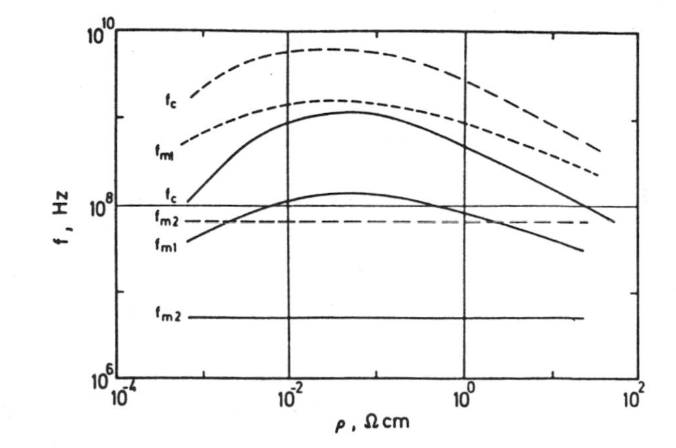

Fig. 5. Frequency limits for slow-mode region versus resistivity of n–Si: $V + V_D = 0.5$ V and $\sigma_m = 3.3 \cdot 10^5\ \Omega^{-1}\ \text{cm}^{-1}$. Parameter is the substrate thickness: —— $l = 0.02$ cm; ---- $l = 0.005$ cm.

For the above-mentioned example, $f_{m2} \approx 5$ MHz if now $l_m \gtrsim 80\ \mu$m.

In Fig. 5 the three frequency limits are given as a function of semiconductor resistivity, when $l_m \geq 2\delta_m$ is assumed. The significance of imperfect conductors is evident. A substrate thickness $l = 50\ \mu$m, for example, results in $f_{m1} \gtrsim 0.3 f_c$ so that the slow-mode region has nearly disappeared.

IV. Experimental Study

In the following, a quantitative comparison is carried out between experimental values and the presented theory; i.e., the equivalent circuit of Fig. 3. Fringing fields will not be considered. This may be done, however, in a quasi-static approximation by introducing spreading factors for the elements of the equivalent circuit [4], similar to W_m from [16].

SCML's with Al and Au Schottky contacts are prepared on n–Si wafers; a sample preparation is described in [4]. The measured reverse voltage-capacitance characteristic agrees well with (1). The values of barrier height as determined from the saturation current density under forward bias correspond with those given in [14], thus confirming the good properties of the rectifying metal–semiconductor transition with diffusion potentials V_D between 0.5 and 0.6 V.

Z and γ are determined by measurements of the open-circuit and short-circuit input impedance of the SCML. An immitance and transfer-function bridge is used. Experimental values of sample SCML-4 from [4], with $\rho = 4\ \Omega \cdot$ cm, $l = 0.11$ mm, $l_3 = 8\ \mu$m, $l_4 = 1.5\ \mu$m, $w = 0.055$ cm, and Al–n–Si Schottky contact, are given in Fig. 4. Bias is 12 V. A comparison with the theoretical curves show only small differences between experimental values and theory. These differences, however, may be traced back to a smaller actual conductivity of the vacuum-evaporated conductors than that of Al bulk material which has been used in the theoretical calculation. Thus the agreement with theory is satisfactory, bearing in mind the large

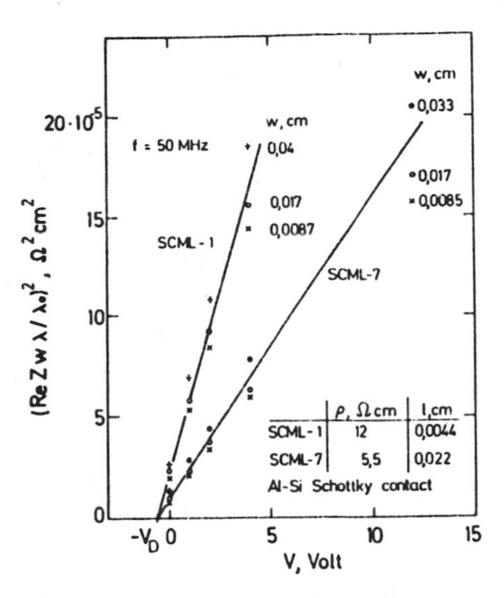

Fig. 6. Experimental determination of n_D and V_D.

number of approximations during analysis and the errors in measurement.

The transmission-line parameters are bias-dependent. Fig. 6 shows the measured reverse-voltage characteristics of six different SCML's within the slow-mode region. From (19) and (20) it is seen that the product

$$(\text{Re } Zw\lambda/\lambda_0)^2 = \frac{2\varepsilon_0 Z_0{}^2}{en_D\varepsilon_r}(V + V_D)$$

is proportional to the sum of reverse bias V and barrier potential V_D of the Schottky barrier. Thus the intercept of the voltage axis yields $V_D = 0.6$ V and the slopes the doping concentrations $n_D(\text{SCML-1}) = 3.8 \cdot 10^{14}$ cm^{-3} and $n_D(\text{SCML-7}) = 8.9 \cdot 10^{14}$ cm^{-3}, in excellent agreement with the resistivities determined by the four-point probe measurements.

In Figs. 7 and 8 the admittance curves of series impedance $W' = \gamma Z$ per unit length and shunt admittance $Y' = \gamma/Z$ are given, respectively. In Fig. 7, W' consists of an inductance in series with a nearly frequency-independent resistance. No bias dependence is observed within experimental error. A quantitative agreement with the equivalent network of Fig. 3 is obtained when L' is calculated from the given parameters on the basis of real microstrip configuration. Re W_m' is then identified with the measured HF resistance of the conductors. It is evident that the internal inductance Im W_m' can be neglected in comparison with the external L' and that the longitudinal losses within the semiconductor may also be ignored: Re $W_m' \gg \omega^2 L'^2/R'$. The values of $-45°$ frequencies determine the lower limit f_{m2}, (23).

The measured shunt admittance Y' (Fig. 8) is that of the Schottky-barrier diode, the depletion-layer capacitance C_2' in series with the bulk conductance G_1'. The values of the 45° frequencies correspond with the upper limit f_c of the slow-mode region. Since l_2 of the SCML is bias-dependent, the frequency markers shift to the origin with increasing bias, resulting in larger values of f_c, (1), (14), and (15). Thus Figs. 7 and 8 show that TM slow-wave propagation along the SCML may be described by the equivalent circuit of Fig. 3 even in the microstrip case.

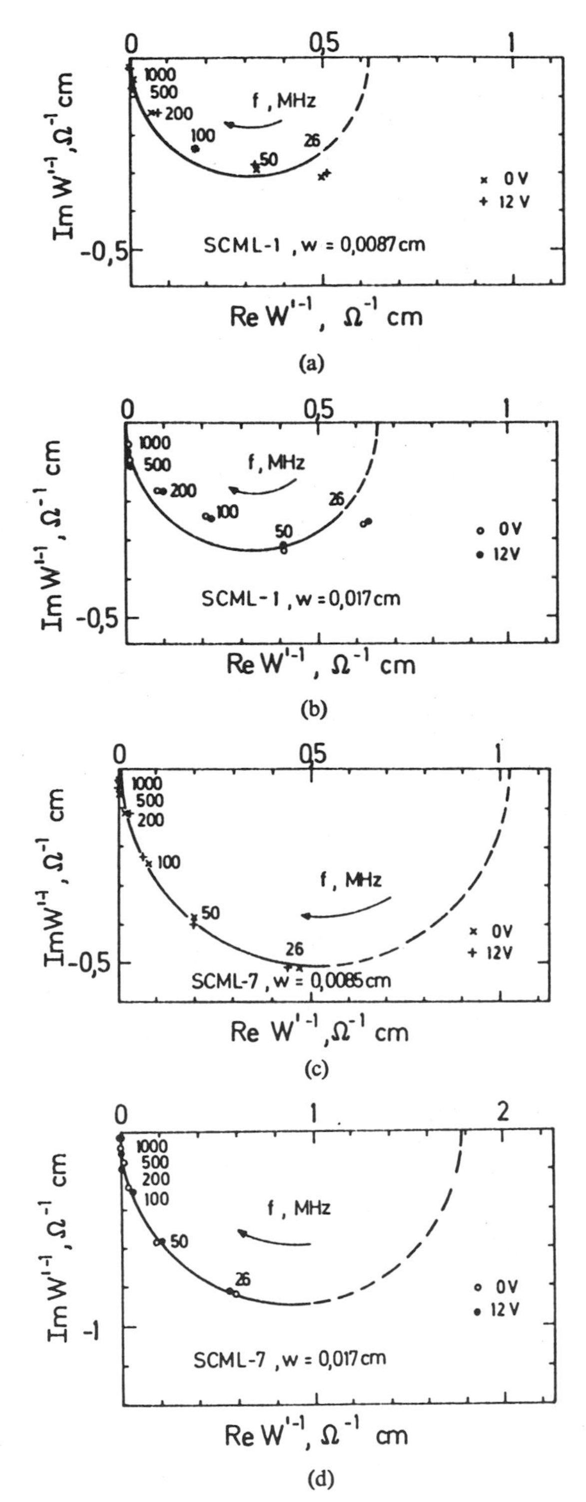

Fig. 7. (a)–(d) Measured admittance curves of series impedance $W' = \gamma Z$; parameter is the bias.

Slow-wave propagation on the Si–SCML's is shown in Fig. 9. Phase velocity and delay time T are calculated from (19). For example, a SCML on an n–Si substrate, of typical thickness 0.02 cm, exhibits phase velocities $v \lesssim 0.01c$ for resistivities $\rho \lesssim 0.1$ Ω · cm. Furthermore, a bias variation from 0 to 12 V changes v by more than a factor of 2. Experimental points are also plotted in Fig. 9.

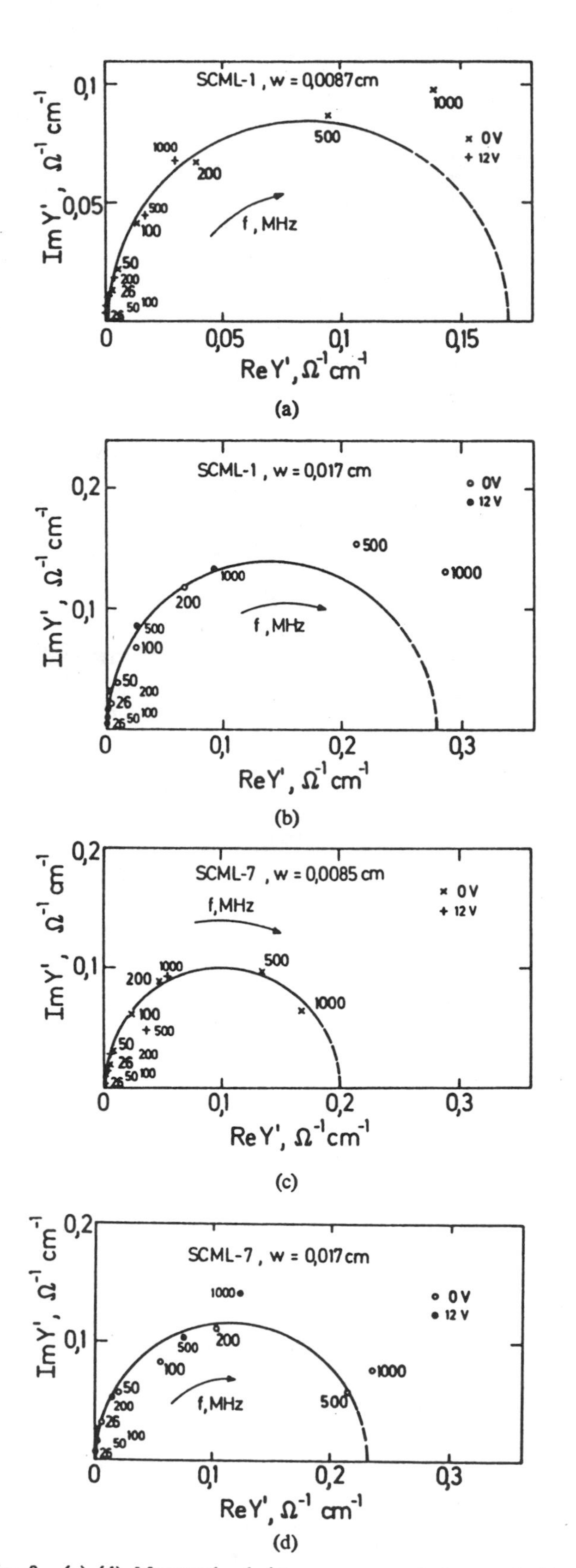

Fig. 8. (a)–(d) Measured admittance curves of shunt admittance $Y' = y/Z$; parameter is the bias.

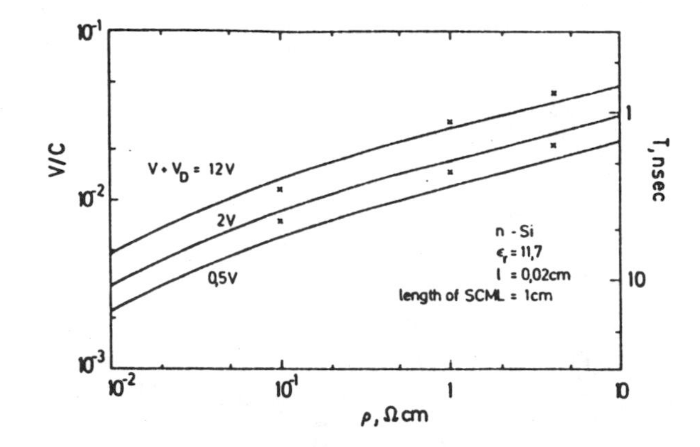

Fig. 9. Phase velocity and delay time of SCML as a function of semiconductor resistivity.

V. Summary and Conclusions

Slow-wave propagation on the SCML's is investigated both theoretically and experimentally, and the frequency limits are given. The upper one results from the frequency limit for interfacial polarization, while the lower one is determined by finite conductivity of the metallic conductors. Formulas for the characteristic impedance and propagation constant are derived together with an equivalent circuit. The analysis is confirmed by experiments performed on the SCML's on the n–Si substrate with Al–Si and Au–Si Schottky contacts.

The fundamental nonlinear properties of the SCML's, together with the very low phase velocity, are the most attractive features for possible applications in IC-microwave components with variable parameters. Such components would be of very small size due to highly reduced guide wavelength. The presented results, however, show clearly the dominating influence of imperfect conductors. This influence cannot be neglected when designing SCML's as well as microstrip lines on MOS systems with optimum performance. Therefore, practical applications are only feasible if optimum stripline parameters are given. These parameters, e.g., substrate thickness, mobility and doping concentration of a suitable semiconductor, Schottky-barrier potential, and strip width, may be calculated from the formulas of Section III. Beyond that, the presented equations and equivalent circuit are also valid if the depletion layer is realized by a p-n-junction, for example, and the theoretical treatment may be employed when similar microstrip lines are analyzed where the cross section equals any other semiconductor diode used in microwave techniques.

Acknowledgment

The author wishes to thank Prof. W. Hampe of the Institute for Applied Physics, University of Münster, for many helpful discussions and his assistance during the course of this work.

References

[1] W. Rabus, "Über die Frequenzvervielfachung längs Schottky-Kontakt-Leitungen," *AEÜ*, vol. 28, pp. 1–11, Jan 1974; also thesis, University of Erlangen-Nürnberg, Germany, 1973.

[2] K. Everszumrode, B. Brockmann, and D. Jäger, "Über den Wirkungsgrad der Frequenzvervielfachung längs Schottky-Kontakt-Leitungen," in preparation.
[3] D. Jäger and W. Rabus, "Bias-dependent phase delay of Schottky contact microstrip line," *Electron. Lett.*, vol. 9, pp. 201–202, May 1973.
[4] D. Jäger, W. Rabus, and W. Eickhoff, "Bias-dependent small-signal parameters of Schottky contact microstrip lines," *Solid-St. Electron.*, vol. 17, pp. 777–783, Aug. 1974.
[5] D. Jäger, "Wellenausbreitung auf Schottky-Kontakt-Leitungen im Kleinsignalbetrieb," thesis, University of Münster, Germany, 1974.
[6] W. Schäfer, J. P. Becker, and D. Jäger, "Variable Tiefpassfilter in Streifenleitungstechnik," in preparation.
[7] D. Jäger and W. Rabus, "Elektrische Leitung auf halbleitendem Substratmaterial," patent pending, 1973.
[8] H. Guckel, P. A. Brennan, and I. Palocz, "A parallel-plate waveguide approach to microminiaturized planar transmission lines for integrated circuits," *IEEE Trans. Microwave Theory Tech.*, vol. MTT-15, pp. 468–476, Aug. 1967.
[9] I. T. Ho and S. K. Mullick, "Analysis of transmission lines on integrated circuit chips," *IEEE J. Solid-St. Circuits*, vol. SC-2, pp. 201–208, Dec. 1967.
[10] H. Hasegawa and H. Yanai, "Characteristics of parallel-plate waveguide filled with a silicon-siliconoxide system," *Electron. and Commun. Jap.*, vol. 53-B, no. 10, pp. 63–73, 1970.
[11] H. Hasegawa, M. Furukawa, and H. Yanai, "Properties of microstrip line on Si–SiO_2 system," *IEEE Trans. Microwave Theory Tech.*, vol. MTT-19, pp. 869–881, Nov. 1971.
[12] J. M. Jaffe, "A high frequency variable delay line," *IEEE Trans. Electron Devices*, vol. ED-19, pp. 1292–1294, Dec. 1972.
[13] U. Günther and E. Voges, "Variable capacitance MIS microstrip lines," *AEÜ*, vol. 27, pp. 131–139, March 1973.
[14] S. M. Sze, *Physics of Semiconductor Devices*. New York: Wiley-Interscience, 1969.
[15] R. E. Collin, *Field Theory of Guided Waves*. New York: McGraw-Hill, 1960.
[16] a) R. A. Pucel, D. J. Masse, and C. P. Hartwig, "Losses in microstrip," *IEEE Trans. Microwave Theory Tech.*, vol. MTT-16, pp. 342–350, June 1968.
b) ——, "Correction to 'Losses in microstrip'," *ibid.* (Corresp.), vol. MTT-16, p. 1064, Dec. 1968.

Characteristics of Metal–Insulator–Semiconductor Coplanar Waveguides for Monolithic Microwave Circuits

ROBERTO SORRENTINO, MEMBER, IEEE, GIORGIO LEUZZI, AND AGNÈS SILBERMANN

***Abstract*—Using a full-wave mode-matching technique, an extensive analysis is presented of the slow-wave factor, attenuation, and characteristic impedance of a metal–insulator–semiconductor coplanar waveguide (MISCPW) as functions of the various structural parameters. Design criteria are given for low-attenuation slow-wave propagation. By a proper optimization of the structure, performances comparable with or even better than those of alternative structures proposed in the literature are theoretically predicted.**

I. Introduction

MONOLITHIC MICROWAVE integrated circuits, using both Si and GaAs technologies, have an increasing impact in a number of applications because of higher reliability, reproducibility, and potentially lower costs [1]. It has already been pointed out that accurate analysis techniques are required in order to reduce necessity for trimming, which is more difficult than for hybrid integrated circuits. Even in this case, however, full-wave analyses are necessary to study propagation effects in active devices [2]. Gigabit logic is another area where propagation effects have to be accounted for through the use of accurate theoretical analyses [3].

Slow-wave propagation in metal–insulator–semiconductor and Schottky-contact planar transmission lines has been both experimentally observed and theoretically explained from different points of view [3]–[10]. The slow-wave properties of such transmission lines can be used to reduce the dimensions and cost of distributed elements to realize delay lines or, when Schottky-contact lines are used, for variable phase shifters, voltage-tunable filters, etc.

A drawback of these slow-wave structures is the loss associated with the semiconducting layer. As an example, the GaAs metal–insulator–semiconductor coplanar waveguide (MISCPW) experimented by Hasegawa and his coworkers [6], [11] presented an attenuation greater than 1 dB/mm, with a slowing factor of about 30 at the frequency of 1 GHz. Since losses and slow-wave effects depend on the distribution of the electromagnetic field inside the various regions of the structure, accurate analyses are required to determine the most favorable conditions for the practical use of such transmission lines.

An extensive study of the properties of MISCPW, based on a full-wave technique, is presented in this paper. The influence of the various structural parameters on the characteristics of the structure is investigated, together with the effect of the addition of a back conducting plane, which

Manuscript received July 22, 1983; revised November 10, 1983. This work was supported in part by the Consiglio Nazionale delle Ricerche (CNR), Italy.

R. Sorrentino is with the Dipartimento di Elettronica, Università di Roma La Sapienza, Via Eudossiana 18, Rome, Italy.

G. Leuzzi is with the Dipartimento di Elettronica, Università di Roma Tor Vergata, Via O. Raimondo, Rome, Italy.

A. Silbermann is with Elettronica S.P.A., Via Tiburtina, Rome, Italy.

Reprinted from *IEEE Trans. Microwave Theory Tech.*, vol. MTT-32, no. 4, pp. 410–416, April 1984.

can be used for increasing the mechanical strength of the circuits [12]. Conductor loss has not been included in the analysis since it is generally negligible with respect to semiconductor loss. The method of analysis is basically a classical mode-matching technique [10]. For clarity of presentation, it is briefly described in the next section, but the analytical details are omitted.

The computed results are presented and discussed in the third section. They indicate that, by properly choosing its parameters, the MISCPW is capable of slow-wave low-loss propagation with characteristics comparable to or better than those of alternative structures proposed for the same type of applications [11].

II. Method of Analysis

Fig. 1 shows a sketch of the MISCPW. The analysis of the structure can be reduced to that of a discontinuity problem in a parallel-plate waveguide by inserting two longitudinal electric or magnetic planes perpendicular to the substrate, sufficiently apart from the slots. The effect of these auxilliary planes is expected to be negligible since the EM field is normally confined to the proximity of the slots. Because of symmetry, a further magnetic longitudinal wall can be placed at the center of the strip conductor for analyzing the dominant even mode of the CPW. The geometry of the reduced structure is shown in Fig. 2. As viewed in the y-direction, it appears as a parallel-plate waveguide (with plates of electric or magnetic type at $x = 0$ and $x = a$) which is loaded with three (lossy) slabs and a metallic iris of finite thickness. The analysis can be performed using a classical mode-matching technique. Assuming a z-dependence as $\exp(-\gamma z)$, the EM field is expanded in each homogeneous section ($i = 1,2,\cdots,6$) in terms of $TE^{(y)}$ and $TM^{(y)}$ modes of the parallel-plate waveguide; the boundary conditions at infinity and at $y = y_i$ ($i = 1,2,\cdots,5$), through the use of the orthogonality properties of the modes, lead to a homogeneous system of equations in the expansion coefficients. For nontrivial solutions the coefficient matrix must be singular; this leads to a transcendental equation in the complex propagation constant.

The system of equations can be manipulated so that the only unknowns are the wave amplitudes in region 5. The result is a small number of equations (typically less than 12), which requires little computing time. The roots of the characteristic equation have been computed using the ZEPLS program [13]. Once a value of γ has been computed, the EM field expansion coefficients are obtained as the eigensolutions of the homogeneous system. From this, any other quantity relevant to the mode of propagation can be computed, such as field distribution, power density, and characteristic impedance. With regard to the last quantity, the following definition has been adopted:

$$Z_0 = |V|^2/(2P^*)$$

where V is the voltage between the strip conductor and ground, and P the complex power flowing through the cross section of the structure.

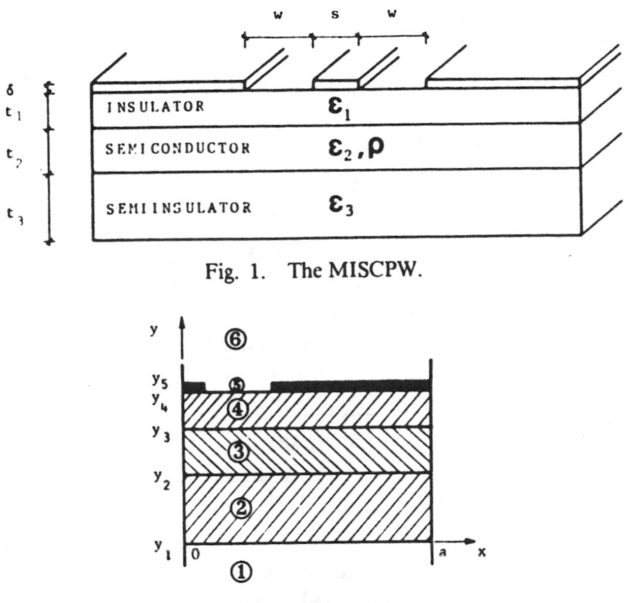

Fig. 1. The MISCPW.

Fig. 2. Reduced geometry of the MISCPW for analysis purposes.

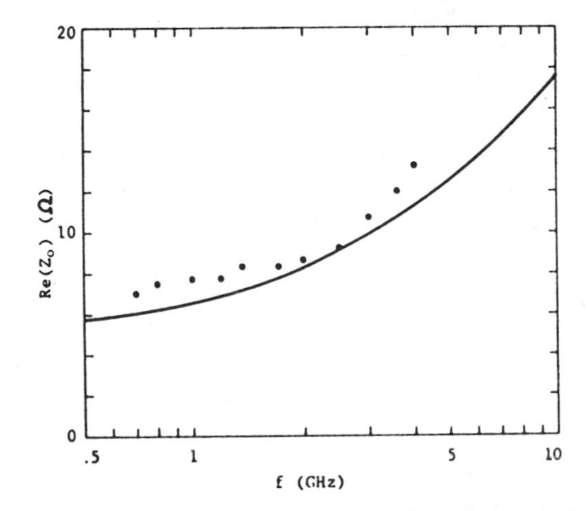

Fig. 3. Frequency behavior of the real part of the characteristic impedance of Hasegawa's MISCPW. Dots represent experiment [6].

III. Results

In a preliminary work [10], the agreement was demonstrated between the slow-wave factors and attenuations computed by the present method and the measurements performed by Hasegawa and coworkers [6], [11] on a specific MISCPW on GaAs substrate. (With reference to Fig. 1, the parameters of this structure were: $s = 0.1$ mm, $w = 0.45$ mm, $t_1 = 0.4$ μm, $t_2 = 3.0$ μm, $\epsilon_1 = 8.5$, $\epsilon_2 = \epsilon_3 = 13.1$, $\rho = 5.5\ 10^{-5}$ $\Omega \cdot$m.) As shown in Fig. 3, a similar agreement has been obtained with regard to the characteristic impedance of the same structure. Such results suggest the suitability of the above-described technique for an extensive analysis of the properties of MISCPW. It should be observed that the behavior of a MISCPW depends on a number of quantities, namely: the frequency f, the width s of the strip conductor, the distance $s + 2w$ between the ground planes, the thickness δ of the metallization, the thicknesses t_i ($i = 1,2,3$) of the substrate layers, the doping level n_d, or the resistivity ρ of the semiconducting layer. $\epsilon_1, \epsilon_2, \epsilon_3$ have not been accounted for, as we suppose the substrate material is given. It is clearly very difficult to work with such a high number (eight) of parameters. We have observed, however, that the thickness of the metallization has a nonsubstantial or even negligible effect; moreover, as long as the semiinsulating layer thickness is large, as in the case previously analyzed, its influence is negligible too. Since we are interested in singling out the conditions for low-loss propagation, the resistivity of the semiconducting layer can be fixed at the value corresponding approximately to the minimum attenuation [10]. We have found, in fact, that this value is slightly sensitive to the structural parameters. Finally, because of the scaling properties of Maxwell's equations, the results obtained for a given structure at a frequency, say, of 1 GHz can be easily extended to another structure with all linear dimensions and semiconductor resistivity scaled by a factor of $1/\kappa$ at the frequency of κ GHz.

We started our computations examining the MISCPW experimented by Hasegawa [6]. As previously shown [10], at $f=1$ GHz, this structure has a minimum theoretical attenuation of about 2 dB/mm for a semiconductor resistivity ρ of about 1.3 10^{-5} $\Omega \cdot$m, corresponding to a doping level $n_d = 6\ 10^{17}$ cm^{-3} (assuming a GaAs electron mobility of 8000 cm^2/Vs). For about the same value of ρ, the slow-wave factor λ_0/λ_g has a maximum of about 40. (λ_0 is the free-space wavelength and λ_g the MISCPW dominant mode wavelength.)

A. Effect of Shape Parameters

Using such an optimum value of ρ, we have computed the results shown in Fig. 4, where α and λ_0/λ_g are plotted at the frequency of 1 GHz against the shape parameters

$$q = t_1/(t_1 + t_2) \qquad r = s/(s+2w)$$

which characterize the substrate and metallization geometries, respectively. The distance $s+2w$ between the ground planes and the thickness t_1+t_2 of the insulating plus semiconducting layers have been kept equal to those of Hasegawa's structure.

It is observed that, while the attenuation has a marked dependence on both q and r, the slow-wave factor is slightly affected by the geometry of the metallization (r), as it is mainly influenced by the geometry of the substrate (q). As the semiconducting layer thickness t_2 is reduced from ~3 μm ($q=0.1$) to ~0.3 μm ($q=0.9$), the corresponding slowing factor is reduced from values greater than 40 to about 15. The attenuation, on the contrary, decreases with q down to a minimum for $q \sim 0.75$; as q approaches unity, α has first a local maximum, which is sharper with the smaller r, then rapidly decreases to zero as the semiconducting layer thickness becomes zero ($q=1$).

Very high slowing factors (> 40) can be obtained for low q's, thus thicker semiconducting layers, but at the price of higher attenuations; it can be observed that the increased attenuation is generally not compensated for by the possible reduction of the dimensions of the circuits due to the higher value of λ_0/λ_g. For example, for $r=0.2$ and q varying from 0.5 to 0.1, λ_0/λ_g increases by a factor of ~2, while α increases by a factor of ~3. In order to get a quantitative comparison between structures with different slowing factors and different attenuations, we can use as a quality factor the parameter

$$Q = (\alpha\lambda_g)^{-1}.$$

The Q behavior computed from the data in Fig. 4 is shown in Fig. 5. These figures indicate that, in order to obtain lower attenuations with still considerable slowing factors, it

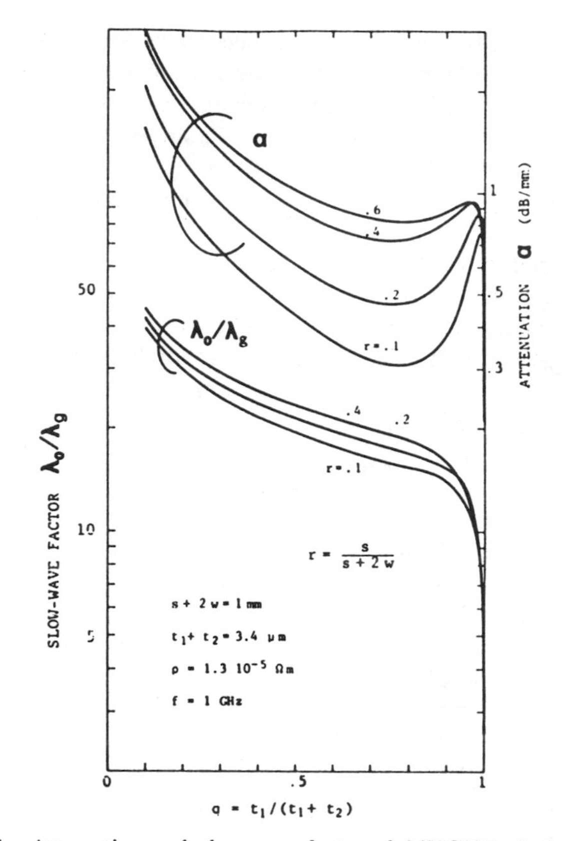

Fig. 4. Attenuation and slow-wave factor of MISCPW. $f=1$ GHz, $\rho=1.3\ 10^{-5}$ $\Omega\cdot$m, $s+2w=1$ mm, $\delta=1$ μm, $t_1+t_2=3.4$ μm, $t_3=0.997$ mm, $\epsilon_1=8.5$, and $\epsilon_2=\epsilon_3=13.1$.

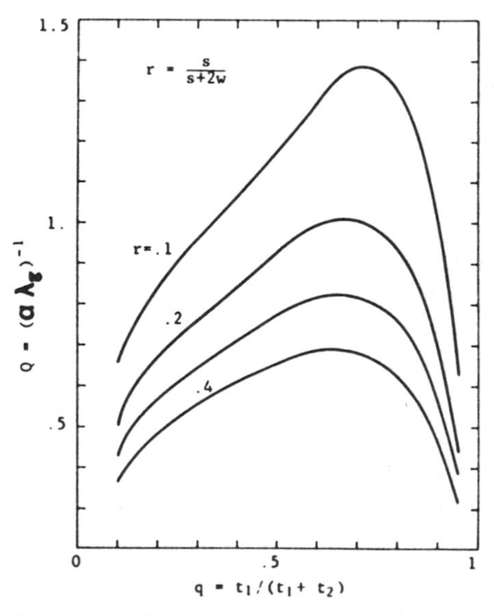

Fig. 5. Quality factor of the MISCPW of Fig. 4.

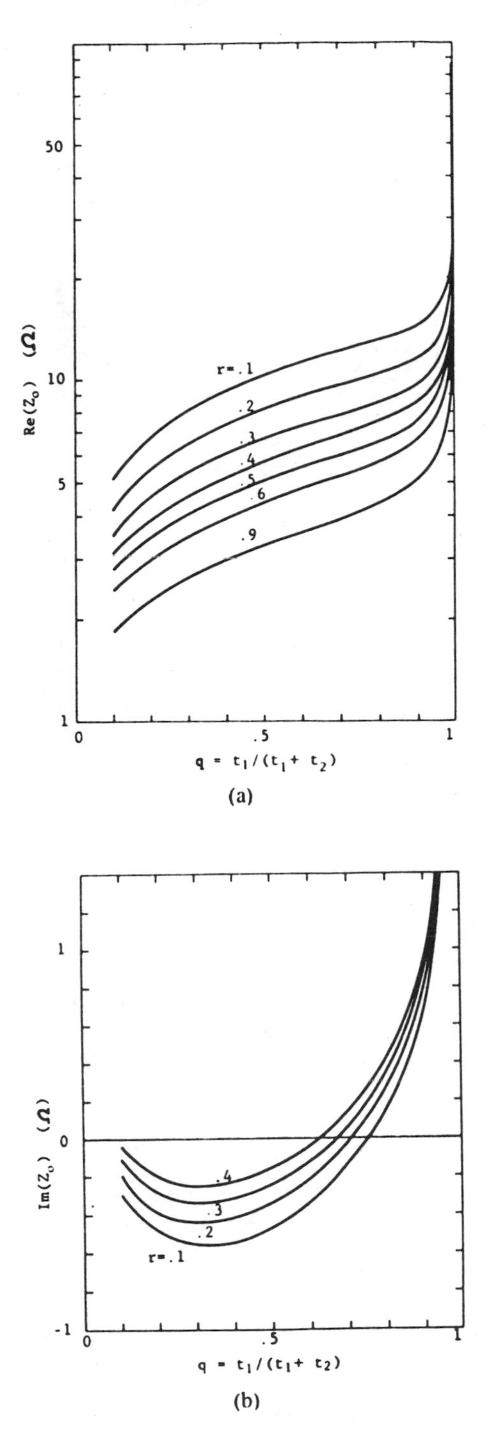

Fig. 6. (a) Real part of the characteristic impedance of the MISCPW of Fig. 4. (b) Imaginary part of the characteristic impedance of the MISCPW of Fig. 4.

is convenient to use narrow strip conductors and proper ratios between insulating and semiconducting layer thicknesses. In practical cases, however, the attenuation cannot be reduced indefinitely by reducing the strip width, since this will also have the effect of increasing the conductor loss. For a given geometry of the metallization, highest Q values are obtained for q ranging from 0.6 (for $r > 0.2$) to 0.7 (for $r < 0.2$). An investigation of the α behavior versus ρ for $q = 0.7$, $r = 0.1$ has shown that the optimum resistivity for minimum attenuation is about the same as the previous one.

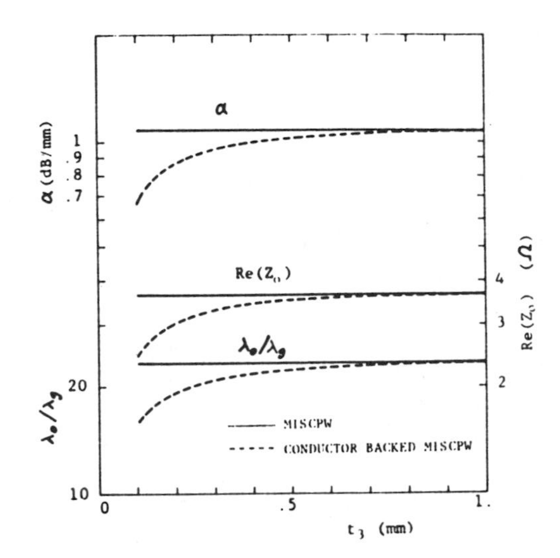

Fig. 7. Slow-wave factor, attenuation, and characteristic impedance as functions of the semi-insulating layer thickness of MISCPW.

The behavior of the complex characteristic impedance as a function of q and r is shown in Fig. 6(a) and (b). The behavior of the real part can be understood from a qualitative point of view in terms of the characteristic impedance of a quasi-TEM transmission line. The presence of the semiconducting layer has the effect of confining the electric field to the insulating layer, thus increasing the capacitance per unit length of the line and decreasing the characteristic impedance. Fig. 6(b) shows that the characteristic impedance has a generally small imaginary part of inductive or capacitive type depending on the geometry of the structure. As the semiconducting layer becomes very thin ($q \sim 1$), both the real and imaginary part of Z_0 undergo a very sharp increase. In the limit for $q = 1$, i.e., when the semiconducting layer is absent, the real part assumes much higher values, while the imaginary part becomes zero (not shown in Fig. (6b)).

B. Effect of Semiinsulating Layer Thickness

The effect of the semi-insulating layer thickness t_3 is illustrated ($q = 0.8$, $r = 0.5$) in Fig. 7 for a MISCPW with and without a ground plane on the back of the substrate. As t_3 decreases from 1.0 to 0.1 mm, the characteristics of the standard MISCPW remain practically unchanged. In the presence of a back conducting plane, on the contrary, the characteristics of the MISCPW are modified for t_3 smaller than 0.5 mm. The increased capacitance per unit length, due to the additional metallic plane, is responsible for the lower characteristic impedance. This figure indicates that the adoption of a back ground plane could be advantageous in reducing attenuation.

C. Effect of Distance Between Ground Planes

It has been already noted (see Fig. 4) that the attenuation can be reduced, for a fixed distance $s + 2w$ between the ground planes, by reducing the strip width s. Even lower attenuations can be obtained by a simultaneous reduction of $s + 2w$ and s. Fig. 8 shows the computed characteristics versus $s + 2w$ of a MISCPW with $q = 0.5$,

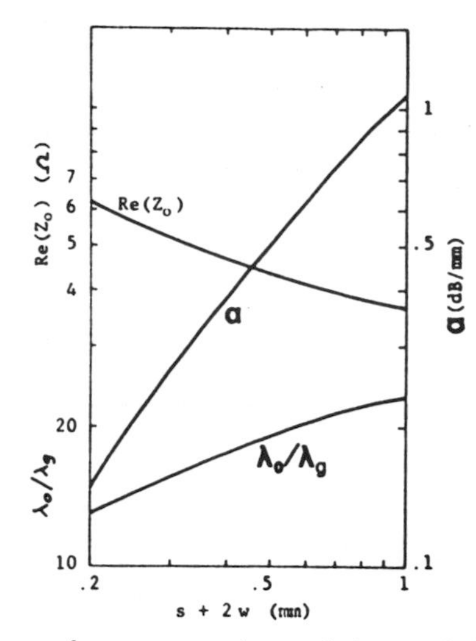

Fig. 8. Slow-wave factor, attenuation, and characteristic impedance as functions of the distance between ground planes.

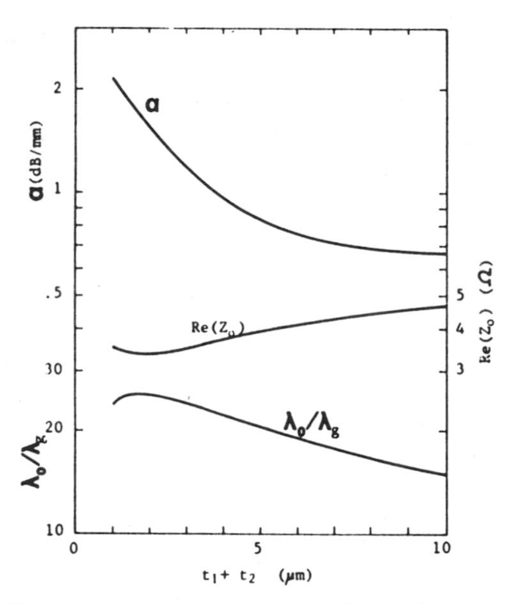

Fig. 9. Slow-wave factor, attenuation, and characteristic impedance as functions of the overall thickness of the insulating and semiconducting layers.

$r = 0.8$. (Since r is kept constant, s varies from 0.8 to 0.16 mm as $w + 2s$ varies from 1.0 to 0.2 mm.) It is seen that the attenuation of the structure can be reduced by about one order of magnitude by reducing the dimensions of the printed circuit; this has also the effect of increasing the Q of the line, since the slow-wave factor undergoes a much smaller reduction. A similar but not so marked effect is obtained by increasing the overall thickness of the insulating and semiconducting layers, as shown in Fig. 9. This way of reducing the attenuation, however, may be impractical because of technological problems.

We have then computed the characteristics of MISCPW having a distance between ground planes reduced with respect to the case of Fig. 4. As shown in Fig. 10, the general behaviors of α and λ_0/λ_g are about the same as the previous ones, but the attenuation is considerably reduced and attains a value lower than 0.1 dB/mm for $q = 0.8$, $r = 0.1$. The slowing factor is also reduced, but to a

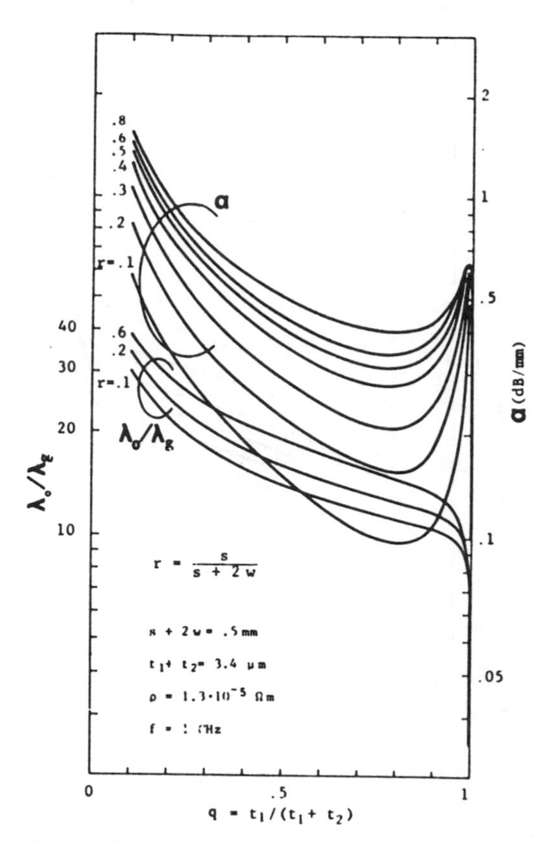

Fig. 10. Same as Fig. 4, except for with $s + 2w = 0.5$ mm and $t_3 = 0.2$ mm.

lesser extent, so that higher values (up to ~ 3.4) of Q are obtained. In any case, except for very small values of the semiconducting layer thickness ($q \sim 1$), slowings better than 10 are obtained. Fig. 11 shows the computed behavior of the real part of the characteristic impedance, which is slightly higher than in the previous case.

D. Comparison between Different Structures

Finally, Fig. 12 shows a comparison between the frequency behaviors of various structures. The a curves represent the computed values of attenuation and slow-wave factor of the original structure tested by Hasegawa: α varies from 2 dB/mm at $f = 1$ GHz up to more than 20 dB/mm at $f = 10$ GHz, while λ_0/λ_g varies from 30 to 11 correspondingly. The b curves represent the computed characteristics of one of the structures of Fig. 10, with $r = 0.1$, $q = 0.8$. In this case, the attenuation is reduced by one order of magnitude, while the slowing factor is about 10 in the whole frequency range. Even lower attenuations can be obtained, as shown by the third structure (c curves) having $s + 2w = 0.25$ mm. The slowing factor, though lower than in the other case, may be still considered as satisfactory. It is interesting to compare these results with those relative to the cross-tie CPW, which has been proposed by Seky and Hasegawa [11] as an alternative to the MISCPW. The Q values of the four structures of Fig. 12 are, at $f = 1$ GHz, 0.46 for a, 3.37 for b, 7.1 for c, 1.4 (CT-CPW). Although conductor loss has not been included in the analysis, these results suggest that the MISCPW is capable of supporting slow-wave propagation with comparable or even better attenuation characteristics with respect to the

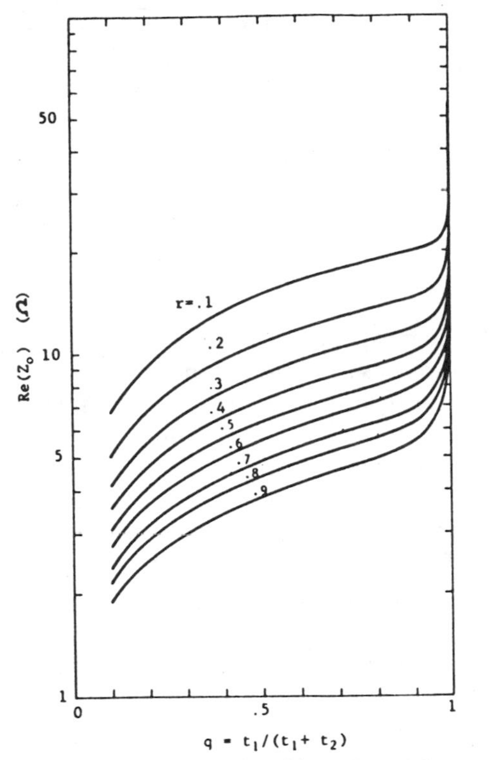

Fig. 11. Same as Fig. 6(a), except for with $s+2w=0.5$ mm and $t_3=0.2$ mm.

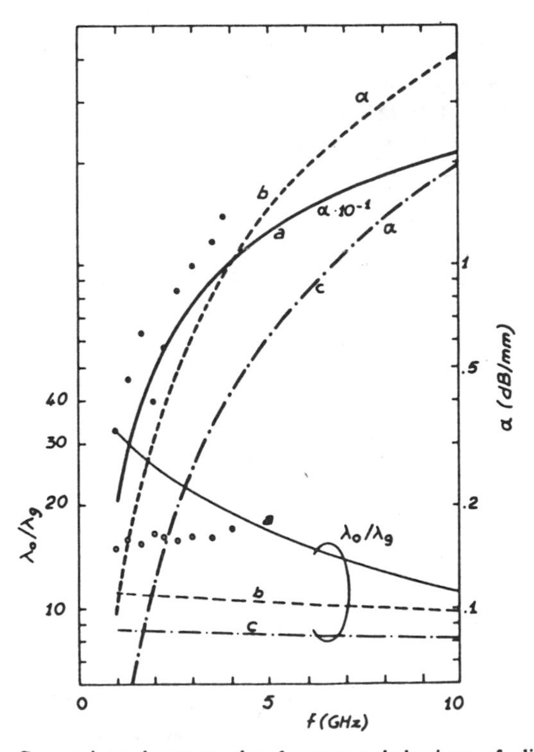

Fig. 12. Comparison between the frequency behaviors of different MISCPW. Dots represent experimental values for CT CPW [11].

cross-tie CPW, provided a suitable optimization of its parameters is made.

IV. Conclusions

A full-wave mode-matching technique has been used for computing the characteristics (slow-wave factor, attenuation, characteristic impedance) of MISCPW in terms of various geometrical parameters and frequency. Theoretical results indicate that low-loss propagation with useful slow-wave factors can be obtained adopting proper shape factors and doping levels of the semiconducting layer. Attenuations lower than 0.1 dB/mm with slowing factors of about 10 at $f=1$ GHz are theoretically predicted. These results render the MISCPW competitive with respect to alternative configurations [11] proposed for applications in the area of MMIC.

References

[1] R. A. Pucel, "Design considerations for monolithic microwave circuits," *IEEE Trans. Microwave Theory Tech.*, vol. MTT-29, pp. 513–534, 1981.

[2] W. Heinrich and H. L. Hartnagel, "Wave-theoretical analysis of signal propagation on FET electrodes," *Electron. Lett.*, pp. 65–67, Jan. 1983.

[3] P. Kennis and L. Faucon, "Rigourous analysis of planar M.I.S. transmission lines," *Electron. Lett.*, vol. 17, no. 13, pp. 454–456, 1981.

[4] H. Hasegawa, M. Furukawa, and H. Yanai, "Properties of microstrip line on Si–SiO_2 system," *IEEE Trans. Microwave Theory Tech.*, vol. MTT-19, pp. 869–881, 1971.

[5] D. Jager and W. Rabus, "Bias dependent phase delay of Schottky contact microstrip lines," *Electron. Lett.*, vol. 9, no. 9, pp. 201–203, 1973.

[6] H. Hasegawa and H. Okizaki, "M.I.S. and Schottky slow-wave coplanar stripline on GaAs substrates," *Electron. Lett.*, vol. 13, no. 22, pp. 663–664, 1977.

[7] G. W. Hughes and R. M. White, "Microwave properties of nonlinear MIS and Schottky-barrier microstrip," *IEEE Trans. Electron Devices*, vol. ED-22, pp. 945–956, 1975.

[8] Y. C. Shih and T. Itoh, "Analysis of printed transmission lines for monolithic integrated circuits," *Electron. Lett.*, vol. 18, no. 14, pp. 585–586, July 1982.

[9] Y. Fukuoka and T. Itoh, "Analysis of slow-wave phenomena in coplanar waveguide on a semiconductor substrate," *Electron. Lett.*, vol. 18, no. 14, pp. 589–590, July 1982.

[10] R. Sorrentino and G. Leuzzi, "Full-wave analysis of integrated transmission lines on layered lossy media," *Electron. Lett.*, vol. 18, no. 14, pp. 607–609, July 1982.

[11] S. Seki and H. Hasegawa, "Cross-tie slow-wave coplanar waveguide on semi-insulating GaAs substrate," *Electron. Lett.*, vol. 17, pp. 940–941, 1981.

[12] Y. C. Shih and T. Itoh, "Analysis of conductor backed coplanar waveguide," *Electron. Lett.*, vol. 18, pp. 538–540, June 1982.

[13] P. Lampariello and R. Sorrentino, "The ZEPLS program for solving characteristic equations of electromagnetic structures," *IEEE Trans. Microwave Theory Tech.*, vol. MTT-23, pp. 457–458, May 1975. See also W. Zieniutycz, "Comments on the ZEPLS program," and Authors' Reply, *IEEE Trans. Microwave Theory Tech.*, vol. MTT-31, p. 420, May 1983.

Analysis of Slow-Wave Coplanar Waveguide for Monolithic Integrated Circuits

YOSHIRO FUKUOKA, STUDENT MEMBER, IEEE, YI-CHI SHIH, MEMBER, IEEE, AND TATSUO ITOH, FELLOW, IEEE

Abstract —The slow-wave characteristics of an MIS coplanar waveguide are analyzed using two different full-wave methods: mode-matching and spectral-domain technique. The theoretical results obtained with them and the experimental values are in good agreement. Several important features of the MIS coplanar waveguide are presented along with some design criteria.

I. Introduction

MONOLITHIC MICROWAVE integrated circuits (MMIC's) based on gallium arsenide (GaAs) technology are being seriously considered as viable candidates for satellite communication systems, airborne radar systems, and other applications [1], [2]. The dielectric properties of semi-insulating GaAs, combined with the excellent microwave performance of GaAs field-effect transistors (GaAs FET's), make it possible to design monolithic microwave integrated circuits. The monolithic approach provides ease in circuit fabrication without the need to wire bond various components. This reduces the cost of manufacturing and improves reliability and reproducibility.

However, as the physical dimensions of the circuit components become smaller, it becomes difficult to trim and trouble-shoot a working circuit. To minimize this need for trimming, and to achieve design goals, a computer-aided design (CAD) procedure is indeed essential. Another potential problem which arises due to the small chip size is the RF coupling within the circuit. To overcome aforesaid problems, it is necessary to acquire a thorough knowledge of the properties of various planar transmission lines on semiconductor substrates.

On a planar substrate there are basically two different types of transmission lines available. They are microstrip lines and several coplanar structures such as slotline and coplanar waveguide. In the past two decades, they have been studied using various analytical and numerical techniques such as conformal mapping, finite difference, finite element, spectral domain, and so on. Based on the field distributions of these lines and their electrical characteristics, microstrip lines and coplanar waveguides are considered to be the most suitable for MMIC's [3].

Manuscript received November, 12, 1982; revised March 1, 1983. This work was supported in part by the Office of Naval Research under Contract N00014-79-0053, in part by the Joint Services Electronics Program under Grant F49620-79-C-0101, and in part by the U.S. Army Research Office under Contract DAAG29-81-K-0053.

Y. Fukuoka and T. Itoh are with the Department of Electrical Engineering, University of Texas at Austin, Austin TX 78712.

Y. C. Shih is with the U.S. Naval Postgraduate School, Monterey, CA 93940.

Planar metal–insulator–semiconductor (MIS) structures have been investigated by many authors. Simplified parallel plate structures were first examined [4], [5], and the existence of slow-wave propagation was experimentally observed for microstrip lines [5], [6] and coplanar waveguides [7]. This characteristic cannot only be used in delay lines, but, if the insulator layer is replaced by a Schottky contact between the metal and the semiconductor substrate, it can also be used as an electronically controllable variable phase shifter [8]. Some other applications employing these structures have also been proposed in [9].

The purpose of this paper is to present detailed analyses of MIS coplanar waveguides. Previous studies have shown the applicability of several techniques to the analysis of MIS microstrip lines [10]–[12], as well as MIS coplanar waveguides [12]–[15]. However, so far no detailed studies have been reported. The methods employed here include the mode-matching method and the spectral-domain technique. These methods have been used to analyze conventional lossless planar structures, and yield accurate results for the frequency-dependent propagation constants. In the present case, the doped region of the semiconductor substrate is treated as a dielectric layer with a finite resistivity, which is included in the analysis by the complex permittivity for the layer. Several basis functions are needed for both methods to obtain accurate results.

The results of both methods are in good agreement. They also agree well with the experimental results reported in [7]. Some important features of MIS transmission lines are presented in this paper.

II. Method of Analysis: Mode-Matching Method

The mode-matching method has been widely used to analyze microstrip lines [16], as well as dielectric waveguides [17], [18]. This method utilizes the eigenfunction expansion of the field. The cross-sectional view of the structure is shown in Fig. 1. Unlike the spectral-domain technique, the thickness of metallization can be taken into account. Since we are interested in the even dominant mode, a magnetic wall can be placed at $x = 0$, and only the right half of the structure is considered. In order to expand the field in terms of the Fourier series, a hypothetical electric wall needs to be placed at the far right side $x = w$ ($w \gg a$). Since the field is expected to be strongly confined

Reprinted from *IEEE Trans. Microwave Theory Tech.*, vol. MTT-31, no. 7, pp. 567–573, July 1983.

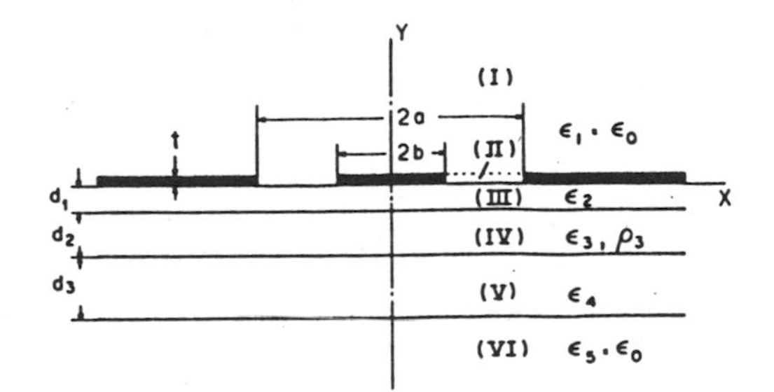

Fig. 1. Cross-sectional view of MIS coplanar waveguide. Region III—insulator layer. Region IV—doped semiconductor layer. Region V—semi-insulating layer. Regions I, II, and VI—air.

in the slot regions, the influence of this hypothetical electric wall is negligible. The cross section is divided into six regions, and the potentials in each region are expanded in terms of eigenfunctions. Since there are many layers in the y-direction, it is preferable to write the electric and magnetic potentials with respect to the y-direction. (This allows the boundary conditions to be easily matched.) The potentials in each region are

$$\psi^1(x,y) = \sum_{n=1}^{M} A_n \cos\beta_n x \exp[-\alpha_{1n}(y-t)]$$
$$\phi^1(x,y) = \sum_{n=1}^{M} B_n \sin\beta_n x \exp[-\alpha_{1n}(y-t)]$$
$$\psi^2(x,y) = \sum_{n=2}^{N} \sin\beta_{pn}(x-a) \cdot \{C_{1n}\sin\alpha_{2n}y + C_{2n}\cos\alpha_{2n}y\}$$
$$\phi^2(x,y) = \sum_{n=1}^{N} \cos\beta_{pn}(x-a) \cdot \{D_{1n}\sin\alpha_{2n}y + D_{2n}\cos\alpha_{2n}y\}$$
$$\vdots$$
$$\psi^6(x,y) = \sum_{n=1}^{M} K_n \cos\beta_n x \exp[\alpha_{6n}(y+d_1+d_2+d_3)]$$
$$\phi^6(x,y) = \sum_{n=1}^{M} L_n \sin\beta_n x \exp[\alpha_{6n}(y+d_1+d_2+d_3)] \tag{1}$$

where ψ's and ϕ's are the magnetic and electric potentials with respect to the y-direction, respectively. $A_n, B_n, \cdots, L_n$ are coefficients to be determined, and N is the number of basis functions in the slot region and M in the other regions. γ is the complex propagation constant in the z-direction which is to be determined, and the other parameters are

$$\beta_n = (n-0.5)\pi/w$$
$$\beta_{pn} = (n-1)\pi/(b-a)$$
$$\alpha_{1n}^2 = \gamma^2 + \beta_n^2 - \omega^2\epsilon_0\mu_0$$
$$\alpha_{2n}^2 = \omega^2\epsilon_0\mu_0 - \gamma^2 - \beta_{pn}^2$$
$$\vdots$$
$$\alpha_{6n}^2 = \gamma^2 + \beta_n^2 - \omega^2\epsilon_0\mu_0. \tag{2}$$

The field components are then calculated in terms of these potentials with the following equations:

$$\vec{E}(x,y) = -\nabla \times \bar{e}_y\phi - j\omega\mu_0\bar{e}_y\psi + \nabla(\nabla\cdot\bar{e}_y\psi)/j\omega\epsilon$$
$$\vec{H}(x,y) = \nabla \times \bar{e}_y\psi - j\omega\epsilon\bar{e}_y\phi + \nabla(\nabla\cdot\bar{e}_y\phi)/j\omega\mu_0 \tag{3}$$

where $\bar{e}_y$ is a unit vector in the y-direction, and ϵ takes a different value in each region and particularly a complex value in region 4. The potential functions in (1) already satisfy the boundary conditions at the magnetic wall at $x=0$ and the electric wall at $x=w$, and $x=a$, $x=b$ in the slot region. The remaining boundary conditions are satisfied to yield the following set of homogeneous equations:

$$\sum_{n=2}^{N}(P_{1mn}C_{1n} + P_{2mn}C_{2n}) + \sum_{n=1}^{N}(P_{3mn}D_{1n} + P_{4mn}D_{2n}) = 0, \quad m=2,3,4,\cdots,N$$
$$\sum_{n=2}^{N}(Q_{1mn}C_{1n} + Q_{2mn}C_{2n}) + \sum_{n=1}^{N}(Q_{3mn}D_{1n} + Q_{4mn}D_{2n}) = 0, \quad m=1,2,3,\cdots,N$$
$$\sum_{n=2}^{N}(R_{1mn}C_{1n} + R_{2mn}C_{2n}) + \sum_{n=1}^{N}(R_{3mn}D_{1n} + R_{4mn}D_{2n}) = 0, \quad m=2,3,4,\cdots,N$$
$$\sum_{n=2}^{N}(S_{1mn}C_{1n} + S_{2mn}C_{2n}) + \sum_{n=1}^{N}(S_{3mn}D_{1n} + S_{4mn}D_{2n}) = 0, \quad m=1,2,3,\cdots,N \tag{4}$$

where P, Q, R, and S are given in the Appendix. Equations (4) can be written in a matrix form

$$ZU = 0 \tag{5}$$

where Z is a square matrix of order $4N-2$, and vector U contains unknown coefficients C_{1n}, C_{2n}, D_{1n}, and D_{2n}. The determinant of the matrix Z is set equal to zero to obtain a nontrivial solution of the vector U, which, at the same time, determines the propagation constant γ.

III. Method of Analysis: Spectral-Domain Technique

The spectral-domain technique has been applied to analyze a number of lossless planar transmission lines [19]–[21]. A simple method for formulating the dyadic Green's functions in the spectral domain, proposed recently by Itoh [19], is based on the transverse equivalent transmission lines. With complex permittivities representing lossy dielectric substrates, this method is followed to analyze the structure (Fig. 1) where the metallic strips are assumed to be infinitesimally thin perfect conductors. The immediate result is the following set of coupled equations:

$$\begin{bmatrix} \tilde{Y}_{xx}(\alpha,\gamma) & \tilde{Y}_{xz}(\alpha,\gamma) \\ \tilde{Y}_{zx}(\alpha,\gamma) & \tilde{Y}_{zz}(\alpha,\gamma) \end{bmatrix} \begin{bmatrix} \tilde{E}_x(\alpha) \\ \tilde{E}_z(\alpha) \end{bmatrix} = \begin{bmatrix} \tilde{J}_x(\alpha) \\ \tilde{J}_z(\alpha) \end{bmatrix} \tag{6}$$

where $\tilde{Y}_{xx}, \tilde{Y}_{xz}, \tilde{Y}_{zx}$, and $\tilde{Y}_{zz}$ are the dyadic Green's functions similar to those derived in [19], and $\tilde{E}_x$, $\tilde{E}_z$, $\tilde{J}_x$, and $\tilde{J}_z$ are the Fourier transform of x and z components of the electric fields and current densities, respectively. α is the

Fourier transform variable and γ is the propagation constant. Following Galerkin's procedure in the spectral domain, we express slot field components $E_x(x)$ and $E_z(x)$ in terms of a complete set of known basis functions

$$E_x(x) = \sum_{n=0}^{N} c_n E_{xn}$$

$$E_z(x) = \sum_{n=0}^{N} d_n E_{zn} \quad (7)$$

where c_n and d_n are unknown coefficients.

The expressions in (7) are Fourier transformed and substituted into (6). The inner products of the resultant equations with each of the basis functions are performed and result in homogeneous linear simultaneous equations. The right-hand side is identically zero by the inner product process [19], [20]. A nontrivial solution may again be obtained by requiring the determinant of the coefficient matrix to be zero; this results in a characteristic equation from which γ is obtained.

Any kind of basis function may be used as long as it is nonzero only in the slot region. However, due to the variational nature of the approach, the efficiency and accuracy of this method depend greatly on the choice of basis functions. In this study, the following sets of basis functions are employed [20]:

$$E_{xn}(x) = \begin{cases} \cos[n\pi x/(b-a)] / \sqrt{\eta^2 - x^2}, & n = 0,2,4,\cdots \\ \sin[n\pi x/(b-a)] / \sqrt{\eta^2 - x^2}, & n = 1,3,5,\cdots \end{cases}$$

$$E_{zn}(x) = \begin{cases} \cos[n\pi x/(b-a)] / \sqrt{\eta^2 - x^2}, & n = 1,3,5,\cdots \\ \sin[n\pi x/(b-a)] / \sqrt{\eta^2 - x^2}, & n = 2,4,6,\cdots \end{cases} \quad (8)$$

where $\eta = (b-a)/2$. Note that these functions are defined only over the slot region. The Fourier transforms of the entire set are

$$\tilde{E}_{xn}(\alpha) = \begin{cases} j\sqrt{\pi/2}\sin[\alpha\xi] \cdot [J_0(r_{n+}) + J_0(r_{n-})], & n = 0,2,4,\cdots \\ -j\sqrt{\pi/2}\cos[\alpha\xi] \cdot [J_0(r_{n+}) - J_0(r_{n-})], & n = 1,3,5,\cdots \end{cases}$$

$$\tilde{E}_{zn}(\alpha) = \begin{cases} \sqrt{\pi/2}\cos[\alpha\xi] \cdot [J_0(r_{n+}) + J_0(r_{n-})], & n = 1,3,5,\cdots \\ \sqrt{\pi/2}\sin[\alpha\xi] \cdot [J_0(r_{n+}) - J_0(r_{n-})], & n = 2,4,6,\cdots \end{cases} \quad (9)$$

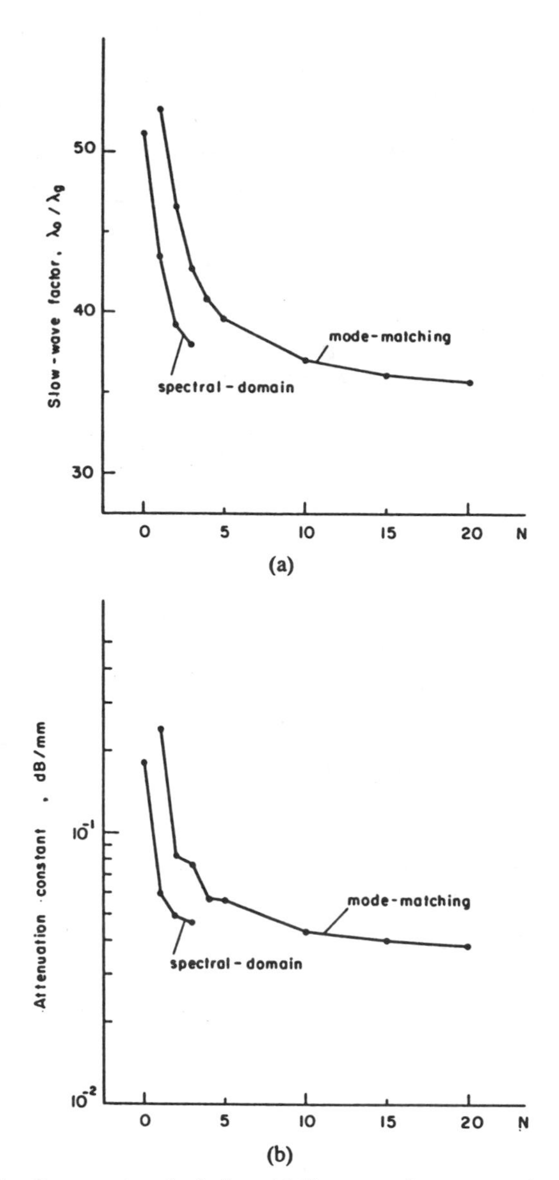

Fig. 2. Convergence of solutions. (a) Slow-wave factor versus frequency. (b) Attenuation constant versus frequency. $a = 50$ μm, $b = 0.5$ mm, $d_1 = 0.4$ μm, $d_2 = 3.0$ μm, $d_3 = 1.0$ mm, $\epsilon_2 = 8.5$, $\epsilon_3 = \epsilon_4 = 13$, $\rho_3 = 0.055$ Ω-mm, $f = 0.1$ GHz, ($t = 1.0$ μm, $w = 1.1$ mm: mode-matching method).

where J_0 denotes the zeroth order Bessel function of the first kind, and

$$\xi = (a+b)/2$$

$$r_{n+} = |(b-a)\alpha + n\pi|/2$$

$$r_{n-} = |(b-a)\alpha - n\pi|/2.$$

IV. Computational Results

In this section, we present numerical results calculated with the analytical procedures described above. The MIS coplanar waveguide structure on GaAs experimentally tested by Hasegawa *et al.* [7] was used to compare the results.

Fig. 2 shows a typical convergence of the numerical results for the slow-wave factors and the attenuation constants. The abscissa is the number of eigenfunctions in region 2 in the mode-matching technique, or the number of basis functions chosen in the spectral-domain method. The rate of convergence strongly depends on the parameters of

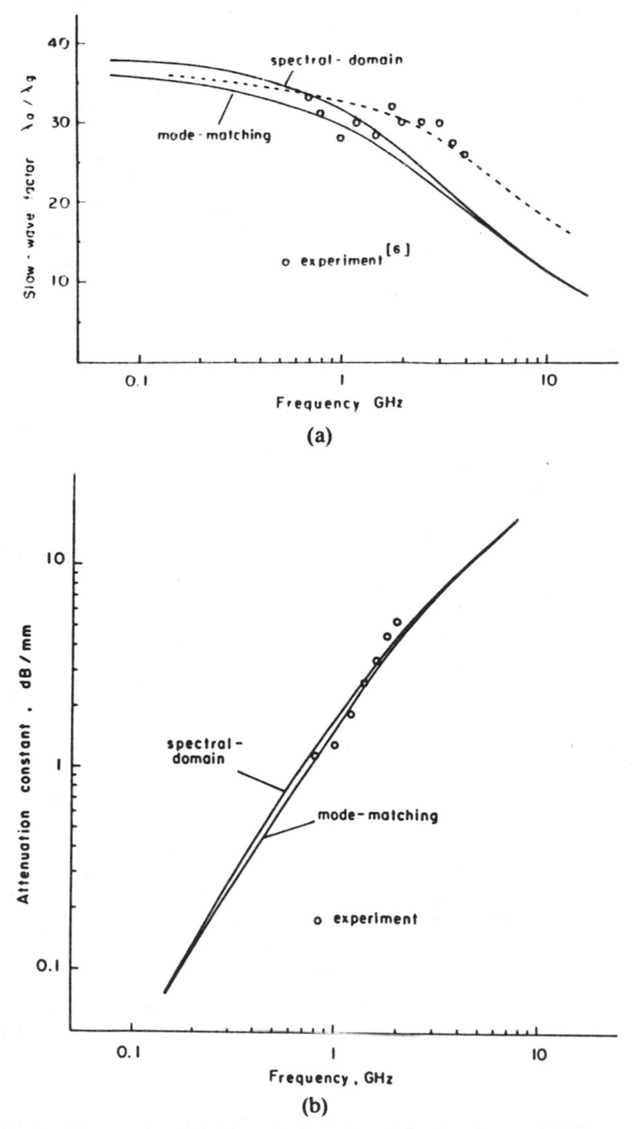

Fig. 3. Comparison of theoretical results with experiment. (a) Slow-wave factor versus frequency. (b) Attenuation constant versus frequency. $a = 50$ μm, $b = 0.5$ mm, $d_1 = 0.4$ μm, $d_2 = 3.0$ μm, $d_3 = 1.0$ mm, $\epsilon_2 = 8.5$, $\epsilon_3 = \epsilon_4 = 13$, $\rho_3 = 0.055$ Ω-mm.

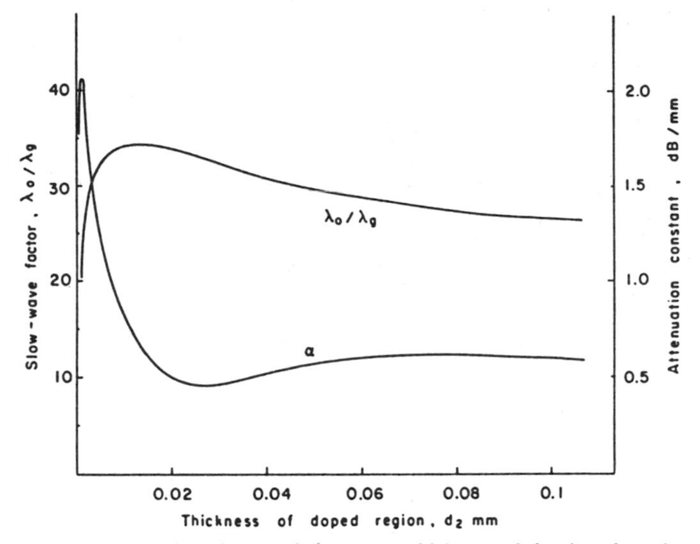

Fig. 4. Propagation characteristics versus thickness of the doped semiconductor region. $a = 50$ μm, $b = 0.5$ mm, $d_1 = 0.4$ μm, $d_2 + d_3 = 1.0$ mm, $\epsilon_2 = 8.5$, $\epsilon_3 = \epsilon_4 = 13$, $\rho_3 = 0.055$ Ω-mm, $f = 1.0$ GHz.

the structure. For this particular structure, it converges slowly. However, faster convergence is achieved with the spectral-domain technique. This is due to the choice of the basis functions for the slot fields. However, for sufficient accuracy, $N = 3$ is not enough for this structure, as seen from Fig. 2. The spectral-domain technique takes up excessive memory size in the computer during numerical integrations involving Bessel functions. Therefore, it is difficult to include more basis functions. On the other hand, the evaluations of the coefficients are fairly easy for the mode-matching method.

Fig. 3 shows the comparison of numerical results obtained by these two methods with the experiment. At higher frequencies, the discrepancy in the results of these two methods is smaller, and the convergence of the solutions is faster. In Fig. 3(a), the experimental results show slower propagation around 2–4 GHz when compared to these theories. A possible reason for this phenomenon is the approximation made to model the doped region (region 4) of the substrate, which is assumed to be uniformly doped through the depth d_2. If we double d_2, we can obtain a better fit between the theoretical and the experimental results. This is shown by dotted line in Fig. 3(a).

Since this is interesting, we show the behavior of the slow-wave factor and the attenuation constant with respect to the thickness of the doped region in Fig. 4. From now on, the mode-matching method is used for calculations. It should be pointed out that the slow-wave factor has a maximum at a certain value of the thickness, while the attenuation constant has a sharp peak at a very small thickness value and a broad minimum at a moderate thickness value. This phenomenon should be understood in relation to the propagation characteristics with respect to the resistivity of the doped region, which will be discussed later. That is, varying the thickness of the doped region has a similar effect as varying the resistivity. However, the attenuation constant should become large at a small thickness value because the current in the doped region is confined in a very small region. This may be an important feature from a design point of view. As one might expect, both slow-wave factor and attenuation constant approach to constant values as the thickness of the doped region increases.

Another interesting parameter is the thickness d_1 of the insulator region (region 3). If a Schottky-contact coplanar waveguide, used as a variable phase shifter, is approximated by a MIS coplanar line, and its electrical tunability is to be evaluated, the most important parameter is d_1, which is analogous to the thickness of the depletion layer. Several curves in Fig. 5 show almost linear relationship between the slow-wave factor and the thickness d_1. The slope of the curve strongly depends on the frequency. At a lower frequency, the propagation constant has greater dependence on d_1. At higher frequency, such as $f = 5$ GHz, the slow-wave factor is not affected by the change in thickness d_1. This is attributed to the fact that the waveguide is not being operated in the slow-wave region.

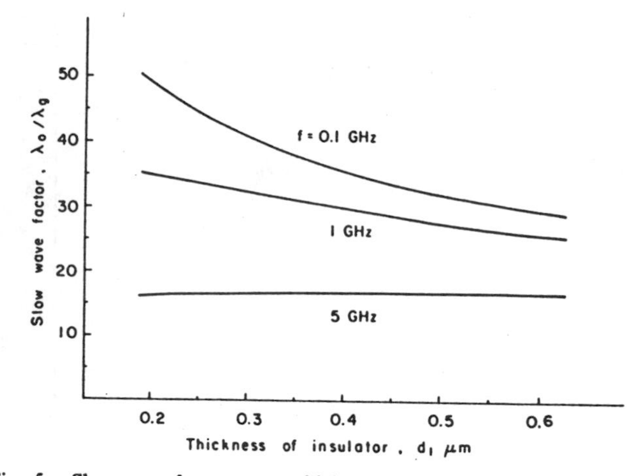

Fig. 5. Slow-wave factor versus thickness of the insulator region. $a = 50$ μm, $b = 0.5$ mm, $d_2 = 3.0$ μm, $d_3 = 1.0$ mm, $\epsilon_2 = 8.5$, $\epsilon_3 = \epsilon_4 = 13$, $\rho_3 = 0.055$ Ω-mm.

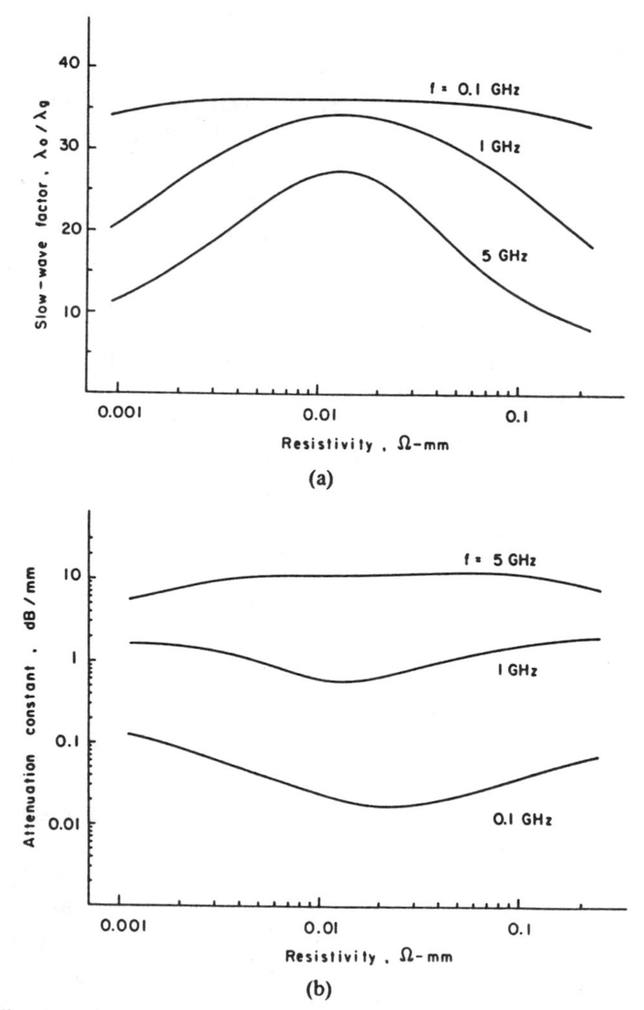

Fig. 6. (a) Slow-wave factor versus resistivity of the doped semiconductor region. (b) Attenuation constant versus resistivity of the doped semiconductor region. $a = 50$ μm, $b = 0.5$ mm, $d_1 = 0.4$ μm, $d_2 = 3.0$ μm, $d_3 = 1.0$ mm, $\epsilon_2 = 8.5$, $\epsilon_3 = \epsilon_4 = 13$.

The propagation characteristics as a function of the resistivity of the doped region shows behavior similar to that predicted for MIS microstrip lines [5]. There are basically three operating regions: the slow-wave region, the skin-effect region, and the lossy-dielectric region. Fig. 6(a) shows a typical behavior of the slow-wave factor for various frequencies. It is observed that there exists an optimum resistivity in the middle of the slow-wave region where the attenuation constant has a local minimum (Fig. 6(b)). The location of this point may be important for applications since the slow-wave factor also has its maximum value at this resistivity value.

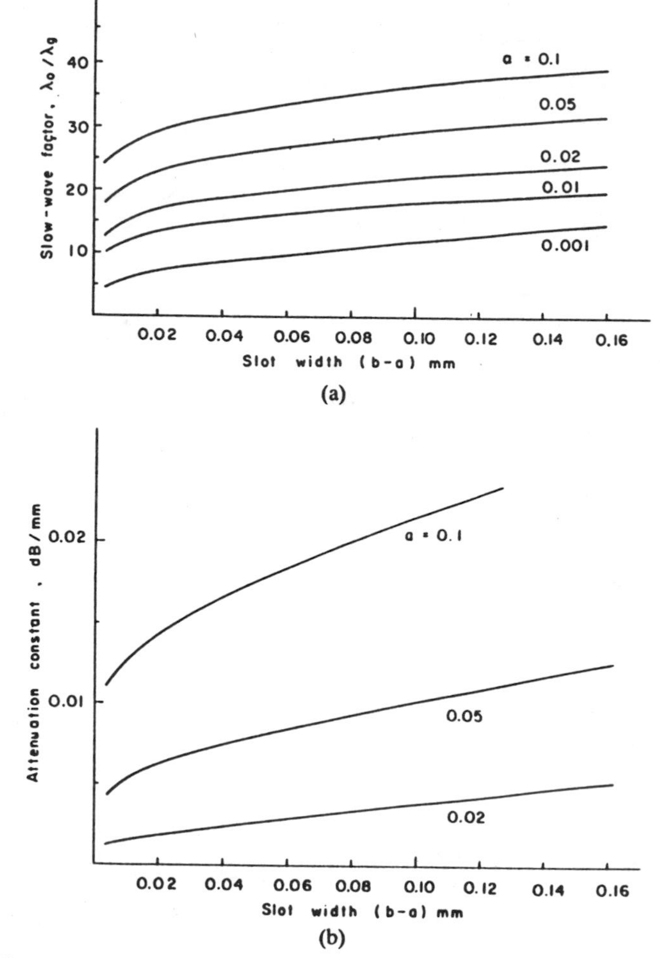

Fig. 7. (a) Slow-wave factor versus slot width. (b) Attenuation constant versus slot width. $d_1 = 0.4$ μm, $d_2 = 3.0$ μm, $d_3 = 1.0$ mm, $\epsilon_2 = 8.5$, $\epsilon_3 = \epsilon_4 = 13$, $\rho_3 = 0.01$ Ω-mm, $f = 0.1$ GHz.

Finally, the behavior of the propagation constant due to variations in some other structural parameters is also investigated. For example, the values of the slow-wave factor and the attenuation constant in the slow-wave mode are plotted as a function of slot width in Fig. 7. $f = 0.1$ GHz and $\rho_3 = 0.01$ Ω-mm were chosen so that the coplanar waveguide operates in a slow-wave region. From Fig. 7(a), we observe that the value of the slow-wave factor increases as the width of either the center strip or the slot increases. This behavior may be understood by considering the capacitance of the waveguide. In the slow-wave region, the electric field of the waveguide sees the doped region as a conductor [5]. Therefore, when the width of the center strip increases, the line capacitance increases. When the slot becomes wider, the electric field near the slot spreads out and causes field lines to enter more into the surface of the doped region. This again increases the capacitance. All these contribute to slower propagation of the wave. Since the wider width of the center strip or slot pushes more energy into the lossy substrate, the attenuation constant shows an increase as shown in Fig. 7(b).

V. Conclusions

MIS coplanar waveguide has been analyzed with two different methods. The conductive semiconductor substrate was treated by a complex permittivity, and the final solution was obtained by searching a complex root of the determinantal equation. The convergence study was successful and the methods employed show good correlation for the propagation constant. The results showed basically similar behavior of the MIS coplanar waveguide to the MIS microstrip line. Some important parameters, such as the thickness and the resistivity of the doped semiconductor layer governing the performance of a MIS coplanar waveguide for practical design of a circuit, were also shown.

Appendix

The coefficients appearing in (4) are given by

$$P_{1mn} = h_{1mn}\cos\alpha_{pn}t - \delta_{mn}[w(b-a)/4]\gamma\sin\alpha_{pm}t$$

$$P_{2mn} = -h_{1mn}\sin\alpha_{pn}t - \delta_{mn}[w(b-a)/4]\gamma\cos\alpha_{pm}t$$

$$P_{3mn} = h_{2mn}\sin\alpha_{pn}t - \delta_{mn}[w(b-a)/4](1/\omega\mu_0)\alpha_{pm}\beta_{pm}\cos\alpha_{pm}t$$

$$P_{4mn} = h_{2mn}\cos\alpha_{pn}t + \delta_{mn}[w(b-a)/4](1/\omega\mu_0)\alpha_{pm}\beta_{pm}\sin\alpha_{pm}t$$

$$Q_{1mn} = k_{1mn}\cos\alpha_{pn}t - \delta_{mn}[w(b-a)/4](1+\delta_{mo})\beta_{pm}\sin\alpha_{pm}t$$

$$Q_{2mn} = -k_{1mn}\sin\alpha_{pn}t - \delta_{mn}[w(b-a)/4](1+\delta_{mo})\beta_{pm}\cos\alpha_{pm}t$$

$$Q_{3mn} = k_{2mn}\sin\alpha_{pn}t + \delta_{mn}[w(b-a)/4](1+\delta_{mo})(\gamma/\omega\mu_0)\alpha_{pm}\cos\alpha_{pm}t$$

$$Q_{4mn} = k_{2mn}\cos\alpha_{pn}t - \delta_{mn}[w(b-a)/4](1+\delta_{mo})(\gamma/\omega\mu_0)\alpha_{pm}\sin\alpha_{pm}t$$

$$R_{1mn} = \sum_{s=1}^{N} [\gamma/(\beta_s^2+\gamma^2)]\{-(\epsilon_2/\epsilon_1\alpha_{2s})(E_{2s}/E_{1s})T_{sn}^1 - (\alpha_{2s}\beta_s/\omega\epsilon_1)(F_{1s}/F_{2s})T_{sn}^2\}A_{ms}^s\alpha_{pn}$$

$$R_{2mn} = -\delta_{mn}[w(b-a)/4]\gamma$$

$$R_{3mn} = -\delta_{mn}[w(b-a)/4](1/\omega\mu_0)\alpha_{pm}\beta_{pm}$$

$$R_{4mn} = \sum_{s=1}^{N} [1/(\beta_s^2+\gamma^2)]\{-(\gamma^2\omega\epsilon_2/\alpha_{2s})(E_{2s}/E_{1s})T_{sn}^3 - \alpha_{2s}\beta_s(F_{1s}/F_{2s})T_{sn}^4\}A_{ms}^s$$

$$S_{1mn} = \sum_{s=1}^{N} [1/(\beta_s^2+\gamma^2)]\{(\epsilon_2\beta_s/\epsilon_1\alpha_{2s})(E_{2s}/E_{1s})T_{sn}^1 - \gamma^2(\alpha_{2s}/\omega\epsilon_1)(F_{1s}/F_{2s})T_{sn}^2\}A_{ms}^s\alpha_{pn}$$

$$S_{2mn} = -\delta_{mn}[w(b-a)/4](1+\delta_{mo})\beta_{pm}$$

$$S_{3mn} = \delta_{mn}[w(b-a)/4](1+\delta_{mo})(\gamma/\omega\mu_0)\alpha_{pm}$$

$$S_{4mn} = \sum_{s=1}^{N} [\gamma/(\beta_s^2+\gamma^2)]\{(\beta_s\omega\epsilon_2/\alpha_{2s})(E_{2s}/E_{1s})T_{sn}^3 - \alpha_{2s}(F_{1s}/F_{2s})T_{sn}^4\}A_{ms}^s$$

where δ_{mn} is the Kronecker delta function, and the ratios of the coefficients E_{2s}/E_{1s} and F_{1s}/F_{2s} are calculated from the boundary conditions at $y=-d_1$, $-d_1-d_2$, and $-d_1-d_2-d_3$. The other coefficients appearing here are given by

$$h_{1mn} = \sum_{s=1}^{N} [\gamma/(\beta_s^2+\gamma^2)] \cdot \{(1/\alpha_{1s})T_{sn}^1 + (\alpha_{1s}\beta_s/\omega\epsilon_1)T_{sn}^2\}A_{ms}^c\alpha_{pn}$$

$$h_{2mn} = \sum_{s=1}^{N} [1/(\beta_s^2+\gamma^2)]\{(\gamma^2\omega\epsilon_1/\alpha_{1s})T_{sn}^3 + \alpha_{1s}\beta_s T_{sn}^4\}A_{ms}^c$$

$$k_{1mn} = \sum_{s=1}^{N} [1/(\beta_s^2+\gamma^2)] \cdot \{-(\beta_s/\alpha_{1s})T_{sn}^1 + (\alpha_{1s}\gamma^2/\omega\epsilon_1)T_{sn}^2\}A_{ms}^s\alpha_{pn}$$

$$k_{2mn} = \sum_{s=1}^{N} [\gamma/(\beta_s^2+\gamma^2)]\{-(\beta_s\omega\epsilon_1/\alpha_{1s})T_{sn}^3 + \alpha_{1s}T_{sn}^4\}A_{ms}^s$$

$$T_{sn}^1 = \beta_s\beta_{pn}A_{ns}^s - \gamma^2 A_{ns}^c$$

$$T_{sn}^2 = (1/\omega\mu_0)\{\beta_{pn}A_{ns}^s + \beta_s A_{ns}^c\}$$

$$T_{sn}^3 = \beta_s A_{ns}^s + \beta_{pn}A_{ns}^c$$

$$T_{sn}^4 = (1/\omega\mu_0)\{\gamma^2 A_{ns}^s - \beta_s\beta_{pn}A_{ns}^c\}$$

$$A_{ns}^s = \int_a^b \cos\beta_{pn}(x-a)\sin\beta_s x\,dx$$

$$A_{ns}^c = \int_a^b \sin\beta_{pn}(x-a)\cos\beta_s x\,dx$$

References

[1] F. J. Moncrief, "Monolithic MIC's gain momentum as GaAs MSI nears," *Microwaves*, vol. 18, no. 7, pp. 42–53, July 1979.

[2] H. Q. Tserng, "Advance in microwave GaAs power FET device and circuit technologies," in *11th Eur. Microwave Conf.*, Sept. 1981, pp. 48–58.

[3] R. A. Pucel, "Design considerations for monolithic microwave circuits," *IEEE Trans. Microwave Theory Tech.*, vol. MTT-29, pp. 513–534, June 1981.

[4] H. Guckel, P. A. Brennan, and I. Palócz, "A parallel-plate waveguide approach to microminiaturized, planar transmission lines for integrated circuits," *IEEE Trans. Microwave Theory Tech.*, vol. MTT-15, pp. 468–476, Aug. 1967.

[5] H. Hasegawa, M. Furukawa, and H. Yanai, "Properties of microstrip line on Si-SiO$_2$ system," *IEEE Trans. Microwave Theory Tech.*, vol. MTT-19, pp. 869–881, Nov. 1971.

[6] D. Jäger and W. Rabus, "Bias dependent phase delay of Schottky-contact microstrip lines," *Electron. Lett.*, vol. 9, no. 9, pp. 201–203, May 1973.

[7] H. Hasegawa and H. Okizaki, "M.I.S. and Schottky slow-wave coplanar striplines on GaAs substrates," *Electron. Lett.*, vol. 13, no. 22, pp. 663–664, Oct. 1977.

[8] J. M. Jaffe, "A high frequency variable delay line," *IEEE Trans. Electron Devices*, vol. ED-19, pp. 1292–1294, Dec. 1972.

[9] G. W. Hughes and R. M. White, "Microwave properties of nonlinear MIS and Schottky barrier microstrip," *IEEE Trans. Electron Devices*, vol. ED-22, pp. 945–956, Oct. 1975.

[10] P. Kennis and L. Faucon, "Rigorous analysis of planar MIS transmission lines," *Electron. Lett.*, vol. 17, no. 13, pp. 454–456, June 1981.

[11] M. Aubourg, J. P. Villotte, F. Codon, and Y. Garault, "Analysis of microstrip line on semiconductor substrate," in *IEEE MTT-S Int. Microwave Symp. Dig.*, June 1981, pp. 495–497.

[12] P. Kennis *et al.*, "Properties of microstrip and coplanar lines on semiconductor substrates," in *Proc. 12th Eur. Microwave Conf.*, Sept. 1982, pp. 328–333.

[13] Y.-C. Shih and T. Itoh, "Analysis of printed transmission lines for monolithic integrated circuits," *Electron. Lett.*, vol. 18, no. 14, pp. 585–586, July 1982.

[14] Y. Fukuoka and T. Itoh, "Analysis of slow-wave phenomena in coplanar waveguide on a semiconductor substrate," *Electron. Lett.*, vol. 18, no. 14, pp. 589–590, July 1982.

[15] R. Sorrentino and G. Leuzzi, "Full-wave analysis of integrated transmission lines on layered lossy media," *Electron. Lett.*, vol. 18, no. 14, pp. 607–609, July 1982.

[16] G. Kowalski and R. Pregla, "Dispersion characteristics of shielded microstrips with finite thickness," *Arch. Elek. Übertragung*, vol. 25, no. 4, pp. 193–196, Apr. 1971.

[17] K. Solbach and I. Wolff, "The electromagnetic fields and the phase constants of dielectric image lines," *IEEE Trans. Microwave Theory Tech.*, vol. MTT-26, pp. 266–274, Apr. 1978.

[18] R. Mittra, Y. L. Hou, and V. Jamnejad, "Analysis of open dielectric waveguides using mode-matching technique and variational methods," *IEEE Trans. Microwave Theory Tech.*, vol. MTT-28, pp. 36–43, Jan. 1980.

[19] T. Itoh, "Spectral-domain immittance approach for dispersion characteristics of generalized printed transmission lines," *IEEE Trans. Microwave Theory Tech.*, vol. MTT-28, pp. 733–736, July 1980.

[20] L.-P. Schmidt and T. Itoh, "Spectral-domain analysis of dominant and higher order modes in fin-lines," *IEEE Trans. Microwave Theory Tech.*, vol. MTT-28, pp. 981–985, Sept. 1980.

[21] J. B. Knorr and K.-D. Kuchler, "Analysis of coupled slots and coplanar strips on dielectric substrate," *IEEE Trans. Microwave Theory Tech.*, vol. MTT-23, pp. 541–548, July 1975.

ANALYTICAL MODEL OF THE SCHOTTKY CONTACT COPLANAR LINE

Ch. SEGUINOT, P. KENNIS, P. PRIBETICH, J.F. LEGIER

ABSTRACT. The purpose of this paper is to present an original model of the Schottky contact coplanar line. Determination of the propagation characteristics of this structure is obtained with a good accuracy in a determined range of conductivity and of geometrical parameters. The comparative study of these lines when DC bias and geometrical parameters are changed is simply performed on a desktop computer.

INTRODUCTION. In recent years growing interest has been paid to transmission lines laid on semiconducting medium. For example, in digital integrated circuits with gigabit rate speed, interconnections must be treated as microwave transmission lines [1]. Furthermore Schottky contact lines exhibit propagation properties which can be modulated by DC bias voltage. Some devices such as switchable attenuators or phase shifters have been reported [2]. Finally the study of Schottky contact coplanar line will be helpful to understand propagation effects in microwave power FETs. In this point of view Schottky contact coplanar lines (fig. 1-a) are more interesting than other planar lines laid on semiconduting medium. Up to now the study of these lines was made by using numerical analysis such as SDA or FEM [3] - [7]. In fact such numerical analysis need very long computer time, so it seemed interesting to develop an analytical model of the Schottky contact coplanar line, based on physical properties of the considered propagated mode. The purpose of this model is to compare potential characteristics of such lines, and also to initialize more powerful numerical studies.

SCHOTTKY CONTACT COPLANAR LINE MODEL. In previous studies [8], [9] relative to a MIS microstrip line a two layers parallel plate waveguide was used. Those works have pointed out the peculiar behaviour of a slow wave propagation mode, which occurs when magnetic energy is distributed in the whole structure while electric one is stored in the insulating layer. Of course for coplanar structures such an analytical frequencial approach can not be carried out. So in our study we suppose that the conditions for slow wave propagation mode are satisfied ; the transmission line model shown Fig. 1-b can be used.

The authors are with "Centre Hyperfréquences et Semiconducteurs"
Université des Sciences et Techniques de Lille
59655 VILLENEUVE D'ASCQ Cédex - FRANCE

Reprinted with permission from *Proc. 14th European Microwave Conference,* pp. 160–165, September 1984.

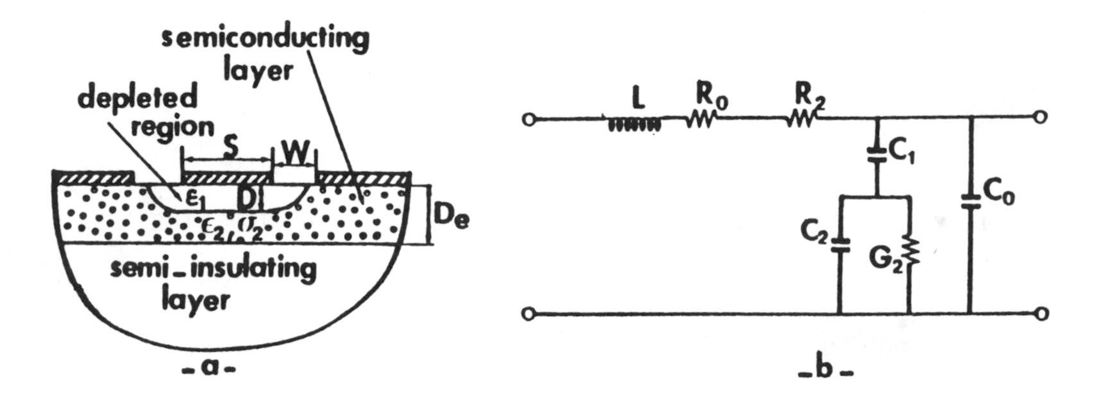

Fig. 1 - Schottky contact coplanar line -
a - cross section
b - Equivalent transmission line model.

In this equivalent circuit L is the inductance of the structure, C_1 the capacitance of the real permittivity layer, and C_0 the capacitance of the half upper plane. Losses in conductors are represented by a series resistance R_0. The determination of those elements is made without any difficulty [2], [10]. For low conductivity values no slow wave mode can propagate and the contributions of the semiconducting medium are represented by the conductance G_2 and the capacitance C_2. The series resistance R_2 represents losses due to skin effect in semiconducting medium. The physical phenomena which lead to R_2, G_2 and C_2 in the equivalent model are quite obvious, however for coplanar structures the determination of those elements can not be rigorous. They are obtained by semi-empirical considerations, the validity of which can only be attested by comparison of analytical model results to more rigorous numerical ones.

DETERMINATION OF THE POTENTIAL SLOW WAVE FACTOR. In the case of slow wave propagation mode, even if the attenuation is important, real elements in the transmission line model can be neglected as compared to the reactive ones, in order to obtain an approximation of the potential slow wave factor. This one is then only relative to L and C_1. This assumption is justified by SDA results [7] which point out that the electric energy is trapped in the insulating layer, under the central conductor, while the magnetic energy spread in the whole structure. So C_1 is only relative to the depleted layer and is calculated using the one dimensional Schockley approximation. As complex permittivity does not affect the inductance calculation, L is obtained by means of conformal mapping as in previous works [10].Some results corresponding to this simplified model are presented fig. 2.

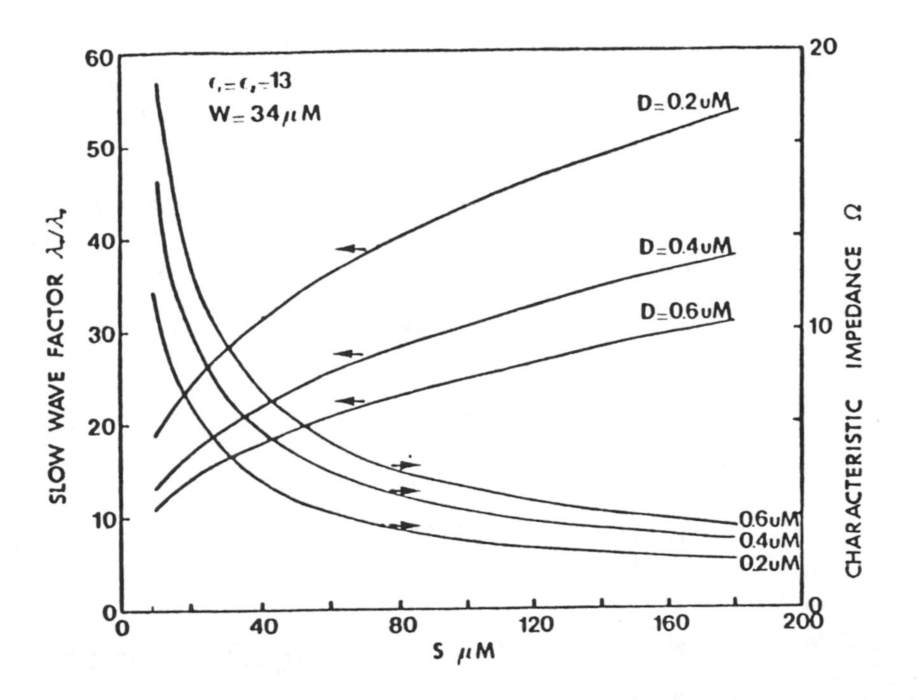

Fig. 2 - Effect of varying D.C. bias on Schottky contact coplanar line characteristics.

They point out the peculiar characteristics of the considered propagated mode which are : - a small guided wavelength corresponding to a high effective relative permittivity, - a low characteristics impedance. We must note that results relative to the slow wave factor are in good agreement with SDA ones.

COMPLETE MODEL : PROPAGATION CHARACTERISTICS. Indeed to obtain the frequency behaviour of the propagated mode we must take into account losses due to the semiconducting layer. Then we use the whole transmission line model. SDA results and analytical ones are compared Fig. 3.

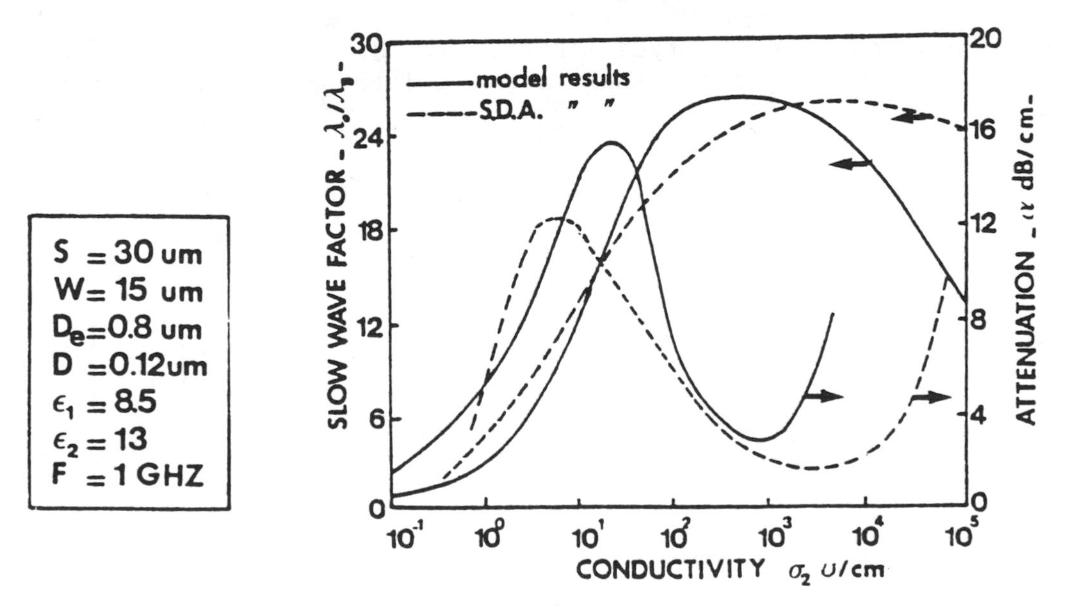

Fig. 3 - Optimal conductivity : comparison of SDA an analytical model results

We note that slowing factor and losses obtained by both methods exhibit quite similar evolution and order of magnitude. However some discrepancies appear for the determination of the optimal conductivity value. This is due to the difficulty to account correctly of, the series and the parallel losses of the semiconductor layer in the transmission line model.Specially it is more difficult to analyse wide structure with high conductivity in which skin effect in the semiconducting layer becomes important. For medium conductivity values corresponding to doping level ranging from 10^{16} to 10^{17} cm^{-3} our model is quite satisfying.

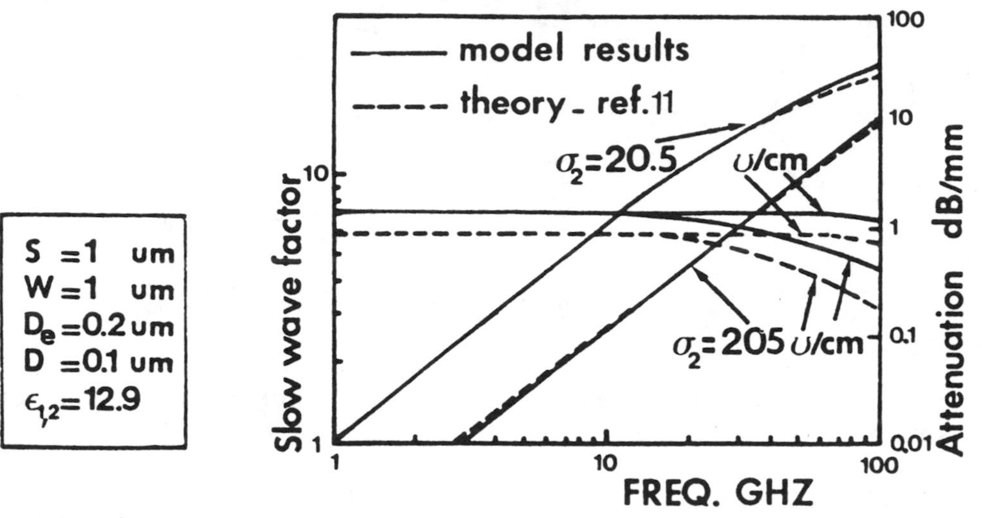

Fig. 4 - Propagation characteristics of a MESFET type structure.

As Shown fig. 4 our results are in good agreement with numerical ones obtained by mode matching method when propagation characteristics in a GaAs MESFET is considered [11]. Note that for the considered structure the frequencial behaviour of the slow wave factor and of the attenuation is well describe by our model up to about 100 GHz.

EFFECTS OF LOSSES IN CONDUCTORS. In the example presented fig. 4 losses in conductors are not taken into account. Unfortunately these ones can no longer be neglected in FET type structures. Thisis due to the small gate length, and to the reduce metallization thickness. Typical results are presented fig. 5.

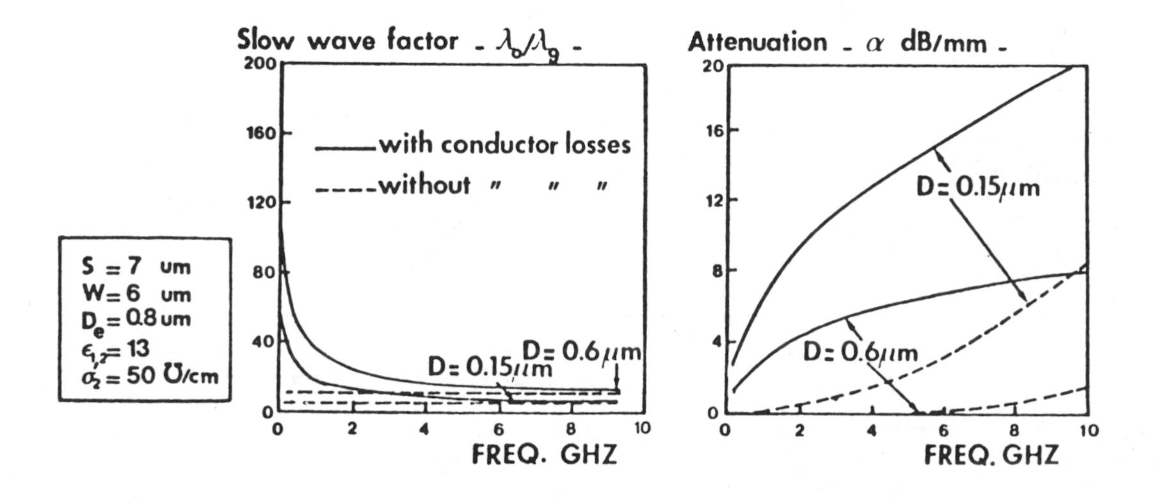

Fig. 5 - Effects of losses in conductors.

For the considered Schottky contact coplanar line we assume a central conductor series resistance of 30 Ω/mm corresponding to a 2000 A thickness gold metallization [2]. For lowest frequencies, losses in conductors are very important and the slowing factor is strongly increased. For highest frequencies semiconductor and conductors losses are in the same order of magnitude, and the influence of metallic losses on the slow wave factor is reduced.

CONCLUSION. We have proposed an analytical model for the Schottky contact coplanar line, the validity of which is attested by comparison with numerical analysis using SDA and Mode Matching technics. With this model a comparative study can be carried out easily on a destop computer for structures having different geometrical parameters under different D.C. bias conditions. The results obtained are useful to initialize numerical methods when analysing Schottly contact and MIS coplanar lines. Our model is suitable for structures with strip and slot widths up to 50 μm with doping level ranging from 10^{16} to 10^{17} cm^{-3}, so it can be helpful for the study of propagation effects in microwave power FETs.

BIBLIOGRAPHY

[1] H. HASEGAWA, S. SEKI
"On chip pulse transmission in very high speed LSI/VLSIS."
IEEE Monolithic Circuits Symposium, San Francisco, May 1984, pp 29-33

[2] PAUL L. FLEMING, T. SMITH, H.E. CARLSON, WILLIAM A. COX
"GaAs SAMP device for KU-Band switching"
I.E.E.E. Trans. on MTT, 27, Dec 1979, pp. 1032-1035

[3] Y.C. SHIH, T. ITOH
"Analysis of printed transmission lines for monolithic integrated circuits"
ELECTRON. LETT., 1982, Vol 18, N°14, pp 585-586

[4] Y.FUKUOKA, T. ITOH
"Analysis of slow-wave phenomena in coplanar waveguide on a semiconductor substrate"
ELECTRON. LETT., 1982, Vol 18, N°14, pp 589-590

[5] P. KENNIS, M. AUBOURG and all
"Properties of microstrip and coplanar lines on semiconductor substrate"
Proc. 12th EuMC, Helsinski, Sept. 1982

[6] M. AUBOURG, J.P. VILOTTE, F. GODON, Y. GARAULT
"Finite element analysis of lossy waveguide"
I.E.E.E. Trans. on MTT, March 1983

[7] M. AUBOURG, P. KENNIS and all
"Analysis of MIS or Schottky contact coplanar lines using the F.E.M. or SDA"
I.E.E.E. MTT-S, International microwave symposium Boston, June 1983.

[8] HIDEKI HASEGAWA, MIEKO FURUKAWA, HISAYOSHI YANAI
"Properties of microstrip lines on Si Si O_2 system"
I.E.E.E. Trans. on MTT, 19 Nov. 1971, pp. 869-881

[9] HENRY GUCKEL, PIERCE A BRENNAN, ISTUAN PALOCZ
"A parallel plate waveguide approach to miniaturized planar transmission lines for integrated circuits"
I.E.E.E. Trans. on MTT, 15 Aug. 1967, pp. 468-476

[10] C.P. WEN
"Coplanar waveguide : a surface strip transmission line suitable for non reciprocal gyromagnetic application"
I.E.E.E. Trans. MTT, 17 Dec. 1979, pp. 1087-1091

[11] W. HEINRICH, H.L. HARTNAGEL
"Wave-theorical analysis of signal propagation on FET electrodes".
ELECTRON. LETT., 1983, Vol. 19, p 65.

Author Index

Subject Index

Editor's Biography

Tatsuo Itoh (S'69–M'69–SM'74–F'82) received the Ph.D. degree in electrical engineering from the University of Illinois, Urbana in 1969.

From September 1966 to April 1976, he was with the Electrical Engineering Department, University of Illinois. From April 1976 to August 1977, he was a Senior Research Engineer in the Radio Physics Laboratory, SRI International, Menlo Park, CA. From August 1977 to June 1978, he was an Associate Professor at the University of Kentucky, Lexington. In July 1978, he joined the faculty at the University of Texas at Austin, where he is now a Professor of Electrical Engineering. During the summer of 1979, he was a guest researcher at AEG-Telefunken, Ulm, West Germany. Since September 1983, he has held the Hayden Head Centennial Professorship of Engineering at the University of Texas and since September 1984, has been the Associate Chairman for Research and Planning of the Electrical and Computer Engineering Department.

Dr. Itoh is a Fellow of the IEEE, a member of the Institute of Electronics and Communication Engineers of Japan, Sigma Xi, and Commissions B and D of USNC/URSI. He served as the Editor of IEEE TRANSACTIONS ON MICROWAVE THEORY AND TECHNIQUES for 1983–1985. He serves on the Administrative Committee of IEEE Microwave Theory and Techniques Society, and is a Professional Engineer registered in the state of Texas.